T0318878

EMERY AND RIMOIN'S PRINCIPLES AND PRACTICE OF MEDICAL GENETICS AND GENOMICS

EMERY AND RIMOIN'S PRINCIPLES AND PRACTICE OF MEDICAL GENETICS AND GENOMICS

Hematologic, Renal, and Immunologic Disorders

Seventh Edition

Edited by

Reed E. Pyeritz
Perelman School of Medicine at the University of Pennsylvania, Philadelphia, PA, United States

Bruce R. Korf
University of Alabama at Birmingham, Birmingham, AL, United States

Wayne W. Grody
UCLA School of Medicine, Los Angeles, CA, United States

Academic Press is an imprint of Elsevier
125 London Wall, London EC2Y 5AS, United Kingdom
525 B Street, Suite 1650, San Diego, CA 92101, United States
50 Hampshire Street, 5th Floor, Cambridge, MA 02139, United States
The Boulevard, Langford Lane, Kidlington, Oxford OX5 1GB, United Kingdom

ISBN: 978-0-12-812534-2

For information on all Academic Press publications visit our website at https://www.elsevier.com/books-and-journals

Publisher: Stacy Masucci
Acquisitions Editor: Peter B Linsley
Editorial Project Manager: Pat Gonzalez
Production Project Manager: Maria Bernard
Cover Designer: Miles Hitchen

Typeset by TNQ Technologies

CONTENTS

PART III Immunologic Disorders

LIST OF CONTRIBUTORS

Merrill D. Benson
Professor of Pathology and Laboratory Medicine, Professor of Medical and Molecular Genetics, and Professor of Medicine, Indiana University School of Medicine, Indianapolis, IN, United States

Dervla M. Connaughton
Schulich School of Medicine & Dentistry, University of Western Ontario, London, ON, Canada
Department of Medicine, Division of Nephrology, London Health Sciences Centre, London, ON, Canada

David J. Friedman
Division of Nephrology, Department of Medicine, Beth Israel Deaconess Medical Center and Harvard Medical School, Boston, MA, United States

David Ginsburg
Howard Hughes Medical Institute and Departments of Internal Medicine, Pediatrics, and Human Genetics, University of Michigan Medical School, Ann Arbor, Michigan, United States

Friedhelm Hildebrandt
Division of Nephrology, Professor of Pediatrics, Harvard Medical School, Department of Pediatrics, Boston Children's Hospital, Harvard Medical School, Boston, MA, United States

Hannu Jalanko
Children's Hospital, University of Helsinki, Helsinki, Finland

Helena Kääriäinen
National Institute for Health and Welfare, Helsinki, Finland

Selina M. Luger
Division of Hematology-Oncology, Department of Medicine, Perelman School of Medicine, University of Pennsylvania, Philadelphia, PA, United States

Jennifer J.D. Morrissette
Division of Precision and Computational Diagnostics, Department of Pathology and Laboratory Medicine, Perelman School of Medicine, University of Pennsylvania, Philadelphia, PA, United States

Michal Mrug
University of Alabama at Birmingham, Birmingham, AL, United States

Scott Peslak
Division of Hematology/Oncology, Department of Medicine, University of Pennsylvania Perelman School of Medicine, Philadelphia, PA, United States
Division of Hematology, Children's Hospital of Philadelphia, Philadelphia, PA, United States

Martin R. Pollak
Division of Nephrology, Department of Medicine, Beth Israel Deaconess Medical Center and Harvard Medical School, Boston, MA, United States

Reed E. Pyeritz
Perelman School of Medicine at the University of Pennsylvania, Philadelphia, PA, United States

Frederic Rahbari Oskoui
Emory University, Atlanta, GA, United States

Dana V. Rizk
University of Alabama at Birmingham, Birmingham, AL, United States

Jacquelyn J. Roth
Division of Precision and Computational Diagnostics, Department of Pathology and Laboratory Medicine, Perelman School of Medicine, University of Pennsylvania, Philadelphia, PA, United States

Takamitsu Saigusa
University of Alabama at Birmingham, Birmingham, AL, United States

Farzana Sayani
Division of Hematology/Oncology, Department of Medicine, University of Pennsylvania Perelman School of Medicine, Philadelphia, PA, United States

Jordan A. Shavit
Departments of Pediatrics and Human Genetics, University of Michigan Medical School, Ann Arbor, Michigan, United States

Edward A. Stadtmauer
Division of Hematology-Oncology, Department of Medicine, Perelman School of Medicine, University of Pennsylvania, Philadelphia, PA, United States

Kathleen E. Sullivan
Division of Allergy Immunology, The Children's Hospital of Philadelphia, University of Pennsylvania Perelman School of Medicine, Philadelphia, PA, United States

Angela C. Weyand
Department of Pediatrics, Division of Hematology/Oncology, University of Michigan Medical School, Ann Arbor, Michigan, United States

PREFACE TO THE SEVENTH EDITION OF *EMERY AND RIMOIN'S PRINCIPLES AND PRACTICE OF MEDICAL GENETICS AND GENOMICS*

The first edition of *Emery and Rimoin's Principles and Practice of Medical Genetics* appeared in 1983. This was several years prior to the start of the Human Genome Project in the early days of molecular genetic testing, a time when linkage analysis was often performed for diagnostic purposes. Medical genetics was not yet a recognized medical specialty in the United States, or anywhere else in the world. Therapy was mostly limited to a number of biochemical genetic conditions, and the underlying pathophysiology of most genetic disorders was unknown. The first edition was nevertheless published in two volumes, reflecting the fact that genetics was relevant to all areas of medical practice.

35 years later we are publishing the seventh edition of *Principles and Practice of Medical Genetics and Genomics*. Adding "genomics" to the title recognizes the pivotal role of genomic approaches in medicine, with the human genome sequence now in hand and exome/genome-level diagnostic sequencing becoming increasingly commonplace. Thousands of genetic disorders have been matched with the underlying genes, often illuminating pathophysiological mechanisms and in some cases enabling targeted therapies. Genetic testing is becoming increasingly incorporated into specialty medical care, though applications of adequate family history, genetic risk assessment, and pharmacogenetic testing are only gradually being integrated into routine medical practice. Sadly, this is the first edition of the book to be produced without the guidance of one of the founding coeditors, Dr. David Rimoin, who passed away just as the previous edition went to press.

The seventh edition incorporates two major changes from previous editions. The first is publication of the text in 11 separate volumes. Over the years, the book had grown from two to three massive volumes, until the electronic version was introduced in the previous edition. The decision to split the book into multiple smaller volumes represents an attempt to divide the content into smaller, more accessible units. Most of these are organized around a unifying theme, for the most part based on specific body systems. This may make the book more useful to specialists who are interested in the application of medical genetics to their area but do not wish to invest in a larger volume that covers all areas of medicine. It also reflects our recognition that genetic concepts and determinants now underpin all medical specialties and subspecialties. The second change might seem on the surface to be a regressive one in today's high-tech world—the publication of the 11 volumes in print rather than strictly electronic form. However, feedback from our readers, as well as the experience of the editors, indicated that access to the web version via a password-protected site was cumbersome and printing a smaller volume with two-page summaries was not useful. We have therefore returned to a full print version, although an eBook is available for those who prefer an electronic version.

One might ask whether there is a need for a comprehensive text in an era of instantaneous Internet searches for virtually any information, including authoritative open sources such as *Online Mendelian Inheritance in Man* and *GeneReviews*. We recognize the value of these and other online resources, but believe that there is still a place for the long-form prose approach of a textbook. Here the authors have the opportunity to tell the story of their area of medical genetics and genomics, including in-depth background about pathophysiology, as well as giving practical advice for medical practice. The willingness of our authors to embrace this approach indicates that there is still enthusiasm for a textbook on medical genetics; we will appreciate feedback from our readers as well.

The realities of editing an 11-volume set have become obvious to the three of us as editors. We are grateful to our authors, many of whom have contributed to multiple past volumes, including some who have updated their contributions from the first or second editions. We are also indebted to staff from Elsevier, particularly Peter Linsley and Pat Gonzalez, who have worked patiently with us in the conception and production of this large project. Finally, we thank our families, who have indulged our occasional disappearances into writing and editing. As always, we look forward to feedback from our readers, as this has played a critical role in shaping the evolution of *Principles and Practice of Medical Genetics and Genomics* in the face of the exponential changes that have occurred in the landscape of our discipline.

PREFACE TO *HEMATOLOGIC, RENAL, AND IMMUNOLOGIC DISORDERS*

This volume of *Principles and Practice of Medical Genetics and Genomics* presents topics focused on three organ systems, renal, immunologic, and hematologic. The latter two interact with each other intimately and both have positive and potentially negative effects on the kidney. Some of the authors have revised their chapters since much earlier editions of this treatise. New authors have brought fresh perspectives to the subjects introduced in the first edition in 1983. Several chapters are entirely original to this 7th edition. The explosive growth in genomics underlies enhanced comprehension of the normal, pathogenic, diagnostic, and therapeutic aspects of the topics in this volume. The fundamental and clinical perspectives discussed in the first two volumes of this 7th edition are well illustrated by the conditions in this volume.

PART I

Renal Disorders

1

Congenital Anomalies of the Kidney and Urinary Tract

Dervla M. Connaughton[1,2], *Friedhelm Hildebrandt*[3]

[1]Schulich School of Medicine & Dentistry, University of Western Ontario, London, ON, Canada
[2]Department of Medicine, Division of Nephrology, London Health Sciences Centre, London, ON, Canada
[3]Division of Nephrology, Professor of Pediatrics, Harvard Medical School, Department of Pediatrics, Boston Children's Hospital, Harvard Medical School, Boston, MA, United States

1.1 INTRODUCTION

Congenital anomalies of the kidney and urinary tract (CAKUT) are defined as any abnormality in structure or function of the kidneys, ureters, bladder, urethra, and/or distal genitourinary tract. CAKUT is a common disease entity detected at a frequency of one in 500 fetal ultrasound examinations [1]. CAKUT account for 20%–30% of all congenital malformations [2–4] and are frequently observed birth defects at a rate of ~3–6/1000 live births [3,5].

CAKUT can have major health implications as they are the most common cause of chronic kidney disease in the first three decades of life. Registry-based data from Europe and North America indicate that CAKUT account for 41% and 39% of pediatric onset end-stage renal disease, respectively [6,7]. The peak incidence of end-stage kidney disease (ESKD) in children with CAKUT is 15–19 years of age with a gradual decline throughout adulthood [8].

In adults, the prevalence of CAKUT is less clear. This is likely related to the fact that many individuals remain asymptomatic, with the diagnosis of CAKUT only made incidentally following routine imaging of the abdomen [9]. Based on epidemiological data, CAKUT has an estimated prevalence of 5%–10% in the general chronic kidney disease population and an estimated incidence of <5% in renal replacement therapy populations; however, these estimates are likely prone to underestimation [8,10–12].

1.2 CLINICAL MANIFESTATIONS OF CAKUT

CAKUT is defined as any abnormality in the number, size, shape, structure or function of the kidney and/or genitourinary tract. CAKUT can be categorized based on anatomical position into phenotypes that involve the upper urinary tract; renal agenesis, renal hypodysplasia and multicystic dysplastic kidney (MCDK), phenotypes involving the ureters; ureter duplication, hydronephrosis, hydroureter or megaureter, ectopic ureter or uretero-pelvic junction obstruction (UPJO) to phenotypes predominately affecting the lower urinary tract; vesicourethral reflux (VUR), ureterovesical junction obstruction (UVJO), and posterior urethral valves (PUV) [2,5,13]. CAKUT can also encompass anomalies in kidney shape or position, namely horseshoe kidney or pelvic kidney, respectively. Pathologies involving bladder innervation resulting in discoordinated detrusor and urethral function can result in secondary urinary tract malformations [14]. For example, myelomeningocele,

Emery and Rimoin's Principles and Practice of Medical Genetics and Genomics. https://doi.org/10.1016/B978-0-12-812534-2.00002-3

which can result in dysfunctional innervation of the bladder, can lead to hydronephrosis and secondary changes in the kidney due to obstruction and urinary reflux. Many of these conditions lead to an increased susceptibility to urinary tract infections due to stasis and obstruction of urine flow. This in turn can lead to parenchymal renal scarring and progressive renal impairment due to so-called "reflux nephropathy" [14].

Clinical heterogenicity is one of the hallmark features of the CAKUT phenotype since different phenotypes from the CAKUT spectrum can occur within the same individual [2,15]. In a study of over 200 families with CAKUT, a total of 546 pathologies were observed in 273 individuals. This included single unilateral pathologies (e.g., unilateral renal agenesis), bilateral concordant disease (e.g., bilateral VUR) to bilateral discordant disease (e.g., right side MCDK and left VUR) (Fig. 1.1) [16]. In a Japanese study, 26% of the study population with a single functioning kidney had a concurrent CAKUT phenotype [17]. Equally, unilateral disease such as single renal agenesis can occur in the setting of an entirely normal contralateral kidney and urinary tract [18].

Long-term renal outcomes are variable again depending on the number and type of CAKUT pathologies.

The median reported age of starting renal replacement therapy in all patients with CAKUT is 31 years, which is significantly younger than patients with other forms of kidney disease [8]. However, there is large variability, with sometimes conflicting data on disease progression and renal outcomes depending on the CAKUT subtype and the study population under study. For example, in a US study of over 300 patients with CAKUT, the presence of a solitary kidney or renal hypodysplasia associated with posterior urethral valves was associated with an increased risk of progression to ESKD, compared to patients with either unilateral or bilateral renal hypodysplasia, or dysplastic or horseshoe kidney [19]. In an European population, patients with isolated renal hypodysplasia required renal replacement therapy at an earlier age (median, 16 years) than those with renal hypoplasia and associated urinary tract disorders (median, 29.5–39.5 years) [8]. Therefore, further studies are required to determine the exact risk associated with each CAKUT subtype.

Interestingly, although many individuals remain asymptomatic, in the KIMONO study, it was found that 32% of children born with a single functioning kidney had evidence of renal injury by 10 years of age [17]. By

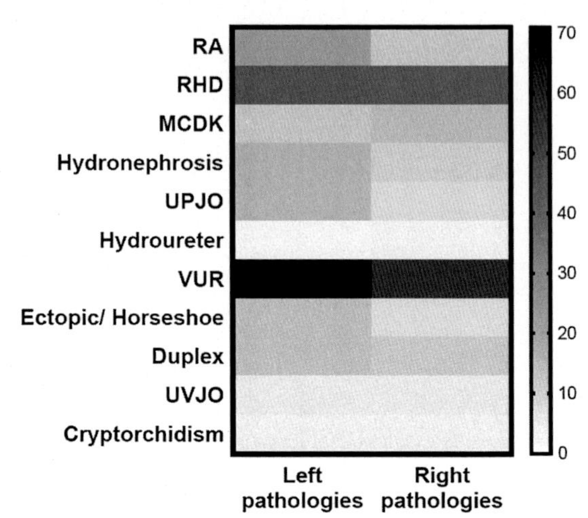

Figure 1.1 Heatmap comparing the distribution of the specific phenotypic pathologies in a cohort of individuals with congenital anomalies of the kidney and urinary tract (CAKUT). A study of 232 families with CAKUT. The absolute numbers of pathologies are graded according to the color chart displayed above on the *right*. Specific CAKUT phenotypic pathologies are listed from cranial to caudal positions in the y-axis. The x-axis is further subdivided into *left* and *right* pathologies. Note, 130 pathologies were present in 130 individuals with unilateral CAKUT. In the 143 individuals with bilateral CAKUT, a total of 286 pathologies were present. A total of 416 pathologies were present in 273 individuals with CAKUT. The bilateral renal pathologies consist of individuals with both bilateral concordant and bilateral discordant CAKUT. The total number of pathologies were calculated independent of both laterality and whether the bilateral pathologies were concordant or discordant in the same individual. Individuals with the CAKUT pathology of posterior urethral value (PUV) or epi/hypospadias ($n = 25$) and individuals in whom the CAKUT phenotype is undefined ($n = 21$) in this analysis, due to the lack and/or inability to determine laterality with these specific pathologies, were excluded from the above analysis. *Duplex*, duplex collecting system, *Ectopic/Horseshoe*, ectopic or horseshoe kidney; *MCDK*, multicystic dysplastic kidney; *RA*, renal agenesis; *RHD*, renal hypoplasia/dysplasia; *UPJO*, ureteropelvic junction obstruction; *UVJO*, ureterovesical junction obstruction; *VUR*, vesicourethral reflux. (Adapted from van der Ven AT, Connaughton DM, Ityel H, Mann N, Nakayama M, Chen J, et al. Whole-exome sequencing identifies causative mutations in families with congenital anomalies of the kidney and urinary tract. *J Am Soc Nephrol* 2018;29(9):2348–2361.)

adulthood (classified as 30 years of age), 18.5% (58 out of 312) of patients with either renal agenesis, renal hypodysplasia with or without posterior urethral valves, multicystic dysplastic kidney, or horseshoe kidney had reached end stage kidney disease requiring renal replacement therapy. This demonstrates that even in individuals with subclinical CAKUT, a poorer outcome in terms of maintaining long-term renal function may occur. Equally there is increasing evidence that even in the asymptomatic patients, a single functional kidney, either due to unilateral renal agenesis, MCDK, or renal hypodysplasia, can lead to glomerular damage with subsequent hypertension, albuminuria, and progression to end-stage renal disease in adulthood [20].

1.3 FAMILIAL CAKUT

Although routine screening in asymptomatic family members is not currently recommended, there is evidence of an increased incidence of CAKUT in first degree relatives of affected individuals. Epidemiological data have revealed that familial clustering occurs in patients with CAKUT with approximately 10%–15% of cases with CAKUT noted to have a familial component [10]. In a Turkish study of 145 families, screening of 412 asymptomatic first-degree relatives revealed a diagnosis of CAKUT in 23 individuals from 21 different families [21]. The most frequently observed phenotypes in asymptomatic family members were renal agenesis, renal hypodysplasia and hydronephrosis. It is important to note that wide variation can also occur within family members where in some, a single pathology is observed, and in others multiple different pathologies are evident. In parents or siblings of 41 patients with either bilateral or unilateral renal agenesis or dysgenesis, there was a 9% risk (10 of 111 asymptomatic family members) of detecting asymptomatic CAKUT by ultrasound examination. The most frequent CAKUT subtype observed in these asymptomatic individuals was again unilateral renal agenesis [22]. In patients with either, bilateral renal agenesis, or bilateral severe dysgenesis, or agenesis of one kidney and dysgenesis of the other kidney, 9% of either parents or siblings had asymptomatic urogenital malformations incidentally detected by ultrasonography. In patients with either unilateral or bilateral bifid or double ureters, there is also an increased incidence of duplex collecting system in first degree relatives of patients [23].

1.4 SPECTRUM OF CAKUT PHENOTYPES

1.4.1 Renal Agenesis

Renal agenesis refers to congenital absence of one of both kidneys. In a study of more than 625,000 consecutive births in British Columbia, Canada, 92 cases of bilateral renal agenesis and 117 cases of unilateral renal agenesis were identified, with a male predominance of 2.45:1 [24]. Bilateral renal agenesis represents one of the most severe phenotypes in the CAKUT spectrum. Presentation generally occurs *in utero* with oligohydramnios evident as early as the second trimester. Owing to primary renal dysfunction, oligohydramnios occurs, leading to pulmonary hypodysplasia. Many infants perish *in utero*, with studies indicating a postnatal mortality of up to 100% [25]. The classic presentation includes a triad of club foot, respiratory compromise, and cranial anomalies referred to as the Potter sequence [26].

On the other hand, unilateral renal agenesis may be entirely subclinical only detected incidentally following routine imaging of the abdomen and pelvis.

Renal aplasia refers to the presence of renal parenchymal tissue without any function. After birth, involution of this rudimentary tissue can occur and often patients present with a clinical picture of renal agenesis.

1.4.2 Renal Hypodysplasia (RHD)

Renal hypodysplasia is characterized by under and/or abnormal development of the kidney usually resulting in reduced size of the kidney. Specifically, hypoplasia refers to small kidneys with reduced nephron numbers while dysplasia refers to abnormal kidney development. In reality, the two processes often occur together and can have a very similar clinical presentation. RHD is a common subtype within in the CAKUT disease spectrum with unilateral RHD occurring at a frequency of one in 1000 and bilateral RHD occurring at a frequency of one in 5000 in the general population [27].

1.4.3 Multicystic Dysplastic Kidney (MCDK)

Multicystic dysplastic kidney describes the condition of multiple irregular cysts of varying sizes that are surrounded by dysplastic renal tissue [28]. Ultrasonography at 20 weeks gestation has a high likelihood of diagnosis. Ultrasonography examination is characterized by multiple thin-walled cysts that do not connect and are randomly distributed throughout the kidney(s) [27].

Schreuder and colleagues performed a meta-analysis of 19 populations encompassing 2500 individuals with unilateral MCDK and demonstrated an incidence of one in 4300. There was a male predominance at 59%

and MCDK was more commonly (53.1%) identified on the left side. In one-third of individuals, an additional subtype from the CAKUT spectrum was observed, most commonly vesico-urethral reflux (VUR) [29].

1.4.4 Horseshoe Kidney

Horseshoe kidney is a renal fusion defect that occurs when both kidneys fuse, usually at the upper pole, resulting in a horseshoe shape. Fusion results in the inability of the kidneys to ascend to the anatomically normal position, rather they remain in the embryonic pelvic position. Fusion therefore results in defects of not only position but also rotation and vascular supply; however, the functional mass of the kidney is generally intact, and the ureters remain uncrossed. The incidence in the general population is approximately one in 500 [30]; however, this may be subject to underestimation as it is believed that up to one-third of patients with a horseshoe kidney remain asymptomatic. Patients usually present if there is an associated CAKUT subtype such as UPJO, hydronephrosis, other genitourinary anomalies such as hypospadias and cryptorchidism or secondary complications related to urinary tract infections or renal stones.

1.4.5 Duplex Kidney

Duplex kidney occurs when more than one kidney forms. Duplex kidney is a common subtype of CAKUT, with an estimated incidence of approximately 1% [31]. In a study of 1716 children and 3480 adults, 79 cases of unilateral duplex and 16 cases of bilateral duplex kidneys were detected [32]. In individuals with unilateral duplex, the "extra" renal tissue can be partial, as seen in 10%, or complete duplication of the kidney resulting in a kidney of equal size in 39%. The duplex portion can be functionally normal but in up to 27%, the renal tissue is defective with evidence of impaired kidney function. In kidney duplication, there appears to be a slight female predominance, but there is no predilection to either the right or left side of the genitourinary tract [32]. Isolated duplex kidney can remain entirely asymptomatic. As with many of the CAKUT phenotypes, other subtypes of CAKUT may occur in conjunction with a duplex kidney such as ectopic ureter, ureterocele or most commonly VUR. When symptoms do occur, they usually result from the presence of these additional CAKUT phenotypes.

1.4.6 Duplex Ureter

Duplex ureter is the formation of more than one ureter, which are usually defective. Complete duplication of the ureters with separate insertion into the bladder is a rare phenomenon. More commonly partial duplication, also known as a bifid ureter, is observed; however, the exact prevalence is difficult to ascertain as many patients remain asymptomatic. The incidence of urinary tract duplication is estimated to be between 0.7% and 4% in the general population with a female predominance [32].

1.4.7 Obstructive Uropathy

Obstructive uropathy is an umbrella term for a diverse range of both acquired and inherited diseases that results in the impedance of normal urinary flow in either the upper or lower urinary tract. The obstruction can be secondary to either an intrinsic obstruction within the genitourinary tract or an extrinsic obstruction such as tumor compression. The clinical manifestation in most cases of obstructive uropathy is ultrasonographic evidence of either hydronephrosis or hydroureter.

1.4.8 Hydronephrosis and Hydroureter

Hydronephrosis is an abnormal distention of the pelvicalyceal region of the kidney, and hydroureter is abnormal distention of the ureter. Both occur usually secondary to distal obstruction and can occur in conjunction with each other. Megaureter is a subtype of hydroureter, where a large increase in ureter size is observed. *In utero*, hydronephrosis is a common clinical manifestation of obstructive uropathy, affecting between 1% and 4.5% of all pregnancies [33]. However, the definitive etiology of the obstruction is often not revealed until the postpartum period. Within the CAKUT spectrum, uretero-pelvic junction obstruction (UPJO) and vesicoureteral reflux (VUR) are the two most common postnatal causes of obstructive uropathy as well as any cause of lower urinary tract obstruction (please see below for further details).

1.4.9 Uretero-Pelvic Junction Obstruction (UPJO)

Uretero-pelvic junction obstruction is an abnormal constriction resulting in blockage of urine at the junction between the renal pelvis and the ureter. It has a reported incidence of one in 500 live births [34]. It occurs at higher frequency in males and has a predilection for the left side of the urinary tract, although can be bilateral. The most common clinical manifestation is ultrasonographic evidence of hydronephrosis without hydroureter [35].

1.4.10 Uretero-Vesical Junction Obstruction (UVJO)

Uretero-vesical junction obstruction is abnormal constriction resulting in blockage of urine at the junction between the ureter the bladder. UVJO is also a cause of antenatal hydronephrosis or megaureter and can occur in conjunction with other CAKUT phenotypes such as UPJO [36].

1.4.11 Vesico-Urethral Reflux (VUR)

Vesico-urethral reflux, also known as vesico-ureteric reflux, is the abnormal passage of urine from the bladder to the ureter or kidney in a retrograde direction. The primary etiology is dysfunction of the vesico-urethral value, which normally allows one-way flow of urine from the ureter to the bladder thus preventing back flow of urine during micturition. Normal valvular function is maintained through a number of coordinated processes and anatomical structures including the length of the submucosal ureter, the width of the ureteric opening, the muscles of the trigone and ureter, and coordinated ureteric peristalsis [37]. Value dysfunction is thought to occur due to a multitude of both genetic and environmental factors interfering with this process. VUR is predominately diagnosed based on voiding cystourography and is graded based on the International Reflux Study system, which is a scale grading system from I to V depending on findings on voiding cystourography [37]. The severity of VUR can range from mild disease that spontaneously resolves in childhood to progressive chronic kidney disease secondary to chronic infections and renal parenchymal scarring, which is commonly referred to as "reflux nephropathy."

1.4.12 Lower Urinary Tract Obstruction (LUTO)

Posterior urethral valve (PUV), urethral anomalies including urethral atresia and the Prune Belly Syndrome (PBS) are the most common causes of LUTO with a combined incidence of 2.2 in 10,000 births [38]. Rarer causes of LUTO include anterior urethral valve, ureterocele, and other causes of external compression such as a urethral diverticulum or hydrocolpos due to cloacal anomalies [39].

1.4.13 Posterior Urethral Valve (PUV)

PUV occurs as a result of an obstructive membrane or "valve" in the posterior segment of the urethra resulting in blockage of urine flow from the bladder to the urethra. This subtype of CAKUT, also referred to as infra-vesical urinary tract obstruction, is the most common cause of lower urinary tract obstruction in childhood [40], with an estimated prevalence of three per 10,000 births [39]. PUV predominantly occurs in males and is a common cause of chronic kidney disease in this population. PUV can occur in isolation or conjunction with other CAKUT pathologies, including hydronephrosis, UVJO, VUR, and RHD. There is also an increased incidence of other genitourinary tract pathologies with an estimated incidence of PUV in male with hypospadias of 1% [41]. Long-term outcomes following PUV include bladder dysfunction in childhood with an elevated risk of lower urinary tract infections in adulthood and progressive kidney disease [42].

The initial presentation of PUV is usually within the first 12 months of life; however, presentation is variable depending on the degree and severity of the obstruction. With the widespread utilization of prenatal ultrasonography, up to 62% of cases are now diagnosed *in utero* [39]. The classic presentation *in utero* is ultrasonographic evidence of a dilated bladder with a thickened bladder wall, bilateral hydronephrosis, dilated ureters, and a dilated posterior urethra, which is commonly referred to as the "keyhole sign" [43]. If severe, oligohydramnios can ensue with subsequent pulmonary hypoplasia.

In the postnatal period, the clinical manifestations include a palpable abdominal mass secondary to a distended bladder, urinary tract infections, poor voiding pressure with a weak urinary stream, bladder dysfunction, and progressive renal impairment. Voiding cystourethrogram followed by urethral endoscopy is generally required to confirm the diagnosis. Endoscopic treatment with primary valve ablation through transurethral fulguration is the treatment of choice for most patients with PUV, although increasingly *in utero* interventions are being explored. For example, first trimester decompression by means of vesico-amniotic shunting or fetal cystoscopic ablation is currently being undertaken at a number of specialized centers [39].

1.4.14 Urethral Agenesis and Atresia

Urethral Agenesis and Atresia are rare subtypes of CAKUT occurring predominantly in males and occur when the urethra either fails to develop or there is abnormal development of the urethra.

1.4.15 Duplication of the Urethra

Duplication of the urethra is another rare subtype of CAKUT demonstrating a male predominance [44]. Urethral duplication is divided into four subclasses:

- Type 1 both bladder and urethral duplication
- Type 2 single bladder with urethral duplication
- Type 3 Y-type duplication, which is characterized by two limbs, a penile limb, which contains the urethral channel, and an ectopic limb, which tends to be a fistula tract to the perineum or anal cancel
- Type 4 miscellaneous type, which include urethral channels, spindle urethra, and other female forms

1.5 SYNDROMIC CAKUT

CAKUT may occur in isolation as a monogenic disorder (Table 1.1) or as part of a syndromic disorder (Table 1.2). It may occur in conjunction with other structural defects (Table 1.3) or chromosomal anomalies (Table 1.4) [13,46–48].

1.6 DIAGNOSIS OF CAKUT

The widespread use of prenatal ultrasonography means than increasingly the diagnosis of CAKUT is established prenatally, with a diagnostic rate of over 85% following fetal screening during pregnancy ultrasonography. In the prenatal period, ultrasonography is the modality of choice for the diagnosis of most subtypes of CAKUT. In severe cases, CAKUT can often be suspected in the presence of oligohydramnios, which can be detected as early as the second trimester, usually at 14–16 weeks gestation. Renal agenesis can in some cases be detected as early as 15 weeks' gestation, but usually a definitive diagnosis can only be confirmed on ultrasonography after 18–19 weeks of gestation.

Although not routinely performed, postnatal ultrasonography has a specificity of 100% and a sensitivity of 92.1% for the detection of single renal agenesis [49]. In a Japanese study of infants age 1 month old, routine screening with ultrasonography revealed a diagnosis of CAKUT in 3.5% (198 positive cases out of 5700 screened infants) [50]. However, not infrequently asymptomatic patients present in adulthood following an incidental finding on imaging of the abdomen or pelvis. Although ultrasonography is the standard method for the diagnosis of CAKUT, occasionally renal anomalies can also be detected following other modalities of abdominal or pelvic imaging such as computed tomography (CT) or magnetic resonance imaging (MRI).

In certain subtypes of CAKUT, additional imaging is required particularly in obstructive lesions or pathologies involving the distal genitourinary tract. Current recommendations from the American Academy of Pediatrics suggest ultrasonography and either voiding cystourethrography or radionuclide cystography for VUR and other potential causes of obstructive uropathy [51]. In lower urinary tract anomalies, ultrasound scanning is often of limited value as it will only reveal severe ureteral anomalies with significant dilatation. Therefore, further imaging modalities are generally warranted including, voiding cystourethrography (VCUG), intravenous pyelography (IVP), computed tomography (CT), dimercaptosuccinic acid (DMSA) renal cortical scintigraphy, ^{99m}technetium-mercaptoacetyltriglycine (^{99m}Tc-MAG-3) renography, and magnetic resonance urography (MRU) [52].

1.7 EMBRYONIC DEVELOPMENT OF THE KIDNEY AND URINARY TRACT

Development of the kidney in humans occurs at day 35–37 of embryonic development. The genitourinary tract arises from two structures; the nephric ducts (also known as the mesonephric ducts or Wolffian ducts) and the nephric cord, both of which arise from the intermediate mesoderm (Fig. 1.2). The nephric duct (ND) is an epithelial tube from which the ureteric bud (UB) arises as an epithelial outgrowth. Ultimately the nephric duct fuses caudally with the cloacal epithelium, which is a precursor of the urinary bladder. The metanephric mesenchyme (MM) arises from mesenchymal cells in the posterior intermediate mesoderm. Renal morphogenesis is initiated and maintained by reciprocal interactions between the epithelial portion of the

TABLE 1.1 Monogenic Causes of Human Isolated CAKUT, if Mutated. (Sorted Alphabetically by Mode of Inheritance)

Gene	Protein	Reference	Mode of Inheritance	Phenotype	OMIM #
ACE	Angiotensin I-converting enzyme	Gribouval *Nat Genet* 37:964, 2005	AR	Renal tubular dysgenesis	# 267430
AGT	Angiotensinogen	Gribouval *Nat Genet* 37:964, 2005	AR	Renal tubular dysgenesis	# 267430
AGTR1	Angiotensin II receptor, type 1	Gribouval *Nat Genet* 37:964, 2005	AR	Renal tubular dysgenesis	# 267430
CHRM3	Muscarinic acetylcholine receptor M3	Weber *AJHG* 19:634, 2011	AR	Prune belly syndrome	# 100100
ETV4	ETS translocation variant 4, E1A enhancer binding protein	Chen *IJPCH* 4:61, 2016	AR	NA	* 600711
FRAS1	Extracellular matrix protein FRAS1	Kohl *JASN* 25:1917, 2014	AR	Fraser syndrome 1	# 219000
FREM1	FRAS1-related extracellular matrix protein 1	Kohl *JASN* 25:1917, 2014	AR	Manitoba oculotrichoanal syndrome	# 248450
FREM2	FRAS1-related extracellular matrix protein 2	Kohl *JASN* 25:1917, 2014	AR	Fraser syndrome 2	# 617666
GRIP1	Glutamate receptor interacting protein 1	Kohl *JASN* 25:1917, 2014	AR	Fraser syndrome 3	# 617667
HPSE2	Heparanase 2 (inactive)	Bulum *Nephron* 130:54, 2015	AR	Urofacial syndrome 1	# 236730
ITGA8	Integrin α8	Humbert *AJHG* 189:1260, 2014	AR	Renal hypodysplasia/aplasia 1	# 191830
REN	Renin	Gribouval *Nat Genet* 37:964, 2005	AR	Renal tubular dysgenesis	# 267430
TRAP1	Heat shock protein 75 (also known as TNF receptor-associated protein 1)	Saisawat *Kid Int* 85:880, 2014	AR	NA	* 606219
FGF20	Fibroblast growth factor 20	Barak *Dev Cell* 22:1191, 2012	AR	Renal hypodysplasia/aplasia 2	# 615721
BMP4	Bone morphogenic protein 4	Weber *JASN* 19:891, 2008	AD	Microphthalmia, syndromic 6	# 607932
CHD1L	Chromodomain helicase DNA binding protein 1-like	Brockschmidt *NDT* 27:2355, 2012	AD	NA	* 613039

Continued

TABLE 1.1 Monogenic Causes of Human Isolated CAKUT, if Mutated. (Sorted Alphabetically by Mode of Inheritance)—cont'd

Gene	Protein	Reference	Mode of Inheritance	Phenotype	OMIM #
CRKL	CRK like proto-oncogene, adaptor protein	Lopez-rivera *NEJM* 376:742, 2017	AD	NA	* 602007
DSTYK	Dual serine/threonine and tyrosine protein kinase	Sanna-cherchi *NEJM* 369:621, 2013	AD	Congenital anomalies of kidney and urinary tract 1	# 610805
EYA1	Eyes absent homolog 1	Abdelhak *Nat Genet* 15:157, 1997	AD	Branchiootorenal syndrome 1, with or without cataracts	# 113650
GATA3	GATA-binding protein 3	Pandolfi *Nat Genet* 11:40, 1995; Van Esch *Nature* 406:419, 2000	AD	Hypoparathyroidism, sensorineural deafness, and renal dysplasia	# 146255
GREB1L	Growth regulation by estrogen in breast cancer 1 like	Brophy *Genetics* 207:215, 2017, Sanna-Cherchi *AJHG* 101:1034, 2017	AD	Renal hypodysplasia/aplasia 3	# 617805
HNF1B	HNF homeobox B	Lindner *Hum Mol Genet* 24:263, 1999	AD	Renal cysts and diabetes syndrome	# 137920
MUC1	Mucin 1	Kirby *Nat Genet* 45:299, 2013	AD	Medullary cystic kidney disease 1	# 174000
NRIP1	Nuclear receptor interacting protein 1	Vivante *JASN* 28:2364, 2107	AD	NA	* 602490
PAX2	Paired box 2	Sanyanusin *Hum Mol Genet* 4:2183, 1995	AD	Papillorenal syndrome	# 120330
PBX1	PBX homeobox 1	Heidet *JASN* 28:2901, 2017	AD	Congenital anomalies of kidney and urinary tract syndrome with or without hearing loss, abnormal ears, or developmental delay	# 617641
RET	Proto-oncogene tyrosine-protein kinase receptor ret	Skinner *AJHG* 82:344, 2008	AD	Multiple OMIM classifications	* 164761
ROBO2	Roundabout, axon guidance receptor, homolog 2 (*Drosophila*)	Hwang *Hum Genet* 134:905, 2015; Lu *AJHG* 80:616, 2007	AD	Vesicoureteral reflux 2	# 610878
SALL1	Sal-like protein 1 (also known as spalt-like transcription factor 1)	Kohlhase *Nat Genet* 18:81, 1998	AD	Townes-brocks syndrome 1	# 107480

Continued

TABLE 1.1 Monogenic Causes of Human Isolated CAKUT, if Mutated. (Sorted Alphabetically by Mode of Inheritance)—cont'd

Gene	Protein	Reference	Mode of Inheritance	Phenotype	OMIM #
SIX1	SIX homeobox 1	Ruf *PNAS* 101: 8090, 2004	AD	Branchio-otic syndrome 3	# 608389
SIX2	SIX homeobox 2	Weber *JASN* 19:891, 2008	AD	NA	* 604994
SIX5	SIX homeobox 5	Hoskins *AJHG* 80:800, 2007	AD	Branchiootorenal syndrome 2	# 610896
SLIT2	Slit homolog 2	Hwang *Hum Genet* 134:905, 2015	AD	NA	* 603746
SOX17	Transcription factor SIX-17	Gimelli *Hum Mut* 31:1352, 2010	AD	Vesicoureteral reflux 3	# 613674
SRGAP1	SLIT-ROBO rho GTPase activating protein 1	Hwang *Hum Genet* 134:905, 2015	AD	NA	* 606523
TBX18	T-box transcription factor	Vivante *AJHG* 97:291, 2015	AD	Congenital anomalies of kidney and urinary tract 2	# 143400
TNXB	Tenascin XB	Gbadegesin *JASN* 24:1313, 2013	AD	Vesicoureteral reflux 8	# 615963
UPK3A	Uroplakin 3A	Jenkins *JASN* 16:2141, 2005	AD	NA	* 611559
WNT4	Protein Wnt-4	Biason-Lauber *NEJM* 351:792, 2004; Mandel *AJHG* 82:39, 2008; Vivante *JASN* 24:550, 2013	AD	Mullerian aplasia and hyperandrogenism	# 158330
KAL1	Anosmin 1	Hardelin *PNAS* 89:8190, 1992	XL	Hypogonadotropic hypogonadism 1 with or without anosmia (kallmann syndrome 1)	# 308700

AD, autosomal dominant; *AR*, autosomal recessive; *NA*, not available; *OMIM*, online mendelian inheritance in man; *XL*, X-linked; *#*, phenotype MIM number; * gene/locus MIM number if not phenotype MIM number available.Adapted from Connaughton DM, Kennedy C, Shril S, Mann N, Murray SL, Williams PA, et al. Monogenic causes of chronic kidney disease in adults. Kidney Int 2019;95(4):914–928.

nephric duct and the MM. Nephrogenesis then arises from this structure due to reciprocal induction between the ureteric bud (UB) and the metanephric mesenchyme (MM). The portion of the MM that comes in closest proximity to the UB condenses and forms the cap mesenchyme (CM). Following induction from the UB, the CM then undergoes mesenchymal-to-epithelial transition (MET) with formation of the renal vesicle. This structure then forms the common-shaped body followed by the S-shaped body, which progressively invades endothelial cells. This is then followed by branching morphogenesis and nephrogenesis through a process called nephron

TABLE 1.2 Monogenic Causes of Human Syndromic CAKUT, if Mutated. (Sorted Alphabetically by Mode of Inheritance and Includes Syndromes That are Presumed Monogenic but Causative Gene Not yet Identified)

Gene	Protein	Reference	Mode of Inheritance	Phenotype MIM Number	Gene/ locus MIM Number
B3GALTL	Beta 3-glucosyltransferase	Lesnik Oberstein *AJHG* 79:562, 2006	AR	Peters-plus syndrome (261540)	# 610308
BSCL2	BSCL2, seipin lipid droplet biogenesis associated	Haghighi *Clin Genet* 89: 434, 2016	AR	Lipodystrophy, congenital generalized, type 2 (296700)	* 606158
CD151	CD151 Molecule (raph blood group)	Karamatic *Blood* 104:2217, 2004	AR	Nephropathy with pretibial epidermolysis bullosa and deafness (609057)	* 602243
CD96	CD96 molecule	Kaname *AJHG* 81:835, 2007	AR	C syndrome (211750)	# 606037
CHRNG	Cholinergic receptor nicotinic gamma subunit	Vogt *J Med Genet* 49:21, 2012	AR	Escobar syndrome (265000)	# 100730
CISD2	CDGSH iron sulfur domain 2	Amr *AJHG* 81:673, 2007	AR	Wolfram syndrome 2 (604928)	# 611507
CTU2	Cytosolic thiouridylase, subunit 2	Shaheen *AJMG* 170:3222, 2016	AR	Microcephaly, facial dysmorphism, renal agenesis, and ambiguous genitalia syndrome (618142)	# 617057
CYP21	Cytochrome P450 family 21	Martul *Arch Dis Child* 55:324, 1980	AR	HyperandrogenisM, nonclassic type, due to 21-hydroxylase deficiency (201910)	# 613815
DACH1	Dachshund family transcription factor 1	Schild *NDT* 28:227, 2013	AR	NA	* 603803
DHCR7	7-Dehydrocholesterol reductase	Löffler *AJHG* 13; 95:174, 2000	AR	Smith-lemli-opitz syndrome (270400)	# 602858
DIS3L2	DIS3 like 3′-5′ exoribonuclease 2	Astuti *Nat Genet* 5; 44:277, 2012	AR	Perlman syndrome (267000)	# 614184
EMG1	EMG1, N1-Specific pseudouridine methyltransferase	Armistead *AJHG* 84:728, 2009	AR	Bowen-conradi syndrome (211180)	# 611531
ERCC8	Excision repair cross-complementing, group 8	Bertola *J Hum Genet* 51:701, 2006	AR	Cockayne syndrome, type a (216400)	# 609412

ESCO2	Establishment of sister chromatid Cohesion N-Acetyltransferase 2	Vega *J Med Genet* 47:30, 2010	AR	Roberts syndrome (268300)	# 609353
ETFA	Electron transfer flavoprotein alpha subunit	Lehnert *Eur J Pediatr* 139:56, 1982	AR	Glutaric acidemia IIA (231680)	# 608053
ETFB	Electron transfer flavoprotein beta subunit	Lehnert *Eur J Pediatr* 139:56, 1982	AR	Glutaric acidemia IIB (231680)	# 130410
ETFDH	Electron transfer flavoprotein dehydrogenase	Lehnert *Eur J Pediatr* 139:56, 1982	AR	Glutaric acidemia IIC (231680)	# 231675
FANCA	Fanconi anemia complementation group A	Joenje and Patel *Nat Rev Genet* 2:466, 2001	AR	Fanconi anemia, complementation group a (227650)	* 607139
FANCB	Fanconi anemia complementation group B	McCauley *AJMG* 155A:2370, 2011	AR	Fanconi anemia, complementation group B (300514)	# 300514
FANCD2	Fanconi anemia complementation group D2	Kalb *AJHG* 80:895, 2007	AR	Fanconi anemia, complementation group D2 (227646)	# 613984
FANCE	Fanconi anemia complementation group E	Wegner *Clin Genet* 50:479, 1996	AR	Fanconi anemia, complementation group E (600901)	# 613976
FANCI	Fanconi anemia complementation group I	Savage *AJMG* 170A:386, 2015	AR	Fanconi anemia, complementation group I (609053)	# 611360
FANCL	Fanconi anemia complementation Group L	Vetro *Hum Mutat* 36:562, 2015	AR	Fanconi anemia, complementation group L (614083)	# 608111
FAT4	FAT atypical cadherin 4	Alders *Hum Genet* 133:1161, 2014	AR	Van maldergem syndrome 2 (615546)	# 612411
FOXP1	Forkhead box P1	Bekheirnia *Genet Med* 19:412, 2017	AR	Mental retardation with language impairment and with or without autistic features (613670)	* 605515
HES7	Hes family BHLH transcription factor 7	Sparrow *Hum Mol Genet* 17:3761, 2008	AR	Spondylocostal dysostosis 4, autosomal recessive (613686)	* 608059
HYLS1	HYLS1, centriolar and ciliogenesis associated	Paetau *J Neuropathol* Exp *Neurol* 67:750, 2008	AR	Hydrolethalus syndrome (236680)	# 610693
ICK	Intertinal cell kinase	Lahiry *AJHG* 84:822, 2009	AR	Endocrine-cerebroosteodysplasia (612651)	* 612325

Continued

TABLE 1.2 Monogenic Causes of Human Syndromic CAKUT, if Mutated. (Sorted Alphabetically by Mode of Inheritance and Includes Syndromes That are Presumed Monogenic but Causative Gene Not yet Identified)—cont'd

Gene	Protein	Reference	Mode of Inheritance	Phenotype MIM Number	Gene/ locus MIM Number
IFT46	Intraflagellar transport 46	Lee *Dev Biol* 400:248, 2015	AR	Short-rib thoracic dysplasia 16 with or without polydactyly (617102)	# 617094
IFT74	Intraflagellar transport 74	Cevik *PLoS GeneT* 9:e1003977, 2013	AR	Bardet-biedl syndrome 20 (617119)	# 608040
ITGA3	Integrin subunit alpha 3	Yalcin *Hum Mol Genet* 24:3679, 2015	AR	Interstitial lung disease, nephrotic syndrome, and epidermolysis bullosa, congenital (614748)	# 605025
JAM3	Junctional adhesion molecule 3	Mochida *AJHG* 10; 87:882, 2010	AR	Hemorrhagic destruction of the brain, subependymal calcification, and cataracts (613730)	# 606871
LRIG2	Leucine-rich repeats and immunoglobulin like domains containing protein 2	Stuart *AJHG* 92:259, 2013	AR	Urofacial syndrome 2 (615112)	# 608869
LRP2	LDL receptor related protein 2	Kantarci *Nat Genet* 39:957, 2007	AR	Donnai-barrow syndrome (222448)	# 600073
LRP4	LDL receptor related protein 4	Li Am *J Hum Genet* 86:696, 2010	AR	Cenani-lenz syndactyly syndrome (212780)	# 604270
MESP2	Mesoderm posterior BHLH transcription factor 2	George-Abraham *AJMG* a 158A:1971, 2012	AR	Spondylocostal dysostosis 2, autosomal recessive (608681)	* 605195
MKKS	McKusick-Kaufman syndrome	Stone *Nature Genet* 25:79, 2000	AR	McKusick–Kaufman syndrome (236700) Bardet-biedl syndrome 6 (605231)	* 604896
MKS3	Meckel syndrome type 3 protein	Baala *AJHG* 80:186, 2007	AR	Meckel syndrome 3 (607361)	# 609884
PEX5	Peroxisomal biogenesis factor 5	Sundaram *Nat Clin Pract Gastroenterol Hepatol* 5:456, 2008	AR	Peroxisome biogenesis disorder 2A (Zellweger) (214110)	# 600414
PMM2	Phosphomannomutase 2	Horslen *Arch Dis Child* 66:1027, 1991	AR	Congenital disorder of glycosylation, type IA (212065)	# 610785

POC1A	POC1 centriolar protein	Shaheen *AJHG* 91:330, 2012	AR	Short stature, onychodysplasia, facial dysmorphism, and hypotrichosis (614813)	# 614783
PROK2	Prokineticin 2	Madan *Mol Genet Metab Rep* 12:57, 2017	AR	Hypogonadotropic hypogonadism 4 with or without anosmia (610628)	# 607002
RECQL4	RecQ like helicase 4	Siitonen *Eur J Hum Genet* 17:151, 2009	AR	Baller-Gerold syndrome (218600)	# 603780
ROR2	Receptor tyrosine kinase like orphan receptor 2	Wiens *Clin Genet* 37:481, 1990	AR	Robinow syndrome (268310)	# 602337
RPS19	Ribosomal protein S19	Hoefele *Pediatr Nephrol* 25:1255, 2010	AR	Diamond-blackfan anemia 1 (105650)	* 603474
SCARF2	Scavenger receptor class F member 2	Anastasio *AJHG* 87:553, 2010	AR	Van den ende-gupta syndrome (600920)	# 613619
STRA6	Stimulated by retinoic acid 6	Golzio *AJHG* 80:1179, 2007	AR	Microphthalmia, syndromic 9 microphthalmia, isolated, with coloboma 8 (601186)	# 610745
TMCO1	Transmembrane and coiled-coil domains 1	Xin *PNAS* 107:258, 2010	AR	Craniofacial dysmorphism, skeletal anomalies, and mental retardation syndrome (213980)	# 614123
UBR1	Ubiquitin protein ligase E3 component N-Recognin 1	Vanlieferinghen *Genet Couns* 14:105, 2003	AR	Johanson-blizzard syndrome (604292)	# 603273
PEX1	Peroxisomal biogenesis factor 1	Crane *Hum Mutat* 26:167, 2005	AR	Peroxisome biogenesis disorder 1A (Zellweger) (214100)	# 602136
PIGL	Phosphatidylinositol glycan anchor biosynthesis Class L	Schnur *AJMG* 72:24, 1997	AR	CHIME syndrome (280000)	# 605947
PIGO	Phosphatidylinositol glycan anchor biosynthesis class O	Krawitz *AJHG* 91:146, 2012	AR	Hyperphosphatasia with mental retardation syndrome 2 (614749)	# 614730
PIGN	Phosphatidylinositol glycan anchor biosynthesis class N	Ohba *Neurogenetics* 15:85, 2014	AR	Multiple congenital anomalies-hypotonia-seizures syndrome 1 (614080)	# 606097
PIGT	Phosphatidylinositol glycan anchor biosynthesis class T	Nakashima *Neurogenetics* 15:193, 2014	AR	Multiple congenital anomalies-hypotonia-seizures syndrome 3 (615398)	# 601272

Continued

TABLE 1.2 **Monogenic Causes of Human Syndromic CAKUT, if Mutated. (Sorted Alphabetically by Mode of Inheritance and Includes Syndromes That are Presumed Monogenic but Causative Gene Not yet Identified)—cont'd**

Gene	Protein	Reference	Mode of Inheritance	Phenotype MIM Number	Gene/ locus MIM Number
PIGV	Phosphatidylinositol glycan anchor biosynthesis class V	Horn *Eur J Hum Genet* 22:762, 2014	AR	Hyperphosphatasia with mental retardation syndrome 1 (239300)	* 610274
PIGY	Phosphatidylinositol glycan anchor biosynthesis class Y	Ilkovski *Hum Mol Genet* 24:6146, 2015	AR	Hyperphosphatasia with mental retardation syndrome 6 (616809)	# 610662
PTF1A	Pancreas specific transcription factor, 1a	Gurung *Mol Med Rep* 12:1579, 2015	AR	Pancreatic agenesis 2 (615935) Pancreatic and cerebellar agenesis (609069)	* 607194
WFS1	Wolframin ER transmembrane glycoprotein	Salih *Acta Paediatr Scand* 80:567, 1991	AR	Wolfram syndrome 1 (222300)	# 606201
WNT3	Wnt family member 3	Niemann *AJHG* 74:558, 2004	AR	Tetra-amelia syndrome 1 (273395)	# 165330
ZMPSTE24	Zinc metallopeptidase STE24	Chen *AJMG A* 149A:1550, 2009	AR	Restrictive dermopathy, lethal (275210)	# 606480
ACTB	Actin beta	Rivière *Nat Genet* 44:440, 2012	AD	Baraitser-Winter syndrome 1 (243310)	# 102630
ACTG1	Actin gamma 1	Rivière *Nat Genet* 44:440, 2012	AD	Baraitser-Winter syndrome 1 (243310)	# 614583
AIFM3	Apoptosis inducing factor, mitochondria associated 3	Lopez-rivera *NEJM* 376:742, 2017	AD	Di george syndrome	* 617298
ARID1B	AT-rich interaction domain 1B	Levy *J Med Genet* 28, 1991	AD	Coffin-Siris syndrome 1 (135900)	# 614556
ATXN10	Ataxin 10	Matsuura *Nat Genet* 26:191, 2000	AD	Spinocerebellar ataxia 10 (603516)	# 611150
BICC1	BicC family RNA binding protein 1	Kraus *Hum Mutat* 33:86, 2012	AD	Renal cystic/dysplasia (601331)	# 614295
BMP7	Bone morphogenetic protein 7	Hwang *Kidney Int* 85:1429, 2014	AD	CAKUT	* 112267
BRAF	B-raf proto-oncogene, serine/Threonine kinase	Sarkozy *Hum Mutat* 30:695, 2009	AD	Cardiofaciocutaneous syndrome	# 115150

CDC5L	Cell division cycle 5 like	Groenen *Genomics* 49:218, 1998	AD	Multicystic renal dysplasia	* 602868
CREBBP	CREB binding protein	Kanjilal *J Med Genet* 29:669, 1992 Bartsch *AJMG*, 152A: 2254, 2010	AD	Rubinstein–Taybi syndrome 1 (180849) Menke-Hennekam syndrome 1 (618332)	* 600140
EEC1	Ectrodactyly, ectodermal dysplasia and cleft lip/palate syndrome 1	Fukushima *Clin Genet* 44:50 only, 1993	AD	Ectrodactyly-ectodermal dysplasia-clefting (129900)	% 129900
EEC3	Ectrodactyly, ectodermal dysplasia and cleft lip/Palate syndrome 3	Celli *Cell* 99:143, 1999	AD	Ectrodactyly-ectodermal dysplasia-clefting (604292)	% 129900
DACT1	Disheveled binding antagonist of beta catenin 1	Webb *Hum Mutat* 38:373, 2017	AD	Townes-brocks syndrome 2 (617466)	# 607861
EP300	E1A binding protein P300	Roelfsema *AJHG* 76:572, 2005	AD	Rubinstein-Taybi syndrome 2 (613684)	# 602700
ESRRG	Estrogen related receptor gamma	Harewood *PLoS One* 5:e12375, 2010	AD	Bilateral renal agenesis/hypoplasia/dysplasia (BRAHD)	* 602969
FBN1	Fibrillin 1	Tokhmafshan *Pediatr Nephrol* 32:565, 2017	AD	Marfan syndrome (154700)	# 134797
FGFR1	Fibroblast growth factor receptor 1	Farrow *AJHG* 140A:537, 2006	11	Hartsfield syndrome (615465) Hypogonadotropic hypogonadism 2 with or without anosmia (147950) Jackson-Weiss syndrome (123150) Osteoglophonic dysplasia (166250) Pfeiffer syndrome (101600) Trigonocephaly 1 (190440)	* 136350
FGFR2	Fibroblast growth factor receptor 2	LeHeup *Eur J Pediatr* 154:130, 1995 Hains *Pediatr Res* 64:592, 2008	AD	Lacrimoauriculodentodigital (levy—Hollister, LADD) (149730)	* 176943
FGFR3	Fibroblast growth factor receptor 3	Rohmann *Nat Genet* 38:495, 2006	AD	Lacrimoauriculodentodigital (levy—Hollister, LADD) (149730)	* 134934

Continued

TABLE 1.2 Monogenic Causes of Human Syndromic CAKUT, if Mutated. (Sorted Alphabetically by Mode of Inheritance and Includes Syndromes That are Presumed Monogenic but Causative Gene Not yet Identified)—cont'd

Gene	Protein	Reference	Mode of Inheritance	Phenotype MIM Number	Gene/ locus MIM Number
FGF10	Fibroblast growth factor 10	Milunsky *Clin Genet* 69:349, 2006; Bamforth *AJMG* 43:932, 1992	AD	Lacrimoauriculodentodigital (levy—Hollister, LADD) (149730)	* 602115
FGF8	Fibroblast growth factor 8	Falardeau *J Clin Invest* 118:2822, 2008	AD	Hypogonadotropic hypogonadism 6 with or without anosmia (612702)	* 600483
FGFRL2	Forkhead box C1	LeHeup *Eur J Pediatr* 154:130, 1995	AD	Antley-bixler syndrome without genital anomalies or disordered steroidogenesis (207410)	# 176943
FMN1	Formin 1	Woychik *Nature* 346:850, 1990	AD	Renal aplasia and limb deformities	* 136535
FOXF1	Forkhead box F1	Hilger *Hum Mutat* 36:1150, 2015	AD	Alveolar capillary dysplasia with misalignment of pulmonary veins (265380)	# 601089
GDF3	Growth differentiation factor 3	Karaca *AJMG A* 167A:2795, 2015	AD	Klippel-Feil syndrome 3 (613702)	# 606552
GDNF	Glial cell line derived neurotrophic factor	Pini prato *Medicine (Baltimore)* 88:83, 2009	AD	Susceptibility hirschsprung disease (613711)	# 600837
GFRA1	GDNF family receptor alpha 1	Shefelbine *Hum Genet* 102:474, 1998	AD	Complex vesicoureteral reflux	* 601496
GLI2	GLI family Zinc finger 2	Carmichael *J Urol* 190:1884, 2013	AD	Culler-Jones syndrome (615849) holoprosencephaly 9 (610829)	* 615230
HOXA13	Homeobox A13	Halal *AJMG* 30:793, 1998	AD	Hand-foot-genital syndrome (142959)	# 140000
HOXD13	Homeobox D13	Garcia-Barceló *AJMG A*146A:3181, 2008	AD	Brachydactyly-syndactyly syndrome (610713) Brachydactyly, type D (113200) Brachydactyly, type D (113300) Syndactyly, type V (186300) Synpolydactyly 1 (186000)	* 142989
JAG1	Jagged 1	Kamath *Nat Rev Nephrol* 9:409, 2013	AD	Alagille syndrome 1 (118450)	# 601920
KAT6B	Lysine acetyltransferase 6B	Campeau *AJMG* 90:282, 2012	AD	Genitopatellar syndrome (606170)	# 605880

KCTD1	Potassium channel tetramerization domain containing 1	Marneros *AJMG* 92:621, 2013	AD	Scalp-ear-nipple syndrome (181270)	# 613420
KCNH2	Potassium voltage-gated channel subfamily H member 2	Caselli *AJMG* 146A:1195, 2008	AD	Scalp-ear-nipple syndrome	* 152427
KRAS	KRAS proto-oncogene, GTPase	Schubbert *Nat Gene* 38:331, 2006	AD	Noonan syndrome 3 (609942)	# 190070
LMX1B	LIM homeobox transcription factor 1 beta	Dreyer *Nat Genet* 19:47, 1998	AD	Nail-patella syndrome (161200)	# 602575
LPP	LIM domain containing preferred translocation partner in lipoma	Hernández-García *AJMG A* 158A:1785, 2012	AD	Esophageal atresia, tracheoesophageal fistula, and VACTERL association	* 600700
MAP2K1	Mitogen-activated protein kinase kinase 1	Schulz *Clin Genet* 73:62, 2007	AD	Cardiofaciocutaneous syndrome 3 (615279)	# 176872
MAP2K2	Mitogen-activated protein kinase kinase 2	Schulz *Clin Genet* 73:62, 2007	AD	Cardiofaciocutaneous syndrome 4 (615280)	# 601263
MLL2/ KMT2D	Myeloid/lymphoid or mixed-lineage leukemia protein 2	Banka *Eur J Hum Genet* 20:381, 2012	AD	Kabuki syndrome 1 (147920)	# 602113
MYCN	Feingold syndrome	Marcelis *Hum Mut* 29:1125, 2006	AD	Feingold syndrome 1 (164280)	# 164840
NOTCH2	Notch 2	Kamath *Nat Rev Nephrol* 9:409, 2013	AD	Alagille syndrome 2 (610205) Hajdu-cheney syndrome (102500)	# 600275
PKD1	Polycystin 1, transient receptor potential channel interacting	Rossetti *JASN* 18:2143, 2007	AD	Polycystic kidney disease 1 (173900)	# 601313
PKD2	Polycystin 2, transient receptor potential cation channel	Rossetti *JASN* 18:2143, 2007	AD	Polycystic kidney disease 2 (613095)	# 173910
PROKR2	Prokineticin receptor 2	Sarfati *Front Horm Res* 39:121, 2010	AD	Hypogonadotropic hypogonadism 3 with or without anosmia (244200)	# 607123
PTPN11	Protein tyrosine phosphatase, non-receptor type 11	Bertola *AJMG* 130A:378, 2004	AD	Leopard syndrome 1 (151100)	# 176876
RAF1	Raf-1 proto-oncogene, serine/Threonine kinase	Razzaque *Nat Genet* 39:1013, 2007	AD	Noonan syndrome 5 (611553)	# 164760

Continued

TABLE 1.2 Monogenic Causes of Human Syndromic CAKUT, if Mutated. (Sorted Alphabetically by Mode of Inheritance and Includes Syndromes That are Presumed Monogenic but Causative Gene Not yet Identified)—cont'd

Gene	Protein	Reference	Mode of Inheritance	Phenotype MIM Number	Gene/ locus MIM Number
RAI1	Retinoic acid induced 1	Vilboux *PLoS One* 6:e22861, 2011	AD	Smith-Magenis syndrome (182290)	# 607642
SALL1	Spalt like transcription factor 1	Kohlhase *Nat Genet* 18:81, 1998	AD	Townes–Brocks syndrome 1/ Townes-Brocks branchiootorenal-like syndrome (107480)	# 602218
SALL4	Spalt like transcription factor 4	Kohlhase *GeneReviews Book Section*, 1993	AD	Duane-radial ray syndrome (607323)	# 607323
SEMA3A	Semaphorin 3A	Young *Hum Reprod* 27:1460, 2012	AD	Hypogonadotropic hypogonadism 16 with or without anosmia (614897)	# 603961
SEMA3E	Semaphorin 3E	Lalani *J Med Genet* 41:e94, 2004	AD	Charge syndrome (214800)	# 608166
SETBP1	SET binding protein 1	Schinzel *AJMG* 1:361, 1978	AD	Schinzel-Giedion midface retraction syndrome (269150)	# 611060
SHH	Sonic hedgehog	Lurie *AJMG* 35:286, 1990	AD	Holoprosencephaly 3 (142945)	# 600725
SF3B4	Splicing factor 3B Sub-unit 4	Bernier *AJMG* 90:925, 2012	AD	Acrofacial dysostosis 1, nager type (154400)	# 605593
SNAP29	Synaptosome associated protein 29	Lopez-rivera *NEJM* 376:742, 2017	AD	Di George syndrome (609523)	* 604202
SOS1	SOS Ras/Rac guanine nucleotide exchange factor 1	Ferrero *Eur J Med Genet* 51:566, 2008	AD	Noonan syndrome 4 (610733)	# 182530
SOX9	SRY-box 9	Airik *Hum Mol Genet* 19:4918, 2010	AD	Campomelic dysplasia (114290)	# 608160
SRCAP	Snf2 related CREBBP activator protein	Hood *AJHG* 90:308, 2012	AD	Floating-Harbor syndrome (136140)	# 611421
TBX1	T-box 1	Kujat *AJMG A* 140:1601, 2006	AD	Di George syndrome (188400)	# 602054
TBX3	T-box 3	Meneghini *Eur J Med Genet* 49:151, 2006	AD	Ulnar-mammary syndrome (181450)	# 601621
TFAP2A	Transcription factor AP-2 alpha	Milunsky *AJHG* 82:1171, 2008	AD	Branchiooculofacial syndrome (113620)	# 107580
TP63	Tumor protein p63	Celli *Cell* 99:143, 1999 Bougeard *EJHG* 11:700, 2003	AD	Ectrodactyly, ectodermal dysplasia, and cleft lip/palate syndrome 3 (604292)	# 603273

TRPS1	Zinc finger transcription factor; trichorhinophalangeal syndrome	Tasic *Ren Fail* 36:619, 2014	AD	Trichorhinophalangeal syndrome (190350) (190351)	# 604386
TSC1	Tuberous sclerosis 1	Curatolo *Lancet* 372:657, 2008	AD	Tuberous sclerosis-1 (191100)	# 605284
TSC2	Tuberous sclerosis 2	Kumar *Hum Mol Genet* 4:1471, 1995	AD	Tuberous sclerosis-2 (613254)	# 191092
TWIST2	Twist family BHLH transcription factor 2	Stevens *AJMG* 107:30, 2002	AD	Ablepharon-macrostomia syndrome (200110)	# 607556
WNT5A	Wnt family member 5A	Roifman *Clin Genet* 87:34, 2015; person *Dev Dyn* 239:327, 2010	AD	Robinow syndrome (180700)	# 164975
ZMYM2	MYM type 2 zinc finger	Connaughton *AJHG* 107:4, 2020	AD	Syndromic CAKUT	NA
GDF6	Growth differentiation factor 6	Tassabehji *Hum Mutat* 29:1017, 2008	AD/AR	Klippel-Feil syndrome 1, autosomal dominant (118100)	* 601147
GLI3	GLI family Zinc finger 3	Cain *PLoS One* 4:e7313, 2009	AD/AR	Pallister-Hall syndrome (146510) Greig cephalopolysyndactyly syndrome (175700)	* 165240
LMNA	Lamin A/C	Klupa *Endocrine* 36:518, 2009	AR/AD	Mandibuloacral dysplasia (248370) Lipodystrophy, familial partial, type 2 (151660) Malouf syndrome (212111)	* 150330
PCSK5	Proprotein convertase subtilisin and kexin type 5	Nakamura *BMC Res Notes* 8:228, 2015	AD/AR	Vacterl association	* 600488
PTEN	Phosphatase and tensin homolog	Reardon *J Med Genet* 38:820, 2001	AD/AR	Cowden syndrome 1 Lhermitte-Duclos syndrome (158350)	* 601728
RPS24	Ribosomal protein S24	Yetgin *Turk J Pediatr* 36:239, 1994	AD/AR	Diamond-blackfan anemia 3 (610629)	* 602412
TBXT	T-box transcription factor T	Postma *J Med Genet* 51:90–97, 2014	AR/AD	Sacral agenesis with vertebral anomalies (615709) Neural tube defects, susceptibility to (182940)	* 601397
VANGL1	Van gogh-like protein 1 VANGL planar cell polarity protein 1	Bartsch *Mol Syndromol* 3:76, 2012	AD/AR	Sacral defect with anterior meningocele (sirenomelia) (600145) Neural tube defects, susceptibility to (182940) Caudal regression syndrome (600145)	# 610132

Continued

TABLE 1.2 Monogenic Causes of Human Syndromic CAKUT, if Mutated. (Sorted Alphabetically by Mode of Inheritance and Includes Syndromes That are Presumed Monogenic but Causative Gene Not yet Identified)—cont'd

Gene	Protein	Reference	Mode of Inheritance	Phenotype MIM Number	Gene/ locus MIM Number
VANGL2	Van gogh-like protein 2	Kibar *Clin Genet* 80:76, 2011	AD	Neural tube defects (182940)	# 600533
AXIN1	Axin 1	Oates *AJHG* 79:155, 2006	De novo	Caudal duplication anomaly (607864)	# 603816
CDKN1C	Cyclin dependent kinase inhibitor 1C	Mussa *Pediatr Nephrol* 27:397, 2012	De novo	Beckwith-Wiedemann syndrome (130650)	# 600856
CHD7	Chromodomain helicase DNA binding protein 7	Janssen *Hum Mutat* 33:1149 2012	De novo	Charge syndrome (214800)	# 214800
H19	H19, imprinted maternally expressed transcript (non-protein coding)	Hur *PNAS* 113:10,938, 2016	De novo	Silver–Russell syndrome phenotypes	* 103280
KCNQ1OT1	KCNQ1 opposite strand and antisense transcript 1 (non-protein coding)	Chiesa *Hum Mol Genet* 21:10, 2012	De novo	Beckwith-Wiedemann syndrome (130650)	# 604115
NIPBL	NIPBL, cohesin loading factor	Rohatgi *AJMG* 152A:1641, 2010	De novo	Cornelia de Lange syndrome 1 (122470)	# 608667
AMER1	APC membrane recruitment protein 1	Pellegrino *AJMG* 16:159, 1997	XL	Osteopathia striata with cranial sclerosis (300373)	# 300647
ATP7A	ATPase copper transporting alpha	Vulpe *Nat Genet* 3:7, 1993	XL	Menkes disease (309400)	# 300011
BCOR	BCL6 corepressor	Ng *Nat Genet* 36:411, 2004	XL	Microphthalmia, syndromic 2 (300166)	# 300485
DLG3	Disc large, drosphilia, homologue of 3	Philips *Orphanet J Rare Dis* 9:49, 2014	XL	Mental retardation, X-linked 90 (300850)	# 300189
FAM58A	Family with sequence similarity 58 member A	Green *J Med Genet* 33:594, 1996; Unger *Nat Genet* 40:287, 2008	XL	Star syndrome (300707)	# 300708
FLNA	Filamin A	Robertson *AJMG* a 140:1726, 2006	XL	Frontometaphyseal dysplasia 1 (305620) Heterotopia, periventricular, 1 (300049)	* 300017

GPC3	Glypican 3	Cottereau *AJMG* C *Semin Med Genet* 163:92, 2013	XL	Simpson-golabi-behmel syndrome, type 1 (312870)	# 300037
MID1	Midline 1	Preiksaitiene *Clin Dysmorphol* 24:7, 2015	XL	Opitz GBBB syndrome, type I (300000)	# 300552
NSDHL	NAD(P) dependent steroid dehydrogenase-like	König *J Am Acad Dermatol* 46:594, 2002	XL	Child syndrome (308050)	# 300275
PIGA	Phosphatidylinositol glycan anchor biosynthesis class A	Johnston *AJHG* 90:295, 2012	XL	Multiple congenital anomalies-hypotonia-seizures syndrome 2 (300868)	# 311770
PORCN	Porcupine O-acyltransferase	Suskan *Pediatr Dermatol* 7:283, 1990	XL	Focal dermal hypoplasia (305600)	# 300651
SMC1A	Structural maintenance of chromosomes 1A	Deardorff *GeneReviews Book Section Seattle(WA)*, 1993	XL	Cornelia de Lange syndrome 2 (300590)	# 300040
UPF3B	UPF3B, regulator of nonsense mediated MRNA decay	Lynch *Eur J Med Genet* 55:476, 2012	XL	Mental retardation, X-linked, syndromic 14 (300676)	* 300298
ZIC3	Zic family member 3	Chung *AJMG* 155:1123, 2011	XL	VACTERL association (314390)	# 300265
OSR1	Odd-Skipped related transcription factor 1	Zhang *Hum Mol Genet* 20:4167, 2011	Unknown	CAKUT	* 608891
SH2B1	SH2B adaptor protein 1	Sampson *AJMG* 152:2618, 2010	Unknown	CAKUT	* 608937
Not identified			?AD ?AR	Acrorenal syndrome (102520) (201310)	Not identified
Not identified			?AR	Acrorenal-mandibular syndrome (200980)	Not identified
Not identified			?AR	AREDYLD (acrorenal field defect, ectodermal dysplasia, and lipoatrophic diabetes; hypotrichosis, absent dentition, aplasia of the breast) (207780)	Not identified
Not identified			?AR	DK phocomelia (223340)	Not identified

Continued

TABLE 1.2 Monogenic Causes of Human Syndromic CAKUT, if Mutated. (Sorted Alphabetically by Mode of Inheritance and Includes Syndromes That are Presumed Monogenic but Causative Gene Not yet Identified)—cont'd

Gene	Protein	Reference	Mode of Inheritance	Phenotype MIM Number	Gene/locus MIM Number
Not identified			AD	Hemifacialmicrosomia Oculoauriculovertebral spectrum Goldenhar syndrome (164210)	Not identified
Not identified			AR	Hirschsprung disease with polydactyly, renal agenesis and deafness (164210)	Not identified
Not identified			Sporadic	MURCS—MUllerian duct aplasia, renal agenesis/ectopia, cervical somite dysplasia % 601076	Not identified
Not identified			?AR, dominance not excluded	Orocraniodigital syndrome Juberg—Hayward syndrome (216100)	Not identified
Not identified			AD	Radial—renal syndrome (179280)	Not identified

AD, autosomal dominant; *AR*, autosomal recessive; *NA*, not available; *OMIM*, online mendelian inheritance in man; *XL*, X-linked; *#*, phenotype MIM number; *Unknown*, mode of inheritance not clearly characterized; * gene/locus MIM number if not phenotype MIM number available.Adapted from Connaughton DM, Kennedy C, Shril S, Mann N, Murray SL, Williams PA, et al. Monogenic causes of chronic kidney disease in adults. Kidney Int 2019;95(4):914–928.

TABLE 1.3 List and Frequency of Extra-renal Structural Defects in Patients With CAKUT

Structural Anomaly/ Pattern	Percentage with Urinary Tract Anomalies	Most Common Urinary Tract Defects (in Relative Order of Frequency)
Absent gall bladder	32	Cystic dysplasia, renal agenesis, horseshoe kidneys
Agenesis of the corpus callosum	45–55	Reflux, ureterocele, unilateral renal agenesis, crossed fused renal ectopia, bladder diverticulae
Anencephaly	5–16	Hydronephrosis, horseshoe kidneys, polycystic kidneys, renal agenesis, renal hypoplasia, urethral atresia
Anorectal malformation	26–58[a]	Hydronephrosis, unilateral renal agenesis, cystic dysplasia, reflux, cystic dysplasia, renal ectopia, cloacal exstrophy
Biliary atresia	3	Double ureter, hydronephrosis, renal cysts
Caudal dysplasia	40	Renal agenesis, hypoplasia, cystic dysplasia; horseshoe kidneys, crossed renal ectopia, urachal anomalies
Charge	42	Unilateral renal agenesis, hydronephrosis, renal hypoplasia
Diaphragmatic hernia	15–18	Renal agenesis, cystic dysplasia, hydronephrosis, ureteropelvic obstruction
Esophageal atresia and tracheoesophageal fistula	33	Unilateral renal agenesis, horseshoe kidneys, reflux
Gastroschisis	15	Unilateral renal agenesis, horseshoe kidneys
Heart defect	5–39[a]	Duplex collecting system, unilateral renal agenesis, renal ectopia
Lateral body wall defect	50–65	Renal agenesis, urethral atresia, hydronephrosis
Limb reduction defects	9	Renal agenesis, hydronephrosis, cystic dysplasia, horseshoe kidney
MURCS association	28–80	Renal agenesis, renal ectopia
Myelomeningocele	9	Renal agenesis, horseshoe kidney
Omphalocele	11–47	Cloacal exstrophy, horseshoe kidneys, patent urachus
Oral clefts	4	Renal agenesis, horseshoe kidney
Penoscrotal transposition	90	Renal agenesis, cystic dysplasia, ectopia, horseshoe kidneys
Persistent cloaca	83	Cloacal and bladder anomalies
Pulmonary hypoplasia	18–21	Cystic dysplasia, renal agenesis, horseshoe kidney, polycystic kidney, urethral atresia
Single umbilical artery (isolated)	26	Dilated renal pelvis, duplicated renal pelvis, reflux, hydronephrosis, horseshoe kidneys, unilateral renal agenesis
Sirenomelia	100	Renal agenesis, cystic dysplasia, urethral atresia
Supernumerary nipples	4	No specific pattern
Tracheal agenesis	38	Renal agenesis, cystic dysplasia, horseshoe kidney
VACTERL association	82–87	Reflux, unilateral renal agenesis, ureteropelvic junction obstruction, crossed fused ectopia
Vertebral defects	27–46	Unilateral renal agenesis, duplicated ureter, renal ectopia, horseshoe kidney

[a]Higher numbers seen in autopsy series and/or those with $\geq$3 congenital anomalies.**Source:** From Stevenson RE. Human malformations and related anomalies. 2nd ed. Oxford, New York: Oxford University Press; 2006. p. 116. Adapted from Chapter 62 Congenital Anomalies of the Kidney and Urinary Tract fourth Edition: Emery and Rimoin's Essential Medical Genetics.

TABLE 1.4 **Common Chromosomal Disorders Associated With CAKUT**

Chromosome Disorder	Urinary Tract Anomaly	Frequency
2q terminal deletion	Wilms tumor, horseshoe kidney, dysplastic kidney, renal hypoplasia, ureteral stenosis	11%
4p	Renal agenesis or hypoplasia, vesicoureteral reflux, hydronephrosis	33%
4q partial duplication	Horseshoe kidney, renal hypoplasia, renal duplication, ectasia of distal tubule	Frequent
5p	Horseshoe kidney, renal agenesis, renal duplication, ectasia of distal tubules	Occasional
6q partial duplication	Unilateral renal agenesis, cystic dysplasia	Uncertain[a]
7 trisomy	Cystic dysplasia, enlarged kidneys	Frequent
7q partial deletion (Williams–Beuren)	Renal aplasia/hypoplasia, duplicated kidney, bladder diverticula	18%
8 trisomy, 8 trisomy mosaicism	Cystic dysplasia, enlarged kidneys, small cortical cysts, hydronephrosis, duplication of kidneys/ureters/pelvis	Frequent
9p partial duplication	Horseshoe kidney, hydronephrosis	Frequent
9 trisomy	Hydronephrosis, cystic dysplasia, duplication kidneys and ureters	Frequent
10p partial duplication	Unilateral agenesis, cystic kidney, renal dysplasia	Uncertain
10q partial duplication	Hypoplastic kidney, hydronephrosis	Frequent
11p13 deletion	Wilms tumor	Frequent
13q	Hydronephrosis, vesicoureteral junction obstruction	Uncommon
13 ring	Renal hypoplasia and ectopy, duplication of kidney and ureter, unilateral renal agenesis, polycystic kidney	Frequent
13 trisomy	Hydronephrosis, cystic dysplasia, micropolycystic dysplasia, hydroureter, horseshoe kidney, ureteral duplication, duplication of renal pelvis, small cortical cysts	60%–70%
17p partial deletion (Smith–Magenis)	Variable; no characteristic structural abnormality	19%
17p13.3 deletion (Miller–Dieker)	Cystic or pelvic kidney	Uncertain
18q	Horseshoe kidney, unilateral agenesis, hydronephrosis	40%
18 ring	Hydronephrosis, tubular dilation	20%
18 trisomy	Horseshoe kidney, ectopia, ureteral duplication, cortical cysts, exstrophy of the cloaca, hydronephrosis	70%
21q	Unilateral renal agenesis, abnormal kidney shape, dilated calyces	Uncertain[a]
21 trisomy	Renal agenesis, hypoplasia, horseshoe kidney, posterior urethral valves, hydronephrosis, ureteropelvic junction obstruction, vesicoureteral reflux, fetal hydronephrosis	3%–7%
22q11 deletion	Renal agenesis, dysplasia, multicystic kidneys	36%
22 partial trisomy	Renal agenesis, horseshoe kidney, hydronephrosis	Frequent
22 partial tetrasomy (cat eye)	Unilateral or bilateral renal agenesis, hydronephrosis, supernumerary kidneys	Occasional
45,X (plus other turner karyotype abnormalities)	Horseshoe kidney, duplication collecting system, abnormal rotation, ureteropelvic junction obstruction, cystic malformation of collecting tubules; cystic, double, and ectopic kidneys, intrarenal vascular changes	60%–80%

Continued

TABLE 1.4 Common Chromosomal Disorders Associated With CAKUT—cont'd

Chromosome Disorder	Urinary Tract Anomaly	Frequency
XXXXY	Hydronephrosis	10%
XXXXX	Renal hypoplasia, dysplasia	Uncertain[a]
XXY (klinefelter)	Renal cysts, hydronephrosis	Uncertain[a]
Triploidy	Hydronephrosis, renal cysts, polycystic kidneys, cystic dysplasia	Frequent

[a]Reported abnormality may not be in excess over background risk for all urinary tract anomalies.**Source:** Chapter 62 Congenital Anomalies of the Kidney and Urinary Tract: Emery and Rimoin's Essential Medical Genetics.

patterning and elongation, which ultimately gives rise to the structures of the nephron (glomerulus, the proximal tubule, and the distal tubule) (Fig. 1.2).

1.8 PATHOGENESIS OF CAKUT

The majority of subtypes of CAKUT are thought to arise following disruption in any of the stages of the normal nephrogenesis outlined above [15,53,54].

Renal agenesis is hypothesized to result from failure of the UB to make contact with the MM either due to complete absence of the nephric duct or MM, failure of induction of the UB, failure of the UB to connect with the MM, or lack of communication between the UB and MM. This process is tightly regulated by the GDNF-RET signaling pathway and disturbances in this process are hypothesized to result in both renal agenesis and renal hypodysplasia (See below for further details) [2]. Renal hypodysplasia is hypothesized to also occur due to abnormal interaction between the UB and MM resulting in defective branching morphogenesis and nephron differentiation. This is thought to arise in part due to imbalances in cell proliferation in the epithelium and uncontrolled apoptosis in the mesenchyme. An increase in unregulated apoptosis can lead to dysplasia and subsequent involution of the dysplastic kidney. This dysregulation in cell death is thought to be partially mediated through abnormalities expression of Pax2 and Pax8 resulting in downregulation of Wnt11, which is a target of RET and GDNF signaling [55]. Indeed, in the North American CKiD cohort, pathogenic variants in *PAX2* and *HNF1B* were found in a significant proportion of patients with RHD [56]. PAX2, also known as paired-box gene 2, is a transcription factor implicated in renal coloboma syndrome (OMIM # 120330) [47]. This condition, which encompasses multiple phenotypes from the CAKUT spectrum, is characterized by optic nerve colobomas, renal hypoplasia, and vesicoureteral reflux.

Hepatocyte nuclear factor 1-beta (HNF1B), also known as also known as transcription factor-2 (TCF2), is a transcription factor expressed in multiple tissues during embryonic development, including the kidneys, pancreas, lung, liver, and gut. Patients carrying heterozygous pathogenic mutations in *HNF1B* develop renal cysts and diabetes syndrome (OMIM # 137920) and maturity-onset diabetes of the young type 5 (OMIM # 606391). Despite the name, *HNF1B* associated disease can present with multiple CAKUT phenotypes including renal agenesis, RHD, and MCDK.

The latter is felt to be related to the abnormal interaction between *HNF1B* and the *PKHD1* gene, which when mutated causes autosomal recessive polycystic kidney disease (OMIM # 263200). The reason for this shared cystic phenotype is because *HNF1B* binds to the proximal promoter of the *PKHD1* to stimulate gene transcription. Mutations in *HNF1B* can therefore lead to renal cyst formation as seen in the MCDK subtype of human CAKUT [57]. Copy number variants, including whole gene deletions, have also been described in patient with CAKUT and diabetes since *HNF1B* is located on chromosome 17, a region that is highly susceptible to genomic rearrangements [58,59].

The UB gives rise to the ureters, and abnormalities at the time of UB formation are thought to result in pathologies of the ureters.

Vesicoureteral reflux occurs due to incompetence of the vesicoureteral junction and is thought to be mediated by a number of interreacting genetic and environmental factors [37]. During development, abnormalities of the UB and/or abnormal interaction between the UB and MM are implicated in the development of VUR [60]. Heterozygous mutations in *ROBO2*, with subsequent reduction in *ROBO2* gene dosage, have been identified both in humans

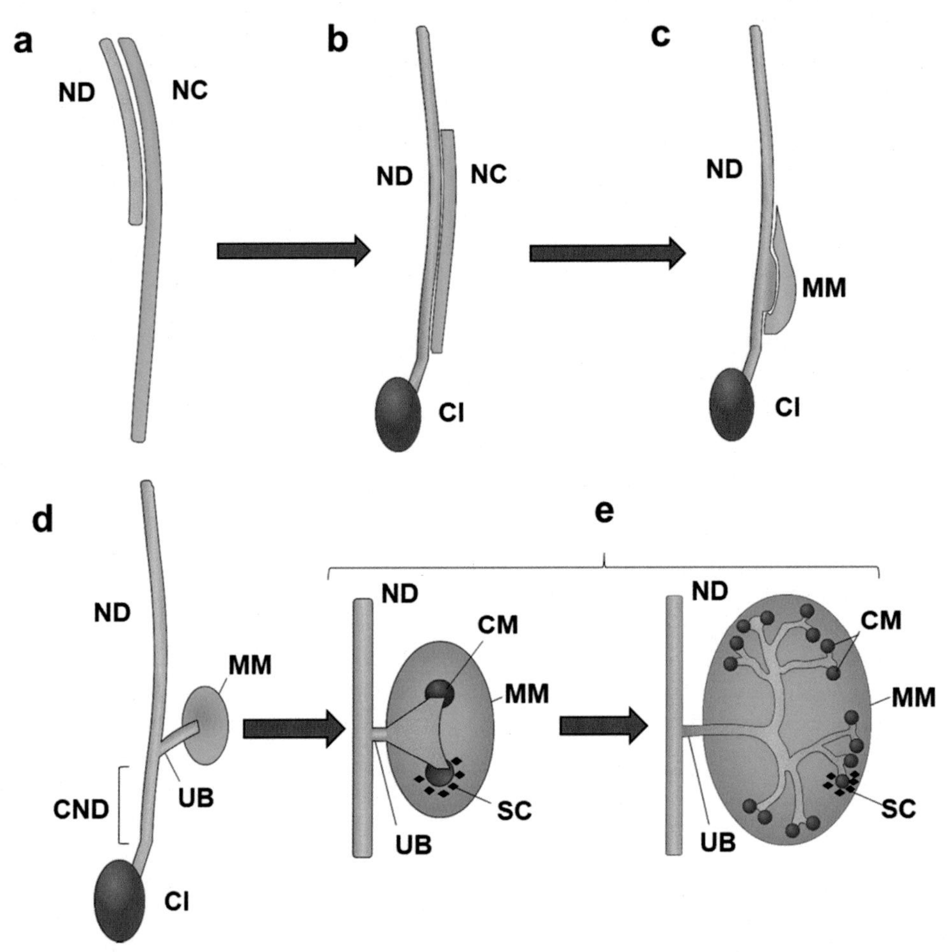

Figure 1.2 Development of the kidneys and urinary tract. (a) The nephric duct (ND) and the nephric cord (NC) are the precursors of the urinary system, arising from the intermediate mesoderm. The cells of the ND undergo an early mesenchymal-to-epithelial transition and assemble into epithelial tube–like structures. (b) As the embryo develops, the ND elongates caudally fusing with the cloacal epithelium (CI), which is the precursor of the urinary bladder. (c) The NC gives rise to the metanephric mesenchyme (MM). Renal morphogenesis is initiated and maintained by reciprocal interactions between the epithelial ND and the MM. (d) The UB arises from the ND and elongates toward the MM. The caudal part of the ND, which is located between the UB and the insertion into the CI, is referred to as the common nephric duct (CND). (e) Stimulated by MM-derived signals, the UB begins to branch repeatedly (branching morphogenesis). Through continuous reciprocal induction, the MM promotes and maintains branching morphogenesis of the UB. Via branching morphogenesis, the UB gives rise to the renal collecting system consisting of collecting ducts and renal pelvis as well as the ureter. Reciprocally, signals from the UB also support development of MM cells. The MM that is in closest proximity to the UB tips condenses and forms the cap mesenchyme (CM). Stimulated by signals from the UB, the CM undergoes a mesenchymal-to-epithelial transition, which subsequently gives rise to structures of the nephron (glomerulus, the proximal tubule, and the distal tubule). (Adapted with permission from van der Ven AT, Vivante A, Hildebrandt F. Novel insights into the pathogenesis of monogenic congenital anomalies of the kidney and urinary tract. *J Am Soc Nephrol* 2018;29(1):36–50.)

and mice with a CAKUT-VUR phenotype [61]. ROBO2 is a transmembrane receptor that binds to its SLIT2 ligand, and they together have been implicated in regulating GDNF, which plays an essential role in UB formation (see below for further details) [62].

Recently, Mann et al. demonstrated that disruption of the neural pathways innervating the bladder can lead to bladder dysfunction ultimately leading to a secondary CAKUT phenotype [14]. A number of genes such as *CHRM3, ACTG2, ACTA2, MYH11, MYLK, HPSE2, LRIG2,* and most recently *CHRNA3* have been implicated in regulation of smooth muscle actin contraction, neuronal patterning, and synaptic neuronal transmission in the bladder, while mutations in these genes have been shown to cause ureteral and bladder dysfunction leading to obstructive changes in the urinary tract.

1.9 DISEASE CAUSATION IN CAKUT

CAKUT is a heterogeneous disease entity with a large degree of variability in genotype and phenotype. It is hypothesized that the molecular basis of CAKUT ranges from monogenic to polygenic in nature. Environmental factors that have been implicated in CAKUT pathogenesis including exposure to toxins during embryogenesis including Vitamin A [63], high maternal body mass index [64], and maternal diabetes mellitus [65].

Increasingly, the role of epigenetic mechanisms involved in posttranslation modification and control of gene expression such as DNA methylation, histone acetylation, and chromatin remodeling are being investigated in CAKUT, although further data are required to definitively elucidate the role in disease pathogenesis [66].

In terms of monogenic causation, several other lines of evidence support the hypothesis that many cases of CAKUT are due to an underlying monogenic disease. This includes the observation that CAKUT is congenital in nature; that CAKUT also occurs in conjunction with other monogenic syndromes; the development of numerous monogenic mouse models that exhibit CAKUT; and the fact that the development of the kidney and urinary tract is governed by distinct developmental genes, and defects likely contribute to disease pathogenesis [13,46,67,68].

Monogenic causation in CAKUT can follow an autosomal dominant, autosomal recessive, and X-linked pattern of inheritance. The most prevalence monogenic disorder is due to pathogenic mutations in the transcription factors *PAX2* or *HNF1B*, both of which follow an autosomal dominant mode of inheritance [56]. With the expansion of next-generation sequencing technology, now 40 monogenic causes of isolated CAKUT (Table 1.1) and over 150 monogenic causes of syndromic CAKUT (Table 1.2) have been described.

These findings have resulting in a current estimated prevalence of monogenic causation between 14% and 20% [47,48,69–99]; however, prevalence can vary depending on the population under study, the CAKUT subtypes included, and the method of analysis. For example, in a population of fetuses with bilateral kidney anomalies, Rasmussen, using a gene-panel approach in combination with whole exome sequencing, identified a deleterious variants in 11 of 56 (20%) [11]. In a pediatric cohort of 100 children with renal hypodysplasia, Weber detected a monogenic cause in 17% of affected individuals [100]. In a heterogeneous pediatric cohort of 232 families encompassing a variety of subtypes of CAKUT, monogenic causation was observed in 14% [16]. In a Korean population of children with CAKUT, targeted exome sequencing identified genetic causes in 13.8% of the 94 recruited patients following detection of pathogenic variants in *HNF1B, PAX2, EYA1, UPK3A,* and *FRAS1.* Interestingly, pathogenic copy number variations in many known CAKUT causing genes regions, such as *HNF1B, EYA1,* and *CHD1L,* were also detected [101]. Overall it is estimated that an additional ~10%–15% of cases of CAKUT are due to copy number variations, the majority of which involve either the *HNFB1* locus on chromosome 17 or the DiGeorge/velocardiofacial syndrome region on chromosome 22 [58].

1.10 GENETIC FEATURES CHARACTERISTIC OF CAKUT

Both the clinical and molecular characteristics of CAKUT are heterogeneous. A number of factors are hypothesized to contribute to this heterogeneity. Significant heterogeneity can be attributed to the observation of variable expressivity and incomplete penetrance, which are frequently encountered in the CAKUT disease spectrum [46]. Variable expressivity occurs when an individual carrying a mutation in a disease-causing gene presents with a phenotype that differs from the phenotypic manifestation of other individuals with an identical mutation, whereas variable expressivity occurs when an individual carrying a mutation does not exhibit the phenotype at all [46].

Along with these features, which are predominately observed in autosomal dominant disease, other theories proposed for the observed phenotypic variability in CAKUT include those proposed by Ichikawa *et al.* This group hypothesize that the both intraindividual and interindividual difference may occur as a result of stochastic spatiotemporal differences during key events of renal morphogenesis; gene dosage effects; and redundancy of genes that belong to functionally related gene families [15,46,68]. These theories contribute to the understanding of the lack of genotype–phenotype correlation in some patients with CAKUT, where identical mutations in the same gene can result is different phenotypes [61].

Increasingly, phenotypic variability in CAKUT can be attributed to allelism. Allelism is defined as the existence of two or more variants of a given gene in an outbred population. An example of allelism in CAKUT is, for instance, seen in the Fraser-complex associated diseases due to mutations in the genes *FRAS1*, *FREM1*, *FREM2*, and *GRIP1*. Truncating disease-causing alleles, which result in loss of function of the corresponding protein, give rise to a severe phenotype observed in patients with Fraser syndrome (OMIM # 219000, # 617666, # 617667), Manitoba-oculo-tricho-anal/MOTA syndrome (OMIM # 248450 and bifid nose with or without anorectal and renal anomalies (OMIM # 608980). On the other hand, missense, hypomorphic alleles cause milder, disease consistent with isolated CAKUT (Fig. 1.3) [90].

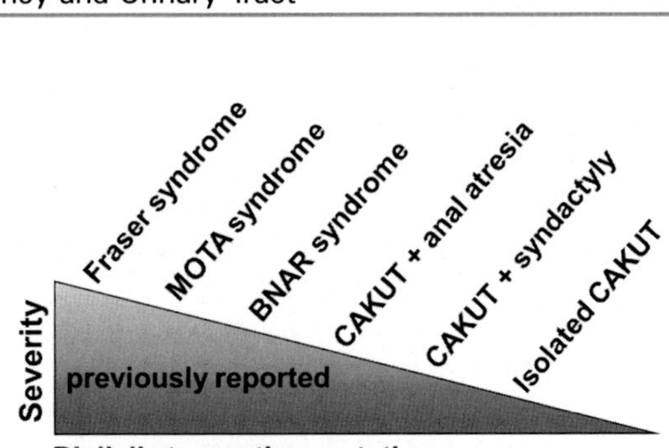

Figure 1.3 Allelism determining syndromic *versus* isolated CAKUT in the example of genes encoding the Fraser-complex. Biallelic recessive truncating mutations in genes encoding members of the Fraser-Complex (***FRAS1, FREM1, FREM2***) and the associated protein ***GRIP1*** are known monogenic causes of the severe, syndromic CAKUT phenotype Fraser syndrome (***FRAS1, FREM2, GRIP1***) and MOTA/BNAR syndrome (Manitoba-oculo-tricho-anal/bifid nose with or without anorectal and renal anomalies; ***FREM1***) [90]. Studies by Kohl *et al.* revealed that missense (rather than truncating) mutations in the same Fraser-complex encoding genes may cause isolated CAKUT in a significant proportion (13/590; ~2.2%) of individuals with isolated CAKUT. This indicates that in the case of the ***FRAS1, FREM2, GRIP1***, and ***FREM1*** genes, a mechanism of "allelism" determines the occurrence of syndromic versus isolated CAKUT, i.e., truncating *versus* missense mutations determine whether a syndromic or isolated CAKUT phenotype results. (Adapted from van der Ven AT, Connaughton DM, Ityel H, Mann N, Nakayama M, Chen J, et al. Whole-exome sequencing identifies causative mutations in families with congenital anomalies of the kidney and urinary tract. *J Am Soc Nephrol* 2018;29(9):2348–2361.)

1.11 MOUSE MODELS OF CAKUT

Contributing to the monogenic hypothesis in CAKUT is the large number of monogenic mouse models that display phenotypes consistent to those identified in humans with CAKUT [16,53]. To date, over 180 mouse models exist representing the diverse spectrum of CAKUT phenotypes including renal aplasia and hypoplasia, renal cystic disease, kidney ectopia, urethral duplication, and horseshoe kidney. Mouse models have also been developed that represent syndromic forms of CAKUT with models for Alagille syndrome (OMIM # 118450, # 610205), Branchio-Oto-Renal (BOR) syndrome (OMIM # 113650), Fraser syndrome (OMIM # 219000, # 617666, # 617667), Kallmann syndrome (OMIM # 308700), Meckel syndrome (OMIM # 249000), Pallister–Hall syndrome (OMIM # 146510), Renal Coloboma syndrome (OMIM # 120330), and Townes–Brocks syndromes (OMIM # 107480, # 617466) available [102]. This in turn has made possible the study of developmental pathways in CAKUT pathogenesis as outlined in the next section.

1.12 MOLECULAR PATHWAYS IN CAKUT PATHOGENESIS

1.12.1 Rearranged in Transfection (RET) Tyrosine Kinase Signaling System

Pathogenic mutations in the gene *RET*, a tyrosine kinase receptor, have been implicated in 5% of patients with CAKUT [103] and up to a third of fetuses with either unilateral or bilateral renal agenesis [76]. RET activation through the GDNF-GFRα1-RET pathway is an important mediator in the developmental pathway of the kidney and urinary tract. Specifically this pathway promotes normal nephrogenesis through regulation of the reciprocal induction between the ureteric bud (UB) and the metanephric mesenchyme (MM) (Fig. 1.2) [3,15,104,105]. In the developing kidney RET, which is expressed in nephric duct derived structures including the ureteric bud, is recruited following binding of its ligand GDNF to two homodimers of the coreceptor GFRα1 [53,104]. Central to this process is the initial release of GDNF from the MM followed by binding of GDNF and its coreceptor GDNF Family Receptor Alpha 1 (GFRα1) and subsequently RET [104,106].This activated GDNF-GFRα1-RET complex then promotes phosphorylation of specific tyrosine residues resulting in activation of downstream signaling cascades. Mutations in genes that function either upstream or downstream of the GDNF-GFRα1-RET pathway have also been implicated in both human and murine CAKUT. For example, mutations in GDNF regulatory genes *EYA1* [107], *SIX1* [98], and *SIX5* [99] have been implicated in CAKUT and have been described in patients with Brachio–Oto–Renal syndrome (OMIM # 113650), which is characterized by defects in branchial arch, ear, and renal development. Transcription factors, known to regulate the level and expression of GDNF, have also been implicated in human and murine CAKUT namely *PAX2, GATA3,* and *SALL1.* Heterozygous mutations in *PAX2,* causing Renal Coloboma Syndrome (OMIM # 120330), are a common monogenic causes of CAKUT [56] having been described in families with optic nerve colobomas, renal anomalies, and vesicoureteral reflux [47]. Heterozygous mutations in *SALL1* are implicated in Townes–Brocks syndrome (OMIM # 107480), characterized by external ear anomalies with sensorineural hearing loss, limb anomalies, and renal and anorectal malformations [97], while heterozygous mutations in *GATA3* have been described in patients with DiGeorge-like phenotype (OMIM # 146255) that includes hypoparathyroidism, heart defects, immune deficiency, deafness, and renal malformations [108].

1.12.2 The Fraser Complex and the Extra-Cellular Matrix Complex

The role of the Fraser Complex (FC) in CAKUT pathogenesis first came to light following the identification of pathogenic mutations in the genes encoding members of the FC in patients with Fraser-syndrome (OMIM # 219000). FS is a rare autosomal recessive form of syndromic CAKUT characterized by renal anomalies with extra-renal features including syndactyly, cryptophthalmos, and abnormalities of the respiratory tract. Truncating mutations in the genes *FRAS1, FREM2, FREM1,* and *GRIP1,* which are involved in either the formation or regulation of the FC, have all been described in patients with CAKUT [89,109–113].

The Fraser complex itself is an extracellular matrix complex that is governed by a *FRAS1, FREM2,* and *FREM1*, which are integral in maintaining the integrity of this complex (Fig. 1.4). In the mouse, Fras1, Frem1, and Frem2 are located in the basement membrane primarily during embryogenesis, but can also be seen to a lesser degree into adulthood [114]. In addition to providing support [114–116], these proteins play a regulatory role in cell–cell and cell–matrix interaction [117]. It is hypothesized that truncating, loss-of function mutations in these genes lead to failure of the assembly of the Fraser complex [109,114,118]. Interestingly this loss of integrity of the Fraser complex has also been shown to decrease, Fraser complex-mediated GDNF expression in the MM, thereby impairing interaction between the UB and MM [119]. GRIP1 is a cytosolic protein, which specifically interacts with FRAS1 to ensure appropriate trafficking and targeting to the basolateral surface of the ureteric bud cells [109]. The FC once assembled at the epithelial–mesenchymal interface of the ureteric bud epithelium directly interacts with nephronectin (Npnt). Npnt is extracellular matrix protein that is expressed throughout the kidney including the ureteric bud epithelium, although to date no mutations in *NPNT* have been described in human with CAKUT. However at the UB, Npnt serves as an adaptor for proteins expressed in the MM such as Integrin Subunit Alpha 8 (ITGA8) and Integrin Subunit Beta 1 (ITGB1) [118,120,121]. Interestingly, ITGA8 is also an upstream activator of GDNF, which, as outlined above, is

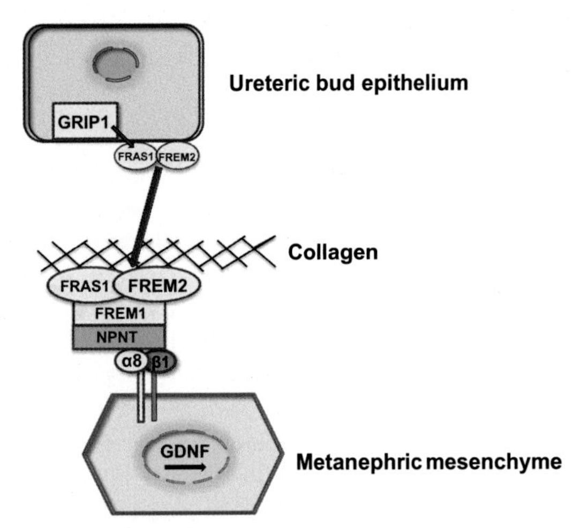

Figure 1.4 Schematic demonstrating the interaction between the Fraser complex with the ureteric bud (UB) and the metanephric mesenchyme (MM) in renal development. Known monogenic causes of CAKUT in humans are indicated in *yellow* while the *red* frames indicate a monogenic cause of CAKUT in murine animal models. FRAS1 and FREM2 are transmembrane proteins that localize to the epithelial cells of the ureteric bud. GRIP1 is an intracellular protein that facilitates targeting of FRAS1 and FREM2 to the basal surface of the UB cells and shedding of FRAS1 from the membrane. FREM1 is produced by the MM and secreted into the extracellular space. FRAS1, FREM2, and FREM1 assemble at the epithelial–mesenchymal interface to form the Fraser-Complex (FC). Npnt functions as an adaptor protein to interconnect the ternary FC with the integrin α8 integrin-β1 heterodimer (ITGA8, ITGB1) on the surface of MM cells. ITGA8/ITGB1 signaling leads to an increased expression of GDNF by the MM, thereby promoting renal morphogenesis. Loss of Fraser Complex integrity results in a significant decrease in GDNF expression in the MM, thereby hampering the interaction between the UB and MM and consequentially impeding renal morphogenesis. (Adapted with permission from van der Ven AT, Vivante A, Hildebrandt F. Novel insights into the pathogenesis of monogenic congenital anomalies of the kidney and urinary tract. J Am Soc Nephrol 2018;29(1):36–50.)

involved in the tyrosine kinase signaling pathway [118,120]. Recessive mutations in *ITGA8* cause renal hypodysplasia/aplasia 1 (OMIM # 191830), which is characterized by multiple renal and extra-renal anomalies including renal dysplasia, aplasia, and agenesis along with ureteric and bladder pathologies [90,92].

Another regulator in the interaction between the UB and MM, which occurs at the level of the extracellular matrix, is the heparin sulfate proteoglycans (HSPGs) [115,122,123]. One of the key regulators in HPSE activity itself is the enzyme protein heparanase 2 (HPSE2) [124]. Urofacial syndrome (also known as Ochoa syndrome, OMIM # 236730) is characterized by syndromic facial features, where there is a crying facial expression when laughing along with features of the CAKUT disease spectrum including hydronephrosis, hydroureter, and PUV [125].

1.12.3 Vitamin A and Retinoic Acid Signaling

Vitamin A exposure during embryogenesis has been implicated as an environmental factor in CAKUT pathogenesis [63], with abnormal levels during embryogenesis associated with a congenital birth defect termed Vitamin-A-deficient or VAD syndrome [126–128]. VAD syndrome is characterized by immune and hematological defects, ocular abnormalities such as xerophthalmia, and night blindness as well as anomalies of the cardiac, respiratory, and urinary systems [127,129,130]. The active metabolite of vitamin A is retinoic acid. In mouse models, retinoic acid has been implicated in kidney development [106,131–135], where it plays a role in the expression of Ret. Moreover, vitamin A and retinoic acid are required both for successful insertion of the ND into the cloaca as well as the branching morphogenesis of the UB [53,106,132]. Mouse models with mutations in genes involved in the regulation of intracellular retinoic acid have also been shown to display phenotypes that resemble VAD syndrome including pathologies within the CAKUT disease spectrum. These genes include *Retinol Dehydrogenase 10* (*Rdh10*) [136]; *Aldehyde Dehydrogenase 1 Family Member A2* (*Aldh1a2*) [133]; *Cytochrome P450 Family 26 Subfamily A Member 1* (*Cyp26a1*) [137,138] as well as double mutants of specific isoforms within the RXR (*Rxrα*) and RAR family of proteins (*Rarα, Rarβ*) [129]. Recently in humans, mutations in a gene related to vitamin A function (*NRIP1, Nuclear Receptor Interacting Protein 1*) were discovered in patients with CAKUT [139]. Vivante et al. described heterozygous truncating mutations in *NRIP1* in large kindred with an isolated CAKUT phenotype [139]. *NRIP1* encodes a nuclear receptor transcriptional cofactor that directly interacts with the retinoic acid receptors (RARs) to modulate retinoic acid transcriptional signaling (Fig. 1.5).

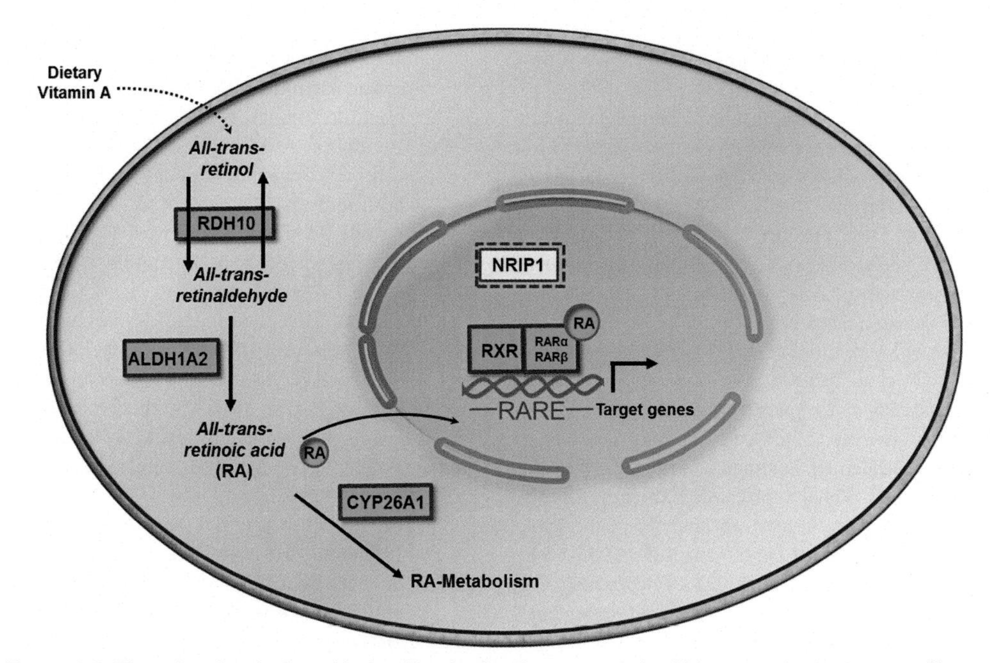

Figure 1.5 The role of retinoic acid signaling in development of the kidneys and urinary tract. Known monogenic causes of CAKUT in humans are indicated in *yellow* while the *red* frames indicate a monogenic cause of CAKUT in murine animal models. The reversible reaction of converting retinol to retinal is catalyzed by retinol dehydrogenase 10 (RDH10). Retinal is then converted into the active metabolite all-trans retinoic acid (RA) via an irreversible reaction catalyzed by the enzyme aldehyde dehydrogenase 1 (ALD1A2). RA then either enters the nucleus or is metabolized via enzymes of the cytochrome P26 family (CYP26) in the endoplasmic reticulum. In the nucleus, RA derivatives can serve as a ligand for two receptors: retinoid X receptor (RXR; 9-*cis*-RA, rexinoids) and retinoic acid receptor (RAR, all-*trans*-RA). Both receptor proteins have at least three subtypes and isoforms each. RXR and RAR heterodimers bind to retinoic-acid responsive elements (RARE) of the cellular DNA. RAREs are predominantly located in promoter regions of target genes. The presence of RA recruit coactivators leads to enhanced binding to RAREs and expression of target genes. The transcriptional control of downstream genes is further modified by the binding of additional coregulators (coactivators and corepressors, including NRIP1) to RXR and RAR. *NRIP1* encodes a nuclear receptor transcriptional cofactor that directly interacts with the retinoic acid receptors (RARs) to modulate retinoic acid transcriptional signaling. (Adapted with permission from van der Ven AT, Vivante A, Hildebrandt F. Novel insights into the pathogenesis of monogenic congenital anomalies of the kidney and urinary tract. J Am Soc Nephrol 2018;29(1):36–50.)

REFERENCES

[1] Nakanish K, Yoshikawa N. Genetic disorders of human congenital anomalies of the kidney and urinary tract (CAKUT). Pediatr Int Off J Japan Pediatr Soc 2003;45(5):610–6.
[2] Nicolaou N, Renkema KY, Bongers EM, Giles RH, Knoers NV. Genetic, environmental, and epigenetic factors involved in CAKUT. Nat Rev Nephrol 2015;11(12):720–31.
[3] Schedl A. Renal abnormalities and their developmental origin. Nat Rev Genet 2007;8(10):791–802.
[4] Loane M, Dolk H, Kelly A, Teljeur C, Greenlees R, Densem J, et al. Paper 4: EUROCAT statistical monitoring: identification and investigation of ten year trends of congenital anomalies in Europe. Birth Def Res A Clin Mol Teratol 2011;91(1):S31–43.
[5] Hildebrandt F. Genetic kidney diseases. Lancet 2010;375(9722):1287–95.
[6] Chesnaye N, Bonthuis M, Schaefer F, Groothoff JW, Verrina E, Heaf JG, et al. Demographics of paediatric renal replacement therapy in Europe: a report of the ESPN/ERA-EDTA registry. Pediatr Nephrol 2014;29(12):2403–10.
[7] North American Pediatric Renal Trials and Collaborative Studies. NAPRTCS 2008 Annual Report. Rockville, MD: The EMMES Corporation; 2008.
[8] Wuhl E, van Stralen KJ, Verrina E, Bjerre A, Wanner C, Heaf JG, et al. Timing and outcome of renal replacement therapy in patients with congenital

malformations of the kidney and urinary tract. Clin J Am Soc Nephrol 2013;8(1):67–74.

[9] Rubenstein M, Meyer R, Bernstein J. Congenital abnormalities of the urinary system. I. A postmortem survey of developmental anomalies and acquired congenital lesions in a children's hospital. J Pediatr 1961;58:356–66.

[10] Connaughton DM, Bukhari S, Conlon P, Cassidy E, O'Toole M, Mohamad M, et al. The Irish kidney gene project—prevalence of family history in patients with kidney disease in Ireland. Nephron 2015;130(4):293–301.

[11] Rasmussen M, Sunde L, Nielsen ML, Ramsing M, Petersen A, Hjortshoj TD, et al. Targeted gene sequencing and whole-exome sequencing in autopsied fetuses with prenatally diagnosed kidney anomalies. Clin Genet 2018;93(4):860–9.

[12] Wuhl E, van Stralen KJ, Wanner C, Ariceta G, Heaf JG, Bjerre AK, et al. Renal replacement therapy for rare diseases affecting the kidney: an analysis of the ERA-EDTA Registry. Nephrol Dial Transpl 2014;29:iv1–8.

[13] Vivante A, Kohl S, Hwang DY, Dworschak GC, Hildebrandt F. Single-gene causes of congenital anomalies of the kidney and urinary tract (CAKUT) in humans. Pediatr Nephrol 2014;29(4):695–704.

[14] Mann N, Kause F, Henze EK, Gharpure A, Shril S, Connaughton DM, et al. CAKUT and autonomic dysfunction caused by acetylcholine receptor mutations. Am J Hum Genet 2019;105(6):1286–93.

[15] Ichikawa I, Kuwayama F, Pope JC, Stephens FD, Miyazaki Y. Paradigm shift from classic anatomic theories to contemporary cell biological views of CAKUT. Kidney Int 2002;61(3):889–98.

[16] van der Ven AT, Connaughton DM, Ityel H, Mann N, Nakayama M, Chen J, et al. Whole-exome sequencing identifies causative mutations in families with congenital anomalies of the kidney and urinary tract. J Am Soc Nephrol 2018;29(9):2348–61.

[17] Westland R, Schreuder MF, Bökenkamp A, Spreeuwenberg, MD, van Wijk JA. Renal injury in children with a solitary functioning kidney—the KIMONO study. Nephrol Dial Transpl Off Publ Eur Dial Transpl Assoc Eur Renal Assoc 2011;26(5):1533–41.

[18] Westland R, Schreuder MF, Ket JCF, van Wijk JA. Unilateral renal agenesis: a systematic review on associated anomalies and renal injury. Nephrol Dial Transpl Off Publ Eur Dial Transpl Assoc Eur Renal Assoc 2013;28(7):1844–55.

[19] Sanna Cherchi S, Ravani P, Corbani V, Parodi S, Haupt R, Piaggio G, et al. Renal outcome in patients with congenital anomalies of the kidney and urinary tract. Kidney Int 2009;76(5):528–33.

[20] Schreuder MF. Life with one kidney. Pediatr Nephrol 2018;33(4):595–604.

[21] Gök ES, Ayvacı A, Ağbaş A, Adaletli İ, Canpolat N, Sever L, et al. The frequency of familial congenital anomalies of the kidney and urinary tract: should we screen asymptomatic first-degree relatives using urinary tract ultrasonography? Nephron 2020;144(4):170–5.

[22] Roodhooft AM, Birnholz JC, Holmes LB. Familial nature of congenital absence and severe dysgenesis of both kidneys. N Engl J Med 1984;310(21):1341–5.

[23] Atwell JD, Cook PL, Howell CJ, Hyde I, Parker BC. Familial incidence of bifid and double ureters. Arch Dis Child 1974;49(5):390–3.

[24] Wilson RD, Baird PA. Renal agenesis in British Columbia. Am J Med Genet 1985;21(1):153–69.

[25] Thomas AN, McCullough LB, Chervenak FA, Placencia FX. Evidence-based, ethically justified counseling for fetal bilateral renal agenesis. J Perinat Med 2017;45(5):585–94.

[26] Potter EL. Bilateral renal agenesis. J Pediatr 1946;29:68–76.

[27] Winyard P, Chitty LS. Dysplastic kidneys. Semin Fetal Neonatal Med 2008;13(3):142–51.

[28] Spence HM. Congenital unilateral multicystic kidney: an entity to be distinguished from polycystic kidney disease and other cystic disorders. J Urol 1955;74(6):693–706.

[29] Schreuder MF, Westland, R, van Wijk JAE. Unilateral multicystic dysplastic kidney: a meta-analysis of observational studies on the incidence, associated urinary tract malformations and the contralateral kidney. Nephrol Dial Transpl Off Publ Eur Dial Transpl Assoc Eur Renal Assoc 2009;24(6):1810–8.

[30] Kirkpatrick JJ, Leslie SW. Horseshoe kidney. StatPearls. StatPearls Publishing; 2020, NBK431105.

[31] Doery AJ, Ang E, Ditchfield MR. Duplex kidney: not just a drooping lily. J Med Imag Rad Oncol 2015;59(2):149–51.

[32] Privett JT, Jeans WD, Roylance J. The incidence and importance of renal duplication. Clin Radiol 1976;27(4):521–30.

[33] Liu DB, Armstrong 3rd WR, Maizels M. Hydronephrosis: prenatal and postnatal evaluation and management. Clin Perinatol 2014;41(3):661–78.

[34] Koff SA, H MK. In: Gillenwater JY, T GJ, Howards SS, Mitchell ME, editors. Anomalies of the kidney: adult and pediatric urology. 4th ed. Philadelphia, Pa, USA: Lippincott Williams and Wilkins; 2002. 2129 p.

[35] Has R, Sarac Sivrikoz T. Prenatal diagnosis and findings in ureteropelvic junction type hydronephrosis. Front Pediatr 2020;8:492.

[36] Lee YS, Im YJ, Lee H, Kim MJ, Lee MJ, Jung HJ, et al. Coexisting ureteropelvic junction obstruction and ureterovesical junction obstruction: is pyeloplasty always the preferred initial surgery? Urology 2014;83(2):443–9.

[37] Willaims G, Fletcher JT, Alexander SI, Craig JC. Vesicoureteral reflux. J Am Soc Nephrol JASN 2008;19(5):847–62.

[38] Anumba DO, Scott JE, Plant ND, Robson SC. Diagnosis and outcome of fetal lower urinary tract

obstruction in the northern region of England. Prenat Diagn 2005;25(1):7–13.
[39] Farrugia MK. Fetal bladder outlet obstruction: embryopathology, in utero intervention and outcome. J Pediatr Urol 2016;12(5):296–303.
[40] Hennus PML, van der Heijden GJMG, Bosch JLHR, de Jong TPVM, de Kort LMO. A systematic review on renal and bladder dysfunction after endoscopic treatment of infravesical obstruction in boys. PLoS One 2012;7(9):e44663.
[41] Lee B, Driver CP, Flett ME, Steven L, Steven M, O'Toole S. PUVs are more common in boys with hypospadias. J Pediatr Urol 2020;16(3):299–303.
[42] Taskinen S, Heikkilä J, Rintala R. Effects of posterior urethral valves on long-term bladder and sexual function. Nat Rev Urol 2012;9(12).
[43] Farrugia MK. Fetal bladder outflow obstruction: interventions, outcomes and management uncertainties. Early Hum Dev 2020;150(05189). https://doi.org/10.1016/j.earlhumdev.2020.105189.
[44] Effmann EL, Lebowitz RL, Colodny AH. Duplication of the urethra. Radiology 1976;119(1):179–85.
[45] Connaughton DM, Kennedy C, Shril S, Mann N, Murray SL, Williams PA, et al. Monogenic causes of chronic kidney disease in adults. Kidney Int 2019;95(4):914–28.
[46] Vivante A, Hildebrandt F. Gentics of congenital anomalies of the kidneys and urinary tract. In: Barakat A, Rushton H, editors. Congenital anomalies of the kidney and urinary tract. 1. Switzerland: Springer Nature; 2016. p. 303–22.
[47] Sanyanusin P, Schimmenti LA, McNoe LA, Ward TA, Pierpont ME, Sullivan MJ, et al. Mutation of the PAX2 gene in a family with optic nerve colobomas, renal anomalies and vesicoureteral reflux. Nat Genet 1995;9(4):358–64.
[48] Lindner TH, Njolstad PR, Horikawa Y, Bostad L, Bell GI, Sovik O. A novel syndrome of diabetes mellitus, renal dysfunction and genital malformation associated with a partial deletion of the pseudo-POU domain of hepatocyte nuclear factor-1beta. Hum Mol Genet 1999;8(11):2001–8.
[49] Urisarri A, Gil M, Mandiá N, Aldamiz-Echevarría L, Iria R, González-Lamuño D, et al. Retrospective study to identify risk factors for chronic kidney disease in children with congenital solitary functioning kidney detected by neonatal renal ultrasound screening. Medicine 2018;97(32):e11819.
[50] Tsuchiya M, Hayashida M, Yanagihara T, Yoshida J, Takeda S, Tatsuma N, et al. Ultrasound screening for renal and urinary tract anomalies in healthy infants. Pediatr Int 2003;45(5):617–23.
[51] Practice parameter: the diagnosis, treatment, and evaluation of the initial urinary tract infection in febrile infants and young children. American Academy of Pediatrics. Committee on Quality Improvement. Subcommittee on urinary tract infection. Pediatrics 1999;103(4 Pt 1):843–52. https://doi.org/10.1542/peds.103.4.843.
[52] Didier RA, Chow JS, Kwatra NS, Retik AB, Lebowitz RL. The duplicated collecting system of the urinary tract: embryology, imaging appearances and clinical considerations. Pediatr Radiol 2017;47(11):1526–38.
[53] van der Ven AT, Vivante A, Hildebrandt F. Novel insights into the pathogenesis of monogenic congenital anomalies of the kidney and urinary tract. J Am Soc Nephrol 2018;29(1):36–50.
[54] Costantini F. Genetic controls and cellular behaviors in branching morphogenesis of the renal collecting system. Wiley Interdiscip Rev Dev Biol 2012;1(5):693–713.
[55] Narlis M, Grote D, Gaitan Y, Boualia SK, Bouchard M. Pax2 and pax8 regulate branching morphogenesis and nephron differentiation in the developing kidney. J Am Soc Nephrol 2007;18(4):1121–9.
[56] Thomas R, Sanna-Cherchi S, Warady BA, Furth SL, Kaskel FJ, Gharavi AG. HNF1B and PAX2 mutations are a common cause of renal hypodysplasia in the CKiD cohort. Pediatr Nephrol 2011;26(6):897–903.
[57] Hiesberger T, Bai Y, Shao X, McNally BT, Sinclair AM, Tian X, et al. Mutation of hepatocyte nuclear factor-1beta inhibits Pkhd1 gene expression and produces renal cysts in mice. J Clin Invest 2004;113(6):814–25.
[58] Sanna-Cherchi S, Kiryluk K, Burgess KE, Bodria M, Sampson MG, Hadley D, et al. Copy-number disorders are a common cause of congenital kidney malformations. Am J Hum Genet 2012;91(6):987–97.
[59] Mefford HC, Clauin S, Sharp AJ, Moller RS, Ullmann R, Kapur R, et al. Recurrent reciprocal genomic rearrangements of 17q12 are associated with renal disease, diabetes, and epilepsy. Am J Hum Genet 2007;81(5):1057–69.
[60] Bertoli-Avella AM, Conte ML, Punzo F, de Graaf BM, Lama G, La Manna A, et al. ROBO2 gene variants are associated with familial vesicoureteral reflux. J Am Soc Nephrol 2008;19(4):825–31.
[61] Hwang DY, Dworschak GC, Kohl S, Saisawat P, Vivante A, Hilger AC, et al. Mutations in 12 known dominant disease-causing genes clarify many congenital anomalies of the kidney and urinary tract. Kidney Int 2014;85(6):1429–33.
[62] Lu W, van Eerde AM, Fan X, Quintero-Rivera F, Kulkarni S, Ferguson H, et al. Disruption of ROBO2 is associated with urinary tract anomalies and confers risk of vesicoureteral reflux. Am J Hum Genet 2007;80(4):616–32.
[63] Lee LMY, Leung CY, Tang WWC, Choi HL, Leung YC, McCaffery PJ, et al. A paradoxical teratogenic mechanism for retinoic acid. Proc Natl Acad Sci USA 2012;109(34):13668–73.
[64] Dart AB, Ruth CA, Sellers EA, Au W, Dean HJ. Maternal diabetes mellitus and congenital anomalies of the kidney and urinary tract (CAKUT) in the child. Am J Kidney Dis 2015;65(5):684–91.
[65] Parikh CR, McCall D, Engelman C, Schrier RW. Congenital renal agenesis: case-control analysis of birth characteristics. Am J Kidney Dis 2002;39(4):689–94.

[66] Patel SR, Kim D, Levitan I, Dressler GR. The BRCT-domain containing protein PTIP links PAX2 to a histone H3, lysine 4 methyltransferase complex. Dev Cell 2007;13(4):580–92.
[67] Barratt TM, Avner ED, Harmon WE. Pediatric nephrology. In: Pine JW, editor. Pediatric nephrology. Media, PN 19063-2043. USA: Lippinkott Williams & Wilkins; 1999.
[68] Davies J. Mesenchyme to epithelium transition during development of the mammalian kidney tubule. Acta Anat Baseline 1996;156(3):187–201.
[69] Weber S, Taylor JC, Winyard P, Baker KF, Sullivan-Brown J, Schild R, et al. SIX2 and BMP4 mutations associate with anomalous kidney development. J Am Soc Nephrol 2008;19(5):891–903.
[70] Brockschmidt A, Chung B, Weber S, Fischer DC, Kolatsi-Joannou M, Christ L, et al. CHD1L: a new candidate gene for congenital anomalies of the kidneys and urinary tract (CAKUT). Nephrol Dial Transpl Off Publ Eur Dial Transpl Assoc Eur Renal Assoc 2012;27(6):2355–64.
[71] Lopez-Rivera E, Liu YP, Verbitsky M, Anderson BR, Capone VP, Otto EA, et al. Genetic drivers of kidney defects in the DiGeorge syndrome. N Engl J Med 2017;376(8):742–54.
[72] Sanna-Cherchi S, Sampogna RV, Papeta N, Burgess KE, Nees SN, Perry BJ, et al. Mutations in DSTYK and dominant urinary tract malformations. N Engl J Med 2013;369(7):621–9.
[73] Abdelhak S, Kalatzis V, Heilig R, Compain S, Samson D, Vincent C, et al. Clustering of mutations responsible for branchio-oto-renal (BOR) syndrome in the eyes absent homologous region (eyaHR) of EYA1. Hum Mol Genet 1997;6(13):2247–55.
[74] Pandolfi PP, Roth ME, Karis A, Leonard MW, Dzierzak E, Grosveld FG, et al. Targeted disruption of the GATA3 gene causes severe abnormalities in the nervous system and in fetal liver haematopoiesis. Nat Genet 1995;11(1):40–4.
[75] Kirby A, Gnirke A, Jaffe DB, Baresova V, Pochet N, Blumenstiel B, et al. Mutations causing medullary cystic kidney disease type 1 lie in a large VNTR in MUC1 missed by massively parallel sequencing. Nat Genet 2013;45(3):299–303.
[76] Skinner MA, Safford SD, Reeves JG, Jackson ME, Freemerman AJ. Renal aplasia in humans is associated with RET mutations. Am J Hum Genet 2008;82(2):344–51.
[77] Hwang DY, Kohl S, Fan X, Vivante A, Chan S, Dworschak GC, et al. Mutations of the SLIT2-ROBO2 pathway genes SLIT2 and SRGAP1 confer risk for congenital anomalies of the kidney and urinary tract. Hum Genet 2015;134(8):905–16.
[78] Gimelli S, Caridi G, Beri S, McCracken K, Bocciardi R, Zordan P, et al. Mutations in SOX17 are associated with congenital anomalies of the kidney and the urinary tract. Hum Mutat 2010;31(12):1352–9.
[79] Vivante A, Kleppa MJ, Schulz J, Kohl S, Sharma A, Chen J, et al. Mutations in TBX18 cause dominant urinary tract malformations via transcriptional dysregulation of ureter development. Am J Hum Genet 2015;97(2):291–301.
[80] Gbadegesin RA, Brophy PD, Adeyemo A, Hall G, Gupta IR, Hains D, et al. TNXB mutations can cause vesicoureteral reflux. J Am Soc Nephrol 2013;24(8):1313–22.
[81] Hart TC, Gorry MC, Hart PS, Woodard AS, Shihabi Z, Sandhu J, et al. Mutations of the UMOD gene are responsible for medullary cystic kidney disease 2 and familial juvenile hyperuricaemic nephropathy. J Med Genet 2002;39(12):882–92.
[82] Jenkins D, Bitner-Glindzicz M, Malcolm S, Hu CC, Allison J, Winyard PJ, et al. De novo Uroplakin IIIa heterozygous mutations cause human renal adysplasia leading to severe kidney failure. J Am Soc Nephrol JASN 2005;16(7):2141–9.
[83] Biason-Lauber A, Konrad D, Navratil F, Schoenle EJ. A WNT4 mutation associated with Mullerian-duct regression and virilization in a 46, XX woman. N Engl J Med 2004;351(8):792–8.
[84] Mandel H, Shemer R, Borochowitz ZU, Okopnik M, Knopf C, Indelman M, et al. SERKAL syndrome: an autosomal-recessive disorder caused by a loss-of-function mutation in WNT4. Am J Hum Genet 2008;82(1):39–47.
[85] Vivante A, Mark-Danieli M, Davidovits M, Harari-Steinberg O, Omer D, Gnatek Y, et al. Renal hypodysplasia associates with a WNT4 variant that causes aberrant canonical WNT signaling. J Am Soc Nephrol JASN 2013;24(4):550–8.
[86] Gribouval O, Gonzales M, Neuhaus T, Aziza J, Bieth E, Laurent N, et al. Mutations in genes in the renin-angiotensin system are associated with autosomal recessive renal tubular dysgenesis. Nat Genet 2005;37(9):964–8.
[87] Weber S, Thiele H, Mir S, Toliat MR, Sozeri B, Reutter H, et al. Muscarinic acetylcholine receptor M3 mutation causes urinary bladder disease and a prune-belly-like syndrome. Am J Hum Genet 2011;89(5):668–74.
[88] Barak H, Huh SH, Chen S, Jeanpierre C, Martinovic J, Parisot M, et al. FGF9 and FGF20 maintain the stemness of nephron progenitors in mice and man. Dev Cell 2012;22(6):1191–207.
[89] McGregor L, Makela V, Darling SM, Vrontou S, Chalepakis G, Roberts C, et al. Fraser syndrome and mouse blebbed phenotype caused by mutations in FRAS1/Fras1 encoding a putative extracellular matrix protein. Nat Genet 2003;34(2):203–8.
[90] Kohl S, Hwang DY, Dworschak GC, Hilger AC, Saisawat P, Vivante A, et al. Mild recessive mutations in six Fraser syndrome-related genes cause

isolated congenital anomalies of the kidney and urinary tract. J Am Soc Nephrol JASN 2014;25(9):1917–22.

[91] Bulum B, Ozcakar ZB, Duman D, Cengiz FB, Kavaz A, Burgu B, et al. HPSE2 mutations in urofacial syndrome, non-neurogenic neurogenic bladder and lower urinary tract dysfunction. Nephron 2015;130(1):54–8.

[92] Humbert C, Silbermann F, Morar B, Parisot M, Zarhrate M, Masson C, et al. Integrin alpha 8 recessive mutations are responsible for bilateral renal agenesis in humans. Am J Hum Genet 2014;94(2):288–94.

[93] Stuart HM, Roberts NA, Burgu B, Daly SB, Urquhart JE, Bhaskar S, et al. LRIG2 mutations cause urofacial syndrome. Am J Hum Genet 2013;92(2):259–64.

[94] Saisawat P, Kohl S, Hilger AC, Hwang DY, Yung Gee H, Dworschak GC, et al. Whole-exome resequencing reveals recessive mutations in TRAP1 in individuals with CAKUT and VACTERL association. Kidney Int 2014;85(6):1310–7.

[95] Hardelin JP, Levilliers J, del Castillo I, Cohen-Salmon M, Legouis R, Blanchard S, et al. X chromosome-linked Kallmann syndrome: stop mutations validate the candidate gene. Proc Natl Acad Sci USA 1992;89(17):8190–4.

[96] Vivante A, Hildebrandt F. Exploring the genetic basis of early-onset chronic kidney disease. Nat Rev Nephrol 2016;12(3):133–46.

[97] Kohlhase J, Wischermann A, Reichenbach H, Froster U, Engel W. Mutations in the SALL1 putative transcription factor gene cause Townes-Brocks syndrome. Nat Genet 1998;18(1):81–3.

[98] Ruf RG, Xu PX, Silvius D, Otto EA, Beekmann F, Muerb UT, et al. SIX1 mutations cause branchio-oto-renal syndrome by disruption of EYA1-SIX1-DNA complexes. Proc Natl Acad Sci U S A 2004;101(21):8090–5.

[99] Hoskins BE, Cramer CH, Silvius D, Zou D, Raymond RM, Orten DJ, et al. Transcription factor SIX5 is mutated in patients with branchio-oto-renal syndrome. Am J Hum Genet 2007;80(4):800–4.

[100] Weber S, Moriniere V, Knuppel T, Charbit M, Dusek J, Ghiggeri GM, et al. Prevalence of mutations in renal developmental genes in children with renal hypodysplasia: results of the ESCAPE study. J Am Soc Nephrol 2006;17(10):2864–70.

[101] Ahn YH, Lee C, Kim NKD, Park E, Kang HG, Ha IS, et al. Targeted exome sequencing provided comprehensive genetic diagnosis of congenital anomalies of the kidney and urinary tract. J Clin Med 2020;9(3):751. https://doi.org/10.3390/jcm9030751.

[102] Kuure S, Sariola H. Mouse models of congenital kidney anomalies. Adv Exp Med Biol 2020;1236:109–36. https://doi.org/10.1007/978-981-15-2389-2_5.

[103] Chatterjee R, Ramos E, Hoffman M, Van Winkle J, Martin DR, Davis TK, et al. Traditional and targeted exome sequencing reveals common, rare and novel functional deleterious variants in RET-signaling complex in a cohort of living US patients with urinary tract malformations. Hum Genet 2012;131(11):1725–38.

[104] Davis TK, Hoshi M, Jain S. To bud or not to bud: the RET perspective in CAKUT. Pediatr Nephrol 2014;29(4):597–608.

[105] Short KM, Smyth IM. The contribution of branching morphogenesis to kidney development and disease. Nat Rev Nephrol 2016;12(12):754–67.

[106] Chia I, Grote D, Marcotte M, Batourina E, Mendelsohn C, Bouchard M. Nephric duct insertion is a crucial step in urinary tract maturation that is regulated by a Gata3-Raldh2-Ret molecular network in mice. Development 2011;138(10):2089–97.

[107] Abdelhak S, Kalatzis V, Heilig R, Compain S, Samson D, Vincent C, et al. A human homologue of the Drosophila eyes absent gene underlies branchio-oto-renal (BOR) syndrome and identifies a novel gene family. Nat Genet 1997;15(2):157–64.

[108] Van Esch H, Groenen P, Nesbit MA, Schuffenhauer S, Lichtner P, Vanderlinden G, et al. GATA3 haplo-insufficiency causes human HDR syndrome. Nature 2000;406(6794):419–22.

[109] Takamiya K, Kostourou V, Adams S, Jadeja S, Chalepakis G, Scambler PJ, et al. A direct functional link between the multi-PDZ domain protein GRIP1 and the Fraser syndrome protein Fras1. Nat Genet 2004;36(2):172–7.

[110] Jadeja S, Smyth I, Pitera JE, Taylor MS, van Haelst M, Bentley E, et al. Identification of a new gene mutated in Fraser syndrome and mouse myelencephalic blebs. Nat Genet 2005;37(5):520–5.

[111] Slavotinek A, Li C, Sherr EH, Chudley AE. Mutation analysis of the FRAS1 gene demonstrates new mutations in a propositus with Fraser syndrome. Am J Med Genet 2006;140(18):1909–14.

[112] van Haelst MM, Maiburg M, Baujat G, Jadeja S, Monti E, Bland E, et al. Molecular study of 33 families with Fraser syndrome new data and mutation review. Am J Med Genet 2008;146A(17):2252–7.

[113] Nathanson J, Swarr DT, Singer A, Liu M, Chinn A, Jones W, et al. Novel FREM1 mutations expand the phenotypic spectrum associated with Manitoba-oculo-tricho-anal (MOTA) syndrome and bifid nose renal agenesis anorectal malformations (BNAR) syndrome. Am J Med Genet 2013;161A(3):473–8.

[114] Pavlakis E, Chiotaki R, Chalepakis G. The role of Fras1/Frem proteins in the structure and function of basement membrane. Int J Biochem Cell Biol 2011;43(4):487–95.

[115] Bonnans C, Chou J, Werb Z. Remodelling the extracellular matrix in development and disease. Nat Rev Mol Cell Biol 2014;15(12):786–801.

[116] Yurchenco PD, Patton BL. Developmental and pathogenic mechanisms of basement membrane assembly. Curr Pharmaceut Des 2009;15(12):1277–94.

[117] Rozario T, DeSimone DW. The extracellular matrix in development and morphogenesis: a dynamic view. Dev Biol 2010;341(1):126–40.

[118] Kiyozumi D, Takeichi M, Nakano I, Sato Y, Fukuda T, Sekiguchi K. Basement membrane assembly of the integrin alpha8beta1 ligand nephronectin requires Fraser syndrome-associated proteins. J Cell Biol 2012;197(5):677–89.

[119] Pitera JE, Scambler PJ, Woolf AS. Fras1, a basement membrane-associated protein mutated in Fraser syndrome, mediates both the initiation of the mammalian kidney and the integrity of renal glomeruli. Hum Mol Genet 2008;17(24):3953–64.

[120] Linton JM, Martin GR, Reichardt LF. The ECM protein nephronectin promotes kidney development via integrin alpha8beta1-mediated stimulation of Gdnf expression. Development 2007;134(13):2501–9.

[121] Brandenberger R, Schmidt A, Linton J, Wang D, Backus C, Denda S, et al. Identification and characterization of a novel extracellular matrix protein nephronectin that is associated with integrin alpha8-beta1 in the embryonic kidney. J Cell Biol 2001;154(2):447–58.

[122] Steer DL, Shah MM, Bush KT, Stuart RO, Sampogna RV, Meyer TN, et al. Regulation of ureteric bud branching morphogenesis by sulfated proteoglycans in the developing kidney. Dev Biol 2004;272(2):310–27.

[123] Patel VN, Pineda DL, Hoffman MP. The function of heparan sulfate during branching morphogenesis. Matrix Biol J Int Soc Matrix Biol 2017;57–58:311–23.

[124] Levy-Adam F, Feld S, Cohen-Kaplan V, Shteingauz A, Gross M, Arvatz G, et al. Heparanase 2 interacts with heparan sulfate with high affinity and inhibits heparanase activity. J Biol Chem 2010;285(36):28010–9.

[125] Vivante A, Hwang DY, Kohl S, Chen J, Shril S, Schulz J, et al. Exome sequencing discerns syndromes in patients from consanguineous families with congenital anomalies of the kidneys and urinary tract. J Am Soc Nephrol 2017;28(1):69–75.

[126] Das BC, Thapa P, Karki R, Das S, Mahapatra S, Liu TC, et al. Retinoic acid signaling pathways in development and diseases. Bioorg Med Chem 2014;22(2):673–83.

[127] Wilson JG, Roth CB, Warkany J. An analysis of the syndrome of malformations induced by maternal vitamin A deficiency. Effects of restoration of vitamin A at various times during gestation. Am J Anat 1953;92(2):189–217.

[128] Shannon SR, Moise AR, Trainor PA. New insights and changing paradigms in the regulation of vitamin A metabolism in development. Wiley Interdiscip Rev Dev Biol 2017;6(3). https://doi.org/10.1002/wdev.264.

[129] Mark M, Ghyselinck NB, Chambon P. Function of retinoic acid receptors during embryonic development. Nucl Recept Signal 2009;7:e002.

[130] Wilson JG, Warkany J. Aortic-arch and cardiac anomalies in the offspring of vitamin A deficient rats. Am J Anat 1949;85(1):113–55.

[131] Mendelsohn C, Batourina E, Fung S, Gilbert T, Dodd J. Stromal cells mediate retinoid-dependent functions essential for renal development. Development 1999;126(6):1139–48.

[132] Batourina E, Gim S, Bello N, Shy M, Clagett-Dame M, Srinivas S, et al. Vitamin A controls epithelial/mesenchymal interactions through Ret expression. Nat Genet 2001;27(1):74–8.

[133] Rosselot C, Spraggon L, Chia I, Batourina E, Riccio P, Lu B, et al. Non-cell-autonomous retinoid signaling is crucial for renal development. Development 2010;137(2):283–92.

[134] Batourina E, Choi C, Paragas N, Bello N, Hensle T, Costantini FD, et al. Distal ureter morphogenesis depends on epithelial cell remodeling mediated by vitamin A and Ret. Nat Genet 2002;32(1):109–15.

[135] Batourina E, Tsai S, Lambert S, Sprenkle P, Viana R, Dutta S, et al. Apoptosis induced by vitamin A signaling is crucial for connecting the ureters to the bladder. Nat Genet 2005;37(10):1082–9.

[136] Rhinn M, Schuhbaur B, Niederreither K, Dolle P. Involvement of retinol dehydrogenase 10 in embryonic patterning and rescue of its loss of function by maternal retinaldehyde treatment. Proc Natl Acad Sci USA 2011;108(40):16687–92.

[137] Abu-Abed S, Dolle P, Metzger D, Beckett B, Chambon P, Petkovich M. The retinoic acid-metabolizing enzyme, CYP26A1, is essential for normal hindbrain patterning, vertebral identity, and development of posterior structures. Gene Dev 2001;15(2):226–40.

[138] Sakai Y, Meno C, Fujii H, Nishino J, Shiratori H, Saijoh Y, et al. The retinoic acid-inactivating enzyme CYP26 is essential for establishing an uneven distribution of retinoic acid along the anterio-posterior axis within the mouse embryo. Gene Dev 2001;15(2):213–25.

[139] Vivante A, Mann N, Yonath H, Weiss AC, Getwan M, Kaminski MM, et al. A dominant mutation in nuclear receptor interacting protein 1 causes urinary tract malformations via dysregulation of retinoic acid signaling. J Am Soc Nephrol 2017;28(8):2364–76.

2

Cystic Diseases of the Kidney

Frederic Rahbari Oskoui[1], Michal Mrug[2], Takamitsu Saigusa[2], Dana V. Rizk[2]

[1]Emory University, Atlanta, GA, United States
[2]University of Alabama at Birmingham, Birmingham, AL, United States

2.1 INTRODUCTION

Genetic cystic diseases of the kidneys are some of the most significant monogenic causes of renal disorders causing significant morbidity and mortality in both pediatric and adult populations. In the past few decades, there have been considerable advances in identifying the genes responsible for these disorders and characterizing their protein products. This allowed the recognition of the primary cilium as a crucial structure in the pathogenesis of these disorders. The genetic aspects, molecular pathogenesis, clinical features, and management of these disorders will be highlighted in this chapter.

2.2 AUTOSOMAL DOMINANT POLYCYSTIC KIDNEY DISEASE (MIM 173900)

2.2.1 Clinical Features and Natural History

Autosomal dominant polycystic kidney disease (ADPKD), once erroneously called "adult polycystic kidney disease," is the most common inherited disease of the kidney, affecting an estimated 12.5 million people worldwide with a prevalence of 1/500−1/1000 live births and accounting for 5%−10% of all cases of adult end-stage kidney disease (ESKD) [1,2]. Two genes, *PKD1* and *PKD2*, have been identified to cause ADPKD encoding two distinct proteins polycystin 1 and polycystin 2 [3,4]. Other genetic factors have been discovered to explain the phenotypic variability. Patients with a *PKD1* mutation have earlier onset of disease and reach ESKD at an earlier age than patients with *PKD2* mutations. However, the annual rate of progression [decline in glomerular filtration rate (GFR) or increase in total kidney volume] is similar in both mutations [5].

ADPKD is often not diagnosed until adulthood. Many patients are incidentally diagnosed while having an imaging study during pregnancy, back pain, or accidents. Hematuria, renal colic, recurrent urinary tract infection (UTI), lower back discomfort, and hypertension are common presenting symptoms. Progressive disease is defined by cyst growth and decline in kidney function, leading to ESKD in majority of patients. Risk factors for rapid progression include hypertension, gross hematuria, nephrolithiasis, microalbuminuria/proteinuria, and *PKD1*mutation [3]. Women with *PKD2* mutations preserve renal function longer than men with the same mutations, but no gender difference is observed with *PKD1* mutations. Median age of onset of ESKD or death was reported at 53 years in *PKD1* patients and 69 years in *PKD2* patients based on the original historical cohorts [6].

Emery and Rimoin's Principles and Practice of Medical Genetics and Genomics. https://doi.org/10.1016/B978-0-12-812534-2.00005-9

Although ADPKD is a completely penetrant disorder, a significant degree of clinical variability exists. Onset of hypertension is typically early: 50% of patients develop hypertension with a mean age of onset of 31 years, long before the development of ESKD, and sometimes before the discovery of renal cysts [7]. Intrarenal activation of the renin–angiotensin–aldosterone (RAA) system occurs early and has been proposed as a cause of hypertension in ADPKD. Cyst enlargement and compression of normal parenchyma are also hypothesized to decrease intrarenal perfusion and cause more tubulointerstitial and glomerular fibrosis [8]. However, this theory does not adequately explain why the age- and gender-adjusted prevalence of hypertension is four times greater in those with *PKD1* compared to *PKD2* mutations. About 25%–30% of patients develop cyst infection, and 18%–25% of patients develop gross hematuria from cyst hemorrhage [9]. Nephrolithiasis, seen in 10%–20% of all patients, occurs with equal frequency in PKD1 and PKD2 and is due mostly to either urate or oxalate crystals. Expanding cysts compressing the collecting system producing urinary stasis, along with low urine pH and hypocitraturia, contribute to stone formation [10].

Hepatic cysts are commonly present in ADPKD patients. The incidence of hepatic cysts increases with age and is higher (up to 83%) if more sensitive imaging modalities such as magnetic resonance imaging (MRI) are used in adults younger than 46 years of age. In comparison, ultrasound imaging in adults older than 60 years of age only reveals cysts in 50%–70% of the cases [11,12]. Hepatic cysts present approximately 10 years after the development of renal cysts. The overall prevalence of hepatic cysts is similar in men and women, but the number and size of these cysts increase with age female gender, and multiparity correlates with earlier onset and increased burden of liver involvement. This phenomenon is thought to be due to exposure to female steroid hormones. Polycystic liver disease (PLD) is rare in children and is typically asymptomatic. Symptoms are being seen more commonly as the life span of ADPKD patients has lengthened with dialysis and transplantation [13]. The primary complication of polycystic liver disease is severe hepatomegaly with abdominal discomfort and shortness of breath; hepatic function is typically preserved. Cyst infection, hemorrhage, and posttraumatic rupture occur much less often. Portal hypertension with ascites, variceal bleeding, and encephalopathy are very rare but are perhaps the most significant hepatic sequela of ADPKD [11].

Cardiac valvulopathies, specifically mitral valve prolapse (MVP), hemodynamically significant mitral regurgitation (MR), or aortic insufficiency (AI), are also seen in patients with ADPKD with respective estimated prevalence of 25%–30%, 13%, and 8% [14–16]. Valvular changes are due to myxomatous degeneration and loss of collagen. Patients with hypertension also develop compensatory left ventricular hypertrophy, but the prevalence of LVH seems to be lower (0.74%–2.23%) since broad utilization of RAAS blockers in ADPKD [17]. Dilatation of the aortic root and thoracic aortic dissection have also been reported. Aortic dissections tend to occur earlier in ADPKD patients compared to other non-PKD populations who are almost invariably hypertensive. Therefore, chest and back pain even in young patients with ADPKD should require a workup to rule out aortic dissection [18]. Although abdominal aortic aneurysms and dissection were thought to occur with increased frequency in ADPKD patients, subsequent studies have shown that this is not the case [19].

Intracranial aneurysms, typically small saccular aneurysms of the circle of Willis, are found in approximately 10% of all ADPKD patients and are less common in those younger than 30 years The risk factors for aneurysms are older age, personal or family history of cerebral aneurysms or subarachnoid hemorrhage, Finnish or Japanese ancestry, smoking, and excessive alcohol intake [20–22]. PKD1 and PKD2 patients are equally at risk [23]. Aneurysms are found in 22%–25% of those with a family history of aneurysms and ADPKD and 6% of those without a family history [24–26]. Most aneurysms are small, with 90% less than 10 mm in diameter and 70% less than 6 mm in diameter [25]. Over 95% of aneurysms identified in asymptomatic screening are less than 6 mm in diameter [27,28]. Aneurysm surveillance studies indicate that aneurysms smaller than 10 mm have a low rupture risk of 0.05% per year, but since the studies were done on the general population, it is uncertain whether this low risk can be applied to ADPKD patients alone [29]. Aneurysmal rupture results in subarachnoid hemorrhage (SAH), perhaps the most sudden and devastating complication of ADPKD. Symptoms of rupture are no different from those in non-ADPKD patients and include sudden-onset severe headache, acute focal neurologic deficit, altered mental status, nausea, vomiting, or photophobia. The estimated immediate mortality of such an event is 10%, with 38% long-term, permanent morbidity in survivors [30]. The mean age of rupture is 40 years, with 65%–75% of all ruptures occurring prior to age 50 [20]. About 10% of aneurysm rupture occurs in ADPKD patients younger than 20 years [31]. Patients with family history of aneurysms or strokes, patients who will undergo major surgery or being considered for chronic anticoagulation or patients with high risk jobs (pilots, drivers, etc.) should be offered an symptomatic screening of cerebral aneurysms by using a time-of-flight magnetic resonance angiography (MRA) without contrast [32]. This modality does not require Gadolinium injection and is safe at any level of estimated GFR (eGFR). In case of a positive MRA, a four-vessel angiography of the cerebral vessels should be considered. Recognition of symptoms of SAH and emergent neurosurgical intervention are crucial for survival and reduction of long-term morbidity.

Colonic diverticular disease has been reported to occur with increased frequency in ADPKD patients with ESKD awaiting transplant, but not in patients without ESKD [33,34]. Diverticulitis, if not recognized and treated promptly, can lead to colonic perforation. Ventral and inguinal hernias have also been documented in ADPKD patients with ESKD in up to 45% of the cases [34,35].

In males, cysts can be found in the seminal vesicles, epididymis, and prostate in up to 40% of the cases. Sperm abnormalities and defective motility can be also seen but rarely cause male infertility [36]. Cysts of the pancreas are rare and usually asymptomatic. However, patients may develop recurrent pancreatitis and intraductal papillary mucinous tumor or carcinoma. Arachnoid cysts may occur and present with headache, diplopia, hearing loss, and ataxia due to cerebrospinal fluid leak [37].

2.2.2 Gene Mapping, Structure, Function, and Genotype–Phenotype Correlations

Historically, pathogenetic variants in two genes (*PKD1* and *PKD2*) have been shown to cause ADPKD. Loss-of-function variants in *PKD1* and *PKD2* account for 75.5% and 18.3% of cases of ADPKD based on the Genkyst cohort in Europe or 74.6% and 16.4% based on the HALT-PKD Cohort in the United States. The remaining 6.2%–9% of patients had no pathogenic variant detected [38,39]. However, recent studies have suggested new causal or modifying genetic variants in GANAB, *DNAJB11*, and *COL4A1* to explain some of the phenotype–genotype variability.

2.2.2.1 PKD1 and PKD2 Variants

Early studies demonstrated linkage between the *PKD* locus and the α-globin locus on chromosome 16p. Subsequent linkage studies refined the locus to a 750 kilobase (kb) segment at 16p13 [40,41]. The presence of several *PKD* pseudogenes made direct positional cloning difficult, but in 1995, sequencing of the *PKD* cDNA showed a 14,148 bp transcript with 46 exons spanning 52 kb of genomic DNA [42]. The *PKD1* protein, called polycystin-1, is a plasma membrane protein containing 4304 amino acids and weighing 463 kDa (Fig. 2.1). The N-terminal extracellular region is composed of an aggregation of multiple domains involved in protein–protein and protein–carbohydrate interactions, while the C-terminal intracellular region ends in a series of protein–protein interaction motifs called coiled-coil regions. In addition to the extracellular and intracellular signaling regions, polycystin-1 contains a sea urchin receptor for egg jelly (suREJ) domain [43]. This domain offers some insight into the function of the protein. The sea urchin homolog is involved in the activation of the sperm's acrosome reaction. Contact of the sperm with the egg's zona pellucida results in calcium ion influx triggered by the suREJ protein, leading to release of proteolytic enzymes and polymerization of globular actin.

PKD2 gene, a 680 kb segment on chromosome 4q21, containing 15 exons spanning 68 kb and encoding a 5.4 kb mRNA, encodes the polycystin-2 protein. It was cloned several years after the discovery of *PKD1*, from a group of patients who were unlinked to the *PKD1* gene. The original idea had come from the observation of two families with PKD where the linkage of PKD to the α-globin locus did not link to chromosome 16p [44–46]. The polycystin-2 translated product, a 110 kDa protein, contains 968 amino acids and 6 transmembrane domains (Fig. 2.1). Polycystin-2 belongs to a subfamily of transient receptor potential channels (TRPCs) and functions as a calcium channel. The C-terminal intracellular region shares homology with the C-terminal region of *PKD1*, and

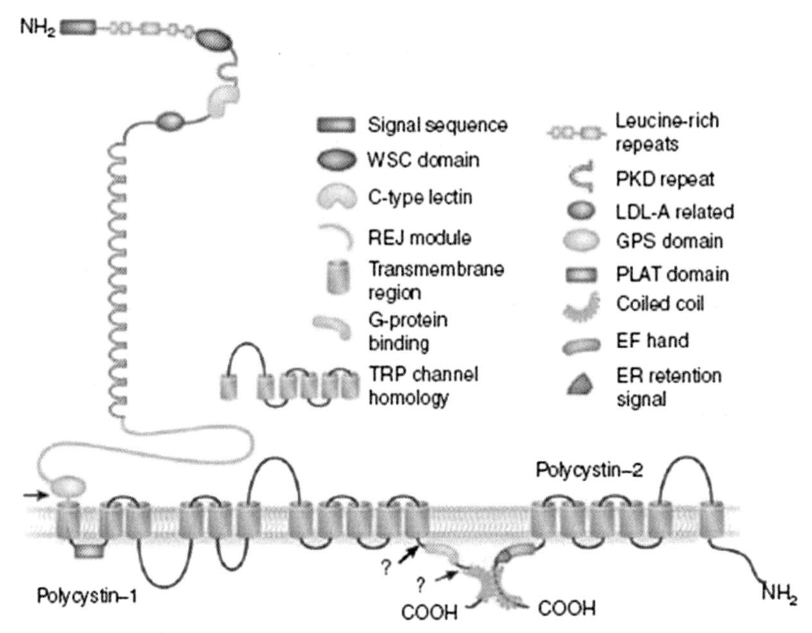

Figure 2.1 The polycystin1—polycystin2 complex. (From Torres VE, Harris PC. Autosomal dominant polycystic kidney disease: The last 3 years. Nature 2009;76(2).)

coimmunoprecipitation studies indicate that *PKD1* and *PKD2* interact with one another via their C-terminal coiled-coil regions [47,48].

More recent studies based on cryogenic electron microscopy (Cryo-EM) generated a high-resolution of the Polycystin-1 plus Polycystin-2 complex showing a 1:3 ration (in favor of PC2) and overall dimensions of 130 Å × 110 Å × 130 Å [49] (Fig. 2.2). The voltage-gated ion channel (VGIC) fold consisting of the last six transmembrane domains of PC1 and a regulating TOP domain (homologous to that of PC2) substitutes for one PC2 molecule in the complex. The other five transmembrane domains and the PLAT (polycystin 1, lipoxygenase, and α-toxin) region of PC1 constitute the amino-terminal domain that associates with the 24 transmembrane domains of the four VGIC regions. The structure of the last transmembrane domain of PC1 is distinct from the corresponding domain of PC2 in that it splits into two segments offset at 120 degrees and lacks a PC2 selective filter and supporting pore helices. This region contains three positively charged residues facing the pore that would likely impede Ca^{2+}permeability. These findings suggested that the PC1-PC2 complex may not function as a Ca^{2+} channel [50].

Polycystin-1 and polycystin-2 are ubiquitously expressed, at higher levels during embryonic development, then at lower levels during postnatal and adult life [51—53]. mRNA expression is observed in the brain, cardiac and skeletal muscles, smooth muscle of blood vessels, vascular endothelium, breast, lung, liver, pancreas, kidney, and reproductive organs [54,55]. Within the fetal kidney, the two proteins are expressed in maturing renal tubular epithelium and at lower levels at the distal ureteric buds. Both proteins are required for normal embryonic development; *PKD1* knockout mice die at birth because of pulmonary hypoplasia from massive cystic enlargement of the kidneys and pancreas while the *PKD2* knockout dies in the late embryonic period with abnormalities in cardiac atrioventricular septation and cysts in the kidneys and pancreas [56,57]. Postnatally, the proteins are seen in the distal convoluted tubular and collecting duct epithelia. To date, 2323 variants in *PKD1* have been reported, most of which result in protein truncation; 868 variants are classified as definitely pathogenic, and 185 are highly likely pathogenic (www.pkdb.mayo.edu *accessed on 5/20/2021*). Of the "definitely pathogenic" cases, 449 (51.7%) are frameshift, 263 (30.3%) nonsense, 45 (10%)

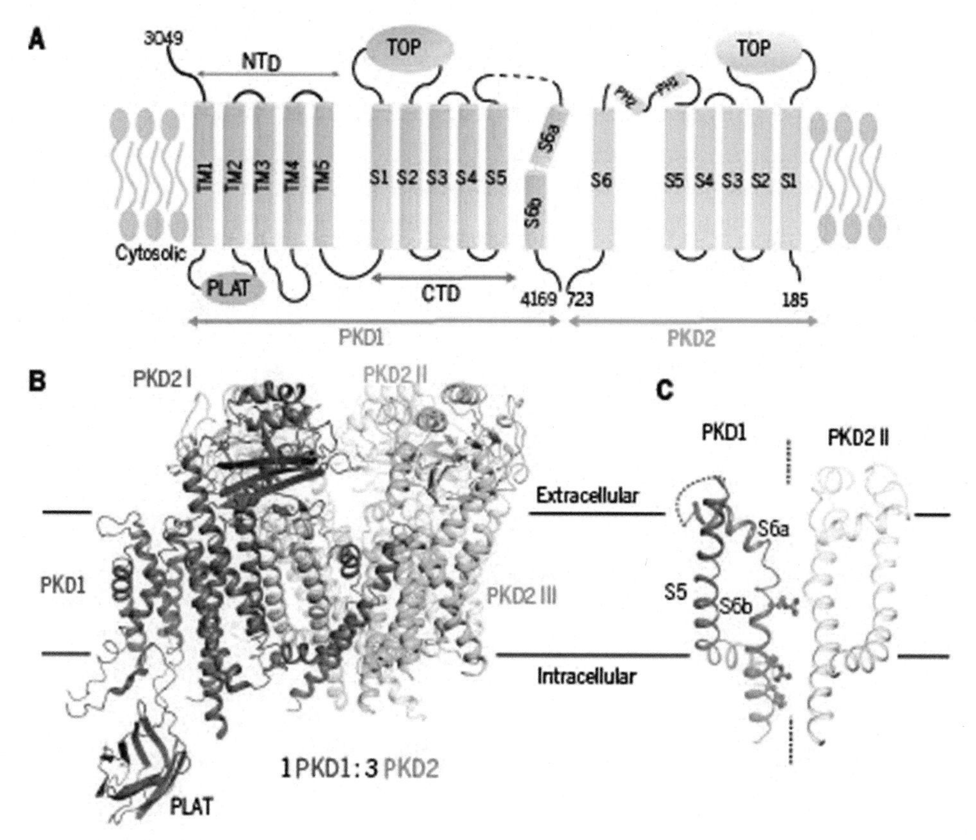

Figure 2.2 The Cryo-EM structure of the truncated human PKD1-PKD2 complex at 3.6-A resolution. A) Topological illustration of PKD1 and PKD2. NTD, N-terminal domain; TOP, aslo known as the polycystin domain; CTD, C-termianl domain (which includes S1 to S6 and the TOP domain). B) The 1:3 organization of the PKD1-PKD2 complex. PKD2 I, II and II are the three PKD2 subunits. C) Unconventional conformation of the S6 segment in PKD1. The sequence between the S5 and S6 segments are flexible and disordered in PKD1. Structure of polycystin 1 and polycystin 2 complex based on Cryo-EM. (Reproduced with permission from Su Q, Hu F, Ge X, Lei J, Yu S, Wang T, Zhou Q, Mei C, Shi Y. Structure of the human PKD1-PKD2 complex. Science 2018;361(6406):eaat9819.)

splice site, 51 (5.9%) deletions, and 12 (1.4%) insertions. The high mutation rate in the *PKD1* gene is thought to be related to the large polypyrimidine tract in intron 21, the longest such tract known in the human genome [58]. The high cytosine and thymine content allows for triple-helix formation, which can result in mutagenesis.

A large majority of germline *PKD2* mutations result in a truncated protein product too. A total of 278 variants have been described, of which 162 have been classified as definitely pathogenic and another 17 as highly likely pathogenic (www.pkdb.mayo.edu *accessed on 5/20/2021).* Of these "definitely pathogenic" cases, 75 (46.3%) are frameshift, 48 (29.6%) nonsense, 31 (19.1%) splice site, and 6 (3.7%) or deletions, and 1 (0.06%) insertion. While the variants are located throughout the PKD2 gene, a slight clustering is seen in exons 4, 5, and 6, which make up 24.2% of the cDNA sequence but account for 38.6% of germline variants. There are several unrelated families (by ethnicity and haplotype analysis) with identical variants, indicating a few mutation-prone "hot spots," but most PKD2 germline variants are unique to the affected family.

Although all cells in an ADPKD kidney inherit the same germinal pathogenic variant, histopathologic studies indicate that only 1%–5% of all nephrons have cysts [59]. Experience gained from *Rb* variants and hereditary retinoblastoma led to the hypothesis that a "second hit," or somatic inactivation of the normal PKD gene, is necessary to trigger cystogenesis in both the kidney and the liver. The second hit hypothesis was confirmed in both *PKD1* and *PKD2* patients. Clonal loss of heterozygosity, specifically, loss of the normal *PKD1* haplotype, was demonstrated in cysts isolated from *PKD1*kidneys and the same was demonstrated for *PKD2* kidney and liver [60–62]. Truncating and missense variants from the normal *PKD1* haplotype were discovered [63]. Cysts from *PKD2* kidneys and livers were also found to have truncating variants in the inherited wild-type *PKD2* gene. These studies indicate that, while inherited in a dominant fashion, ADPKD is recessive at the cellular level. In addition, a transheterozygous state can arise when compound heterozygous variants occur in both *PKD1* and *PKD2* [63–65].

Regarding genotype–phenotype correlations, it is well known that mutations in *PKD1* are associated with more severe disease than those in *PKD2*. Individuals with *PKD1* pathogenic variants are diagnosed at an earlier age, have a higher incidence of hypertension and hematuria, and progress to ESKD on average 20 years earlier [6,66]. PKD1 kidneys are significantly larger than age-matched *PKD2* kidneys because of a greater number of cysts rather than the rate of cystic expansion [5,67]. This finding is consistent with the two-hit model of cystogenesis as *PKD1* is more prone to mutation.

Variants of any type at the 5′ end of *PKD1* are correlated with decreased renal survival, earlier onset of ESKD, and increased risk of intracranial aneurysms compared to variants at the 3′ end of the gene. There appears to be no significant difference in the severity of disease between families with truncating, in-frame, or missense deletions [23,68]. With *PKD2*, splice site variants appear to produce milder renal symptoms compared with other types of variants [69]. Excepting this association, no other genotype–phenotype correlations have been determined from genomic analysis [70,71]. There is, however, a gender effect observed with *PKD2* disease, with females having milder renal involvement than males [69]. Furthermore, truncating variants carry a worse prognosis: truncating *PKD1* variants (65% of cases) reach ESKD at a mean age of 55.6 years compared to 67.9 years for nontruncating *PKD1* variants (35% of cases), and 79.7 years for PKD 2 [38].

In addition to the phenotypic variation observed in families with different variants, individuals within a family can manifest with differing degrees of clinical severity [71]. Genetic modifiers have been purported to play a role in this intrafamilial variability, and their effect has been estimated to be between 18% and 59% in those with *PKD1* variants. Thus far, association studies of the angiotensin-1-converting enzyme (*ACE*) gene and the endothelial nitric oxide synthase (*ENOS*) gene have shown equivocal results [72–75]. One clear example of genetic background affecting phenotype expression is the contiguous gene deletion syndrome involving the tuberin (*TSC2*) and *PKD1* genes, both located on chromosome 16p13.3 [76] (Fig. 2.3). Affected individuals suffer from tuberous sclerosis and have severe polycystic kidney and liver disease typically arising in infancy to early childhood

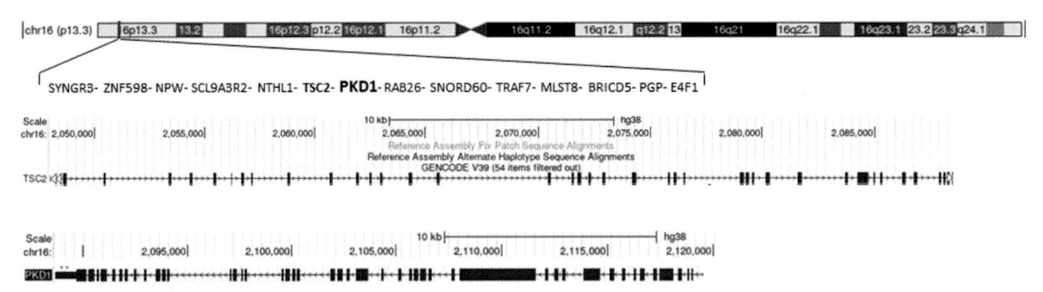

Figure 2.3 Schematic representation of PKD1 and TSC2 loci and their surrounding genes on the short arm of Chromosome 16. *BRICD5*, BRICHOS domain containing 50; *CASKIN1*, CASK interacting protein 1; *E4F1*, E4F transcription factor 1; *MLST8*, mammalian lethal with SEC13 protein 8; *NPW*, neuropeptide W protein; *NTHL1*, endonuclease III-like protein 1; *PGP*, phosphoglycolate phosphatase; *PKD1*, polycystic kidney disease 1; *RAB26*, Ras-related protein Rab-26; *SCL9A3R2*, SLC9A3 regulator 2; *SNORD60*, small nucleolar RNA, C/D Box 60; *SYNGR3*, synaptogyrin 3; *TRAF7*, TNF receptor associated factor 7; *TSC2*, tuberous sclerosis complex 2; *ZNF598*, zinc finger protein 598. (Taken from the University of California Santa Cruz Genome Browser at http://www.genome.ucsc.edu/.)

with rapid progression to ESKD. In the past, tuberin was thought to function in localizing polycystin-1 to the plasma membrane [77]. Furthermore, the cytoplasmic tail of polycystin-1 directly interacts with tuberin in the mammalian target of rapamycin (mTOR) pathway, and inhibition of this pathway reverses renal cystogenesis [78].

The immune response may also play a role in pathogenesis of ADPKD. One of the strongly suggestive mediators is monocyte chemoattractant protein 1 (MCP1), which recruits monocytes into the sites of inflammation and controls T Helper lymphocyte differentiation TH1 or TH2 phenotypes. MCP1 has shown a pathogenic role in PKD by promoting macrophage-dependent cyst expansion [50,79]. Urine MCP1 levels have been correlated with kidney function decline and proposed as a novel biomarker of ADPKD [80].

2.2.2.2 Other Genetic Mutations (Table 2.1)

GANAB: The missense variant in neutral α-glucosidase AB *(GANAB)* encoding glucosidase II subunit a has been reported to cause both polycystic kidney disease and polycystic liver disease. *GANAB* (located in chromosomal region 11q12.3; genomic size 21.9 kb) has two splice forms shown by in silico and RT-PCR analysis to be approximately equally expressed in the human kidneys and liver: isoform 3 (GenBank: NM_198335.3) has 966 aa (~110 kDa), 25 exons, and 2898 bp of coding sequence, and isoform 2 (GenBank: NM_198334.2) has 944 aa (~107 kDa), 24 exons (in-frame skipping of exon 6), and 2832 bp of coding sequence [81–83]. This variant has been reportedly found in 3% of the genotypically unresolved cases, which represents 0.3% of all ADPKD patients [84]. Patients with GANAB variant present with a milder renal phenotype (probably even milder than PKD2 individuals) with absence of renomegaly, typically few large cysts, and preserved renal function. The liver phenotype can be variable ranging from no liver cysts to severe polycystic, liver.

DNAJB11: variants in *DNAJB11*, discovered in few families of patients with atypical ADPKD, can lead to an ADPKD-like phenotype [85]. DNAJB11 has a 1698 bp mRNA (GenBank: NP_016306.5) with a 1077 bp coding segment, extending over 15,124 bp of genomic DNA (186,570,676–186,585,800 nt) as 10 exons in chromosome region 3q27. DNAJB11 encodes a soluble glycoprotein (358 aa, MW 1⁄4 40.5 kDa; GenBank: NP_057390.1) of the endoplasmic reticulum (ER) lumen, one of the most abundant cochaperones of binding immunoglobulin protein (BiP, also known as GRP78), a heat shock protein chaperone required for the proper folding and assembly of proteins in the ER [86,87]. All affected individuals presented with non-enlarged polycystic kidneys, preserved renal function, and some patients had multiple liver cysts. The *DNAJB11* variants were found in 0.2% of the genetically unresolved cases, which represents around 0.02% of all APKD cases [83].

COL4A1: Isolated or syndromic form of ADPKD was identified by observation in a single family of

TABLE 2.1 Main Pathogenic Variants in ADPKD

Variant Type	Gene Location	Phenotype	Proportion of ADPKD	Age of Onset of ESKD (years)
PKD1 - Truncating	16p13.3	Most severe Early onset of renomegaly	49%	55.6
- Non-truncating		Intermediate severity	26%	67.9
PKD2	4q21.2	Less severe Later onset renomegaly	16%–18%	79.7
GANAB	11q12.3	Least severe renal phenotype Polycystic liver disease +++	0.03%	Variable 70–80s (if ever)
DNAJB11	3q27.3	Normal size kidneys	Extremely rare	60s
COL4A1/PKD2	13q34	ADPKD/HANAC	Extremely rare	50s

digenic *PKD2* and *COL4A1* variants that led to a more severe phenotype. Several individuals reached ESKD before or around age of 50. However, other individuals in the same pedigree did not have PKD2 variant but presented with the hereditary angiopathy with nephropathy, aneurysms, and muscle cramps (HANAC) syndrome and PKD-like renal cysts [88]. A cytosine to thymine substitution at nucleotide 739 of the coding sequence (c.739C>T) in exon 13 of COL4A1 was identified and predicted to lead to a nonsense variant (introduction of a premature stop codon) at the glutamine at amino acid 247 (p.Gln247*).

2.2.2.3 Nongenetic Factors

Nongenetic factors also contribute to phenotypic variation in ADPKD. Hormonal effects have been described, including the predominance of cystic liver disease in females due to the effects of estrogen [89]. Smoking and other premutagenic factors may confer higher risk by increasing the chance for somatic mutations, thereby providing the "second hit" [90]. More recently, the concept of a "third hit" that can cause accelerated and extensive cystic proliferation was introduced base on the observations of renal ischemia–reperfusion injury models in Pkd1 IKO mice that caused extensive tubular damage [91]. But human studies to evaluate the role of ischemic tubular necrosis and injury in ADPKD are lacking currently.

2.2.3 Laboratory, Imaging, and Histopathologic Findings

Serum electrolyte studies are normal unless there is advanced chronic kidney disease. Urinalysis reveals microscopic or gross hematuria in about 40%–45%, especially if there is cyst hemorrhage [92]. Dipstick proteinuria is present in 20% of patients and microalbuminuria in about 35% [93]. Liver transaminases, alkaline phosphatase, and markers of hepatic synthetic function such as prothrombin time and partial thromboplastin time are usually normal even with severe cystic disease. Gamma-glutamyltransferase levels correlate with hepatic cyst burden [11]. Abdominal ultrasound has become the modality of choice for diagnosis of ADPKD due to its wide spread availability, low cost, and lack of exposure to radiation [94]. The original Ravine diagnostic criteria, where the number of cysts was stratified based on three age categories (younger than 30 years, between 31 and 59 years and greater than 60 years of age), were used to confirm ADPKD [95]. Those criteria are highly specific (near 100%), but sensitivity is age-dependent, suggesting that the longer an individual remains negative, the lower the likelihood of disease [96]. An international collaborative effort, where both genetic mutation analysis (*PKD1 and PKD2*) and ultrasound imaging studies were used, established the unified diagnostic criteria for ADPKD in 2009 [97]. In individuals aged 15–39, the presence of at least three (unilateral or bilateral) kidney cysts is sufficient to establish a diagnosis of ADPKD. In those individuals 40–59 years of age, two cysts in each kidney are required, and in those older than 60, in whom acquired cystic disease is common, four or more cysts in each kidney are required for diagnosis. For patients with no family history, the diagnostic criteria are more stringent with at least five cysts bilaterally by the age of 30 and a phenotype consistent with ADPKD required. Furthermore, inclusion and exclusion criteria were defined, the latter being particularly useful to clear a family member of an affected ADPKD patient who wants to be considered for kidney donation [90]. The performance characteristics of these criteria are presented in Table 2.2.

Computerized tomography: Computerized tomography (CT) and magnetic resonance imaging (MRI) can be used in diagnosis of ADPKD and its potential complications [98]. They are more sensitive to diagnose cysts and can detect cysts as small as 1 mm. Therefore, the total number of cysts is typically higher compared to ultrasound imaging. Enlarged kidneys with multiple anechoic cysts of varying sizes throughout the cortex, distorting normal pelvicalyceal architecture are commonly observed. Cyst hemorrhage or infection may be visualized as echogenic debris within the cyst. Hepatic, pancreatic, and splenic cysts can be visualized with ultrasound as well. CT with intravenous contrast or MRI can be used for diagnosis of ADPKD when higher-resolution imaging is required.

Renal biopsy of patients with ADPKD is not recommended because of risk of bleeding. Gross renal specimens are enlarged with cysts present over the surface of the kidney. Cut specimens reveal cysts throughout the thickness of the kidney, filled with straw-colored, serosanguinous, or brown fluid. Infected cysts are filled with purulent material. Microscopically, interstitial fibrosis and intact nephrons can be seen in

TABLE 2.2 **Performance Characteristics of Ultrasound-Based Cyst Number Criteria for the Diagnosis and Exclusion of ADPKD [90]**

Purpose of Imaging Testing	Number of Cysts	Age of Diagnosis	Performance Characteristics: Known PKD1 Mutation	Performance Characteristics: Known PKD2 Mutation	Performance Characteristics: Unknown Family Mutation
Confirmation of diagnosis	≥3 cysts bilaterally	15–30	PPV: 100% Sens: 94.3%	PPV: 100% Sens: 69.5%	PPV: 100% Sens: 81.7%
	≥3 cysts bilaterally	30–39	PPV: 100% Sens: 96.6%	PPV: 100% Sens: 94.9%	PPV: 100% Sens: 95.5%
	≥2 cysts in each kidney	40–50	PPV: 100% Sens: 92.6%	PPV: 100% Sens: 88.8%	PPV: 100% Sens: 90%
Exclusion of disease	≥1cyst	15–30	NPV: 99.1% Spec: 97.6%	NPV: 83.5% Spec: 96.6%	NPV: 90.8% Spec: 97.1%
	≥1 cyst	30–39	NPV: 100% Spec: 96%	NPV: 96.8% Spec: 93.8%	NPV: 98.3% Spec: 94.8%
	≥1 cyst	40–59	NPV: 100% Spec: 93.9%	NPV: 100% Spec: 93.7%	NPV: 100% Spec: 93.9%

Reproduced with permission from Pei Y. Practical genetics for autosomal dominant polycystic kidney disease. Nephron Clin Pract 2011;118:c19-c30. https://doi.org/10.1159/000320887.

the parenchyma between cysts. Cysts are lined with renal epithelium originating primarily from the proximal convoluted tubule and collecting [99]. Hepatic cysts are lined with biliary duct epithelium.

2.2.4 Molecular and Prenatal Diagnosis/ Preimplantation Genetic Diagnosis

As mentioned earlier, abdominal ultrasound is the first line diagnostic tool of choice. This point is especially more relevant to individuals at risk for PKD1. CT scan or MRI may be used as an adjunct if ultrasound is equivocal, higher resolution is needed, or if total kidney volume needs to be calculated. Molecular diagnostic testing is commercially available for both *PKD1* and *PKD2*, however, the high cost and restricted insurance coverage have limited their clinical use in the United States. Sequence analysis has a variant detection rate of approximately 90% for *PKD1* [100]. An additional 4% of patients may have gross rearrangements, which are detectable using multiplex ligation-dependent probe amplification [101]. For *PKD2*, over 90% of variants are identified with direct sequence analysis. There are limitations to molecular testing, however. Most *PKD1* and *PKD2* mutations are "private" with very few recurrent variants. Consequently, sequencing results must be interpreted with caution, as some sequence variations may represent polymorphisms that are not pathogenic. In fact, each individual is known to carry more than 10 polymorphic variants in the *PKD1* gene [102]. In addition, exon sequencing does not detect variants in the noncoding regions of the genes. Extensive and updated libraries of pathogenic and likely pathogenic variants should be routinely consulted when interpreting the results of a genetic testing for inherited renal cystic diseases and ADPKD.

Preimplantation genetic diagnosis (PDG) may be considered in families with or without infertility to minimize the risk of transmitting the PKD mutations to their offspring. IVF with PGD has been reported to successfully prevent transmission of ADPKD to a child [103]. The chance of having a child with a monogenic disease (and more particularly in ADPKD) after PGD + IVF procedure is around 4%–5%. But those odds are primarily based on the prerequisite of identifying a pathogenic variant in the parent. If direct mutation analysis fails to identify a single pathogenic variant, either because multiple potential pathogenic variants are identified or no mutation is detected, the use of family-based linkage analysis may be required to determine the locus of the pathogenic PKD variant [104]. Without proper identification of the locus, the success rate of the procedure is expected to be significantly lower. The benefits of PGD need to be carefully weighed against the risks of hyperstimulation syndrome and the high cost of these procedures (IVF + PDG).

Prenatal diagnosis can be performed via analysis of cells obtained from chorionic villus sampling or amniocentesis if linkage to a specific haplotype or disease-causing variants has already been established. However, due to the availability of treatment, lack of cognitive impairment, and adult onset of disease, prenatal diagnosis is usually not performed unless there is a family history of a severely affected infant.

2.2.5 Total Kidney Volume: A Novel Biomarker for Disease Progression

Given the limited utility of serum creatinine in defining disease progression in ADPKD patients with preserved renal function, The Consortium for Radiologic Imaging Studies in Polycystic kidney disease (CRISP) was funded in late 1990 to develop and validate new imaging markers for polycystic kidney disease. Through an impressive and persistently innovative effort over two decades, CRISP established the value of volume progression in ADPKD. The study validated the relationship between the total kidney volume (TKV) measured by MRI and iothalamate clearance and estimated GFR. Increase in TKV is shown to precede the decline in kidney function, and TKV is a better marker for disease progression than serum creatinine [5]. Longitudinal follow-up of this cohort, at 8 years, established the prognostic value of TKV (assessed by MRI) to predict progression to chronic kidney disease (CKD)-stage 3 [105]. Furthermore, a classification tool, named the Irazabal or the Mayo Classification, was developed in 2014 to categorize patients with typical ADPKD (called class 1) into five distinct progression groups (1A, 1B, 1C, 1D, 1E) based on their height-adjusted TKV and age [106] (Fig. 2.4).

This tool was originally intended to help with selection of patients with fast progression to enhance recruitment for clinical trials but rapidly became a clinical tool for selection of patients for tolvaptan after this drug became available on the market. It also gave birth to an online calculator that could predict the age of onset of ESKD in these typical individuals. ADPKD is the only kidney disease where such a tool has been validated (https://www.mayo.edu/research/documents/classification-of-typical-adpkd-calculator/doc-20094754). One needs to remember that the Mayo classification tool is currently only validated for typical ADPKD (Class 1 cases) and not for patients with atypical APDKD (Class 2) who present with asymmetric, lop-sided, unilateral, segmental, unilaterally or bilaterally atrophic disease, and constitute around 10% of all ADPKD patients [107].

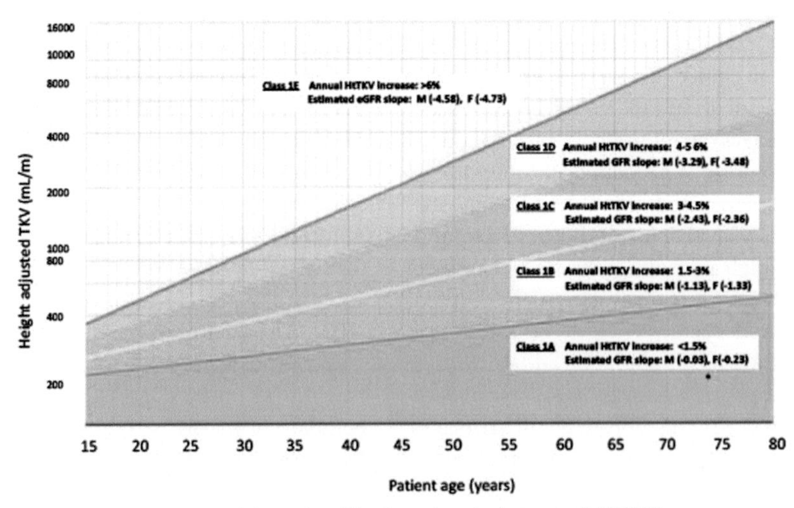

Figure 2.4 Mayo classification of typical cases of ADPKD.

2.2.6 Management

Since there is no definitive treatment for the underlying genetic defect of ADPKD, management had historically focused on surveillance for and control of the disease complications. More recently, the approval of the first disease-modifying agent (tolvaptan) has opened another exciting door to slow down disease progression in ADPKD. A few novel therapies are in the clinical trial pipelines that could hopefully be used in the management of these patients in the future.

2.2.6.1 Hypertension

Since hypertension and proteinuria are predictors of earlier onset of ESKD, and the RAA system is the key mechanism underlying hypertension in ADPKD, angiotensin-converting enzyme inhibitors (ACE-i), which target the RAA system, decrease blood pressure, and reduce proteinuria, would seem to be the treatment of choice. There have been epidemiologic and non-randomized studies suggesting that ACE-i can slow the progression of renal insufficiency and reduce proteinuria [108–110]. However, the few randomized, prospective studies on ACE-i have indicated otherwise. The Modification of Diet in Renal Disease (MDRD) Study Group was a randomized, prospective trial that included 200 patients with ADPKD and found that treatment with low-protein diet or ACE-i did not slow the loss of glomerular filtration rate (GFR) during the 3 years of the trial [111]. However, the 12 year follow-up of the MDRD study after completion of the randomized part of the study suggested a beneficial role of the low-protein diet [112]. Therefore, moderate protein restriction (0.8 mg/kg of body weight/day) is often recommended in most CKDs including ADPKD. The ACEi in Progressive Renal Insufficiency Study included 64 patients with ADPKD and found that treatment with benazepril did not decrease the number of patients who experienced doubling of serum creatinine or required initiation of dialysis [113,114]. Finally, no difference in GFR decline or level of microalbuminuria was noted in hypertensive ADPKD patients treated with either atenolol (a β-blocker) or enalapril. Normotensive ADPKD patients treated with enalapril did not have any significant decrease in GFR decline compared to those not treated [115]. There are no data to suggest that antihypertensive therapy reduces the risk of intracranial hemorrhage or cardiac disease in ADPKD patients. The optimal blood pressure target in ADPKD has been evaluated in two randomized trials: First, a study that compared a blood pressure target of less than 135–140/85–90 mmHg versus 120/80 mmHg found significantly decreased left ventricular mass index, a cardiovascular disease risk factor, in patients with strict control [116]. Another large-scale randomized trial, the HALT-PKD trial randomized 558 hypertensive ADPKD patients (15–49 years of age), who were early in the course of their disease (eGFR >60 mL/min/1.73 m^2), to a standard blood pressure target (120/70–130/80 mmHg) or to a low blood pressure target (95/60–110/75 mmHg) and to either an ACE inhibitor (lisinopril) plus an ARB (telmisartan) or lisinopril plus placebo [39]. Compared with the standard-target group, the annual percentage increase in TKV was lower in the low-target group (6.6 vs. 5.6%, respectively, with an overall reduction in TKV growth of 14.2%). In addition, compared with the standard-target group, the left ventricular mass index decreased more in the low-target group (−0.57 vs. −1.17 g per m^2 per year, respectively). Albumin excretion decreased by 3.9% in the low-target group and increased by 2.4% in the standard-target group. There was no statistically significant difference in eGFR decline between the two groups (−2.71 vs. −3.0 mL/min/year, respectively), but this apparently negative result was confounded by an initial rapid decline in eGFR that was seen in the low-target arm (possibly a salvatory hemodynamic effect), followed by a slower slope of progression during the rest of the study. The HALT-PKD-study A cohort is currently being followed for an additional 8 years after the randomized trial to assess the long-term effects of an unusually low blood pressure target. Results should be available by end of 2022. Based on the HALT-PKD trial results, the current recommendation in early ADPKD is to target a blood pressure goal of less than 110/75 mmHg if patients are not having hypotensive symptoms and are functional. In more advanced ADPKD, a blood pressure goal of 120–130/70–80 mmHg is desired.

2.2.6.2 Vasopressin-2 Receptor Antagonist: Tolvaptan

Arginine-Vasopressin (AVP) increases intracellular cyclic adenosine monophosphate (cAMP) in distal nephron segments and collecting ducts, promoting chloride-driven fluid secretion. cAMP also stimulates B-RAF/MEK/extracellular signal–regulated signaling, mitogenesis, and proliferation of polycystic kidney epithelial cells or wild-type kidney epithelial cells under experimental conditions of calcium deprivation [117]. Interestingly, V2 receptors (site of action of vasopressin) are localized in the distal nephron and collecting duct, which happen to be the main site of cystogenesis in autosomal recessive polycystic kidney disease (ARPKD) and arguably in ADPKD. Since increased circulating levels of vasopressin had been shown in both animal and human models of ADPKD, experimental studies, using a vasopressin-2 receptor antagonist (tolvaptan), were conducted in rodent models of polycystic kidney disease [118]. Given the beneficial role of this agent in those models, two large-scale randomized trials (TEMPO 3:4 and REPRISE) were conducted in patients with ADPKD, from 2010 to 2017 [119,120]. Those efforts culminated in the approval of tolvaptan in rapidly progressive ADPKD by various regulatory agencies in Europe, Japan, and the United States. The TEMPO 3:4 trail was conducted in patients age 18–50 with eGFR of >60 mL/min/m2 and a TKV of >750 mL REPRISE was performed in patients age 18–65 with eGFR 25–60 mL/min regardless of their TKV. These complementary trails showed an impressive 31% reduction in the rate of growth of kidneys (TEMPO 3:4) and rate of decline of eGFR in a large population of ADPKD patients, with a broad spectrum of renal function. The main side effects of the drug were polyuria and polydipsia, which led to discontinuation of medication in 10%–15% of patients. There was also a 4% risk of hepatocellular toxicity and transaminitis, which seemed to be reversible after discontinuation of the drug. Due to this risk, the US Food and Drug Administration (FDA) approved the marketing of tolvaptan under the Risk, Evaluation, Mitigation, and Strategy (REMS) program in April 2018. Patients and providers are required to monitor liver function tests on a monthly basis, during 18 months, and quarterly after that. At this point, patients at high risk of progression with ADPKD are recommended to be considered for tolvaptan. Patients at Mayo classes IC, ID, or IE are considered the best candidates even at earlier stages with preserved renal function. Alternatively, if MRI-measured TKV is not feasible or available, a renal length (by ultrasound) of greater than 16.5 cm before age of 50 can be used [121]. Following the strict REPRISE trial criteria, an eGFR of less than 65 mL/min at age younger than 55 (or eGFR of less than 44 mL/min at age younger than 65) could also be used.

2.2.6.3 High Water Intake

High water intake could conceptually mimic the results of tolvaptan on polycystic kidneys by suppressing the production of vasopressin. Rodent models of ADPKD confirmed those suspicions [122]. The only trial of high water intake in ADPKD, conducted in 30 Japanese patients, did not show any significant difference in clinical renal outcomes based on the water intake, over 1 year even though high water intake effectively reduced AVP and serum copeptin levels. However, the study suffered from confounding due to relatively high water intake even at baseline, short period of follow-up, and small sample size [123]. A larger randomized trial of high versus standard water intake, with a longer follow-up period, is under way in Australia with expected results in early 2022.

2.2.6.4 Low-Salt Diet

Salt restriction (sodium intake of <2 g/day or salt intake of <4 g/day) is a common recommendation in hypertensive and chronic kidney disease patients. However, more recent evidence suggests an independent renoprotective effect of low-salt diet, beyond the pure benefits on blood pressure control. In fact, in HALT-PKD trials, where all patients were advised to limit their intake of sodium to less than 2400 mg/day (or 5 g of salt/day), a high urinary sodium intake as reflected by a higher urinary sodium excretion was independently associated with increased growth of the kidneys and faster decline of eGFR [124]. Therefore, low-salt diet should be recommended in all ADPKD patients, not only to improve blood pressure control, but maybe also to give them additional and independent kidney protection.

2.2.6.5 Other Lifestyle Modifications

Several lifestyle-related factors have been evaluated in progression of ADPKD. These include cigarette smoking, alcohol, caffeine intake, and obesity.

Smoking has been associated with increased microalbuminuria and proteinuria [93] and higher odds ratios of ESKD (3.5- and 5.8-fold increase for 5–15 and greater than 15 pack-years of smoking compared to less than 5 pack-years) in a small cohort of ADPKD patients [125].

It is difficult to make an evidence-based recommendation about consumption of caffeine in ADPKD. Caffeine intake has been shown to increase cAMP accumulation and activated transepithelial fluid secretion in ADPKD cystic epithelium of the rodent models of ADPKD [126], and an original study has suggested higher blood pressure levels but no difference in kidney function in caffeine consumers, compared to no caffeine consumers [127]. However, larger studies (such as the Swiss cohort and CRISP cohort) do not support a significant difference in TKV or eGFR based on caffeine intake [128,129]. It is important to note that those studies were not designed to evaluate the effect of caffeine intake as a primary outcome.

Maintaining an ideal body mass index (BMI) is recommend in ADPKD. Overweight and obesity have been linked to faster progression in ADPKD in patients. In a posthoc analysis of the HALT-PKD trial, the annual percent change in TKV and the slope of the eGFR decline were both greater in patients who were overweight/obese [129].

2.2.6.6 Dialysis and Renal Transplantation

At onset of ESRD, renal replacement therapy with dialysis should be initiated or, if living donors are available, patients should be evaluated for preemptive kidney transplantation. Assessment for ventral hernias, which interfere with peritoneal dialysis, should be performed prior to consideration for peritoneal catheter placement. Outcomes of renal transplantation are comparable with all nondiabetic renal transplant patients, with 73% 5-year and 67% 10-year graft survival and 84% 5-year and 73% 10-year patient survival rate [130,131]. Malignancy due to chronic immunosuppression is the most common cause of graft failure and death. Patients with ADPKD are excellent candidates for transplantation but living-related donor transplantation must be carefully assessed because of the high risk of disease in relatives and late onset of cysts in some of them. Owing to these concerns, potential living-related donors should undergo genetic testing or be screened carefully for evidence of ADPKD before proceeding to living donation.

2.2.6.7 Polycystic Liver Disease (PLD)

Most patients have mild hepatic cyst burden and do not experience any significant symptoms related to their liver cysts and therefore, do not need any treatment for their polycystic liver disease. However, in patients with massive (and or symptomatic) polycystic liver disease, several therapeutic avenues have been investigated. The observation that postmenopausal women with ADPKD taking hormone replacement therapy have higher burden of hepatic cysts prompted a recommendation to avoid estrogen, but there are no studies that indicate that avoidance of estrogen or estrogen receptor blockade is effective [11]. Patients with symptomatic liver disease and a few dominant cysts may be candidates for CT-guided aspiration/sclerosis or laparoscopic cyst fenestration, but most symptomatic patients have too many cysts for these interventions to improve symptoms. There is a risk of bleeding, bile leak, ascites, or recurrent symptoms with these techniques, and some patients may require conversion to open laparotomy. Partial liver resection has been attempted in patients with severe symptoms and hepatomegaly with relief of symptoms, but there is a high risk of perioperative morbidity. Orthotopic liver transplantation (OLT) or combined liver and kidney transplantation is reserved for those patients who have ESKD and severe refractory liver disease or for patients with symptomatic portal hypertension, Budd–Chiari syndrome, or intrahepatic biliary obstruction. Five-year survival is 69% for OLT alone and 76% for combined liver and kidney transplantation [11].

More recently, data from randomized trials of somatostatin analogues (octreotide, lanreotide, pasireotide) have shown promising results in reducing the rate of growth of liver cysts in ADPKD patients with massive PLD. The 2-year follow-up of the randomized clinical trial of long-acting octreotide repeatable depot (OctLAR) in 40 patients with ADPKD and PLD showed an average difference in the total liver volume of −7.99% in favor of the OctLAR group [132]. Another study showed a 1.99% reduction in height adjusted total liver volume in the intervention group compared to a 3.92% increase in the control group when lanreotide was used [133]. However, they can be associated with severe side effects. More specifically, pasireotide should probably be avoided due to the high incidence of hyperglycemia and new onset diabetes mellitus in 79% and 59% of cases, respectively [134].

2.2.6.8 Abdominal Pain

Abdominal pain, both acute and chronic, affects the majority of ADPKD patients at some point in their lives. Acute pain can be due to renal conditions (nephrolithiasis, renal cyst rupture, cyst infection, pyelonephritis) or extrarenal conditions (diverticulitis, hepatic cyst rupture and infection, biliary and gallbladder issues, peptic ulcer disease, irritable bowel syndrome, and inflammatory bowel disease). Chronic pain is typically due to the extensive nephromegaly or hepatomegaly, but some patients could have chronic severe pain without massive organomegaly.

Nephrolithiasis should be treated with analgesia, bed rest, and hydration. Lithotripsy and urolithotomy can be considered if the calculi are obstructing the urinary tract. Renal cyst hemorrhage is usually self-limited resolving over few days or weeks and responds to analgesics. The possibility of concomitant infection should always be considered and treated with antibiotics. Retroperitoneal hemorrhage with severe blood loss and hypovolemia has been reported in rare cases. Diverticulitis in ADPKD patients must be recognized promptly and treated with antibiotics and bowel rest. Lower UTI presents in the usual manner with frequency, urgency, and dysuria. Oral antibiotics that cover Gram-negative Enterobacteriaceae should be administered for the standard duration.

Cyst infections, however, are more difficult to recognize and treat. Renal cyst infections are likely due to ascending infection, as suggested from the observation that 92% of cyst infections occur in females [135]. Renal cyst infection classically presents with fever and focal tenderness overlying the affected kidney, hepatic cyst infection with fever, and right upper quadrant tenderness. However, some patients have more subtle clinical signs. Not all patients with renal cyst infection have pyuria or positive urine cultures. Leukocytosis and elevated erythrocyte sedimentation rate aid somewhat in diagnosis, but blood cultures have been shown to be the most effective diagnostic test for both renal and hepatic cyst infection [135,136]. 18F-fluorodeoxyglucose positron emission tomography (18-FDG PET-scan) has become a promising tool for detection of infected cysts but can be expensive [137,138]. Empiric therapy with an antibiotic that penetrates the cyst, such as ciprofloxacin, trimethoprim-sulfamethoxazole, or vancomycin, should be initiated in patients with suspected cyst infection. Enterobacteriaceae are the usual pathogens. If an organism is identified, the patient should receive an extended course of treatment lasting a minimum of 4 weeks to ensure adequate cyst penetration; CT-guided drainage may be necessary with cyst infections that do not respond.

2.2.6.9 Cardiac Disease

Periodic echocardiography should be performed to assess for valvular disease, left ventricular hypertrophy, and dilatation of the aortic root. Subacute bacterial endocarditis prophylaxis should be given prior to dental and gastrointestinal procedures if there is hemodynamically significant valvular insufficiency. Valve replacement is indicated for severe, symptomatic regurgitation.

Semiannual monitoring of the aortic root should be performed if the aortic root reaches 4 cm, and aortic root replacement is indicated if the diameter of the aortic root exceeds 5 cm. This recommendation, however, is based on data from patients with Marfan syndrome. There are no ADPKD patient studies that follow the velocity of aortic root dilatation or assess the risk of aortic dissection in relation to aortic root diameter. It is also not known whether β-blockade in ADPKD patients with aortic root dilatation can successfully slow the rate of dilatation as in Marfan syndrome. Finally, patients with mitral valve prolapse should be considered for treatment with beta-blockers especially if they present with palpitations.

2.2.6.10 Screening of Intracranial Aneurysm

Prior to the development of MRI and magnetic resonance angiography (MRA), routine screening of all ADPKD patients was not recommended because the screening test, intraarterial angiography, held a significant risk of complications including permanent neurologic deficit [139]. High-resolution time-of-flight MRA (without injection of Gadolinium) is now the modality of choice for aneurysm screening. While the technique is highly specific and sensitive for aneurysms 5 mm in diameter and above, the sensitivity drops considerably for smaller aneurysms [140]. CT angiography has been shown to have better sensitivity, but the requirement of intravenous contrast may render this study unfeasible for patients with CKD or ESKD [141]. While screening relatives of patients with subarachnoid hemorrhage (SAH) successfully identified small, asymptomatic aneurysms and resulted in more surgical interventions performed, the resultant operative complications, postoperative morbidity, and long-term reduction in function outweighed any benefits gained from preventing SAH [29,142]. Caution, however, must be exercised in extrapolating the data to ADPKD patients, where only theoretical cost—benefit analyses have been performed [143]. The current recommendation is to screen only ADPKD patients with a history of SAH or a family history of SAH or aneurysm, beginning at age 20.

Since aneurysm rupture has been observed following renal transplantation, other groups have suggested screening of those planning to undergo major elective surgery. Individuals in "high-risk" professions such as airline pilots have also been identified as candidates for screening [31,144].

Very few individuals (2.4%) without aneurysm identified at initial screening developed an aneurysm in a 10-year longitudinal study [145]. Studies of those with an aneurysm found at screening have shown that aneurysmal enlargement or development of new aneurysms occurs slowly. In one study of 18 individuals with an aneurysm found at screening, one developed a new aneurysm and one experienced enlargement of the existing aneurysm [28]. Both events occurred in the group of 10 patients imaged over an average period of 10 years and previously reported in a separate study [146]. None of the patients followed for less than 10 years developed a new aneurysm. In another study, 25% (5 of 20) of individuals with previously identified aneurysms developed new aneurysms over an average 15-year follow-up period. Four of those five were initially discovered to have intracranial aneurysm when they experienced an intracranial bleed; none had recurrent intracranial bleeding [147]. These findings suggest that those who have experienced SAH have a high incidence of developing new aneurysms. Other studies have also shown that they are at higher risk of aneurysm rupture [148]. Although there are no specific recommendations for imaging intervals, these data indicate that those who screened negative should have the longest interval for reimaging (5—10 years), those with asymptomatic aneurysms an intermediate interval (3—5 years), and those with prior history of SAH the shortest (1—3 years). Clipping of an aneurysm or intraarterial coiling should be performed in patients with aneurysms that are symptomatic or larger than 10 mm.

2.2.7 Novel Therapies

Better understanding of the pathophysiology, availability of ADPKD animal models, new discoveries of promising pharmacological agents, and the validation of TKV as an enrichment tool to identify high-risk patients for disease progression have led to the identification of several promising interventions that are in clinical trials. Currently several dietary trials trying to elucidate the role of ketogenic diet (NCT04680780) and high water intake in ADPKD are ongoing. New pharmacological agents such as metformin, Venglustat (a glucosylceramide synthase (GCS) inhibitor), Tesevatinib (a new molecule that inhibits multiple receptor tyrosine kinases (RTKs)) (ErbB family members [EGFR and HER2/ErbB2], Src, and vascular endothelial growth factor receptor [VEGFR2/KDR], Bardoxolone methyl (a semisynthetic triterpenoid, acting as an activator of the Nrf2 pathway and an inhibitor of the NF-κB pathway), and RGLS4326 (an anti-miRNA17) are in clinical trials (1 — NCT02656017, 2 — NCT04705051 ,3 — NCT032 03642, 4 — NCT03918447 and 5 — NCT04536688).

2.3 AUTOSOMAL RECESSIVE POLYCYSTIC KIDNEY DISEASE (MIM 263200)

2.3.1 Clinical Features and Natural History

Hepatorenal fibrocystic diseases (HRFDs) are characterized by fibrocystic disease of the kidney and dysgenesis of the portobiliary tract (Fig. 2.5) (reviewed in Ref. [149]). ARPKD, a hallmark disorder representing this phenotypic classification, is broadly known as a severe, early-onset disorder characterized by renal cystic disease, renal function loss, and variable involvement of the portobiliary tract. ARPKD is among the leading causes of pediatric end-stage kidney disease (ESKD), as well as the requirement for kidney, liver, and combined kidney and liver transplantation in children and young adults [150]. However, its phenotypic manifestations and the age when ARPKD is diagnosed are variable [151]. The ages at which ARPKD patients' diagnoses are established can be distributed into three approximately equally sized groups: under 1 year, between ages 1 and 20, and above 20 years of age [152].

The reported incidence of ARPKD is 1 in 20,000 live births [153]. However, it can be higher since severely affected infants may die as newborns without a diagnosis, and patients with mild disease may not be diagnosed until later in life [154]. The estimated neonatal mortality in ARPKD is 30% with children surviving the first month of life likely to be alive at the end of the first year of life. The reported survival rates are 92%—95% at 1 year, 50%—87% at 5 years [151,155,156], 87% at 10 years, and 67%—79% at 15 years of age [157]. The surviving patients may develop multiple complications, including growth retardation (in 20%—25% of cases) and systemic hypertension (in 65% of cases) [151,158]. The median onset of systemic hypertension is 7 days of life for ventilated and 70 days of life for nonventilated patients [151]. Approximately 40% of the surviving ARPKD patients develop chronic renal

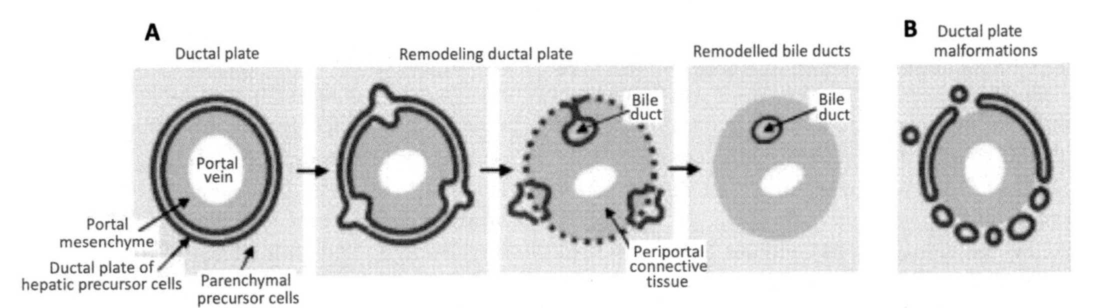

Figure 2.5 Embryogenesis of bile ducts. (A) Schematic diagrams of normal embryogenesis, progressing from the ductal plate stage (gestational weeks 9 to 12) to remodelled bile ducts of the portal tract (gestational weeks 18 to 40). Remodelling is thought to be the result of epithelial-mesenchyme inductive interactions. The embryonic structures are in cross section, with a branch of the portal vein (lumen in white) and a cuff of surrounding mesenchyme (dense red dots) at the central axis. The red lines represent the ductal plate. Refer to the text for further details. (B) Examples of biliary dysgenesis, resulting in ductal plate malformations owing to incomplete remodelling. The ductal plate (red lines) is in the form of either an interrupted circle or peripheral tubular structures. The ductal plate malformation is a characteristic manifestation of a number of congenital fibrocystic syndromes that are inherited in an autosomal recessive manner. (From Johnson, C.A.; Gissen, P.; Sergi, C. Molecular pathology and genetics of congenital hepatorenal fibrocystic syndromes. J Med Genet 2003;40:311–319.)

insufficiency. Its median onset is 1 day of life for ventilated and 1 year of life for nonventilated patients [151]. Other renal manifestations include urine concentrating defects and the ensuing polydipsia and polyuria. ARPKD patients may also develop a urine acidification defect that leads to metabolic acidosis. Additional ARPKD-related urine abnormalities include mild proteinuria, glucosuria, hyperphosphaturia, and an increased urinary excretion of magnesium [159]. In 10%–30% of ARPKD patients, advanced chronic kidney disease progresses to end-stage kidney disease (ESKD), usually in the first decade of life [151].

Phenotypic manifestations of ARPKD are highly variable. While the most severe forms of ARPKD lead to perinatal mortality (30% of cases), the vast majority (92%–95%) of less severely affected ARPKD patients who survive the first month are alive at the 1-year mark [151]. Furthermore, many of them retain stable renal and hepatic function into young adulthood [160]. In addition, in perhaps a third of ARPKD patients, the symptoms are so minimal during childhood that the ARPKD diagnosis is established after they reach 20 years of age [152]. Also, the age at ARPKD diagnosis is associated with different clinical manifestations and outcomes. For example, most ARPKD patients identified in utero or at birth have massively enlarged kidneys. The oligohydramnios onset in the second trimester may be associated with the 'Potter sequence that includes pulmonary hypoplasia. The ensuing respiratory insufficiency with air leaks (pneumothorax or pneumomediastinum) leads to death in the most severe cases. Oligohydramnios may also lead to the development of the Potter facies (hypertelorism, prominent infraorbital creases, flattened nasal bridge and tip, low-set flattened and fleshy pinnae, and micrognathia) and positional limb deformities (*talipes equinovarus*) due to extrinsic compression. In addition, hyponatremia may occur during the first few weeks of life due to the inability to dilute urine [161]. While renal dysfunction is a common complication in this age group, it has become a rare cause of neonatal death. In those who survive the perinatal period, ARPKD is often complicated by systemic hypertension and renal function decline that progresses to ESKD at highly variable, patient-specific rates [159]. Systemic hypertension usually develops within the first 6 months of life and is often associated with a transient improvement in renal function, e.g., glomerular filtration rate (GFR) attributed to renal maturation. However, this initial improvement is often followed by a progressive but variable GFR decline [149].

In addition, ARPKD patients surviving the perinatal period may also develop portal hypertension as a complication of congenital hepatic fibrosis (CHF; their hepatocellular function is usually preserved). In 15% –34% of these ARPKD children, the median age for portal hypertension diagnosis is 7 months for ventilated and 4.6 years for nonventilated patients; 2.4%–7% of these ARPKD patients require portocaval shunting or liver transplantation. The severity of renal disease and the presence of systemic hypertension did not correlate with the severity of hepatic disease [151]. In older children, portal hypertension leads to gastric or esophageal varices and portal vein thrombosis. Often, it also leads to hepatosplenomegaly, which may cause thrombocytopenia, anemia, and leukopenia. Children with ARPKD may also develop growth retardation [151]. In addition, some ARPKD children may have increased susceptibility to urinary tract infections and recurrent bacteremia with enteric pathogens [162]. Moreover, they may develop ascending suppurative cholangitis that may lead to fulminant hepatic failure [163]. Advanced renal dysfunction in this age group may be complicated by anemia and renal osteodystrophy. However, there is also a subset of patients with a late onset of ARPKD manifestations that develop mild or no renal problems [152]. Their predominant symptoms are instead related to hepatobiliary manifestations such as portal hypertension and its complications, as described above. Rarely, ARPKD patients (and especially these older patients) may also develop intracranial aneurysms (summarized in [164]).

The predicted carrier frequency of pathogenic variants in the causative ARPKD gene *PKHD1* is 1 in 70 individuals in nonisolated populations. Until recently, these individuals were considered unaffected carriers of mutations in this gene. However, recent data indicate that carrier status for ARPKD (heterozygosity for a pathogenic *PKHD1* variant) predisposes to renal abnormalities such as increased medullary echogenicity on ultrasound (possibly medullary sponge kidney). It may also lead to liver abnormalities, including asymptomatic polycystic liver disease, and in some cases, CHF. For example, out of 85 parents of an offspring with a known pathogenic variant in *PKHD1* gene, 17 (20%) (age range 30–62 years; average 42.6 years) had cystic liver and/or kidney abnormalities [165]. In these parents, the phenotypic manifestations included liver and kidney cysts, hyperechoic liver, gallbladder polyps, increased renal medullary echogenicity resembling nephrocalcinosis, and splenomegaly. However, the phenotypic manifestations were highly variable and were not associated with any symptoms at the time of the evaluation. Another study found adult carriers of pathogenic *PKHD1* variants among patients with polycystic liver disease (PLD) [166]. Most of these patients had innumerable small liver cysts (>10 cm cysts in one patient), 30% had kidney cysts, and 70% were females.

Finally, pathogenic variants in the *DZIP1L* gene were reported in seven patients with ARPKD-like phenotype from four unrelated consanguineous families [167]. All these patients had enlarged hyperechogenic kidneys and systemic hypertension, while four progressed to ESKD.

2.3.2 Gene Mapping, Structure, Function, and Genotype–Phenotype Correlations

The principal gene affected in ARPKD is *PKHD1* (polycystic kidney and hepatic disease 1). The ARPKD locus that harbors this gene was initially mapped to the short arm of chromosome 6, and this locus was eventually refined to a 3.8 centiMorgan (cM) interval on chromosome 6 (6p21.1.12) [168]. Later, *PKHD1* was identified by three groups as the ARPKD gene [169–171]. The first group succeeded using a syntenic mapping between the ARPKD locus and mutation in PCK rat, a model of PKD that mapped to rat chromosome 9 (currently designated as $Pkhd1^{pck/pck}$). This research group named the *PKHD1*-encoded protein product fibrocystin. Independently, two other groups reported the same gene product, the second group named it polyductin. The *PKHD1* gene is a large gene spanning 470 kilobase pairs (kb) of genomic DNA. It contains at least 86 exons [170]. Among these exons, 71 are nonoverlapping, and 15 exons have alternative splicing boundaries. While *PKHD1* exons generate multiple alternatively spliced transcripts, 67 exons form the longest open reading frame. It encodes a 16.2 kb mRNA and is translated into a 447 kDa protein with 4074 amino acids (Fig. 2.6). The fibrocystin/polyductin has an extracellular domain that extends over 3835 amino acids (amino acid positions 12–3858), 21 amino acid transmembrane domain (position 3859–3879), and a 195 amino acid cytoplasmic tail (position 3880–4074). The *PKHD1*-encoded protein product is highly glycosylated, and its N-terminal segment contains multiple immunoglobulin-like and plexin domains. The *PKHD1* protein product localizes to primary cilia and basal bodies in renal tubular epithelial cells [151,172]. It is also found in a subset of PKD1-positive urine exosomes together with other cystoproteins [173]. The function of fibrocytin/polyductin is not well understood, but domain analysis suggests that the protein may play a role in the regulation of cellular adhesion, repulsion, and proliferation, and/or the regulation and maintenance of renal collecting tubules and hepatic bile ducts [174]. *PKHD1* truncating variants also alter mitochondrial metabolism and morphology, resembling the effects of polycystin 1 deficiency [175].

To date, a total of 748 unique *PKHD1* variants have been reported in the Mutation Database for Autosomal Recessive Polycystic Kidney Disease (ARPKD/*PKHD1*) hosted by the Aachen University, Germany (http://www.humgen.rwth-aachen.de). Of these, 313 were expected to be pathogenic or likely pathogenic. Of these

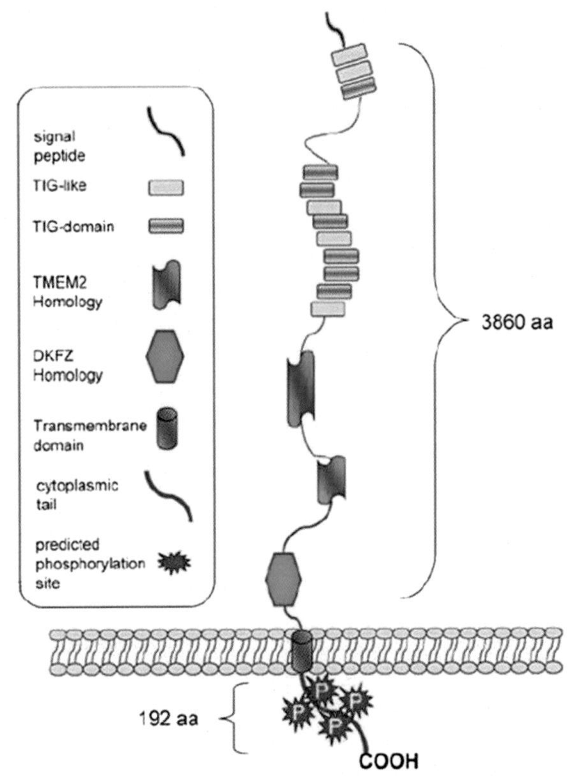

Figure 2.6 Predicted structure of fibrocystin/polyductin, a large integral membrane protein, with the largest known open reading frame, translating a 4074-amino-acid novel peptide (4059 amino acids in mouse). Fibrocystin has homologies to several known proteins, including TIG/IPT (immunoglobulin-like fold shared by plexins and transcription factors) or TIG-like domains in the extracellular region. This domain has 80–100 amino acids and is found in several receptor molecules, including Met, various plexins, and Ron, although the large number of domains seems unique to the fibrocystins. The structure of fibrocystin as an integral membrane protein with a large extracellular portion containing multiple glycosylation sites, and a short intracellular C-terminal domain containing potential protein kinase A and PCK phosphorylation sites suggest that this protein acts as a transducer of extracellular information into the cell by eliciting signal transduction cascades resulting in the modulation of gene transcription (aa amino acids). (From Sweeney, WE, Jr.; Avner, ED. Molecular and Cellular Pathophysiology of Autosomal Recessive Polycystic Kidney Disease (ARPKD). Cell and Tissue Research 2006, 326 (3), 675.)

313 variants, 183 (58%) are missense mutations, 63 (20%) are frameshift variants, 38 (12%) are nonsense variants, and 27 (9%) are splice site variants. Large deletions in *PKHD1* have also been documented and make up a significantly smaller proportion of cases [176]. Pathogenic variants are mostly scattered throughout the *PKHD1* gene without evidence of clustering or a "mutation hot spot" (Fig. 2.7). As with *PKD1* and *PKD2*, many pathogenic variants result in a truncated protein product unique to each family. One of them, 107C>T (p.T36M), has been observed in unrelated patients of different ethnic origins [177]. Due to the large number of pathogenic *PKHD1* variants, ARPKD patients are nearly always compound heterozygotes for two different germline variants of the *PKHD1* gene. Homozygosity for the same pathogenic variant is seen mostly when parents have the same common ancestor. For example, due to the founder effect, 67% of Afrikaner ARPKD patients are homozygotes for the p.M627K substitution [178]. The p.R496X and p.V3471G are founder mutations, in the Finnish population [177]. The observed clinical variability seen in homozygotes suggests that ARPKD phenotypic manifestations are modulated by other genetic loci or by gene–environment interactions. Such effects are also supported by animal studies [179–181].

While intrafamilial phenotypic heterogeneity and high rates of compound heterozygosity make genotype–phenotype correlation in ARPKD unreliable, the following trends were observed. Patients carrying truncating pathogenic variants on both parental alleles died shortly after birth regardless of the site of truncation [177]. Analysis of families classified as "severe" (at least one child presenting with perinatal disease and neonatal demise) showed that 57% of the detected pathogenic *PKHD1* variants were truncating. Families classified as "moderate" (survival or diagnosis beyond 1 month) had 70% missense variants. However, seven missense variants were correlated with perinatal demise [182]. Only one allele with a pathogenic variant was identified in some ARPKD patients, implying the existence of mutations within introns, regulatory elements, or larger genomic rearrangements undetectable by gene sequencing. No specific pathogenic variants in the *PKHD1* gene were associated with severe portobiliary ARPKD manifestations [71].

In support of the functional relevance of carrier status for a pathogenic ARPKD allele heterozygosity, an enrichment of pathogenic variants in the *PKHD1* gene was observed on a genome-wide basis among a cohort of 102 unrelated patients with dominantly inherited isolated polycystic liver disease that was not associated with pathogenic variants in PRKCSH and SEC63. These two genes are mutated in autosomal dominant polycystic liver disease (MIM 174050) [166].

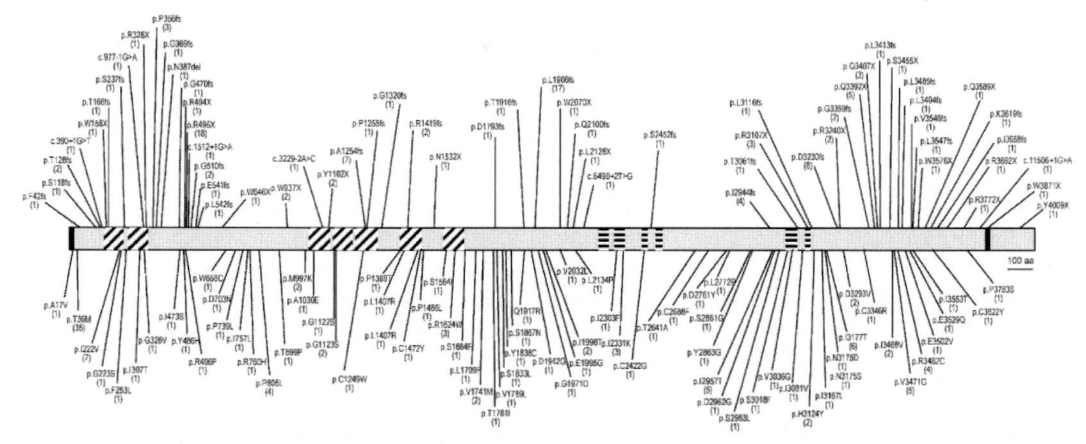

Figure 2.7 Simplified linear representation of fibrocystin/polyductin encoded by the longest potential open reading frame. The sites of PKHD1 mutations (denoted by trivial names) detected so far are indicated with the number of occurrences of each mutation shown in parentheses below the mutation. Nucleotide numbering was done according to GenBank NM_138694.2 with nucleotide +1 being the A of the ATG-translation initiation codon. Truncating mutations are shown above, missense mutations below the bar. Putative protein do- mains are symbolized (black box at the N-term, signal peptide; cross-hatched boxes, IPTdomains; transverse hatched boxes, PbH1 repeats; black box near the C-term, transmembrane domain). (From Bergmann, C, Senderek, Jan, et al. PKHD1 Mutations in Autosomal Recessive Polycystic Kidney Disease (ARPKD). Human Mutation 2004, 23 (5), 453–463.)

While the expected number of loss-of-function variants in *PKHD1* gene among the 1266 total variants was 0.48, the observed number among the studied PLD cohort was 9 (19-fold enrichment; $P = 2.31 \times 10^{-9}$). Eight out of 10 (80%) of identified *PKHD1* abnormalities were truncating variants. These associations were confirmed in an aged animal model where heterozygosity for a truncating *PKHD1* pathogenic variants was associated with renal and liver cystic phenotypes [183]. Finally, recent studies identified a second ARPKD locus on 3q22.3 chromosome. It harbors a *DZIP1L* gene that encodes the DAZ-interacting zinc finger protein 1-like, a ciliary transition zone protein. This protein appears to be involved in the function of primary apical cilia, likely by regulating ciliary membrane translocation of polycystin-1 and polycystin-2 (the two principal proteins that are affected in patients with ADPKD) [167].

2.3.3 Laboratory, Imaging and Histopathologic Findings

Historically, the kidney and the hepatobiliary tract were recognized as the two organ systems primarily affected in most ARPKD patients. However, more recent data suggest that these associations might be more complex and include additional manifestations (e.g., in some patients, the ARPKD-associated intracranial aneurysms). The kidneys from ARPKD patients typically harbor non-obstructive fusiform dilatations of collecting ducts. These dilatations mostly radiate in corticomedullary orientation. Since the size of these dilatations in ARPKD children is mostly 1–2 mm, the surface of their kidneys typically had a sponge-like appearance. Histologic examination reveals dilated collecting ducts with flattened epithelium [184]. Microdissection studies and scanning electron microscopy suggest that while these dilated collecting ducts have a cyst-like appearance, there is no obstruction of the urinary flow [184]. However, as noted above, there is a prominent variability in renal manifestations of ARPKD, and the severity of the renal manifestations is in part proportional to the percentage of affected nephrons. In patients with milder disease, the ectasia of the collecting tubules is less prominent and irregularly distributed. Over time, these patients develop larger renal cysts (up to 1 cm) and interstitial fibrosis contributing to the progressive deterioration of renal function seen in patients who survive beyond the neonatal period. ARPKD is also associated with enhanced renal innate immune responses that include complement system activation [185] and interstitial renal infiltrates of repair and fibrosis-associated CD163+ macrophages [185,186]. However, it remains to be determined if ARPKD kidneys also contain increased numbers of resident macrophages [187,188] that promote renal cystogenesis in animal models [189,190].

Most patients with ARPKD develop a urine concentrating defect with maximal urine osmolality below 500 mOsmol/kg. Additional abnormalities include hyponatremia (25%), elevated blood urea nitrogen (BUN) and serum creatinine (60%–67%), and proteinuria (10%–55%). Since renal manifestations of ARPKD are variable (see details in the above Section 2.3.1 Clinical Features and Natural History), this variability is also reflected in imaging studies. Frequently reported findings include nephromegaly, increased echogenicity, loss of corticomedullary differentiation, and multiple small cysts.

Hepatobiliary tract changes, the hallmark manifestation of ARPKD, stem from the biliary ductal plate formation defect (Fig. 2.5). Histologic examination demonstrates the increased numbers of hyperplastic, ectatic biliary ducts surrounded by fibrosis confound strictly to portal fields; liver parenchyma is normal. However, the severity of the intrahepatic bile duct dilatations (Caroli disease) is variable [191,192]. Some patients may have macroscopic dilations of the intrahepatic bile ducts in addition to congenital fibrosis, a combination of findings referred to as Caroli syndrome [193]. The degree of liver involvement varies in ARPKD, but hepatomegaly and portal hypertension develop in most patients over time (See "Caroli disease").

The renal and biliary manifestations were used to establish the diagnosis of ARPKD. However, there is high variability in ARPKD manifestations that also overlap with phenotypic manifestations of other disorders from the hepatorenal fibrocystic disease spectrum. Due to recent breakthrough advances in technologies allowing affordable and accessible identification of pathogenic gene variants in hundreds of genes arranged within a diagnostic panel (e.g., Next Generation Sequencing), genetic testing is becoming more important for establishing and confirming the diagnosis of ARPKD. The individuals that benefit from such evaluation include siblings of patients with established diagnosis of ARPKD and those with a history of parental consanguinity. While the genetic testing identifies the underlying genetic defect in most patients with phenotypic manifestations of ARPKD, in a small subset of patients with negative genetic tests, the diagnoses have to be established based on the above-stated phenotypic manifestations.

2.3.4 Molecular and Prenatal Diagnosis

Most often, the ARPKD diagnosis is based on imaging studies. While molecular genetic testing is not required to establish the diagnosis in ARPKD patients who fulfill the clinical criteria, genetic testing may confirm the diagnosis in those with the uncertain outcome of routine diagnostic assessment. For the prenatal diagnosis, standard second-trimester ultrasound imaging is usually sufficient to suggest the diagnosis of ARPKD, mainly if findings include bilateral large hyperechogenic kidneys with poor corticomedullary differentiation [194]. At this age, bilateral 5–7 mm large cysts were observed in 29% of ARPKD patients (larger cysts >10 mm in diameter suggest multicystic dysplasia) [195]. However, since ARPKD manifestations are variable, normal antenatal ultrasound does not exclude a diagnosis of ARPKD [196].

Fetal magnetic resonance imaging may better delineate the renal anatomy [197] in the case of severe oligohydramnios. Oligohydramnios together with increased renal size >4 standard deviations (SD) was associated with 100% perinatal mortality [198]. However, since kidney volume in ARPKD does not increase over time, kidney volume-based indices cannot be used as biomarkers or predictors of outcomes in ARPKD in a fashion that resembles their use in ADPKD. Genetic testing for ARPKD is often performed by direct sequencing of the *PKHD1* gene. However, thanks to recent advances in the Next-Generation Sequencing that enhanced the throughput and reduced the cost, the genetic analysis is shifting from a single gene evaluation to analyses of panels containing hundreds of genes. Such panels include the most known genes relevant to disorders from the hepatorenal fibrocystic disease spectrum, including ARPKD. Some commercial providers of genetic testing with the Next-Generation Sequencing panels even offer a complimentary information session with a board-certified genetic counselor to facilitate the review of the results and their interpretation. Yet, in some cases, the interpretation of sequence results may be complicated by a large number of private pathogenic missense variants. This may occur despite the availability of advanced tools for sequence interpretation and reference resources such as the Human Genome Mutation Database (http://www.hgmd.cf.ac.uk) or the Broad Institute gnomAD database (https://gnomad.broadinstitute.org).

Also, in a subset of patients with no genetic mutation identified by exome sequencing-based panels, whole-genome sequencing may be considered. The screening for large deletions or gross rearrangements in *PKHD1* may further increase the diagnostic yield [176] that approaches 85% for the entire spectrum of ARPKD [156,199].

The early and reliable prenatal diagnosis is possible using single-gene testing methodologies. The indirect, haplotype-based linkage analyses are not anymore considered for a typical diagnostic genetic evaluation. From imaging analyses, a standard second-trimester ultrasound demonstrating bilateral large hyperechogenic kidneys with poor corticomedullary differentiation can provide sufficient support for the diagnosis of ARPKD. A high-resolution ultrasound may improve diagnostic sensitivity. In patients with severe oligohydramnios, structural abnormalities of fetal kidneys may be better revealed by magnetic resonance imaging. However, the prenatal imaging-based assessment of ARPKD severity does not accurately predict lethal neonatal pulmonary hypoplasia.

2.3.5 Management

There is no curative therapy for ARPKD. Therefore, the therapeutic management of these patients is focused on supportive care [194]. Since ARPKD manifestations are highly variable, individual patients may benefit from engagement with different multidisciplinary medical teams that best meet their needs. Also, some families pursue preimplantation genetic diagnosis (PGD), which is also available for ARPKD. In the prenatal period, the monitoring is focused on the kidney size and the volume of amniotic fluid. However, such monitoring may not accurately predict the perinatal ARPKD outcomes, including pulmonary hypoplasia and death. Since 40% of perinatally diagnosed ARPKD patients require pulmonary ventilation, the delivery should be performed at a center that offers neonatal intensive care with mechanical pulmonary ventilation and dialysis. Cesarean delivery may be considered, especially in cases of fetal abdominal dystocia due to markedly enlarged kidneys.

Since the perinatal mortality for diagnosed ARPKD patients is high (approximately 30%), family preferences should be established before the delivery. This approach assures good alignment between family wishes and medical interventions during and after delivery (e.g., resuscitation and withholding of specific therapeutic options such as mechanical pulmonary ventilation or dialysis). Neonatal management of ARPKD is centered on stabilization or respiratory status in newborns with respiratory distress. This approach reflects the fact that in severe cases, pulmonary hypoplasia leads to critical respiratory insufficiency and perinatal death in 30%–40% of patients with a presumptive diagnosis of ARPKD [200]. However, in its milder forms, pulmonary hypoplasia and its adverse consequences are treatable with high-frequency ventilation and the ensuing pulmonary hypertension with inhaled nitric oxide. Respiratory failure due to causes unrelated to the pulmonary hypoplasia may respond to extracorporeal membrane oxygenation [194]. In addition, the survival of infants with severe renal ARPKD manifestations may improve with dialysis [201].

Therefore, renal function and electrolyte status should be closely monitored and hemodialysis or peritoneal dialysis should be initiated as indicated. Nephrectomy is not recommended because the renal function of even substantially enlarged ARPKD kidneys may remain preserved over the years. For patients that develop ESKD, the preferred method for managing their advanced chronic kidney disease is kidney transplantation. Hypertension is usually managed with angiotensin-converting enzyme inhibitors and hyponatremia with fluid restriction.

During their childhood, ARPKD patients mostly require monitoring and management of renal and biliary ARPKD complications. In addition to monitoring with ultrasound, these studies may include liver function tests and complete blood count. A reduction of platelet counts is a surrogate marker for the presence and severity of portal hypertension in ARPKD. Its complications, such as esophageal varices, may be treated with a nonselective beta-blocker or with endoscopic banding or sclerotherapy. In some patients who still have a well-maintained liver function, portocaval shunting may be considered.

Among hepatobiliary complications, bacterial cholangitis is a major cause of death among patients with ARPKD. In most cases, initial treatment with empiric antibiotics is appropriately modified as soon as microbiology sensitivities are available. The risk of bacterial cholangitis is further increased in ARPKD patients that are immunosuppressed (e.g., after kidney transplantation). In addition, ARPKD patients with hypersplenism have an increased risk for infections with encapsulated bacteria (*Streptococcus pneumoniae, Neisseria meningitis, Haemophilus influenzae,* and *Streptococcus agalactiae* – Group B Streptococcus) and should follow prophylactic immunizations protocols for asplenic patients. Also, due to the ARPKD-related malabsorption of fat-soluble vitamins (e.g., vitamin D), ARPKD patients may benefit from their supplementation.

Finally, weight and growth changes should be used to guide nutritional support and the use of recombinant growth hormone [202].

2.4 FAMILIAL NEPHRONOPHTHISIS

2.4.1 Clinical Features and Natural History

Nephronophthisis is the most common genetic cause of kidney failure in the first three decades of life [203]. First described by Fanconi in 1950, nephronophthisis was previously classified together with medullary cystic kidney disease (MCKD) since both had similar presenting symptoms and renal histopathologic appearance. However, MCKD was later placed into its own category because of its autosomal dominant inheritance and later age of onset than familial nephronophthisis. Nephronophthisis is an autosomal recessive disorder with variable presentation and considerable locus heterogeneity [204]. Three different forms have been identified and are classified according to age of onset (juvenile, infantile, and adolescent). Juvenile nephronophthisis usually presents by age 6 years, with development of ESKD by a median age of 13 years and generally between the first and second decades of life [205]. Infantile nephronophthisis begins in infancy when mothers notice their babies producing an excessive number of wet diapers. The adolescent variant presents with ESKD at a median age of 19 years. In all types, progressive kidney dysfunction is invariable, and anemia develops out of proportion to the degree of CKD. Nephronophthisis typically progresses to ESKD, and if untreated, death occurs by uremia.

2.4.1.1 Juvenile Nephronophthisis (MIM 256100)

Juvenile nephronophthisis is the most common of the three forms. Patients present with polyuria and secondary enuresis at the age of 4–6. This is due to tubular dysfunction and a decrease in urinary concentrating ability. Despite fluid deprivation, urine osmolality is low in these children and is not increased with the administration of vasopressin, mimicking nephrogenic diabetes insipidus. Hypertension is rare because of sodium wasting [206].

2.4.1.2 Infantile Nephronophthisis (MIM 602088)

Infantile nephronophthisis is rare and the most severe type among nephronophthisis [207]. Some infants present prenatally with an ARPKD-like phenotype of oligohydramnios, Potter facies, and severe respiratory failure without hepatic fibrosis. Patients presenting in infancy and early childhood do not have the typical polyuria and polydipsia, but instead suffer from hypertension and hyperkalemia. *Situs inversus totalis* and cardiac defects may be present. The renal histological phenotype differs from that of juvenile nephronophthisis in that features of ARPKD (e.g., nephromegaly, microcystic cortical dilatation) are seen along with the characteristic tubular cell atrophy, interstitial cell infiltration, and fibrosis seen in nephronophthisis. Affected children develop ESKD by the age of 1 year.

In most cases, mutations in the *NPHP2* gene cause infantile nephronophthisis. Mutations in *NPHP3* and *NPHP9* have also been reported in patients with very-early-onset nephronophthisis, thus establishing locus heterogeneity for infantile nephronophthisis [208,209].

2.4.1.3 Adolescent Nephronophthisis (MIM 604387)

In adolescent nephronophthisis, ESKD develops in late adolescence and young adulthood, with a median onset of age 19 years-old (range 4–37 years) [210]. *NPHP3* was the first gene identified to cause adolescent nephronophthisis. A recent study from 5606 kidney transplant recipients revealed that 26 patients had homozygous *NPHP1* deletion leading to adult-onset ESKD [211]. In children with variants in *NPHP1–4*, retinitis pigmentosa occurs in about 10% of families. However, all patients with *NPHP5* variants have retinitis pigmentosa [204,206].

2.4.2 Syndromic Forms of Nephronophthisis

An additional characteristic of nephronophthisis that distinguishes the condition from MCKD is its association with extrarenal malformations. Optic nerve colobomas, cone-shaped epiphyses, and liver fibrosis have been reported in separate kindreds. Other forms of nephronophthisis with specific extrarenal malformations are associated with ciliary dysfunction and are mentioned here.

2.4.2.1 Joubert Syndrome (MIM 213300)

Joubert syndrome is an autosomal recessive disorder characterized by congenital hypotonia, psychomotor retardation, an abnormal breathing pattern of tachypnea alternating with apnea, cerebellar ataxia, oculomotor apraxia, and most importantly, aplasia or hypoplasia of the cerebellar vermis giving the classic "molar tooth sign" on brain neuroimaging (Fig. 2.8). Kidney disease is present in 25% of patients with

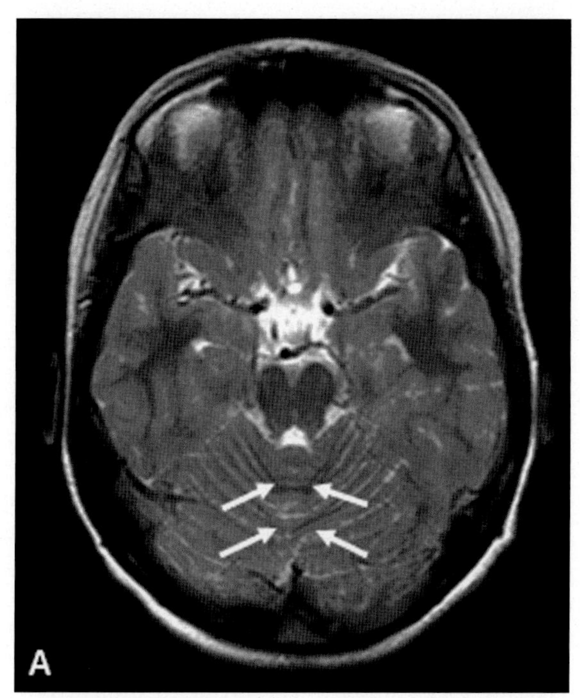

Figure 2.8 Molar tooth sign in Joubert syndrome Axial MRI image through the cerebellum and brain stem of a normal individual showing intact cerebellar vermis (outlined by white arrows). (Reproduced with permission Parisi M, Glass I, Joubert syndrome. 2003 Jul 9 [Updated 2017 Jun 29]. In: Adam MP, Ardinger HH, Pagon RA et al., editors. GeneReviews [Internet]. Seattle (WA): University of Washington, Seattle; 1993–2021. Available from: https://www.ncbi.nlm.nih.gov/books/NBK1325/).

Joubert syndrome and most often manifests as nephronophthisis [212]. Children typically present in the first decade of life with polyuria and polydipsia followed by CKD and eventually ESKD by the second decade. Juvenile nephronophthisis is the most common form associated with Joubert syndrome, but infantile and adolescent onset nephronophthisis can occur.

There are at least 35 ciliopathy-related genes known to cause Joubert syndrome. Pathogenic variants in *CEP290* accounted for approximately 38% [213], *AHI1*, *NPHP6*, *MKS3*, and *CC2D2A* each account for approximately 10% of cases [214–217]. About 1%–2% of patients have a homozygous *NPHP1* deletion [216].

2.4.2.2 Senior–Løken Syndrome (MIM 266900)

Also known as Senior syndrome or renal–retinal syndrome, this disorder is a combination of nephronophthisis and retinitis pigmentosa and accounts for approximately 10%–15% of all cases of nephronophthisis. Renal symptoms and histopathology are identical to that of isolated nephronophthisis, including the variability in age of onset. Similarly, onset of ocular pathology varies from birth to childhood. Newborns present with Leber's congenital amaurosis, characterized by visual inattention, nystagmus, near-absent pupillary reaction, and blindness or severe visual impairment. Older children present with nyctalopia (night blindness), then develop progressive loss of bilateral peripheral vision. Color perception and central visual acuity are preserved until end-stage disease, which typically occurs in adolescence, although there is one report of an adult patient with some preservation of visual fields and acuity [218]. Fundoscopic examination reveals attenuation of blood vessels and pallor of the optic disk without the "bone spicule" retinal pigmentary abnormalities typically seen in retinitis pigmentosa.

In the majority of cases, Senior–Løken syndrome is caused by variants in *NPHP5*, but variants in *NPHP1*, including the classic 290 kb deletion, *NPHP2*, *NPHP3*, and *NPHP4* have been reported [203,206,210,219–222]. In addition, variants in *NPHP6* cause 20% of cases of Leber congenital amaurosis, which is characterized by early onset of severe vision impairment from retinitis pigmentosa, without renal disease [223,224]. The frequency of retinitis pigmentosa in Senior–Løken syndrome varies with the gene involved, from 6% with *NPHP1* to 100% with *NPHP5* and *NPHP6* [206].

2.4.2.3 Meckel–Gruber Syndrome (MIM 249000)

Meckel–Gruber syndrome is a severe, perinatally lethal, multiple congenital anomaly syndrome with an estimated incidence of 1 in 140,000 to 1 in 13,250 births [225]. Infants have bilaterally enlarged, cystic kidneys and hepatic ductal plate dysplasia, biliary proliferation, and fibrosis similar to those with ARPKD. Renal cysts involve both the cortex and the medulla. Cortical cysts tend to be small, thin-walled, and surrounded by normal glomeruli, while medullary cysts are larger, up to several millimeters in diameter, and separated by thick fibromuscular walls [226]. Central nervous system malformations are severe. Posterior encephalocele is the most common malformation, while aplasia or hypoplasia of various central nervous system (CNS) structures, such as the cerebellum, cerebrum, corpus callosum, and optic or olfactory tracts, is present in other patients. Preaxial or postaxial polydactyly is another cardinal feature of the syndrome. Prenatal diagnosis of Meckel–Gruber syndrome can be made with transvaginal ultrasound as early as 10–12 weeks gestational age. The mode of inheritance is autosomal recessive, and locus heterogeneity has been well established. Pathogenic variants in *MKS1*, *TMEM216*, *MKS3/TMEM67*, *NPHP6/CEP290*, *NPHP8/RGRIP1L*, *NPHP3*, and *CC2D2A* all have been reported to cause Meckel–Gruber syndrome. Many of these genes overlap with those causing Joubert syndrome, suggesting that these disorders form a continuum of phenotypic variation rather than separate entities [214,227].

2.4.2.4 Cogan-type Oculomotor Apraxia (MIM 257550)

Patients with Cogan-type oculomotor apraxia suffer from an inability to make horizontal eye movements. In order to bring objects into view, they must make quick, jerking movements of their heads. Involuntary horizontal eye movements are not impaired, indicating that the oculomotor nerves function normally. Nephronophthisis has been reported in association with this disorder, but unlike Joubert syndrome, these patients do not have the molar tooth sign or retinitis pigmentosa. Pathogenic variants in *NPHP1* and *NPHP4* have been found in patients with Cogan-type oculomotor apraxia and nephronophthisis. Homozygosity of the classic *NPHP1* deletion has been reported in one patient, while another patient was found to have the classic deletion of one allele and an inactivating splice junction point variant of the other [147]. Homozygosity for a frameshifting 3272delT variant in *NPHP4* has also been documented [228].

2.4.2.5 Mainzer–Saldino Syndrome (MIM266920)

Mainzer–Saldino syndrome or conorenal syndrome is a rare autosomal recessive disease defined by phalangeal cone-shaped epiphyses, chronic renal disease, nearly constant retinal dystrophy, and mild radiographic abnormality of the proximal femur [229,230]. The association with nephronophthisis has been described, and it is considered a ciliopathy from *IFT140* gene mutation [231].

2.4.3 Gene Mapping, Structure, Function, and Genotype–Phenotype Correlations

At least 20 different genes are known to be associated with nephronophthisis and other nephronophthisis-associated syndromes [232] (Table 2.3). The most common variant in patients with nephronophthisis is a homozygous deletion of *NPHP1*, which accounts for 21% of cases [206]. The remaining genes each comprise <3% of cases, and the variant is unknown in 30% [233].

2.4.3.1 NPHP1

The first gene locus for nephronophthisis was mapped to 2p by linkage analysis of 22 multiplex families [234]. The locus was refined first to 2q13, then with fine mapping techniques to a 2 Mb critical region [235–239]. An approximately 250 kb deletion was discovered in 80% of familial cases of nephronophthisis mapping to the 2q13 locus [238]. Two ORFs were identified within the deleted region, one encoding a small 148 amino acid protein involved in cellular trafficking and the other encoding a then-uncharacterized protein that was eventually shown to be mutated in patients with nephronophthisis [240,241].

The *NPHP1* protein, named nephrocystin-1, contains 20 exons spanning 83 kb of the genome, and encodes a 4.5 kb mRNA. The predicted protein is 732 amino acids long and weighs 83 kDa. It contains coiled-coil domains and an *Src*-homology 3 domain (SH3), both of which are protein–protein interaction motifs. SH3 domains are typically found on "adapter" proteins that bring other proteins together. Through the yeast 2-hybrid system, nephrocystin-1 was shown to bind the focal adhesion complex protein p130cas. Immunofluorescence demonstrated that nephrocystin-1 and p130cas colocalize in vivo at focal adhesions of the basolateral cell membrane [242,243]. p130cas is a cytosolic protein involved in the focal adhesion complex, which mediates signal transduction from the extracellular matrix to the nucleus [244]. Nephrocystin-1 was also found to colocalize with E-cadherin at adherens junctions and with β-tubulin in the primary cilia of renal cells, specifically, at the transition zone [245–247]. In addition, it interacts with other nephrocystin proteins and Jouberin to form a multiprotein complex [210,228,247,248]. These findings, along with similar results from PKD1, PKD2, and PKHD1 localization, indicate the important role of cilia, cell–cell, and cell–matrix interactions in inherited kidney disease [206,249,250]. As stated earlier, more than 80% of patients with juvenile nephronophthisis have a homozygous 290 kb deletion of *NPHP1* [205,251]. The gene is flanked by two 330 kb inverted repeats, with two nearby

TABLE 2.3 Genes Associated with Nephronophthisis and Associated Syndromes

Gene (Protein)	Chromosome	Phenotype (Median Age at ESRD)	Extrarenal Symptoms	Variant Frequency (%)	Interaction Partners
NPHP1 (nephrocystin-1)	2q13	NPHP (13 years)	RP (10%), OMA (2%), JS (rarely)	~20%	Inversion, nephro cystin-3, nephrocystin-4, filamin A and B, tensin, β-tubulin, PTK2B
NPHP2/INVS (inversin)	9q31	Infantile NPHP (<5 years)	RP (10%), LF, *situs inversus*, VSD	1%–2%	Nephrocystin-1, calmodulin, catenins, β-tubulin, APC2
NPHP3 (nephrocystin-3)	3q22	Infantile and adolescent NPHP	LF, RP (10%), *situs inversus*, MKS	1%–2%	Nephrocystin-1
NPHP4 (nephrocystin-4)	1p36	NPHP (21 years)	RP (10%), OMA, LF	3%–4%	Nephrocystin-1, BCAR1, PTK2B
NPHP5/IQCB 1 (nephrocystin-5)	3q21	NPHP (13 years)	Early-onset RP	3%	Calmodulin, RPGR, nephrocystin-6
NPHP6/CEP290 (nephrocystin-6/CEP290)	12q21	NPHP	JS, MKS	3%–5%	ATF4, nephro cystin-5, CC2D2A
NPHP7/GLIS2 (nephrocystin-7/GLIS2)	16p	NPHP	–	<1%	–
NPHP8/RPGRIPIL (nephrocystin-8/RPGRIPIL)	16q	NPHP	JS, MKS	<1%	Nephrocystin-1
NPHP9/NEK8(nephrocystin-8/NEK8)	17q11	Infantile NPHP	–	<1%	–
NPHP10/SDCCAG8 (nephrocystin-10/SDCCAG8 or CCCAP)	1q43-q44	NPHP	SLSN/BBS	<1%	OFD1
NPHP11/TMEM67/MKS3 (nephrocystin-11/transmemberane protein-67/Meckelin)	8q22.1	MKS, JS, NPHP(10 years) +	JS, MKS, COACH, LF	~3%	
NPHP12/TTC21B (nephrocystin12/Tetratricopeptide repeat domain 21B)	2q24	NPHP, LF	JS, ATD	1%	

Continued

TABLE 2.3 Genes Associated with Nephronophthisis and Associated Syndromes—cont'd

Gene (Protein)	Chromosome	Phenotype (Median Age at ESRD)	Extrarenal Symptoms	Variant Frequency (%)	Interaction Partners
NPHP13/WDR19 (nephrocystin13/WD repeat-containing protein19)	4q14	NPHP	SLSN, Caroli, Sensenbrenner, Jeune syndrome	1%	
NPHP14/ZNF423 (nephrocystin14/zinc finger protein 423)	16q12.1		JS	<1%	PARP1, CEP290
NPHP15/CEP164 (nephrocystin15/centrosomal protein 164)	11q23.3	NPHP	SLSN	<1%	
NPHP16/ANKS6 (nephrocystin16/ankyrin repeat and sterile alpha motif domain containing 6)	9q22.33	NPHP		<1%	Inversin, NPHP3, NEK8
NPHP17/IFT172 (nephrocystin17/intraflagellar transport protein 172)	2q23.3	NPHP	Jeune, Saldino—Mainzer syndrome, BBS	<1%	
NPHP18/CEP83 (nephrocystin18/centrosomal protein 83)	12q22	NPHP	RP, LF	<1%	
NPHP19/DCDC2 (nephrocystin19/doublecortin domain-containing protein 2)	6q22.3	NPHP	Sclerosing cholangitis, LF, deafness	<1%	
NPHP20/MAPKBP1 (nephrocystin20/mitogen-activated protein kinase-binding protein 1)	15q15.1	NPHP		<1%	
NPHP1L/XPNPEP3 (nephrocystin-1 L/XPNPEP3)	22q13	NPHP	Cardiomyopathy, seizures	<1%	
AHI1/CC2D2A	6q23.3	NPHP	JS, RP	1%	CEP290

ATF4, Activating transcription factor 4; APC2, anaphase-promoting complex 2; BCAR 1, breast cancer antiestrogen resistance 1; CC2D2A, coiled-coil and C2 domain containing 2A; ESRD, end-stage renal disease; JS, Joubert syndrome; LF, liver fibrosis; MKS, Meckel–Gruber syndrome; SLSN, Senior- Løken syndrome; BBS, Bardet–Biedl syndrome (BBS); ATD, Asphyxiating thoracic dystrophy; OMA, oculomotor apraxia; PTK2B, protein tyrosine kinase 2B; RP, retinitis pigmentosa; RPGR, retinitis pigmentosa GTPase regulator; VSD, ventricular septal defect.

45 kb repeats, making the segment prone to unequal homologous recombination and intrachromosomal loop excision, creating an interstitial deletion [252]. Nonsense, frameshift, and splice site variants in *NPHP1* have also been documented, and all likely result in loss of function of nephrocystin-1 [253] Only one missense variant, 1024G > A, has been reported, leading to a nonconservative amino acid change in a residue that is conserved in mouse and dog *NPHP1* [254].

Why some patients with the classic *NPHP1* deletion have only renal manifestations while others have Senior—Løken syndrome, Joubert syndrome, or Cogan-type oculomotor apraxia is not clear. There may be other genes within the deleted region or around the deletion boundaries that are responsible for extrarenal disease. Another possible explanation is triallelic inheritance, in which a variant in another gene, in addition to deletion of *NPHP*, is required for multisystemic disease. Other theories include subtle rearrangements of the break points undetectable by haplotype analysis, or disruption of enhancer sequences altering expression of genes outside the deletion region.

2.4.3.2 NPHP2

Linkage analysis in an inbred Bedouin family with infantile nephronophthisis localized the gene locus to a 12.9 cM region at 9q22–q31 [255]. The discovery of variants of *inversin* in the *inv*/*inv* mouse, which has enlarged, cystic kidneys, *situs inversus*, and histopathologic findings similar to nephronophthisis, and the location of human *inversin* within the NPHP2 critical region, strongly suggested that *inversin* was the candidate gene. Confirmation came when mutational analysis in seven families with infantile nephronophthisis showed recessive variants in the gene, which was then named *NPHP2* or *INVS* [247].

NPHP2 is a 17-exon gene spanning 100 kb of genomic DNA, encoding a 1065 amino acid protein, nephrocystin-2/inversin, which is conserved from the zebrafish to the mouse. It has a highly conserved N-terminal domain containing 16 tandem ankyrin repeats, a split nuclear localization signal, two calmodulin-binding IQ domains, and two "destruction" or D boxes that bind a protein involved in the anaphase-promoting complex, a cell-cycle regulatory ubiquitin ligase that regulates cellular transition from metaphase to anaphase and mitosis to G_1. Inversin colocalizes with nephrocystin-1 in the primary cilia of renal tubular epithelium, connecting it with other proteins that have a role in cilia formation, maintenance, and intraflagellar transport [247,256] (Fig. 2.9). In addition, cell-cycle-dependent expression of *NPHP2* and its role in the Wnt signaling pathway support the hypothesis that nephrocystin-2/inversin is involved in the maintenance of normal planar cell polarity [257–259].

Frameshift, missense, and splice site variants have all been reported in *NPHP2*, and most variants result in protein truncation. Five recurrent variants have been found in unrelated families: R899X, R907X, Q485fsX509, Q671X, and E970fsX971 [209,253]. No clear genotype–phenotype correlations have been established. In a cohort of seven families with *NPHP2* variants, one patient with homozygous missense variants did not have onset of ESKD until age 5 years, but neither did children of a family with compound heterozygous truncating variants. The patient with *situs*

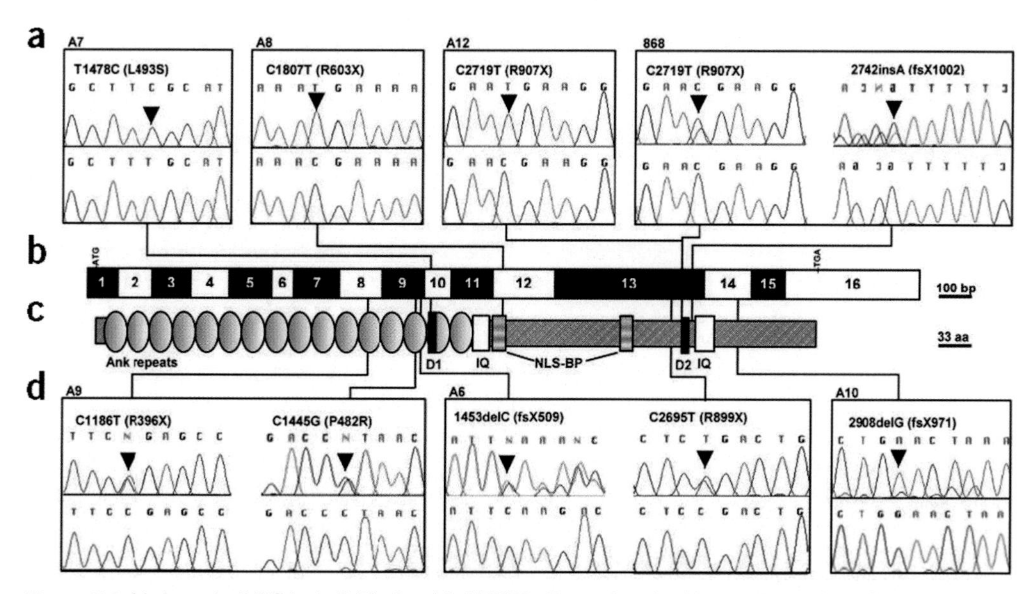

Figure 2.9 Variants in INVS in individuals with NPHP2. (Reproduced with permission from Springer Nature, Nature Genetics 34,413-420 (2003), Mutations in INVS encoding inversin cause nephronophthisis type 2, linking renal cystic disease to the function of primary cilia and left-right axis determination. Otto E., Schermer B., Obara T. et al. (2003))

inversus totalis had homozygous truncating R603X variants and was the only patient with variants in both alleles that removed all domains C-terminal to the ankyrin repeats, including the nuclear localization domains [247].

2.4.3.3 NPHP3

Examination of a large, inbred Venezuelan family with nephronophthisis indicated, at the time, that their disease was distinct from juvenile and infantile nephronophthisis on the basis of both the later age of onset and the gene locus. A total genome scan followed by higher resolution haplotype analysis mapped the family's locus to a 3.3 Mb interval at 3q22 [260,261]. Seven genes with known function and eight expressed sequence tags (ESTs) were found within this interval.

The mapping of several families with Senior—Løken syndrome to the same region narrowed the candidate genes to five ESTs, and sequencing revealed variants in a 40.5 kb gene that was named *NPHP3*. *NPHP3* is composed of 27 exons and encodes a 3990 nucleotide mRNA, which is translated into a 1330 amino acid cytosolic protein product called nephrocystin-3. The protein has an N-terminal coiled-coil region and a tubulin-tyrosine ligase domain and interacts with nephrocystin-1 and nephrocystin-2/inversin [210,262] (Fig. 2.10). In addition, ANKS6 were found to be linked with proteins inversin, nephrocystin-3 and NEK8, which are localized at the proximal part of the primary cilia known as the inversion compartment [263]. This may explain why patents with *INVS*, *NPHP3*, and *NEK8* variants share similar phenotype. In embryonic mice, the mRNA is expressed first in the node during gastrulation, then in the brain, retina, liver and biliary tract, renal tubules, and respiratory epithelium. Variants in the murine ortholog *Nphp3* produce the *pcy* mouse variant, which has cystic kidney disease due to a hypomorphic *Nphp3* allele [210]. Interestingly, treating the *pcy* mouse with a vasopressin-2 receptor antagonist has been shown to ameliorate disease by lowering renal cAMP levels [264]. This discovery may lead to future therapeutic interventions.

The phenotypic spectrum associated with *NPHP3* variants has broadened as more cases are amassed and now includes infantile and adolescent nephronophthisis, nephronophthisis with liver fibrosis, nephronophthisis with retinitis pigmentosa, and embryonic patterning defects. With regard to genotype—phenotype correlations, truncating and splice site variants in *NPHP3* produce a more severe phenotype similar to

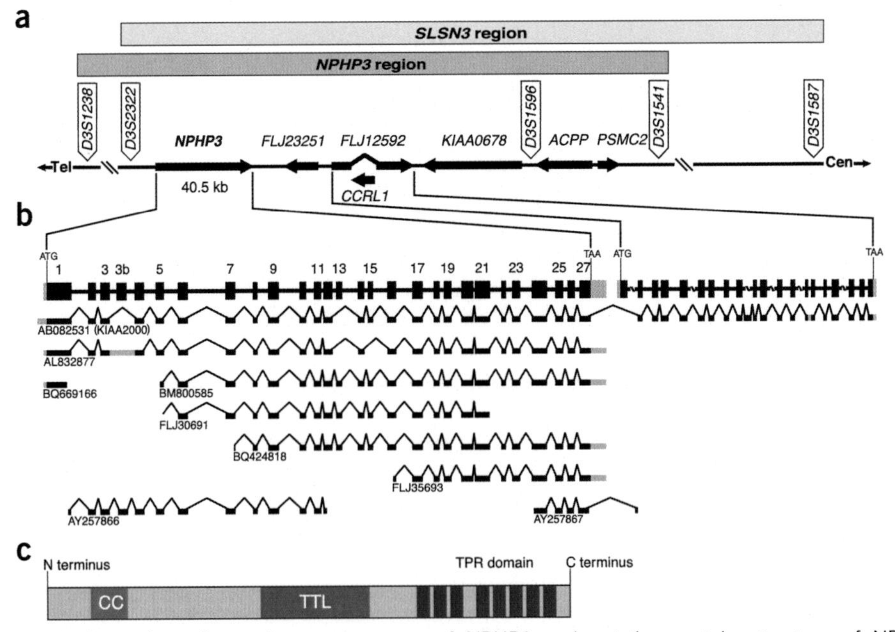

Figure 2.10 Genomic region and gene structure of NPHP3 and putative protein structure of NPHP3. (Reproduced with permission from Olbrich H, Fliegauf M, Hoefele J. et al. Mutations in a novel gene, NPHP3, cause adolescent nephronophthisis, tapeto-retinal degeneration and hepatic fibrosis. Nat Genet 2003;34:455—459.)

Meckel–Gruber syndrome, whereas missense and other nontruncating variants are associated with milder and later-onset disease [203,209].

2.4.3.4 NPHP4

After the discovery of *NPHP1*, *NPHP2*, and *NPHP3*, there continued to be kindreds that did not map to any of these loci [265]. Those families were gathered, and linkage analysis mapped them to a 2 Mb region within 1p36.31 [221]. Higher-resolution haplotype mapping narrowed the region to a 700 kb interval containing six genes. Two groups identified variants in the same gene and named it *NPHP4* [220,228].

NPHP4 is a large, 30 exon gene that spans 130 kb and encodes nephrocystin-4/nephroretinin, a 1426 amino acid protein product that has no known homologs in the human genome (Fig. 2.11). It is a highly conserved protein, with orthologs found in the mouse, cow, pig, zebrafish, *Xenopus*, *Ascaris*, and *Caenorhabditis elegans*. The mouse ortholog shares 78% sequence identity, while the *C. elegans* ortholog has 24% identity with the human sequence. Domain analysis shows a putative nuclear localization signal, an acidic domain, and an SH3 domain, the latter two also being features of nephrocystin-1 [220]. Nephrocystin-4 forms a complex with nephrocystin-1 and other proteins including p130Cas and Pyk2. It localizes to primary cilia, basal bodies, and centrosomes, consistent with the pathogenic mechanism of cystogenesis ascribed to other nephrocystin proteins [266]. In addition, nephrocystin-4 localizes to the connecting cilium of photoreceptor cells where it interacts with RPGRIP1 (retinitis pigmentosa GTPase regulator interacting protein 1), deficiency of which is responsible for Leber congenital amaurosis [267]. *NPHP4* has been identified as a negative regulator of the Hippo pathway, which is involved in tumor suppression and the control of cell proliferation [268].

A recent mutational analysis of *NPHP4* in 250 patients with nephronophthisis identified 23 novel sequence variants in 26 (10%) different patients [269]. Of the six patients who had either homozygous or compound heterozygous variants, 5/8 variants (63%) were thought to be loss-of-function or truncating variants. In the 20 patients with only one sequence variant detected, only 1 was a likely loss-of-function mutation. No genotype–phenotype correlations were recognized.

Previously reported variants in *NPHP4* have included nonsense, missense, frameshift, and splice site variants, most resulting in a truncated protein [220,228]. Three kindreds with no extrarenal involvement were found to have homozygous missense variants, G754R, R848W, and F991S, indicating a possible genotype–phenotype correlation between isolated nephronophthisis and missense variants [220,228].

2.4.3.5 NPHP5

NPHP5, also called *IQCB1*, spans 65.7 kb and contains 15 exons. The first two exons are not translated. *NPHP5* encodes nephrocystin-5, a 598 amino acid protein that is ubiquitously expressed in fetal and adult tissues (*193*). It

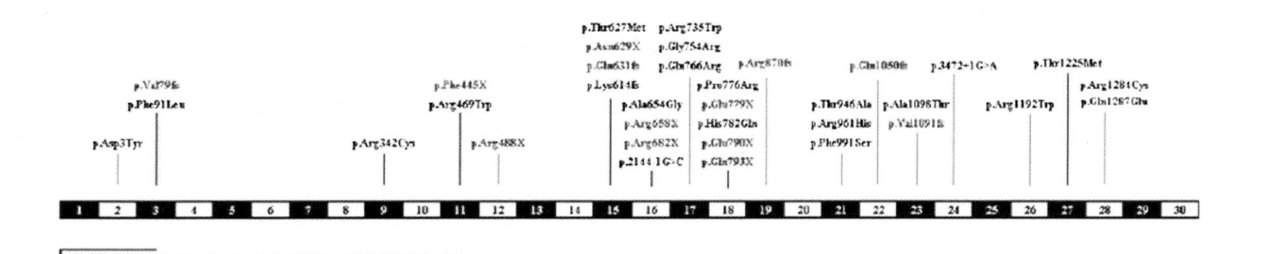

Figure 2.11 Linear representation of all sequence variants detected in the NPHP4 gene. Numbering based on cDNA position +1 corresponds to te A of the ATG translation initiation codon in the reference sequence NM_015102. *Black and white boxes* represent the 30 exons encoding nephrocystin-4/nephroretinin. Positions of novel sequence variants detected in this study and mutations described in the literature are indicated. Truncating mutations are shown in red. The extent of the truncation is shown beneath the exon structure. (Reproduced with permission from Hoefele J, et al. Mutational analysis of the NPHP4 gene in 250 patients with nephronophthisis. Human Mutation 2005;25(4): 411.)

has a coiled-coil domain and two IQ calmodulin-binding regions. Nephrocystin-5 directly interacts with calmodulin and forms a complex with retinitis pigmentosa GTPase regulator (RPGR) that localizes to primary cilia in renal epithelial cells and their homologous structures in the retina, photoreceptor connecting cilia (*193*). Mutations in *NPHP5* cause Senior–Løken syndrome, which is characterized by nephronophthisis and early-onset retinitis pigmentosa. All the eight variants originally described resulted in protein truncation; no missense variants were found (*193*). Most of the affected individuals had homozygous variants.

2.4.3.6 NPHP6

NPHP6, also known as *CEP290*, spans 93.2 kb and contains 55 exons. It encodes nephrocystin-6, a 290 kDa protein of 2479 amino acids with 13 putative coiled-coil domains, a region with homology to structural maintenance of chromosomes (SMC) chromosome segregation ATPases [14], a bipartite nuclear localization signal (NLS_BP), six RepA/Rep$^+$ protein KID motifs (KID), three tropomyosin homology domains, and an ATP/GTP-binding site motif A (P-loop) [217]. Nephrocystin-6 localizes to centrosomes in renal epithelial cells during interphase and to connecting cilia of photoreceptors. It interacts with nephrocystin-5 and ATF4, a transcription factor involved in the regulation of the cell cycle and in cAMP-dependent renal cyst formation [217,227,264]. Variants in *NPHP6* along with interaction with proteins TMEM67 [270] and CC2D2A [271] lead to ciliopathy and cause cystic renal disease, syndromic forms of nephronophthisis such as Joubert syndrome and Meckel–Gruber syndrome. The nine initially reported variants were all nonsense or frameshift [217].

2.4.3.7 NPHP7

NPHP7 is also known as *GLIS2* and contains six exons spanning more than 7.5 kb. The protein nephrocystin-7 or "Gli-similar protein 2" is highly expressed in the kidney and localizes to primary cilia of renal epithelial cells and nuclei [272,273]. The GLIS2 protein is related to the Gli transcription factor, suggesting a possible link between cyst formation and the sonic hedgehog signaling network [206]. Loss of Glis2 increases cell senescence and suppresses kidney cyst growth [274]. Mutations in *NPHP7* were first reported in a consanguineous Oji-Cree kindred [272]. Affected members had a homozygous transversion (IVS5 + 1G > T) affecting a splice donor site. Other than this index family, mutations in *NPHP7* appear to be a rare cause of nephronophthisis.

2.4.3.8 NPHP8

NPHP8, or *RPGRIP1L* as it is also known, is composed of 27 exons, the last of which is noncoding, and extends over 103.2 kb. The encoded protein, nephrocystin-8/RPGRIP1L (retinitis pigmentosa GTPase regulator interacting protein 1-like), is ubiquitously expressed in human embryonic and fetal tissues. It contains five coiled-coil domains, a C-terminal region homologous to the RPGR-interacting domain of RPGRIP1, and a central region with two protein kinase C conserved region 2 (C2) motifs. Nephrocystin-8 colocalizes with nephrocystin-4 and nephrocystin-6 to basal bodies, centrosomes, and primary cilia of renal tubular cells [275,276]. The spectrum of phenotypic variability observed in individuals with *NPHP8* variants can be explained in part by the type of variant. Homozygous truncating variants of *NPHP8* produce Meckel–Gruber syndrome or similar severe phenotypes, whereas the presence of at least one missense variant causes milder and later-onset disease [264,276]. Heterozygous and homozygous missense variants as well as frameshift and splice site variants have all been reported in individuals with Joubert syndrome [275–277]. Other extrarenal features may include polydactyly, liver fibrosis, pituitary agenesis, and partial growth deficiency.

2.4.3.9 NPHP9

NPHP9, also known as *NEK8*, is an extremely rare cause of nephronophthisis. In a study of 588 patients with nephronophthisis, only three missense variants were identified in *NPHP9*, and in one of these individuals, a homozygous variant in *NPHP5* was also present, suggesting possible oligogenic inheritance [208]. Like all known nephrocystin proteins, nephrocystin-9 localizes to primary cilia of renal tubular epithelial cells [208,278]. It has been shown to interact with polycystin-2, and in mouse models, mutations in *Nek8* cause increased expression of polycystin-1 and polycystin-2 [278]. Nephrocystin-9/NEK8 is also thought to play a role in cell cycle regulation [206]. *NEK8* mutation increases DNA damage [279] and associated with dysregulated Hippo pathway and may have role in severe renal cystic dysplasia [280].

2.4.3.10 NPHP10

NHPH10 encodes for a protein SDCCAG8 known as colon cancer antigen 8 or centrosomal colon cancer autoantigen protein (CCCAP), which has been identified in families with retinitis pigmentosa and NPHP [281]. Patients with this gene mutation has also been reported in Senior–Løken syndrome and may have features of BBS [282]. SDCCAG8 localizes in the ciliary centrioles and transgenic mice with *SDCCAG8* mutation exhibited phenotypes resembling human retinal degeneration and nephronophthisis.

2.4.3.11 NPHP11

The *NPHP11* gene at 8q22 is also known as the *MKS3* and *TMEM67* gene, encodes for a ciliary protein meckelin. Hypomorphic *TMEM67* gene variant was found in 8% of patients who had NPHP and liver fibrosis [283]. *TMEM67* has been implicated in the pathogenesis of Meckel–Gruber syndrome, Joubert syndrome, and COACH syndrome (cerebellar vermis hypo/aplasia, oligophrenia, congenital ataxia, coloboma. and congenital hepatic fibrosis). Patients with Joubert syndrome who have *TMEM67* variant were frequently associated with kidney disease [284].

2.4.3.12 NPHP12

The *NPHP12* gene, also known as the *TTC21B* gene, on chromosome 2 encodes IFT139, a subunit of the intraflagellar transport-A complex, which regulates the retrograde trafficking in the primary cilia [285]. Mutation in *TTC21B* has been reported in patients with NPHP, hepatic fibrosis, or bone anomalies and Jeune asphyxiating thoracic dystrophy [286]. Loss of *TTC21B* in mice were shown to downregulate Hedgehog signaling and promote cystogenesis [287]. A homozygous missense variant of *TTC21B* is reported to be associated with onset of familial FSGS [288].

2.4.3.13 NPHP13

The *NPHP13* gene at 4q14, also known as *WDR19*, encodes for a ciliary protein IFT144, a subunit of the intraflagellar transport-A complex essential for ciliogenesis. Mutation of *WDR19* have been identified in patients with Senior–Loken syndrome, Caroli disease, Sensenbrenner syndrome, Joubert syndrome, and Jeune syndrome [289–291]. Biallelic mutations in the *WDR19* gene in a patient with nephronophthisis also exhibited dilatation of the intrahepatic bile ducts [291].

2.4.3.14 NPHP14

Gene mutation of *NPHP14* gene also known as *ZNF423* on chromosome 16 have been reported in patient with Joubert syndrome. ZNF423 is a nuclear protein and functions as a DNA binding transcription factor and interacts with DNA repair gene *PARP1* and *CEP290* [292].

2.4.3.15 NPHP15

Mutation of *NPHP15* or *CEP164* at 11q23 may lead to NPHP and Senior–Loken syndrome [292]. CEP164 is a protein involved in ciliogenesis and formation of distal appendage of the centriole [293]. CEP164 is involved in the DNA repair response and defects in the DNA repair response signaling pathway may contribute to the pathogenesis of NPHP and related ciliopathies.

2.4.3.16 NPHP16

NPHP16 or *ANKS6S* mutation is reported as a cause of NPHP [294]. *ANKS6S* on chromosome 9 is localized in the inversion compartment of the cilium and links to the gene *NPHP2* (Inversin) and *NPHP3* to *NPHP9* (NEK8) [263].

2.4.3.17 NPHP17

NPHP17 or *IFT172* mutation is reported in patients with Jeune syndrome, Saldino–Mainzer syndrome [233]. It has been reported in patients with retinal degeneration and BBS [295]. IFT172 is a gene encoding one of the intraflagellar transport complex B proteins, which are involved in anterograde protein transport in the primary cilia.

2.4.3.18 NPHP18

The gene mutation of *NPHP18/CEP83* located at 12q22 have been reported in seven families with infantile NPHP and cause intellectual disability [296]. *CEP83* gene encodes for a ciliary protein CEP83 or CCDC41 localized in the distal appendage of the centriole, which is involved in docking and anchoring of the mother centriole [297].

2.4.3.19 NPHP19

Pathogenic variants of *NPHP19* or *DCDC2* at chromosome 6 were detected in patients with ESKD and liver fibrosis and sclerosing cholangitis [298,299]. DCDC2 is localized in the ciliary axoneme, mitotic spindles and is involved in Wnt signaling. Knockdown of *DCDC2* suppresses kidney cystogenesis [298].

2.4.3.20 NPHP20

Gene variants of *NPHP20* or *MAPKBP1* at 15q.15 have been reported in eight patients with juvenile or late onset NPHP with severe kidney fibrosis. MAPKBP1 is a scaffolding protein for JNK signaling. This protein is not localized to the primary cilia but it is in the mitotic spindle pole, indicating a role of non-ciliary protein to the development of NPHP [300].

2.4.3.20.1 Other genes. Genes that encode for proteins that are part of the primary cilia or ciliary function are associated with nephronophthisis. Variants in genes encoding intraflagellar transport (IFT) complex A (retrograde transport in the primary cilia) and B (anterograde transport) result in NPHP phenotype. IFT-A complex gene *SRDT9* encoding protein IFT140 is associated with Mainzer–Saldino Syndrome, a rare disorder characterized by phalangeal cone-shaped epiphyses, CKD, and severe retinal dystrophy [231,301]. IFT-B complex gene: Variants in *IFT27, IFT80, IFT172* are reported in patients with Jeune syndrome, Saldino–Mainzer syndrome, and BBS [233,302,303]. *IFT54* gene variant is linked to development of tubulointerstitial nephritis and ESKD in those who have Senior–Loken syndrome [304]. *IFT81* is a gene encoding for IFT-B core protein and linked to a very rare cause of ciliopathy [305].

Pathogenic variants in genes *AHI1* and *CC2D2A* were both initially reported in patients with Joubert syndrome without cystic kidney disease [271,306,307]. Both AHI1 and CC2D2A localize in the basal body. CC2D2A colocalizes with NPHP6 or CEP290 and mutation of these genes are associated with development of cystic kidney disease [308,309]. R830W allele of the *AHI1* gene was found more frequently in patients with nephronophthisis and retinal degeneration than in those with nephronophthisis without retinal degeneration (25% compared to 1.8%, $P = 5.36 \times 10{-}^{6}$), equaling a relative risk of 7.5 [310]. This association was true regardless of the primary mutation. Similarly, the Thr229 variant of the *RPGRIP1L* gene was shown to be associated with retinitis pigmentosa in a small cohort with nephronophthisis [311]. Other factors may become apparent as the mechanism of disease is further elucidated.

2.4.4 Laboratory, Imaging, and Pathologic Findings

A complete blood count reveals severe microcytic, hypochromic, hypoproliferative anemia due to low levels of iron and erythropoietin. There is progressive decline in GFR. Urine osmolality following water deprivation/dDAVP administration is inappropriately low compared to serum osmolality and reflects loss of urinary concentrating ability (also known as nephrogenic diabetes insipidus). Proteinuria and pyuria are rare. Kidney ultrasound shows normal to decreased kidney size, increased echogenicity, and diminished corticomedullary differentiation. Although corticomedullary cysts may be present, it is important to note that their presence is not necessary for diagnosis. Kidney biopsy shows a characteristic set of histopathologic findings: disorganized tubular basement membranes, tubular atrophy with cyst formation, interstitial lymphohistiocytic infiltration, and periglomerular and interstitial fibrosis. Patients with nephronophthisis should also be screened for ocular involvement with fundoscopic examination and an electroretinogram if Senior–Løken syndrome is suspected. A brain MRI may show the molar tooth sign in patients with Joubert syndrome.

2.4.5 Molecular and Prenatal Diagnosis

Molecular genetic screening allows gene identification in 70% of NPHP cases. Given the high degree of genetic heterogeneity and phenotypic variation, molecular genetic testing using a whole-exome sequencing panels is recommended for patients with clinical diagnosis of NPHP. If whole-exome sequencing is not available, screening for homozygous or heterozygous variants of *NPHP1* should be performed since this is the most common variant in NPHP. Other gene variants are responsible for less than 3% of the cases. Based on the clinical manifestation of the individual patients, specific gene testing may be considered. In children less than 5 years of age with early onset of CKD, testing for variants in *NPHP2* and *NPHP3* should be sought [209]. In those with severe retinitis pigmentosa, *NPHP5* should be tested. For patients with neurological symptoms, *NPHP6* and *NPHP8* may be screened. If no variants are detected, testing of other genes is dictated by the nature of the extrarenal complications. In all cases, a careful family history should be elicited, looking in particular for consanguinity or a history of early death or renal disease. More information regarding genetic testing is available at www.kidneygenes.org.

2.4.6 Management

There is currently no therapy for the underlying disease or the ophthalmologic and neurologic manifestations. Anemia responds to administration of erythropoietin and iron. Supportive care should be given for complications of CKD such as metabolic acidosis, electrolyte disorder, secondary hyperparathyroidism, and hypovolemia. Periodic monitoring for kidney function is necessary so that dialysis and enrollment in transplantation registries can be initiated before reaching ESKD. Analysis of the North American Pediatric Renal Trials and Collaborative Studies database demonstrates that nephronophthisis transplant recipients have excellent clinical outcomes that are better than those of the general pediatric transplant population [312].

2.5 AUTOSOMAL DOMINANT TUBULOINTERSTITIAL KIDNEY/ MEDULLARY CYSTIC KIDNEY DISEASE

2.5.1 Clinical Features and Natural History

Patients with medullary cystic kidney diseases (MCKD) typically present with polyuria, polydipsia, or isosthenuria along with a bland urine sediment and absent to mild proteinuria. These clinical features are indistinguishable from juvenile nephronophthisis, however, unlike nephronophthisis, MCKD is dominantly inherited and has adult onset of disease. Recent advances in the understanding of these disorders and the identification of some of their underlying disease-causing variants led to the proposal of a new terminology to describe them. The KDIGO consensus 2015 report suggested using the term "Autosomal Dominant Tubulointerstitial Kidney Disease" (ADTKD) appended by a gene-based subclassification to collectively describe these rare kidney disorders. This highlighted the inheritance of these disorders and the fact that an interstitial fibrotic rather than cystic phenotype dominates their presentation. To date four major genes with disease causing variants have been identified: Uromodulin (UMOD), renin (REN), hepatocyte nuclear factor 1β (HNF1B), and mucin-1 (MUC1). We now know that previously coined MCKD type 1 is due to *MUC1* variants while MCKD type 2 is caused by *UMOD* variants [313].

2.5.2 Gene Mapping, Structure, Function, and Genotype—Phenotype Correlations

2.5.2.1 ADTKD-UMOD (OMIM 191845) Previously Called MCKD2

The uromodulin gene (16p12.3) has 12 exons (10 coding) spanning 18.6 kb of the human genome. It is transcribed into a 2.4 kb mRNA, which is translated into a 640 amino acid protein that is also known as Tamm—Horsfall protein. Uromodulin/Tamm—Horsfall protein (UMOD/THP) is produced almost exclusively by the epithelial cells lining the thick ascending loop (TAL) of Henle and to a lesser extent by the early distal convoluted tubules [314]. UMOD is a heavily glycosylated protein with a leader peptide at its N terminal allowing its insertion into the endoplasmic reticulum and a C—terminal hydrophobic stretch that directs it toward specialized membrane domains called lipid rafts [314]. Although not fully elucidated, the physiologic functions of UMOD include protection against urinary tract infections as well as kidney stone formation [314]. UMOD also enhances sodium retention and a genetic mutation causing a decreased production of UMOD leads to mild natriuresis and subsequent increase in sodium and uric acid reabsorption proximally, which explains the hypouricosuric hyperuricemia characteristic of ADTKD-*UMOD* [315]. To date, more than 120 *UMOD* variants have been identified with most being missense and clustering in exons 3 and 4 [316]. Additionally, more than 50% of variants affect the conserved cysteine residues. *UMOD* variants lead to protein misfolding, which results in defective proteins being trapped in the endoplasmic reticulum leading to ER stress and cellular apoptosis [316]. No clear genotype—phenotype associations have been established to date.

Hyperuricemia and gout are present early in life. Chronic kidney disease is often diagnosed in childhood with a median age of 54 years at the time of kidney failure [315]. A family with a rare homozygous *UMOD* variants in the setting of consanguinity was observed to survive to adulthood but have more severe hyperuricemia and faster progression to ESKD [317]. No disease specific treatments are currently available; however, early medical treatment of hyperuricemia may play a protective role [315].

2.5.2.2 ADTKD-MUC1 (OMIM 158340)

Pathogenic variants in *MUC1* located on chromosome 1 (1q22) have recently been identified to cause a subset of ADTKD cases formerly referred to as MCKD type 1. Sanger sequencing of *MUC1* to establish the diagnosis is particularly challenging as variants reside in a coding variable number of tandem repeats (VNTR) in exon 2 that is rich in guanosine and cytosine. The most common reported variation leads to the addition of a cytosine residue to a seven residue tract in the VNTR leading to a frameshift mutation [315]. Additional reported variants include the addition of a guanosine or loss of two cytosine residues, all of which result in a similar frameshift protein [315]. Because of the challenge in identifying variants on genetic testing, other immunohistochemical examination of urinary cells are being evaluated [318]. MUC1 protein is a transmembrane protein with a cytoplasmic tail that is responsible for intracellular signaling. It is expressed in a variety of epithelial cells and in the kidneys. UMOD and MUC1 form glycoprotein polymers that protect the uroepithelium [319]. More recently, MUC1 was also found to regulate the renal calcium channel TRPV5 decreasing the risk of calcium based nephrolithiasis by reducing calciuresis [320].

Mutated proteins accumulate in the ER and lead to tubular cell death, nephron dropout and ultimately CKD [315]. Based on a large international cohort study, *MUC1* variants account for the second most common cause of ADTKD following *UMOD* variants [321]. Patients affected by ADTKD-MUC1 appear to have a more severe kidney phenotype with a higher prevalence of ESKD occurring at a younger age. On the other hand, gout was much less common in that population [321]. To date, there is no disease specific treatment for this condition; however, recent studies in a murine model of ADTKD revealed promising results for a small molecule (BRD4780) that has the ability to clear the misfolded protein out of the ER. Further studies translating these early exciting results into future human therapeutic trials are needed [322].

2.5.2.3 ADTKD- REN (OMIM 179820)

REN (located on chromosome 1q32.1) variants are rare causes of ADTKD. Heterozygous mutations lead to reduced synthesis of renin and prorenin and subsequently result in mild hyperkalemia, anemia, hyperuricemia, and mild hypotension with an increased risk of acute kidney injury [323]. *REN* variants have been reported in the signal peptide (which allows the processing of the preprorenin protein in the ER), the mature renin protein, and more recently the prosegment (a segment of the gene that assists in protein folding) [323]. All types of variants ultimately result in renin deficiency and ER stress. Patients with variants in the signal peptide have the most severe phenotype followed by variants in the prosegment, with variants in the mature renin peptide imparting the mildest course. Retrospective clinical data from a large cohort of affected patients suggest that treatment with fludrocortisone can reverse some the disease manifestations but more information needed [323].

2.5.2.4 ADTKD- HNF1B (OMIM 189907)

Pathogenic variants in the hepatocyte nuclear factor 1β (*HNF1β*) gene (located on chromosome 17q12) typically lead to the development of ADTKD as well as maturity-onset diabetes of the young (MODY), hypomagnesemia, hypokalemia, and hypolcaciuria and abnormal liver function tests [324]. There is however significant variability in the disease presentation, and a recent kindred was described whose phenotype included medullary sponge kidney disease [325].

HNF1β is a transcription factor that regulates multiple genes expressed in the kidneys, pancreas, as well as the liver, which explains the multiorgan manifestations of the disease. Studies in mouse derived kidney cell lines and mouse models suggest that *HNF1β* variants lead to reduced activity of the potassium channel Kir5.1 expressed in the distal convoluted tubule with subsequent magnesium and potassium wasting [324]. Disease diagnosis requires direct sequencing for point variants and multiplex ligation-dependent probe amplification with quantitative PCR for deletions and duplications [313].

2.5.3 Imaging and Histopathologic Findings

Renal ultrasonography shows normal to decreased renal size and reduced cortical thickness. Although bilateral corticomedullary or medullary cysts may be present (40%), they are not necessary for diagnosis, nor are they pathognomonic for ADTKD [326]. Histopathologic findings of renal biopsies are similar for all genetic causes of ADTKD and typically reveal interstitial fibrosis with tubular atrophy but intact glomeruli. Thickening and lamellation of tubular basement membranes can be seen with possible tubular microcystic dilatation. Immunofluorescence studies on kidney biospecimens are negative of any complement of immunoglobulin staining [313].

2.6 MULTICYSTIC DYSPLASTIC KIDNEY DISEASE (MCDK)

MCDK is a form of congenital anomaly of the kidney and urinary tract (CAKUT). Most cases are unilateral although bilateral MCDK has been described and in those cases can lead to Potter syndrome [327]. When unilateral, they can be associated with contralateral anomalies in 30%—40% of cases [328]. By ultrasound, congenital MCDK consists of cysts of various sizes with loss of lobar organization. Microscopically, there is loss of normal structure, with the presence of primitive ductules and cartilage, suggesting aberrations in renal differentiation. Dysplasia can result in cysts that are solid or cystic, large or small. Function is variable [329]. MCDK is usually detected prenatally by ultrasound and presents as an abdominal mass in the newborn period. Other significant manifestations are uncommon and include flank pain, urinary frequency, dysuria, failure to thrive, and UTI. Most cases of MCDK occur sporadically; however, a number of syndromes can have MCDK as their renal manifestation. These include Meckel—Gruber syndrome, Jeune asphyxiating thoracic dystrophy, Zellweger syndrome, VACTERL association, brachiootorenal syndrome, Williams syndrome, Beckwith—Wiedemann syndrome, and certain chromosomal trisomies. The best diagnostic test is high-resolution ultrasonography. Chromosome microarray analysis can also be helpful to detect submicroscopic copy number variations, which has been reported in about 15% of isolated MCDK cases and around 20% of those with extrarenal abnormalities [328]. Children with MCDK should be prospectively monitored for the development of hypertension and CKD [328].

2.7 GENETIC SYNDROMES WITH CYSTIC RENAL DISEASE AS A MAJOR COMPONENT

2.7.1 Tuberous Sclerosis Complex (MIM 191100, 613254)

Tuberous sclerosis complex (TSC) is an autosomal dominant neurocutaneous disorder with an incidence of 1 in 6000 to 10,000 live births [330]. It is characterized by the involvement of many organ systems including benign hamartomas of the brain, eye, heart, lung, liver, kidney, and skin. Two genes are responsible for the disease pathogenesis: *TSC1*, located on chromosome 9q34, and *TSC2* located on chromosome 16p13.3, which lies in close proximity to the *PKD1* gene. Deletions in both *TSC2* and *PKD1* result in the contiguous gene syndrome characterized by severe early onset of features of polycystic phenotype with cutaneous and neurologic manifestations of TSC. The disease can be familial with an autosomal dominant inheritance in about 2/3 of patients, or sporadic in about 1/3 of patients, with around 1%—2% revealing germline mosaicism. Definitive genotype—phenotype correlations have not been well established; however, *TSC1* variants, which account for about 25% of cases, typically portend a milder phenotype. *TSC2* variants account for about 70% of cases (mostly sporadic ones) and are associated with a more severe phenotype including earlier onset of seizures along with more neurologic and renal manifestations [331]. The protein products of *TSC1* and *TSC2,* hamartin and tuberin, respectively, interact together and along with a third protein TBC1D7, form a complex that serves as a negative regulator of the mTOR pathway, controlling cell growth and proliferation [332]. Patients with TSC have a germline mutation in either *TSC1* or *TSC2,* which is followed by somatic mutation in the unaffected allele resulting in the disruption of the tuberin—hamartin complex and subsequent unabated activation of the mTOR pathways. This results in the formation and growth of hamartomas throughout the body. which are the hallmark of the disease and, although benign, lead to organ dysfunction. The diagnostic criteria for TSC were initially put forth during a National Institutes of Health sponsored conference in 1998 and updated in 2012 [333]. Accordingly, a pathogenic variant identified on genetic testing is sufficient to establish the diagnosis.

Clinical diagnostic criteria are also available and include 11 major criteria (≥3 hypomelanotic macules at least 5 mm in size, ≥3 angiofibromas or fibrous cephalic plaque, ≥2 ungual fibromas, shagreen patch, multiple retinal hamartomas, cortical dysplasias, subependymal nodules, subependymal giant cell astrocytoma, cardiac rhabdomyoma, lymphangioleiomyomatosis, and angiomyolipomas) and six minor criteria ("confetti" skin lesions, >3 dental enamel pits, ≥2 intraoral fibromas, retinal achromic patch, multiple kidney cysts, nonkidney hamartomas). The clinical diagnosis of TSC requires the presence of two major criteria or one major and two minor criteria [333] (Table 2.4).

Approximately 85% of affected patients experience CNS complications including epilepsy and cognitive impairment, collectively referred to as TSC-associated neuropsychiatric disorders or TAND. Based on consensus conference recommendations, all TSC patients should undergo baseline brain MRI [to assess for the presence of cortical/subcortical tubers, subependymal nodules, and subependymal giant cell astrocytomas (SEGA)] as well as electroencephalogram (EEG). Subsequent brain MRI every 1–3 years at least should be performed for follow-up [334]. Epilepsy occurs in 70%–90% of children with TSC and is often difficult to treat. Vigabatrin is first line agent for the treatment of seizures, and recent data show that preventive therapy with vigabatrin in children with abnormal EEG (prior to seizure development) is safe and reduces the risk of clinical seizures as well as drug-resistant seizures [335]. Everolimus is indicated for the treatment of SEGA when tumors cannot be curatively resected and results in successful size reduction followed by stabilization [336]. More recent studies suggest that long-term maintenance can be successfully achieved on a lower dose of everolimus [337].

Kidney manifestations are the second most common feature with 80% of patients having renal angiomyolipomas (AML) and 50% having cystic kidney disease. MRI is recommended for screening and for follow-up every 1–3 years [334]. AML can grow with time and their abnormal vascular structure is prone to aneurysmal dilation. The risk of bleeding increases significantly once AML enlarges to more than 4 cm or their aneurysmal component reaches 5 mm in diameter. Currently, the standard of care to control active bleeding is arterial embolization. Post-embolectomy syndrome is common within 48 h of the procedure and manifests as nausea, pain, fever, and hemodynamic instability and should be treated with corticosteroids. For AML larger than 3 cm, the first-line treatment is an mTOR inhibitor. Results from the EXIST-2 trial (Everolimus for angiomyolipoma associated with tuberous sclerosis complex or sporadic lymphangioleiomyomatosis) showed that everolimus treatment was safe and effective in reducing the volume of AML and helped to preserve eGFR in most patients, hence supporting its use for asymptomatic growing AML [338,339].

Lymphangioleiomyomatosis (LAM) is a devastating pulmonary complication that occurs almost exclusively in women and can lead to cystic and interstitial lung

TABLE 2.4 Diagnostic criteria for tuberous sclerosis complex

CLINICAL DIAGNOSTIC CRITERIA	
Major features	**Minor features**
Hypomelanotic macules (≥3, at least 5-mm in diameter)	"Confetti" skin lesions
Angiofibromas (≥3) or fibrous cephalic plaque	Dental enamel pits (>3)
Ungual fibromas (≥2)	Intraoral fibromas (≥2)
Shagreen patch	Retinal achromic patch
Multiple retinal hamartomas	Multiple renal cysts
	Nonrenal hamartomas
Cortical dysplasias	
Subependymal nodules	
Subependymal giant cell astrocytoma	
Cardiac rhabdomyoma	
Lymphangioleiomyomatosis (LAM)	
Angiomyolipomas (≥2)	

Definite diagnosis: Two major features or one major feature with ≥2 minor features
Possible diagnosis: Either one major feature or ≥2 minor features

disease, pneumothoraces, and chylous pleural effusions. Screening high-resolution CT scans are recommended every 5–10 years for asymptomatic at-risk patients, but the frequency of imaging studies should increase once symptoms arise. In cases of moderate to severe lung disease, mTOR inhibitors can be used [340].

2.7.2 Bardet–Biedl Syndrome (MIM 209900)

BBS also sometimes known as Laurence–Moon–Bardet–Biedl syndrome is a genetically heterogeneous disorder with 2 gene loci identified to date (Table 2.5). It is usually inherited in an autosomal recessive manner. Owing to considerable phenotypic heterogeneity, a classification system has been proposed in which a minimum of four major criteria are required for diagnosis. The major diagnostic criteria are learning disability (62%–87%), rod–cone dystrophy (92%–100%), polydactyly (58%–74%), truncal obesity (72%–96%), genital anomalies (59%–98%), and renal anomalies (46%–95%). Secondary features include speech delay (54%–81%); developmental delay (50%–91%), diabetes mellitus (6%–48%), ataxia (40%–86%); dental anomalies (51%), congenital heart disease (7%), anosmia and hyposmia (60%), brachydactyly (46%–100%), and syndactyly (8%–95%) [341]. Molecular diagnostic testing is commercially available and molecular confirmation of the disease is currently estimated at 40%–80% [341]. The diagnosis of BBS is typically triggered by the symptomatic rod-cone dystrophy and the onset of night blindness. BBS shares some clinical features of nephronophthisis, with polyuria and polydipsia, and the renal pathology is sometimes indistinguishable between the two diseases. Progression to ESKD, however, is rare estimated at around 8% based on large BBS cohort [342]. Structural renal anomalies include cortical cysts, urinary tract malformation, vesicoureteral reflux with recurrent UTI, renal dysplasia, and fetal lobulation [343].

The BBS proteins that have been evolutionary conserved localize to the primary cilium. Collectively, they maintain, stabilize, and assemble the intraflagellar transport system that allows the trafficking of proteins in and out of the primary cilium [344]. The exact mechanism for the development of renal cystic disease remains unelucidated but suggests possible contributions of abnormal mTOR signaling or noncanonical Wnt signaling and planar cell polarity [345–347].

TABLE 2.5 Genetic Loci Associated Bardet-Biedl Syndrome

Gene	Locus
BBS1	11q13
BBS2	16q21
BBS3/ARL6	3p12–q13
BBS4	15q22.3–q23
BBS5	2q31
BBS6/MKKS	20p12
BBS7	4q27
BBS8/TTC8	14q32.1
BBS9/B1	7p14
BBS10	12q21.2
BBS11/TRIM32	9q31–34.1
BBS12	4q27
BBS13/MKS1	17q23
BBS14/CEP290	12q21.3

2.7.3 Jeune Asphyxiating Thoracic Dystrophy (MIM 208500) and the Short-Rib Polydactyly Syndromes

Patients with asphyxiating thoracic dystrophy (ATD) have extremely narrow, constricted thoracic cages caused by shortened, horizontal ribs. Death often occurs in infancy because of severe pulmonary hypoplasia as a consequence of the small chest cavity. Survivors face recurrent pneumonia and the possibility of respiratory failure incurred by restrictive lung disease. Long bones and phalanges are shortened, and some patients have postaxial polydactyly of the hands and/or feet. Radiographs show short ribs with irregular costochondral junctions, hypoplastic iliac wings with trident-shaped acetabulae, transient irregularity of epiphyses and metaphyses, and cone epiphyses of the phalanges. Short stature and relative size of the thoracic cage improve with age, but children also develop biliary ductal plate malformations and hepatic fibrosis, pancreatic cysts and fibrosis, and polycystic kidneys. Renal histopathology shows cystic renal tubular dysplasia and possibly disorganized tubular basement membranes, tubular atrophy, and interstitial infiltration and fibrosis seen in nephronophthisis. Interestingly, some patients with ATD are affected with tapetoretinal degeneration and situs inversus, which are also seen in patients with nephronophthisis [348,349]. Currently, there are four genes that are associated with ATD, *IFT80* and *DYNC2H1*, *TTC21B*, and *WDR19*. Like the nephronophthisis and BBS genes, all four of these genes play a role in ciliary and/or microtubular function.

Short rib polydactyly syndromes (SRPS) are rare, lethal skeletal dysplasias, which is inherited by autosomal recessive fashion. SRPS is categorized into four groups: Type 1 (Saldino Noonan), Type 2 (Majewski), Type 3 (Verma Naumoff) and Type 4 (Beemer-Langer) [350]. There is overlap between Type 1 and Type 3 and ATD [351]. Type 1 is an extremely rare SRPS with neonatal onset characterized by polydactyly, hydropic appearance, and small thorax with short horizontal ribs causing fatal cardiorespiratory distress. Patients have micromelia, pointed metaphyses, and ossification defects involving vertebrae, calvaria, pelvis, hand, and foot. Polycystic kidneys along with gastrointestinal and genitourinary abnormalities are reported. Type 2 may have atrial septal defect, brachycephaly, cerebellar vermis hypoplasia, dilation of lateral ventricles, and hepatic fibrosis. Type 3 is characterized by short limb dwarfism, short ribs with thoracic dysplasia, postaxial polydactyly, and protuberant abdomen. Patients may have cardiovascular defects, renal agenesis/hypoplasia, abnormal cloacal development (ambiguous genitalia and anal atresia), and cerebellar hypoplasia. Type 4 is a rare form of SRPS developing prenatally or immediately after birth and characterized by short and narrow thorax with horizontally oriented ribs. Infants with this syndrome are hydropic and die in the perinatal period. Patients have micromelia and small thoracic circumference. They have normal cranial size and bone mineralization. Polydacyly is rare, but commonly associated with brain defect, and some cases have absent genitalia, renal agenesis, and pancreatic cyst.

2.7.4 Hajdu–Cheney Syndrome (MIM 102500)

This autosomal dominant skeletal dysplasia is an acroosteolytic condition with progressive, slow, centrifugal resorption of the distal and middle phalanges of the hands and feet beginning in adolescence. The initial abnormality in affected infants is widening of the sagittal and/or lambdoid sutures. Other craniofacial abnormalities that develop with age are persistently open sutures, wormian bones, elongation of the sella turcica, coarse hair and facies, broad forehead, hypertelorism with downslanting palpebral fissures, broad nasal bridge, and premature loss of teeth. Short stature, osteoporosis, and joint laxity are also observed. Polycystic kidneys similar to ADPKD were noted in approximately 10%–14% of affected children, often with concomitant hypertension and CKD [352–354]. Other renal anomalies, such as unilateral multicystic dysplastic kidney, vesicoureteral reflux, and glomerulonephritis, have been noted. Hajdu–Cheney Syndrome is associated with variants in exon 34 of *NOTCH2* gene [355]. Notch regulates skeletal development and bone remodeling, and gain-of-function mutation of *NOTCH2* is associated with this syndrome.

2.7.5 Campomelia, Cumming Type (MIM 211890)

Campomelia, Cumming type is characterized by the association of limb defects and multivisceral anomalies. This autosomal recessive, skeletal dysplasia is characterized by premature stillbirth, cervical hygromas with hydrops fetalis, narrow chest, shortened limbs with bowing of all long bones, especially the tibiae and ulnae, and cloverleaf skull [356–358] Biliary ductal plate malformation and hepatic fibrosis, pancreatic fibrosis, and renal cortical and medullary cysts are seen [357,359]. Abnormalities of situs, specifically dextrocardia, left-sided superior vena cava, right-sided aortic arch, total anomalous pulmonary venous return, and polysplenia have been reported in several fetuses with this disorder [356,359,360].

2.7.6 Glutaric Acidemia Type II (MIM 231680)

Glutaric acidemia type II, also known as multiple acyl-CoA dehydrogenase deficiency, is an autosomal recessive disorder caused by variants in *ETFA*, *ETFB*, or *ETFDH*, genes that encode the electron transport flavoprotein (ETF) and ETF ubiquinone oxidoreductase (ETF:QO). ETF is the ultimate electron acceptor in branched-chain amino acid, long-chain fatty acid, and choline metabolism. ETF:QO transfers electrons from ETF to the electron transport chain. Deficiency in either enzyme causes a phenotype that is a combination of fatty acid oxidation disorders and isovaleric acidemia, with metabolic acidosis, hypoketotic hypoglycemia, hepatomegaly, hypotonia, and a "sweaty foot" odor. Malformations such as enlarged anterior fontanel, high forehead, flat nasal bridge, telecanthus, and ear anomalies may be present. Bilaterally enlarged, polycystic kidneys with medullary dysplasia can be seen, especially in patients with more severe disease. Presumptive diagnosis can be made from plasma acylcarnitine and urine organic acid analysis. Fibroblast assay of ETF and ETF:QO activity and molecular genetic testing of *ETFA*, *ETFB*, and *ETFDH* confirm the diagnosis.

2.7.7 Carnitine Palmitoyltransferase II Deficiency, Neonatal Lethal Form (MIM 608836)

The carnitine palmitoyltransferase II (CPT II) enzyme catalyzes the reconjugation of long and very-long-chain acylcarnitines to coenzyme A following translocation into the mitochondrion. Patients with CPT II deficiency are consequently unable to metabolize long-chain fatty acids and become energy deficient during periods of fasting. CPT II deficiency can present at any time, but severely affected infants who present in the first few days of life with lethargy, poor feeding, tachypnea, metabolic acidosis, hepatomegaly, hypothermia, and hypoglycemia have been noted to have enlarged, polycystic kidneys with parenchymal dysplasia. Presumptive diagnosis can be made from free and total carnitine levels, plasma acylcarnitine profile, and urine organic acid analysis. The diagnosis is confirmed by measuring CPT II activity in fibroblasts or molecular testing of the *CPT2* gene [361].

2.7.8 Zellweger Syndrome (MIM 214100)

Also known as cerebrohepatorenal syndrome, Zellweger syndrome is a disorder of peroxisome biogenesis that lies at the severe end of a clinical spectrum that includes neonatal adrenoleukodystrophy and infantile Refsum disease. This group of disorders is genetically heterogeneous. Pathogenic variants have been found in 12 different genes, all encoding proteins necessary for normal peroxisome assembly. *PEX1* is the most commonly implicated, and the mode of inheritance for all types is autosomal recessive. Patients present in early infancy with severe hypotonia, hyporeflexia, and poor feeding. Many are initially suspected of having a type of myopathy. Dysmorphologic examination shows macrocephaly with an extremely large anterior fontanelle, high forehead, flat facies, hypertelorism with upslanting palpebral fissures, and hepatomegaly. Plain-film X-rays reveal stippling of the patellae and long bone epiphyses. Most patients have brain malformations and suffer from myoclonic or generalized tonic—clonic seizures. A vast majority die before the first birthday; survivors are severely mentally retarded. Polycystic kidneys are present in most patients. Zellweger syndrome can be screened for by measuring plasma very-long-chain fatty acids, which are elevated in this condition. Additional testing may include analysis of plasmalogens in erythrocytes, bile acid intermediates, and measurement of phytanic acid and pristanic acids in plasma. Complementation studies can be performed on fibroblasts to identify which peroxisomal gene is deficient. Molecular diagnostic testing is commercially available for many of the *PEX* genes. There are other genetic syndromes associated with renal cysts. These syndromes and their mode of inheritance are shown in Table 2.6.

TABLE 2.6 Genetic Syndromes Associated with Renal Cysts and Their Mode of Inheritance

Autosomal Recessive Inheritance
Meckel syndrome
Kaufman—McKusick syndrome
Retina-renal dysplasia syndromes
Ivemark syndrome
Fryns syndrome
Autosomal Dominant Inheritance
Brachio—oto—renal syndrome
von Hippel Lindau syndrome
Townes—Brocks syndrome
X-Linked
Oro—facial—digital syndrome
Chromosomal
Trisomy 18
Trisomy 13
Trisomy 9
Triploidy
Deletion 3p
Inheritance Variable
VATER association
Proteus syndrome
Prune-belly syndrome

2.8 MECHANISMS OF CYSTOGENESIS

How do variants in the polycystins, fibrocystin, and nephrocystins result in cyst formation? While all the pathogenic processes are not entirely elucidated, increasing understanding of the role of primary cilia on renal tubular epithelial cells has shed light on some of the mechanisms. In 1999, the discovery that the protein products of *LOV1* and *PKD-2* (the *C Elegans* gene homologues of *PKD1* and *PKD2*) localized to the flagella was the first suggestion that cilia may play a role in cystic phenotype. Subsequently, fibrocystin was also found to be localized in the cilia. Since these initial observations, data has accumulated showing that proteins associated with cystic kidney diseases all localize to the cilia or their basal bodies [362]. Cilia are classified as motile or nonmotile with the latter being characterized by a 9 + 0 microtubular arrangement that lacks the central microtubule doublet. These nonmotile or primary cilia are ubiquitously present in all organ system and were once thought to be vestigial remnants, but now recognized to be sensing and transduction organelles. As such, nonmotile cilia transduce extracellular information into physiologic responses through complex intracellular signaling pathways [344]. Cilia rely on intraflagellar antegrade and retrograde transport system (IFT) for the trafficking of proteins required for ciliary maintenance, growth, and function. A transition zone (TZ) separates the cilia from the rest of the cell membrane and controls the entry and exit of proteins into the ciliary structure. Cilia sense extracellular signals (both chemical and mechanical) and through a variety of signaling pathways convert these stimuli into downstream effects on cell growth and proliferation, as well as organ development and function. Cilia have also been involved in the planar polarity regulation, which is lost in cystogenesis. Mutations affecting ciliary function result in disturbed signaling pathways that control oriented cell division crucial for the integrity of renal tubular structure and function [344]. The majority of genes mutated in renal ciliopathies encode proteins that have been localized to various parts of the primary cilium (Fig. 2.12) and result in dysregulation of this complex ciliary machinery [344].

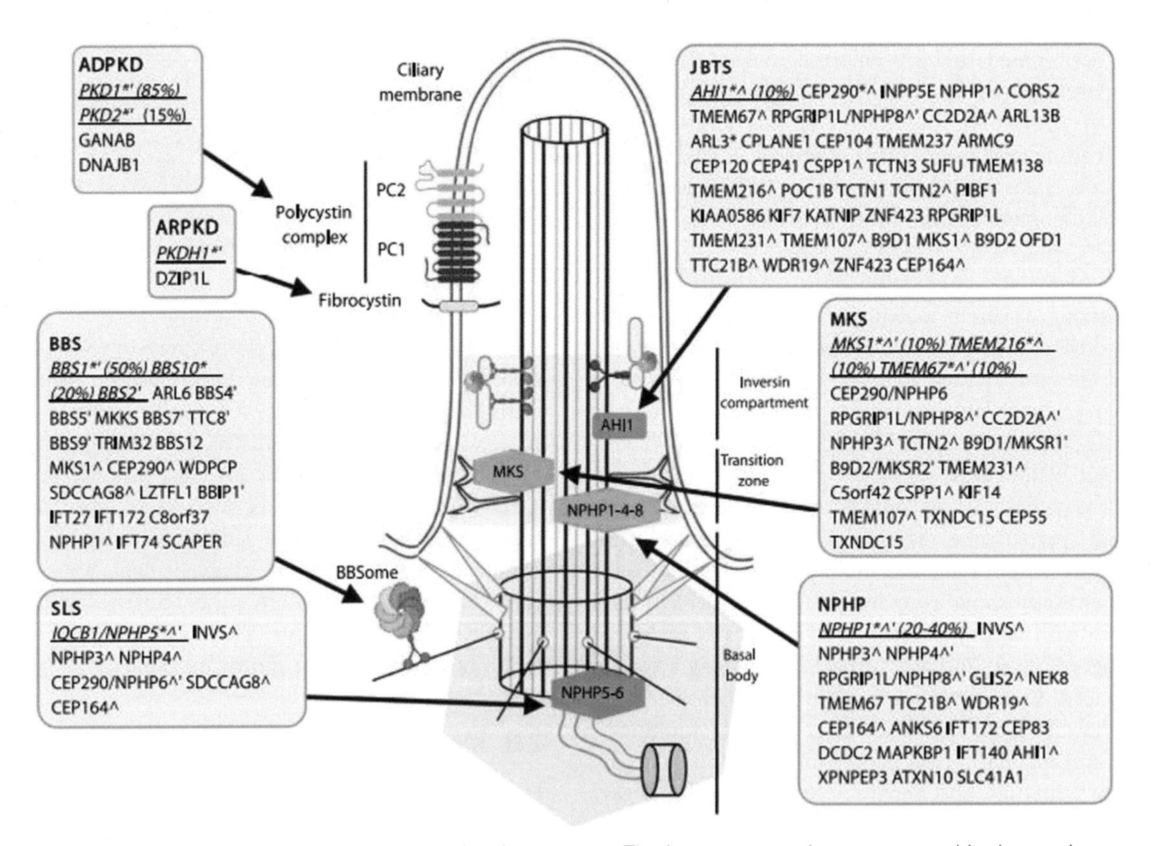

Figure 2.12 Renal ciliopathies and associated genotypes. The known reported genes mutated in the renal ciliopathies outlined in this review overlaid on a primary cilium schematic *Main associated genes. ^Mutated in another renal ciliopathy. 'Part of displayed protein/protein complex. Abbreviations: ADPKD, autosomal dominant polycystic kidney disease; ARPKD, autosomal recessive polycystic kidney disease; BBS, Bardet–Biedl syndrome; JBTS, Joubert syndrome; MKS, Meckel–Gruber syndrome; NPHP, nephronophthisis; PC1(2), polycystin 1(2); SLS, Senior–Loken syndrome. (Reproduced with permission from McConnachie et al. Ciliopathies and the kidney: a review. Am J Kidney Dis. 77(3):410–419. Published online October 9, 2020.)

Much progress has been made in our understanding of renal cystogenesis. Although questions remain to be answered, our increasing knowledge of the genetic and molecular basis of cystic kidney diseases will enhance insight into the pathogenesis, clinical consequences, and ultimately, the treatment of these disorders.

REFERENCES

[1] Gabow PA. Autosomal dominant polycystic kidney disease. N Engl J Med 1993;329:332–42.

[2] Rizk D, Chapman AB. Cystic and inherited kidney diseases. Am J Kidney Dis 2003;42:1305–17.

[3] Grantham JJ. Clinical practice. Autosomal dominant polycystic kidney disease. N Engl J Med 2008;359:1477–85.

[4] Bergmann C, Guay-Woodford LM, Harris PC, Horie S, Peters DJM, Torres VE. Polycystic kidney disease. Nat Rev Dis Prim 2018;4:50.

[5] Grantham JJ, Torres VE, Chapman AB, Guay-Woodford LM, Bae KT, King Jr BF, et al. Volume progression in polycystic kidney disease. N Engl J Med 2006;354:2122–30.

[6] Hateboer N, v Dijk MA, Bogdanova N, Coto E, Saggar-Malik AK, San Millan JL, et al. Comparison of phenotypes of polycystic kidney disease types 1 and 2. European PKD1-PKD2 Study Group. Lancet 1999;353:103–7.

[7] Rahbari-Oskoui F, Williams O, Chapman A. Mechanisms and management of hypertension in autosomal dominant polycystic kidney disease. Nephrol Dial Transplant 2014;29:2194–201.

[8] Ecder T, Schrier RW. Hypertension in autosomal-dominant polycystic kidney disease: early occurrence and unique aspects. J Am Soc Nephrol 2001;12:194–200.

[9] Chebib FT, Torres VE. Autosomal dominant polycystic kidney disease: core curriculum 2016. Am J Kidney Dis 2016;67:792–810.

[10] Nishiura JL, Neves RF, Eloi SR, Cintra SM, Ajzen SA, Heilberg IP. Evaluation of nephrolithiasis in autosomal dominant polycystic kidney disease patients. Clin J Am Soc Nephrol 2009;4:838–44.

[11] Everson GT. Hepatic cysts in autosomal dominant polycystic kidney disease. Am J Kidney Dis 1993;22:520–5.

[12] Gabow PA, Johnson AM, Kaehny WD, Manco-Johnson ML, Duley IT, Everson GT. Risk factors for the development of hepatic cysts in autosomal dominant polycystic kidney disease. Hepatology 1990;11:1033–7.

[13] Torres VE, Harris PC. Autosomal dominant polycystic kidney disease: the last 3 years. Kidney Int 2009;76:149–68.

[14] Bardaji A, Martinez-Vea A, Valero A, Gutierrez C, Garcia C, Ridao C, et al. Cardiac involvement in autosomal-dominant polycystic kidney disease: a hypertensive heart disease. Clin Nephrol 2001;56:211–20.

[15] Lumiaho A, Ikaheimo R, Miettinen R, Niemitukia L, Laitinen T, Rantala A, et al. Mitral valve prolapse and mitral regurgitation are common in patients with polycystic kidney disease type 1. Am J Kidney Dis 2001;38:1208–16.

[16] Timio M, Monarca C, Pede S, Gentili S, Verdura C, Lolli S. The spectrum of cardiovascular abnormalities in autosomal dominant polycystic kidney disease: a 10-year follow-up in a five-generation kindred. Clin Nephrol 1992;37:245–51.

[17] Perrone RD, Abebe KZ, Schrier RW, Chapman AB, Torres VE, Bost J, et al. Cardiac magnetic resonance assessment of left ventricular mass in autosomal dominant polycystic kidney disease. Clin J Am Soc Nephrol 2011;6:2508–15.

[18] Silverio A, Prota C, Di Maio M, Polito MV, Cogliani FM, Citro R, et al. Aortic dissection in patients with autosomal dominant polycystic kidney disease: a series of two cases and a review of the literature. Nephrology 2015;20:229–35.

[19] Torra R, Nicolau C, Badenas C, Bru C, Perez L, Estivill X, et al. Abdominal aortic aneurysms and autosomal dominant polycystic kidney disease. J Am Soc Nephrol 1996;7:2483–6.

[20] Gieteling EW, Rinkel GJ. Characteristics of intracranial aneurysms and subarachnoid haemorrhage in patients with polycystic kidney disease. J Neurol 2003;250:418–23.

[21] Schievink WI, Torres VE, Piepgras DG, Wiebers DO. Saccular intracranial aneurysms in autosomal dominant polycystic kidney disease. J Am Soc Nephrol 1992;3:88–95.

[22] Xu HW, Yu SQ, Mei CL, Li MH. Screening for intracranial aneurysm in 355 patients with autosomal-dominant polycystic kidney disease. Stroke 2011;42:204–6.

[23] Rossetti S, Chauveau D, Kubly V, Slezak JM, Saggar-Malik AK, Pei Y, et al. Association of mutation position in polycystic kidney disease 1 (PKD1) gene and

development of a vascular phenotype. Lancet 2003;361:2196–201.

[24] Huston 3rd J, Torres VE, Sulivan PP, Offord KP, Wiebers DO. Value of magnetic resonance angiography for the detection of intracranial aneurysms in autosomal dominant polycystic kidney disease. J Am Soc Nephrol 1993;3:1871–7.

[25] Rinkel GJ, Djibuti M, Algra A, van Gijn J. Prevalence and risk of rupture of intracranial aneurysms: a systematic review. Stroke 1998;29:251–6.

[26] Ruggieri PM, Poulos N, Masaryk TJ, Ross JS, Obuchowski NA, Awad IA, et al. Occult intracranial aneurysms in polycystic kidney disease: screening with MR angiography. Radiology 1994;191:33–9.

[27] Chapman AB, Johnson AM, Gabow PA. Intracranial aneurysms in patients with autosomal dominant polycystic kidney disease: how to diagnose and who to screen. Am J Kidney Dis 1993;22:526–31.

[28] Gibbs GF, Huston 3rd J, Qian Q, Kubly V, Harris PC, Brown Jr RD, et al. Follow-up of intracranial aneurysms in autosomal-dominant polycystic kidney disease. Kidney Int 2004;65:1621–7.

[29] International Study of Unruptured Intracranial Aneurysms I: Unruptured intracranial aneurysms–risk of rupture and risks of surgical intervention. N Engl J Med 1998;339:1725–33.

[30] Chauveau D, Pirson Y, Verellen-Dumoulin C, Macnicol A, Gonzalo A, Grunfeld JP. Intracranial aneurysms in autosomal dominant polycystic kidney disease. Kidney Int 1994;45:1140–6.

[31] Pirson Y, Chauveau D, Torres V. Management of cerebral aneurysms in autosomal dominant polycystic kidney disease. J Am Soc Nephrol 2002;13:269–76.

[32] Wiebers DO, Whisnant JP, Huston 3rd J, Meissner I, Brown Jr RD, Piepgras DG, et al. Unruptured intracranial aneurysms: natural history, clinical outcome, and risks of surgical and endovascular treatment. Lancet 2003;362:103–10.

[33] Dominguez Fernandez E, Albrecht KH, Heemann U, Kohnle M, Erhard J, Stoblen F, et al. Prevalence of diverticulosis and incidence of bowel perforation after kidney transplantation in patients with polycystic kidney disease. Transpl Int 1998;11:28–31.

[34] Hadimeri H, Norden G, Friman S, Nyberg G. Autosomal dominant polycystic kidney disease in a kidney transplant population. Nephrol Dial Transplant 1997;12:1431–6.

[35] Morris-Stiff G, Coles G, Moore R, Jurewicz A, Lord R. Abdominal wall hernia in autosomal dominant polycystic kidney disease. Br J Surg 1997;84:615–7.

[36] Torra R, Sarquella J, Calabia J, Marti J, Ars E, Fernandez-Llama P, et al. Prevalence of cysts in seminal tract and abnormal semen parameters in patients with autosomal dominant polycystic kidney disease. Clin J Am Soc Nephrol 2008;3:790–3.

[37] Schievink WI, Palestrant D, Maya MM, Rappard G. Spontaneous spinal cerebrospinal fluid leak as a cause of coma after craniotomy for clipping of an unruptured intracranial aneurysm. J Neurosurg 2009;110:521–4.

[38] Cornec-Le Gall E, Audrezet MP, Chen JM, Hourmant M, Morin MP, Perrichot R, et al. Type of PKD1 mutation influences renal outcome in ADPKD. J Am Soc Nephrol 2013;24:1006–13.

[39] Schrier RW, Abebe KZ, Perrone RD, Torres VE, Braun WE, Steinman TI, et al. Blood pressure in early autosomal dominant polycystic kidney disease. N Engl J Med 2014;371:2255–66.

[40] Germino GG, Weinstat-Saslow D, Himmelbauer H, Gillespie GA, Somlo S, Wirth B, et al. The gene for autosomal dominant polycystic kidney disease lies in a 750-kb CpG-rich region. Genomics 1992;13:144–51.

[41] Reeders ST, Breuning MH, Davies KE, Nicholls RD, Jarman AP, Higgs DR, et al. A highly polymorphic DNA marker linked to adult polycystic kidney disease on chromosome 16. Nature 1985;317:542–4.

[42] Hughes J, Ward CJ, Peral B, Aspinwall R, Clark K, San Millan JL, et al. The polycystic kidney disease 1 (PKD1) gene encodes a novel protein with multiple cell recognition domains. Nat Genet 1995;10:151–60.

[43] Moy GW, Mendoza LM, Schulz JR, Swanson WJ, Glabe CG, Vacquier VD. The sea urchin sperm receptor for egg jelly is a modular protein with extensive homology to the human polycystic kidney disease protein, PKD1. J Cell Biol 1996;133:809–17.

[44] Kimberling WJ, Fain PR, Kenyon JB, Goldgar D, Sujansky E, Gabow PA. Linkage heterogeneity of autosomal dominant polycystic kidney disease. N Engl J Med 1988;319:913–8.

[45] Mochizuki T, Wu G, Hayashi T, Xenophontos SL, Veldhuisen B, Saris JJ, et al. PKD2, a gene for polycystic kidney disease that encodes an integral membrane protein. Science 1996;272:1339–42.

[46] Romeo G, Devoto M, Costa G, Roncuzzi L, Catizone L, Zucchelli P, et al. A second genetic locus for autosomal dominant polycystic kidney disease. Lancet 1988;2:8–11.

[47] Qian F, Germino FJ, Cai Y, Zhang X, Somlo S, Germino GG. PKD1 interacts with PKD2 through a probable coiled-coil domain. Nat Genet 1997;16:179–83.

[48] Tsiokas L, Kim E, Arnould T, Sukhatme VP, Walz G. Homo- and heterodimeric interactions between the gene products of PKD1 and PKD2. Proc Natl Acad Sci U S A 1997;94:6965–70.

[49] Ward CJ, Turley H, Ong AC, Comley M, Biddolph S, Chetty R, et al. Polycystin, the polycystic kidney disease 1 protein, is expressed by epithelial cells in fetal, adult, and polycystic kidney. Proc Natl Acad Sci U S A 1996;93:1524–8.

[50] Van Adelsberg J, Chamberlain S, D'Agati V. Polycystin expression is temporally and spatially regulated during renal development. Am J Physiol 1997;272:F602–9.

[51] Ong AC, Ward CJ, Butler RJ, Biddolph S, Bowker C, Torra R, et al. Coordinate expression of the autosomal dominant polycystic kidney disease proteins, polycystin-2 and polycystin-1, in normal and cystic tissue. Am J Pathol 1999;154:1721–9.

[52] Palsson R, Sharma CP, Kim K, McLaughlin M, Brown D, Arnaout MA. Characterization and cell distribution of polycystin, the product of autosomal dominant polycystic kidney disease gene 1. Mol Med 1996;2:702–11.

[53] Peters DJ, Spruit L, Klingel R, Prins F, Baelde HJ, Giordano PC, et al. Adult, fetal, and polycystic kidney expression of polycystin, the polycystic kidney disease-1 gene product. Lab Invest 1996;75:221–30.

[54] Lu W, Peissel B, Babakhanlou H, Pavlova A, Geng L, Fan X, et al. Perinatal lethality with kidney and pancreas defects in mice with a targetted Pkd1 mutation. Nat Genet 1997;17:179–81.

[55] Somlo S, Markowitz GS. The pathogenesis of autosomal dominant polycystic kidney disease: an update. Curr Opin Nephrol Hypertens 2000;9:385–94.

[56] Germino GG. Autosomal dominant polycystic kidney disease: a two-hit model. Hosp Pract 1997;32:81–2. 85-88, 91-82 passim.

[57] Grantham JJ, Geiser JL, Evan AP. Cyst formation and growth in autosomal dominant polycystic kidney disease. Kidney Int 1987;31:1145–52.

[58] Qian F, Watnick TJ, Onuchic LF, Germino GG. The molecular basis of focal cyst formation in human autosomal dominant polycystic kidney disease type I. Cell 1996;87:979–87.

[59] Koptides M, Constantinides R, Kyriakides G, Hadjigavriel M, Patsalis PC, Pierides A, et al. Loss of heterozygosity in polycystic kidney disease with a missense mutation in the repeated region of PKD1. Hum Genet 1998;103:709–17.

[60] Koptides M, Hadjimichael C, Koupepidou P, Pierides A. Constantinou Deltas C: germinal and somatic mutations in the PKD2 gene of renal cysts in autosomal dominant polycystic kidney disease. Hum Mol Genet 1999;8:509–13.

[61] Koptides M, Mean R, Demetriou K, Pierides A, Deltas CC. Genetic evidence for a trans-heterozygous model for cystogenesis in autosomal dominant polycystic kidney disease. Hum Mol Genet 2000;9:447–52.

[62] Pei Y, Watnick T, He N, Wang K, Liang Y, Parfrey P, et al. Somatic PKD2 mutations in individual kidney and liver cysts support a "two-hit" model of cystogenesis in type 2 autosomal dominant polycystic kidney disease. J Am Soc Nephrol 1999;10:1524–9.

[63] Watnick T, He N, Wang K, Liang Y, Parfrey P, Hefferton D, et al. Mutations of PKD1 in ADPKD2 cysts suggest a pathogenic effect of trans-heterozygous mutations. Nat Genet 2000;25:143–4.

[64] Su Q, Hu F, Liu Y, Ge X, Mei C, Yu S, et al. Cryo-EM structure of the polycystic kidney disease-like channel PKD2L1. Nat Commun 2018;9:1192.

[65] Torres VE, Harris PC. Progress in the understanding of polycystic kidney disease. Nat Rev Nephrol 2019;15:70–2.

[66] Torra R, Badenas C, Darnell A, Nicolau C, Volpini V, Revert L, et al. Linkage, clinical features, and prognosis of autosomal dominant polycystic kidney disease types 1 and 2. J Am Soc Nephrol 1996;7:2142–51.

[67] Harris PC, Bae KT, Rossetti S, Torres VE, Grantham JJ, Chapman AB, et al. Cyst number but not the rate of cystic growth is associated with the mutated gene in autosomal dominant polycystic kidney disease. J Am Soc Nephrol 2006;17:3013–9.

[68] Rossetti S, Burton S, Strmecki L, Pond GR, San Millan JL, Zerres K, et al. The position of the polycystic kidney disease 1 (PKD1) gene mutation correlates with the severity of renal disease. J Am Soc Nephrol 2002;13:1230–7.

[69] Magistroni R, He N, Wang K, Andrew R, Johnson A, Gabow P, et al. Genotype-renal function correlation in type 2 autosomal dominant polycystic kidney disease. J Am Soc Nephrol 2003;14:1164–74.

[70] Deltas CC. Mutations of the human polycystic kidney disease 2 (PKD2) gene. Hum Mutat 2001;18:13–24.

[71] Rossetti S, Harris PC. Genotype-phenotype correlations in autosomal dominant and autosomal recessive polycystic kidney disease. J Am Soc Nephrol 2007;18:1374–80.

[72] Baboolal K, Ravine D, Daniels J, Williams N, Holmans P, Coles GA, et al. Association of the angiotensin I converting enzyme gene deletion polymorphism with early onset of ESRF in PKD1 adult polycystic kidney disease. Kidney Int 1997;52:607–13.

[73] Pereira TV, Nunes AC, Rudnicki M, Magistroni R, Albertazzi A, Pereira AC, et al. Influence of ACE I/D gene polymorphism in the progression of renal failure in autosomal dominant polycystic kidney disease: a meta-analysis. Nephrol Dial Transplant 2006;21:3155–63.
[74] Persu A, Stoenoiu MS, Messiaen T, Davila S, Robino C, El-Khattabi O, et al. Modifier effect of ENOS in autosomal dominant polycystic kidney disease. Hum Mol Genet 2002;11:229–41.
[75] Walker D, Consugar M, Slezak J, Rossetti S, Torres VE, Winearls CG, et al. The ENOS polymorphism is not associated with severity of renal disease in polycystic kidney disease 1. Am J Kidney Dis 2003;41:90–4.
[76] Brook-Carter PT, Peral B, Ward CJ, Thompson P, Hughes J, Maheshwar MM, et al. Deletion of the TSC2 and PKD1 genes associated with severe infantile polycystic kidney disease–a contiguous gene syndrome. Nat Genet 1994;8:328–32.
[77] Kleymenova E, Ibraghimov-Beskrovnaya O, Kugoh H, Everitt J, Xu H, Kiguchi K, et al. Tuberin-dependent membrane localization of polycystin-1: a functional link between polycystic kidney disease and the TSC2 tumor suppressor gene. Mol Cell 2001;7:823–32.
[78] Shillingford JM, Murcia NS, Larson CH, Low SH, Hedgepeth R, Brown N, et al. The mTOR pathway is regulated by polycystin-1, and its inhibition reverses renal cystogenesis in polycystic kidney disease. Proc Natl Acad Sci U S A 2006;103:5466–71.
[79] Grantham JJ, Chapman AB, Blais J, Czerwiec FS, Devuyst O, Gansevoort RT, et al. Tolvaptan suppresses monocyte chemotactic protein-1 excretion in autosomal-dominant polycystic kidney disease. Nephrol Dial Transplant 2017;32:969–75.
[80] Cassini MF, Kakade VR, Kurtz E, Sulkowski P, Glazer P, Torres R, et al. Mcp1 promotes macrophage-dependent cyst expansion in autosomal dominant polycystic kidney disease. J Am Soc Nephrol 2018;29:2471–81.
[81] Pelletier MF, Marcil A, Sevigny G, Jakob CA, Tessier DC, Chevet E, et al. The heterodimeric structure of glucosidase II is required for its activity, solubility, and localization in vivo. Glycobiology 2000;10:815–27.
[82] Tannous A, Pisoni GB, Hebert DN, Molinari M. N-linked sugar-regulated protein folding and quality control in the ER. Semin Cell Dev Biol 2015;41:79–89.
[83] Treml K, Meimaroglou D, Hentges A, Bause E. The alpha- and beta-subunits are required for expression of catalytic activity in the hetero-dimeric glucosidase II complex from human liver. Glycobiology 2000;10:493–502.
[84] Porath B, Gainullin VG, Cornec-Le Gall E, Dillinger EK, Heyer CM, Hopp K, et al. Mutations in GANAB, encoding the glucosidase IIalpha subunit, cause autosomal-dominant polycystic kidney and liver disease. Am J Hum Genet 2016;98:1193–207.
[85] Cornec-Le Gall E, Olson RJ, Besse W, Heyer CM, Gainullin VG, Smith JM, et al. Monoallelic mutations to DNAJB11 cause atypical autosomal-dominant polycystic kidney disease. Am J Hum Genet 2018;102:832–44.
[86] Shen Y, Hendershot LM. ERdj3, a stress-inducible endoplasmic reticulum DnaJ homologue, serves as a cofactor for BiP's interactions with unfolded substrates. Mol Biol Cell 2005;16:40–50.
[87] Shen Y, Meunier L, Hendershot LM. Identification and characterization of a novel endoplasmic reticulum (ER) DnaJ homologue, which stimulates ATPase activity of BiP in vitro and is induced by ER stress. J Biol Chem 2002;277:15947–56.
[88] Cornec-Le Gall E, Chebib FT, Madsen CD, Senum SR, Heyer CM, Lanpher BC, et al. The value of genetic testing in polycystic kidney diseases illustrated by a family with PKD2 and COL4A1 mutations. Am J Kidney Dis 2018;72:302–8.
[89] Sherstha R, McKinley C, Russ P, Scherzinger A, Bronner T, Showalter R, et al. Postmenopausal estrogen therapy selectively stimulates hepatic enlargement in women with autosomal dominant polycystic kidney disease. Hepatology 1997;26:1282–6.
[90] Pei Y. Practical genetics for autosomal dominant polycystic kidney disease. Nephron Clin Pract 2011;118:c19–30.
[91] Takakura A, Contrino L, Zhou X, Bonventre JV, Sun Y, Humphreys BD, et al. Renal injury is a third hit promoting rapid development of adult polycystic kidney disease. Hum Mol Genet 2009;18:2523–31.
[92] Hateboer N, Lazarou LP, Williams AJ, Holmans P, Ravine D. Familial phenotype differences in PKD11. Kidney Int 1999;56:34–40.
[93] Chapman AB, Johnson AM, Gabow PA, Schrier RW. Overt proteinuria and microalbuminuria in autosomal dominant polycystic kidney disease. J Am Soc Nephrol 1994;5:1349–54.
[94] Bear JC, McManamon P, Morgan J, Payne RH, Lewis H, Gault MH, et al. Age at clinical onset and at ultrasonographic detection of adult polycystic kidney disease: data for genetic counselling. Am J Med Genet 1984;18:45–53.

[95] Ravine D, Gibson RN, Walker RG, Sheffield LJ, Kincaid-Smith P, Danks DM. Evaluation of ultrasonographic diagnostic criteria for autosomal dominant polycystic kidney disease 1. Lancet 1994;343:824–7.
[96] Nicolau C, Torra R, Badenas C, Vilana R, Bianchi L, Gilabert R, et al. Autosomal dominant polycystic kidney disease types 1 and 2: assessment of US sensitivity for diagnosis. Radiology 1999;213:273–6.
[97] Pei Y, Obaji J, Dupuis A, Paterson AD, Magistroni R, Dicks E, et al. Unified criteria for ultrasonographic diagnosis of ADPKD. J Am Soc Nephrol 2009;20:205–12.
[98] Rahbari-Oskoui F, Mittal A, Mittal P, Chapman A. Renal relevant radiology: radiologic imaging in autosomal dominant polycystic kidney disease. Clin J Am Soc Nephrol 2014;9:406–15.
[99] Bisceglia M, Galliani CA, Senger C, Stallone C, Sessa A. Renal cystic diseases: a review. Adv Anat Pathol 2006;13:26–56.
[100] Rossetti S, Consugar MB, Chapman AB, Torres VE, Guay-Woodford LM, Grantham JJ, et al. Comprehensive molecular diagnostics in autosomal dominant polycystic kidney disease. J Am Soc Nephrol 2007;18:2143–60.
[101] Consugar MB, Wong WC, Lundquist PA, Rossetti S, Kubly VJ, Walker DL, et al. Characterization of large rearrangements in autosomal dominant polycystic kidney disease and the PKD1/TSC2 contiguous gene syndrome. Kidney Int 2008;74:1468–79.
[102] Deltas C, Papagregoriou G. Cystic diseases of the kidney: molecular biology and genetics. Arch Pathol Lab Med 2010;134:569–82.
[103] De Rycke M, Georgiou I, Sermon K, Lissens W, Henderix P, Joris H, et al. PGD for autosomal dominant polycystic kidney disease type 1. Mol Hum Reprod 2005;11:65–71.
[104] Murphy EL, Droher ML, DiMaio MS, Dahl NK. Preimplantation genetic diagnosis counseling in autosomal dominant polycystic kidney disease. Am J Kidney Dis 2018;72:866–72.
[105] Chapman AB, Bost JE, Torres VE, Guay-Woodford L, Bae KT, Landsittel D, et al. Kidney volume and functional outcomes in autosomal dominant polycystic kidney disease. Clin J Am Soc Nephrol 2012;7:479–86.
[106] Irazabal MV, Rangel LJ, Bergstralh EJ, Osborn SL, Harmon AJ, Sundsbak JL, et al. Imaging classification of autosomal dominant polycystic kidney disease: a simple model for selecting patients for clinical trials. J Am Soc Nephrol 2015;26:160–72.
[107] Iliuta IA, Kalatharan V, Wang K, Cornec-Le Gall E, Conklin J, Pourafkari M, et al. Polycystic kidney disease without an apparent family history. J Am Soc Nephrol 2017;28:2768–76.
[108] Ecder T, Edelstein CL, Fick-Brosnahan GM, Johnson AM, Chapman AB, Gabow PA, et al. Diuretics versus angiotensin-converting enzyme inhibitors in autosomal dominant polycystic kidney disease. Am J Nephrol 2001;21:98–103.
[109] Ecder T, Chapman AB, Brosnahan GM, Edelstein CL, Johnson AM, Schrier RW. Effect of antihypertensive therapy on renal function and urinary albumin excretion in hypertensive patients with autosomal dominant polycystic kidney disease. Am J Kidney Dis 2000;35:427–32.
[110] Schrier RW, Johnson AM, McFann K, Chapman AB. The role of parental hypertension in the frequency and age of diagnosis of hypertension in offspring with autosomal-dominant polycystic kidney disease. Kidney Int 2003;64:1792–9.
[111] Klahr S, Levey AS, Beck GJ, Caggiula AW, Hunsicker L, Kusek JW, et al. The effects of dietary protein restriction and blood-pressure control on the progression of chronic renal disease. Modification of diet irenal disease study group. N Engl J Med 1994;330:877–84.
[112] Levey AS, Greene T, Sarnak MJ, Wang X, Beck GJ, Kusek JW, et al. Effect of dietary protein restriction on the progression of kidney disease: long-term follow-up of the Modification of diet in renal disease (MDRD) study. Am J Kidney Dis 2006;48:879–88.
[113] Maschio G, Alberti D, Locatelli F, Mann JF, Motolese M, Ponticelli C, et al. Angiotensin-converting enzyme inhibitors and kidney protection: the AIPRI trial. The ACE inhibition in progressive renal insufficiency (AIPRI) study group. J Cardiovasc Pharmacol 1999;33(Suppl. 1):S16–20. discussion S41-13.
[114] Maschio G, Alberti D, Janin G, Locatelli F, Mann JF, Motolese M, et al. Effect of the angiotensin-converting-enzyme inhibitor benazepril on the progression of chronic renal insufficiency. The angiotensin-converting-enzyme inhibition in progressive renal insufficiency study group. N Engl J Med 1996;334:939–45.
[115] van Dijk MA, Breuning MH, Duiser R, van Es LA, Westendorp RG. No effect of enalapril on progression in autosomal dominant polycystic kidney disease. Nephrol Dial Transplant 2003;18:2314–20.
[116] Schrier R, McFann K, Johnson A, Chapman A, Edelstein C, Brosnahan G, et al. Cardiac and renal effects of standard versus rigorous blood pressure control in autosomal-dominant polycystic kidney disease: results of a seven-year prospective randomized study. J Am Soc Nephrol 2002;13:1733–9.

[117] Grantham JJ. Lillian Jean Kaplan International Prize for advancement in the understanding of polycystic kidney disease. Understanding polycystic kidney disease: a systems biology approach. Kidney Int 2003;64:1157–62.
[118] Torres VE, Meijer E, Bae KT, Chapman AB, Devuyst O, Gansevoort RT, et al. Rationale and design of the TEMPO (tolvaptan efficacy and safety in management of autosomal dominant polycystic kidney disease and its outcomes) 3-4 study. Am J Kidney Dis 2011;57:692–9.
[119] Torres VE, Chapman AB, Devuyst O, Gansevoort RT, Grantham JJ, Higashihara E, et al. Tolvaptan in patients with autosomal dominant polycystic kidney disease. N Engl J Med 2012;367:2407–18.
[120] Torres VE, Chapman AB, Devuyst O, Gansevoort RT, Perrone RD, Koch G, et al. Tolvaptan in later-stage autosomal dominant polycystic kidney disease. N Engl J Med 2017;377:1930–42.
[121] Bhutani H, Smith V, Rahbari-Oskoui F, Mittal A, Grantham JJ, Torres VE, et al. A comparison of ultrasound and magnetic resonance imaging shows that kidney length predicts chronic kidney disease in autosomal dominant polycystic kidney disease. Kidney Int 2015;88:146–51.
[122] Nagao S, Nishii K, Katsuyama M, Kurahashi H, Marunouchi T, Takahashi H, et al. Increased water intake decreases progression of polycystic kidney disease in the PCK rat. J Am Soc Nephrol 2006;17:2220–7.
[123] Higashihara E, Nutahara K, Tanbo M, Hara H, Miyazaki I, Kobayashi K, et al. Does increased water intake prevent disease progression in autosomal dominant polycystic kidney disease? Nephrol Dial Transplant 2014;29:1710–9.
[124] Torres VE, Abebe KZ, Schrier RW, Perrone RD, Chapman AB, Yu AS, et al. Dietary salt restriction is beneficial to the management of autosomal dominant polycystic kidney disease. Kidney Int 2017;91:493–500.
[125] Orth SR, Stockmann A, Conradt C, Ritz E, Ferro M, Kreusser W, et al. Smoking as a risk factor for end-stage renal failure in men with primary renal disease. Kidney Int 1998;54:926–31.
[126] Belibi FA, Wallace DP, Yamaguchi T, Christensen M, Reif G, Grantham JJ. The effect of caffeine on renal epithelial cells from patients with autosomal dominant polycystic kidney disease. J Am Soc Nephrol 2002;13:2723–9.
[127] Tanner GA, Tanner JA. Chronic caffeine consumption exacerbates hypertension in rats with polycystic kidney disease. Am J Kidney Dis 2001;38:1089–95.
[128] Girardat-Rotar L, Puhan MA, Braun J, Serra AL. Long-term effect of coffee consumption on autosomal dominant polycystic kidneys disease progression: results from the Suisse ADPKD, a prospective longitudinal cohort study. J Nephrol 2018;31:87–94.
[129] Nowak KL, You Z, Gitomer B, Brosnahan G, Torres VE, Chapman AB, et al. Overweight and obesity are predictors of progression in early autosomal dominant polycystic kidney disease. J Am Soc Nephrol 2018;29:571–8.
[130] Stiasny B, Ziebell D, Graf S, Hauser IA, Schulze BD. Clinical aspects of renal transplantation in polycystic kidney disease. Clin Nephrol 2002;58:16–24.
[131] Fitzpatrick PM, Torres VE, Charboneau JW, Offord KP, Holley KE, Zincke H. Long-term outcome of renal transplantation in autosomal dominant polycystic kidney disease. Am J Kidney Dis 1990;15:535–43.
[132] Hogan MC, Masyuk TV, Page L, Holmes 3rd DR, Li X, Bergstralh EJ, et al. Somatostatin analog therapy for severe polycystic liver disease: results after 2 years. Nephrol Dial Transplant 2012;27:3532–9.
[133] van Aerts RMM, Kievit W, D'Agnolo HMA, Blijdorp CJ, Casteleijn NF, Dekker SEI, et al. Lanreotide reduces liver growth in patients with autosomal dominant polycystic liver and kidney disease. Gastroenterology 2019;157:481–491 e487.
[134] Hogan MC, Chamberlin JA, Vaughan LE, Waits AL, Banks C, Leistikow K, et al. Pansomatostatin agonist pasireotide long-acting release for patients with autosomal dominant polycystic kidney or liver disease with severe liver involvement: a randomized clinical trial. Clin J Am Soc Nephrol 2020;15:1267–78.
[135] Schwab SJ, Bander SJ, Klahr S. Renal infection in autosomal dominant polycystic kidney disease. Am J Med 1987;82:714–8.
[136] Telenti A, Torres VE, Gross Jr JB, Van Scoy RE, Brown ML, Hattery RR. Hepatic cyst infection in autosomal dominant polycystic kidney disease. Mayo Clin Proc 1990;65:933–42.
[137] Bleeker-Rovers CP, Vos FJ, Corstens FH, Oyen WJ. Imaging of infectious diseases using [18F] fluorodeoxyglucose PET. Q J Nucl Med Mol Imaging 2008;52:17–29.
[138] Soussan M, Sberro R, Wartski M, Fakhouri F, Pecking AP, Alberini JL. Diagnosis and localization of renal cyst infection by 18F-fluorodeoxyglucose PET/CT in polycystic kidney disease. Ann Nucl Med 2008;22:529–31.
[139] Cloft HJ, Joseph GJ, Dion JE. Risk of cerebral angiography in patients with subarachnoid hemorrhage, cerebral aneurysm, and arteriovenous malformation: a meta-analysis. Stroke 1999;30:317–20.
[140] Huston 3rd J, Nichols DA, Luetmer PH, Goodwin JT, Meyer FB, Wiebers DO, et al. Blinded prospective evaluation of sensitivity of MR angiography to known

intracranial aneurysms: importance of aneurysm size. AJNR Am J Neuroradiol 1994;15:1607–14.
[141] White PM, Teasdale EM, Wardlaw JM, Easton V. Intracranial aneurysms: CT angiography and MR angiography for detection prospective blinded comparison in a large patient cohort. Radiology 2001;219:739–49.
[142] Raaymakers TW. Aneurysms in relatives of patients with subarachnoid hemorrhage: frequency and risk factors. MARS study group. Magnetic resonance angiography in relatives of patients with subarachnoid hemorrhage. Neurology 1999;53:982–8.
[143] Crawley F, Clifton A, Brown MM. Should we screen for familial intracranial aneurysm? Stroke 1999;30:312–6.
[144] Hughes PD, Becker GJ. Screening for intracranial aneurysms in autosomal dominant polycystic kidney disease. Nephrology 2003;8:163–70.
[145] Schrier RW, Belz MM, Johnson AM, Kaehny WD, Hughes RL, Rubinstein D, et al. Repeat imaging for intracranial aneurysms in patients with autosomal dominant polycystic kidney disease with initially negative studies: a prospective ten-year follow-up. J Am Soc Nephrol 2004;15:1023–8.
[146] Huston 3rd J, Torres VE, Wiebers DO, Schievink WI. Follow-up of intracranial aneurysms in autosomal dominant polycystic kidney disease by magnetic resonance angiography. J Am Soc Nephrol 1996;7:2135–41.
[147] Belz MM, Fick-Brosnahan GM, Hughes RL, Rubinstein D, Chapman AB, Johnson AM, et al. Recurrence of intracranial aneurysms in autosomal-dominant polycystic kidney disease. Kidney Int 2003;63:1824–30.
[148] David CA, Vishteh AG, Spetzler RF, Lemole M, Lawton MT, Partovi S. Late angiographic follow-up review of surgically treated aneurysms. J Neurosurg 1999;91:396–401.
[149] O'Connor AK, Guay-Woodford LM. Chapter 20 - Polycystic kidney diseases and other hepatorenal fibrocystic diseases: Clinical phenotypes, molecular pathobiology, and variation between mouse and man. In: Litle MH, editor. Kidney Development, Disease, Repair and Regeneration. Academic Press; 2016. p. 241–64. ISBN 9780128001028. https://doi.org/10.1016/B978-0-12-800102-8.00020-5.
[150] Mekahli D, van Stralen KJ, Bonthuis M, Jager KJ, Balat A, Benetti E, et al. Kidney versus combined kidney and liver transplantation in young people with autosomal recessive polycystic kidney disease: data from the European society for pediatric nephrology/European renal association-European dialysis and transplant (ESPN/ERA-EDTA) registry. Am J Kidney Dis 2016;68:782–8.
[151] Guay-Woodford LM, Desmond RA. Autosomal recessive polycystic kidney disease (ARPKD): the clinical experience in north America. Pediatrics 2003;111:1072–80.
[152] Adeva M, El-Youssef M, Rossetti S, Kamath PS, Kubly V, Consugar MB, et al. Clinical and molecular characterization defines a broadened spectrum of autosomal recessive polycystic kidney disease (ARPKD). Medicine (Baltim) 2006;85:1–21.
[153] Zerres K, Rudnik-Schoneborn S, Steinkamm C, Becker J, Mucher G. Autosomal recessive polycystic kidney disease. J Mol Med (Berl) 1998;76:303–9.
[154] Shaikewitz ST, Chapman A. Autosomal recessive polycystic kidney disease: issues regarding the variability of clinical presentation. J Am Soc Nephrol 1993;3:1858–62.
[155] Capisonda R, Phan V, Traubuci J, Daneman A, Balfe JW, Guay-Woodford LM. Autosomal recessive polycystic kidney disease: outcomes from a single-center experience. Pediatr Nephrol 2003;18:119–26.
[156] Bergmann C, Senderek J, Windelen E, Kupper F, Middeldorf I, Schneider F, et al. Clinical consequences of PKHD1 mutations in 164 patients with autosomal-recessive polycystic kidney disease (ARPKD). Kidney Int 2005;67:829–48.
[157] Roy S, Dillon MJ, Trompeter RS, Barratt TM. Autosomal recessive polycystic kidney disease: long-term outcome of neonatal survivors. Pediatr Nephrol 1997;11:302–6.
[158] Zerres K, Rudnik-Schoneborn S, Senderek J, Eggermann T, Bergmann C. Autosomal recessive polycystic kidney disease (ARPKD). J Nephrol 2003;16:453–8.
[159] Gunay-Aygun M, Font-Montgomery E, Lukose L, Tuchman M, Graf J, Bryant JC, et al. Correlation of kidney function, volume and imaging findings, and PKHD1 mutations in 73 patients with autosomal recessive polycystic kidney disease. Clin J Am Soc Nephrol 2010;5:972–84.
[160] Burgmaier K, Kilian S, Bammens B, Benzing T, Billing H, Buscher A, et al. Clinical courses and complications of young adults with autosomal recessive polycystic kidney disease (ARPKD). Sci Rep 2019;9:7919.
[161] Kaplan BS, Fay J, Shah V, Dillon MJ, Barratt TM. Autosomal recessive polycystic kidney disease. Pediatr Nephrol 1989;3:43–9.
[162] Kashtan CE, Primack WA, Kainer G, Rosenberg AR, McDonald RA, Warady BA. Recurrent bacteremia with

enteric pathogens in recessive polycystic kidney disease. Pediatr Nephrol 1999;13:678–82.
[163] Davis ID, Ho M, Hupertz V, Avner ED. Survival of childhood polycystic kidney disease following renal transplantation: the impact of advanced hepatobiliary disease. Pediatr Transplant 2003;7:364–9.
[164] Gately R, Lock G, Patel C, Clouston J, Hawley C, Mallett A. Multiple cerebral aneurysms in an adult with autosomal recessive polycystic kidney disease. Kidney Int Rep 2021;6:219–23.
[165] Gunay-Aygun M, Turkbey BI, Bryant J, Daryanani KT, Gerstein MT, Piwnica-Worms K, et al. Hepatorenal findings in obligate heterozygotes for autosomal recessive polycystic kidney disease. Mol Genet Metabol 2011;104:677–81.
[166] Besse W, Dong K, Choi J, Punia S, Fedeles SV, Choi M, et al. Isolated polycystic liver disease genes define effectors of polycystin-1 function. J Clin Invest 2017;127:1772–85.
[167] Lu H, Galeano MCR, Ott E, Kaeslin G, Kausalya PJ, Kramer C, et al. Mutations in DZIP1L, which encodes a ciliary-transition-zone protein, cause autosomal recessive polycystic kidney disease. Nat Genet 2017;49:1025–34.
[168] Harris PC, Rossetti S. Molecular genetics of autosomal recessive polycystic kidney disease. Mol Genet Metabol 2004;81:75–85.
[169] Ward CJ, Hogan MC, Rossetti S, Walker D, Sneddon T, Wang X, et al. The gene mutated in autosomal recessive polycystic kidney disease encodes a large, receptor-like protein. Nat Genet 2002;30:259–69.
[170] Onuchic LF, Furu L, Nagasawa Y, Hou X, Eggermann T, Ren Z, et al. PKHD1, the polycystic kidney and hepatic disease 1 gene, encodes a novel large protein containing multiple immunoglobulin-like plexin-transcription-factor domains and parallel beta-helix 1 repeats. Am J Hum Genet 2002;70:1305–17.
[171] Xiong H, Chen Y, Yi Y, Tsuchiya K, Moeckel G, Cheung J, et al. A novel gene encoding a TIG multiple domain protein is a positional candidate for autosomal recessive polycystic kidney disease. Genomics 2002;80:96–104.
[172] Wang S, Luo Y, Wilson PD, Witman GB, Zhou J. The autosomal recessive polycystic kidney disease protein is localized to primary cilia, with concentration in the basal body area. J Am Soc Nephrol 2004;15:592–602.
[173] Hogan MC, Manganelli L, Woollard JR, Masyuk AI, Masyuk TV, Tammachote R, et al. Characterization of PKD protein-positive exosome-like vesicles. J Am Soc Nephrol 2009;20:278–88.
[174] Sweeney Jr WE, Avner ED. Molecular and cellular pathophysiology of autosomal recessive polycystic kidney disease (ARPKD). Cell Tissue Res 2006;326:671–85.
[175] Chumley P, Zhou J, Mrug S, Chacko B, Parant JM, Challa AK, et al. Truncating PKHD1 and PKD2 mutations alter energy metabolism. Am J Physiol Ren Physiol 2019;316:F414–25.
[176] Bergmann C, Kupper F, Schmitt CP, Vester U, Neuhaus TJ, Senderek J, et al. Multi-exon deletions of the PKHD1 gene cause autosomal recessive polycystic kidney disease (ARPKD). J Med Genet 2005;42:e63.
[177] Bergmann C, Senderek J, Sedlacek B, Pegiazoglou I, Puglia P, Eggermann T, et al. Spectrum of mutations in the gene for autosomal recessive polycystic kidney disease (ARPKD/PKHD1). J Am Soc Nephrol 2003;14:76–89.
[178] Lambie L, Amin R, Essop F, Cnaan A, Krause A, Guay-Woodford LM. Clinical and genetic characterization of a founder PKHD1 mutation in Afrikaners with ARPKD. Pediatr Nephrol 2015;30(2):273–9.
[179] Mrug M, Li R, Cui X, Schoeb TR, Churchill GA, Guay-Woodford LM. Kinesin family member 12 is a candidate polycystic kidney disease modifier in the cpk mouse. J Am Soc Nephrol 2005;16(4):905–16.
[180] Mrug M, Zhou J, Yang C, Arrow BJ, Cui X, Schoeb TR, et al. Genetic and informatic analyses implicate *Kif12* as a candidate gene within the *Mpkd2* locus that modulates renal cystic disease severity in the *Cys1*^{cpk} mouse. PLoS One 2015;10(8):e0135678.
[181] Olson RJ, Hopp K, Wells H, Smith JM, Furtado J, Constans MM, et al. Synergistic genetic interactions between Pkhd1 and Pkd1 result in an ARPKD-like phenotype in murine models. J Am Soc Nephrol 2019;30:2113–27.
[182] Bergmann C, Senderek J, Schneider F, Dornia C, Kupper F, Eggermann T, et al. PKHD1 mutations in families requesting prenatal diagnosis for autosomal recessive polycystic kidney disease (ARPKD). Hum Mutat 2004;23:487–95.
[183] Shan D, Rezonzew G, Mullen S, Roye R, Zhou J, Chumley P, et al. Heterozygous Pkhd1(C642*) mice develop cystic liver disease and proximal tubule ectasia that mimics radiographic signs of medullary sponge kidney. Am J Physiol Ren Physiol 2019;316:F463–72.
[184] Kissane JM. Renal cysts in pediatric patients. A classification and overview. Pediatr Nephrol 1990;4:69–77.
[185] Mrug M, Zhou J, Woo Y, Cui X, Szalai AJ, Novak J, et al. Overexpression of innate immune response genes in a model of recessive polycystic kidney disease. Kidney Int 2008;73:63–76.
[186] Swenson-Fields KI, Vivian CJ, Salah SM, Peda JD, Davis BM, van Rooijen N, et al. Macrophages promote polycystic kidney disease progression. Kidney Int 2013;83:855–64.

[187] Zimmerman KA, Hopp K, Mrug M. Role of chemokines, innate and adaptive immunity. Cell Signal 2020;73:109647.
[188] Zimmerman KA, Bentley MR, Lever JM, Li Z, Crossman DK, Song CJ, et al. Single-cell RNA sequencing identifies candidate renal resident macrophage gene expression signatures across species. J Am Soc Nephrol 2019;30:767–81.
[189] Zimmerman KA, Huang J, He L, Revell DZ, Li Z, Hsu J-S, et al. Interferon regulatory factor 5 in resident macrophage promotes polycystic kidney disease. Kidney 2020. https://doi.org/10.34067/KID.0001052019.
[190] Zimmerman K, Song C, Lever J, Crossman D, Li Z, Zhou J, et al. Tissue resident macrophages promote renal cystic disease. J Am Soc Nehrol 2019. Accepted for publication 5/26/2019.
[191] Turkbey B, Ocak I, Daryanani K, Font-Montgomery E, Lukose L, Bryant J, et al. Autosomal recessive polycystic kidney disease and congenital hepatic fibrosis (ARPKD/CHF). Pediatr Radiol 2009;39:100–11.
[192] Kamath BM, Piccoli DA. Heritable disorders of the bile ducts. Gastroenterol Clin N Am 2003;32:857–75. vi.
[193] Sgro M, Rossetti S, Barozzino T, Toi A, Langer J, Harris PC, et al. Caroli's disease: prenatal diagnosis, postnatal outcome and genetic analysis. Ultrasound Obstet Gynecol 2004;23:73–6.
[194] Guay-Woodford LM, Bissler JJ, Braun MC, Bockenhauer D, Cadnapaphornchai MA, Dell KM, et al. Consensus expert recommendations for the diagnosis and management of autosomal recessive polycystic kidney disease: report of an international conference. J Pediatr 2014;165:611–7.
[195] Chaumoitre K, Brun M, Cassart M, Maugey-Laulom B, Eurin D, Didier F, et al. Differential diagnosis of fetal hyperechogenic cystic kidneys unrelated to renal tract anomalies: a multicenter study. Ultrasound Obstet Gynecol 2006;28:911–7.
[196] Hartung EA, Guay-Woodford LM. Autosomal recessive polycystic kidney disease: a hepatorenal fibrocystic disorder with pleiotropic effects. Pediatrics 2014;134:e833–845.
[197] Hawkins JS, Dashe JS, Twickler DM. Magnetic resonance imaging diagnosis of severe fetal renal anomalies. Am J Obstet Gynecol 2008;198. 328 e321-325.
[198] Tsatsaris V, Gagnadoux MF, Aubry MC, Gubler MC, Dumez Y, Dommergues M. Prenatal diagnosis of bilateral isolated fetal hyperechogenic kidneys. Is it possible to predict long term outcome? BJOG 2002;109:1388–93.
[199] Sharp AM, Messiaen LM, Page G, Antignac C, Gubler MC, Onuchic LF, et al. Comprehensive genomic analysis of PKHD1 mutations in ARPKD cohorts. J Med Genet 2005;42:336–49.
[200] Guay-Woodford L. Other cystic diseases. In: Floege J, Johnson R, Feehally J, editors. Comprehensive clinical nephrology. 4th ed. London: Mosby; 2010. p. 543–59.
[201] Rheault MN, Rajpal J, Chavers B, Nevins TE. Outcomes of infants <28 days old treated with peritoneal dialysis for end-stage renal disease. Pediatr Nephrol 2009;24:2035–9.
[202] Hartung EA, Dell KM, Matheson M, Warady BA, Furth SL. Growth in children with autosomal recessive polycystic kidney disease in the CKiD cohort study. Front Pediatr 2016;4:82.
[203] Wolf MT, Hildebrandt F. Nephronophthisis. Pediatr Nephrol 2011;26:181–94.
[204] Caridi G, Dagnino M, Gusmano R, Ginevri F, Murer L, Ghio L, et al. Clinical and molecular heterogeneity of juvenile nephronophthisis in Italy: insights from molecular screening. Am J Kidney Dis 2000;35:44–51.
[205] Hildebrandt F, Strahm B, Nothwang HG, Gretz N, Schnieders B, Singh-Sawhney I, et al. Molecular genetic identification of families with juvenile nephronophthisis type 1: rate of progression to renal failure. APN Study Group. Arbeitsgemeinschaft Fur Padiatrische Nephrologie. Kidney Int 1997;51:261–9.
[206] Hildebrandt F, Attanasio M, Otto E. Nephronophthisis: disease mechanisms of a ciliopathy. J Am Soc Nephrol 2009;20:23–35.
[207] Srivastava S, Sayer JA. Nephronophthisis. J Pediatr Genet 2014;3:103–14.
[208] Otto EA, Trapp ML, Schultheiss UT, Helou J, Quarmby LM, Hildebrandt F. NEK8 mutations affect ciliary and centrosomal localization and may cause nephronophthisis. J Am Soc Nephrol 2008;19:587–92.
[209] Tory K, Rousset-Rouviere C, Gubler MC, Moriniere V, Pawtowski A, Becker C, et al. Mutations of NPHP2 and NPHP3 in infantile nephronophthisis. Kidney Int 2009;75:839–47.
[210] Olbrich H, Fliegauf M, Hoefele J, Kispert A, Otto E, Volz A, et al. Mutations in a novel gene, NPHP3, cause adolescent nephronophthisis, tapeto-retinal degeneration and hepatic fibrosis. Nat Genet 2003;34:455–9.
[211] Snoek R, van Setten J, Keating BJ, Israni AK, Jacobson PA, Oetting WS, et al. (Nephrocystin-1) gene deletions cause adult-onset ESRD. J Am Soc Nephrol 2018;29:1772–9.
[212] Brancati F, Dallapiccola B, Valente EM. Joubert Syndrome and related disorders. Orphanet J Rare Dis 2010;5:20.
[213] Radha Rama Devi A, Naushad SM, Lingappa L. Clinical and molecular diagnosis of Joubert syndrome and related disorders. Pediatr Neurol 2020;106:43–9.
[214] Baala L, Romano S, Khaddour R, Saunier S, Smith UM, Audollent S, et al. The Meckel-Gruber syndrome gene,

MKS3, is mutated in Joubert syndrome. Am J Hum Genet 2007;80:186–94.
[215] Parisi MA. Clinical and molecular features of Joubert syndrome and related disorders. Am J Med Genet C Semin Med Genet 2009;151C:326–40.
[216] Parisi MA, Doherty D, Eckert ML, Shaw DW, Ozyurek H, Aysun S, et al. AHI1 mutations cause both retinal dystrophy and renal cystic disease in Joubert syndrome. J Med Genet 2006;43:334–9.
[217] Sayer JA, Otto EA, O'Toole JF, Nurnberg G, Kennedy MA, Becker C, et al. The centrosomal protein nephrocystin-6 is mutated in Joubert syndrome and activates transcription factor ATF4. Nat Genet 2006;38:674–81.
[218] Sarangapani S, Chang L, Gregory-Evans K. Cataract surgery in Senior-Loken syndrome is beneficial despite severe retinopathy. Eye 2002;16:782–5.
[219] Caridi G, Murer L, Bellantuono R, Sorino P, Caringella DA, Gusmano R, et al. Renal-retinal syndromes: association of retinal anomalies and recessive nephronophthisis in patients with homozygous deletion of the NPH1 locus. Am J Kidney Dis 1998;32:1059–62.
[220] Otto E, Hoefele J, Ruf R, Mueller AM, Hiller KS, Wolf MT, et al. A gene mutated in nephronophthisis and retinitis pigmentosa encodes a novel protein, nephroretinin, conserved in evolution. Am J Hum Genet 2002;71:1161–7.
[221] Schuermann MJ, Otto E, Becker A, Saar K, Ruschendorf F, Polak BC, et al. Mapping of gene loci for nephronophthisis type 4 and Senior-Loken syndrome, to chromosome 1p36. Am J Hum Genet 2002;70:1240–6.
[222] Otto EA, Loeys B, Khanna H, Hellemans J, Sudbrak R, Fan S, et al. Nephrocystin-5, a ciliary IQ domain protein, is mutated in Senior-Loken syndrome and interacts with RPGR and calmodulin. Nat Genet 2005;37:282–8.
[223] den Hollander AI, Koenekoop RK, Yzer S, Lopez I, Arends ML, Voesenek KE, et al. Mutations in the CEP290 (NPHP6) gene are a frequent cause of Leber congenital amaurosis. Am J Hum Genet 2006;79:556–61.
[224] Perrault I, Delphin N, Hanein S, Gerber S, Dufier JL, Roche O, et al. Spectrum of NPHP6/CEP290 mutations in Leber congenital amaurosis and delineation of the associated phenotype. Hum Mutat 2007;28:416.
[225] Salonen R, Norio R. The Meckel syndrome in Finland: epidemiologic and genetic aspects. Am J Med Genet 1984;18:691–8.
[226] Johnson CA, Gissen P, Sergi C. Molecular pathology and genetics of congenital hepatorenal fibrocystic syndromes. J Med Genet 2003;40:311–9.
[227] Salomon R, Saunier S, Niaudet P. Nephronophthisis. Pediatr Nephrol 2009;24:2333–44.
[228] Mollet G, Salomon R, Gribouval O, Silbermann F, Bacq D, Landthaler G, et al. The gene mutated in juvenile nephronophthisis type 4 encodes a novel protein that interacts with nephrocystin. Nat Genet 2002;32:300–5.
[229] Beals RK, Weleber RG. Conorenal dysplasia: a syndrome of cone-shaped epiphysis, renal disease in childhood, retinitis pigmentosa and abnormality of the proximal femur. Am J Med Genet 2007;143A:2444–7.
[230] Mainzer F, Saldino RM, Ozonoff MB, Minagi H. Familial nephropathy associatdd with retinitis pigmentosa, cerebellar ataxia and skeletal abnormalities. Am J Med 1970;49:556–62.
[231] Perrault I, Saunier S, Hanein S, Filhol E, Bizet AA, Collins F, et al. Mainzer-Saldino syndrome is a ciliopathy caused by IFT140 mutations. Am J Hum Genet 2012;90:864–70.
[232] Srivastava S, Molinari E, Raman S, Sayer JA. Many genes-one disease? Genetics of nephronophthisis (NPHP) and NPHP-associated disorders. Front Pediatr 2017;5:287.
[233] Halbritter J, Bizet AA, Schmidts M, Porath JD, Braun DA, Gee HY, et al. Defects in the IFT-B component IFT172 cause Jeune and Mainzer-Saldino syndromes in humans. Am J Hum Genet 2013;93:915–25.
[234] Antignac C, Arduy CH, Beckmann JS, Benessy F, Gros F, Medhioub M, et al. A gene for familial juvenile nephronophthisis (recessive medullary cystic kidney disease) maps to chromosome 2p. Nat Genet 1993;3:342–5.
[235] Hildebrandt F, Cybulla M, Strahm B, Nothwang HG, Singh-Sawhney I, Berz K, et al. Physical mapping of the gene for juvenile nephronophthisis (NPH1) by construction of a complete YAC contig of 7 Mb on chromosome 2q13. Cytogenet Cell Genet 1996;73:235–9.
[236] Hildebrandt F, Singh-Sawhney I, Schnieders B, Centofante L, Omran H, Pohlmann A, et al. Mapping of a gene for familial juvenile nephronophthisis: refining the map and defining flanking markers on chromosome 2. APN Study Group. Am J Hum Genet 1993;53:1256–61.
[237] Hildebrandt F, Singh-Sawhney I, Schnieders B, Papenfuss T, Brandis M. Refined genetic mapping of a gene for familial juvenile nephronophthisis (NPH1) and physical mapping of linked markers. APN Study Group. Genomics 1995;25:360–4.
[238] Konrad M, Saunier S, Heidet L, Silbermann F, Benessy F, Calado J, et al. Large homozygous deletions of the 2q13 region are a major cause of juvenile nephronophthisis. Hum Mol Genet 1996;5:367–71.

[239] Konrad M, Saunier S, Silbermann F, Benessy F, Le Paslier D, Weissenbach J, et al. A 11 Mb YAC-based contig spanning the familial juvenile nephronophthisis region (NPH1) located on chromosome 2q. Genomics 1995;30:514–20.
[240] Hildebrandt F, Otto E, Rensing C, Nothwang HG, Vollmer M, Adolphs J, et al. A novel gene encoding an SH3 domain protein is mutated in nephronophthisis type 1. Nat Genet 1997;17:149–53.
[241] Saunier S, Calado J, Heilig R, Silbermann F, Benessy F, Morin G, et al. A novel gene that encodes a protein with a putative src homology 3 domain is a candidate gene for familial juvenile nephronophthisis. Hum Mol Genet 1997;6:2317–23.
[242] Benzing T, Gerke P, Hopker K, Hildebrandt F, Kim E, Walz G. Nephrocystin interacts with Pyk2, p130(Cas), and tensin and triggers phosphorylation of Pyk2. Proc Natl Acad Sci U S A 2001;98:9784–9.
[243] Hildebrandt F, Otto E. Molecular genetics of nephronophthisis and medullary cystic kidney disease. J Am Soc Nephrol 2000;11:1753–61.
[244] Brugge JS. Casting light on focal adhesions. Nat Genet 1998;19:309–11.
[245] Donaldson JC, Dempsey PJ, Reddy S, Bouton AH, Coffey RJ, Hanks SK. Crk-associated substrate p130(Cas) interacts with nephrocystin and both proteins localize to cell-cell contacts of polarized epithelial cells. Exp Cell Res 2000;256:168–78.
[246] Fliegauf M, Horvath J, von Schnakenburg C, Olbrich H, Muller D, Thumfart J, et al. Nephrocystin specifically localizes to the transition zone of renal and respiratory cilia and photoreceptor connecting cilia. J Am Soc Nephrol 2006;17:2424–33.
[247] Otto EA, Schermer B, Obara T, O'Toole JF, Hiller KS, Mueller AM, et al. Mutations in INVS encoding inversin cause nephronophthisis type 2, linking renal cystic disease to the function of primary cilia and left-right axis determination. Nat Genet 2003;34:413–20.
[248] Eley L, Gabrielides C, Adams M, Johnson CA, Hildebrandt F, Sayer JA. Jouberin localizes to collecting ducts and interacts with nephrocystin-1. Kidney Int 2008;74:1139–49.
[249] Huan Y, van Adelsberg J. Polycystin-1, the PKD1 gene product, is in a complex containing E-cadherin and the catenins. J Clin Invest 1999;104:1459–68.
[250] Wilson PD, Geng L, Li X, Burrow CR. The PKD1 gene product, "polycystin-1," is a tyrosine-phosphorylated protein that colocalizes with alpha2beta1-integrin in focal clusters in adherent renal epithelia. Lab Invest 1999;79:1311–23.
[251] Otto E, Betz R, Rensing C, Schatzle S, Kuntzen T, Vetsi T, et al. A deletion distinct from the classical homologous recombination of juvenile nephronophthisis type 1 (NPH1) allows exact molecular definition of deletion breakpoints. Hum Mutat 2000;16:211–23.
[252] Saunier S, Calado J, Benessy F, Silbermann F, Heilig R, Weissenbach J, et al. Characterization of the NPHP1 locus: mutational mechanism involved in deletions in familial juvenile nephronophthisis. Am J Hum Genet 2000;66:778–89.
[253] Otto EA, Helou J, Allen SJ, O'Toole JF, Wise EL, Ashraf S, et al. Mutation analysis in nephronophthisis using a combined approach of homozygosity mapping, CEL I endonuclease cleavage, and direct sequencing. Hum Mutat 2008;29:418–26.
[254] Hildebrandt F, Rensing C, Betz R, Sommer U, Birnbaum S, Imm A, et al. Establishing an algorithm for molecular genetic diagnostics in 127 families with juvenile nephronophthisis. Kidney Int 2001;59:434–45.
[255] Haider NB, Carmi R, Shalev H, Sheffield VC, Landau D. A Bedouin kindred with infantile nephronophthisis demonstrates linkage to chromosome 9 by homozygosity mapping. Am J Hum Genet 1998;63:1404–10.
[256] Tsuji T, Matsuo K, Nakahari T, Marunaka Y, Yokoyama T. Structural basis of the Inv compartment and ciliary abnormalities in Inv/nphp2 mutant mice. Cytoskeleton (Hoboken) 2016;73:45–56.
[257] Morgan D, Eley L, Sayer J, Strachan T, Yates LM, Craighead AS, et al. Expression analyses and interaction with the anaphase promoting complex protein Apc2 suggest a role for inversin in primary cilia and involvement in the cell cycle. Hum Mol Genet 2002;11:3345–50.
[258] Simons M, Gloy J, Ganner A, Bullerkotte A, Bashkurov M, Kronig C, et al. Inversin, the gene product mutated in nephronophthisis type II, functions as a molecular switch between Wnt signaling pathways. Nat Genet 2005;37:537–43.
[259] Werner ME, Ward HH, Phillips CL, Miller C, Gattone VH, Bacallao RL. Inversin modulates the cortical actin network during mitosis. Am J Physiol Cell Physiol 2013;305:C36–47.
[260] Omran H, Fernandez C, Jung M, Haffner K, Fargier B, Villaquiran A, et al. Identification of a new gene locus for adolescent nephronophthisis, on chromosome 3q22 in a large Venezuelan pedigree. Am J Hum Genet 2000;66:118–27.
[261] Omran H, Haffner K, Burth S, Fernandez C, Fargier B, Villaquiran A, et al. Human adolescent nephronophthisis: gene locus synteny with polycystic kidney disease in pcy mice. J Am Soc Nephrol 2001;12:107–13.
[262] Bergmann C, Fliegauf M, Bruchle NO, Frank V, Olbrich H, Kirschner J, et al. Loss of nephrocystin-3

function can cause embryonic lethality, Meckel-Gruber-like syndrome, situs inversus, and renal-hepatic-pancreatic dysplasia. Am J Hum Genet 2008;82:959–70.
[263] Hoff S, Halbritter J, Epting D, Frank V, Nguyen TM, van Reeuwijk J, et al. ANKS6 is a central component of a nephronophthisis module linking NEK8 to INVS and NPHP3. Nat Genet 2013;45:951–6.
[264] Gattone 2nd VH, Wang X, Harris PC, Torres VE. Inhibition of renal cystic disease development and progression by a vasopressin V2 receptor antagonist. Nat Med 2003;9:1323–6.
[265] Omran H, Haffner K, Burth S, Ala-Mello S, Antignac C, Hildebrandt F. Evidence for further genetic heterogeneity in nephronophthisis. Nephrol Dial Transplant 2001;16:755–8.
[266] Mollet G, Silbermann F, Delous M, Salomon R, Antignac C, Saunier S. Characterization of the nephrocystin/nephrocystin-4 complex and subcellular localization of nephrocystin-4 to primary cilia and centrosomes. Hum Mol Genet 2005;14:645–56.
[267] Roepman R, Letteboer SJ, Arts HH, van Beersum SE, Lu X, Krieger E, et al. Interaction of nephrocystin-4 and RPGRIP1 is disrupted by nephronophthisis or Leber congenital amaurosis-associated mutations. Proc Natl Acad Sci U S A 2005;102:18520–5.
[268] Habbig S, Bartram MP, Muller RU, Schwarz R, Andriopoulos N, Chen S, et al. NPHP4, a cilia-associated protein, negatively regulates the Hippo pathway. J Cell Biol 2011;193:633–42.
[269] Hoefele J, Sudbrak R, Reinhardt R, Lehrack S, Hennig S, Imm A, et al. Mutational analysis of the NPHP4 gene in 250 patients with nephronophthisis. Hum Mutat 2005;25:411.
[270] Brancati F, Iannicelli M, Travaglini L, Mazzotta A, Bertini E, Boltshauser E, et al. MKS3/TMEM67 mutations are a major cause of COACH Syndrome, a Joubert Syndrome related disorder with liver involvement. Hum Mutat 2009;30:E432–42.
[271] Gorden NT, Arts HH, Parisi MA, Coene KL, Letteboer SJ, van Beersum SE, et al. CC2D2A is mutated in Joubert syndrome and interacts with the ciliopathy-associated basal body protein CEP290. Am J Hum Genet 2008;83:559–71.
[272] Attanasio M, Uhlenhaut NH, Sousa VH, O'Toole JF, Otto E, Anlag K, et al. Loss of GLIS2 causes nephronophthisis in humans and mice by increased apoptosis and fibrosis. Nat Genet 2007;39:1018–24.
[273] Zhang F, Jetten AM. Genomic structure of the gene encoding the human GLI-related, Kruppel-like zinc finger protein GLIS2. Gene 2001;280:49–57.
[274] Lu D, Rauhauser A, Li B, Ren C, McEnery K, Zhu J, et al. Loss of Glis2/NPHP7 causes kidney epithelial cell senescence and suppresses cyst growth in the Kif3a mouse model of cystic kidney disease. Kidney Int 2016;89:1307–23.
[275] Arts HH, Doherty D, van Beersum SE, Parisi MA, Letteboer SJ, Gorden NT, et al. Mutations in the gene encoding the basal body protein RPGRIP1L, a nephrocystin-4 interactor, cause Joubert syndrome. Nat Genet 2007;39:882–8.
[276] Delous M, Baala L, Salomon R, Laclef C, Vierkotten J, Tory K, et al. The ciliary gene RPGRIP1L is mutated in cerebello-oculo-renal syndrome (Joubert syndrome type B) and Meckel syndrome. Nat Genet 2007;39:875–81.
[277] Wolf MT, Saunier S, O'Toole JF, Wanner N, Groshong T, Attanasio M, et al. Mutational analysis of the RPGRIP1L gene in patients with Joubert syndrome and nephronophthisis. Kidney Int 2007;72:1520–6.
[278] Sohara E, Luo Y, Zhang J, Manning DK, Beier DR, Zhou J. Nek8 regulates the expression and localization of polycystin-1 and polycystin-2. J Am Soc Nephrol 2008;19:469–76.
[279] Choi HJ, Lin JR, Vannier JB, Slaats GG, Kile AC, Paulsen RD, et al. NEK8 links the ATR-regulated replication stress response and S phase CDK activity to renal ciliopathies. Mol Cell 2013;51:423–39.
[280] Grampa V, Delous M, Zaidan M, Odye G, Thomas S, Elkhartoufi N, et al. Novel NEK8 mutations cause severe syndromic renal cystic dysplasia through YAP dysregulation. PLoS Genet 2016;12:e1005894.
[281] Otto EA, Hurd TW, Airik R, Chaki M, Zhou W, Stoetzel C, et al. Candidate exome capture identifies mutation of SDCCAG8 as the cause of a retinal-renal ciliopathy. Nat Genet 2010;42:840–50.
[282] Schaefer E, Zaloszyc A, Lauer J, Durand M, Stutzmann F, Perdomo-Trujillo Y, et al. Mutations in SDCCAG8/NPHP10 cause bardet-biedl syndrome and are associated with penetrant renal disease and absent polydactyly. Mol Syndromol 2011;1:273–81.
[283] Otto EA, Tory K, Attanasio M, Zhou W, Chaki M, Paruchuri Y, et al. Hypomorphic mutations in meckelin (MKS3/TMEM67) cause nephronophthisis with liver fibrosis (NPHP11). J Med Genet 2009;46:663–70.
[284] Vilboux T, Doherty DA, Glass IA, Parisi MA, Phelps IG, Cullinane AR, et al. Molecular genetic findings and clinical correlations in 100 patients with Joubert syndrome and related disorders prospectively evaluated at a single center. Genet Med 2017;19:875–82.
[285] Taschner M, Bhogaraju S, Lorentzen E. Architecture and function of IFT complex proteins in ciliogenesis. Differentiation 2012;83:S12–22.
[286] Davis EE, Zhang Q, Liu Q, Diplas BH, Davey LM, Hartley J, et al. TTC21B contributes both causal and

modifying alleles across the ciliopathy spectrum. Nat Genet 2011;43:189–96.
[287] Tran PV, Talbott GC, Turbe-Doan A, Jacobs DT, Schonfeld MP, Silva LM, et al. Downregulating hedgehog signaling reduces renal cystogenic potential of mouse models. J Am Soc Nephrol 2014;25:2201–12.
[288] Huynh Cong E, Bizet AA, Boyer O, Woerner S, Gribouval O, Filhol E, et al. A homozygous missense mutation in the ciliary gene TTC21B causes familial FSGS. J Am Soc Nephrol 2014;25:2435–43.
[289] Coussa RG, Otto EA, Gee HY, Arthurs P, Ren H, Lopez I, et al. WDR19: an ancient, retrograde, intraflagellar ciliary protein is mutated in autosomal recessive retinitis pigmentosa and in Senior-Loken syndrome. Clin Genet 2013;84:150–9.
[290] Bredrup C, Saunier S, Oud MM, Fiskerstrand T, Hoischen A, Brackman D, et al. Ciliopathies with skeletal anomalies and renal insufficiency due to mutations in the IFT-A gene WDR19. Am J Hum Genet 2011;89:634–43.
[291] Halbritter J, Porath JD, Diaz KA, Braun DA, Kohl S, Chaki M, et al. Identification of 99 novel mutations in a worldwide cohort of 1,056 patients with a nephronophthisis-related ciliopathy. Hum Genet 2013;132:865–84.
[292] Chaki M, Airik R, Ghosh AK, Giles RH, Chen R, Slaats GG, et al. Exome capture reveals ZNF423 and CEP164 mutations, linking renal ciliopathies to DNA damage response signaling. Cell 2012;150:533–48.
[293] Graser S, Stierhof YD, Lavoie SB, Gassner OS, Lamla S, Le Clech M, et al. Cep164, a novel centriole appendage protein required for primary cilium formation. J Cell Biol 2007;179:321–30.
[294] Taskiran EZ, Korkmaz E, Gucer S, Kosukcu C, Kaymaz F, Koyunlar C, et al. Mutations in ANKS6 cause a nephronophthisis-like phenotype with ESRD. J Am Soc Nephrol 2014;25:1653–61.
[295] Bujakowska KM, Zhang Q, Siemiatkowska AM, Liu Q, Place E, Falk MJ, et al. Mutations in IFT172 cause isolated retinal degeneration and Bardet-Biedl syndrome. Hum Mol Genet 2015;24:230–42.
[296] Failler M, Gee HY, Krug P, Joo K, Halbritter J, Belkacem L, et al. Mutations of CEP83 cause infantile nephronophthisis and intellectual disability. Am J Hum Genet 2014;94:905–14.
[297] Joo K, Kim CG, Lee MS, Moon HY, Lee SH, Kim MJ, et al. CCDC41 is required for ciliary vesicle docking to the mother centriole. Proc Natl Acad Sci U S A 2013;110:5987–92.
[298] Schueler M, Braun DA, Chandrasekar G, Gee HY, Klasson TD, Halbritter J, et al. DCDC2 mutations cause a renal-hepatic ciliopathy by disrupting Wnt signaling. Am J Hum Genet 2015;96:81–92.
[299] Girard M, Bizet AA, Lachaux A, Gonzales E, Filhol E, Collardeau-Frachon S, et al. DCDC2 mutations cause neonatal sclerosing cholangitis. Hum Mutat 2016;37:1025–9.
[300] Macia MS, Halbritter J, Delous M, Bredrup C, Gutter A, Filhol E, et al. Mutations in MAPKBP1 cause juvenile or late-onset cilia-independent nephronophthisis. Am J Hum Genet 2017;100:372.
[301] Schmidts M, Frank V, Eisenberger T, Al Turki S, Bizet AA, Antony D, et al. Combined NGS approaches identify mutations in the intraflagellar transport gene IFT140 in skeletal ciliopathies with early progressive kidney Disease. Hum Mutat 2013;34:714–24.
[302] Aldahmesh MA, Li Y, Alhashem A, Anazi S, Alkuraya H, Hashem M, et al. IFT27, encoding a small GTPase component of IFT particles, is mutated in a consanguineous family with Bardet-Biedl syndrome. Hum Mol Genet 2014;23:3307–15.
[303] Beales PL, Bland E, Tobin JL, Bacchelli C, Tuysuz B, Hill J, et al. IFT80, which encodes a conserved intraflagellar transport protein, is mutated in Jeune asphyxiating thoracic dystrophy. Nat Genet 2007;39:727–9.
[304] Bizet AA, Becker-Heck A, Ryan R, Weber K, Filhol E, Krug P, et al. Mutations in TRAF3IP1/IFT54 reveal a new role for IFT proteins in microtubule stabilization. Nat Commun 2015;6:8666.
[305] Perrault I, Halbritter J, Porath JD, Gerard X, Braun DA, Gee HY, et al. IFT81, encoding an IFT-B core protein, as a very rare cause of a ciliopathy phenotype. J Med Genet 2015;52:657–65.
[306] Ferland RJ, Eyaid W, Collura RV, Tully LD, Hill RS, Al-Nouri D, et al. Abnormal cerebellar development and axonal decussation due to mutations in AHI1 in Joubert syndrome. Nat Genet 2004;36:1008–13.
[307] Dixon-Salazar T, Silhavy JL, Marsh SE, Louie CM, Scott LC, Gururaj A, et al. Mutations in the AHI1 gene, encoding jouberin, cause Joubert syndrome with cortical polymicrogyria. Am J Hum Genet 2004;75:979–87.
[308] Utsch B, Sayer JA, Attanasio M, Pereira RR, Eccles M, Hennies HC, et al. Identification of the first AHI1 gene mutations in nephronophthisis-associated Joubert syndrome. Pediatr Nephrol 2006;21:32–5.
[309] Al-Hamed MH, Kurdi W, Alsahan N, Alabdullah Z, Abudraz R, Tulbah M, et al. Genetic spectrum of Saudi Arabian patients with antenatal cystic kidney disease and ciliopathy phenotypes using a targeted renal gene panel. J Med Genet 2016;53:338–47.

[310] Louie CM, Caridi G, Lopes VS, Brancati F, Kispert A, Lancaster MA, et al. AHI1 is required for photoreceptor outer segment development and is a modifier for retinal degeneration in nephronophthisis. Nat Genet 2010;42:175–80.
[311] Khanna H, Davis EE, Murga-Zamalloa CA, Estrada-Cuzcano A, Lopez I, den Hollander AI, et al. A common allele in RPGRIP1L is a modifier of retinal degeneration in ciliopathies. Nat Genet 2009;41:739–45.
[312] Hamiwka LA, Midgley JP, Wade AW, Martz KL, Grisaru S. Outcomes of kidney transplantation in children with nephronophthisis: an analysis of the North American pediatric renal trials and collaborative studies (NAPRTCS) registry. Pediatr Transplant 2008;12:878–82.
[313] Eckardt KU, Alper SL, Antignac C, Bleyer AJ, Chauveau D, Dahan K, et al. Autosomal dominant tubulointerstitial kidney disease: diagnosis, classification, and management–A KDIGO consensus report. Kidney Int 2015;88:676–83.
[314] Schaeffer C, Devuyst O, Rampoldi L. Uromodulin: roles in Health and disease. Annu Rev Physiol 2021;83:477–501.
[315] Bleyer AJ, Wolf MT, Kidd KO, Zivna M, Kmoch S. Autosomal dominant tubulointerstitial kidney disease: more than just HNF1beta. Pediatr Nephrol 2021.
[316] Rampoldi L, Scolari F, Amoroso A, Ghiggeri G, Devuyst O. The rediscovery of uromodulin (Tamm-Horsfall protein): from tubulointerstitial nephropathy to chronic kidney disease. Kidney Int 2011;80:338–47.
[317] Rezende-Lima W, Parreira KS, Garcia-Gonzalez M, Riveira E, Banet JF, Lens XM. Homozygosity for uromodulin disorders: FJHN and MCKD-type 2. Kidney Int 2004;66:558–63.
[318] Zivna M, Kidd K, Pristoupilova A, Baresova V, DeFelice M, Blumenstiel B, et al. Noninvasive immunohistochemical diagnosis and novel MUC1 mutations causing autosomal dominant tubulointerstitial kidney disease. J Am Soc Nephrol 2018;29:2418–31.
[319] Wenzel A, Altmueller J, Ekici AB, Popp B, Stueber K, Thiele H, et al. Single molecule real time sequencing in ADTKD-MUC1 allows complete assembly of the VNTR and exact positioning of causative mutations. Sci Rep 2018;8:4170.
[320] Nie M, Bal MS, Yang Z, Liu J, Rivera C, Wenzel A, et al. Mucin-1 increases renal TRPV5 activity in vitro, and urinary level associates with calcium nephrolithiasis in patients. J Am Soc Nephrol 2016;27:3447–58.
[321] Olinger E, Hofmann P, Kidd K, Dufour I, Belge H, Schaeffer C, et al. Clinical and genetic spectra of autosomal dominant tubulointerstitial kidney disease due to mutations in UMOD and MUC1. Kidney Int 2020;98:717–31.
[322] Dvela-Levitt M, Kost-Alimova M, Emani M, Kohnert E, Thompson R, Sidhom EH, et al. Small molecule targets TMED9 and promotes lysosomal degradation to reverse proteinopathy. Cell 2019;178. 521-535 e523.
[323] Zivna M, Kidd K, Zaidan M, Vyletal P, Baresova V, Hodanova K, et al. An international cohort study of autosomal dominant tubulointerstitial kidney disease due to REN mutations identifies distinct clinical subtypes. Kidney Int 2020;98:1589–604.
[324] Kompatscher A, de Baaij JHF, Aboudehen K, Farahani S, van Son LHJ, Milatz S, et al. Transcription factor HNF1beta regulates expression of the calcium-sensing receptor in the thick ascending limb of the kidney. Am J Physiol Ren Physiol 2018;315:F27–35.
[325] Izzi C, Dordoni C, Econimo L, Delbarba E, Grati FR, Martin E, et al. Variable expressivity of HNF1B nephropathy, from renal cysts and diabetes to medullary sponge kidney through tubulo-interstitial kidney disease. Kidney Int Rep 2020;5:2341–50.
[326] Stavrou C, Koptides M, Tombazos C, Psara E, Patsias C, Zouvani I, et al. Autosomal-dominant medullary cystic kidney disease type 1: clinical and molecular findings in six large Cypriot families. Kidney Int 2002;62:1385–94.
[327] Chen TJ, Song R, Janssen A, Yosypiv IV. Cytogenomic aberrations in isolated multicystic dysplastic kidney in children. Pediatr Res 2021.
[328] Raina R, Chakraborty R, Sethi SK, Kumar D, Gibson K, Bergmann C. Diagnosis and management of renal cystic disease of the newborn: core curriculum 2021. Am J Kidney Dis 2021.
[329] Ericsson NO, Ivemark BI. Renal dysplasia and pyelonephritis in infants and children. II. Primitive ductules and abnormal glomeruli. AMA Arch Pathol 1958;66:264–9.
[330] Crino PB, Nathanson KL, Henske EP. The tuberous sclerosis complex. N Engl J Med 2006;355:1345–56.
[331] Marom D. Genetics of tuberous sclerosis complex: an update. Childs Nerv Syst 2020;36:2489–96.
[332] Caban C, Khan N, Hasbani DM, Crino PB. Genetics of tuberous sclerosis complex: implications for clinical practice. Appl Clin Genet 2017;10:1–8.
[333] Northrup H, Krueger DA. International tuberous sclerosis complex consensus G: tuberous sclerosis complex diagnostic criteria update: recommendations of the 2012 iinternational tuberous sclerosis complex consensus conference. Pediatr Neurol 2013;49:243–54.
[334] Krueger DA, Northrup H. International tuberous sclerosis complex consensus G: tuberous sclerosis

complex surveillance and management: recommendations of the 2012 international tuberous sclerosis complex consensus conference. Pediatr Neurol 2013;49:255–65.

[335] Kotulska K, Kwiatkowski DJ, Curatolo P, Weschke B, Riney K, Jansen F, et al. Prevention of epilepsy in infants with tuberous sclerosis complex in the EPISTOP trial. Ann Neurol 2021;89:304–14.

[336] Franz DN, Belousova E, Sparagana S, Bebin EM, Frost M, Kuperman R, et al. Efficacy and safety of everolimus for subependymal giant cell astrocytomas associated with tuberous sclerosis complex (EXIST-1): a multicentre, randomised, placebo-controlled phase 3 trial. Lancet 2013;381:125–32.

[337] Bobeff K, Krajewska K, Baranska D, Kotulska K, Jozwiak S, Mlynarski W, et al. Maintenance therapy with everolimus for subependymal giant cell astrocytoma in patients with tuberous sclerosis - final results from the EMINENTS study. Front Neurol 2021;12:581102.

[338] Bissler JJ, Kingswood JC, Radzikowska E, Zonnenberg BA, Frost M, Belousova E, et al. Everolimus for angiomyolipoma associated with tuberous sclerosis complex or sporadic lymphangioleiomyomatosis (EXIST-2): a multicentre, randomised, double-blind, placebo-controlled trial. Lancet 2013;381:817–24.

[339] Bissler JJ, Kingswood JC, Radzikowska E, Zonnenberg BA, Frost M, Belousova E, et al. Everolimus for renal angiomyolipoma in patients with tuberous sclerosis complex or sporadic lymphangioleiomyomatosis: extension of a randomized controlled trial. Nephrol Dial Transplant 2016;31:111–9.

[340] Bissler JJ, McCormack FX, Young LR, Elwing JM, Chuck G, Leonard JM, et al. Sirolimus for angiomyolipoma in tuberous sclerosis complex or lymphangioleiomyomatosis. N Engl J Med 2008;358:140–51.

[341] Forsythe E, Kenny J, Bacchelli C, Beales PL. Managing bardet-biedl syndrome-now and in the future. Front Pediatr 2018;6:23.

[342] Forsythe E, Sparks K, Best S, Borrows S, Hoskins B, Sabir A, et al. Risk factors for severe renal disease in bardet-biedl syndrome. J Am Soc Nephrol 2017;28:963–70.

[343] Bergmann C. Educational paper: ciliopathies. Eur J Pediatr 2012;171:1285–300.

[344] McConnachie DJ, Stow JL, Mallett AJ. Ciliopathies and the kidney: a review. Am J Kidney Dis 2021;77:410–9.

[345] Marion V, Schlicht D, Mockel A, Caillard S, Imhoff O, Stoetzel C, et al. Bardet-Biedl syndrome highlights the major role of the primary cilium in efficient water reabsorption. Kidney Int 2011;79:1013–25.

[346] Putoux A, Attie-Bitach T, Martinovic J, Gubler MC. Phenotypic variability of Bardet-Biedl syndrome: focusing on the kidney. Pediatr Nephrol 2012;27:7–15.

[347] Tobin JL, Beales PL. Bardet-Biedl syndrome: beyond the cilium. Pediatr Nephrol 2007;22:926–36.

[348] Majewski E, Ozturk B, Gillessen-Kaesbach G. Jeune syndrome with tongue lobulation and preaxial polydactyly, and Jeune syndrome with situs inversus and asplenia: compound heterozygosity Jeune-Mohr and Jeune-Ivemark? Am J Med Genet 1996;63:74–9.

[349] Wilson DJ, Weleber RG, Beals RK. Retinal dystrophy in Jeune's syndrome. Arch Ophthalmol 1987;105:651–7.

[350] Wu MH, Kuo PL, Lin SJ. Prenatal diagnosis of recurrence of short rib-polydactyly syndrome. Am J Med Genet 1995;55:279–84.

[351] Bonafe L, Cormier-Daire V, Hall C, Lachman R, Mortier G, Mundlos S, et al. Nosology and classification of genetic skeletal disorders: 2015 revision. Am J Med Genet 2015;167A:2869–92.

[352] Barakat AJ, Saba C, Rennert OM. Kidney abnormalities in Hajdu-Cheney syndrome. Pediatr Nephrol 1996;10:712–5.

[353] Brennan AM, Pauli RM. Hajdu–Cheney syndrome: evolution of phenotype and clinical problems. Am J Med Genet 2001;100:292–310.

[354] Kaplan P, Ramos F, Zackai EH, Bellah RD, Kaplan BS. Cystic kidney disease in Hajdu-Cheney syndrome. Am J Med Genet 1995;56:25–30.

[355] Canalis E, Zanotti S. Hajdu-Cheney syndrome: a review. Orphanet J Rare Dis 2014;9:200.

[356] Cumming WA, Ohlsson A, Ali A. Campomelia, cervical lymphocele, polycystic dysplasia, short gut, polysplenia. Am J Med Genet 1986;25:783–90.

[357] Perez del Rio MJ, Fernandez-Toral J, Madrigal B, Gonzalez-Gonzalez M, Ablanedo P, Herrero A. Two new cases of Cumming syndrome confirming autosomal recessive inheritance. Am J Med Genet 1999;82:340–3.

[358] Urioste M, Arroyo A, Martinez-Frias ML. Campomelia, polycystic dysplasia, and cervical lymphocele in two sibs. Am J Med Genet 1991;41:475–7.

[359] Ming JE, McDonald-McGinn DM, Markowitz RI, Ruchelli E, Zackai EH. Heterotaxia in a fetus with campomelia, cervical lymphocele, polysplenia, and multicystic dysplastic kidneys: expanding the phenotype of Cumming syndrome. Am J Med Genet 1997;73:419–24.

[360] Bedeschi MF, Spaccini L, Rizzuti T, Coviello DA, Castorina P, Natacci F, et al. Cumming syndrome with heterotaxia, campomelia and absent uterus/fallopian tubes. Am J Med Genet 2005;132A:329–30.

[361] Wieser T, Deschauer M, Olek K, Hermann T, Zierz S. Carnitine palmitoyltransferase II deficiency: molecular and biochemical analysis of 32 patients. Neurology 2003;60:1351–3.
[362] Dell KM. The role of cilia in the pathogenesis of cystic kidney disease. Curr Opin Pediatr 2015;27:212–8.

FURTHER READING

Fain PR, McFann KK, Taylor MR, Tison M, Johnson AM, Reed B, et al. Modifier genes play a significant role in the phenotypic expression of PKD1. Kidney Int 2005;67:1256–67.

3

Nephrotic Disorders

Hannu Jalanko[1], Helena Kääriäinen[2]

[1]Children's Hospital, University of Helsinki, Helsinki, Finland
[2]National Institute for Health and Welfare, Helsinki, Finland

3.1 INTRODUCTION

Nephrotic syndrome (NS) is a clinical diagnosis characterized by heavy proteinuria, hypoproteinemia, and edema. It occurs in various forms of acquired renal diseases, such as glomerulonephritides and membranous nephropathy, or may be part of systemic diseases, such as vasculitis, lupus erythematosus, and amyloidosis. NS may also be caused by infections, including HIV, maternal syphilis, toxoplasmosis, parvovirus B19, and malaria. Moreover, intoxications caused by mercury, heroin, lithium, interferon-α, and pamidronate are also associated with NS [1].

The most common form of primary NS is steroid-sensitive NS (SSNS) accounting for 80%−90% of the pediatric cases. Practically all these patients show minimal change histology (MCNS). Twenty percent have steroid resistant NS (SRNS), and in many of them histological diagnosis is focal segmental glomerulosclerosis (FSGS). Overall, MCNS and FSGS account for over 90% and 5% of NS in the pediatric age group, respectively. FSGS is especially common in African Americans, and the incidence of FSGS in children has increased. In adults the situation is different, and FSGS is the most common type of NS. Typically, FSGS manifests as NS but progresses to renal failure within months or years. Overall, FSGS is a major cause of end-stage renal disease (ESRD), comprising up to 5% of adults and 20% of children with ESRD. The pathogenesis of SSNS and nongenetic SRNS is mostly unknown, but T-lymphocyte dysfunction has been regarded as the basic event. Also, circulating permeability factors, such as soluble urokinase-type plasminogen activator receptor (suPAR), interleukin-13, and hemopexin, have been suggested as causative factors for SSNS and SRNS, but their pathogenic role has not been verified [2].

During the past two decades, increasing number of pathogenic gene defects both in autosomal recessive (AR) and dominant (AD) forms of SRNS have been identified. Causative mutations have so far been reported in over 50 genes. Gene defects are found in 10%−20% of sporadic and 30%−40% of familial cases of SRNS, respectively [3]. Gene mutations are more probable in pediatric (especially in infants) than adult patients [4]. Autosomal recessive inheritance is typical for NS in small children while dominant forms are more likely in adults [5].

Many of the gene defects are rare and reported in only few patients. There are, however, around 10 genes responsible for the majority of genetic SRNS. These include genes for nephrin (*NPHS1*), podocin (*NPHS2*), Wilms' tumor factor 1 (*WT1*), phospholipase Ce1 (*PLCE1*), laminin-β2 (*LAMB2*), transient receptor potential C6 ion channel (*TRPC6*), CD2-associated

Emery and Rimoin's Principles and Practice of Medical Genetics and Genomics. https://doi.org/10.1016/B978-0-12-812534-2.00004-7

protein (*CD2AP*), alpha-actinin-4 (*ACTN4*), inverted formin 2 (*INF2*), and collagen IV alpha 3−5 (*COL4A3-5*) [6−10]. While gene defects are detected in about two-thirds of small infants with NS, this is the case in only 10%−20% of the adolescent and adult patients with SRNS [5].

The clinical classification of NS into congenital (CNS; onset <3 months of age), infantile (INS, onset at 4−12 months), childhood/adolescence (1−18 years), and adult forms is relevant, as it gives a clue for the etiology of the disease. NS caused by a particular gene defect can, however, manifest at various ages. Also, the histologic features of the kidney overlap in the different genetic and nongenetic entities, so that the microscopic findings do not tell the specific diagnosis. The clinical renal and extrarenal manifestations, kidney histology, and the age at onset, however, help in searching for the possible gene defect involved. The development of next-generation sequencing (NGS) during the last 10 years has greatly increased the probabilities for specific genetic diagnosis, which helps in the management of NS patients [11]. It can be expected that new but possibly very rare monogenic entities will still be found in the coming years as searching for variants from the whole exome is becoming more common. In addition to the monogenic causes of NS, multifactorial genetic variants predisposing to or protecting from NS are being investigated and may in the future provide a way to predict the occurrence as well as prognosis of NS.

3.2 GLOMERULAR FILTRATION BARRIER

In kidney, filtration of plasma takes place in the glomerular capillary tuft surrounded by the urinary space and Bowman's capsule (Fig. 3.1). The ultrafiltration barrier in the capillary wall comprises three layers: fenestrated endothelium, glomerular basement membrane (GBM), and epithelial cell (podocyte) layer with distal foot processes and interposed slit diaphragms (SDs). The barrier is a highly sophisticated size- and charge-selective molecular sieve, and normally only water, electrolytes, and small plasma solutes pass through this barrier [12−14]. The flow of glomerular filtrate follows the extracellular route, passing across the GBM and then across SDs, which bridge adjacent foot processes just above the GBM. (Fig. 3.2).

Studies performed during the past two decades have clearly indicated that podocytes and the SD are essential in restricting the passage of plasma proteins into urine [15]. The precise molecular structure of SD is still unresolved, but podocyte proteins nephrin, Neph1, Neph2, Fat1, Fat2, cadherins, and dendrin seem to be

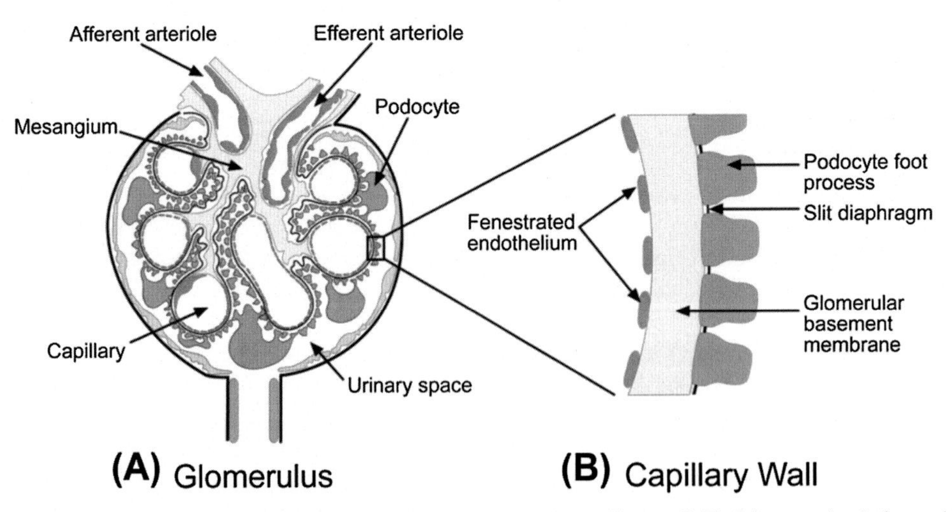

Figure 3.1 The structure of kidney glomerulus (A) and glomerular capillary wall (B). Primary urine is formed in the glomerulus by passing of plasma water and small-molecular-weight molecules through the capillary wall (glomerular filtration barrier) into the urinary space. In nephrosis plasma, proteins also leak through this barrier, which is composed of three layers: fenestrated endothelium, glomerular basement membrane, and podocyte foot processes. These are connected by the slit diaphragm, which is important for the proper function of the filtration barrier. (Modified from the original figure by Professor Wilhelm Kriz.)

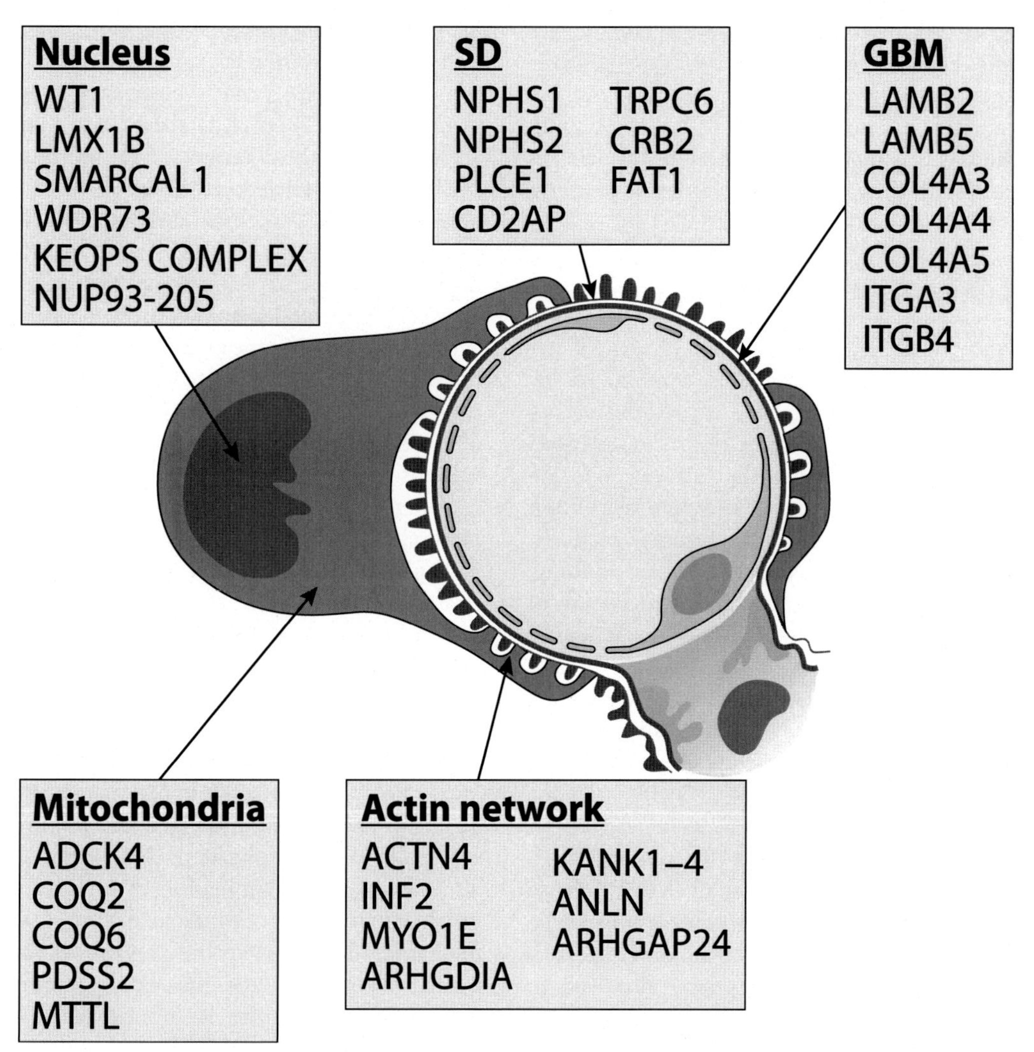

Figure 3.2 Glomerular epithelial cell, podocyte, is composed of cell body, primary and secondary processes, and distal foot processes, which are interposed by slit diaphragms (SD). Podocyte foot processes are tightly attached to the glomerular basement membrane (GBM). Mutations in over 50 podocyte genes have been identified in patients with nephrotic syndrome. Some of them are listed in boxes.

its essential components [16,17]. Nephrin and Neph1 associate with each other extracellularly, and they form the backbone of the SD [18,19]. The SD proteins interact with the adapter proteins podocin, CD2AP, ZO-1, CASK, IQGAP1, MAG1, catenins, and spectrins that are located in the cytosolic part of the podocyte foot process [20–22]. The adapter proteins connect the SD with the actin cytoskeleton of the podocyte foot process and take part in the signal transduction from the SD into podocyte [20,21,23].

The actin network and the interacting proteins are critical for the maintenance of the complex structure of the podocyte foot process [24]. Actin network and the interacting proteins, such as α-actinin-4, inverted formin-2, myosin-9, myosin 1E, and ARHGDIA, maintain the architecture of the foot processes and respond to signals from outside modifying the cellular functions accordingly. Practically all forms of NS lead to podocyte injury with foot process flattening (effacement) and distortion of normal SD-structure [25]. This is believed to reflect a defective interplay of the SD components, adapter proteins, and the actin network itself. The discovery of NS caused by mutations in the GBM components (laminin-β2 and collagen IV alpha 3–5) highlights the importance of the GBM, podocyte foot processes, and the SD in maintaining a properly functioning glomerular filtrations barrier [14].

3.3 NEPHRIN GENE (*NPHS1*) MUTATIONS

Nephrin is a transmembrane adhesion protein of the immunoglobulin family. Nephrin is synthesized by glomerular podocytes and localized at the SD area between the podocyte foot processes [15]. The extracellular part binds to nephrin from the adjacent podocyte foot process as well as NEPH1-3 and ephrin-β1 [26]. The intracellular domain has nine tyrosine residues, which become phosphorylated during ligand binding. Phosphorylated nephrin is involved in the regulation of multiple intracellular signaling pathways that influence actin polymerization and dynamics [27]. Phosphorylation-mediated increase in PI3K enzymatic activity subsequently increases activity of small GTPases that play a pivotal role in membrane ruffling, cell motility, and actin organization.

Nephrin is encoded by the *NPHS1* gene, located on chromosome 19q13.1. It consists of 29 exons and has a size of 26 kb. Pathogenic variants in *NPHS1* cause congenital nephrotic syndrome of the Finnish type (CNF, NPHS1), which is a prototype of CNS [28]. "Mild" *NPHS1* pathogenic variants may also lead to SRNS manifesting later in childhood or even in adults.

3.3.1 Congenital Nephrotic Syndrome of the Finnish Type (CNF, NPHS1)

CNF (OMIM 602716) is an autosomal recessive disease first described in Finland. The incidence in Finland is 1 in 8000 live births. CNF has been reported from all over the world in different ethnic groups [29,30]. A very high incidence of CNF has been observed among the Old Order Mennonites in Lancaster County, Pennsylvania. Exon sequencing analyses of the Finnish patients revealed the presence of two truncating mutations. One or the other of these founder mutations are found in 94% of *NPHS1* chromosomes (Fin-major and Fin-minor) [28]. The uniform mutation pattern seen in the Finnish population can be explained by the founder effect. Most non-Finns have individual mutations. Almost 280 NPHS1 variants associating to NS are listed in ClinVar and HGMD databases, and many of them are unique. Of these variants, 54% are missense, 21% small insertions/deletions/indels, 14% splicing, 9% nonsense, and 1% gross deletions. Most missense pathogenic variants in *NPHS1* lead to the misfolding of nephrin molecule and a defective intracellular nephrin transport in the podocyte [31]. In reports of pathogenic variants in the European and worldwide cohorts of CNF, *NPHS1* mutations account for 39%–80% of the congenital NS cases [8,32–34]. Overall, about one-half of CNF seems to be caused by *NPHS1* pathogenic variants.

The majority of typical CNF children are born prematurely with a birth weight ranging between 1500 and 3500 g [35]. The index of placental weight/birth weight (ISP) is over 25% in practically all newborns. The reason for this is not known. Proteinuria begins in utero and is thus detectable in the first urine sample tested. Fetal proteinuria also leads to high level of α-fetoprotein (AFP) in the amniotic fluid and maternal serum, but an increased AFP is not specific for the disease. Heavy protein losses result in hypogammaglobulinemia as well as low antithrombin III levels and increased risk for thrombotic complications [36]. Hyperlipidemia is also present, as in other nephroses. Infants with CNF do not have any major extrarenal malformations. Minor functional disorders in the central nervous system, cardiac hypertrophy, and muscular hypotonia, however, are common during the nephrotic stage.

No single histological finding in a kidney biopsy is pathognomonic for CNF. In the early stage, proliferation of mesangial cells and increase of mesangial matrix are seen [37]. Dilations of the proximal and distal tubules are the most characteristic findings. Their amount varies greatly, from an occasional dilation to a universal dispersion throughout the renal cortex. In the interstitium, fibrosis and inflammatory infiltrates increase with age. In electron microscopy, the principal finding is the fusion and effacement of podocyte foot processes, seen in many nephrotic kidney diseases [25]. The Fin-major and Fin-minor mutations of *NPHS1* lead to a complete absence of nephrin in the kidney glomerulus [35]. These kidneys also lack the filamentous image of podocyte slit diaphragms as studied by electron microscopy, indicating that the absence of nephrin leads to distortion of the SD and the leakage of plasma proteins into urine through the "empty" podocyte pores [18].

3.3.2 Childhood-Onset NS

Pathogenic variants in *NPHS1* are quite rare in cases manifesting after the first weeks of life [33]. Putative disease-causing *NPHS1* pathogenic variants were found in 10 out of 142 families, with the age of onset ranging from 6 months to 8 years [38]. Biopsy findings showed minimal changes in six of the patients and FSGS in three patients. ESRD had developed in five patients at the age of 6–25 years of age. Santin et al. report a similar study of 97 FSGS patients from 89 unrelated families. Disease-causing *NPHS1* defects were observed in seven sporadic and five familial cases. Three of the individuals were 1 year of age or older, and one was 27 years old at the time of disease onset [34]. In a study of 160 patients from 142 unrelated families, compound heterozygous *NPHS1* pathogenic variants were found in one familial case and nine sporadic pediatric cases [39]. In a large worldwide study, *NPHS1* pathogenic variants were diagnosed in 10% of the genetic SRNS patients at the age group of 4–12 months but in only solitary cases thereafter [8]. The results, however, broaden the spectrum of renal disease related to *NPHS1* pathogenic variants into older age groups.

3.4 PODOCIN GENE (*NPHS2*) PATHOGENIC VARIANTS

Podocin is a hairpin-like protein of the stomatin family and is exclusively expressed in podocytes, where it is important component of the lipid rafts of the filtration slit [40]. Podocin plays a major role in the structural integrity and function of the SD. It is encoded by the *NPHS2* gene composed of eight exons and located on chromosome 1q25-q31. Genetic defects in *NPHS2* (OMIM 604766) can lead to NS starting at any age. To date, more than 200 different *NPHS2* pathogenic variants have been identified in patients with SRNS. The mutations are distributed throughout the entire gene. Most pathogenic variants cause a severe disease with an early onset and development of ESRD within a few years. Many of the disease-causing pathogenic variants in podocin seem to disrupt nephrin trafficking to the plasma membrane of podocytes, which severely interferes with the SD structure and function [20,21,41].

3.4.1 Congenital and Childhood SRNS

Homozygous *NPHS2* pathogenic variants were first reported in two small infants [42]. Thereafter recessive pathogenic variants have abundantly been reported in CNS patients. In a large European cohort, *NPHS1* and *NPHS2* pathogenic variants accounted for an equal amount (39.1%) of CNS cases [33]. On the other hand, *NPHS2* pathogenic variants were clearly less common in CNS patients in a more recent worldwide survey (*NPHS2* 10%, *NPHS1* 40%) [8]. The severity of proteinuria and the clinical findings are more variable in CNS patients with *NPHS2* pathogenic variants as compared to NPHS1 patients. While nephrotic range proteinuria is typically detected in the first few days of life in NPHS1 patients, this often occurs at age of a few weeks in infants with *NPHS2* mutations [43]. Nonrenal manifestations are uncommon, as is the case with NPHS1. CNS patients with *NPHS2* mutations often show minimal histological changes or NPHS1-type histology, and FSGS is seen in less than half of the patients. CNS patients with *NPHS2* pathogenic variants develop ESRD on average 6 years after diagnosis [33].

NPHS2 pathogenic variants are common cause of SRNS throughout the childhood [44,45]. They were found in 18% of over 400 SRNS cases from a worldwide cohort, with the disease onset ranging from birth to 21 years [46]. Truncating pathogenic variants (nonsense or frame shift) resulted in early onset of NS (from birth to 9.1 years of age). This was also true for homozygous R138Q mutation (from birth to 5.4 years), which is common in European patients. In a study of 338 patients, the pathogenic variant detection rate in *NPHS2* was 43% for the familial cases [47]. NS in these patients manifested at the mean age of 3.4 years. In another study, homozygous or compound heterozygous pathogenic variants in *NPHS2* were detected in a third of the patients. Half of them represented familial cases. The age of the patients varied from birth to 24 years, with the median of 3.5 years. Kidney biopsy revealed FSGS histology in a majority of the patients.

3.4.2 Adult-Onset SRNS

NPHS2 pathogenic variants have also been described in SRNS patients with an adult onset. Many are compound heterozygotes, with one allele harboring a R229Q mutation [46,48,49]. The R229Q variant is present in heterozygous state in approximately 4% of Western populations. Heterozygous or homozygous R229Q genotype alone does not cause SRNS. This variant, however, is "non-neutral" and R229Q allele leads to a disease phenotype when it is associated with certain 3′ *NPHS2* pathogenic variants. The disease-associated mutation exerts a dominant-negative effect on R229Q podocin but otherwise behave as a recessive allele [50]. Analysis of *NPHS2* in 546 patients from 455 families with SRNS revealed 36 patients from 27 families with compound heterozygosity for the p.R229Q variant and one pathogenetic variant [51]. These patients had significantly later onset of NS than patients with two pathogenic variants. One pathogenic variant, p.R229Q, however, has also been reported in children with SRNS.

NPHS2 pathogenic variants are the most common podocyte gene defects. Yet, pathogenic variants are found only in a minority of SRNS patients. No homozygotes or compound heterozygotes were observed in 377 biopsy confirmed late-onset FSGS [52]. In three large worldwide analyses, *NPHS2* pathogenic variants were found in roughly 10% of the families and patients with SRNS (Table 3.1).

3.5 WILMS TUMOR SUPPRESSOR GENE (WT1) PATHOGENIC VARIANTS

WT1 gene encodes for a nuclear WT1 protein, which is a transcription factor of the zinc finger family. WT1 regulates a broad set of podocyte genes and plays a crucial role in the embryonic development of the kidney and genitalia. In the mature kidney, WT1 is expressed in podocytes and epithelial cells of Bowman's capsule. *WT1* contains 10 exons, the first six of which encode a proline/glutamine-rich transcriptional regulatory region. Exons 7–10 encode the four zinc fingers of the DNA-binding domain. Up to 24 different isoforms of WT1 may result from the combination of alternative translations sites, alternative RNA splicing, and RNA editing. The biological role of all these isoforms is not known [53].

A variety of WT1 variants, which either affect development or induce tumor formation, have been identified. Developmental defects include the WAGR syndrome, Denys–Drash syndrome (DDS), and Frasier syndrome (FS). Mutations in *WT1* can also cause an isolated kidney disease with NS. The histopathological

TABLE 3.1 Percentage (%) of Families/Patients Screened Positive for SRNS in Three Worldwide Studies of Over 2000 Individuals. Study 1, [8]; Study 2, [9]; Study 3, [10]

Gene	Mode of Inheritance	Study 1%	Study 2%	Study 3%
NPHS1	AR	7.3	18	19.7
NPHS2	AR	9.9	11	12.7
WT1	AD	4.8	3	5.3
PLCE1	AR	2.1	15	13.3
LAMB2	AR	1.1	8	6.0
SMARCAL1	AR	1.1	11	17.6
INF2	AD	0.5	1	3.6
TRPC6	AD	0.5	1	1.0
COQ6	AR	0.5	1	1.8
ITGA3	AR	0.3	3	–
MYO1E	AR	0.3	3	4.2
QOQ2	AR	0.2	1	1.8
LMX1B	AD	0.2	3	3.7
ADCK4	AR	0.2	–	3.7
DGKE	AR	0.1	1	-
PDSS2	AR	0.1	1	0
ARHGAP24	AD	0.6	–	–
ARHGDIA	AR	0.1	–	–
ITGB4	AR	0.1	–	–

AR, Autosomal Recessive; AD, Autosomal Dominant.

diagnosis in these cases may be diffuse mesangial sclerosis (DMS) or FSGS [54]. Loss of WT1 in mature podocytes modulates podocyte Notch activation, which could mediate early events in WT1-related glomerulosclerosis [55].

3.5.1 Denys–Drash Syndrome

DDS (OMIM 194080) is a combination of NS showing a histopathologic picture of DMS, male pseudohermaphroditism, and Wilms tumor. Male pseudohermaphroditism in most published cases refers to an XY karyotype, some testicular tissue elsewhere than in the scrotum, variable degrees of ambiguous genitalia, such as hypospadia and cryptorchidism in children appearing like boys, or hypertrophy of the clitoris in phenotypic girls. Some cases with an XX karyotype and female phenotype have been reported. Three clinical categories of DDS have been noted: genotypic males with all three abnormalities, genotypic males with nephropathy and ambiguous external and/or internal genitalia only, and genotypic females with nephropathy and Wilms' tumor only [53].

DDS is caused by dominant, heterozygous pathogenic variants in *WT1*. Both familial and de novo germline pathogenic variants have been described in DDS patients. Most are missense pathogenic variants within exon 8 and 9, coding for zinc finger domains 2 and 3, which leads to alteration in the DNA-binding capacity of WT1. It is believed that the mutant protein actively suppresses the normal allele, which explains the more severe phenotype seen in DDS compared with children with complete deletion of one *WT1* allele. In most patients, the nuclear expression of WTI is absent or reduced in podocytes. The podocyte function is affected and, at the molecular level, upregulation of PAX2 and downregulation of nephrin have been reported [56]. Also, the expression of growth factors that regulate glomerular capillary development is affected in DDS [57,58].

WT1 analysis is important in young patients with NS for early detection and tumor prophylaxis. The nephropathy is usually discovered at the age of a few months, sometimes at birth. To avoid the development of Wilms tumor, bilateral nephrectomy at the onset of terminal renal failure is recommended. Removal of native kidneys has been suggested for all patients with nephropathy caused by *WT1* mutations [59].

3.5.2 Frasier Syndrome

FS (OMIM 136680) is characterized by the association of male pseudohermaphroditism and glomerulopathy. There is complete male-to-female gender reversal in 46,XY patients. FS is associated with gonadoblastomas, but not with Wilms tumor. Proteinuria is detected in childhood, usually between 2 and 6 years of age, and kidney biopsy reveals FSGS. The renal disease does not respond to medical therapy, but it has a slower progressive course to renal failure than DDS.

FS is caused by point mutations in intron 9 in *WT1*, which interferes with the recognition of the second splice donor site of *WT1*. This results in loss of the lysine-threonine-serine (KTS) containing isoforms of WT1. How this change in the relative expression of +KTS and −KTS isoforms results in the severe developmental defects in FS is not completely understood. Also, pathogenic variants in exon 9 that do not alter the +KTS isoform expression have been reported in FS patients. On the other hand, typical FS pathogenic variants have been observed in DDS patients. Based on this overlap, both diseases should be considered as part of a spectrum of *WT1* gene pathogenic variants, rather than as separate entities. Chernin et al. followed 19 patients with mutations in intron 9 splice site (KTS mutations), 27 patients with missense pathogenic variants, and six patients with other variants. Totally 24 different *WT1* pathogenic variants were detected. Sixteen of the 19 patients with KTS mutations were females. The results showed that KTS pathogenic variants cause isolated NS with absence of Wilms tumor in 46,XX females. On the other hand, these pathogenic variants cause FS with gonadoblastoma risk in 46,XY phenotypic females [60].

3.5.3 Isolated Kidney Disease

Originally, three studies reported *WT1* pathogenic variants in patients with isolated DMS. Since then, the incidence of WT1 mutations was evaluated in a large cohort of sporadic and familiar SRNS and SSNS [61]. Pathogenic variants in exons 6–9 of *WT1* were identified in 8 of the 115 SRNS patients (7%). In two females, pathogenic variants in exon 9 consistent with the diagnosis of DDS were identified; one of them had Wilms tumor. In three male and three female patients, splice site variants in exon 9 consistent with the diagnosis of FS were found. Two of the male patients presented with urinary or genital malformations, one of them with sexual reversal and bilateral gonadoblastoma. All three female patients presented with isolated FSGS. No pathogenic variants were found in 110 patients with sporadic SSNS.

WT1 pathogenic variants were found in around 5% of SRNS patients in three large cohorts (Table 3.1). Most patients were small children at the onset, but pathogenic variants were found at any age. In a European registry data on SRNS, WT1 patients more frequently exhibited rapid disease progression than patients with SRNS from other causes [62]. FSGS was equally prevalent histological diagnosis, but DMS was quite specific for WT1 disorder and was present in a third of the cases. Sex reversal and/or urogenital abnormalities (52%), Wilms tumor (38%), and gonadoblastoma (5%) were almost exclusive to WT1 disease. Patients with truncating mutations seem to have a milder renal phenotype and develop proteinuria and ESRD later, although the risk for bilateral Wilms tumor is higher compared with missense variants. KTS pathogenic variants are most likely to present as isolated SRNS, FSGS on biopsy, and slow progression. Missense substitutions affecting DNA-binding residues are associated with DMS histology (74%), early STNS onset, and rapid progression to ESRD. Truncating variants confer the highest Wilms tumor risk but typically late-onset SRNS.

3.6 PHOSPHOLIPASE ε1GENE (*PLCE1*) MUTATIONS

PLCε1 belongs to the phospholipase family of proteins that catalyze hydrolysis of phosphoinositides, generating messenger such as IP3 and DAG, which are involved in a wide spectrum of cellular functions [63]. Phospholipase C isoenzymes can be activated by G-protein-coupled receptors. The main action of IP3 is the stimulation of calcium release from intracellular storage pools, which could be involved in cytoskeletal reorganization and modulate the function of other signal transduction cascades. PLCε1 was found to associate with IQGAP1, a protein that has been shown to interact with nephrin [64]. Moreover, PLCe1 seems to activate TRPC6 and interact with actin binding protein advillin. Defects in both molecules may cause SRNS [65].

In 2006, Hinkes et al. identified pathogenic variants in *PLCE1* as causing early onset NS (OMIM 610725). The age at onset in the 12 patients varied from 2 months to 4 years and the age at ESRD from five to over 13 years. No extrarenal manifestations were observed. Kidney histology showed DMS in most patients. Two siblings with a missense variants had characteristics of FSGS. Two other affected individuals responded to immunosuppressive therapy, making this the first report of a molecular cause of NS that may resolve after therapy. The study showed that PLCε1 is expressed in developing and mature podocytes, and the DMS histology possibly represents developmental arrest [66].

A subsequent study identified 40 children from 35 families with idiopathic DMS from a worldwide cohort of 1368 children with NS [67]. The age of onset varied from 1 month to 6 years. Truncating variants in *PLCE1* were detected in 28.6% of the families and, interestingly, *WT1* variants in only 8.5% of the families. Except in one family, all the mutations detected were homozygous loss of function variants. One child had two compound heterozygous mutations. Fourteen children were placed on corticosteroid or cyclosporine therapy and no one responded. Age at ESRD varied from 8 months to 5 years.

So far, the results show that *PLCE1* is a major gene causing isolated DMS, and subjects with this type of histological lesion should be screened for *PLCE1* mutations. The mutations may also lead to FSGS with a relatively late onset of proteinuria. An infant with rapidly progressive FSGS has, however, also been reported. In the three recent worldwide cohorts, *PLCE1* pathogenic variants were found in 2.1%–15% of patients and families with SRNS (Table 3.1). Most patients were infants at the time of diagnosis. Interestingly, some individuals with *PLCE1* pathogenic variants may also remain asymptomatic, implying that there may be modifier genes that interact with *PLCE1* to cause DMS/FSGS.

3.7 LAMININ-β2 GENE (*LAMB2*) MUTATIONS

Laminin-β2 is one of the laminin chains and component of laminin-521. It is specifically expressed in the GBM and at some other sites such as intraocular muscles and neuromuscular synapses. Laminin is able to polymerize in a regular manner and has a critical role in maintaining the structural properties of the GBM.

Laminin-521 also interacts with its receptor integrin α3β1, which links the GBM to the actin cytoskeleton of the podocyte [68].

In 2004, Zenker et al. reported that pathogenic variants in the *LAMB2* gene, encoding the laminin-β2 chain, are associated with Pierson syndrome [69]. This syndrome (OMIM 609049) is a rare autosomal recessive disorder characterized by early-onset NS with variable ocular and neurologic defects. The typical ophthalmic sign is a fixed narrowing of pupils (microcoria), which is due to a defect of the dilatator pupillae.

After the original observation, milder variants of the syndrome have been reported with less prominent extrarenal abnormalities [70]. Matejas et al. reviewed the findings in 51 patients from 39 families with *LAMB2* mutations [71]. The majority (71%) of the mutations was truncating and was evenly distributed along *LAMB2*. The rest (29%) were missense variants causing amino acid changes in the N-terminal region of laminin-β2, which is critical area for interacting with neighboring laminin. The age of the patients at diagnosis was mostly less than 3 months and was over 1 year in only four patients. The age at onset of ESRD was mainly under 1 year (from 1 week to 16 years). Renal biopsy showed DMS histology in 73%, FSGS in 14%, and minimal changes in 8% of the patients. Neurodevelopmental deficits were observed in 82% of the patients, and all but two patients had some ocular abnormalities.

Among patients with isolated NS, *LAMB2* mutations are apparently rare. They were observed in only 2.5% (2/80) of the European patients with congenital or infantile NS [33]. Similarly, no *LAMB2* pathogenic variant was observed in 33 patients with SRNS and FSGS histology and ocular involvement in eight patients [72]. On the other, *LAMB2* pathogenic variants were found in 1.2%–8.0% of patients in three large cohorts (Table 3.1). Most patients were less than 3 months of age.

3.8 INVERTED FORMIN 2 GENE (*INF2*) PATHOGENIC VARIANTS

INF2 is a member of formins that accelerate cytoplasmic filament nucleation and elongation. Formins are widely expressed proteins governing several dynamic events that require remodeling of the actin cytoskeleton such as cell polarity, morphogenesis, and cytokinesis. INF2 has the unique ability to accelerate both actin polymerization and depolymerization [73]. The regulatory mechanisms controlling INF2 and its cellular function are not well known. INF2 is a widely expressed protein with high expression in the kidney, including glomerular podocytes.

In 2010, Brown et al. detected nine independent nonconservative missense *INF2* variants in a total of 72 individuals with autosomal dominant form of SRNS from 11 families coming from United States, Canada, and Mexico [74]. Patients with *INF2* pathogenic variants presented proteinuria in adolescence or adulthood (age range 11–67 years), typically with moderate or nephrotic range proteinuria, but not with NS. The disease was progressive, often leading to ESRD (age range 13–67 years). All pathogenic variants altered highly conserved amino acid residues in diaphanous inhibitory domain of the IFN2 protein. More recently, the prevalence of *INF2* pathogenic variants was analyzed in a cohort of 54 families (78 patients) with proteinuric disorder of apparent autosomal dominant inheritance [75]. Seven missense variants were found in nine families (28 patients), which translates to a detection rate of 16.7%. Median age at onset of proteinuria was 27 years, and NS was noted in four patients, and ESRD developed at the median age of 36 years (20–70 years). Significant intrafamilial phenotypic variability was evident with a wide range of age at presentation and ESRD.

INF2 pathogenic variants have been identified in both isolated FSGS and Charcot–Marie–Tooth disease, which is a hereditary motor and sensory neuropathy characterized by peripheral nerve demyelination [76]. *INF2* pathogenic variants were identified in 32 patients from 28 families with dominant SRNS [77]. Clinical features included mild proteinuria (range 1–2.5 g/L) and hematuria as the first recognized sign at the age of 8–30 years. Four patients had symptoms of Charcot–Marie–Tooth neuropathy. Eighteen patients developed ESRD in their third decade. While *INF2* pathogenic variants are rarely detected in small children, they are quite common in adolescents and adults with a dominant SRNS (Table 3.1). In a large registry study, *INF2* pathogenic variants were detected in 20% of dominant form of SRNS patients. The age range at the time of diagnosis was 10–16 years [78].

3.9 TRANSIENT RECEPTOR POTENTIAL C6 ION CHANNEL GENE (*TRPC6*) PATHOGENIC VARIANTS

TRPC6 is a receptor-operated cation channel that contributes to changes in the cytosolic free Ca^{2+}-concentration. TRPC6 is expressed in the podocyte and clustered in the podocin–nephrin–lipid complex. It has been speculated that TRPC6 could be involved in monitoring the integrity of the SD [79].

The alteration of TRPC6 function may impair the cytoskeletal adaptive response of podocytes to injury, eventually leading to progressive damage. Nephrin binds to phosphorylated TRPC6 via its cytoplasmic domain, competitively inhibiting TRPC6 expression and activation [80]. FSGS-associated gain-of-function variants render the mutated TRPC6 insensitive to nephrin suppression, thereby promoting their surface expression and channel activation.

TRPC6 pathogenic variants were originally reported a subset of patients with autosomal dominant form of FSGS [81,82]. A missense variant in *TRPC6* was reported to be a cause of familial FSGS in a large New Zealand family with FSGS. Subsequently, other families with different *TRPC6* pathogenic variants have been reported in both adults and children [83,84]. *TRPC6* variants has also be detected in children with early-onset and sporadic SRNS [85]. In recent worldwide surveys, *TRPC6* pathogenic variants were found in 0.5%–1.0% of the patients with SRNS (Table 3.1). The affected individuals are typically young adults with high-grade proteinuria. The incomplete penetrance of *TRPC6* variants to some extent limits its diagnostic value.

3.10 TYPE IV COLLAGEN GENE (*COL4A3-5*) PATHOGENIC VARIANTS

Type IV collagen forms the backbone of the glomerular basement membrane (GBM). Each collagen IV molecule is a heterotrimer, which in the mature GBM contains α-chains α3:α4:α5. While *COL4A5* gene encoding for α5-chain is located in chromosome X, *COL4A3* and *COL4A4* are mapped to chromosome 2. Pathogenic variants in any of the *COL4A3-5* genes are responsible for Alport syndrome characterized by hematuria, renal failure, hearing loss, eye changes, and sometimes leiomyomatosis.

Somewhat surprisingly, the NGS technology has revealed that *COLA3-5* pathogenic variants are also important cause of adult onset SRNS with FSGS histology. A series of 70 families with FSGS revealed recessive and dominant *COL4A3/4* variants in 10% of [86]. The age range was 10–64 years at the time of diagnosis, but most were adults. The patients developed ESRD without recurrence after kidney transplantation. *COL4A3* pathogenic variants have occurred in 5 of 40 Chinese families with FSGS (12.5%), and in one of 50 sporadic cases [87]. In a more recent analysis, collagen pathogenic variants were identified in 38% of families with familial FSGS, and 3% with sporadic FSGS, with over half the variants occurring in *COL4A5* [88]. Patients with pathogenic variants both in *COL4A3* and *NPHS1* or *NPHS2* genes have also been reported [89]. Overall, *COL4A3-5* pathogenic variants are an important cause of isolated SRNS especially in adult patients, and the inheritance can be recessive or dominant and as well as X-linked or autosomal.

3.11 MUTATIONS IN GENES ENCODING MITOCHONDRIAL PROTEINS

The development of NS has been found in rare cases of mitochondrial disorders. Genetic defects of coenzyme Q_{10} biosynthesis can cause SRNS as part of multiorgan involvement but may cause isolated SRNS. Mutations in *COQ2*, located on chromosome 4q21, are associated with early onset of collapsing glomerulopathy-associated SRNS, with or without extrarenal symptoms (OMIM 607426) [90,91]. Also, variants in decaprenyl diphosphate synthase subunit 2 (*PDSS2*), which is located on chromosome 6q21, led to both CoQ10 deficiency and NS with Leigh syndrome (OMIM 607426) [92]. Pathogenic variants in mitochondrially encoded tRNA leucine 1(MT-TL1) cause MELAS syndrome, which is sometimes associated with SRNS, characterized histologically by FSGS and extensive obliteration of podocyte foot processes [93].

COQ8B (AarF domain containing kinase, ADCK4) is a mitochondrial enzyme also taking part in coenzyme Q10 (CoQ10) synthesis. Recent studies show that *ADCK4* pathogenic variants can cause glomerulopathy especially in adolescents with SRNS and renal failure [94,95]. Recessive variants in *ADCK4* occurred in 26 patients from 12 families. They mostly had isolated kidney disorder, with three subjects exhibiting mild neurological symptoms (occasional seizures, mental retardation, retinitis pigmentosa). Interestingly, six Korean children (5–14 years) with *ADCK4* pathogenic variants and moderate proteinuria showed medullary nephrocalcinosis, but no extrarenal manifestations. ESRD developed in 4 years [96]. *ADCK4* pathogenic variants have also been reported in 15 patients with SRNS from eight unrelated families. These patients showed decreased mitochondrial activity, which in one patient could be reversed by ubiquinone therapy [97].

3.12 RARE GENETIC DISORDERS OF THE SD COMPLEX

CD2-associated protein (CD2AP) is an adapter protein, which in the kidney localizes to the slit diaphragm of the podocyte, where it links podocin and nehrin to the phosphoinositide 3-OH kinase to form a signaling complex [22]. CD2AP-deficient mice develop severe proteinuria and renal dysfunction shortly after birth [98]. Individuals with homozygous mutations in *CD2AP* are present with early onset FSGS (OMIM 607832) [99]. A homozygous variant (p.R612X) was identified in an infant who presented with NS and severe glomerular sclerosis on renal biopsy. The truncated protein displayed a dramatic reduction of actin-binding efficiency in vitro [100]. Homozygous *CD2AP* variants, however, are rare and not reported in the large international cohorts (Table 3.1). A heterozygous pathogenic variant in *CD2AP* is a predisposing factor toward developing a late-onset FSGS. However, heterozygous *CD2AP* pathogenic variants have been found in unaffected individuals and the causal link between heterozygous *CD2AP* variants and FSGS is not yet clear.

Nonmuscle class I myosin, myosin 1E (*MYO1E*), represents a component of the SD complex and may contribute to regulating junctional integrity in podocytes [101]. Myosin 1E expression is believed to be necessary for normal glomerular filtration and *MYO1E* mutations have been reported in solitary cases with SRNS [102]. described two such patients. The first patient presented with nephritic-range proteinuria, microhematuria, hypoalbuminemia, and edema at 9 years, and the other had edema, nephrotic-range proteinuria hematuria, and hypoalbuminemia at 1 year of age. A kidney biopsy at 4 years showed 20% of glomeruli with FSGS. Also, an 8-year-old girl presented with ESRD and an infant brother presented with nephrotic syndrome, progressing to ESRD by 3 years of age. Both were subsequently found to have homozygous variants of MYO1E [103].

Membrane-associated guanylate kinase, WW, and PDZ domain-containing 2 (MAGI2) interact with nephrin and regulates podocyte cytoskeleton and SD dynamics. Two siblings and one sporadic case with autosomal recessive pathogenic variants were detected in a large pediatric SRNS cohort from the United Kingdom. All three patients had congenital NS. One patient had no extrarenal manifestations, the second patient had minor cardiac abnormalities, and the third one had polydactyly and pyloric stenosis without other syndromic features [6].

3.13 RARE GENETIC DISORDERS OF THE ACTIN NETWORK

Alpha-actinin-4 is an actin filament cross-linking protein important for the integrity of the cytoskeleton in podocyte foot process. Pathogenic variants in *ACTN4* are associated with an autosomal dominant form of familial FSGS (OMIM 603278) [104]. The mutated ACTN4 proteins showed higher binding affinity to F-actin, which may change the mechanical characteristics of the podocyte. By now, only a few *ACTN4* missense variants have been described [105]. Most reported patients show a mild to moderate degree of proteinuria during adolescence or later, and some patients gradually progress to ESRD. In contrast to this, Choi et al. reported a familial case of FSGS in which two affected siblings showed rapidly progressing NS in early childhood [106]. Renal pathological findings were of an FSGS collapsing variant.

ARHGDIA encodes Rho (guanosine diphosphate) dissociation inhibitor (RhoGDia, OMIM601925), which interacts with RHO GTPases that are critical regulators of the actin cytoskeleton. The protein is highly expressed in podocytes and depletion of *ARHGDIA* causes early-onset NS in mice. Gee et al. described four patients with recessive variants in *ARHDGIA*. The patients were from 2 weeks to 2.4 years of age. Two had intellectual disability. Microscopy showed DMS histology, and all patients developed ESRD in 6 weeks to 2.4 years [107].

Similarly, Gupta reported a homozygous in-frame deletion in two sisters with CNS and heavy proteinuria already during the neonatal period. Both showed DMS histology in kidney biopsy and fast development of ESRD [108].

3.14 MUTATIONS IN PODOCYTE NUCLEAR PROTEINS

In the nail-patella syndrome, NS occurs in a minority of patients and is accompanied by changes in the expression of the GBM components and some podocyte proteins. This syndrome is caused by pathogenic variants in *LMX1B*, which codes for a transcription factor strongly expressed in podocytes [109]. The link between *LMX1B* variants and the GBM alterations is not completely resolved [110].

Galloway–Mowat syndrome (GMS) is characterized by NS with central nervous system anomalies including microcephaly, psychomotor retardation, and macroscopic and microscopic brain anomalies [111]. Other extrarenal features have also been reported, such as hiatus hernia, congenital spondylorhizomelic shortness, and diaphragmatic defects. NS appears usually at the age of a few months (0–34 months), and kidney biopsy may show MCNS, FSGS, DMS, or mesangioproliferative glomerulonephritis. Recently, homozygous truncating as well as missense variants in *WDR73* were identified in families with GMS [112]. *WDR73* encodes a 378 amino acid protein with six WD40 repeats. *WDR73* is expressed in the brain and kidney, and the protein is predicted to function as scaffolds for the assembly of protein complexes. *WDR73* mutations are limited to those with classical GMS features [113].

More recently, pathogenic variants in KEOPS complex have also been associated with GMS. The KEOPS complex contains four subunits LAGE3, OSGEP, TP53RK, and TPRKB. They regulate chemical modification of tRNAs necessary for translational accuracy and efficiency. Other known functions relate to control of telomere length, genome maintenance, and regulation of gene transcription. Individuals with pathogenic variants in any of the four KEOPS genes had primary microcephaly, developmental delay, propensity for seizures, and NS of early onset. Most patients died in early childhood. Several individuals were noted to have facial dysmorphism sometimes with features of progeria and skeletal abnormalities such as arachnodactyly. The most frequently observed brain anomalies included a spectrum of gyration abnormalities ranging from lissencephaly to pachygyria and polymicrogyria and cerebellar hypoplasia [114].

Schimke immuno-osseous dysplasia (SIOD, MIM 242900) is an autosomal-recessive multisystem disorder with growth failure, immunodeficiency, cerebral vascular complications, and progressive SRNS with FSGS histology. Approximately half of patients with SIOD have biallelic pathogenic variants in *SMARCAL1* (SWI/SNF-related matrix-associated actin-dependent regulator of chromatin, subfamily a-like 1), which encodes the DNA translocase SMARCAL1 [115]. Many homozygous/heterozygous missense variants lead to a severe phenotype. These children develop nephrotic syndrome and renal insufficiency by preschool age and suffer from life-threatening neurologic complications such as transient ischemic attacks or cerebral infarctions. On the other hand, large number of biallelic missense variants are associated with mild clinical symptoms. These patients do not develop cerebral vascular complications until at least early adulthood [116,117]. Interestingly, a 5-year-old girl's nephrotic range proteinuria decreased after commencement of ACE inhibitors and sartans [118].

3.15 POLYMORPHIC GENE VARIANTS IN NS

SRNS with FSGS histology is more common in Blacks than in other ethnic groups. Forty percent of FSGS occurs in Blacks and over 70% is associated with genetic variants of apolipoprotein 1 (*APOL1*) gene in the United States. *APOL1*-associated FSGS is a major form of FSGS in subjects of sub-Saharan African descent [119]. APOL1 risk allele G1 comprises two missense variants (S342G and I384M), and risk allele G2 has six base pair in-frame deletion (N3988del:Y389del). GI and G2 are mutually exclusive and do not occur in the same chromosome. While about half of the African Americans with FSGS have the G1 risk allele, it can be found 18%–23% of those without FSGS. The prevalence of the G2 risk allele in African Americans with FSGS is 23% and 15% in those without FSGS [120]. Overall, individuals with two *APOL1* risk alleles have an estimated 4% lifetime risk for developing FSGS. *APOL1* variants conferred 17-fold higher odds for FSGS in an American cohort. *APOL1* variants also have a tendency to manifest FSGS at relatively young age so that FSGS begins between the ages of 15–39 years in 70% of individuals with *APOL1* risk alleles and 42% of those with 0 or 1 risk alleles. *APOL1*-associated disorder can present as a primary FSGS or as an NS recurrence after kidney transplantation [121].

In steroid sensitive NS (SSNS) as well as in a majority of SRNS cases, monogenic pathogenic variants have not been identified in any gene. The possible association of polymorphic gene variants with SSNS has been analyzed in several studies. These genes include glucocorticoid-induced transcript 1 gene (*GLCCI*), soluble urokinase plasminogen activator receptor (suPAR), Fc gamma receptors (*FCGR2A, FCGR2B, FCGR3A, FCGR3B*), HLA-DR-allotypes (e.g., *DQB1*02*), multidrug resistance-1 (*MDR1 or ABCB1*) gene, angiopoietin-like 4 (*ANGPTL4*), glypican 5 (*GPC5*), interleukin-13 (*IL-13*), macrophage migration inhibitory factor (*MIF*), neural nitric oxide synthetase (*nNOS*), and nuclear receptor subfamily-3 (*NR3C1*) gene. No clinically relevant associations, however, have so far been observed [122].

3.16 DIAGNOSIS OF NS

Heavy proteinuria, hypoproteinemia, hyperlipidemia, and edema are the cardinal signs of NS. The serum albumin level is typically <10–20 g/L and urinary protein 2–100 g/L (after serum albumin has been corrected >15 g/L). The severity of proteinuria, however, shows variation and can be quite modest in the beginning. Urine erythrocyte count in NS patients is mostly low, but there is variation depending on the etiology of NS. Serum creatinine and urea can be normal in the beginning but both show elevation with time as the renal failure develops. SRNS associated with DMS or FSGS often progresses to ESRD quite rapidly [13].

Diagnostics of NS should happen in a clinic with ample experience on NS and rare syndromes, if possible. In principle, two approaches lead toward as exact diagnosis as possible. One is to do comprehensive clinical investigations, including search for extrarenal findings to detect the syndromic forms of NS and histologic investigation of kidney biopsy. When the most likely entity has thus been identified, a specific genetic test could be performed. Another approach is to use a comprehensive NGS gene panel with genes known to cause NS and hope for exact etiologic diagnosis in this way. Both approaches can also be taken simultaneously. At present, the best approach depends on the traditions and resources of the clinic concerned but it can be foreseen that starting diagnostics with a gene panel or an exome wide test are becoming the first choice [10].

Genetic NS is often an isolated kidney disease with no extrarenal manifestations (i.e., *NPHS1, NPHS2, PLCe1, MYO1E, ARHGDIA, TRPC6,* and many *WT1, LAMB2,* and *COL4A3-5* gene disorders). Searching for abnormalities in genitalia (karyotype), urinary tract, central nervous system, eyes, ears, heart, bone, growth, bones, and heart is, however, important for the management of the patient. In newborns, placental weight >25% of birth weight is typically present in classic CNF, but may be seen in other entities with congenital NS. The indications for renal biopsy are not clear. Histological lesions, such as FSGS, MCNS, and DMS, are not specific for the gene defect involved. On the other hand, histology tells about the severity of the histological lesions that helps in deciding the treatment strategies. The diagnosis is helpful in assessing the therapy and prognosis and in genetic counseling of the family. Unnecessary treatment of SRNS patients with immunosuppressive drugs may be avoided if a genetic defect is found.

3.17 MANAGEMENT OF PATIENTS WITH NEPHROTIC SYNDROME

3.17.1 Supportive Therapy

The magnitude of the protein losses into urine is crucial for therapeutic decision-making in patients with NS. Heavy and constant proteinuria (5–100 gr/L) inevitably leads to life-threatening edema, and protein substitution by parenteral albumin infusions is mandatory [36]. On the other hand, patients with moderate proteinuria (1–5 gr/L) may manage without albumin substitute, especially if proteinuria is associated with renal failure and reduced urinary output. Patients with constant heavy proteinuria often have low levels of serum thyroid-binding globulin and thyroxine substitution is recommended. Imbalance of plasma coagulation factor levels contribute to hypercoagulability and risk for thrombosis, and the use of warfarin, aspirin, and dipyridamole therapy has been recommended. Antithrombin III (50 U/kg) may be given before surgical procedures. Because of urinary losses of immunoglobulin, prophylactic use of antibiotics and immunoglobulin has been recommended, but in our experience, they are not helpful and may induce resistant bacterial strains. A high degree of suspicion for septic infections is warranted in NS patients. Infants with severe NS have traditionally been treated with a high-energy (130 kcal/kg per day) and a high-protein (3–4 g/kg per day) diet. Glucose polymers are given to increase energy intake, and a mixture of rapeseed and sunflower oil is given to balance lipid levels. The children also receive vitamin D_2 (400 IU/day), multivitamin preparations, magnesium (50 mg/day), and calcium (500–1000 mg/day). Most infants need a nasogastric tube to guarantee the energy intake.

3.17.2 Medication

In nongenetic forms of NS (SSNS and SRNS) immunosuppressive medication with prednisone, calcineurin inhibitor (cyclosporine A, tacrolimus), antimetabolite (cyclophosphamide, chlorambucil, mycophenolic acid), or anti-CD20 antibodies (rituximab) often bring the disease in remission. Use of immunosuppressive drugs in hereditary forms of NS has traditionally been regarded unhelpful [123,67]. Few patients with proven hereditary SRNS reached partial remission by cyclosporine A (CsA) [66,124]. In one study seven patients with genetic SRNS received CyA and two went into partial remission. Both had pathogenic variants in *WT1* [125]. In a more recent study of 32 children with genetic SRNS, complete remission was achieved by CsA in one patient (3%) but no response was seen in 26 patients (81%). Five patients (16%) responded partially to CsA but developed rapidly renal failure [126]. CsA and perhaps also glucocorticoids have a direct effect on podocyte cell signaling and maintenance of actin network, which may explain the few cases with beneficial effect [127,128].

Reduction of protein losses by antiproteinuric drugs, ACE inhibitors, and indomethacin has been reported [36]. These drugs lower the perfusion pressure in the glomerulus and have direct effects on the podocyte functions, e.g., nephrin expression [129], which lead to reduced protein leakage. In our experience, patients who have severe *NPHS1* mutations (such as Fin-major and Fin-minor) do not respond to this medication. On the other hand, patients with mild missense variants in podocyte genes may show reduced proteinuria and treatment with "antiproteinuric therapy" with an ACE-inhibitor or/and AT2 blocker is worth trying. Calcium signaling is important for the podocyte functions, and its manipulation with new drugs may in the future be a part of the therapy [130].

Pathogenic variants in genes associated with mitochondrial function (*COQ2, COQ6, ADCK4,* or *PDSS2*) lead to coenzyme Q10 deficiency, abnormalities in podocytes and proteinuria. Coenzyme Q10 is synthesized de novo in the mitochondria and plays a crucial role in the generation of ATP. Several small studies suggest that CoQ10 supplementation in such patients at an early stage may ameliorate proteinuria [97,131].

3.17.3 Kidney Transplantation

Renal transplantation is an established mode of therapy for children with NS not responding to medical therapy. Overall, the results of kidney transplantation in NS are quite similar to those obtained in other etiologies. Patient survival at 5 years is over 90% and graft survival over 80% in registry databases and in single centers [132]. Recurrence of NS in the graft is rare but has occurred in 30% of the CNF children homozygous for Fin-major mutation. Antinephrin antibodies have been observed in these children after transplantation. Treatment of the recurrence with cyclophosphamide, plasmapheresis, and anti-CD20 often leads to remission [133,134]. The recurrence rate of NS in FSGS is in general about 30%. However, in genetic FSGS, this proportion is much lower, which is an important aspect when planning renal replacement therapy.

3.18 CONCLUSIONS

Our knowledge on the genetic basis of NS has greatly increased during the past decade. Over 50 genetic defects responsible for isolated and syndromic steroid resistant nephrotic syndrome have been detected. Podocyte proteins play an important role in the glomerular sieving and their genetic defects result in proteinuria and NS. Pathogenic variants in *NPHS1* lead especially to congenital NS, whereas NPHS2 variants can cause NS at any age. The third important gene is *WT1*, which may cause syndromic and isolated NS. In addition, several other podocyte genes are responsible for rare cases of early and late-onset NS. Gene diagnostics is crucial in diagnostics of NS cases not responding to medical therapy.

REFERENCES

[1] Chen Y, Liapis H. Focal segmental glomerulosclerosis: molecular genetics and targeted therapies. BMC Nephrol 2015;16:101–11.
[2] Bierzynska A, Saleem M. Recent advances in understanding and treating nephrotic syndrome. F100Research 2017;6:121.
[3] Liu J, Wang W. Genetic basis of adult-onset nephrotic syndrome and focal segmental glomerulosclerosis. Front Med 2017;11:333–9.
[4] Boyer O, Dorval G, Servais A. Hereditary podocytopathies in adults: the next generation. Kidney Dis 2017;3:50–6.
[5] Rosenberg A, Kopp J. Focal segmental glomerulosclerosis. Clin J Am Soc Nephrol 2017;12:502–17.
[6] Bierzynska A, McCarthy H, Soderquest K, et al. Genomic and clinical profiling of a national nephrotic syndrome cohort advocates a precision medicine approach to disease management. Kidney Int 2017;91:937–47.
[7] Bierzynska A, Soderquest K, Dean P, et al. MAGI2 mutations cause congenital nephrotic syndrome. J Am Soc Nephrol 2017;28:1614–21.
[8] Sadowski C, Lovric S, Ashraf Sal. A single-gene cause in 29.5% of cases of steroid –resistant nephrotic syndrome. J Am Soc Nephrol 2015;26:1279–89.
[9] Trautman A, Bodria M, Ozaltin F, et al. Spectrum of steroid-resistant and congenital nephrotic syndrome in children: the podoNet Registry Cohort. Clin J Am Soc Nephrol 2015;10:592–600.
[10] Warejko J, Tan W, Daga A, et al. Whole exome sequencing of patients with steroid-resistant nephrotic syndrome. Clin J Am Soc Nephrol 2018;13:53–62.
[11] Cil O, Perward F. Monogenic causes of proteinuria in children. Front Med 2018;5:55–60.
[12] Dámico G, Bazi C. Pathophysiology of proteinuria. Kidney Int 2003;63:809–25.
[13] Jalanko H. Pathogenesis of proteinuria: lessons learned from nephrin and podocin. Pediatr Nephrol 2003:487–91.
[14] Suh J, Miner J. The glomerular basement membrane as a barrier to albumin. Nat Rev Nephrol 2013;9:470–7.
[15] Tryggvason K, Patrakka J, Wartiovaara J. Hereditary proteinuria syndromes and mechanisms of proteinuria. N Engl J Med 2006;354:1387–401.
[16] Inoue T, Yaoita E, Kurihara H, et al. FAT is a component of glomerular slit diaphragms. Kidney Int 2001;59:1003–12.
[17] Patrakka J, Tryggvason K. Molecular make-up of the glomerular filtration barrier. Biochem Biophys Res Commun 2010;396:164–79.
[18] Wartiovaara J, Ofverstedt L, Khoshnoodi J, et al. Nephrin strands contribute to a porous slit diaphragm scaffold as revealed by electron tomography. J Clin Invest 2004;114:1475–83.
[19] Gerke P, Huber T, Sellin L, Benzing T, Walz G. Homodimerization and heterodimerization of the glomerular podocyte proteins nephrin anf NEPH1. J Am Soc Nephrol 2003;14:918–26.
[20] Huber T, Harteleben B, Kim J, et al. Nephrin and CD2AP associate with phosphoinositide 3-OH kinase and stimulate AKT dependent signaling. Mol Cell Biol 2003;23:4917–28.
[21] Huber T, Simons M, Hartleben B, et al. Molecular basis of the functional podocin-nephrin complex: mutations in the NPHS2 gene disrupt nephrin targeting to lipid raft microdomains. Hum Mol Genet 2003;12:3397–405.
[22] Schwarz K, Simons M, Reiser J, et al. Podocin, a raft-associated component of the glomerular slit diaphragm, interacts with CD2AP and nephrin. J Clin Invest 2001;108:1621–9.
[23] Huber T, Schermer B, Benzing T. Podocin organizes ion channel–lipid supercomplexes: implications for mechanosensation at the slit diaphragm. Nephrol Exp Nephrol 2007;106:e27–31.
[24] Yao J, Le T, Kos C, Henderson J, Allen P, Denker B, Pollak M. α-actinin-4-mediated FSGS: an inherited kidney disease caused by an aggregated and rapidly degraded cytoskeletal protein. PLoS Biol 2004:787–94.
[25] Lahdenkari A, Lounatmaa K, Patrakka J, et al. Podocytes are firmly attached to glomerular basement membrane in kidneys with heavy proteinuria. J Am Soc Nephrol 2004;15:2611–8.
[26] Fukusumi Y, Zhang Y, Yamagishi R, et al. Nephrin-binding ephrin-B1 at the slit diaphragm controls podocyte function through the JNK pathway. J Am Soc Nephrol 2018;29(5):1462–74.
[27] New L, Martin C, Scott R, et al. Nephrin tyrosine phosphorylation is required to stabilize and restore podocyte foot process architecture. J Am Soc Nephrol 2016;27:2422–35.
[28] Kestilä M, Lenkkeri U, Männikkö M, et al. Positionally cloned gene for novel glomerular protein - nephrin-is mutated in congenital nephrotic syndrome. Mol Cell 1998;1:575–82.
[29] Beltcheva O, Lenkkeri U, Kestilä M, Tryggvason K. Mutation spectrum in the nephrin gene (NPHS1) in congenital nephrotic syndrome. Hum Mutat 2001;17:368–73.
[30] Koziell A, Grech V, Hussain S, Lee G, Lenkkeri U, Tryggvason K, Scambler P. Genotype/phenotype correlations of NPHS1 and NPHS2 mutations in

nephrotic syndrome advocate a functional inter-relationship in glomerular filtration. Hum Mol Genet 2002;11:379–88.

[31] Liu L, Done S, Khoshnoodi J, Bertorello A, Wartiovaara J, Berggren P, Tryggvason K. Defective nephrin trafficking caused by missense mutations in the NPHS1 gene: insight in to the mechanisms of congenital nephrotic syndrome. Hum Mol Genet 2001;10:2637–44.

[32] Heeringa S, Vlangos C, Chernin G, et al. Thirteen novel NPHS1 mutations in a large cohort of children with congenital nephrotic syndrome. Nephrol Dial Transplant 2008;23:3527–33.

[33] Hinkes B, Mucha B, Vlangos C, et al. Nephrotic syndrome of the first year of life: two thirds of cases are caused by mutations in 4 genes (NPHS1,NPHS2, WT1, and LAMB2). Pediatrics 2007;119:e907–19.

[34] Santin S, Garcia-Maset R, Ruiz P, et al. Nephrin mutations cause childhood- and adult onset focal segmental glomerulosclerosis. Kidney Int 2009;76:1268–76.

[35] Patrakka J, Kestilä M, Wartiovaara J, et al. Congenital nephrotic syndrome (NPHS1): features resulting from different mutations in Finnish patients. Kidney Int 2000;58:972–80.

[36] Jalanko H. Congenital nephrotic syndrome. Pediatr Nephrol 2009;24:2121–8.

[37] Kuusniemi AM, Merenmies J, Lahdenkari A, et al. Glomerular sclerosis in kidney with congenital nephrotic syndrome (NPHS1). Kidney Int 2006;70:1423–31.

[38] Schoeb D, Chernin G, Heeringa S, et al. Nineteen novel NPHS1 mutations in a worldwide cohort of patients with congenital nephrotic syndrome (CNS). Nephrol Dial Transplant 2010;25:2970–6.

[39] Philippe A, Nevo F, Esquivel E, et al. Nephrin mutations can cause childhood-onset steroid resistant nephrotic syndrome. J Am Soc Nephrol 2008;19:1871–8.

[40] Roselli S, Gribouval O, Boute N, et al. Podocin localizes in the kidney to the slit diaphragm area. Am J Pathol 2002;160:131–9.

[41] Nishibori Y, Liu L, Hosoyamada M, et al. Disease-causing missense mutations in NPHS2 gene alter normal nephrin trafficking to the plasma membrane. Kidney Int 2004;66:1755–65.

[42] Boute N, Gribouval O, Roselli S, et al. NPHS2, encoding the glomerular protein podocin, is mutated in autosomal recessive steroid-resistant nephrotic syndrome. Nat Genet 2000;24:349–54.

[43] Machuca E, Benoit G, Nevo F, et al. Genotype-phenotype correlations in non-Finnish congenital nephrotic syndrome. J Am Soc Nephrol 2010;21:1209–17.

[44] Caridi G, Bertelli R, Carrea A, et al. Prevalence, genetics, and clinical features of patients carrying podocin mutations in steroid-resistant nonfamilial focal segmental glomerulosclerosis. J Am Soc Nephrol 2001;12:2742–6.

[45] Frishberg Y, Rinat C, Megged O, Shapira E, Feinstein S, Raas-Rotschild A. Mutations in NPHS2 encoding podocin are a prevalent cause of steroid-resistant nephrotic syndrome among Israeli-Arab children. J Am Soc Nephrol 2002;13:400–5.

[46] Hinkes B, Vlangos C, Heeringa S, et al. Specific podocin mutations correlate with age of onset in steroid-resistant nephrotic syndrome. J Am Soc Nephrol 2008;19:365–71.

[47] Weber S, Gribouval O, Esquivel E, et al. NPHS2 mutation analysis shows genetic heterogeneity of steroid-resistant nephrotic syndrome and low post-transplant recurrence. Kidney Int 2004;66:571–9.

[48] Phelan P, Hall G, Wigfall D, et al. Variability in phenotype induced by the podocin variant R229Q plus a single pathogenic mutation. Clin Kidney J 2015;8:538–42.

[49] Tsukaguchi H, Sudhakar A, Le T, et al. NPHS2 mutations in late-onset focal segmental glomerulosclerosis: R229Q is a common disease-associated allele. J Clin Invest 2002;110:1659–66.

[50] Tory K, Menyhard D, Woerner S, et al. Mutation-dependent recessive inheritance of NPHS2-associated steroid- resistant nephrotic syndrome. Nat Genet 2014;46:299–304.

[51] Machuca E, Hummel A, Nevo F, et al. Clinical and epidemiological assessment of steroid-resistant nephrotic syndrome associated with the NPHS2 R229Q variant. Kidney Int 2009;75:727–35.

[52] McKenzie L, Hendrickson S, Briggs W, et al. NPHS2 variation in sporadic focal segmental glomerulosclerosis. J Am Soc Nephrol 2007;18:2987–95.

[53] Niaudet P. Genetic forms of nephrotic syndrome. Pediatr Nephrol 2004;19:1313–8.

[54] Ito S, Takata A, Hataya H, et al. Isolated diffuse mesangial sclerosis and Wilms' tumor suppressor gene. J Pediatr 2001;138:425–8.

[55] Asfahani R, Tahoun M, Miller-Hodges E, et al. Activation of podocyte Notch mediates early Wt1 glomerulopathy. Kidney Int 2018;93:903–20.

[56] Natoli T, Liu J, Eremina V, et al. A mutant form of the Wilms tumor suppressor gene *WT1* observed in Denys-Drash syndrome interferes with glomerular capillary development. J Am Soc Nephrol 2002;13:2058–67.

[57] Guo JK, Menke A, Gubler M. WT1 is a key regulator of podocyte function: reduced expression levels cause crescentic glomerulonephritis and mesangial sclerosis. Hum Mol Genet 2002;11:651–9.
[58] Lipska B, Ranchin B, Iatropoulos P, et al. Genotype-phenotype associations in WT1 glomerulopathy. Kidney Int 2014;85:1169–78.
[59] Hu M, Zhang G, Arbuckle S, et al. Prophylactic bilateral nephrectomies in two paediatric patients with missense mutations in the WT1 gene. Nephrol Dial Transplant 2004;19:223–6.
[60] Chernin G, Vega-Warner V, Schoeb D, et al. Genotype/phenotype correlation in nephrotic syndrome caused by WT1 mutations. Clin J Am Soc Nephrol 2010;5:1655–62.
[61] Ruf R, Schultheiss M, Lichtenberger A. Prevalence of WT1 mutations in a large cohort of patients with steroid-resistant and steroid-sensitive nephrotic syndrome. Kidney Int 2004;66:564–70.
[62] Lehnhardt A, Karnatz C, Ahlenstiel-Grunow T, et al. Clinical and molecular charazterization of patients with heterozygous mutations in Wilms Tumor Suppressor Gene 1. Clin J Am Soc Nephrol 2015;10:825–31.
[63] Zenker M, Machuca E, Antignac C. Genetics of nephrotic syndrome: new insights into molecules acting at the glomerular filtration barrier. J Mol Med 2009;87:849–57.
[64] Chaib H, Hoskins B, Asheaf S, Goyal M, Wiggins R, Hildebrandt F. Identification of BRAF as a new interactor of PLCepsilon1, the mutated in nephrotic syndrome type 3. Am J Physiol Ren Physiol 2008;294:F93–9.
[65] Rao J, Ashraf S, Tan W, et al. Advillin acts upstream of phospholipase C e 1 in steroid –resistant nephrotic syndrome. J Clin Invest 2017;127:4257–69.
[66] Hinkes B, Wiggins R, Gbadegesin R, et al. Positional cloning uncovers mutations in PLCE1 responsible for a nephrotic syndrome variant that may be reversible. Nat Genet 2006;38:1397–405.
[67] Gbadegesin R, Batkowiat B, Lavin P, et al. Exclusion of homozygous PLCE1 NPHS39 mutations in 69 families with idiopathic and hereditary FSGS. Pediatr Nephrol 2009;24:281–5.
[68] Miner J. Glomerular basement membrane composition and the filtration barrier. Pediatr Nephrol 2011;26:1413–7.
[69] Zenker M, Aigier T, Tralau T. Human laminin beta-2 deficiency causes congenital nephrosis with mesangial sclerosis and distinct eye abnormalities Hum. Mol Genet 2004;13:2625–32.
[70] Hasselbacher K, Wiggins R, Matejas V, et al. Recessive missense mutations in LAMB2 expand the clinical spectrum of LAMB2 associated disorders. Kidney Int 2006;70:1008–12.
[71] Matejas V, Hinkes B, Alkandari F, et al. Mutations in the human laminin beta2 (LAMB2) gene and the associated phenotypic spectrum. Hum Mutat 2010;31:992–1002.
[72] Ozaltin F, Heeringa S, Poyraz C, et al. Eye involvement in children with primary focal segmental glomerulosclerosis. Pediatr Nephrol 2008;23:421–7.
[73] Chhabra E, Higgs H. INF2 is a WASP homology 2 motif-containing formin that severs actin filaments and accelerate both polymerization and depolymerization. J Biol Chem 2006;281:26754–67.
[74] Brown E, Schlöndorff J, Becker D, et al. Mutations in the formin gene IFN2 cause focal segmental glomerulosclerosis. Nat Genet 2010;42:72–6.
[75] Boyer O, Benoit G, Gribouval O, et al. Mutations in INF2 are a major cause of autosomal dominant focal segmental glomerulosclerosis. J Am Soc Nephrol 2011;22:239–45.
[76] Lee H, Han K, Jung Y, Kang H, Moon K, Ha H, Choi Y, Cheong H. Variable renal phenotype in a family with INF2 mutation. Pediatr Nephrol 2011;26:73–6.
[77] Caridi G, Lugani F, Dagnino M, et al. Novel INF2 mutations in an Italian cohort of patients with focal segmental glomerulosclerosis, renal failure and Charcot-Marie-Tooth neuropathy. Nephrol Dial Transplant 2014;29:iv80–6.
[78] Lipska B, Iatropoulos P, Maranta R, et al. Genetic screening in adolescents with steroid-resistant nephrotic syndrome. Kidney Int 2013;84:206–13.
[79] Möller C, Flesche J, Reiser J. Sensitizing the slit diaphragm with TRPC6 ion channels. J Am Soc Nephrol 2009;20:950–3.
[80] Kanda S, Harita Y, Shibagaki Y, et al. Tyrosine phosphorylation-dependent activation of TRPC6 regulated by PLC(gamma)1 and nephrin: effect of mutations associated with focal segmental glomerulosclerosis. Mol Biol Cell 2011;22:1824–35.
[81] Reiser J, Polu K, Moller C, et al. TRPC6 is a glomerular slit diaphragm –associate channel required for normal renal function. Nat Genet 2005;37:739–44.
[82] Winn M, Conlon P, Lynn K, et al. A mutation in the TRPC6 cation channel causes familial focal segmental glomerulosclerosis. Science 2005;308:1801–4.
[83] Heeringa S, Möller C, Du J, et al. A novel TRPC6 mutation that causes childhood FSGS. PLoS One 2009;10:4e7771.

[84] Zhu B, Chen N, Wang Z, Pan XX, Ren H, Zhang W, Wang W. Identification and functional analysis of a novel TRPC6 mutation associate with late onset familial focal segmental glomerulosclerosis in Chinese patients. Mutat Res 2009;664:84–90.
[85] Gicante M, Caridi G, Montenurno E, et al. TRPC6 mutations in children with steroid-resistant nephrotic syndrome and atypical phenotype. Clin J Am Soc Nephrol 2011;6:1626–34.
[86] Malone A, Phelan P, Hall G, et al. Rare hereditary COL4A3/COL4A4 variants may be mistaken for familial focal segmental glomerulosclerosis. Kidney Int 2014;86:1253–9.
[87] Xie J, Wu X, Ren H, et al. COL4A3 mutations cause focal segmental glomerulosclerosis. J Mol Cell Biol 2014;6:498–505.
[88] Gast C, Pengelly R, Lyon M, et al. Collagen (COL4A) mutations are the most frequent mutations underlying adult focal segmental glomerulosclerosis. Nephrol Dial Transplant 2016:961–70.
[89] Bullich G, Trujillano D, Santin S, et al. Targeted sequencing in steroid-resistant nephrotic syndrome: mutations in multiple glomerular genes may influence disease severity. Eur J Hum Genet 2015;23:1192–9.
[90] Diomedi-Camassei F, Di Giandomenico S, Santorelli F, et al. COQ2 nephropathy: a newly described inherited mitochondriopathy with primary renal involvement. J Am Soc Nephrol 2007;18:2773–80.
[91] Salviati L, Sacconi S, Murer L, et al. Infantile encephalomyopathy and neuropathy with CoQ10 deficiency: a CoQ10-responsive condition. Neurology 2005;65:606–8.
[92] Lopez L, Schuelke M, Quinzii C, et al. Leigh syndrome with neuropathy and CoQ10 deficiency due to decaprenyl diphosphate synthase subunit2 (PDSS2) mutations. Am J Hum Genet 2006;79:1125–9.
[93] Löwik M, Hol F, Steenbergen E, Wetzels J, van den Heuvel L. tRNALeu(UUR) mutation in a patient with steroid-resistant nephrotic syndrome and focal segmental glomerulosclerosis. Nephrol Dial Transplant 2005;20:336–41.
[94] Fonseca L, Doimo M, Caldren C, et al. Mutations in COQ8B (ADCK4) found in patients with steroid-resistant nephrotic syndrome alter COQ8B function. Hum Mutat 2017;39:406–14.
[95] Korkmaz E, Lipska-Zietkiewicz B, Boyer O, et al. ADCK4-associated glomerulopathy causes adolescence-onset FSGS. J Am Soc Nephrol 2016;27:63–8.
[96] Park E, Kang H, Choi Y, et al. Focal segmental glomerulosclerosis and medullary nephrocalcinosis in children with ADCK4 mutations. Pediatr Nephrol 2017;32:1547–54.
[97] Ashraf S, Gee H, Woerner S, et al. J Clin Invest 2013;123:5179–89.
[98] Kim JM, Wu H, Green G, et al. CD2-associated protein haploinsufficiency is linked to glomerular disease susceptibility. Science 2003;300:251–6.
[99] Gigante M, Pontrelli P, Montemurno E, et al. CD2AP mutations are associated with sporadic nephrotic syndrome and focal segmental glomerulosclerosis (FSGS). Nephrol Dial Transplant 2009;24:1653–60.
[100] Löwik M, Groenen P, Pronk I, et al. Focal segmental glomerulosclerosis in a patient homozygous for a CD2AP mutation. Kidney Int 2007;72:1198–203.
[101] Sanna-Cherchi S, Burgess K, Nees S, et al. Exome sequencing identified MYO1E and NEIL1 as candidate genes for human autosomal recessive steroid-resistant nephrotic syndrome. Kidney Int 2011;80:389–96.
[102] Mele C, Iatropoulos P, Donadelli R, et al. MYO1E mutations and childhood familial focal segmental glomerulosclerosis. N Engl J Med 2011;3655:295–306.
[103] Lennon R, Stuart H, Bierzynska A, Randels M, et al. Coinheritance of col4A5 and MYO1E mutations accentuate the severity of kidney disease. Pediatr Nephrol 2015;30:1459–65.
[104] Kaplan J, Kim S, North K, et al. Mutations in ACTN, encoding alpha-actinin-4, cause familial focal segmental glomerulosclerosis. Nat Genet 2000;24:251–6.
[105] Weins A, Kenlan P, Herbert S, et al. Mutational and biological analysis of alpha-actin-4 in focal segmental glomerulosclerosis. J Am Soc Nephrol 2005;16:3694–701.
[106] Choi HJ, Lee B, Cho H, Moon K, Ha I, Nagata M, Chioi Y, Cheong H. Familial focal segmental glomerulosclerosis with ACTN4 mutation and paternal germline mosaicism. Am J Kidney Dis 2008;51:834–8.
[107] Gee H, Saisawat P, Ashraf S, et al. ARDGDIA mutations cause nephrotic syndrome via defective RHOGTPase signaling. J Clin Invest 2013;123:3243–53.
[108] Gupta I, Baldwin C, Auguste D, et al. ARHGDIA: a novel gene implicated in nephrotic syndrome. J Med Genet 2013;50:330–8.
[109] Bongers E, Gubler M, Knoers N. Nail-patella syndrome. Overview on clinical and molecular findings. Pediatr Nephrol 2002;17:703–12.
[110] Heidet L, Bongers E, Sich M. In vivo expression of putative LMX1B targets in nail-patella syndrome kidneys. Am J Pathol 2003;163:145–55.

[111] Srivastava T, Whiting J, Garola R. Podocyte proteins in Galloway–Mowat syndrome. Pediatr Nephrol 2001;16:1022–9.

[112] Colin E, Huynh C, Mollet G, et al. Loss-of-function mutations in WDR73 are responsible for microcephaly and steroid-resistant nephrotic syndrome: Galloway-Mowat syndrome. Am J Hum Genet 2014;95:637–48.

[113] Rosti R, Dikoglu E, Zaki M, et al. Extending the mutation spectrum for Galloway-Mowat syndrome to include homozygous missense mutations in the WDR73 gene. Am J Med Genet 2016;170A:992–8.

[114] Braun D, Rao J, Mollet G, Schapiro D, Daugeron M, Tan W, Gribouval O, Hildebrandt F. Mutations in the evolutionary highly conserved KEOPS complex genes cause nephrotic syndrome with microcephaly. Nat Genet 2017;49:1529–38.

[115] Zivicnjak M, Franke D, Zenker M, Hoyer J, Lucke T, Pape L, Ehrich H. Smarcall mutations: a cause of prepubertal idiopathic steroid-resistant nephrotic syndrome. Pediatr Res 2009;65:564–8.

[116] Carroll C, Hunley T, Guo Y, Cortez D. A novel spice site mutation in SMATRCAL1 results in aberrant exon definition in a child with Schimke immunoosseous dysplasia. Am J Genet 2015;5:2260–4.

[117] Liu S, Zhang M, Ni M, Zhu P, Xia X. A novel heterozygous mutation of Smarcal 1 gene leading to mild Schimke immune-osseous dysplasia: a case report. BMC Pediatr 2017;17:217–23.

[118] Santangelo L, Gicante M, Stafano G, et al. A novel SMARCAL1 mutation associated with a mild phenotype of Schimke immuno-osseous dysplasia (SIOD). BMC Nephrol 2014;15:41.

[119] Genovese G, Friedman D, et al. Association of trypanolytic ApolL 1 variants with kidney disease in African-Americans. Science 2010;329:841–5.

[120] Kopp J, Nelson G, Sampath K, et al. APOL1 genetic variants in focal segmental glomerulosclerosis and HIV-associated nephropathy. J Am Soc Nephrol 2011;22:2129–37.

[121] Kopp J, Winkler C, Zhao X, et al. Clinical features and histology of apolipoprotein L 1-associate nephropathy in the FSGS clinical Trial. J Am Soc Nephrol 2015;26:1443–8.

[122] Suvanto M, Jahnukainen T, Kestilä M, Jalanko H. Single nucleotide polymorphism in pediatric idiopathic nephrotic syndrome. Int J Nephrol 2016;2016:1417456.

[123] Van Husen M, Kemper M. New therapies in steroid-sensitive and steroid-resistant idiopathic nephrotic syndrome. Pediatr Nephrol 2011;6:881–92.

[124] Gellermann J, Stefanidis C, Mitsioni A, Querfeld I. Successful treatment of steroid-resistant nephrotic syndrome associated with WT1mutations. Pediatr Nephrol 2010;25:1285–9.

[125] Buscher A, Kranz B, Buscher R, et al. Immunosuppression and renal outcome in congenital and pediatric steroid-resistant nephrotic syndrome. Clin J Am Soc Nephrol 2010;5:2075–84.

[126] Buscher A, Beck B, Melk A, , et alfor the German Pediatric Nephrology Association (GPN). Rapid response to cyclosporine A and favorable renal outcome in nongenetic versus genetic steroid-resistant nephrotic syndrome. Clin J Am Soc Nephrol 2016;11:245–53.

[127] Faul C, Donelly M, Merscher-Gomez S, et al. The actin cytoskeleton of kidney podocytes is a direct target of antiproteinuric effect of cyclosporine A. Nat Med 2008;14:931–8.

[128] Schönenberger E, Ehrich J, Haller H, Schiffer M. The podocyte as a direct target of immunosuppressive agents. Nephrol Dial Transplant 2010;26:18–24.

[129] Patrakka J, Martin P, Salonen R, et al. Proteinuria and prenatal diagnosis of congenital nephrosis in fetal carriers of nephrin gene mutations. Lancet 2002;359:1575–7.

[130] Wieder N, Greka A. Calcium, TRPC channels, and regulation of the actin cytoskeleton in podocytes: towards a future of targeted therapies. Pediatr Nephol 2016;31:1047–54.

[131] Heeringa S, Chernin G, Chaki M, Zhou W, Sloan A, et al. COQ6 mutations in human patients produce nephrotic syndrome with sensorineural deafness. J Clin Invest 2011;121:2013–24.

[132] Jalanko H, Mattila I, Holmberg C. Renal transplantation in infants. Pediatr Nephrol 2016;31:725–35.

[133] Kuusniemi A, Qvist E, Sun Y, Patrakka J, Rönnholm K, Karikoski R, Jalanko H. Plasma exchange and retransplantation in recurrent nephrosis of patients with congenital nephrotic syndrome of the Finnish type (NPHS1). Transplantation 2007;83:1316–23.

[134] Holmberg C, Jalanko H. Long-term effects of paediatric kidney transplantation. Nat Rev Nephrol 2016;12:301–11.

4

Renal Tubular Disorders

Reed E. Pyeritz

Perelman School of Medicine at the University of Pennsylvania, Philadelphia, PA, United States

4.1 INTRODUCTION

All small molecules not tightly bound to protein are filtered by the glomerulus. The renal tubules are the primary mechanism by which the body can reabsorb these compounds. The proximal tubules (and, for bicarbonate and water, the distal tubules and collecting ducts) are responsible for reabsorbing 80%–98% of the solutes filtered by the glomerulus. Although a few substances are actively secreted by the kidney, most disorders of tubular function lead to decreased reabsorption and thus increased loss in the urine.

Disorders of renal tubular function are divided by the segment of the tubule involved and by the substrate(s) whose transport is affected. Molecules are reabsorbed by mechanisms that are both specific and saturable. In addition, the entry of substrates into the renal tubule from the brush border side (the lumen) is under different genetic control from the exit step through the basolateral membranes into the blood.

Most of the transport proteins involved in specific transport systems have been identified and mapped [1,2]. However, the identification of the transport proteins has still left much to be explained. For example, it is unclear why a mutation in a transport protein may be expressed only in the kidney, only in the gut, or in both, with differing clinical manifestations. There is not yet a good correlation with specific mutations and clinical severity in most cases. A few defects have yet to be identified.

Transport in the proximal tubule is largely energy dependent and is driven by oxidative metabolism. Experimentally it can be supported by acetate, lactate, and a variety of other organic acids. The distal tubules, and particularly the collecting ducts, are more dependent on energy derived from glycolysis. Many substrates are cotransported with sodium, and the driving force for movement across membranes is provided by the electrochemical potential difference generated by Na^+ and H^+ movements. Thus, inherited defects in energy metabolism, such as those in cytochrome C oxidase, and compounds such as heavy metals, which interfere with energy metabolism, can have effects on many different transport systems in the tubule. Several of the specific defects in energy metabolism, particularly the mitochondrial defects, occur [3].

Most disorders of transport involve defects in the proximal tubule. Clinically important systemic disease is produced by generalized transport defects and defects in bicarbonate, phosphate, and L-carnitine reabsorption. Defects in cystine reabsorption cause renal stones. Defects in neutral amino acid transport, when also present in the intestine, cause Hartnup disease. However, many of the defects in transport, although genetically well defined, while of little clinical importance may be important to the nutrition of seniors.

Renal tubular acidosis (RTA) is a generic term that is a feature of numerous hereditary renal syndromes. The unifying feature is acidosis and may include a

Emery and Rimoin's Principles and Practice of Medical Genetics and Genomics. https://doi.org/10.1016/B978-0-12-812534-2.00011-4

predisposition to nephrocalcinosis. The proximal tubules, distal tubules, or both may be the primary site of the defect. Inheritance may be autosomal dominant or recessive.

4.2 GENERALIZED DISORDERS OF TUBULAR FUNCTION (FANCONI SYNDROME)

The Fanconi syndrome is a generalized tubular disorder that leads to multiple transport abnormalities, including increased urinary loss of several organic substrates—amino acids, glucose, bicarbonate, and organic acids—and the loss of inorganic ions essential for mineral homeostasis—calcium, magnesium, sodium, potassium, and most important clinically, phosphorus [4]. Clinically, the syndrome is usually defined by the combination of glucosuria, phosphaturia, generalized aminoaciduria, and RTA. In addition, increased renal excretion of small-molecular-weight proteins typically occurs. These metabolic abnormalities can lead to metabolic bone disease (rickets or osteomalacia) and stunted growth. Water-soluble vitamins are also lost into urine.

The Fanconi syndrome can be caused by a variety of inherited diseases and environmental factors that are toxic to the transport mechanisms in the proximal tubule. Inherited causes are familial idiopathic cystinosis; disorders of phosphorylated sugar metabolism, including galactosemia, glycogen storage disease, and hereditary fructose intolerance; Lowe syndrome; Wilson disease; tyrosinemia; medullary cystic disease; and rickets with secondary hyperparathyroidism. In the absence of other known diseases, it can also exist in an isolated form that is inherited as an autosomal recessive or dominant trait and in an X-linked recessive form with nephrolithiasis and renal failure, which, although discussed here, could be considered a form of RTA. In the general population, the most common causes of childhood Fanconi syndrome are cystinosis and galactosemia [5]. In the inner cities, lead poisoning may be the leading noninherited form.

The clinical presentation of the Fanconi syndrome depends on the age of the patient and the other manifestations of the primary disease. Idiopathic Fanconi syndrome may manifest itself primarily as bone disease or as a gastrointestinal problem with vomiting, anorexia, constipation probably due to volume depletion, and chronic acidosis, or can be found incidentally on routine urine analysis. The diagnosis of idiopathic Fanconi syndrome can only be made after an exhaustive search for other primary causes.

Families with Fanconi syndrome may have autosomal dominant inheritance [6–8]. All these families with dominant inheritance were reported to have mild disease and were termed the adult Fanconi syndrome. However, in a follow-up of Luder and Sheldon's family that three members had developed renal failure [8]. Interfamilial and intrafamilial clinical variations are considerable. Some patients have only had the urinary findings without overt clinical expression, whereas others have had severe rickets. Acquired causes of Fanconi syndrome are often drug-induced, with or without RTA [9]. The gene, *GATM*, for adult Fanconi syndrome maps to chromosome 15q15.3 [10] and encodes L-argine:glycine amidinotransferase [11].

At least one autosomal recessive form occurs. In the severe childhood form, the picture is much like cystinosis, and some questions must be raised whether the early reports of this mode of inheritance might not represent undiagnosed secondary causes of this syndrome [12]. The most convincing pedigree is that of Klajman and Arber, who described a consanguineous Iraqi Jewish family with six affected siblings [13]. The existence of an autosomal recessive adult form of the disease has been called into question.

A huge family, of Irish descent, with X-linked nephrolithiasis, nephrocalcinosis, renal tubular dysfunction, and renal insufficiency was described by Frymoyer and colleagues [14]. Five additional families were reported [15]. Linkage analysis on 102 members from five generations (LOD score 5.91) established the locus to Xp11.22 [16]. In five unrelated British families, the locus mapped to the same position [17]. The gene is *CLCN5*, which encodes a chloride channel, and is also the gene that causes the similar Dent disease [18].

4.3 DISORDERS OF AMINO ACID TRANSPORT

Except for tryptophan and perhaps homocysteine, which are tightly bound to protein, amino acids are filtered freely by the glomerulus. Most amino acids are normally reabsorbed in the proximal tubule with only very small amounts remaining in the urine. Glycine and histidine, which are less efficiently reabsorbed than the other amino acids, account for most of the α-amino nitrogen found in the urine. Only very small infants, particularly premature babies, excrete significant quantities of the other amino acids. In these infants alanine, proline, hydroxyproline, serine, and threonine are commonly present in the urine. Cystine is also seen in premature and occasionally in other newborns. It is now a more common finding because cystine is added to infant formulas.

Amino acids are reabsorbed by energy-dependent mechanisms of high specificity [19,20]. In most cases, transport is coupled to the movement of sodium ions. Essentially, the transport systems can be considered group-specific. The amino acid transport systems are as follows:

Neutral amino acid system L (leucine preferring)
Neutral amino acid system A (alanine preferring)
Iminoglycine system
Acidic amino acid system
Dibasic amino acid system (includes cystine)

At least two transport systems are present for neutral amino acids with differing affinities. At least one transport system is present for amino acids containing two ammonium groups (the dibasic amino acids plus cystine). This transport system is the first in which a definite membrane-associated protein was cloned and mapped. One system is present for acidic amino acids and another for the combination of glycine, proline, and hydroxyproline (the iminoglycine system). Studies in animals suggest that secondary systems with lower affinity are also present for the same substrates [21,22]. This concept has been used to explain some of the variant amino acid disorders in humans, in which the urine pattern does not correspond to the listed systems. The transport systems are present in the brush border of the tubule and the gut and have been observed in a variety of cell types. A defect in basic amino acid transport in the basolateral membrane of the cell has been described for the dibasic amino acids, lysinuric protein intolerance [23]. Pathogenic variants in *SLC7A7* cause this condition [24]. This system has a different inheritance and different affinities for amino acids than the dibasic system found in brush borders.

The genetics of amino acid transport are not always clear. In some families, defects are found only in kidney tubules, in some only in gut, and in others in both.

4.4 GLYCINE AND THE IMINO ACIDS

Some children have marked increases in the urinary excretion of proline, hydroxyproline, and glycine, which share renotubular resporptive mechanisms. Newborn screening studies have reported an incidence as high as 1 in 20,000 in the general population [25]. Initial association of iminoglycinuria with a variety of neurological disorders (e.g., deafness, blindness, mental retardation) was reported. However, family studies and population screening suggest that this is a benign condition inherited as an autosomal recessive trait [26]. Variants in three genes, *SLC6A20, SLC6A19,* and *SLC36A2*) produce this syndrome, which has a much higher affinity for the imino acids than for glycine [27]. Heterozygotes may excrete glycine in the urine. This probably explains the dominant glycinuria reported in three generations [28]. Three different forms of hyperglycinuria exist [29]. This transport system is quite late developing in humans, and these amino acids are commonly elevated in the urine of newborn infants, particularly premature ones. Like many of the other disorders of amino acid transport in the kidney, a defect in the intestine may or may not also be present [30]. The primary importance of this hereditary condition is that it can be mistaken for a primary defect in glycine metabolism or a secondary defect due to organic aciduria.

4.5 DIBASIC AMINO ACIDS AND CYSTINE

4.5.1 Classic Cystinuria

Cystinuria is not only a clinically significant disease but also important historically because it was one of the original inborn errors reported by Garrod in his famous Croonian lectures [31]. McKusick [32] quotes Marcet as suggesting in 1817 that the disease was familial. In classic cystinuria, the urinary excretion of lysine, ornithine, arginine, and cystine is greatly increased compared to normal [33]. The absolute increase in cystine is less than that of the other amino acids. Because it is far less soluble, especially at pH values greater than 7, it forms stones. Of all renal stones, cysteine stones represent only 1%–2%, but 3–4 times that in children. Many cysteine calculi also contain calcium oxalate, calcium phosphate, or struvite. Serum amino acids are normal, and no nutritional deficiency in patients has been demonstrated. Cystinuria is the first amino acid transport disease in which the molecular defect was identified and mapped. Pathogenic variants in two genes, *SLC3A1* and *SLC7A*, cause cystinuria.

Three hereditary patterns of excretion have been described in heterozygotes for classic cystinuria [34]. Homozygotes, including compound heterozygotes between two types, cannot be distinguished. Heterozygotes can be separated based on their pattern of urinary amino acids and whether or not they also have an intestinal defect. Type I heterozygotes have normal excretion of cystine in their urine. Types II and III heterozygotes have increased excretion in their urine. They can only be distinguished by studying intestinal transport. Type II heterozygotes have a transport defect in the gut, but type III do not.

The prevalence of clinical cystinuria varies widely in different populations: 1 in 7000 in the United States; 1 in 18,000 in Japan, 1 in 2500 in Israel; and only 1 in 100,00 in Sweden [35,36]. These figures must be accepted very conditionally because of the increased secretion normally noted in newborns and particularly because newborn heterozygotes are difficult to distinguish from homozygotes. The prevalence is particularly high in the Old Order Mennonites, both in Pennsylvania and Missouri.

Men are affected by cysteine calculi twice as often as women, with the first stone appearing on average by the 20s. One-quarter of stones appear in children. Other stone-related chemical imbalance is common, including hypocitraturia, hypercalcuria, and hyperuircosuria.

Cystine excretion in the absence of other dibasic amino acids in two siblings, and there were without stones but were only 2 and 4 years old when reported [37].

Plain radiographs cannot detect pure cysteine calculi. Ultrasonography can detect stones greater than about 4 mm in diameter and, importantly, determine if hydronephrosis is present. Computed tomography without contrast is the commonly available diagnostic procedure of highest resolution.

Treatment by modification of diet has limited benefit. Reduction of protein, especially methionine, and relatively higher intake of sodium and alkalinizing beverages (e.g., orange juice and mineral water with high bicarbonate levels) are recommended [33]. Along with these dietary measures, increased hydration is essential. The goal is a urinary level of cysteine of 250 mg/L or, preferably, less. Increasing the urine pH to greater than 7.5 can be achieved by intake of citrus juices or a combination of potassium citrate, sodium bicarbonate, and magnesium citrate. At pH levels above 7.5, the risk of calcium phosphate stone formation increases, so hypercalcuria must be controlled. Thiol-based drugs (e.g., captopril, tiopronin, bucillamine) have a role because of their competition for the disulfide bond between cysteine molecules that forms cystine. Extracorporeal shock wave lithotripsy is less effective for cystine stones (especially larger ones) than for stones of other composition.

Ultimately, treatment is often ineffective over a lifetime, renal damage is likely, and quality of life is reduced [38].

4.6 CYSTINOSIS

Cystinosis is classified as a lysosomal transport disorder and is a cause of the renal Fanconi syndrome [5]. Inheritance is autosomal recessive with pathogenic variants in *CTNS*, the gene encoding cystinosin [12]. Allelic variation results in disease of pediatric, adolescent, and adult onset [39]. Cystine accumulates in lysosomes because of the inability to transport to the cytoplasm [12,40].

Organs affected include the kidneys, eyes (cornea), bone, bone marrow, thyroid, muscle, gastrointestinal system, and peripheral nerves [41]. Renal involvement typically determines clinical outcome [5]. Bone disease is not solely due to renal osteodystrophy [42]. For those patients with milder disease or who respond partially to treatment, adult-onset problems including neuropsychologic issues can arise [43,44]. Treatment consists or oral immediate- or delayed-release cysteamine. Cysteamine eye drops are used to manage crystal deposition in the cornea.

4.7 OTHER FORMS OF DIBASIC AMINOACIDURIA

The description of lysinuric protein intolerance has led to the first documentation of a basolateral transport defect for dibasic amino acids (but not cystine) [45]. This disorder leads to a marked protein intolerance because of hyperammonemia and suggests that the urea cycle is not perfect but requires the intake of the dibasic amino acids. The defect can be partially overcome by citrulline, which is not transported by the same system. Dibasic aminoaciduria in the absence of cystine excretion was reported in 13 asymptomatic members of a family [46].

The disease has been reported throughout the world but has a particularly high incidence in Finland (one in 60,000).

4.8 NEUTRAL AMINO ACIDS

Hartnup disorder is primarily a neutral amino acid transport defect across epithelial cells in renal proximal tubules and intestinal mucosa. However, histidine, glutamine, and asparagine are also increased in the urine. It can be distinguished from a generalized aminoaciduria because the other amino acids are not increased. Often the disease is diagnosed from urine because of the presence of indoles and tryptamine produced by bacteria, owing to malabsorption of tryptophan by the gut. Stool amino acids, when measured, are increased. The clinical manifestations of the disease are very variable and include any combination of cerebellar ataxia, emotional instability, delayed development, severe retardation, and a pellagra-like rash [47]. However, the renal defect in the absence of an intestinal transport defect is of less significance than the combined defect or an intestinal defect alone [48]. Many of the clinical manifestations have been believed to be due to tryptophan malabsorption, leading to niacin deficiency. Patients who have clinical manifestations other than rash (primarily developmental delay and ataxia) continue to have slow development even when receiving large amounts of niacin and a good protein intake. Magnetic resonance imaging in these patients leads one to speculate that a transport defect may also be present in the brain.

Autosomal recessive inheritance is due to homozygosity or compound heterozygosity at the *SLC6A19* locus [49].

4.9 RENAL TUBULAR ACIDOSIS

Several hereditary disorders cause impaired transport of transport of various substrates and are classified according to which portion of the tubule is most affected [50,51].

4.10 PROXIMAL RENAL TUBULAR ACIDOSIS (TYPE 2 RTA)

Proximal RTA is caused by bicarbonate loss only in the proximal tubule. Because most bicarbonate is reabsorbed more distally, when the filtered load is decreased by a fall in serum bicarbonate, normal acidification mechanisms in the distal tubule can still lower the urine pH. The serum electrolytes demonstrate a hyperchloremic acidosis [52].

RTA type II occurs as part of the Fanconi syndrome. Nearly all isolated cases have been in young boys, most before 2 years of age [3,53]. The great predominance of males (4:1) suggests an X-linked recessive form of inheritance. This mode of inheritance has not been proven, however, because no family history can be demonstrated in a majority of cases.

4.11 LOWE OCULOCEREBRORENAL SYNDROME

In this X-linked disorder, congenital cataracts, impaired intellectual development, and growth retardation are accompanied by proximal renal tubular dysfunction causing metabolic acidosis and renal failure. There is phenotypic overlap with Dent syndrome, both being due to mutations in *OCRL*, which encodes an inositol polyphosphate 5-phosphatase [54–56].

4.12 DISTAL RENAL TUBULAR ACIDOSIS (TYPE I RTA)

In type I RTA, bicarbonate is lost in the distal tubules. This disease presents with nephrocalcinosis, fixed urinary specific gravity, a low serum bicarbonate, and hypocalcemia. The bone manifestations are very variable but may be severe. Chaabani and colleagues [57] reported a large pedigree in which 28 members had RTA I, the clinical presentation being quite variable. Some had nephrocalcinosis and growth retardation, whereas others were clinically unaffected.

RTA I has autosomal dominant inheritance and is due to pathogenic variants in *ATP6B1*, which encodes the B subunit of the apical proton pump mediating distal nephron acid secretion [58].

4.13 DISTAL RENAL TUBULAR ACIDOSIS WITH NEURAL HEARING LOSS (TYPE 2 RTA)

Most patients are diagnosed in childhood, some in infancy when they present acutely with dehydration and vomiting, or with failure to thrive and/or growth impairment. The diagnosis can be made based on inappropriately alkaline urine (pH greater than 5.5) and the presence of systemic metabolic acidosis with normal anion gap, evidence of renal potassium wasting, and no evidence of secondary causes of distal RTA [59].

This disorder is caused by homozygous or compound heterozygous pathogenic variants in *ATP6V1B*, which encodes the B subunit of the apical proton pump mediating distal nephron acid secretion.

4.14 DISTAL RENAL TUBULAR ACIDOSIS WITH OR WITHOUT DEAFNESS (TYPE 3 DRTA)

The phenotype is similar Type 2 RTA, but with variable effect on hearing. The defect is in *ATP6N1B*, which encodes a kidney-specific proton pump [60].

4.15 CARBONIC ANHYDRASE II DEFICIENCY

RTA is primarily distal but also proximal occurred in three sisters with osteopetrosis and cerebral calcifications [61]. Carbonic anhydrase is found in both the kidney and brain and was deficient in the patients and decreased in the obligate heterozygotes [62]. About half of all the subsequently reported cases have occurred in families of Arab descent in the Middle East or North Africa [63]. The same novel splicing mutation was present in six unrelated Arab families The gene involved is *CA2*, localized to chromosome 8q22, which encodes one of the three carbonic anhydrase proteins expressed in the kidney and brain.

4.16 DISORDERS OF SUGAR TRANSPORT

At least two transport systems mediate the absorption in the gut or the reabsorption of sugars in the proximal tubule. The most important is the system for glucose and galactose. Also present in the proximal tubule are systems for the reabsorption of other sugars. Fructose is not actively transported by the tubule, and the condition known as fructosuria is due to overflow rather than a transport defect. As with the amino acid transport disorders, defects have been described that affect the intestine as well as the kidney. Because heterozygotes for these disorders may spill small amounts of sugar into the urine, these disorders have been described as having dominant modes of inheritance [64]. However, heavy excretion is seen only in homozygotes, and the conditions are considered autosomal recessive.

4.17 RENAL GLYCOSURIA

Clinically, renal glycosuria may be considered as two disorders. First, patients may have a low threshold for glucose but a normal total capacity to reabsorb it. Thus, these patients will spill glucose into their urine at normal serum concentrations, but their total daily loss is not great. This disorder is important mainly because it may be confused with diabetes mellitus and because a low renal threshold for glucose may make urine monitoring of a diabetic difficult. Second, patients may have a decreased capacity to reabsorb glucose with or without a decreased threshold. These patients may spill very large amounts of glucose in a day, sometimes as much as 100 g. These two defects appear to be allelic. Elsas and Rosenberg [65] showed that both types of defects can appear in the same family and that compound heterozygotes have clinical glycosuria. The defect is in sodium glucose cotransporter encoded by *SGLT2*. Some of these patients have polydipsia and polyuria, and many of the women have yeast infections of the vagina. The defect in these patients is in the low-affinity Na^+ glucose cotransporter in the proximal tubule encoded by *SLC5A2*. Heterozygotes often have less severe glycosuria [66].

4.18 FRUCTOSURIA

Fructosuria in the absence of fructose intolerance (fructose phosphate aldolase deficiency) is not a renal tubular disease. Instead, this benign condition is the result of a defect in hepatic fructokinase. It is inherited as an autosomal recessive trait with pathogenic variants in *KHK* [67].

Elevated fructose has been implicated in the development of chronic kidney disease.

4.19 PENTOSURIA

Pentosuria was also one of the original inborn errors described in the Garrod lectures [31]. Patients excrete 1–4 g of pentose, primarily l-xylulose, daily [68]. It is a benign condition that occurs primarily in Ashkenazi Jews of Polish origin and in Lebanese. This condition, like fructosuria, is usually a disorder of metabolism rather than transport. The defect is in *DCXR*, which encodes the enzyme NADP-xylitol dehydrogenase [69]. Heterozygotes have intermediate levels of the enzyme, confirming its autosomal recessive mode of inheritance.

4.20 HYPOPHOSPHATEMIC RICKETS

Hypophosphatemic rickets behaves as an X-linked dominant trait in most families. Defects in *PHEX*, which encodes phosphate regulating endopeptidase homolog X-linked, lead to abnormal expression of fibroblast growth factor 23 (*FGF23*). The net result is high FGF23 that impairs renal resorption of phosphate leading to hypophosphatemia [70]. Treatment with vitamin D is typically ineffective and can lead to hypercalcuria and nephrolithiasis.

Hereditary hypophosphatemic rickets with hypercalcuria is an autosomal recessive condition due to pathogenic variants in *SLC34A3*, which encodes a sodium-potassium 2c transporter in the proximal renal tubule [71]. One complication is nephrolithiasis, which may be the presenting symptom in some patients.

4.21 IMPORTANT AREAS OF CURRENT AND FUTURE RESEARCH

Investigations of the renal tubules will remain clinically important for the disorders described in this chapter and perhaps disorders awaiting description. Some areas that will be particularly relevant include newborn screening [72], understanding risks of drug toxicity [73], the role of mitochondrial defects in renal diseases [3,74,75], and the importance of sodium-glucose cotransport mechanisms in diabetes and cardiovascular disease [76–81].

4.22 CONCLUSION

The renal tubules reabsorb 80%–98% of filtered small molecules by energy-dependent mechanisms. Disorders, primary or secondary, that interfere with these mechanisms can cause loss in the urine of sugars, amino acids, bicarbonate, phosphorus, and other cations, as well as a variety of organic acids. Specific transport systems are under genetic control. Thus, inherited disorders can affect the reabsorption of only one or a small group of compounds. Defects altering energy metabolism can produce a more generalized loss of renal tubular mechanisms. With the notable exceptions of bicarbonate and phosphorus, unless an intestinal defect is also present, these disorders produce little systemic disease and have not yet been implicated in nutritional deficiencies, although there is now some suggestion that they may be import in seniors. Cystine transport deficiency produces problems primarily because of the low solubility of this compound. Nonetheless, these disorders are important to recognize because they cause confusion with defects in metabolism, where increased serum levels of a compound cause the filtered load to exceed the tubules' capacity for reabsorption.

REFERENCES

[1] Verrer F, Singer D, Ramadan T, et al. Kidney amino acid transport. Pfugers Arch 2009;458:53–60.
[2] Kandasamy P, Gyimesi G, Kanai Y, Hediger MA. Amino acid transporters revisited: new views in health and disease. Trends Biochem Sci 2018;43(10):752–89.
[3] Govers LP, Toka HR, Walsh SB, Bockenhauer D. Mitochondrial DNA mutations in renal disease: an overview. Pediatr Nephrol 2021;36:9–17.
[4] Lemaire M. Novel Fanconi renotubular syndromes provide insights in proximal tubule pathophysiology. Am J Physiol Ren Physiol 2021;320:F145–60.
[5] Cherqui S, Courtoy PJ. The renal Fanconi syndrome in cystinosis: pathogenic insights and therapeutic perspectives. Nat Rev Nephrol 2017;13(2):115–31.
[6] Hunt DD, Stearns G, Froning EC, et al. Long term study of a family with the Fanconi syndrome and cystinuria. Surg Forum 1965;16:462–4.
[7] Smith R, Lindenbaum RH, Walton RJ. Hypophosphataemic osteomalacia and Fanconi syndrome of adult onset with dominant inheritance: possible

relationship with diabetes mellitus. Q J Med 1976;45(79):387–400.

[8] Patrick A, Cameron JS, Ogg CS. A family with a dominant form of idiopathic Fanconi syndrome leading to renal failure in adult life. Clin Nephrol 1981;16:289–92.

[9] Tolaymat A, Sakarcan A, Neiberger R. Idiopathic Fanconi syndrome in a family. Part I. Clinical aspects. J Am Soc Nephrol 1992;2:1310–7.

[10] Lichter-Konecki U, Broman KW, Blau EB, Konecki DS. Genetic and physical mapping of the locus for autosomal dominant renal Fanconi syndrome on chromosome 15q15.3. Am J Med 2001;68:264–8.

[11] Reichold M, Klootwijk ED, Reinders J, et al. Glycine amidinotransferase (GATM), renal Fanconi syndrome, and kidney failure. J Am Soc Nephrol 2018;29:1849–58.

[12] Town M, Jean G, Cherqui S, et al. A novel gene encoding an integral membrane protein is mutated in nephropathic cystinosis. Nat Genet 1998;18:319–24.

[13] Klajman A, Arber I. Familial glycosuria and aminoaciduria associated with a low serum alkaline phosphatase. Isr J Med Sci 1967;3(3):392–6.

[14] Frymoyer PA, Scheinman SJ, Dunham PB. X-linked recessive nephrolithiasis with renal failure. N Engl J Med 1991;325:681–6.

[15] Wong O, Norden AGW, Feest TG. X-linked recessive nephrolithiasis with renal failure. N Engl J Med 1992;326:1029.

[16] Scheinman S, Pook MA, Wooding C, et al. Mapping the gene causing X-linked recessive nephrolithiasis to Xp11.22 by linkage studies. J Clin Invest 1993;91:2351–7.

[17] Pook MA, Wong O, Wooding C, et al. Dent's disease, a renal Fanconi syndrome with nephrocalcinosis and kidney stones, is associated with a microdeletion involving DXS255 and maps to Xp11.22. Hum Mol Genet 1993;2:2129–34.

[18] Schurman SJ, Norden AGW, Scheinman SJ. X-linked recessive nephrolithiasis: presentation and diagnosis in children. J Pediatr 1998;132(5):859–62.

[19] Hillman RE, Albrecht I, Rosenberg LE. Identification and analysis of multiple glycine transport systems in isolated mammalian renal tubules. J Biol Chem 1968;243:5566–71.

[20] Fleck M, Schwertfeger M, Taylor PM. Regulation of renal amino acid (AA) transport by hormones, drugs and xenobiotics - a review. Amino Acids 2003;24(4):347–74.

[21] Scriver CR. Renal tubular transport of proline, hydroxyproline and glycine III. genetic basis for more than one mode of transport in human kidney. J Clin Invest 1968;47:823–35.

[22] Scriver CR, Wilson OH. Amino acid transport in human kidney: evidence for genetic control of two types. Science 1967;155(3768):1428–30.

[23] Noguchi A, Takahashi T. Overview of symptoms and treatment for lysinuric protein intolerance. J Hum Genet 2019;64(9):849–58.

[24] Sperandeo MP, Andria G, Sebastio G. Lysinuric protein intolerance: update and extended mutation analysis of the SLC7A7 gene. Hum Mutat 2008;29:14–21.

[25] Levy HL, Madigan PM, Shih VE. Massachusetts Metabolic Screening Program. I: technique and results of urine screening. Pediatrics 1972;49:825–36.

[26] Rosenberg LE, Durant JL, Elsas LJ. Familial iminoglycinuria: an inborn error of renal tubular transport. N Engl J Med 1968;278:1407–13.

[27] Hillman RE, Rosenberg LE. Amino acid transport by isolated mammalian renal tubules. II: proline transport systems. J Biol Chem 1969;244:4494–8.

[28] Greene ML, Lietman PS, Rosenberg LE, Seegmiller JE. Familial hyperglycinuria: new defect in renal tubular transport of glycine and minoacids. Am J Med 1973;54:265.

[29] Broer S, Baile CG, Kowalczuk S, et al. Iminoglycinuria and hyperglycinuria are discrete human phenotypes resulting from complex mutations in proline and glycine transporters. J Clin Invest 2008;118:3881–92.

[30] Goodman SI, McIntyre CA, O'Brien D. Impaired intestinal transport of proline in a patient with familial iminoaciduria. J Pediatr 1967;71:246–9.

[31] Garrod AE. The croonian lectures. Lancet 1908;2(173):142–214.

[32] McKusick V. Mendelian inheritance in man. Baltimore, MD: Johns Hopkins University Press; 1994.

[33] D'Ambrosio V, Capolongo G, Goldfarb D. Cystinuria: an update on pathophysiology, genetics, and clinical management. Pediatr Nephrol 2021. https://doi.org/10.1007/s00467-021-05342-y.

[34] Servais A, Thomas K, Dello Strologo L, et al. Cystinuria: clinical practice recommendation. Kidney Int 2021;99(1):48–58.

[35] Weinberger A, Sperling O, Rabinovitz M, et al. High frequency of cystinuria among Jews of Libyan origin. Hum Hered 1974;24(5–6):568–72.

[36] Bostrom H, Tottie K. Cystinuria in Sweden. II: the incidence of homozygous cystinuria in Swedish school children. Acta Paediatr 1959;48:345–52.

[37] Brodehl J, Gallissen K, Kowalewski S. Isolated cystinuria (without lysine-ornithine-arginuria) in a family with hypocalcemic tetany. Klin Wochenschr 1967;115(4):317–20.

[38] Modersitzki F, Pizzi L, Grasso M, Goldfarb DS. Natural history and quality of life in patients with cystine

urolithiasis: a single centre study. BJU Int 2015;116(Suppl. 3):31–5.
[39] Huizing M, Gahl WA. Inherited disorders of lysosomal membrane transporters. Biochim Biophys Acta Biomembr 2020;1862(12):183336.
[40] Topaloglu R. Nephropathic cystinosis: an update of genetic conditioning. Pediatr Nephrol 2021;36:1347–52.
[41] Topaloglu R, Keser AG, Gulhan B, et al. Cystinosis beyond kidneys: gastrointestinal system and muscle involvement. BMC Gatroenterol 2020;20:242–9.
[42] Machuca-Gayet I, Quinaux T, Bertholet-Thomas A, et al. Bone disease in nephropathic cystinosis: beyond renal osteodystrophy. Int J Mol Sci 2020;21:3109. https://doi.org/10.3390/ijma221093109.
[43] Kasimer RE, Langman CB. Adult complications of nephropathic cystinosis: a systematic review. Pediatr Nephrol 2021;36:223–36.
[44] Curie A, Touil N, Gaillard S, et al. Neuropsychological and neuroanatomical phenotype in 17 patients with cystinosis. Orphanet J Rare Dis 2020;15(1):59.
[45] Rajantie J, Simell O, Rapola J, et al. Lysinuric protein intolerance: a two year trial of dietary supplementation therapy with citrulline and lysine. J Pediatr 1980;97(6):927–32.
[46] Whelan DT, Scriver CR. Hyperdibasic aminoaciduria: an inherited disorder of amino acid transport. Pediatr Res 1968;2(6):525–34.
[47] Scriver CR, Mahon B, Levy HL, et al. The Hartnup phenotype: Mendelian transport disorder, multifactorial disease. Am J Hum Genet 1987;40(5):401–12.
[48] Wilcken B, Yu JS, Brown DA. Natural history of Hartnup disease. Arch Dis Child 1977;52:38–40.
[49] Kleta R, Romeo E, Risti Z, et al. Mutations in SLC6A19, encoding BoAT1, cause Hartnup disorder. Nat Genet 2004;36(9):999–1002.
[50] Bagga A, Sinha A. Renal tubular acidosis. Indian J Pediatr 2020;87(9):733–44.
[51] Scriver CR, MacDonald W, Reade TM, et al. Hypophosphatemic non-rachitic bone disease. Am J Med Genet 1977;1:101–17.
[52] Kashoor I, Batile D. Proximal renal tubular acidosis with and without Fanconi syndrome. Kidney Res Clin Pract 2019;38(3):267–81.
[53] Brenes LG, Brenes JN, Hernandez MM. Familial renal tubular acidosis: a distinct clinical entity. Am J Med 1977;63:244–52.
[54] Sakakibara N, Ijuin T, Horinouchi T, et al. Identification of novel OCRL isoforms associated with phenotypic differences between Dent disease-2 and Lowe syndrome. Nephrol Dial Transplant 2021. https://doi.org/10.1093/ndt/gfab274.
[55] Gianesello L, Arroyo J, Del Prete D, et al. Genotype phenotype correlation in Dent Disease 2 and review of the literature: OCRL gene pleiotropism or extreme phenotypic variability of Lowe syndrome? Genes 2021;12(10):1597. https://doi.org/10.3390/genes12101597.
[56] David S, De Waele K, De Wilde B, et al. Hypotonia and delayed motor development as an early presentation of Lowe syndrome: case report and literature review. Acta Clin Belg 2019;74(6):460–4.
[57] Chaabani H, Hadj-Khlil A, Ben-Dhia N. The primary hereditary form of distal renal tubular acidosis: clinical and genetic studies in 60-member kindred. Clin Genet 1994;45:194–9.
[58] Naveen PS, Srikanth L, Venkatesh K, et al. Distal renal tubular acidosis with nerve deafness secondary to ATP6B1 gene mutation. Saudi J Kidney Dis Transpl 2015;26(1):119–21. https://doi.org/10.4103/1319-2442.148757.
[59] Nance WE, Sweeney A. Evidence for autosomal recessive inheritance of the syndrome of renal tubular acidosis with deafness. Birth Defects Orig Artic Ser 1971:770–2.
[60] Giglio S, Montini G, Trepiccione F, et al. Distal renal tubular acidosis: a systematic approach from diagnosis to treatment. J Nephrol 2021;34(6):2073–83. https://doi.org/10.1007/s40620-021-01032-y.
[61] Sly WS, Whyte MP, Sundaram V, et al. Carbonic anhydrase II deficiency in 12 families with the autosomal recessive syndrome of osteopetrosis with renal tubular acidosis and cerebral calcification. N Engl J Med 1985;313:139–45.
[62] van Karnebeek C, Häberle J. Carbonic anhydrase VA deficiency. In: Adam MP, Ardinger HH, Pagon RA, Wallace SE, Bean LJH, Mirzaa G, Amemiya A, editors. Gene reviews® [internet]. Seattle (WA): University of Washington; 2021. p. 1993–2021.
[63] Fathallah DM, Bejaoui M, Lepaslier D, et al. Carbonic anhydrase II (CA II) deficiency in Maghrebian patients: evidence for founder effect and genomic recombination at the CA II locus. Hum Genet 1997;99:634–7.
[64] Hjarne V. Study of orthoglycemic glycosuria with particular reference to its hereditability. Acta Med Scand 1927;67:422.
[65] Elsas LJ, Busse D, Rosenberg LE. Autosomal recessive inheritance of renal glycosuria. Metabolism 1971;20:968–75.
[66] Kanai Y, Lee WS, You G, et al. The human kidney low affinity Na+/glucose cotransporter SGLT2. Delineation of the major renal reabsorptive mechanism for D-glucose. J Clin Invest 1994;93(1):397–404.
[67] Tran C. Inborn errors of fructose metabolism. What can we learn from them? Nutrients 2017;9:356.

[68] Khachadurian AK. Essential pentosuria. Am J Hum Genet 1962;14:249–56.
[69] Pierce SB, Spurrell CH, Mandell JB, et al. Garrod's fourth inborn error of metabolism solved by the identification of mutations causing pentosuria. Proc Natl Acad Sci USA 2011;108:18313–7.
[70] Baroncelli GI, Mora S. X-linked hypophosphatemic rickets: multisystem disorder in children requiring multidisciplinary management. Front Endocrinol 2021:12. https://doi.org/10.3389/fendo.2021.688309.
[71] Christensen S, Tebben PJ, Sas D, Creo AL. Variable clinical presentation of children with hereditary hypophosphatemic rickets with hypercalcuria. Hum Res Paediatr 2021. https://doi.org/10.1159/000520299.
[72] Arnold GL. Inborn errors of metabolism in the 21st century: past to present. Ann Transl Med 2018;6(24):467–75.
[73] Hall AM, Trepiccione F, Unwin RJ. Drug toxicity in the proximal tubule: new models, methods and mechanisms. Pediar Nephrol 2021. org/101007/s00467-021-05121-9.
[74] Forst A-L, Reichold M, Kleta R, Warth R. Distinct mitochondrial pathologies caused by mutations of the proximal tubular enzymes EHHADH and GATM. Front Physiol 2021;12:1–10.
[75] Heidari R. The footprints of mitochondrial impairment and cellular energy crisis in the pathogenesis of xenobiotic-induced nephrotoxicity, serum electrolytes imbalance, and Fanconi's syndrome. Toxicology 2019;423:1–31.
[76] Abdlmasih R, Abdelmaseih R, Thakker R, et al. Update on the cardiovascular benefits of sodium-glucose co-transporter-2 inhibitors: mechanism of action, available agents and comprehensive review of the literature. Cardiol Res 2021;12(4):210–8.
[77] Shaffner J, Chen B, Malhotra DK, et al. Therapeutic targeting of SGL2: a new era in the treatment of diabetes and diabetic kidney disease. Front Endocrinol 2021;12:749010.
[78] Yamazaki T, Mimura I, Tanka T, Nangaku M. Treatment of diabetic kidney disease: current and future. Diabetes Metab J 2021;45:11–26.
[79] Schernthaner G, Shehadeh N, Armetov AS, et al. Worldwide inertia to the use of cardiorenal protective glucose-lowering drugs (SGLT2i and GLP-1 RA) in high risk patients with type 2 diabetes. Cardiovasc Diabetol 2020;19:185–202.
[80] Gronda E, Jessup M, Iacoviello M, et al. Glucose metabolism in the kidney: neurohormonal activation and heart failure development. J Am Heart Assoc 2020;9:e018889. https://doi.org/10.1161/JAHA.120.018889.
[81] Prattichizzo F, de Candia P, Ceriello A. Diabetes and kidney disease: emphasis on treatment with SGLT-2 inhibitors and GLP-1 receptor agonists. Metab Clin Exp 2021;120:154799. https://doi.org/10.1016/j.metabol.2021.154799.

5

APOL1-Associated Kidney Disease

Martin R. Pollak, David J. Friedman

Division of Nephrology, Department of Medicine, Beth Israel Deaconess Medical Center and Harvard Medical School, Boston, MA, United States

5.1 INTRODUCTION

Two coding variants in the Apolipoprotein-L1 gene APOL1 are strongly associated with risk of a variety of kidney disease phenotypes under a recessive pattern of inheritance. APOL1 is an atypical disease gene: its disease-associated genotypes are both fairly common and have a large effect on disease risk. There is not universal agreement on how the various APOL1-associated phenotypes should best be named—lumped together under some category such as "APOL1-associated kidney disease" or split up by the various presentations influenced by APOL1 genotype, such as focal segmental glomerulosclerosis (FSGS) and hypertension-attributed end-stage kidney disease (H-ESKD). Here, we will largely follow the lumping strategy unless pointing out specific differences in clinical presentation.

Worldwide, more than 2 million people receive some form of dialysis or have a kidney transplant [1]. This number would be significantly larger if so-called "renal replacement therapy" (specifically dialysis or kidney transplantation) was more widely available and able to extend the lives of these people. In the United States, there is a large racial disparity in rates of kidney disease, with much greater rates in people who identify as Black or African American. Although only about 12% of the US population identifies as African American, this group comprises about 40% of the ESKD patients. A major contributor to this disparity is a pair of specific variants in APOL1, which originated in western sub-Saharan Africa [2,3].

A single, very strong kidney disease locus on chromosome 22 was identified through genetic studies of African American individuals with and without kidney disease [4,5]. These admixture mapping-based genetic association studies observed a large excess of recent African ancestry among African Americans with FSGS or H-ESKD at the chromosome 22 locus. We now know that two coding sequence variants in APOL1 drive this association. The first of these, known as the G1 allele, codes for two amino acid substitutions near the C terminus that nearly always occur together, S342G and I384M (dbSNP numbers rs73885319 and rs60910145). The second, known as G2 (rs71785313), is a six-base-pair deletion leading to loss of amino acid residues 388N and 389Y (Fig. 5.1).

Inheritance of kidney disease risk follows an essentially recessive pattern. Inheritance of two risk alleles (one from each parent) markedly increases the risk of kidney disease, while inheritance of one risk allele is associated with only a minimal increase. These basic findings have been widely replicated [6]. Although the

Emery and Rimoin's Principles and Practice of Medical Genetics and Genomics. https://doi.org/10.1016/B978-0-12-812534-2.00007-2

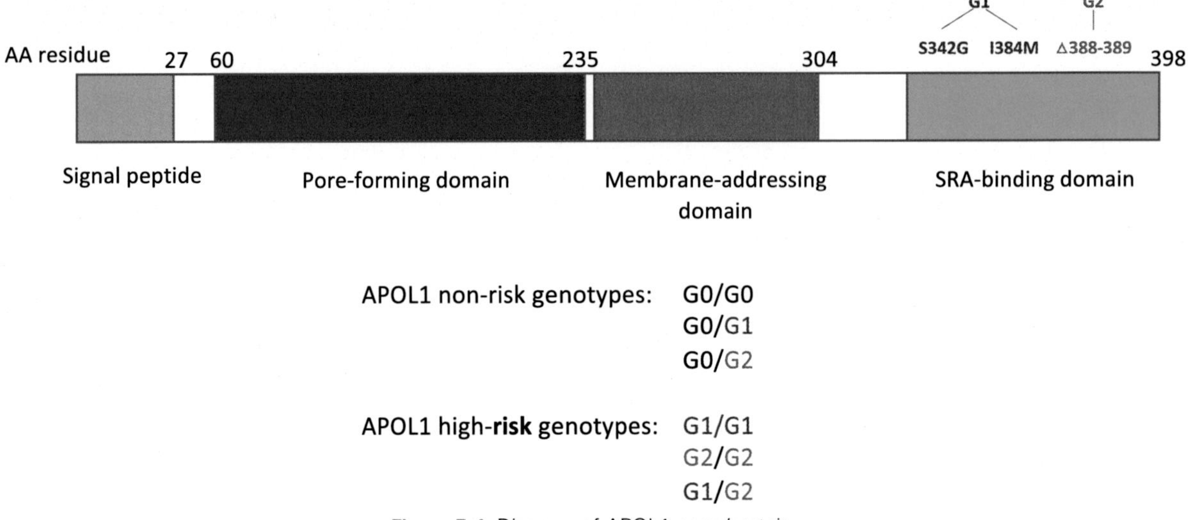

Figure 5.1 Diagram of APOL1 gene/protein.

APOL1 locus was initially identified based on studies in African Americans, we now know that it is ancestry driving this particular genetic risk factor. Studies in Africa and South America show that this is not a geographically limited association [7,8].

It is important to note that most individuals who identify as Black or African American do not have high-risk APOL1 genotypes. Conversely, not everyone with a high-risk APOL1 genotype identifies as Black or African American. The presence of APOL1 risk alleles, and therefore APOL1-associated kidney disease, is associated with ancestry, which only roughly correlates with racial identity.

Case-control odds ratios for APOL1-associated kidney disease differ somewhat by phenotypic classification. Odds ratios are ~7–10 for H-ESKD, 10–20 for FSGS, and have been reported to be as high as 89 for HIVAN [9,10]. These alleles are common in people of recent African ancestry. Among African Americans, G1 frequency is 23% and the G2 frequency 15%. Most people now refer to any allele lacking G1 or G2 as G0. The SNPs that define G1 are in just under 100% linkage disequilibrium. Approximately 1% of haplotypes with the G342 allele do not also have the M384 allele [11]. In the absence of M384, the G342 allele seems to have the same effect on disease risk as these alleles do together [12]. There are six basic APOL1 genotypes that an individual can have: the three nonrisk genotypes G0/G0, G0/G1, and G0/G2, and the three high-risk genotypes G1/G1, G2/G2, and G1/G2. About 12% of African Americans have a high-risk genotype, while about 50% of African Americans carry either one or two copies of risk variant APOL.

Although G1 is defined by the presence of two specific SNP alleles and G2 is defined by a six base-pair deletion, the genomic context (i.e., haplotype) on which these G1 or G2 defining variants reside affects their pathogenicity in experimental systems [13]. There are several common APOL1 haplotypes that lack G1 or G2. Thus G0, as commonly used, refers to any haplotype that is not G1 or G2. Whether there are biologically important differences between the naturally occurring G0 haplotypes is not known.

5.2 APOL1-ASSOCIATED NEPHROPATHIES

The *APOL1* risk locus and genotypes were originally identified in African Americans with FSGS, HIV-associated nephropathy (HIVAN), and H-ESKD [4,5]. FSGS is a histologically defined entity, referring to sclerosis in parts of some of the glomeruli. Typically, this histologic pattern is observed in individuals with abnormally high urine protein excretion and can present in childhood or adulthood. Close to 75% of African Americans with FSGS have a high-risk APOL1 genotype [9]. Individuals with FSGS often progress to ESKD. Those with a high-risk genotype typically have a more

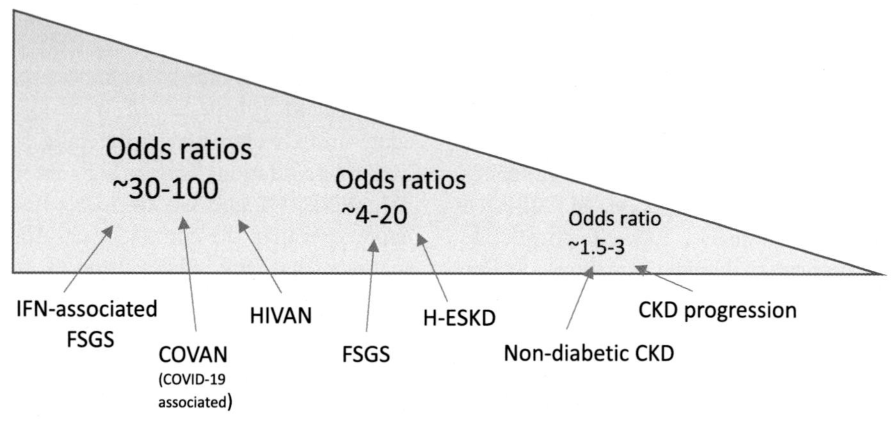

Figure 5.2 Spectrum of phenotypes picture.

rapidly progressive course of deteriorating kidney function [14,15] (Fig. 5.2).

Hypertension-attributed ESKD is strongly associated with *APOL1* risk variants, as is nondiabetic chronic kidney disease (CKD) [16]. HIV nephropathy (or HIVAN) is an infection-driven kidney disease that is morphologically similar to the aggressive form of FSGS known as collapsing nephropathy [10]. Aggressive forms of lupus-associated kidney disease [17,18] and membranous nephropathy are also associated with *APOL1* genotype [19]. Most recently, it has been observed that collapsing nephropathy developing acutely in the setting of severe COVID-19 infection is overwhelmingly observed in individuals with high-risk APOL1 genotypes [20,21]. Preeclampsia, a disorder of pregnancy characterized by proteinuria and hypertension, was reported to be associated with a high-risk *APOL1* genotype in the fetus and, in some but not all studies, the mother [22–24]. Most of these disorders share evidence of an inflammatory trigger to the development of overt kidney disease (see more discussion below).

APOL1 genotype and diabetic nephropathy (DN) do not share a simple association. The APOL1 risk genotype does not appear to increase the risk of incident nephropathy in the setting of preexisting diabetes [25]. However, APOL1 genotype does associate with progression of kidney dysfunction in individuals with already established DN [26]. Both APOL1 kidney disease and DN are common; likely many individuals who receive the clinical diagnosis of DN in fact have APOL1-associated kidney disease.

A large fraction of ESKD in African Americans has been traditionally attributed to the effects of hypertension. In fact, so-called H-ESKD is likely driven by APOL1-associated kidney disease, with hypertension occurring secondary to CKD or in parallel with CKD [27]. It is noteworthy that *APOL1* genotype is a stronger predictor of declining kidney function than is the intensity of blood pressure control [26].

Thus, what we refer to as APOL1-associated kidney disease spans a wide phenotypic spectrum, including presentations that range from indolent to aggressive, nonproteinuric to highly proteinuric, and presenting in both inflammatory and noninflammatory settings. Lumping these entities together under the rubric of "APOL1-associated kidney disease" emphasizes the shared strong genetic risk factor and, in all likelihood, shared mechanisms of disease development.

The association between APOL1 and kidney disease is seen in all age groups, including children [28]. Children with nephrotic syndrome and high-risk APOL1 genotypes have a more aggressive disease with faster decline in kidney function [29]. This association is not just seen in idiopathic kidney disease in children, but also in children with HIV-associated nephropathy [30]. There are major geographic differences in the genetic factors underlying childhood kidney disease and nephrotic syndrome. For example, one study in South Africa found that a specific mutation in the podocin gene NPHS2 to be a major contributor to nephrotic syndrome in children, but no association with APOL1 genotype [31].

In addition to viral infections such as HIV and COVID-19, other inflammatory processes lead to more frequent and more aggressive kidney disease in high-risk APOL1 individuals. This is now well documented with SLE-associated kidney disease (i.e., lupus nephritis). While APOL1 genotype does not influence the incidence or activity of SLE per se, SLE-affected individuals have more frequency lupus nephrotic, as well as more aggressive kidney function decline [17,32–34].

5.3 HUMAN GENETICS, TRYPANOLYSIS, AND APOL1

APOL1 is an unusual human disease gene. The disease-associated variants G1 and G2 are common in people of recent west African ancestry. Approximately 13% of African Americans have a high-risk APOL1 genotype [9]. The frequencies of high-risk genotypes are even higher in parts of Africa. The effect size is also high—these individuals have a markedly increased risk of kidney disease (from 1.5- to 89-fold, depending on the study and kidney phenotype in question). The high frequency of such deleterious genetic variants is unusual. This combination of strong effect and high frequency is related to the origin and history of these variants.

The APOL1 gene is one of six members of a family of APOL genes on human chromosome 22 [35]. APOL1 is absent from the genomes of all nonprimates. It is present only in a few primate species and has disappeared from the genome of our close relative, the chimpanzee [35,36]. APOL1's role as the trypanolytic factor of human serum was understood before its role in kidney disease. The circulating APOL1 protein protects humans, gorillas, baboons, and some Old World monkey species against common African trypanosomes [37–39]. APOL1 protein is taken up by trypanosomes, whereupon APOL1 causes trypanosome lysis.

The two coding sequence variants known as G1 and G2 arose in humans in Africa 5,000–10,000 years ago. Because these risk variants arose after the Out-of-Africa expansion, these two variants have been observed only in individuals with recent African ancestry [40,41]. The high frequency of the deleterious *APOL1* kidney risk alleles strongly suggests that they have beneficial as well as deleterious properties. It is believed that the frequency of G1 and G2 rose quickly in Africa because these variants conferred enhanced protection against the virulent subspecies of trypanosomes that cause acute and chronic African sleeping sickness [2,40]. In human populations with a high frequency of the two risk alleles, G1 and G2 are located on larger genomic haplotypes than the nonrisk or G0 *APOL1*. This finding suggests that risk alleles rose in frequency quickly within the last several thousand years. (In general, common variants are old; because of recombination, old variants are located on smaller shared haplotypes than recent variants. Variants under positive selection pressure may come to high frequencies more quickly and therefore exist on larger conserved haplotypes.)

T. br. Rhodesiense is the subspecies of African trypanosome responsible for acute African sleeping sickness in humans. Both G1 and G2 variants are more effective at killing pathogenic *T. br. Rhodesiense* than G0 variants in experimental models. This supports the notion of a selective sweep by the risk variants. More recently, investigators have confirmed a protective effect of G2 against *T. br. rhodesiense* infection in case-control studies, consistent with trypanolysis studies in culture and in mice (but not for G1) [42,43]. While G2 *predisposes* to symptomatic infection with *T. br. Gambiense,* the chronic sleeping sickness agent, G1 does not prevent infection per se but does seem to protect against *symptomatic* infection. Thus, it appears that G2 is older than G1 and protects against *T. br. rhodesiense*, while G1 arose more recently and spread quickly because of its ability to protect humans from *T. br. Gambiense.*

5.4 THE APOL1 GENE AND PROTEIN

APOL1 differs from other APOL proteins by the presence of a signal peptide in some splice isoforms, which enable export from the cell. The most abundant transcripts encode this signal peptide, while it is spliced out of some transcripts, leading to intracellularly forms of APOL1. The protein has traditionally been organized into three major domains, as defined by their putative roles in trypanolysis [38,42]. The N-terminal portion is referred to as the colicin-like domain, named for putative homology to bacterial colicins that can perforate cell membranes. More recent studies suggest that it is instead the C-terminus that is the main pore-forming domain, and structural studies have not supported the existence of a colicin-like domain at the N-terminus, though the nomenclature has lagged behind the science. The central part of APOL1 is a membrane-addressing domain, which provides a pH-dependent mechanism for the activation of APOL1 activity in acid environments. The APOL1 kidney risk variants are within the C terminal SRA-binding domain, the target of the *T. br. rhodesiense* SRA (Serum Resistance Associated) virulence factor. The APOL1 kidney risk variants alter binding to SRA. This C-terminal SRA-binding domain includes a coiled-coil region with a leucine zipper motif, creating an amphipathic helix that is required for SRA binding and possibly for its interaction with lipids [44].

APOL1 circulates at high levels on HDL3 molecules or in a lipid-poor state complexed to immunoglobulin M (IgM) [37,45,46]. Circulating APOL1 is generated by the liver [47]. APOL1 expression is increased in the setting of inflammation, supporting its known role as an immune defense molecule. *APOL1* responds to a variety of cytokines, including interferons, lipopolysaccharide, toll-like receptor (TLR) agonists, tumor necrosis factor (TNF) [12,48]. The APOL1 promotor contains canonical interferon response elements [49,50].

5.5 APOL1 FUNCTION AND EFFECT OF VARIANTS

APOL1 leads to trypanosome lysis. The precise molecular mechanisms leading to trypanosomal killing are still debated [51]. It is thought that APOL1 leads to trypanosome lysis by inserting into trypanosome membranes and acting as an ion channel. In human cells, APOL1 is able to act as a channel as well, leading to speculation that APOL1-associated kidney disease is related to its channel function [52,53]. APOL1 expression in xenopus oocytes has been found to increase ion permeability and lead to severe BH3-independent toxicity [54] (Fig. 5.3).

Using lipid bilayer systems and recombinant APOL1 protein, investigators have shown that APOL1-induced

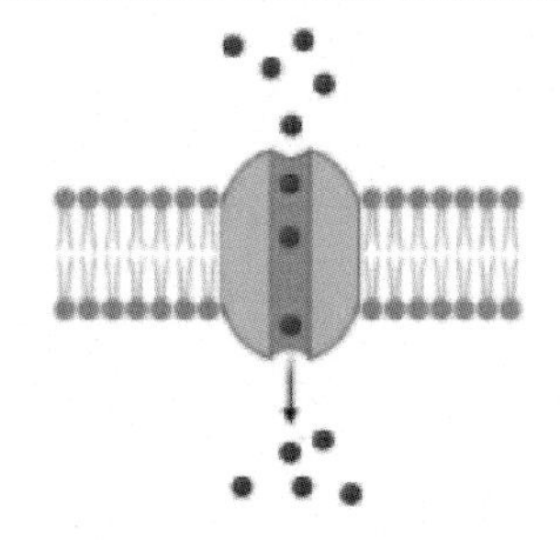

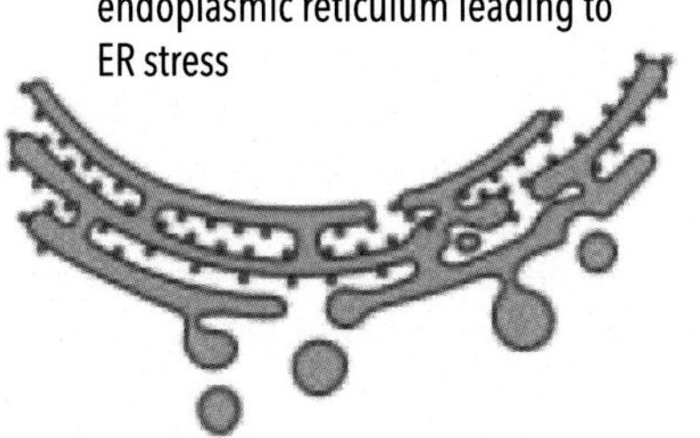

Figure 5.3 Mechanism cartoon.

cation conductances require an initial step at acidic pH, but that the magnitude of these conductances could be increased thousands-fold upon alkalinization [55,56]. This two-step model may be relevant to APOL1 activity and pathogenicity in human cells, where the pH of different cellular compartments may affect its channel activity. In cell-free systems, the G0, G1, and G2 forms of APOL1 have similar ion conductances [55]. While some mammalian cell experiments have not observed differences in APOL1 channel activity based on genotype, there does appear to be a growing consensus that, in some settings, the G1 and G2 forms of APOL1 can have gain-of-function channel effects [57–60]. Both the channel activity and pH gating appear to be controlled by pore-lining residues within the C-terminal domain [56].

The absence of APOL1 from almost all other mammalian genomes suggests that APOL1 does not play an essential role in kidney function. The existence of an APOL1-deficient human being with normal kidney function similarly suggests that APOL1 pays no essential role in human kidney function or development [61].

Overexpression of the *APOL1* G1 or G2 variants causes increased cell toxicity and death in a variety of experimental cell systems [58,60,62–65]. G0 is associated with some toxicity in some studies, but typically much less than that seen with G1 or G2. Patient-derived podocyte cell lines have shown similar differences [66].

Some early studies that suggested that autophagic cell death was important in APOL1-mediated toxicity, consistent with the presence of a BH3-only domain [67]. Enhanced apoptosis has been reported in some APOL1-overexpressing cell systems, with variant-specific differences [48,54,58,68]. Effects on cell swelling have suggested a pore-forming activity of APOL1 [52,58].

Increased ER stress has been suggested to pay a causal role in APOL1 risk-variant-induced toxicity [62,69,70]. Several recent studies have implicated the cGAS-STING pathway, important in the inflammatory response, in mediating APOL1 toxicity [71–73]. Yet another set of studies have suggested that APOL1 C-terminal variants may cause kidney cell injury disease by preventing its homolog APOL3 from activating PI4KB, leading to reorganization of the actin cytoskeleton [74,75].

While there is fairly widespread agreement that the APOL1 expressed in the kidney contributes to the etiology of kidney disease, a role for circulating and extrarenal APOL1 has not been excluded. For example, a possible role of APOL1 expressed in lymphocytes or other immune cells has been recently suggested [76].

5.6 RECESSIVE BUT GAIN OF FUNCTION

It is puzzling that *APOL1* kidney risk variants confer risk in an essentially recessive inheritance pattern if they do in fact cause kidney injury through a gain-of-function mechanism. This is atypical. One possibility is dose dependence. More variant APOL1 is expected to be expressed when there are two risk variant copies of APOL1 compared to one. However, the fact that APOL1 expression can be upregulated 100-fold with inflammatory factors such as interferons argues against this model as normal differences in *APOL1* gene expression at baseline and in inflammatory states would likely dwarf the effect of a twofold difference in gene dosage [12]. APOL1 multimerization could also help explain recessive inheritance, with toxic properties seen if the multimer is made from risk-variant APOL1 subunits only [44,77]. Another plausible hypothesis is that it is not so much the extra "dose" of risk-variant APOL1 as it is the presence of a G0 APOL1 that nullifies risk variant toxicity in heterozygotes as opposed to homozygotes. Recent studies in APOL1 transgenic mice of various genotype combinations and in cells in culture have not been consistent with this "G0 rescue" hypothesis [68,78].

5.7 MODELS OF APOL1-ASSOCIATED DISEASE

5.7.1 Mice

The development of animal models of *APOL1*-associated disease is nontrivial. An APOL1 gene is not present in any of the animals commonly used to model human disease. Bruggeman and colleagues [79] developed a mouse model in which a podocyte-specific promoter drives expression of G0 or G2 forms of APOL1. The G2 transgenic mice developed a preeclampsia-like phenotype. The G0 mice showed a milder version of this phenotype. Beckerman et al. [80] developed differently a mouse model that shows podocyte-specific inducible expression of the G0, G1, or G2 form of APOL1. Podocyte-specific expression of G1 or G2 led to proteinuria and glomerulosclerosis. The degree of kidney damage was found to correlate with APOL1 expression. Ryu et al. [81] described the development of BAC transgenic mice, which carry a bacterial artificial chromosome that contains the entire human *APOL1* genomic region. At 6 months of age, these mice did not demonstrate an overt kidney phenotype.

McCarthy et al. developed coisogenic bacterial artificial chromosome (BAC) transgenic mice harboring the G0, G1, or G2 form of human APOL1. Expression of interferon gamma (IFN-gamma) via plasmid tail vein injection resulted in upregulation of APOL1 protein levels together with robust induction of heavy proteinuria and glomerulosclerosis in G1/G1 and G2/G2 but not G0/G0 mice. Neither heterozygous (G1/G0 or G2/G0) risk-variant mice nor hemizygous (G1/-, G2/-) mice had significant kidney injury in response to IFN-gamma, suggesting that the lack of significant disease in humans heterozygous for G1 or G2 is not due to G0 rescue of G1 or G2 toxicity. Studies using additional mice (multicopy G2 and a nonisogenic G0 mouse) supported the hypothesis that disease is at least in part a function of the level of APOL1 expression [78].

5.7.2 Other Organisms

Other model organisms have also been used to study APOL1-associated kidney disease. Zebrafish have one *APOL* gene, *zAPOL1*. Knockdown of zAPOL1 was found to cause defects in the function of the larval pronephric glomerulus [82,83]. Olabisi et al. expressed G0 or risk variant APOL1 in a podocyte-specific or endothelium-specific manner. The risk-variant expressing fish showed mild ultrastructural changes on electron microscopy that were absent from G0 fish phenotype [84].

Two independent studies reported similar findings when APOL1 was overexpressed in the *Drosophila* nephrocyte, a podocyte-like cell [85,86]. Expression of risk-variant APOL1 in nephrocytes led to much greater hypertrophy, increased cell death, and defective organelle acidification than what was seen with the G0 form [85,86]. Similar effects were observed in a yeast model [85].

5.8 NONKIDNEY PHENOTYPES

APOL1 variants do not seem to have a major effect on most extrarenal phenotypes but various associations (with smaller effect size than for kidney disease) have been reported. These effects, if real, are small; the primary pathologic effect of *APOL1* high-risk genotype is the kidney [87–89]. One explanation for the predominantly kidney-limited nature of APOL1-associated disease may be related to the nature of the kidney's podocytes. These cells are terminally differentiated and have an elaborate structure maintained by an intricate actin cytoskeleton. While most studies of APOL1-associated disease have focused on the podocyte, the wide variety of clinical phenotypes observed may be due to effects of APOL1 in different cell types, depending on the setting [90–92].

There have been conflicting reports on the kidney-independent effects of APOL1 genotype on cardiovascular phenotypes such as heart disease and blood pressure [93–95]. There have also been hard-to-interpret reports of paradoxical effect of APOL1 genotype on mortality [96]. One recent study reported an association between high-risk APOL1 variants and obesity under an additive genetic model [97]. A recent phenome-wide analysis suggested that the nonrenal associations with APOL1 are largely driven by its effects on the kidney [87].

APOL1 is quite widely expressed and circulates at high levels in the bloodstream. Most of the circulating APOL1 is made by the liver [47]. However, the growing consensus is that the kidney-expressed APOL1, rather than circulating APOL1, is responsible for kidney disease. One recent paper reported that APOL1 risk variants specifically present in individuals of African ancestry were associated with increased incidence of sepsis. Data in a mouse model suggested that endothelial expression of APOL1 might mediate this altered susceptibility.

Another recent report found, in contrast to an older study, that kidney recipient APOL1 genotype did alter the posttransplant course, with the number of high-risk alleles in the recipient associated with an increased risk of death-censored allograft loss (i.e., an additive effect) [76]. High expression of APOL1 in activated CD4+/CD8+ T cells and natural killer cells raised the possibility that altered immune function may contribute to posttransplant course.

Several groups have looked at the relationships between APOL1 genotype and preeclampsia, the hypertensive and proteinuric disorder of late pregnancy [24]. Reidy et al. reported APOL1 genotype of the fetus but not the mother fetal nor maternal with preeclampsia [98]. Another study conducted in South African women showed an association of maternal G1 but not G2 allele with early onset [22]. A recent single center study in the United States reported that infant APOL1 genotype was significantly associated with preeclampsia [99]. Most recently, Hong et al. reported the surprising finding that both fetal and maternal APOL1 genotypes associated with greater risk of preeclampsia in African American women but not Haitian American women [23]. Still, while preeclampsia is more common in Black women than white women in the United States, the data do not support an effect for APOL1 in preeclampsia of anywhere near the magnitude as its role in kidney disease.

5.9 APOL1 SECOND HITS: GENES AND/OR ENVIRONMENT

Most individuals with an *APOL1* high-risk genotype never develop overt kidney disease. This indicates that other factors must also be present to drive *APOL1* kidney disease in high-risk genotype individuals with disease. Genetics and environmental factors, or more likely a combination of both, must be contributors to disease presentation (or its absence).

Studies looking for genetic modifiers of the *APOL1* risk genotype have not yet identified genetic factors with major disease-modifying effects. In studies of ESKD subjects, several candidate gene studies have shown some association (e.g., GSTM1, APOL3, or the HGB allele HbS [100–103]. In people with FSGS, where the effect size of *APOL1* risk variants is large, one study found evidence for a modifier locus on chromosome 6 at near the HLA locus [101].

Environmental triggers are probably more important. The odds ratio for the development of HIV-associated nephropathy (HIVAN) and a high-risk *APOL1* is on the order of 29–89. It is thought that in the pre-HAART era, 50% of individuals with a high-risk genotype would develop HIVAN [10,104,105]. In a series of people who developed the severe form of glomerular injury known as collapsing glomerulopathy after treatment with interferon, all were found to have a high-risk *APOL1* genotype [49,106]. Interferons can upregulate APOL1 expression 100-fold in cellular systems. Thus, one model for the triggering of overt disease is that both a high-risk *APOL1* genotype and high APOL1 expression combine to cause kidney cell toxicity and kidney disease. Complex associations between viruria, APOL1 genotype, and kidney disease in African Americans also suggest a role for viruses [107,108].

Over the past 2 years, a new APOL1-associated entity has been described, referred to as COVAN, or COVID-19 -associated nephropathy, as evidenced by numerous recent case series and case reports [21,109,110]. Over 90% of people with COVID-19-associated collapsing nephropathy have a high-risk APOL1 genotype [20]. Presumably the highly inflammatory state associated with severe COVID-19 disease leads to massive upregulation of APOL1 expression.

The nongenetic environmental factors that can trigger or exacerbate kidney disease in high-risk APOL1 individuals are not limited to infectious agents. Environmental toxins and pollutants appear to contribute as well, although the details are not well understood, and confounding is hard to rule out [111,112].

5.10 CLINICAL IMPLICATIONS

It is not yet routine to perform genetic testing in individuals with kidney disease. It is becoming more commonplace, due in part to the increasing ease of DNA sequencing and the availability of large kidney disease gene panels [113]. Similarly, APOL1 genotyping is not a routine part of clinical practice, although the details of how APOL1 genotyping should be used in the clinic are the subject of active debate [114].

5.11 APOL1 AND KIDNEY TRANSPLANTATION

APOL1 influences kidney survival in kidney transplant recipients. The *APOL1* genotype of a deceased-donor kidney impacts its fate, as kidneys with high-risk *APOL1* genotype tend to fail earlier than kidneys with zero or one of these high-risk alleles [115]. This suggests that kidney-expressed APOL1 is a major driver of *APOL1* kidney disease. By contrast, the effect of a high-risk *APOL1* genotype in a transplant recipient is less clear. An early report suggested that a high-risk *APOL1* genotype in a transplant recipient does not alter the survival of the transplanted kidney [116], though this has been called into question by a recent study [76]. The utilization of *APOL1* genotyping in the clinical evaluation of kidney transplant donors remains a matter of some debate, though it may lead to more efficient allocation of kidneys by more accurately assessing allograft failure risk {27862962} (Fig. 5.4).

One clinical setting in which APOL1 genotyping may have immediate implications is in decision-making about kidney transplantation. There are two main issues: does the APOL1 genotype of the donor kidney have an impact on the posttransplant course? And does the APOL1 genotype of a living donor affect his or her risk of kidney damage after kidney donation? [117]. There is fairly wide agreement that transplant physicians should discuss APOL1 genotyping with potential living kidney donors who self-report recent African ancestry or identify as Black or African-American. Because living kidney donors must be free of hypertension, kidney disease, and proteinuria, many candidate donors with a high-risk genotype will be excluded from donation based on clinical criteria.

Some studies have suggested that kidney donors with high-risk APOL1 genotypes are at increased risk of future kidney disease [118–120]. Given the existing racial disparities in access to kidney transplantation and the more general racial disparities in medical care, the clinical implementation of these observations is complex [121,122]. On the one hand, there is a desire to help kidney failure patients avoid chronic dialysis by having a kidney transplant. On the other hand, there is a justifiable wariness among many physicians to allow individuals with a high-risk APOL1 genotype to potentially increase their risk of kidney failure [123,124]. A large ongoing US National Institutes of Health sponsored trial known as APOLLO may help answer these questions [125].

5.12 APOL1 IN THE CLINIC

Is APOL1 genotyping useful in the evaluation and care of patients with new or established kidney disease, such as FSGS? Patients with either CKD or FSGS have a less favorable prognosis as a function of APOL1 genotype [15,26]. Although initial response to standard immunosuppressive does not appear to vary by genotype, long-term risk of progression to kidney failure does [15,126]. While there are some subtle correlations between histologic features and APOL1 genotype, genotype cannot be inferred from histological findings [127].

Nevertheless, knowing the genotype of a patient with FSGS has implications for the risk of kidney disease in family members and also for prognosis. At least two pharmaceutical companies are developing targeted therapies for individuals with APOL1-associated kidney

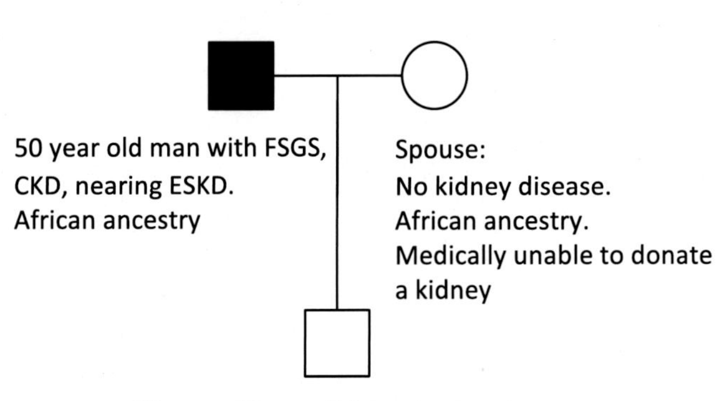

SOME OF THE CLINICAL ISSUES TO CONSIDER:

1. What is the likelihood that the father has a high-risk APOL1 genotype? The spouse? The son?
2. Is donating a kidney to his father likely to cause harm to the son?
3. Is it likely that the donated kidney will be suboptimal in terms of long term prognosis?
4. Have the family members been accurately counseled about APOL1 and do they understand the information?

Figure 5.4 Pedigree w transplant issue.

disease (https://clinicaltrials.gov). If such therapies prove to be effective, this could justify more widespread APOL1 testing. APOL1 testing can be done as a standalone test, part of larger panels of genes, or genome-wide approaches. APOL1 is now a standard part of many of the kidney disease genetic panels that are in common clinical use. Since standard APOL1 genotyping involves genotyping just to determine the presence or absence of two alleles, we can anticipate that commonly used APOL1 tests will be accurate and inexpensive.

In the absence of a histologic evaluation of the kidney, Black people with chronic kidney disease and a high-risk APOL1 genotype progress more quickly to end-stage renal disease than do Black people with low-risk genotypes [26]. APOL1 genotype does not explain all of the differences in rate of kidney disease progression seen in Black individuals, only a fraction of whom have a high-risk genotype [128]. Although APOL1 genotype does not appear to increase the incidence of kidney disease among people with diabetes, individuals with diabetes CKD, and a high-risk APOL1 genotype do progress more quickly to kidney failure than those with a low-risk genotype [25,26]. Strict blood pressure control in individuals with CKD associates with a lower risk of death with a high-risk APOL1 genotype, though its effect on CKD progression, while present, is not large [26,129]. It is possible that APOL1 genotype could inform the selection of optimal blood pressure treatment targets.

5.13 RACIAL DISPARITIES IN KIDNEY DISEASE

High-risk APOL1 genotypes are present in about 13% of Black individuals in the United States [2,3]. APOL1 genotype is a function of ancestry and thus indirectly associated with race in the United States largely as a consequence of the transatlantic slave trade. The high-risk alleles have much greater frequencies in western sub-Saharan Africa than in the east. Thus, while it is certainly the case that APOL1 risk variants are more common in people who identify as Black or African-American, these variants are seen in other groups as well [130,131]. There are geographical differences that vary both on the basis of ancestry and on varying definitions of race—high-risk APOL1 genotypes are quite rare in the non-Black population in the United States. This is quite different from Brazil, where a recent study found that 47% of the high-risk genotype pediatric nephrotic syndrome patients self-reported as white [126].

APOL1 high-risk alleles have highly variable frequencies in different LatinX populations as a function of African ancestry [7,130,132]. While it is certainly appropriate for clinicians to be more suspicious of the presence of APOL1-associated kidney disease in their Black patients, it is important to appreciate that only a portion of the racial disparities in rates of kidney disease are attributable to APOL1 genotype. Nongenetic factors play a significant role. It is important for clinicians to understand that the presence of a high-risk APOL1 genotype does not exclude the possibility of other causes of kidney disease. It is still unclear how the ancestrally driven associations between race and APOL1 should factor into clinical decision-making. A strong case can be made that clinical transplantation algorithms that incorporate race should be replaced by genetics [133—135].

REFERENCES

[1] Thurlow JS, et al. Global epidemiology of end-stage kidney disease and disparities in kidney replacement therapy. Am J Nephrol 2021;52(2):98—107.

[2] Genovese G, et al. Association of trypanolytic ApoL1 variants with kidney disease in African Americans. Science 2010;329(5993):841—5.

[3] Tzur S, et al. Missense mutations in the APOL1 gene are highly associated with end stage kidney disease risk previously attributed to the MYH9 gene. Hum Genet 2010;128(3):345—50.

[4] Kopp JB, et al. MYH9 is a major-effect risk gene for focal segmental glomerulosclerosis. Nat Genet 2008;40(10):1175—84.

[5] Kao WH, et al. MYH9 is associated with nondiabetic end-stage renal disease in African Americans. Nat Genet 2008;40(10):1185—92.

[6] Friedman DJ, Pollak MR. APOL1 nephropathy: from genetics to clinical applications. Clin J Am Soc Nephrol 2021;16(2):294—303.

[7] Riella C, et al. APOL1-Associated kidney disease in Brazil. Kidney Int Rep 2019;4(7):923—9.

[8] Ulasi II, et al. High population frequencies of APOL1 risk variants are associated with increased prevalence of

non-diabetic chronic kidney disease in the Igbo people from south-eastern Nigeria. Nephron Clin Pract 2013;123(1−2):123−8.
[9] Kopp JB, et al. APOL1 genetic variants in focal segmental glomerulosclerosis and HIV-associated nephropathy. J Am Soc Nephrol 2011;22(11):2129−37.
[10] Kasembeli AN, et al. APOL1 risk variants are strongly associated with HIV-associated nephropathy in Black South Africans. J Am Soc Nephrol 2015;26(11):2882−90.
[11] Limou S, et al. Sequencing rare and common APOL1 coding variants to determine kidney disease risk. Kidney Int 2015;88(4):754−63.
[12] Thomson R, et al. Evolution of the primate trypanolytic factor APOL1. Proc Natl Acad Sci U S A 2014;111(20):E2130−9.
[13] Lannon H, et al. Apolipoprotein L1 (APOL1) risk variant toxicity depends on the haplotype background. Kidney Int 2019;96(6):1303−7.
[14] Sampson MG, et al. Integrative genomics identifies novel associations with APOL1 risk genotypes in Black NEPTUNE subjects. J Am Soc Nephrol 2016;27(3):814−23.
[15] Kopp JB, et al. Clinical features and histology of apolipoprotein L1-associated nephropathy in the FSGS clinical trial. J Am Soc Nephrol 2015;26(6):1443−8.
[16] Lipkowitz MS, et al. Apolipoprotein L1 gene variants associate with hypertension-attributed nephropathy and the rate of kidney function decline in African Americans. Kidney Int 2013;83(1):114−20.
[17] Larsen CP, et al. Apolipoprotein L1 risk variants associate with systemic lupus erythematosus-associated collapsing glomerulopathy. J Am Soc Nephrol 2013;24(5):722−5.
[18] Blazer AD, Clancy RM. ApoL1 and the immune response of patients with systemic lupus erythematosus. Curr Rheumatol Rep 2017;19(3):13.
[19] Larsen CP, et al. Histopathologic effect of APOL1 risk alleles in PLA2R-associated membranous glomerulopathy. Am J Kidney Dis 2014;64(1):161−3.
[20] May RM, et al. A multi-center retrospective cohort study defines the spectrum of kidney pathology in coronavirus 2019 disease (COVID-19). Kidney Int 2021;100(6):1303−15.
[21] Shetty AA, et al. COVID-19-associated glomerular disease. J Am Soc Nephrol 2021;32(1):33−40.
[22] Thakoordeen-Reddy S, et al. Maternal variants within the apolipoprotein L1 gene are associated with preeclampsia in a South African cohort of African ancestry. Eur J Obstet Gynecol Reprod Biol 2020;246:129−33.
[23] Hong X, et al. Joint associations of maternal-fetal APOL1 genotypes and maternal country of origin with preeclampsia risk. Am J Kidney Dis 2021;77(6):879−888 e1.
[24] Sedor JR, Bruggeman LA, O'Toole JF. APOL1 and preeclampsia: intriguing links, uncertain causality, troubling implications. Am J Kidney Dis 2021;77(6):863−5.
[25] Friedman DJ, et al. Population-based risk assessment of APOL1 on renal disease. J Am Soc Nephrol 2011;22(11):2098−105.
[26] Parsa A, et al. APOL1 risk variants, race, and progression of chronic kidney disease. N Engl J Med 2013;369(23):2183−96.
[27] Freedman BI, Murea M. Target organ damage in African American hypertension: role of APOL1. Curr Hypertens Rep 2012;14(1):21−8.
[28] Ekulu PM, et al. A focus on the association of Apol1 with kidney disease in children. Pediatr Nephrol 2021;36(4):777−88.
[29] Ng DK, et al. APOL1-associated glomerular disease among African-American children: a collaboration of the chronic kidney disease in children (CKiD) and nephrotic syndrome study network (NEPTUNE) cohorts. Nephrol Dial Transplant 2017;32(6):983−90.
[30] Purswani MU, et al. Brief report: APOL1 renal risk variants are associated with chronic kidney disease in children and youth with perinatal HIV infection. J Acquir Immune Defic Syndr 2016;73(1):63−8.
[31] Asharam K, et al. NPHS2 V260E is a frequent cause of steroid-resistant nephrotic syndrome in Black South African children. Kidney Int Rep 2018;3(6):1354−62.
[32] Blazer A, et al. Apolipoprotein L1 risk genotypes in Ghanaian patients with systemic lupus erythematosus: a prospective cohort study. Lupus Sci Med 2021;8(1).
[33] Vajgel G, et al. Effect of a single apolipoprotein L1 gene nephropathy variant on the risk of advanced lupus nephritis in Brazilians. J Rheumatol 2020;47(8):1209−17.
[34] Freedman BI, et al. End-stage renal disease in African Americans with lupus nephritis is associated with APOL1. Arthritis Rheumatol 2014;66(2):390−6.
[35] Smith EE, Malik HS. The apolipoprotein L family of programmed cell death and immunity genes rapidly evolved in primates at discrete sites of host-pathogen interactions. Genome Res 2009;19(5):850−8.
[36] Friedman DJ. A brief history of APOL1: a gene evolving. Semin Nephrol 2017;37(6):508−13.
[37] Vanhamme L, et al. Apolipoprotein L-I is the trypanosome lytic factor of human serum. Nature 2003;422(6927):83−7.
[38] Pays E, Vanhollebeke B. Human innate immunity against African trypanosomes. Curr Opin Immunol 2009;21(5):493−8.

[39] Molina-Portela Mdel P, et al. Trypanosome lytic factor, a subclass of high-density lipoprotein, forms cation-selective pores in membranes. Mol Biochem Parasitol 2005;144(2):218–26.
[40] Friedman DJ, Pollak MR. Genetics of kidney failure and the evolving story of APOL1. J Clin Invest 2011;121(9):3367–74.
[41] Genovese G, Friedman DJ, Pollak MR. APOL1 variants and kidney disease in people of recent African ancestry. Nat Rev Nephrol 2013;9(4):240–4.
[42] Cooper A, et al. APOL1 renal risk variants have contrasting resistance and susceptibility associations with African trypanosomiasis. Elife 2017;6.
[43] Kamoto K, et al. Association of APOL1 renal disease risk alleles with Trypanosoma brucei rhodesiense infection outcomes in the northern part of Malawi. PLoS Neglected Trop Dis 2019;13(8):e0007603.
[44] Schaub C, et al. Coiled-coil binding of the leucine zipper domains of APOL1 is necessary for the open cation channel conformation. J Biol Chem 2021;297(3):101009.
[45] Shiflett AM, et al. Human high density lipoproteins are platforms for the assembly of multi-component innate immune complexes. J Biol Chem 2005;280(38):32578–85.
[46] Lugli EB, et al. Characterization of primate trypanosome lytic factors. Mol Biochem Parasitol 2004;138(1):9–20.
[47] Shukha K, et al. Most ApoL1 is secreted by the liver. J Am Soc Nephrol 2017;28(4):1079–83.
[48] Zhaorigetu S, et al. ApoL1, a BH3-only lipid-binding protein, induces autophagic cell death. Autophagy 2008;4(8):1079–82.
[49] Nichols B, et al. Innate immunity pathways regulate the nephropathy gene Apolipoprotein L1. Kidney Int 2015;87(2):332–42.
[50] Wang DP, et al. Apolipoprotein L1 is transcriptionally regulated by SP1, IRF1 and IRF2 in hepatoma cells. FEBS Lett 2020;594(19):3108–21.
[51] Vanhollebeke B, Pays E. The trypanolytic factor of human serum: many ways to enter the parasite, a single way to kill. Mol Microbiol 2010;76(4):806–14.
[52] Lan X, et al. APOL1 risk variants enhance podocyte necrosis through compromising lysosomal membrane permeability. Am J Physiol Ren Physiol 2014;307(3):F326–36.
[53] Olabisi OA, Heneghan JF. APOL1 nephrotoxicity: what does ion transport have to do with it? Semin Nephrol 2017;37(6):546–51.
[54] Heneghan JF, et al. BH3 domain-independent apolipoprotein L1 toxicity rescued by BCL2 prosurvival proteins. Am J Physiol Cell Physiol 2015;309(5):C332–47.
[55] Thomson R, Finkelstein A. Human trypanolytic factor APOL1 forms pH-gated cation-selective channels in planar lipid bilayers: relevance to trypanosome lysis. Proc Natl Acad Sci U S A 2015;112(9):2894–9.
[56] Schaub C, et al. Cation channel conductance and pH gating of the innate immunity factor APOL1 are governed by pore-lining residues within the C-terminal domain. J Biol Chem 2020;295(38):13138–49.
[57] O'Toole JF, et al. ApoL1 overexpression drives variant-independent cytotoxicity. J Am Soc Nephrol 2018;29(3):869–79.
[58] Olabisi OA, et al. APOL1 kidney disease risk variants cause cytotoxicity by depleting cellular potassium and inducing stress-activated protein kinases. Proc Natl Acad Sci U S A 2016;113(4):830–7.
[59] Bruno J, Edwards JC. Kidney-disease-associated variants of Apolipoprotein L1 show gain of function in cation channel activity. J Biol Chem 2021;296:100238.
[60] Giovinazzo JA, et al. Apolipoprotein L-1 renal risk variants form active channels at the plasma membrane driving cytotoxicity. Elife 2020;9.
[61] Johnstone DB, et al. APOL1 null alleles from a rural village in India do not correlate with glomerulosclerosis. PLoS One 2012;7(12):e51546.
[62] Granado D, et al. Intracellular APOL1 risk variants cause cytotoxicity accompanied by energy depletion. J Am Soc Nephrol 2017;28(11):3227–38.
[63] Ma L, et al. APOL1 renal-risk variants induce mitochondrial dysfunction. J Am Soc Nephrol 2017;28(4):1093–105.
[64] Cheng D, et al. Biogenesis and cytotoxicity of APOL1 renal risk variant proteins in hepatocytes and hepatoma cells. J Lipid Res 2015;56(8):1583–93.
[65] Lan X, et al. Vascular smooth muscle cells contribute to APOL1-induced podocyte injury in HIV milieu. Exp Mol Pathol 2015;98(3):491–501.
[66] Ekulu PM, et al. Novel human podocyte cell model carrying G2/G2 APOL1 high-risk genotype. Cells 2021;10(8).
[67] Wan G, et al. Apolipoprotein L1, a novel Bcl-2 homology domain 3-only lipid-binding protein, induces autophagic cell death. J Biol Chem 2008;283(31):21540–9.
[68] Datta S, et al. Kidney disease-associated APOL1 variants have dose-dependent, dominant toxic gain-of-function. J Am Soc Nephrol 2020;31(9):2083–96.
[69] Haque S, et al. Effect of APOL1 disease risk variants on APOL1 gene product. Biosci Rep 2017;37(2).
[70] Scales SJ, et al. Apolipoprotein L1-specific antibodies detect endogenous APOL1 inside the endoplasmic reticulum and on the plasma membrane of podocytes. J Am Soc Nephrol 2020;31(9):2044–64.

[71] Wu J, et al. The key role of NLRP3 and STING in APOL1-associated podocytopathy. J Clin Invest 2021;131(20).
[72] Davis SE, Khatua AK, Popik W. Nucleosomal dsDNA stimulates APOL1 expression in human cultured podocytes by activating the cGAS/IFI16-STING signaling pathway. Sci Rep 2019;9(1):15485.
[73] Abid Q, et al. APOL1-Associated collapsing focal segmental glomerulosclerosis in a patient with stimulator of interferon genes (STING)-Associated vasculopathy with onset in infancy (SAVI). Am J Kidney Dis 2020;75(2):287–90.
[74] Pays E. The mechanism of kidney disease due to APOL1 risk variants. J Am Soc Nephrol 2020;31(11):2502–5.
[75] Uzureau S, et al. APOL1 C-terminal variants may trigger kidney disease through interference with APOL3 control of actomyosin. Cell Rep 2020;30(11). 3821–3836 e13.
[76] Zhang Z, et al. Recipient APOL1 risk alleles associate with death-censored renal allograft survival and rejection episodes. J Clin Invest 2021;131(22).
[77] Limou S, et al. APOL1 toxin, innate immunity, and kidney injury. Kidney Int 2015;88(1):28–34.
[78] McCarthy GM, et al. Recessive, gain-of-function toxicity in an APOL1 BAC transgenic mouse model mirrors human APOL1 kidney disease. Dis Model Mech 2021;14(8).
[79] Bruggeman LA, et al. APOL1-G0 or APOL1-G2 transgenic models develop preeclampsia but not kidney disease. J Am Soc Nephrol 2016;27(12):3600–10.
[80] Beckerman P, et al. Transgenic expression of human APOL1 risk variants in podocytes induces kidney disease in mice. Nat Med 2017;23(4):429–38.
[81] Ryu JH, et al. APOL1 renal risk variants promote cholesterol accumulation in tissues and cultured macrophages from APOL1 transgenic mice. PLoS One 2019;14(4):e0211559.
[82] Kotb AM, et al. Knockdown of ApoL1 in zebrafish larvae affects the glomerular filtration barrier and the expression of nephrin. PLoS One 2016;11(5):e0153768.
[83] Anderson BR, et al. In vivo modeling implicates APOL1 in nephropathy: evidence for dominant negative effects and epistasis under anemic stress. PLoS Genet 2015;11(7):e1005349.
[84] Olabisi O, et al. From man to fish: what can Zebrafish tell us about ApoL1 nephropathy? Clin Nephrol 2016;86(13):114–8.
[85] Kruzel-Davila E, et al. APOL1-Mediated cell injury involves disruption of conserved trafficking processes. J Am Soc Nephrol 2017;28(4):1117–30.
[86] Fu Y, et al. APOL1-G1 in nephrocytes induces hypertrophy and accelerates cell death. J Am Soc Nephrol 2017;28(4):1106–16.
[87] Bajaj A, et al. Phenome-wide association analysis suggests the APOL1 linked disease spectrum primarily drives kidney-specific pathways. Kidney Int 2020;97(5):1032–41.
[88] Nadkarni GN, et al. Apolipoprotein L1 variants and blood pressure traits in African Americans. J Am Coll Cardiol 2017;69(12):1564–74.
[89] Wu J, et al. APOL1 risk variants in individuals of African genetic ancestry drive endothelial cell defects that exacerbate sepsis. Immunity 2021;54(11):2632–49.
[90] Chen TK, et al. Association of APOL1 genotypes with measures of microvascular and endothelial function, and blood pressure in MESA. J Am Heart Assoc 2020;9(17):e017039.
[91] Ma L, et al. Localization of APOL1 protein and mRNA in the human kidney: nondiseased tissue, primary cells, and immortalized cell lines. J Am Soc Nephrol 2015;26(2):339–48.
[92] Blessing NA, et al. Lack of APOL1 in proximal tubules of normal human kidneys and proteinuric APOL1 transgenic mouse kidneys. PLoS One 2021;16(6):e0253197.
[93] Ito K, et al. Increased burden of cardiovascular disease in carriers of APOL1 genetic variants. Circ Res 2014;114(5):845–50.
[94] McLean NO, Robinson TW, Freedman BI. APOL1 gene kidney risk variants and cardiovascular disease: getting to the heart of the matter. Am J Kidney Dis 2017;70(2):281–9.
[95] Bick AG, et al. Association of APOL1 risk alleles with cardiovascular disease in blacks in the million veteran program. Circulation 2019;140(12):1031–40.
[96] Gutierrez OM, et al. APOL1 nephropathy risk alleles and mortality in African American adults: a cohort study. Am J Kidney Dis 2020;75(1):54–60.
[97] Nadkarni GN, et al. APOL1 renal risk variants are associated with obesity and body composition in African ancestry adults: an observational genotype-phenotype association study. Medicine 2021;100(45):e27785.
[98] Reidy KJ, et al. Fetal-not maternal-APOL1 genotype Associated with risk for preeclampsia in those with African ancestry. Am J Hum Genet 2018;103(3):367–76.
[99] Miller AK, et al. Association of preeclampsia with infant APOL1 genotype in African Americans. BMC Med Genet 2020;21(1):110.

[100] Skorecki KL, et al. A null variant in the apolipoprotein L3 gene is associated with non-diabetic nephropathy. Nephrol Dial Transplant 2018;33(2):323–30.

[101] Zhang JY, et al. UBD modifies APOL1-induced kidney disease risk. Proc Natl Acad Sci U S A 2018;115(13):3446–51.

[102] Ashley-Koch AE, et al. MYH9 and APOL1 are both associated with sickle cell disease nephropathy. Br J Haematol 2011;155(3):386–94.

[103] Bodonyi-Kovacs G, et al. Combined effects of GSTM1 null allele and APOL1 renal risk alleles in CKD progression in the African American study of kidney disease and hypertension trial. J Am Soc Nephrol 2016;27(10):3140–52.

[104] Rosenberg AZ, et al. HIV-associated nephropathies: epidemiology, pathology, mechanisms and treatment. Nat Rev Nephrol 2015;11(3):150–60.

[105] Goyal R, Singhal PC. APOL1 risk variants and the development of HIV-associated nephropathy. FEBS J 2021;288(19):5586–97.

[106] Markowitz GS, et al. Treatment with IFN-{alpha}, -{beta}, or -{gamma} is associated with collapsing focal segmental glomerulosclerosis. Clin J Am Soc Nephrol 2010;5(4):607–15.

[107] Divers J, et al. JC polyoma virus interacts with APOL1 in African Americans with nondiabetic nephropathy. Kidney Int 2013;84(6):1207–13.

[108] Freedman BI, et al. JC polyoma viruria associates with protection from chronic kidney disease independently from apolipoprotein L1 genotype in African Americans. Nephrol Dial Transplant 2018;33(11):1960–7.

[109] Velez JCQ, Caza T, Larsen CP. COVAN is the new HIVAN: the re-emergence of collapsing glomerulopathy with COVID-19. Nat Rev Nephrol 2020;16(10):565–7.

[110] Friedman DJ. COVID-19 and APOL1: understanding disease mechanisms through clinical observation. J Am Soc Nephrol 2021;32(1):1–2.

[111] Paranjpe I, et al. Association of APOL1 risk genotype and air pollution for kidney disease. Clin J Am Soc Nephrol 2020;15(3):401–3.

[112] Lee DE, Qamar M, Wilke RA. Relative contribution of genetic and environmental factors in CKD. S D Med 2021;74(7):306–9.

[113] Groopman EE, et al. Diagnostic utility of exome sequencing for kidney disease. N Engl J Med 2019;380(2):142–51.

[114] Freedman BI, et al. Diagnosis, education, and care of patients with APOL1-associated nephropathy: a delphi consensus and systematic review. J Am Soc Nephrol 2021;32(7):1765–78.

[115] Reeves-Daniel AM, et al. The APOL1 gene and allograft survival after kidney transplantation. Am J Transplant 2011;11(5):1025–30.

[116] Lee BT, et al. The APOL1 genotype of African American kidney transplant recipients does not impact 5-year allograft survival. Am J Transplant 2012;12(7):1924–8.

[117] Mena-Gutierrez AM, et al. Practical considerations for APOL1 genotyping in the living kidney donor evaluation. Transplantation 2020;104(1):27–32.

[118] Lentine KL, Mannon RB. Apolipoprotein L1: role in the evaluation of kidney transplant donors. Curr Opin Nephrol Hypertens 2020;29(6):645–55.

[119] Doshi MD, et al. APOL1 genotype and renal function of Black living donors. J Am Soc Nephrol 2018;29(4):1309–16.

[120] Locke JE, et al. Apolipoprotein L1 and chronic kidney disease risk in young potential living kidney donors. Ann Surg 2017;267(6):1161–8.

[121] Berrigan M, et al. Opinions of African American adults about the use of apolipoprotein L1 (ApoL1) genetic testing in living kidney donation and transplantation. Am J Transplant 2021;21(3):1197–205.

[122] Asgari E, Hilton RM. One size does not fit all: understanding individual living kidney donor risk. Pediatr Nephrol 2021;36(2):259–69.

[123] Kumar V, Locke JE. APOL1 genotyping in kidney transplantation: to do or not to do, that is the question? (contra). Kidney Int 2021;100(1):30–2.

[124] Freedman BI, Poggio ED. APOL1 genotyping in kidney transplantation: to do or not to do, that is the question? (pro). Kidney Int 2021;100(1):27–30.

[125] Freedman BI, et al. APOL1 long-term kidney transplantation outcomes network (APOLLO): design and rationale. Kidney Int Rep 2020;5(3):278–88.

[126] Watanabe A, et al. APOL1 in an ethnically diverse pediatric population with nephrotic syndrome: implications in focal segmental glomerulosclerosis and other diagnoses. Pediatr Nephrol 2021;36(8):2327–36.

[127] Zee J, et al. APOL1 genotype-associated morphologic changes among patients with focal segmental glomerulosclerosis. Pediatr Nephrol 2021;36(9):2747–57.

[128] Hannan M, et al. Risk factors for CKD progression: overview of findings from the CRIC study. Clin J Am Soc Nephrol 2021;16(4):648–59.

[129] Ku E, et al. Strict blood pressure control associates with decreased mortality risk by APOL1 genotype. Kidney Int 2017;91(2):443–50.

[130] Lin BM, et al. Genetics of chronic kidney disease stages across ancestries: the PAGE study. Front Genet 2019;10:494.

[131] Nadkarni GN, et al. Worldwide frequencies of APOL1 renal risk variants. N Engl J Med 2018;379(26):2571–2.
[132] Kramer HJ, et al. African ancestry-specific alleles and kidney disease risk in hispanics/latinos. J Am Soc Nephrol 2017;28(3):915–22.
[133] Nadkarni GN, et al. APOL1: a case in point for replacing race with genetics. Kidney Int 2017;91(4):768–70.
[134] Freedman BI, Julian BA. Evaluation of potential living kidney donors in the APOL1 era. J Am Soc Nephrol 2018;29(4):1079–81.
[135] Gordon EJ, et al. A national survey of transplant surgeons and nephrologists on implementing apolipoprotein L1 (APOL1) genetic testing into clinical practice. Prog Transplant 2019;29(1):26–35.

PART II

Hematologic Disorders

6

Hemoglobinopathies and Thalassemias

Scott Peslak[1,2], *Farzana Sayani*[1]

[1]Division of Hematology/Oncology, Department of Medicine, University of Pennsylvania Perelman School of Medicine, Philadelphia, PA, United States
[2]Division of Hematology, Children's Hospital of Philadelphia, Philadelphia, PA, United States

6.1 INTRODUCTION

The hemoglobinopathies and the thalassemia syndromes are a diverse group of inherited disorders of hemoglobin synthesis that result from the qualitative defects (hemoglobinopathies) or quantitative defects (thalassemia syndromes) in globin synthesis. Taken together, they are the most common and clinically significant single gene disorder in the world and pose a serious health problem in many countries. The methods of clinical management have improved considerably during the past few years and the life expectancy of affected individuals has significantly increased with patients now surviving into the fifth to sixth decades of life. Although the only definitive cure for the hemoglobinopathies is allogeneic stem cell transplantation, there is promise of gene therapy in both the hemoglobinopathies and the thalassemia syndromes. However, the treatment required is very expensive and is not a realistic means of controlling the disorders especially for many developing countries where the diseases are prevalent. Therefore, standard of care involves employing pharmacologic-based approaches (e.g., hydroxyurea in sickle cell disease), blood transfusions and chelation (thalassemia), or alternative methods of control that involves screening the population for carriers, identifying couples at risk, and providing prenatal diagnosis, particularly in countries with a high incidence of β-thalassemia. This chapter outlines the structure, function, and biosynthesis of normal hemoglobin, then discusses the current knowledge of the clinical diseases associated with the defects in globin synthesis and their molecular pathology, diagnosis, and treatment. A more detailed description of the thalassemia syndromes may be found in the comprehensive book by Weatherall and Clegg [1], and for hemoglobin variants, in the book edited by Steinberg and Nagel [2].

6.2 HEMOGLOBIN GENETICS

In humans, there are eight different genetic loci that code for the six types of globin chains. In addition, there are at least four pseudogenes that have sequences similar to other globin genes but differ in that they are not expressed into globin proteins. A schematic representation of the interaction of the products of these genes is shown in Fig. 6.1. Normally, globin tetramers are formed of two α or α-like chains and two β or β-like chains. As there are two loci encoding the structure of the α-globin chain, there are four α-chain genes. In contrast, there is only a single β-globin locus and,

Emery and Rimoin's Principles and Practice of Medical Genetics and Genomics. https://doi.org/10.1016/B978-0-12-812534-2.00009-6

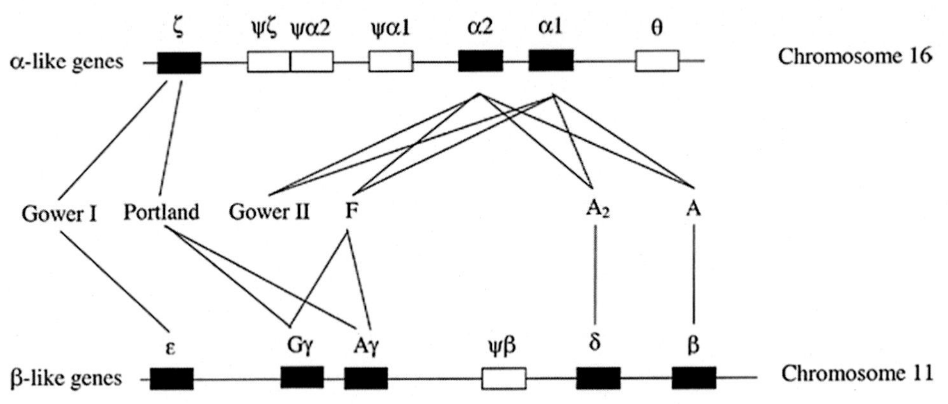

Figure 6.1 Various globin genes and their products.

therefore, two β-genes (Fig. 6.1). The relative numbers of α- and β-loci are important in understanding the different inheritance patterns of α- and β-thalassemias, as well as the different relative amounts of variant hemoglobins in individuals carrying a variant α- or β-globin gene. These quantitative differences correlate directly with the clinical severity of the various disorders.

The α-like genes form a cluster on the tip of chromosome 16 at position 16p13.3 and are arranged in the order 5′-ε-ζ-ψζ-ψα2-ψα1-α2-α1-θ-3′. The cluster contains three functional genes (two α-genes that are 3.6 kb apart and one ζ-gene), three pseudogenes (ψζ, ψα1 and ψα2), and one gene of undetermined function (θ) as depicted in Fig. 6.2. Unlike the pseudogenes, the θ-gene does not contain any apparent defects in its sequence that would prevent its expression. In each case about 4 kb separates the ζ1, ψα1, α2, and α1 loci, suggesting the existence of discrete duplication units in the DNA. There is an α-globin control region called HS-40 for the α-globin gene cluster that is a nuclease-hypersensitive site analogous to LCR-β, although its mechanism of action is different. Deletion of the LCR-α causes α-thalassemia by inactivation of the adjacent intact α-globin genes.

The β-like globin genes form a linked cluster on the short arm of chromosome 11 (11p15.5), spread over a region of approximately 60 kilobases (kb), as depicted in Fig. 6.2. There are five functional genes, arranged in

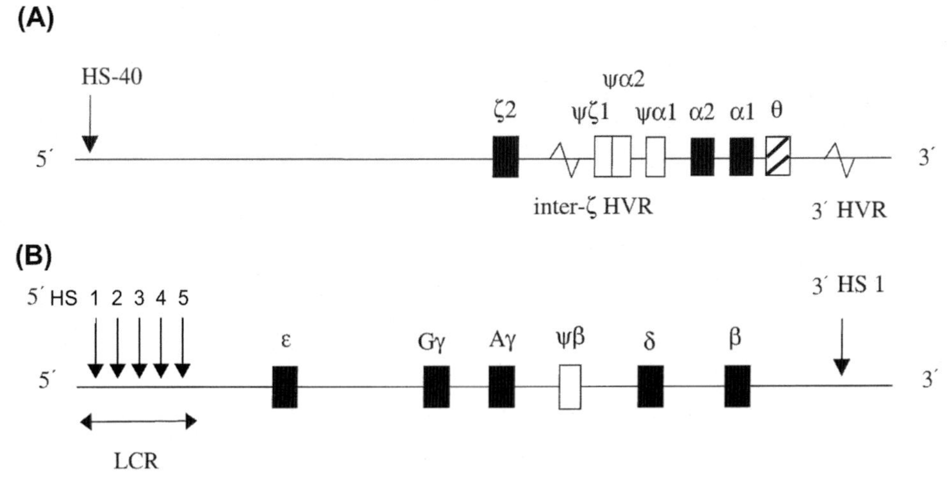

Figure 6.2 Globin gene complexes. (A) α-globin gene complex on chromosome 16. (B) β-globin gene complex on chromosome 11.

the order of their developmental expression: 5′-ε-^{G}γ-^{A}γ-ψβ-δ-β-3′. There are two types of γ-globin genes—those that contain glycine at position 136 (^{G}γ-chains) and those that contain alanine at this position (^{A}γ-chains). The pseudogene (ψβ) has sequences similar to the β-gene but differs in having altered sequences that prevent its expression and the production of functional globin chains. The globin gene sequences comprise only about 7 kb of the 60 kb of DNA in the β-gene region, while the remaining 53 kb are flanking sequences that contain sequences with specific regulatory roles. These include the locus control region, enhancer sequences, and the promoter regions of the globin genes.

The β-locus control region or LCR-β is found in a cluster of DNase I hypersensitive sites 6–18 kb 5′ to the ε-globin gene. The LCR-β establishes a transcriptionally active domain spanning the entire β-globin gene cluster, allowing a high level of erythroid-specific expression of the genes. Transcription is dependent on the promoter regions of these genes becoming opposed to the LCR-β in association with the binding of a variety of regulatory proteins consisting of both tissue-specific and ubiquitous transcription factors. For instance, the two primary repressors of γ-globin transcription, BCL11A and ZBTB7A (LRF), bind specific sites in the γ-globin locus and recruit repressive chromatin modifiers [3–8]. Furthermore, erythroid cells have been shown to contain a variety of erythroid-specific transcription factors, the most important being GATA-1 and NF-E2. GATA-1 belongs to a family of proteins recognizing the consensus sequence (T/A)GATA(A/G) and acts as a tissue-specific activator of transcription with binding sites both in the locus control regions and in the promoters and flanking regions of the globin genes. NF-E2-binding sites have been identified in both the β- and α-locus control regions. The precise mechanism of the developmental switches of gene expression from ε to γ and from γ to β remains an area of active investigation but is known to be complex with respect to both gene silencing and gene competition for the LCR sequences.

6.2.1 Ontogeny of Globin Expression

The globin genes are expressed at different times and in different relative amounts during human development (Fig. 6.3). Hemoglobin (Hb) Gower I ($\zeta_2\varepsilon_2$), Gower II ($\alpha_2\varepsilon_2$), and Portland ($\zeta_2\gamma_2$) are embryonic hemoglobins synthesized in the yolk sac before 8 weeks of gestation. At 4–5 weeks of gestation, a decrease in ζ- and ε-chain production occurs, together with a compensating increase in α- and γ-chain production. Then, at approximately 8 weeks of gestation, the site of hemoglobin synthesis changes to the fetal liver, which produces predominantly HbF ($\alpha_2\gamma_2$), also known as fetal hemoglobin. Starting at 18 weeks gestation, the liver is progressively replaced by bone marrow as the major site of red cell production. In addition, after birth a

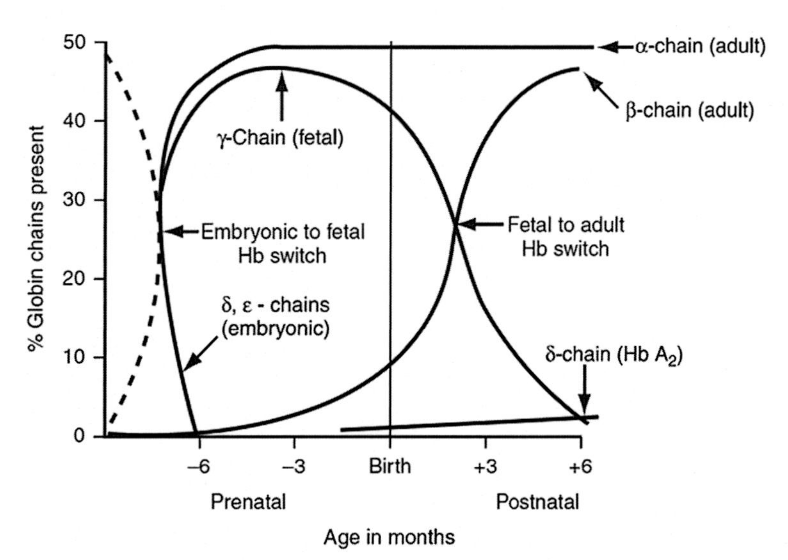

Figure 6.3 Qualitative and quantitative changes in globin chains during human development.

developmental switch occurs during which the γ-globin genes are transcriptionally silenced and the adult counterparts activated to produce adult hemoglobin (HbA, $\alpha_2\beta_2$). Thus, although HbF comprises the vast majority of hemoglobin during the fetal stage of life and in human newborns, HbF levels decline rapidly after birth, reaching concentrations of 10%–15% by 4 months of age and less than 1% by 3–4 years of age.

The sequence of appearance of the various globin chains is helpful in understanding the timing of onset of clinical manifestations of the hemoglobinopathies and thalassemias. For example, a deficiency of α- or γ-chain synthesis or α- or γ-chain variants with abnormal functions would be observed at birth, while a deficiency of β-chains may not cause symptoms until several months of age. Notably, HbF may be increased in β- and δβ-thalassemia, hereditary persistence of fetal hemoglobin (HPFH), some cases of thyrotoxicosis, megaloblastic and aplastic anemias, leukemia, and various malignancies involving marrow, sickle cell disease, and during pregnancy. The mechanism for the switch from HbF to HbA has been the subject of intense research for a number of years because of the therapeutic implications of increasing the amount of HbF in patients with β-thalassemia and SCD.

6.3 NORMAL HUMAN HEMOGLOBIN

Hemoglobin is a tetramer that consists of two α and two non-α globin polypeptide chains, each of which has a single covalently bound heme group. Each of the four heme groups is made up of an iron atom bound within a protoporphyrin IX ring. In humans, there are six known different globin polypeptide chains that form a number of hemoglobin tetramers. The ε-, γ-, and δ-chains are more similar to β-chains than to α-chains, differing from β at 36, 39, and 10 positions, respectively, and are designated β-like chains. The ζ chain is an α-like globin and is found in embryonic erythrocytes together with ε-globin.

Historically, the hemoglobin composition of erythrocytes was quantified by cellulose acetate gel electrophoresis or by isoelectric focusing. More recently, the clinical standard by which hemoglobins are identified and quantified has been high-performance liquid chromatography (HPLC), which provides a rapid and accurate chromatographic method with a degree of automation. Different hemoglobin tetramers, their structure, percentage in normal adult lysate, and conditions in which levels are increased are seen in Table 6.1.

HbA ($\alpha_2\beta_2$) is the major component of hemoglobin in normal adults, usually comprising about 97% of the total hemoglobin. The remainder is HbA_2 ($\alpha_2\delta_2$), which usually constitutes about 2%–3% in normal individuals. HbA_2 is increased in the majority of the β-thalassemias and megaloblastic anemias and decreased in iron deficiency and sideroblastic anemias. HbA1C differs from HbA by the posttranslational addition of a glucose at the NH_2 terminus of the β-chain; hence the tetramer's structure is α_2(β-N-glucose)$_2$. The percentage of HbA1C (~4–5% in normal adults)

TABLE 6.1 Human Hemoglobins

Hemoglobin	Stage of Development	Structure	Percentage in Adults	Conditions in Which Increased
Gower I	Embryonic	$\zeta_2\varepsilon_2$	0	Early embryos (<8 weeks)
Gower II	Embryonic	$\alpha_2\varepsilon_2$	0	Early embryos (>8 weeks)
Portland	Embryonic	$\zeta_2\gamma_2$	0	(<8 weeks) and α^o-thalassemia (hydrops fetalis)
F	Fetal	$\alpha_2\gamma_2$	<1	Newborn, δβ-, β-thalassemia, HPFH and marrow stress
A	Adult	$\alpha_2\beta_2$	97	
A_{1c}		α_2(β-N-glucose)$_2$	5	Diabetes mellitus
A_2		$\alpha_2\delta_2$	2–3	β-Thalassemia
H		β_4	0	Some α-thalassemias
Bart's		γ_4	0	Some α-thalassemias

is related to the intracellular concentration of glucose and the red cell life span. In diabetic patients, the concentration of HbA1C is increased due to the elevated glucose concentration in their red cells. HbH and Hb Bart's are tetramers of β- and γ-chains respectively, and both function very poorly in transporting oxygen. These two hemoglobins may be increased in more severe forms of α-thalassemia in which the deficiency of α-globin chains is sufficient to allow the formation of unpaired β-like globin chains into soluble tetramers.

6.3.1 Hemoglobin Protein Structure

The primary structure of each globin chain is its amino acid sequence; it is composed of 141 amino acids in ζ- and α-chains and 146 amino acids in ε-, γ-, δ-, and β-chains. The relationship between adjacent amino acids along the chain enables interactions that can result in one of two basic configurations of secondary structure: the α-helix or β-pleated sheet. About 75% of hemoglobin in its native state is in the α-helix form. However, at specific locations in the hemoglobin subunits, the rod-like α-helix is interrupted by nonhelical segments that allow folding of the amino acid chain. Tertiary structure refers to the configuration of a protein subunit in three-dimensional space, while quaternary structure refers to the relationships of the four subunits of hemoglobin to each other. The four subunits forming HbA tetramers are labeled α_1, α_2, β_1, and β_2. While there is no contact between the two β-chains, each α-chain touches both β-chains. Bonds across the $\alpha_1\beta_1$ interface are firmer than those at the $\alpha_1\beta_2$ interface, and changes from oxy-to deoxyhemoglobin involve more extensive movement at the $\alpha_1\beta_2$ interface. The quaternary structure changes markedly in going from oxy- to deoxyhemoglobin, and this accounts for many of the observed changes in physical properties. Hemoglobin mutations resulting in amino acid substitutions at key contact points can markedly alter specific functional properties.

6.3.2 Functional Properties

For hemoglobin to fulfill its physiologic role, it must bind oxygen with a certain affinity. One measure of oxygen affinity is P_{50}, or the partial pressure of oxygen in millimeters of mercury (mm Hg) that is required for 50% saturation of hemoglobin: a hemoglobin with increased P_{50} has decreased oxygen affinity (Fig. 6.4). Oxygen affinity is also affected by a number of environmental factors including temperature, pH, organic phosphate concentration, and pCO_2 (Fig. 6.4).

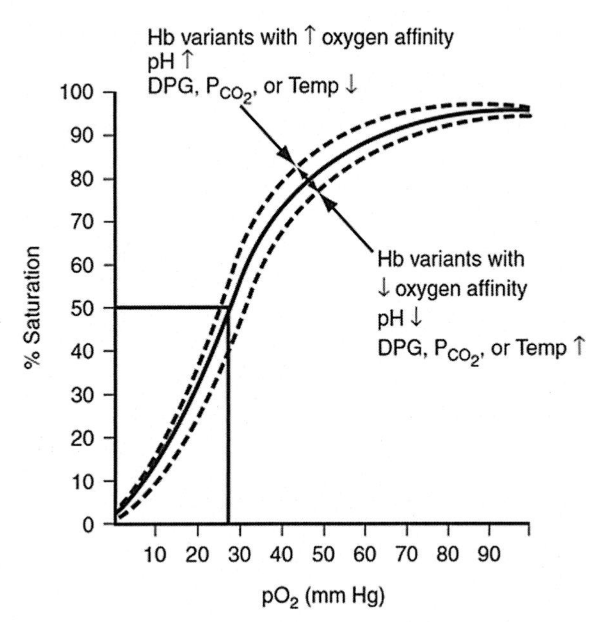

Figure 6.4 The oxyhemoglobin dissociation curve and effect of different factors on oxygen affinity.

The sigmoidal shape of the oxyhemoglobin dissociation curve reflects heme—heme interaction; that is, successive oxygenation of each heme group in the tetramer increases the oxygen affinity of the remaining unoxygenated heme groups. This alteration is amplified by a series of conformational changes that affect the other heme groups. The resulting sigmoidal oxyhemoglobin dissociation curve has great physiologic importance because it enables large amounts of oxygen to be bound or released with a small increase or decrease in oxygen tension. For instance, the Bohr effect is a change in oxygen affinity of hemoglobin with a change in pH. This effect is beneficial at the tissue level where the lower pH decreases oxygen affinity and promotes oxygen release (Fig. 6.4). Oxygen uptake in the lungs is enhanced by the opposite changes in pH and pCO_2.

Red cells have unusually high concentrations of 2,3-diphosphoglycerate (2,3-DPG). One molecule of 2,3-DPG sits in a pocket in deoxyhemoglobin bound to specific β-chain residues (1, 2, 82, and 143 of both β-chains). The importance of the binding is that 2,3-DPG stabilizes the deoxy form of hemoglobin in preference to the oxy form, thereby lowering the oxygen affinity of the molecule. The γ-chain of HbF lacks the β^{143} histidine residue, and the resultant decrease in binding of 2,3-DPG to HbF accounts for the increased oxygen affinity of fetal red cells compared with that of adult red cells.

6.4 HUMAN HEMOGLOBIN VARIANTS

Abnormal hemoglobins result from mutations that change the sequence or number of nucleotides within the globin gene involved, or more rarely, from mispairing and crossover between two like genes during meiosis, creating a fusion protein of both gene sequences. Mutation can cause substitution, addition, or deletion of one or more amino acids in the polypeptide sequence of the affected globin (Table 6.2). Single base changes can result in single amino acid substitutions (e.g., HbS (β6 Glu → Val), HbC (β6 Glu → Lys), and HbE (β26 Glu → Lys)), shortened chains due to premature termination of translation (e.g., Hb McKees Rocks (β145 Tyr → Termination)), or elongated chains due to a mutation in a 'stop' codon such as TAA → CAA (α142 Term → Gln) in Hb Constant Spring. Single base deletions or additions can cause a frameshift in the normal reading process. For example, in the

TABLE 6.2 Molecular Basis of the Hb Variants

Mutation	Example Hb		Clinical Manifestation	Molecular Basis (Presumed)
Nucleotide Base Substitutions				
One amino acid	HbS	β6Glu → Val	Sickling	β:Cd 6 GAG → GTG
	HbC	β6Glu → Lys	Normal	β:Cd 6 GAG → AAG
	HbE	β26 Glu → Lys	Normal	β:Cd 26 GAG → AAG
Two amino acids	Hb C-Harlem	β6Glu → Val + β73Asp → Asn	Sickling	β:Cd 6 GAG → GTG & β:Cd 73 GAT → AAT
Termination	Hb McKees Rocks	β145 Tyr → termination	Increased oxygen affinity and polycythemia	β:Cd 145 TAT → TAA
Amino acid instead of termination	Hb Constant Spring	α2:142 termination → Gln	Decreased synthesis (thalassemia-like)	α2:Cd 142 TAA → CAA
Nucleotide Base Deletions				
Single base deletion → frameshift	Hb Wayne	α2:139–146 Lys-Tyr-Arg → Asn+ 7 residues	Normal	α2:Cd 139 (-A)
Triplet deletion → single amino acid	Hb Leiden	β6 or 7 Glu → 0	Unstable	β:Cd 6 or 7 (–GAG)
Multiple codon	Hb Gun Hill	β91–95 Leu-His-Cys-Asp-Lys → 0)	Unstable	β:Cd 91–95 (–15 bp)
Crossover	Hb Lepore	δβ-fusion with segments of δ and β lost	Decreased synthesis (thalassemia-like)	δβ:7.4 kb deletion
Nucleotide Base Additions				
Two bases added → frameshift	Hb Cranston	β144 Tyr-His → Ser-Ile-Thr	Unstable	β:Cd 144/145 +CT
Multiple codon	Hb Grady	α118 (+Glu-Phe-Thr)	Normal	α2 or α1:Cd 118/119 (+9 bp)

variant Hb Wayne, a single base deletion (-A) at codon α139 causes the subsequent sequence of two codons, the termination codon TAA and the 3′ UTR to be read out of phase for seven triplets until a new terminating codon (TAG) is encountered. Deletions of three, or multiples of three, nucleotides in the DNA can cause deletions of one or more amino acids, such as Hb Leiden (β6 or 7 Glu → 0) and Hb Gun Hill (β91–95 [Leu-His-Cvs-Asp-Lvs] → 0). Deletions of segments of genes may be due to nonhomologous crossing over after mispairing in meiosis. This mechanism accounts for the Hb Lepore globins (δβ fusion chains), the anti-Lepore globins (βδ fusion chains), and Hb Kenya globin (γβ fusion chain).

An updated listing of known hemoglobin variants is available through the globin gene server Web site (HbVar): http://globin.bx.psu.edu [9,10]. The vast majority of these variants arise from a single base substitution, which results in a single amino acid substitution. Many of these substitutions are clinically silent, including some of those that produce abnormal physical properties in the variant hemoglobin, and have been detected only through population screening. The amino acid substitutions can cause a number of abnormal physical properties. These include instability of the tetramer; deformity of the three-dimensional structure; inhibition of ferric iron reduction; alteration of the residues that interact with heme, 2,3-DPG, or the α–β subunit contact site; or abnormality of other properties of the molecule. The varieties of clinical phenotypes that arise from these abnormal physical properties are listed in Table 6.3.

Although most hemoglobin variants are synthesized at a normal rate, several are associated with quantitative as well as qualitative abnormalities. One example, HbE (β26 Glu → Lys), is the second most common hemoglobinopathy worldwide and is associated predominantly with Southeast Asian populations. Both the heterozygous and homozygous states are associated with red cell microcytosis and hypochromia, and the thalassemic phenotype of the βE gene is linked to activation of a cryptic donor splice site by the codon 26 mutation. Competition between the normal and abnormal alternative splice sites reduces βE-mRNA production, resulting in a very mild β-thalassemia phenotype. Individuals homozygous for HbE have a very mild disorder, being only slightly anemic with red cell indices similar to those of β-thalassemia trait. However, for reasons still not fully understood, HbE trait combines with β-thalassemia trait to produce a serious thalassemic disorder, compound heterozygotes having a variable clinical picture ranging from thalassemia intermedia to transfusion-dependent thalassemia major.

Heterozygotes for a hemoglobin containing an abnormal β-globin chain have an abnormal as well as a normal β-gene at that locus, and their status is often described by the term "trait." Since most variants are rare, they usually occur in the heterozygous state and, if they cause clinical symptoms in the heterozygous state, are examples of autosomal dominant conditions. When both alleles code for the same common β-variant, the individual is then homozygous and is said to have the "disease" state. The most commonly encountered variants in the homozygous state are HbS, C, and E and their phenotypes are listed in Table 6.4.

6.5 SICKLE CELL DISEASE AND RELATED DISORDERS

6.5.1 Molecular Basis and Geographical Distribution

The sickle cell gene results from a point mutation of A to T at the second nucleotide of codon 6 of the β-globin

TABLE 6.3 Clinical Manifestations of Hemoglobin Mutants

Type	Example	Clinical Manifestation
Sickling	Hb S	Sickling due to decreased solubility
Unstable	Hb Bristol	Anemia with Heinz body formation
Decreased oxygen affinity	Hb Kansas	Mild anemia possible
Increased oxygen affinity	Hb Chesapeake	Polycythemia due to decreased oxygen transport
M Hemoglobin	Hb M-Boston	Cyanosis due to ferric hemoglobin
Decreased synthesis	Hb Lepore	Thalassemia

TABLE 6.4 Phenotypes of Thalassemias, Sickle Cell Disease, and Common Hb Variant/Thalassemia Interactions

Type	Phenotype
Silent α thalassemia (-α/αα)	None
α thalassemia minor/trait (-α/-α or –/αα)	Mild microcytic anemia
α^+/α^0 thalassemia (-α/–)	Hb H disease
α^0/α^0 thalassemia (–/–)	Hb Bart's (hydrops fetalis)
β-thalassemia minor (β/β^0)	Mild microcytic anemia
β^+/β^+	Thalassemia intermedia
β^0-thalassemia/mild β^+-thalassemia	Thalassemia intermedia
β^0-thalassemia/severe β^+-thalassemia	Thalassemia major
β^0/β^0	Thalassemia major
Homozygous HPFH (hereditary persistence of fetal hemoglobin)	No clinical problems
Homozygous Hb Lepore (δβ fusion chains)	Variable: intermedia to major
HbSS (β6 Glu → Val)	Sickle cell disease
HbCC (β6 Glu → Lys)	Moderate anemia and splenomegaly
HbEE (β26 Glu → Lys)	Mild anemia
HbS/mild β^+-thalassemia	Mild sickle cell disease
HbS/HbC	Sickle cell disease, variable severity
HbS/β^0 or severe β^+-thalassemia	Sickle cell disease
HbS/HbD-Punjab (β121, Glu → Gln)	Sickle cell disease
HbS/HbO-Arab (β121, Glu → Lys)	Sickle cell disease
HbS/C-Harlem (β6 Glu → Val and β87 Asp → Asn)	Severe sickle cell disease
HbS-Antilles (β6 Glu → Val and β23 Val → Ile)	Severe sickle cell disease

gene, causing the amino acid substitution of valine for glutamic acid (i.e., GAG to GTG) and leading to production of sickle hemoglobin (HbS; $\alpha^2\beta_2^S$). The frequency of sickle trait (HbAS) among those of African ancestry in the United States at birth is about 8%, and the incidence of sickle cell disease at birth is approximately 1 in 625 births. This contrasts with the higher carrier frequencies seen in some areas of equatorial Africa because of the protective advantage conferred by the carrier state against *P. falciparum* malaria. HbS is known to occur at frequencies up to 20% in the Cameroon, Guinea, Democratic Republic of Congo, Uganda, Kenya, eastern Saudi Arabia, and parts of India. The sickle cell gene has been reported at lower gene frequencies of up to 5%, in Nepal, in regions around the Mediterranean (e.g., Turkey, Lebanon, Syria, Greece, Portugal, and the coast of North Africa), in the Middle East, and in Iran.

6.5.2 Pathophysiology of Sickling

Substitution of valine for glutamic acid at the β6 residue causes a change on the surface of the deoxygenated βS chain, which results in the formation by $\alpha_2\beta_2^S$ tetramers of a 14-stranded helical polymer of diameter 15–17 nm. The parallel alignment of these rod-like polymers, in turn, causes the sickle-shaped deformation of the erythrocytes. In sickle cell disease, the sickling process begins when the oxygen saturation of HbS in the microvasculature is decreased to 85%. In addition to a decrease in oxygen tension, a reduction in pH or an increase in 2,3-DPG also promotes sickling. These factors probably interact in patients with sickle cell disease, since their blood normally has an increased 2,3-DPG concentration.

The viscosity of oxygenated sickle cell blood is increased, primarily because of irreversibly sickled red cells. When the blood becomes deoxygenated, viscosity increases further because of the cellular rigidity that occurs with sickling. This, in turn, increases the exposure time of erythrocytes to a hypoxic environment, and the lower tissue pH decreases oxygen affinity, which further promotes sickling. The end result is occlusion of capillaries and small arteries and infarction of surrounding tissues. Deformed sickle cells also have a shorter survival time because of their increased mechanical fragility and damaged membranes, resulting in a more rapid red cell turnover and consequent anemia.

6.5.3 Clinical Aspects of Sickle Cell Disease

As can be seen from Fig. 6.3, β-chain production does not usually reach sufficient levels to cause symptoms until the second half of the first year of life. As higher concentrations of HbS are reached in erythrocytes, the cells become susceptible to hemolysis and a progressive hemolytic anemia with splenomegaly is seen. The increased rate of erythropoiesis leads to erythroid marrow expansion and increased folic acid requirements. However, the two major problems for young children with SS disease are infections and vaso-occlusive episodes.

Children with sickle cell disease have increased susceptibility to potentially life-threatening bacterial infections including sepsis and meningitis caused by *Streptococcus pneumoniae* and *Haemophilus influenzae.* These patients are also susceptible to bacterial pneumonia (often *Pneumococcus*), osteomyelitis (*Salmonella* and *Staphylococcus*) (Fig. 6.5), and urinary tract infections (*Escherichia coli* and *Klebsiella*). Increased susceptibility is also seen for *Shigella* and *Mycoplasma* pneumonia. Several factors that contribute to this susceptibility are functional hyposplenism, impaired antibody response, decreased opsonization, impaired complement activation in the properdin pathway, and abnormal chemotaxis. Serious bacterial infections are seen in approximately one-third of children with sickle cell disease before 4 years of age. Infection is the most common cause of death in these children, although infections can often precipitate vaso-occlusive pain.

Vaso-occlusive episodes (VOE) begin in infancy with dactylitis or hand-and-foot syndrome (Fig. 6.5). Later VOE may involve the periosteum, bones, or joints, resulting in infarction that must be differentiated from osteomyelitis and septic arthritis. Pulmonary symptoms such as pleural pain and dyspnea may be due to infection, in situ thrombosis, or embolism. Other clinical manifestations include splenic sequestration, abdominal and aplastic episodes, cholelithiasis, hepatic infarcts, occlusion of cerebral vessels (stroke), ocular changes (sickle retinopathy), hematuria, hyposthenuria, hyponatremia, priapism, and skin ulcers.

Sickle cell disease has a broad spectrum of clinical severity. The most severe forms of SCD (HbSS and HbS/β^0−thalassemia) are associated with significant

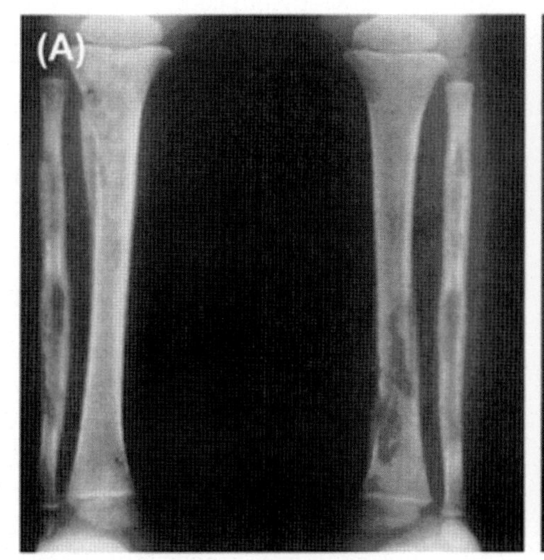

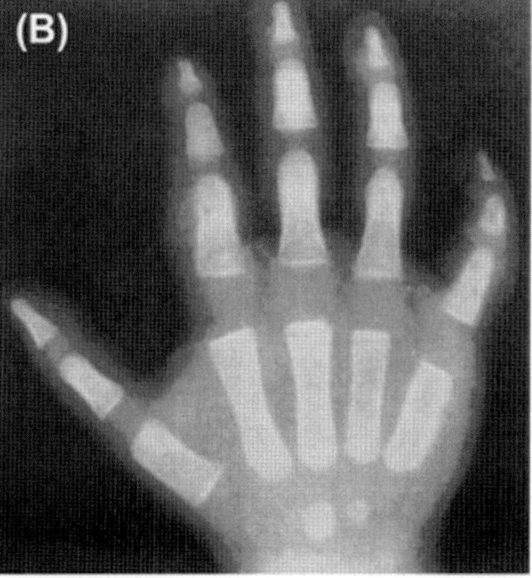

Figure 6.5 Radiographic changes in sickle cell disease. (A) Changes in the tibias and fibulas secondary to *Salmonella osteomyelitis.* (B) Hand-and-foot syndrome with soft-tissue swelling and focal areas of cortical destruction and periosteal new bone formation.

morbidity and mortality, with the overall lifespan of patients with these forms of SCD being shortened by two or three decades compared to the general population [11,12]. Other compound heterozygous forms of SCD (HbSC, HbD-Punjab, HbO-Arab, and HbS/β^+−thalassemia) are usually less severe phenotypes but can still lead to significant SCD-related complications, including retinopathy, nephropathy, and avascular necrosis of bone.

The effect of α-thalassemia on the clinical picture of sickle cell disease is complex. α-thalassemia lowers the mean corpuscular volume (MCV) and the mean corpuscular hemoglobin concentration (MCHC). As the latter is one of the factors that determine the extent of polymerization of HbS, the result of α-thalassemia is a reduction in polymerization, with fewer sickling-related manifestations such as leg ulceration and acute chest syndrome. However, the advantages of the inhibition of sickling are balanced by disadvantages of α-thalassemia in causing a reduction in hemolysis and an increase in total hemoglobin. These effects result in an increase in blood viscosity, making patients more prone to vaso-occlusion.

The persistence of HbF in adults inhibits the polymerization of HbS as well as reduces the HbS concentration. Indeed, patients doubly heterozygous for HbS and the deletion form of HPFH (S/HPFH) have no clinically significant hematologic abnormalities despite having only HbS and HbF. Such compound heterozygotes have nearly normal red cells, each with 20%−30% HbF in a pancellular distribution.

6.5.4 Diagnosis

The peripheral blood smear of patients with sickle cell disease may have normal appearing, irreversibly sickled, target, and nucleated red cells. Howell−Jolly bodies and red cell fragments are also present, especially after functional asplenia develops. The clinical history of vaso-occlusive episodes or severe infections with anemia, abnormal red cell morphology on peripheral smear with a normal or elevated mean corpuscular volume (MCV), positive sickling test, and HbS (greater than 80%) and HbF on electrophoresis or HPLC, makes the diagnosis of sickle cell disease probable. Family studies indicating that both parents have the sickle cell trait are helpful to exclude S/β^0-thalassemia and S/HPFH, although a definitive diagnosis is best made by DNA analysis for the HbS mutation and the other possible interacting genotypes. In addition, siblings should also be tested to identify and treat previously undiagnosed cases.

6.5.5 Treatment

A number of antisickling agents have been developed to ameliorate sickle cell disease, and a variety of new therapies have evolved from our improved understanding of the pathophysiology of SCD [13]. HbF concentrations are inversely related to morbidity in sickle cell disease, and the use of drugs that diminish HbS polymerization by increasing the level of HbF has been the most promising approach. Hydroxyurea (HU, also known as hydroxycarbamide) is the only FDA-approved drug aimed at raising HbF for patients with SCD. HU is a ribonucleotide reductase inhibitor and in addition to raising HbF levels, also lowers the number of circulating leukocytes and reticulocytes to decrease vaso-occlusion, improves cellular deformability and rheology, and increases local vasodilation [14]. HU was approved for use in SCD patients following a Phase III trial in 1995 that showed significant reduction in VOE, time to first painful episode, and incidence of acute chest syndrome [15]. More recently, HU has been shown to have the ability to be safely administered in a global setting [16] and is routinely initiated in patients as young as 2 years old to reduce the risk of both acute and chronic complications [17]. Notably, only limited data from observational studies are available on hydroxyurea therapy in people with genotypes other than HbSS or HbSβ^0-thalassemia, although HU is often trialed in patients with more severe forms of HbSC and HbS/β^+-thalassemia [14].

Despite the benefits of HU in many SCD patients, there remains a strong need to develop more effective pharmacologic HbF inducers. Numerous small molecule compounds have been shown to induce HbF in vitro in recent years, including pomalidomide [18,19], 5-azacytidine [20], the EHMT1/2 inhibitor UNC0638 [21,22], lysine-specific demethylase 1 (LSD1) inhibitors (e.g., RN-1 [23] and tranylcypromine [24]), the histone deacetylase inhibitor entinostat [25], the PDE9 inhibitor PF-04447943 [26,27], and the FOXO3-inducer metformin [28]. Furthermore, in addition to the primary HbF repressors BCL11A and LRF, 3−8 numerous additional genetic targets have recently been identified in HbF regulation, including heme-regulated inhibitor (HRI) [29−31], the E3 ligase adaptor SPOP

[32], the mRNA-binding protein IGF2BP1 [33], members of the NuRD subcomplex such as MBD2 [34] and CHD4 [35], and the CHD4 activator ZNF410 [36]. Ongoing preclinical and clinical studies targeting these specific pathways will likely lead to additional HbF-inducing therapies for the treatment of SCD in the coming years.

In the United States, three additional medications have been approved for the treatment of sickle cell disease—L-glutamine, crizanlizumab, and voxelotor. L-glutamine, an essential amino acid, is a precursor for nicotinamide adenine dinucleotide (NAD) and improves NAD redox potential. Twice daily oral L-glutamine administration was shown to significantly reduce the number of pain episodes, hospitalizations, and frequency of acute chest syndrome in HbS and HbS/β-thalassemia patients [37]. Crizanlizumab is a humanized monoclonal anti-P-selectin antibody that blocks binding of sickled cells to the endothelium, leading to reduction in microvascular vaso-occlusion. Monthly intravenous infusion of crizanlizumab in SCD patients led to significantly lower rate of sickle cell—related pain episodes [38] and was shown to have an additive effect in patients also taking hydroxyurea [39]. Voxelotor functions by binding to α-globin chains of hemoglobin, increasing hemoglobin oxygen affinity, and stabilizing the oxyhemoglobin state. In the HOPE trial, once daily voxelotor showed a significantly higher percentage of patients with a hemoglobin increase of 1.0 g/dL following 24 weeks of therapy, consistent with inhibition of HbS polymerization [40]. Further studies are ongoing to determine the disease-modifying potential of these therapies to reduce acute and chronic complications of SCD.

The clinical management of acute VOE is primarily supportive and prophylactic. Infections should be treated promptly with antibiotics. VOE should be managed with vigorous hydration because of the patient's inability to concentrate urine and the increased blood viscosity. In addition, rapid initiation of opioids is critical to reduce SCD-related complications [14]. Acute management of more severe complications of SCD, including stroke, acute chest syndrome, and multiorgan system failure, may require emergent red cell exchange transfusion and subspeciality involvement [41,42]. Currently, the only effective long-term management options for SCD-related complications such as secondary stroke prevention and treatment of pulmonary hypertension is chronic monthly red cell exchange to maintain HbS levels less than 30%, although curative therapies such as allogeneic stem cell transplant have also shown significant promise.

Major advances have recently been made in prevention of acute and chronic complications of SCD in the past several decades. Newborn screening and close follow-up, especially early in life, have significantly improved survival by allowing vaccination against pneumococcal and other encapsulated pathogens and reinforcing the need for urgent medical care in the event of fever [14,43]. Furthermore, twice daily prophylactic penicillin is initiated in early infancy and continued through at least age 5 to reduce infection risk [14]. The associated anemia in SCD is usually well-tolerated, but if folate deficiency occurs, the anemia becomes more severe and is associated with macrocytosis, hypersegmented granulocytes, and a decrease in the percentage of reticulocytes. Folate deficiency is prevented easily by daily folic acid supplement. Transcranial Doppler (TCD) imaging of large intracranial blood vessels to detect increased velocities secondary to stenosis can predict risk of stroke in children with SCD [44]. In patients with elevated TCD velocities (greater than 200 cm/s), the STOP trial showed that regular blood transfusions can serve as effective primary stroke prevention and has resulted in a declining incidence of primary overt stroke in children with SCD [45,46].

Although hypertransfusion of packed red cells and drugs such as hydroxyurea, L-glutamine, crizanlizumab, and voxelotor may ameliorate the symptoms of sickle cell disease, only allogeneic stem cell transplantation is truly curative at the present time. This approach has been successfully applied to treat sickle cell disease in children and is increasingly being used for patients with less severe but symptomatic SCD. Recent experience with HLA-matched allogeneic stem cell transplantation of SCD patients has been promising, with the 5-year event-free survival and overall survival of 91.4% and 92.9%, respectively [47], and several trials are ongoing studying the role of haploidentical transplantation in SCD patients. Cord blood transplantation has also been performed successfully when an HLA-matched sibling without sickle cell disease has been born. Stem cell transplantation is more challenging in adults who already have irreversible organ damage, and important barriers to more widespread use of bone marrow transplantation remain. These include limited availability of HLA-matched related donors, a short-term mortality of about 10% from the transplant procedure, and uncertainty as to which individuals will have severe symptomatic disease warranting such an intervention.

The concept of gene therapy in sickle cell disease holds promise for a curative approach in the near future. Early preclinical studies demonstrated cure in both mouse thalassemia and SCD models [48,49]. However, it took many years of optimization in the field to bring this therapy successfully to patients utilizing lentiviral gene addition modalities [50,51]. These trials used a vector expressing an antisickling β-globin variant, T87Q, distinguishable from endogenous HbA or HbF by HPLC [52]. Autologous hematopoietic stem cells (HSCs) were harvested in SCD patients utilizing the CXCR4 antagonist plerixafor, allowing for safe mobilization, followed by ex vivo HSC lentiviral gene addition, myeloablative conditioning, and reinfusion of the modified gene product. The ongoing Phase 1/2 HGB-206 trial shows promising early results with high levels of HbA^{T87Q} in a pancellular distribution and significant reduction of vaso-occlusive events [53]. Additionally, the ongoing MOMENTUM trial utilizing a modified gamma-globin gene addition approach (ARU-1801; HbF^{G16D}) has also showed promising early results [54].

In addition to the gene addition approaches above, multiple ongoing trials are currently testing whether the HbF repressor BCL11A can be successfully targeted for the treatment of SCD. One approach is via a short hairpin RNA (shRNA) targeting BCL11A mRNA embedded in a microRNA (shmiR), allowing erythroid lineage−specific knockdown (BCH-BB694 lentiviral vector) [55,56]. Recently published data utilizing this vector showed robust and stable HbF induction of 20%−40% by HPLC, with HbF broadly distributed and highly expressed in red cells and a concomitant reduction in severe sickle cell events [57]. A second approach involves CRISPR-Cas9-based gene editing of the BCL11A + 58 enhancer to achieve erythroid-specific depletion of BCL11A and subsequent increases in HbF levels. Early results from the CLIMB-SCD-121 trial utilizing the CTX001 lentiviral vector showed HbF induction of >40% at 6−12 months posttransplant with durable BCL11A enhancer editing, pancellular HbF expression, and significant reduction in vaso-occlusive episodes [58]. Future studies will continue to explore BCL11A-targeted gene editing approaches as well as direct gene correction of the sickle hemoglobin mutation for the treatment of SCD.

6.5.6 Prevention

During genetic counseling, a couple where each individual has sickle trait is advised of their 25% risk for having a child with SCD-HbSS during each pregnancy. In addition, couples with sickle cell trait and HbC trait or sickle trait and beta-thalassemia trait should also be counseled on the risk of SCD-HbSC and SCD-HbS-β-thalassemia, respectively. The spectrum of additional compound heterozygous states and their clinical severity are outlined in Table 6.4. However, the utilization of prenatal diagnosis for sickle cell disease appears to be very low compared with that for β-thalassemia because of the variable clinical course of the disease. An audit of more than 20-year experience of prenatal diagnosis for the hemoglobin disorders in the United Kingdom showed that only 13% of all couples at risk of having a child with sickle cell disease underwent prenatal diagnosis. This rate has improved in recent years in the United Kingdom and United States [59]; nevertheless, significant work remains to fully incorporate genetic counseling for sickle cell trait parents into standard preventive care.

6.5.7 Interactions With Sickle Hemoglobin

Sickle cell disease comprises a number of compound heterozygous genotypes. Genotypes that are frequently encountered are HbS/β-thalassemia or HbS/δβ-thalassemia, HbS/C, HbS/D-Punjab, and HbS/O-Arab. There are also some rare interactions that have been reported, including HbS/C-Harlem, HbS/S-Antilles, HbS/S-Oman, and the compound heterozygosity for HbS and some unstable β-chain variants. The most common of these genotypes and their clinical severity are summarized in Table 6.4.

6.6 UNSTABLE HEMOGLOBIN VARIANTS

6.6.1 Molecular Basis

At least 300 unstable hemoglobin variants are known in the literature. Among these, β-variants are two times more frequent than α-variants, a discrepancy that may be due to the smaller percentage of unstable hemoglobin and hence milder clinical symptoms associated with the α-chain variants. An individual with a single variant α-gene has three normal α-genes, so that the percentage of unstable hemoglobin in the red cells is very small (5%−20%). In contrast, an individual with a variant β-gene has only a single normal β-gene; so the unstable hemoglobin containing the variant β-chain makes up a greater proportion of the total cellular hemoglobin synthesized (20%−40%). Because the gene frequencies for these variants are extremely low, almost all affected individuals seen are heterozygotes.

The increased propensity of unstable hemoglobins to denature can result from several types of mutations. For instance, the α-helix of α or β-globin can be disrupted by proline replacing another amino acid within the helix. There are at least 10 examples of this type of disruption of primary and secondary structures, including Hb-Bibba (α136 Leu → Pro) and Hb-Genova (β28 Leu → Pro). Deletions of amino acid residues alter primary and secondary structures, as well as the conformation of the hemoglobin molecules, and 8 of the 10 variants of this type are unstable, for example, Hb-Leiden (β6 or 7 Glu → 0) and Hb-Gun Hill (β91−95 (Leu-His-Cys-Asp-Lys) → 0). Interference with interchain contacts permits the αβ dimers to dissociate into monomers, for example, Hb-Philly (β35 Tyr → Phe) and Hb-Tacoma (β30 Arg → Ser) lack hydrogen bonds normally linking the α and β subunits. Substitutions that affect heme binding or disturb the hydrophobic heme pocket decrease the molecule's stability. There are over 30 such mutations, and most result in unstable hemoglobins, such as Hb Bristol (β67 Val → Asp) and Hb-Köln (β98 Val → Mer). Finally, globin chain elongation can result in instability due to hydrophobic properties of the extended chain, for example, Hb Cranston (β144−151).

These variant hemoglobins tend to denature spontaneously; the globin subunits precipitate in the red cell, forming aggregates or Heinz bodies. The Heinz bodies adhere to the red cell membrane and result in decreased pliability of the cell. Inflexible erythrocytes are then selectively trapped by the reticuloendothelial system.

6.6.2 Clinical Aspects of Unstable Hemoglobins

Patients often present in infancy or early childhood with a hemolytic anemia, jaundice, and splenomegaly, or later with cholelithiasis. Some variants also cause cyanosis because of their abnormal properties, that is, propensity to form methemoglobin or decreased oxygen affinity. Clinical severity varies with different unstable variants; for β-variants, symptoms appear after the γ to β transition in hemoglobin synthesis (Fig. 6.3). A number of Hb variants are associated with a thalassemic condition rather than a severe hemolytic anemia. These variants, such as Hb-Agrinio, α 29 Leu → Pro, and Hb Quon Sze, α 125 Leu → Pro, are so unstable that their presence is difficult to detect except by DNA sequence analysis of the appropriate gene.

6.6.3 Diagnosis

The peripheral smear may be normal or hypochromic. Staining with a supravital stain, such as 1% methyl violet, demonstrates preformed Heinz bodies (Fig. 6.6B). Heat instability of the variant hemoglobin is demonstrated by the formation of a hemoglobin precipitate when a hemolysate is incubated at 50 °C or higher or at 37 °C in 17% isopropanol. Hemoglobin electrophoresis by usual methods may detect only about half of unstable variants since the charge of these variants is often unaltered by the substitutions. Oxygen saturation curves of whole blood may indicate normal (20% of the unstable variants), decreased (30%), or increased (50%) oxygen affinity.

6.6.4 Treatment

Treatment is generally supportive. If hemolysis is severe, prophylactic folate may be indicated. Oxidant drugs, such as sulfonamides, increase hemolysis in some patients and should be avoided. Transfusions are indicated in the treatment of aplastic crises as well as symptomatic anemia. While splenectomy may result in improvement of the anemia, it also increases the risk of septicemia, especially in young patients. Because of the mortality associated with septicemia in splenectomized patients, the physician should reserve splenectomy for selected patients. Splenectomy should be postponed until the patient is at least 6 years of age, and the administration of pneumococcal vaccine and prophylactic antibiotics should be considered. However, the risk of pulmonary hypertension and thrombosis in splenectomized patients is significantly elevated, so risks and benefits should always be individually discussed.

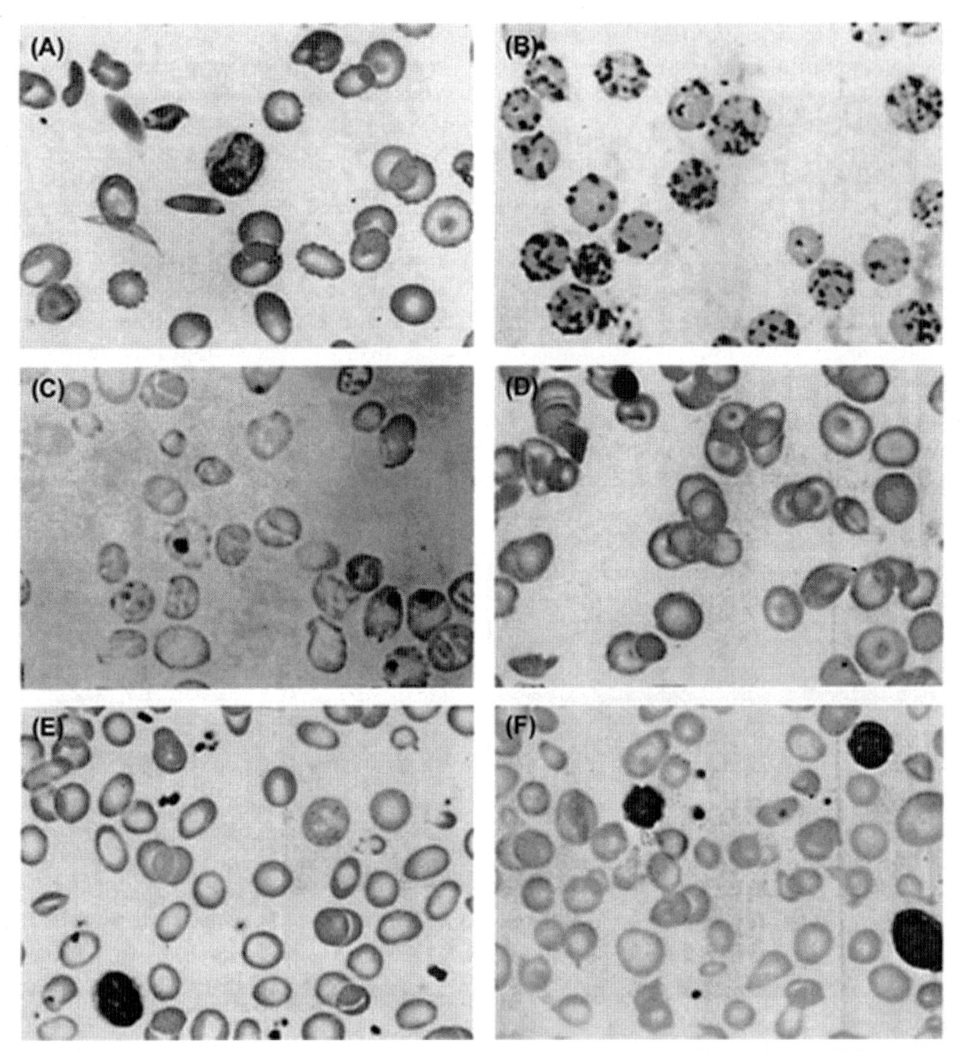

Figure 6.6 Peripheral blood smears from patients with various disorders of globin synthesis. (A) Homozygous sickle cell disease. (B) Unstable Hb Zurich with Heinz bodies. (C) HbH disease. (D) Sickle/β-thalassemia. (E) β-thalassemia trait. (F) Homozygous β-thalassemia.

6.7 HEMOGLOBIN VARIANTS WITH ALTERED OXYGEN AFFINITY

6.7.1 Molecular Basis

The oxygen dissociation curve shown in Fig. 6.4 is sigmoid shaped because of heme–heme interactions. Mutations that affect the heme–heme interaction, the Bohr effect, or the deoxyhemoglobin-2,3-DPG interaction can change the shape or position of the oxygen dissociation curve. Mutations affecting the $\alpha_1\beta_2$ subunit contact point can alter heme–heme interaction by causing the deoxyhemoglobin conformation to be less stable. These mutations result in increased stability of the oxyhemoglobin conformation and increased oxygen affinity (e.g., Hb Kempsey (β99 Asp → Asn)). Alternatively, the oxyhemoglobin conformation can be destabilized by mutations affecting the α94β102 contact point, resulting in decreased oxygen affinity (e.g., Hb Kansas (β102 Asn → Thr)). Substitutions at the COOH-terminal ends of globin chains can lead to instability of deoxyhemoglobin conformations and increased oxygen affinity (e.g., Hb Bethesda (β145 Tyr → His)) as well as a reduction in the Bohr effect. 2,3-DPG binds to residues β1, 2, 82, and 143 in the deoxygenated form. Substitutions altering these residues tend to have increased oxygen affinity (e.g., HbF γ-globin has a serine for histidine substitution at position 143)).

Variants with increased oxygen affinity cause a shift to the left of the oxygen dissociation curve (Fig. 6.4), resulting in less oxygen delivery per gram of hemoglobin. To compensate, hemoglobin concentration or blood flow increases to partially restore oxygen delivery to the tissues. Some variants with increased oxygen affinity do not cause polycythemia because of the small fraction of the total hemoglobin they comprise or the compensatory changes in the shape of the oxygen dissociation curve. Variants with decreased oxygen affinity have a shift to the right and increased oxygen delivery per gram of hemoglobin. As a result, the hemoglobin concentration is normal or decreased (e.g., Hb Beth Israel (β102 Asn → Ser)).

6.7.2 Clinical Aspects and Diagnosis

Because the gene frequencies for nearly all variants of hemoglobins are very low, patients are nearly always heterozygotes. The great majority of patients are asymptomatic, and when oxygen affinity is increased the major finding is polycythemia with erythrocytosis, normal white blood cell and platelet counts, and absence of splenomegaly. Since about half of these variants cannot be detected on routine electrophoresis, whole blood oxygen affinity studies are required for diagnosis. Some concern has been raised regarding the risk to fetuses of mothers who have variants with increased oxygen affinity. The few data available regarding the outcome of such pregnancies do not in general seem to indicate increased fetal mortality.

6.7.3 Treatment

The condition is generally considered benign. It is important to avoid chemical treatment of the compensatory polycythemia unless hematocrit levels are high enough to cause increased viscosity.

6.8 THALASSEMIAS

6.8.1 Quantitative Disorders of Globin Synthesis

The thalassemia syndromes are defined as a group of inherited disorders characterized by the absence or reduced synthesis of one or more of the normal globin chains of hemoglobin. They are subclassified according to the particular globin genes affected and also according to the effect on gene expression. For example, the β-thalassemias are disorders of β-globin gene expression, whereas the δβ-thalassemias are disorders of both δ- and β-gene expression. Absent globin synthesis is designated with an "0" superscript, for example, β^0-thalassemia, while the presence of some normal but insufficient gene product is noted by a "+" superscript, for example, β^+-thalassemia. The lack of synthesis of the affected globin chain results in an unbalanced α/β globin chain ratio, and therefore, the defect is a quantitative one. This is in contrast with the hemoglobinopathies in which the variant hemoglobins are qualitatively or structurally abnormal.

6.8.2 Geographical Basis

Thalassemia is distributed primarily among people of Mediterranean, African, Middle Eastern, Asian Indian, Chinese, and Southeast Asian descent, but sporadic cases have been reported in many ethnic groups. The thalassemias are most common in the Mediterranean region, Africa, and Asia, but have now become a global problem through migration of populations throughout Europe, the Americas, and Australia. The high carrier rates observed in many populations are thought to have resulted from a positive selection pressure due to *Falciparum* malaria. The mechanism by which the thalassemia trait provides protection against malaria is less clear than that with the sickle cell trait, but there is a strong geographic correlation of gene frequencies with the incidence of malaria.

The thalassemias and the hemoglobin variants are regionally specific, with each local population having its own characteristic spectrum of mutations. For most of these populations, the range of mutations has been identified by molecular analysis and the frequency of each mutation relative to the others has been determined. This information is the first step required for the control of the thalassemias through an integrated program of carrier screening, genetic counseling, and prenatal diagnosis.

6.8.3 Molecular Basis of α-Thalassemia

α-Thalassemia is characterized by a deficiency of α-globin chain synthesis, and defective gene expression may occur in either one globin gene or in both. Most of the common α-thalassemia alleles result from the deletion of gene sequences in the α-globin gene cluster, as depicted in Fig. 6.7. An updated listing of all α-thalassemia alleles is available through the HbVar globin gene server website (http://globin.bx.psu.edu) [9,10].

6.8.3.1 α^+-Thalassemia

α^+-Thalassemia is most commonly caused by the deletion of one of the two α-globin genes. Although many different deletions have been identified, only two are commonly encountered in practice. These are the 3.7 kb deletion ($-\alpha^{3.7}$), which has reached high frequencies in the populations of Africa, the Mediterranean area, the Middle East, the Indian subcontinent and Melanesia, and the 4.2 kb deletion ($-\alpha^{4.2}$), which is commonly found in Southeast Asian and Pacific populations. These deletions were created by unequal crossing over between homologous sequences in the α-globin gene cluster, resulting in one chromosome with only one α-gene (-α) and the other chromosome with three α-genes (ααα). Further recombination events between the resulting chromosomes have given rise to a very rare quadruplicated α-gene allele (αααα). Individuals with five or six α-globin genes are hematologically normal. However, the excess α-globin chains may be sufficient to cause thalassemia intermedia when these alleles are coinherited with the β-thalassemia trait.

Various nondeletion defects have also been found to cause α^+-thalassemia, mostly in populations from the Mediterranean area, Africa, and Southeast Asia. Although nondeletion mutations are generally much less frequent than deletion mutations, among Southeast Asians, the nondeletion α^+-thalassemia allele for Hb Constant Spring gene has reached a reasonably high gene frequency. This nondeletion α^+-thalassemia gene encodes an abnormal α-chain that has 31 additional amino acids at the COOH-terminal end and is synthesized at about 3% of the rate of normal α-chains.

6.8.3.2 α^0 -Thalassemia

α^0-Thalassemia results from deletions that involve both α-globin genes in the α-globin gene cluster. At least 34 different such deletions have been described. The deletions that have attained the highest gene frequencies are found in individuals from Southeast Asia and South China ($-^{SEA}$), the Philippine Islands ($-^{FIL}$), Thailand ($-^{THAI}$), and a few Mediterranean countries such as Greece and Cyprus ($-^{MED}$ and $-(\alpha)^{20.5}$). Although one α^0-thalassemia mutation ($-^{SA}$) has been described in individuals of Indian descent, it is extremely uncommon in this ethnic group. No α^0-thalassemia deletions

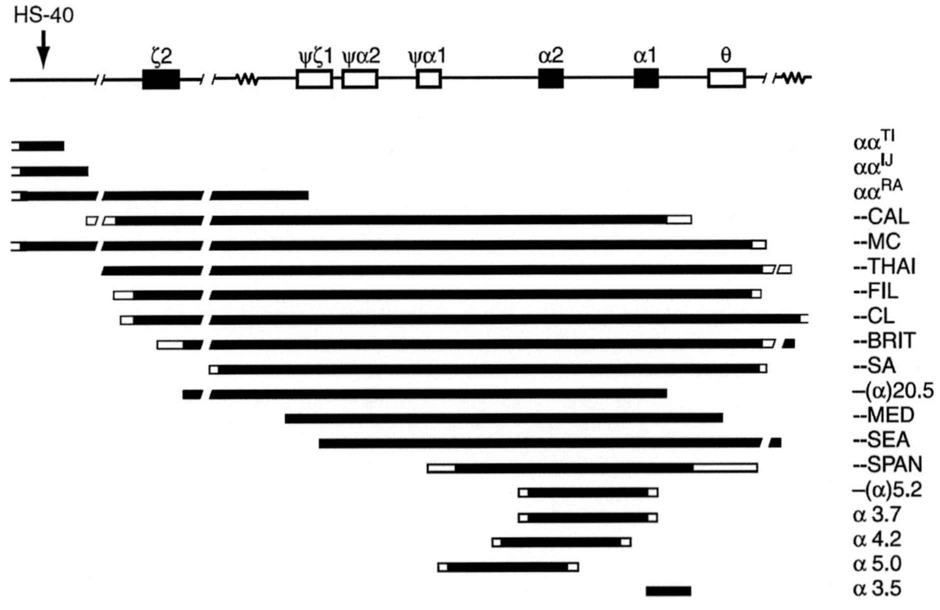

Figure 6.7 Deletions in the α-globin gene cluster.

have been reported in individuals from sub-Saharan Africa, although two very rare α^0-thalassemia alleles involving the combination of the -$\alpha^{3.7}$ deletion and a nondeletion mutation (e.g., the -$\alpha^{3.7}$ CD 30–31 ($-^{AG}$) allele) have been described. In Northern Europe, α-thalassemia only occurs sporadically because of the lack of natural selection, although one particular α^0-thalassemia mutation ($-^{BRIT}$) has been reported in a number of British families. Finally, in extremely rare instances, α^0-thalassemia can also result from deletions that remove the α-globin gene regulatory element about 62 kb upstream of the α-globin gene complex but leave the α-globin genes intact.

6.8.4 Molecular Basis of β-Thalassemia

β-Thalassemia is caused by at least 200 different point mutations or small insertions/deletions of DNA sequence in and around the β-globin gene, together with a much smaller number of gene deletions ranging from 25 bp to 67 kb. An updated listing of all β-thalassemia alleles is available through the HbVar Web site https://globin.bx.psu.edu/) [9,10]. Although more than 230 different β-thalassemia alleles have been characterized, only approximately 30 mutations are found in at-risk groups at a frequency of 1% or greater, and thus just a small number account for the majority of the mutations worldwide. All mutations are regionally specific, and the spectrum of mutations has now been determined for most at-risk populations.

The mutations either reduce the expression of the β-globin gene (β^+-type) or result in the complete absence of β-globin (β^0-type). A few of the β^+-types are associated with an unusually mild phenotype and are sometimes designated β^{++}-type. The mutations may affect globin gene transcription, RNA processing or translation, RNA cleavage, and polyadenylation or result in a highly unstable globin chain. Frameshift and nonsense codon mutations have been observed in all three exons and RNA processing mutations have been found in both introns and the four splice junctions.

6.8.5 Pathophysiology

In the thalassemia syndromes, there is reduced or absent synthesis of the affected globin chain; the unaffected chain continues to be synthesized at relatively normal levels. The result is an imbalance that causes aggregation and precipitation of excess unpaired chains. In β-thalassemia, free α-chains aggregate; the aggregates are highly insoluble and form inclusions in nucleated erythroid precursors in the bone marrow. These inclusion bodies cause intramedullary hemolysis (ineffective erythropoiesis). In contrast, in α-thalassemia the γ_4 (Hb Bart's) and β_4 (Hb H) tetramers that form are more soluble. Thus, in severe α-thalassemias, inclusions are seen in mature erythrocytes and the ineffective erythropoiesis of β-thalassemia is absent. In any severe thalassemia, removal of these inclusions from erythrocytes by the reticuloendothelial system damages the cells and produces "teardrop" forms. Splenomegaly can be secondary to splenic congestion or ineffective erythropoiesis. After the spleen is removed, cell destruction continues at a decreased rate in the liver, and the number of red cell inclusions may increase greatly. The large number of erythroid precursors expands the marrow cavities, and bone deformities, thinning, and occasional pathologic fractures result (Fig. 6.8).

Iron accumulation results from increased gastrointestinal absorption stimulated by the anemia, blood transfusions, and decreased utilization for hemoglobin synthesis. The deposition of excess iron causes damage to the liver, heart, pancreas, and other endocrine tissues. Chelation therapy is essential to remove excess iron stores. Folic acid requirements are increased in thalassemia. If deficiencies develop, they may worsen the anemia.

6.8.6 Clinical Features

The hematological characteristics of heterozygotes and homozygotes for the main types of globin gene disorders are listed in Table 6.5 and described in detail below.

6.8.6.1 α-Thalassemia

Four clinical types are seen, depending on the number of α-genes affected. In order of increasing severity, these are heterozygous α^+-thalassemia (one gene affected), heterozygous α^0-thalassemia and homozygous α^+-thalassemia (two genes affected), HbH disease (three genes affected), and Hb Bart's hydrops fetalis syndrome (all four genes affected).

- α^+-Thalassemia. Heterozygous α^+-thalassemia (genotype: α-/$\alpha\alpha$) is effectively a silent carrier state because the reduction of an mRNA is insufficient to produce significant globin chain imbalance ($\alpha/\beta = 0.8–0.9$) (see Fig. 6.9). The hematological

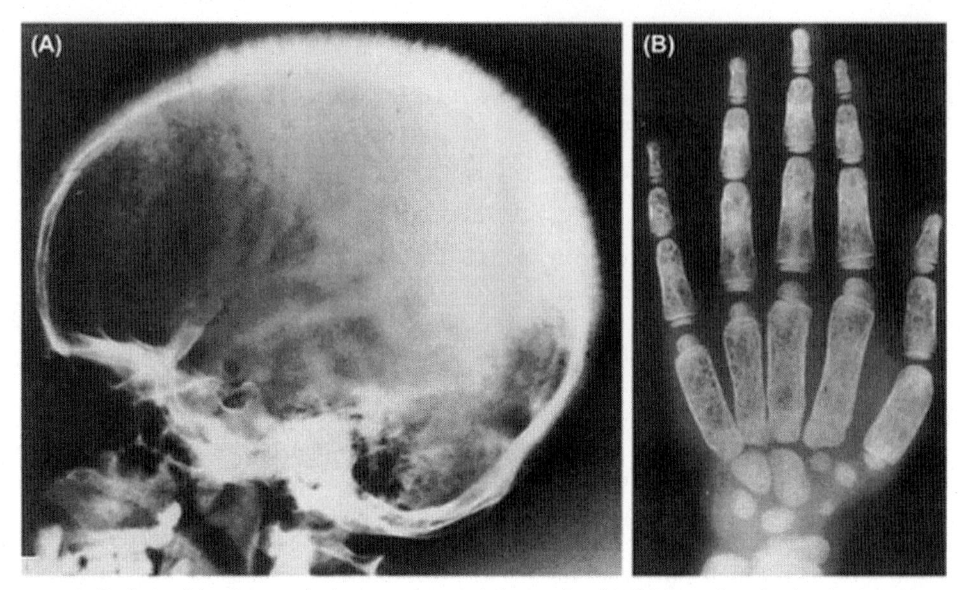

Figure 6.8 Radiographic changes in homozygous β-thalassemia. (A) Thickened parietal calvaria with outer table destruction and "hair-on-end" appearance. (B) Widened medullary cavities, cortical thinning, and coarse trabeculation secondary to intramedullary hyperplasia.

indices in carriers are only slightly altered and overlap considerably with those of normal individuals. A reliable diagnosis can only be made by DNA analysis.

- α^0-Thalassemia. Heterozygous α^0 -thalassemia individuals (genotype: –/αα) are usually of Asian or Mediterranean descent. They are relatively asymptomatic, but have a mild microcytic anemia (hemoglobin of 10–12 g/dL) and mild poikilocytosis and anisocytosis. At birth, Hb Bart's may reach 5% in cord blood. The α/β synthesis ratio is 0.6–0.75 (Fig. 6.9). The diagnosis of α^0-thalassemia trait should be considered when the MCV and MCH are low, the MCHC is relatively normal, the patient is not iron deficient, and the HbA_2 is normal. This phenotype is also associated with that of homozygous α^+-thalassemia (genotype: -α/-α) and the two conditions can only be distinguished by DNA analysis techniques.
- Hb H disease. Hb H disease (genotypes: –/-α) is mostly found in people of Southeast Asian, Greek, and Italian descent. The anemia varies with an average hemoglobin range of 8–10 g/dL, and reticulocytes make up 5%–10% of red cells. Splenomegaly and occasionally hepatomegaly are found. The red cells are microcytic (decreased MCV) and their hemoglobin content is decreased [decreased mean corpuscular hemoglobin (MCH)], but the concentration of hemoglobin per cell is normal (normal MCHC). On the peripheral smear, poikilocytosis, polychromasia, and target cells are seen. The β4 tetramer (HbH) inclusions are seen easily following incubation with 1% brilliant cresyl blue, or after splenectomy they can be seen occasionally with methylene blue reticulocyte stain or Wright's stain (Fig. 6.6C). The α/β ratio is 0.3–0.4. This imbalance and Hb H levels of 4%–30% occur after the switch from γ-to β-chain synthesis is complete. Both tetramers precipitate, causing inclusion body hemolytic anemia. Deficient α-chain synthesis causes a drop in HbA_2 levels to 1%–1.5%. Deficient α-chain synthesis is secondary to a deficiency of α-globin mRNA caused by deletion of three of the four α-genes. A subset of Hb H patients with nondeletional α^+-thalassemia exhibit symptoms more severe than those with deletional α^+-thalassemia, and may require recurrent blood transfusions.
- Hb Bart's hydrops fetalis syndrome. This is the most severe clinical form of α-thalassemia, resulting from homozygous α^0-thalassemia (genotype: –/–), sometimes called hydrops fetalis with Hb Bart's. In the usual case, over 80% of the hemoglobin in the fetal

TABLE 6.5 **Characteristics of the Main Types of Globin Gene Disorders**

Disorders	Heterozygotes	Homozygous State
α^+-thalassemia Deletion (-α)	(α-/αα) 0%–2% Hb Bart's at birth, minimal hematologic changes	(α-/α-) 5%–10% Hb Bart's at birth low MCH and MCV
α^+-thalassemia Nondeletional (αCSα)	0%–2% Hb Bart's at birth, 0.5%–1% Hb Constant Spring	Slightly more severe than heterozygous α^0-thalassemia
α^0 -thalassemia (–)	(αα/–) 5%–10% Hb Bart's at birth, low MCH and MCV; normal HbA_2	(–/–) Hb Bart's hydrops fetalis; 80% Hb Bart's, 20% Hb Portland at birth
α^+-thalassemia/α^0-thalassemia		(α-/–) Hb H disease 20%–40% Hb Bart's, at birth
β^0-thalassemia	Low MCH and MCV HbA_2 3.5%–7.0%	Thalassemia major; HbF 98%; HbA_2 2%
Severe β^+-thalassemia	Low MCH and MCV HbA_2 4.5%–7.0%	Thalassemia major; HbF 70%–95%
Mild β^+-thalassemia	Low MCH and MCV HbA_2 3.5%–7.0%	Thalassemia intermedia; HbF 20%–40%
Hb Lepore	Low MCH and MCV Hb Lepore 8%–20%; low HbA_2	Thalassemia major/intermedia; HbF 80%, Hb Lepore 20%
δβ-thalassemia	Low MCH and MCV Normal HbA_2, HbF 4%–18% (heterocellular)	Thalassemia intermedia; 100% Hb F
Deletional HPFH	Normal indices HbF 15%–35% (pancellular)	Normal 100% HbF
Nondeletional HPFH	Normal indices; HbF 1%–20%	Not described
HbS	HbA, HbA_2 (3%–4%), HbS (35%–40% for αα/αα; 30%–35% for -α/αα; 24%–28% for -α/-α)	HbS, HbF (1%–15%) HbA_2
Hb S/β-thalassemia		β^0: severe sickle cell disease; β^+: less severe sickle cell disease β^{++}: mild sickle cell disease
Hb E/β-thalassemia		Thalassemia major or intermedia; HbE 60%–70%, HbF 30%–40%

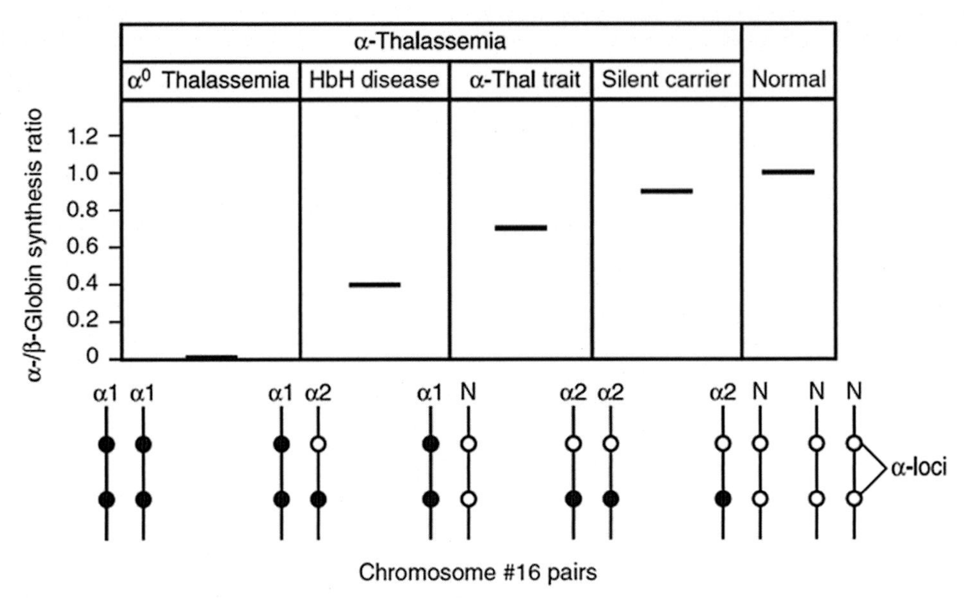

Figure 6.9 Chain synthesis ratios and proposed genotypes in the different α-thalassemia states. α-Genes: o, normal; •, abnormal. Genotypes: α_1,α^0 -thalassemia; α_2,α^+ -thalassemia, N, normal.

blood is Hb Bart's (γ_4), which has a very high oxygen affinity, causing severe tissue hypoxia; the remainder Hb H (β_4) and Hb Portland ($\zeta_2\gamma_2$), the only functional hemoglobin keeping the fetus alive. The resulting severe progressive anemia leads to massive organomegaly and heart failure. Prenatal diagnosis is always indicated because of the serious maternal complications that may occur during the pregnancy. This condition is found usually in Asian infants who are spontaneously aborted or die of severe hydrops shortly after birth, although prenatal diagnosis and intrauterine transfusion therapy is more common in recent years[60]. Such patients require lifetime transfusion therapy unless they can be cured by stem cell transplantation [61]. Ongoing clinical trials exploring the role of in utero hematopoietic stem cell transplantation may offer additional therapeutic options in the future.

6.8.6.2 β-Thalassemia

In contrast to the α-thalassemia states in which there are four levels of severity, the β-thalassemias can be considered as having 3 levels of severity. β-thalassemia trait results from a single β-thalassemia gene mutation; β-thalassemia intermedia results from heterogeneous genotypic combinations; and β-thalassemia major (Cooley's anemia) results from two significant β-thalassemia gene mutations at the β-globin locus, with resultant broad spectrum of intermediate to severe anemia and sequelae.

- β-thalassemia trait. Individuals with β-thalassemia trait (heterozygous β-thalassemia) are usually asymptomatic. The classical picture is mild anemia (hemoglobin of 10–11 g/dL) with decreased MCV (55–70 fL) and HbA_2 raised above 3.5%. Microcytosis, anisocytosis, poikilocytosis, and targeting and stippling of the red cells can be seen on the blood smear (Fig. 6.6E). It should be noted that a small number of β-thalassemia heterozygotes may have normal or borderline-raised HbA_2 levels.
- β-thalassemia intermedia. Patients with β-thalassemia who have a milder clinical course and are not fully transfusion dependent but have a more severe condition than the carrier states for β-thalassemia are classified as having β-thalassemia intermedia. Transfusions are not usually required in these patients early in life, however, may be indicated as complications of chronic hemolysis and ineffective erythropoiesis accumulate. The genetic interactions leading to the phenotype of β-thalassemia intermedia are very heterogeneous and include homozygosity or compound heterozygosity for mild β-thalassemia mutations, coinheritance of α^0

-thalassemia trait or homozygous α^+-thalassemia, and the inheritance of HPFH determinants. Inheritance of single β-thalassemia genes can also cause thalassemia intermedia. In the majority of cases this is related to an increased α-globin production from the coinheritance of one (ααα/αα), or in most cases, two extra α-genes, either as two triplicated α-genes alleles (ααα/ααα), or more rarely, as a quadrupled α-gene allele (αααα/αα).

- β-thalassemia major. β-thalassemia major is a severe disease. At birth, affected infants are relatively normal because the change from γ-chain synthesis to β-chain synthesis has not been completed (Fig. 6.3). However, by 6 months of age the infant develops a severely microcytic hemolytic anemia with aniso- and poikilocytosis, polychromasia, and teardrop red cells (Fig. 6.6F). The failure in β-globin production due to absent or greatly decreased β-globin mRNA leads to imbalance in α— and β-globin synthesis. Subsequent precipitation of free α-chains results in inclusion bodies that damage the erythrocyte membrane and lead to destruction of nucleated red cells in the marrow. The reticulocyte count is usually no greater than 5%—10% because of massive destruction of erythroid precursors in the marrow. To maintain an adequate hemoglobin level, transfusions are usually required every 3—4 weeks. Affected children develop hepatosplenomegaly secondary to extramedullary hematopoiesis, and may also develop characteristic facies due to excessive intramedullary hematopoiesis. The bones have expanded marrow cavities resulting in pathologic fractures and a "hair-on-end" appearance on skull films (Fig. 6.8). Other complications include cholelithiasis, susceptibility to infections, secondary hypersplenism, and delayed growth and maturation. The major causes of mortality are hemochromatosis and overwhelming infections following splenectomy, the former due to excessive iron deposition as a result of blood transfusions and increased gastrointestinal absorption. Excess iron deposited in the heart, pancreas, liver, and other organs damages tissue and leads to cardiac failure, arrhythmias, diabetes mellitus, and liver failure. Without regular transfusion, patients usually die before the age of 20 years. Given optimal transfusion and iron chelation therapy, many patients now survive into the fifth and sixth decades of life.

6.8.7 Differential Diagnosis

The basic hematological tests required for carrier detection are the measurement of the MCV, the MCH value, and the quantity of HbA_2 and HbF. In addition, the hemoglobin pattern needs to be examined, and traditionally, electrophoresis methods have been used for this purpose. However, if HPLC is used to quantitate the HbA_2 and HbF level, it will also detect most of the common, clinically relevant hemoglobin variants, such as HbS, HbC, HbD-Punjab, HbO-Arab, and HbE at the same time. In the general practice of medicine many patients present with a mild microcytic anemia. Nearly all have iron deficiency anemia or a type of thalassemia trait. In heterozygous thalassemia, the peripheral smear may be more abnormal than that of iron deficiency, and the MCV and MCH are decreased; but the MCHC is normal in contrast to the decreased MCHC seen in advanced iron deficiency anemia. The MCV in thalassemia traits also tends to be lower in relation to the red cell count than the MCV in iron deficiency. This difference is the basis of the Mentzer index (MCV/red cell count (RBC)). MCV/RBC values of less than 11.5 suggest thalassemia trait, while values greater than 13.5 suggest iron deficiency anemia. A much more definitive measurement of iron deficiency is to determine the ferritin level. Iron deficiency is indicated by a lower-than-normal ferritin level.

The first step in carrier identification is to measure the HbA_2 level in patients with microcytosis (Table 6.5). Patients with microcytosis and normal HbA_2 should have serum iron or ferritin determinations; a low value suggests iron deficiency anemia, while a normal iron or ferritin value suggests α-thalassemia trait. When microcytosis and an increased HbA_2 (3.5%—6%) with a normal or slightly increased HbF (2%—5%) are found, a diagnosis of β-thalassemia trait is made. Confirmation of the diagnosis is obtained by family studies and DNA analysis. Note that it is common for patients with a clearly raised HbA_2 level to coinherit α-thalassemia, and thus antenatal patients with β-thalassemia trait should be screened for α^0-thalassemia trait by DNA analysis when they are from Southeast Asia or Mediterranean regions where α^0-thalassemia is common.

In the rare cases showing normal or low MCH and MCV values, normal or reduced HbA_2 levels, and a high HbF (5%—30%), δβ-thalassemia trait or HPFH is indicated. These two conditions are distinguished by analyzing the HbF distribution in the red cells and/or by DNA analysis. The δβ-thalassemia trait usually has a heterogeneous distribution of HbF (generally between 4% and 18%) while deletional HPFH typically has a homogenous (pancellular) distribution (15%–35%; Table 6.5).

6.8.8 Molecular Diagnosis

Mutation identification by DNA analysis techniques is required after differential diagnosis in several instances, for example, to distinguish between homozygous α^+-thalassemia and α^0-thalassemia trait, or to distinguish between δβ-thalassemia trait and HPFH with coinherited α-thalassemia. Genotype analysis is also necessary for prenatal diagnosis by fetal DNA analysis and for solving complex difficult diagnostic cases due to the interaction of several alleles or mutations with atypical phenotypes.

6.8.8.1 δβ-Thalassemia

Homozygous δβ-thalassemia is a mild disorder and one of the many genotypes that result in thalassemia intermedia. Homozygotes have 100% HbF and lack HbA and HbA_2. The mild anemia and hemolysis are due to increased γ-chain synthesis, which makes the imbalance between synthesis of α-chains and non-α chains less than that seen in other β-thalassemias. Patients with heterozygous δβ-thalassemias have mild microcytosis and 4%–18% Hb F on electrophoresis see (Table 6.5). They are classified into two groups: the $(\delta\beta)^0$-thalassemias in which the HbF is composed of both $^G\gamma$ and $^A\gamma$ chains (13 deletion mutations), and the $(^A\gamma\delta\beta)^0$-thalassemias in which the HbF contains only $^G\gamma$ chains due to $^A\gamma$-globin gene being deleted.

6.8.8.2 Hb Lepore Thalassemia

As previously mentioned, Hb Lepore is a variant hemoglobin containing a δβ-fusion chain. Four different Hb Lepore variants have been described (Hollandia, Boston/Washington, Baltimore and Leiden), differing in the point at which the δβ-fusion occurs. Hb Lepore has an electrophoretic mobility similar to that of HbS, and it forms 5%–15% of the total hemoglobin of heterozygotes (Table 6.5). It can be easily identified by iso-electric focusing gel electrophoresis in which it focuses to a characteristic position between HbA and S. Decreased Hb Lepore synthesis may be secondary to instability of the δβ-fusion mRNA. Heterozygotes are clinically similar to β^0-thalassemia heterozygotes, and Hb Lepore homozygotes or Lepore/β^0-thalassemia genetic compounds are similar to β^0-thalassemia homozygotes with a severe thalassemic picture ranging from transfusion-dependent disease to thalassemia intermedia when ameliorated by other factors such as α-thalassemia.

6.8.8.3 εγδβ-Thalassemia

Infants heterozygous for this rare condition are born with a severe hemolytic, hypochromic anemia and microcytosis and may require blood transfusions. The condition improves at 3–6 months after birth and adults have a phenotype similar to heterozygous β-thalassemia but with a normal HbA_2 level. Nearly twenty large deletion mutations have been described, which delete all the functional globin genes (ε, $^G\gamma$, $^A\gamma$, δ, and β) and several have been described in which the deletion leaves the β-globin gene intact but not expressed. Mapping of the deleted DNA of the latter cases shows that in all three cases the locus control region (LCR) located 5′ to the ε-gene has been removed. The absence of this segment inactivates all the intact globin genes following the 3′ end of the deletion. The homozygous condition is presumed to be incompatible with fetal survival.

6.8.8.4 δ-Thalassemia

Individuals with heterozygous and homozygous δ^0-thalassemia have, respectively, decreased and absent HbA_2. However, anemia and changes in peripheral smears are not seen because of the normal low level of δ-chain production. The only clinical significance of δ-thalassemia is when it is coinherited with heterozygous β-thalassemia and complicates the diagnosis because of a normal HbA_2 level of 2%–2.5%.

6.8.8.5 Hereditary Persistence of Fetal Hemoglobin

HPFH is a disorder in which HbF is increased above the normal adult level and there are no morphological changes to the red cells. The disorder is caused by at least 25 different mutations, either large deletions in the globin gene cluster or point mutations in the γ-gene promoter regions. HPFH heterozygotes differ from thalassemia heterozygotes in that they have no imbalance between the synthesis of α and non α-chains (i.e., γ and β-chains) and thus are characterized by an asymptomatic heterozygous state without microcytosis. The elevated HbF ranges from 3% to 35%, depending on the type of the mutation. The proportion of γ-chain type ($^{G}\gamma$ vs. $^{A}\gamma$) varies among patients with different HPFHs and usually, but not always, the HbF is homogeneously distributed within red cells. In a few cases, HPFH heterozygotes have two populations of cells: one contains HbF; the other lacks HbF. These patients are said to have heterocellular HPFH, as opposed to the bulk of patients who have pancellular HPFH, which is more common in deletional HPFH variants.

6.8.9 Interaction of Thalassemia With Hemoglobin Variants

6.8.9.1 β-Thalassemia/HbS

Hb S/β-thalassemia is characterized by microcytic red and target cells with occasionally sickled forms. Hemoglobin electrophoresis reveals 60%–90% HbS, 0% –30% HbA, 1%–20% HbF, and an increased HbA_2 level. The percentages of HbS and HbA vary depending on whether the β-thalassemia gene is β^{+} or β^{0} type. Coexisting α-thalassemia increases the Hb concentration, the MCV, and MCH (Fig. 6.6D).

The clinical course of sickle cell β-thalassemia is very variable, ranging from a disorder identical with sickle cell disease to a completely asymptomatic condition. The Hb concentration varies from 5 g/dL to a level within the normal range. The heterogeneity is mostly due to the type of β-thalassemia mutation that is co-inherited. It tends to be very mild in Africans because of the likelihood of the coinheritance of one of the three mild β^{+} mutations commonly found in this racial group (−88, C → T; −29, A → G; CD24, T → A). However, those patients who inherit a β^{0}-thalassemia allele exhibit a clinical disorder very similar to sickle cell disease.

6.8.9.2 β-Thalassemia/HbC

β-Thalassemia/HbC is mostly seen among individuals of African origin. The disorder is clinically heterogeneous because of the different types of β-thalassemia alleles involved. Hb C/β^{+}-thalassemia in Africans has hematologic features similar to those of β-thalassemia trait because β^{+}-thalassemia mutations are the mild types. However, HbC/β^{+}-thalassemia with a severe type of mutation and HbC/β^{0}-thalassemia produce a more severe phenotype, characterized by anemia (hemoglobin of 8–12 g/dL) and splenomegaly. The clinical picture is a slightly more severe form of HbC disease (homozygous HbC), with which individuals have a mild hemolytic anemia (hemoglobin of 10–15 g/dL) and moderate splenic enlargement. HbC is less soluble than HbA in both the oxygenated and deoxy forms, resulting in the formation of crystals in high concentrations. HbCC red cells become dehydrated, dense, and abnormally rigid, resulting in a shortened life span. AC individuals have no clinical manifestations and normal hematological values, although their red cells are slightly dehydrated and dense. Diagnostically, Hb C/β^{0}-thalassemia can be difficult to distinguish from HbC disease (HbCC), although this has become less problematic as HPLC as become standard diagnostic tool in recent years. In addition, raised HbA_2 levels, family studies, or DNA analysis can be diagnostically useful.

6.8.9.3 β-Thalassemia/HbE

HbE/β-thalassemia is a common disease in Thailand and parts of Southeast Asia. It results in a variable clinical picture similar to that of homozygous β-thalassemia, ranging from a condition indistinguishable from thalassemia major to a mild form of thalassemia intermedia. The severest conditions are found in individuals with HbE and β^{0}-thalassemia, who usually have about 60%–70% HbE, the remainder being HbF. Hemoglobin levels may be as low as 4–5 g/dL, and the clinical management of these patients is similar to that for those with thalassemia major. Compound heterozygotes for HbE and β^{+}-thalassemia usually have a milder disorder and produce variable amounts of HbA. As with homozygous β-thalassemia, the genetic factors that account for a mild phenotype in some, but not all patients, are mild β^{+}-type mutations, the coinheritance of α-thalassemia, and the homozygosity for the XmnI restriction site due to the C → T polymorphism at position −158 5′ to the $^{G}\gamma$-globin gene.

6.8.9.4 δβ-Thalassemia/HbS

HbS/δβ-thalassemia is a milder form of sickle cell disease because the high percentage of HbF (15%–25%) produced by the δβ-thalassemia allele protects against red cell sickling by reducing the HbS concentration and inhibiting its polymerization. HbS/δβ-thalassemia has been described in Sicilian, Italian, Greek, Arab, and African-American individuals. Patients have a mild anemia with a Hb concentration in the range of 10–12 g/dL, a significantly reduced MCH and MCV, HbS, HbF, and a normal or low HbA_2 level.

6.8.9.5 HPFH/HbS

Patients doubly heterozygous for HbS and HPFH are either asymptomatic or have an extremely mild form of sickle cell disease. There is usually no anemia and patients have very few VOE, although occasional mild bone pain has been reported. The condition is found in individuals of African origin (with either the black HPFH1 or Ghanaian HPFH2 deletions), and also in Asian Indians (with the Indian HPFH3 deletion). Patients have near-normal red cells with 20%–30% HbF in a pancellular distribution. The condition is difficult to distinguish diagnostically from Hb S/δβ-thalassemia without DNA analysis because many patients have reduced red cell indices due to coinherited α^+-thalassemia.

6.8.10 Management of Thalassemia

6.8.10.1 Prevention

The hemoglobin disorders are the most common clinically serious single-gene disorders in the world. About 250 million people are carriers and about 300,000 affected homozygotes for thalassemia and sickle cell disease are born each year, the majority in countries with limited resources. Since the treatment for affected individuals is still not perfect and is very expensive, from a public health perspective the most important means of controlling thalassemia is prevention by antenatal screening and prenatal diagnosis. Most countries have implemented prevention programs, based on the identification of couples at risk by carrier screening, the provision of information by nondirective counseling, and the option of prenatal diagnosis to allow the termination of an affected fetus if desired. Such programs have been extremely successful in the populations at risk in the Mediterranean regions, leading to a marked decline in the incidence of thalassemia major.

Recent developments in prenatal diagnosis have been directed toward noninvasive methods of fetal sampling and preimplantation diagnosis (PGD). The discovery of fetal DNA in maternal plasma offers a much simpler and more robust approach to noninvasive prenatal diagnosis. PGD represents a "state-of-the-art" procedure that potentially avoids the need to terminate affected pregnancies. PGD is a technically challenging, multistep, and expensive procedure, and is not likely to become a routine alternative to conventional prenatal diagnosis for couples in most countries. PGD has been used successfully for both α-thalassemia [62] and β-thalassemia [63] and is now more widely available around the world. One specific use of this approach is to allow the birth of a normal child that is HLA identical to a sibling affected with β-thalassemia, thus permitting a possible cure by stem cell transplantation [64]. For α-thalassemia, PGD can also allow for intrauterine transfusions followed by unrelated donor stem cell transplantation after birth [61]. Ongoing clinical trials exploring the role of in utero hematopoietic stem cell transplantation may offer additional therapeutic options in the future.

6.8.10.2 Therapy

Advances in the management of thalassemia major have greatly improved the prognosis for patients with this disease, although its management remains complex, difficult, and very expensive. Regular transfusion and optimum iron chelation therapy enables patients to survive at least to the fourth decade with a normal lifestyle in countries able to afford the treatment program, and survival to the fifth to sixth decades is now common. However, in the third world countries where thalassemia reaches its highest frequency, widespread adoption of this expensive treatment is not possible and the problem of thalassemia remains almost untouched. Revised guidelines for the clinical management of thalassemia have recently been published and can be downloaded from the Thalassemia International Federation Web site [65].

The major complications of β-thalassemia major are due primarily to severe anemia and ineffective erythropoiesis. Treatment for β-thalassemia major includes regular red cell transfusions starting at a young age to allow for normal growth and development, and for prevention of complication of ineffective erythropoiesis including bone deformities, paraspinal masses, splenomegaly, hepatosplenomegaly, and osteoporosis. The goal of regular transfusions, approximately every 2–4 weeks, is to maintain a pretransfusion hemoglobin greater than 9.5 g/dL (usually between 9.5 and 10.5 g/dL). In adults, this is approximately 2–3 units every 2–4 weeks. If started early in life this hypertransfusion regimen allows for normal growth and development and prevents many of the bone malformations and abnormalities characteristic of untransfused thalassemia major.

Blood transfusions are associated with iron loading since the body has no mechanism to remove excess iron contained in transfused red cells. Excess iron accumulates in the liver, heart, and endocrine organs, leading to eventual organ damage and dysfunction from free iron-associated reactive oxygen species. Serum ferritin, and liver and cardiac MRI are used to monitor iron burden. Iron chelation therapy is generally started when the serum ferritin is greater than 1000–1500 ng/mL or after receiving approximately 20 units of blood in their lifetime. In poorly chelated individuals, hypogonadotropic hypogonadism due to pituitary iron overload is the most common complication; however, iron-overload associated cardiac failure is the leading cause of death.

There are currently three drugs available for iron chelation therapy. Deferoxamine is the oldest chelator available and is administered via subcutaneous injection for 5–7 days a week, over 8–12 hours per night. It is effective at reducing body iron burden, especially chelating liver and cardiac iron; however, it is cumbersome and thus poor adherence is a major barrier to use. Deferiprone and deferasirox are oral iron chelators now in widespread clinical use following extensive clinical trials. Deferiprone is generally well tolerated, except for the rare but significant side effect of neutropenia and agranulocytosis [66]. Deferiprone reduces serum ferritin, removes liver iron, and most importantly, is the most effective chelator at removing myocardial iron [67] resulting in reduced cardiac morbidity [68]. Combination therapy with deferoxamine and deferiprone significantly reduces cardiac iron, and should be considered for patients with high myocardial iron (low T2*) or with very high levels of iron burden [69]. Deferasirox is an oral iron chelator able to remove both liver and cardiac iron [70], and is convenient as a once a day dosing. Key side effects of deferasirox include gastrointestinal symptoms, rash, and elevations in creatinine and liver enzymes.

Recently, disease-modifying therapies that aim to reduce the ineffective erythropoiesis associated with thalassemia have been developed. Luspatercept is a recombinant fusion protein that binds to select transforming growth factor β (TGF-β) superfamily ligands to enhance erythroid maturation. In the Phase III BELIEVE trial, treatment with luspatercept every 3 weeks led to a significant reduction in transfusional burden (the total number of red-cell units transfused) leading to its approval for transfusion-dependent β-thalassemia [71]. In addition, as in sickle cell disease, a number of pharmacological agents (most commonly hydroxyurea) are being used to elevate hemoglobin F in patients with homozygous β-thalassemia with the aim of increasing HbF to levels high enough to prevent all complications of the disease. Although a significant amelioration of the anemia has been achieved in some thalassemia patients, the effects of hydroxyurea treatment have been more modest compared to the treatment of sickle cell disease, and large randomized trials are still needed to measure the effectiveness of hydroxyurea and other HbF-inducing agents in β-thalassemia.

Splenomegaly is a complication of anemia and ineffective erythropoiesis and can be prevented by adequately maintaining transfusion targets as described above. Splenectomy is not recommended in transfusion or nontransfusion dependent thalassemia patients, due to the increased risk of infection, thrombosis, and pulmonary hypertension postsplenectomy. Instead, a hypertransfusion regimen to maintain a pretransfusion hemoglobin over 10 g/dL is recommended to help shrink the spleen.

In thalassemia patients, transfusion-associated iron overload and ineffective erythropoiesis affects multiple organs including the liver, heart, and endocrine system. A comprehensive approach across the life-span, including clinical care with a specialist hematologist, specialized nursing care, subspecialists, social workers, psychologists and other allied health care workers is key to ensuring improved quality of life and survival.

Individuals with beta thalassemia intermedia generally do not require transfusions in infancy, however, depending on the severity of their disease, may require the initiation of transfusions later in life. Indications for transfusions include failure to thrive, poor growth and development, bony changes or masses due to extramedullary hematopoiesis, significant hepatomegaly or splenomegaly, pulmonary hypertension, and severe osteoporosis, among others [72]. Once red cell transfusions are initiated, the same principles of monitoring iron burden, chelation and comprehensive care apply as for beta thalassemia major.

6.8.10.3 Allogeneic Stem Cell Transplantation and Gene Therapy

Bone marrow transplantation offers the only cure for β-thalassemia major. Allogeneic bone marrow transplantation using related donors has been used in a large number of patients, predominantly in Europe, with studies showing approximately 80% of patients surviving long term and of these, nearly 90% being cured [73]. Better outcomes are associated with adequate iron chelation and the absence of significant hepatic disease, although survival and disease-free survival rates still vary considerably between different centers. The major limitation of bone marrow transplantation is the lack of an HLA-identical sibling donor for most affected patients. Future and experimental approaches to marrow transplantation in patients with thalassemia include transplantation from matched unrelated donors, cord blood transplantation, and in utero transplantation. Good success rates have been reported for a limited number of transplantations from carefully matched unrelated donors [74] and umbilical cord blood transplantation from a related donor offers a good possibility of success with a lowered risk of graft-versus-host disease [75].

Gene therapy using autologous bone marrow has the potential to cure thalassemia major permanently without the limitations of finding a matched donor or the risk of graft-versus-host disease. Similar to the approaches described in Section 6.5.5 for the treatment of SCD, lentiviral gene addition approaches have been employed for the treatment of β-thalassemia utilizing vectors expressing the β-globin variant T87Q (betibeglogene autotemcel). The Phase 1/2 studies HGB-204 and HGB-205 trials showed significant reduction or even elimination of the need for long-term red-cell transfusions in 22 patients with severe β-thalassemia, particularly in those patients with non- β^0/β^0 genotypes [51]. The ongoing Phase 3 HGB-207 and HGB-212 studies showed that 85% of patients achieved transfusion independence following betibeglogene autotemcel infusion with significant reduction in ineffective erythropoiesis. These and other gene therapy approaches are expected to significantly change the management of β-thalassemia in the coming years.

REFERENCES

[1] Weatherall DJ, Clegg JB. The thalassemia syndromes. 4th ed. Oxford: Blackwell Scientific; 2001.

[2] Steinberg MH, Nagel RI. Unstable hemoglobins, hemoglobins with altered oxygen affinity, hemoglobin M, and other variants of clinical and biological interest. 2nd ed. Cambridge University Press; 2009.

[3] Bauer DE, Kamran SC, Lessard S, Xu J, Fujiwara Y, Lin C, Shao Z, Canver MC, Smith EC, Pinello L, Sabo PJ, Vierstra J, Voit RA, Yuan GC, Porteus MH, Stamatoyannopoulos JA, Lettre G, Orkin SH. An erythroid enhancer of BCL11A subject to genetic variation determines fetal hemoglobin level. Science 2013;342(6155):253–7.

[4] Lettre G, Sankaran VG, Bezerra MA, Araujo AS, Uda M, Sanna S, Cao A, Schlessinger D, Costa FF, Hirschhorn JN, Orkin SH. DNA polymorphisms at the BCL11A, HBS1L-MYB, and beta-globin loci associate with fetal hemoglobin levels and pain crises in sickle cell disease. Proc Natl Acad Sci U S A 2008;105(33):11869–74.

[5] Liu N, Hargreaves VV, Zhu Q, Kurland JV, Hong J, Kim W, Sher F, Macias-Trevino C, Rogers JM, Kurita R, Nakamura Y, Yuan GC, Bauer DE, Xu J, Bulyk ML, Orkin SH. Direct promoter repression by BCL11A controls the fetal to adult hemoglobin switch. Cell 2018;173(2):430–442.e17.

[6] Masuda T, Wang X, Maeda M, Canver MC, Sher F, Funnell AP, Fisher C, Suciu M, Martyn GE, Norton LJ, Zhu C, Kurita R, Nakamura Y, Xu J, Higgs DR, Crossley M, Bauer DE, Orkin SH, Kharchenko PV, Maeda T. Transcription factors LRF and BCL11A independently repress expression of fetal hemoglobin. Science 2016;351(6270):285–9.

[7] Sankaran VG, Menne TF, Xu J, Akie TE, Lettre G, Van Handel B, Mikkola HK, Hirschhorn JN, Cantor AB, Orkin SH. Human fetal hemoglobin expression is regulated by the developmental stage-specific repressor BCL11A. Science 2008;322(5909):1839–42.

[8] Uda M, Galanello R, Sanna S, Lettre G, Sankaran VG, Chen W, Usala G, Busonero F, Maschio A, Albai G, Piras MG, Sestu N, Lai S, Dei M, Mulas A, Crisponi L, Naitza S, Asunis I, Deiana M, Nagaraja R, Perseu L, Satta S, Cipollina MD, Sollaino C, Moi P, Hirschhorn JN, Orkin SH, Abecasis GR, Schlessinger D, Cao A. Genome-wide association study shows BCL11A associated with persistent fetal hemoglobin and amelioration of the phenotype of beta-thalassemia. Proc Natl Acad Sci U S A 2008;105(5):1620–5.

[9] Hardison RC, Chui DH, Giardine B, Riemer C, Patrinos GP, Anagnou N, Miller W, Wajcman H. HbVar: a relational database of human hemoglobin variants and thalassemia mutations at the globin gene server. Hum Mutat 2002;19(3):225–33.

[10] Giardine BM, Joly P, Pissard S, Wajcman H, DH KC, Hardison RC, et al. Clinically relevant updates of the HbVar database of human hemoglobin variants and thalassemia mutations. Nucleic Acids Res 2020;49(D1):D1192–6.

[11] Platt OS, Brambilla DJ, Rosse WF, Milner PF, Castro O, Steinberg MH, Klug PP. Mortality in sickle cell disease. Life expectancy and risk factors for early death. N Engl J Med 1994;330(23):1639–44.
[12] Powars DR, Chan LS, Hiti A, Ramicone E, Johnson C. Outcome of sickle cell anemia: a 4-decade observational study of 1056 patients. Medicine 2005;84(6):363–76.
[13] Salinas Cisneros G, Thein SL. Recent advances in the treatment of sickle cell disease. Front Physiol 2020;11:435.
[14] Yawn BP, Buchanan GR, Afenyi-Annan AN, Ballas SK, Hassell KL, James AH, Jordan L, Lanzkron SM, Lottenberg R, Savage WJ, Tanabe PJ, Ware RE, Murad MH, Goldsmith JC, Ortiz E, Fulwood R, Horton A, John-Sowah J. Management of sickle cell disease: summary of the 2014 evidence-based report by expert panel members. JAMA 2014;312(10):1033–48.
[15] Charache S, Terrin ML, Moore RD, Dover GJ, Barton FB, Eckert SV, McMahon RP, Bonds DR. Effect of hydroxyurea on the frequency of painful crises in sickle cell anemia. Investigators of the Multicenter Study of Hydroxyurea in Sickle Cell Anemia. N Engl J Med 1995;332(20):1317–22.
[16] Tshilolo L, Tomlinson G, Williams TN, Santos B, Olupot-Olupot P, Lane A, Aygun B, Stuber SE, Latham TS, McGann PT, Ware RE, Investigators R. Hydroxyurea for children with sickle cell anemia in sub-Saharan Africa. N Engl J Med 2019;380(2):121–31.
[17] Thornburg CD, Files BA, Luo Z, Miller ST, Kalpatthi R, Iyer R, Seaman P, Lebensburger J, Alvarez O, Thompson B, Ware RE, Wang WC, Investigators BH. Impact of hydroxyurea on clinical events in the BABY HUG trial. Blood 2012;120(22):4304–10. quiz 4448.
[18] Moutouh-de Parseval LA, Verhelle D, Glezer E, Jensen-Pergakes K, Ferguson GD, Corral LG, Morris CL, Muller G, Brady H, Chan K. Pomalidomide and lenalidomide regulate erythropoiesis and fetal hemoglobin production in human $CD34^+$ cells. J Clin Invest 2008;118(1):248–58.
[19] Dulmovits BM, Appiah-Kubi AO, Papoin J, Hale J, He M, Al-Abed Y, Didier S, Gould M, Husain-Krautter S, Singh SA, Chan KW, Vlachos A, Allen SL, Taylor N, Marambaud P, An X, Gallagher PG, Mohandas N, Lipton JM, Liu JM, Blanc L. Pomalidomide reverses gamma-globin silencing through the transcriptional reprogramming of adult hematopoietic progenitors. Blood 2016;127(11):1481–92.
[20] Humphries RK, Dover G, Young NS, Moore JG, Charache S, Ley T, Nienhuis AW. 5-Azacytidine acts directly on both erythroid precursors and progenitors to increase production of fetal hemoglobin. J Clin Invest 1985;75(2):547–57.
[21] Renneville A, Van Galen P, Canver MC, McConkey M, Krill-Burger JM, Dorfman DM, Holson EB, Bernstein BE, Orkin SH, Bauer DE, Ebert BL. EHMT1 and EHMT2 inhibition induces fetal hemoglobin expression. Blood 2015;126(16):1930–9.
[22] Krivega I, Byrnes C, de Vasconcellos JF, Lee YT, Kaushal M, Dean A, Miller JL. Inhibition of G9a methyltransferase stimulates fetal hemoglobin production by facilitating LCR/gamma-globin looping. Blood 2015;126(5):665–72.
[23] Cui S, Lim KC, Shi L, Lee M, Jearawiriyapaisarn N, Myers G, Campbell A, Harro D, Iwase S, Trievel RC, Rivers A, DeSimone J, Lavelle D, Saunthararajah Y, Engel JD. The LSD1 inhibitor RN-1 induces fetal hemoglobin synthesis and reduces disease pathology in sickle cell mice. Blood 2015;126(3):386–96.
[24] Sun Q, Ding D, Liu X, Guo SW. Tranylcypromine, a lysine-specific demethylase 1 (LSD1) inhibitor, suppresses lesion growth and improves generalized hyperalgesia in mouse with induced endometriosis. Reprod Biol Endocrinol 2016;14:17.
[25] Bradner JE, Mak R, Tanguturi SK, Mazitschek R, Haggarty SJ, Ross K, Chang CY, Bosco J, West N, Morse E, Lin K, Shen JP, Kwiatkowski NP, Gheldof N, Dekker J, DeAngelo DJ, Carr SA, Schreiber SL, Golub TR, Ebert BL. Chemical genetic strategy identifies histone deacetylase 1 (HDAC1) and HDAC2 as therapeutic targets in sickle cell disease. Proc Natl Acad Sci U S A 2010;107(28):12617–22.
[26] McArthur JG, Svenstrup N, Chen C, Fricot A, Carvalho C, Nguyen J, Nguyen P, Parachikova A, Abdulla F, Vercellotti GM, Hermine O, Edwards D, Ribeil JA, Belcher JD, Maciel TT. A novel, highly potent and selective phosphodiesterase-9 inhibitor for the treatment of sickle cell disease. Haematologica 2020;105(3):623–31.
[27] Charnigo RJ, Beidler D, Rybin D, Pittman DD, Tan B, Howard J, Michelson AD, Frelinger III Al, Clarke N. PF-04447943, a phosphodiesterase 9A inhibitor, in stable sickle cell disease patients: a phase Ib randomized, placebo-controlled study. Clin Transl Sci 2019;12(2):180–8.
[28] Zhang Y, Paikari A, Sumazin P, Ginter Summarell CC, Crosby JR, Boerwinkle E, Weiss MJ, Sheehan VA. Metformin induces FOXO3-dependent fetal hemoglobin production in human primary erythroid cells. Blood 2018;132(3):321–33.
[29] Grevet JD, Lan X, Hamagami N, Edwards CR, Sankaranarayanan L, Ji X, Bhardwaj SK, Face CJ, Posocco DF, Abdulmalik O, Keller CA, Giardine B, Sidoli S, Garcia BA, Chou ST, Liebhaber SA, Hardison RC, Shi J, Blobel GA. Domain-focused

CRISPR screen identifies HRI as a fetal hemoglobin regulator in human erythroid cells. Science 2018;361(6399):285–90.

[30] Huang P, Peslak SA, Lan X, Khandros E, Yano JA, Sharma M, Keller CA, Giardine B, Qin K, Abdulmalik O, Hardison RC, Shi J, Blobel GA. The HRI-regulated transcription factor ATF4 activates BCL11A transcription to silence fetal hemoglobin expression. Blood 2020;135(24):2121–32.

[31] Peslak SA, Khandros E, Huang P, Lan X, Geronimo CL, Grevet JD, Abdulmalik O, Zhang Z, Giardine BM, Keller CA, Shi J, Hardison RC, Blobel GA. HRI depletion cooperates with pharmacologic inducers to elevate fetal hemoglobin and reduce sickle cell formation. Blood Adv 2020;4(18):4560–72.

[32] Lan X, Khandros E, Huang P, Peslak SA, Bhardwaj SK, Grevet JD, Abdulmalik O, Wang H, Keller CA, Giardine B, Baeza J, Duffner ER, El Demerdash O, Wu XS, Vakoc CR, Garcia BA, Hardison RC, Shi J, Blobel GA. The E3 ligase adaptor molecule SPOP regulates fetal hemoglobin levels in adult erythroid cells. Blood Adv 2019;3(10):1586–97.

[33] Chambers CB, Gross J, Pratt K, Guo X, Byrnes C, Lee YT, Lavelle D, Dean A, Miller JL, Wilber A. The mRNA-binding protein IGF2BP1 restores fetal hemoglobin in cultured erythroid cells from patients with beta-hemoglobin disorders. Mol Ther Methods Clin Dev 2020;17:429–40.

[34] Yu X, Azzo A, Bilinovich SM, Li X, Dozmorov M, Kurita R, Nakamura Y, Williams Jr DC, Ginder GD. Disruption of the MBD2-NuRD complex but not MBD3-NuRD induces high level HbF expression in human adult erythroid cells. Haematologica 2019;104(12):2361–71.

[35] Sher F, Hossain M, Seruggia D, Schoonenberg VAC, Yao Q, Cifani P, Dassama LMK, Cole MA, Ren C, Vinjamur DS, Macias-Trevino C, Luk K, McGuckin C, Schupp PG, Canver MC, Kurita R, Nakamura Y, Fujiwara Y, Wolfe SA, Pinello L, Maeda T, Kentsis A, Orkin SH, Bauer DE. Rational targeting of a NuRD subcomplex guided by comprehensive *in situ* mutagenesis. Nat Genet 2019;51(7):1149–59.

[36] Lan X, Ren R, Feng R, Ly LC, Lan Y, Zhang Z, et al. ZNF410 uniquely activates the NuRD component CHD4 to silence fetal hemoglobin expression. Mol Cell 2020;81(2):239–54.

[37] Niihara Y, Smith WR, Stark CW. A phase 3 trial of l-glutamine in sickle cell disease. N Engl J Med 2018;379(19):1880.

[38] Ataga KI, Kutlar A, Kanter J, Liles D, Cancado R, Friedrisch J, Guthrie TH, Knight-Madden J, Alvarez OA, Gordeuk VR, Gualandro S, Colella MP, Smith WR, Rollins SA, Stocker JW, Rother RP. Crizanlizumab for the prevention of pain crises in sickle cell disease. N Engl J Med 2017;376(5):429–39.

[39] Kutlar A, Kanter J, Liles DK, Alvarez OA, Cancado RD, Friedrisch JR, Knight-Madden JM, Bruederle A, Shi M, Zhu Z, Ataga KI. Effect of crizanlizumab on pain crises in subgroups of patients with sickle cell disease: a SUSTAIN study analysis. Am J Hematol 2019;94(1):55–61.

[40] Vichinsky E, Hoppe CC, Ataga KI, Ware RE, Nduba V, El-Beshlawy A, Hassab H, Achebe MM, Alkindi S, Brown RC, Diuguid DL, Telfer P, Tsitsikas DA, Elghandour A, Gordeuk VR, Kanter J, Abboud MR, Lehrer-Graiwer J, Tonda M, Intondi A, Tong B, Howard J, Investigators HT. A phase 3 randomized trial of voxelotor in sickle cell disease. N Engl J Med 2019;381(6):509–19.

[41] DeBaun MR, Jordan LC, King AA, Schatz J, Vichinsky E, Fox CK, McKinstry RC, Telfer P, Kraut MA, Daraz L, Kirkham FJ, Murad MH. American Society of Hematology 2020 guidelines for sickle cell disease: prevention, diagnosis, and treatment of cerebrovascular disease in children and adults. Blood Adv 2020;4(8):1554–88.

[42] Chou ST, Alsawas M, Fasano RM, Field JJ, Hendrickson JE, Howard J, Kameka M, Kwiatkowski JL, Pirenne F, Shi PA, Stowell SR, Thein SL, Westhoff CM, Wong TE, Akl EA. American Society of Hematology 2020 guidelines for sickle cell disease: transfusion support. Blood Adv 2020;4(2):327–55.

[43] Quinn CT, Rogers ZR, McCavit TL, Buchanan GR. Improved survival of children and adolescents with sickle cell disease. Blood 2010;115(17):3447–52.

[44] Adams R, McKie V, Nichols F, Carl E, Zhang DL, McKie K, Figueroa R, Litaker M, Thompson W, Hess D. The use of transcranial ultrasonography to predict stroke in sickle cell disease. N Engl J Med 1992;326(9):605–10.

[45] Adams RJ, McKie VC, Hsu L, Files B, Vichinsky E, Pegelow C, Abboud M, Gallagher D, Kutlar A, Nichols FT, Bonds DR, Brambilla D. Prevention of a first stroke by transfusions in children with sickle cell anemia and abnormal results on transcranial Doppler ultrasonography. N Engl J Med 1998;339(1):5–11.

[46] Enninful-Eghan H, Moore RH, Ichord R, Smith-Whitley K, Kwiatkowski JL. Transcranial Doppler ultrasonography and prophylactic transfusion program is effective in preventing overt stroke in children with sickle cell disease. J Pediatr 2010;157(3):479–84.

[47] Gluckman E, Cappelli B, Bernaudin F, Labopin M, Volt F, Carreras J, Pinto Simoes B, Ferster A, Dupont S, de la Fuente J, Dalle JH, Zecca M, Walters MC, Krishnamurti L, Bhatia M, Leung K, Yanik G, Kurtzberg J, Dhedin N, Kuentz M, Michel G, Apperley J,

Lutz P, Neven B, Bertrand Y, Vannier JP, Ayas M, Cavazzana M, Matthes-Martin S, Rocha V, Elayoubi H, Kenzey C, Bader P, Locatelli F, Ruggeri A, Eapen M, Eurocord, the Pediatric Working Party of the European Society for Blood and Marrow Transplantation, The Center for International Blood and Marrow Transplant Research. Sickle cell disease: an international survey of results of HLA-identical sibling hematopoietic stem cell transplantation. Blood 2017;129(11):1548–56.

[48] May C, Rivella S, Callegari J, Heller G, Gaensler KM, Luzzatto L, Sadelain M. Therapeutic haemoglobin synthesis in beta-thalassaemic mice expressing lentivirus-encoded human beta-globin. Nature 2000;406(6791):82–6.

[49] Pawliuk R, Westerman KA, Fabry ME, Payen E, Tighe R, Bouhassira EE, Acharya SA, Ellis J, London IM, Eaves CJ, Humphries RK, Beuzard Y, Nagel RL, Leboulch P. Correction of sickle cell disease in transgenic mouse models by gene therapy. Science 2001;294(5550):2368–71.

[50] Ribeil JA, Hacein-Bey-Abina S, Payen E, Magnani A, Semeraro M, Magrin E, Caccavelli L, Neven B, Bourget P, El Nemer W, Bartolucci P, Weber L, Puy H, Meritet JF, Grevent D, Beuzard Y, Chretien S, Lefebvre T, Ross RW, Negre O, Veres G, Sandler L, Soni S, de Montalembert M, Blanche S, Leboulch P, Cavazzana M. Gene therapy in a patient with sickle cell disease. N Engl J Med 2017;376(9):848–55.

[51] Thompson AA, Walters MC, Kwiatkowski J, Rasko JEJ, Ribeil JA, Hongeng S, Magrin E, Schiller GJ, Payen E, Semeraro M, Moshous D, Lefrere F, Puy H, Bourget P, Magnani A, Caccavelli L, Diana JS, Suarez F, Monpoux F, Brousse V, Poirot C, Brouzes C, Meritet JF, Pondarre C, Beuzard Y, Chretien S, Lefebvre T, Teachey DT, Anurathapan U, Ho PJ, von Kalle C, Kletzel M, Vichinsky E, Soni S, Veres G, Negre O, Ross RW, Davidson D, Petrusich A, Sandler L, Asmal M, Hermine O, De Montalembert M, Hacein-Bey-Abina S, Blanche S, Leboulch P, Cavazzana M. Gene therapy in patients with transfusion-dependent beta-thalassemia. N Engl J Med 2018;378(16):1479–93.

[52] Nagel RL, Bookchin RM, Johnson J, Labie D, Wajcman H, Isaac-Sodeye WA, Honig GR, Schiliro G, Crookston JH, Matsutomo K. Structural bases of the inhibitory effects of hemoglobin F and hemoglobin A2 on the polymerization of hemoglobin S. Proc Natl Acad Sci U S A 1979;76(2):670–2.

[53] Thompson AA, Walters MC, Mapara MY, Kwiatkowski JL, Krishnamurti L, Aygun B, Kasow KA, Rifkin-Zenenberg S, Schmidt M, DelCarpini J, Pierciey FJ, Miller AL, Gallagher ME, Chen R, Goyal S, Kanter J, Tisdale JF. Resolution of serious vaso-occlusive pain crises and reduction in patient-reported pain intensity: results from the ongoing phase 1/2 HGB-206 group C study of LentiGlobin for sickle cell disease (bb1111) gene therapy. Blood 2020;136(Suppl. 1):16–7.

[54] Grimley M, Asnani M, Shrestha A, Felker S, Lutzko C, Arumugam PI, Witting S, Knight-Madden J, Niss O, Quinn CT, Lo C, Little CR, McIntosh JW, Malik P. Early results from a phase 1/2 study of Aru-1801 gene therapy for sickle cell disease (SCD): manufacturing process enhancements improve efficacy of a modified gamma globin lentivirus vector and reduced intensity conditioning transplant. Blood 2020;136(Suppl. 1):20–1.

[55] Guda S, Brendel C, Renella R, Du P, Bauer DE, Canver MC, Grenier JK, Grimson AW, Kamran SC, Thornton J, de Boer H, Root DE, Milsom MD, Orkin SH, Gregory RI, Williams DA. miRNA-embedded shRNAs for lineage-specific BCL11A knockdown and hemoglobin F induction. Mol Ther 2015;23(9):1465–74.

[56] Brendel C, Guda S, Renella R, Bauer DE, Canver MC, Kim YJ, Heeney MM, Klatt D, Fogel J, Milsom MD, Orkin SH, Gregory RI, Williams DA. Lineage-specific BCL11A knockdown circumvents toxicities and reverses sickle phenotype. J Clin Invest 2016;126(10):3868–78.

[57] Esrick EB, Lehmann LE, Biffi A, Achebe M, Brendel C, Ciuculescu MF, et al. Post-Transcriptional genetic silencing of BCL11A to treat sickle cell disease. N Engl J Med 2021;384(3):205–15.

[58] Frangoul H, Altshuler D, Cappellini MD, Chen YS, Domm J, Eustace BK, et al. CRISPR-Cas9 gene editing for sickle cell disease and beta-thalassemia. N Engl J Med 2021;384(3):252–60.

[59] Weil LG, Charlton MR, Coppinger C, Daniel Y, Streetly A. Sickle cell disease and thalassaemia antenatal screening programme in England over 10 years: a review from 2007/2008 to 2016/2017. J Clin Pathol 2020;73(4):183–90.

[60] Vichinsky EP. Alpha thalassemia major–new mutations, intrauterine management, and outcomes. Hematology Am Soc Hematol Educ Prog 2009:35–41.

[61] Yi JS, Moertel CL, Baker KS. Homozygous alpha-thalassemia treated with intrauterine transfusions and unrelated donor hematopoietic cell transplantation. J Pediatr 2009;154(5):766–8.

[62] Chan V, Ng EH, Yam I, Yeung WS, Ho PC, Chan TK. Experience in preimplantation genetic diagnosis for exclusion of homozygous alpha degrees thalassemia. Prenat Diagn 2006;26(11):1029–36.

[63] Monni G, Cau G, Usai V, Perra G, Lai R, Ibba G, Faa V, Incani F, Rosatelli MC. Preimplantation genetic

diagnosis for beta-thalassaemia: the Sardinian experience. Prenat Diagn 2004;24(12):949–54.

[64] Kuliev A, Rechitsky S, Verlinsky O, Tur-Kaspa I, Kalakoutis G, Angastiniotis M, Verlinsky Y. Preimplantation diagnosis and HLA typing for haemoglobin disorders. Reprod Biomed Online 2005;11(3):362–70.

[65] Cappellini MD, Cohen A, Eleftheriou A, Piga A, Porter J, Taher A. In: nd R, editor. Guidelines for the clinical management of thalassaemia; 2008. Nicosia (CY).

[66] Galanello R. Deferiprone in the treatment of transfusion-dependent thalassemia: a review and perspective. Therapeut Clin Risk Manag 2007;3(5):795–805.

[67] Anderson LJ, Wonke B, Prescott E, Holden S, Walker JM, Pennell DJ. Comparison of effects of oral deferiprone and subcutaneous desferrioxamine on myocardial iron concentrations and ventricular function in beta-thalassaemia. Lancet 2002;360(9332):516–20.

[68] Piga A, Gaglioti C, Fogliacco E, Tricta F. Comparative effects of deferiprone and deferoxamine on survival and cardiac disease in patients with thalassemia major: a retrospective analysis. Haematologica 2003;88(5):489–96.

[69] Tanner MA, Galanello R, Dessi C, Smith GC, Westwood MA, Agus A, Roughton M, Assomull R, Nair SV, Walker JM, Pennell DJ. A randomized, placebo-controlled, double-blind trial of the effect of combined therapy with deferoxamine and deferiprone on myocardial iron in thalassemia major using cardiovascular magnetic resonance. Circulation 2007;115(14):1876–84.

[70] Cappellini MD, Cohen A, Piga A, Bejaoui M, Perrotta S, Agaoglu L, Aydinok Y, Kattamis A, Kilinc Y, Porter J, Capra M, Galanello R, Fattoum S, Drelichman G, Magnano C, Verissimo M, Athanassiou-Metaxa M, Giardina P, Kourakli-Symeonidis A, Janka-Schaub G, Coates T, Vermylen C, Olivieri N, Thuret I, Opitz H, Ressayre-Djaffer C, Marks P, Alberti D. A phase 3 study of deferasirox (ICL670), a once-daily oral iron chelator, in patients with beta-thalassemia. Blood 2006;107(9):3455–62.

[71] Cappellini MD, Viprakasit V, Taher AT, Georgiev P, Kuo KHM, Coates T, Voskaridou E, Liew HK, Pazgal-Kobrowski I, Forni GL, Perrotta S, Khelif A, Lal A, Kattamis A, Vlachaki E, Origa R, Aydinok Y, Bejaoui M, Ho PJ, Chew LP, Bee PC, Lim SM, Lu MY, Tantiworawit A, Ganeva P, Gercheva L, Shah F, Neufeld EJ, Thompson A, Laadem A, Shetty JK, Zou J, Zhang J, Miteva D, Zinger T, Linde PG, Sherman ML, Hermine O, Porter J, Piga A, BELIEVE Investigators. A phase 3 trial of luspatercept in patients with transfusion-dependent beta-thalassemia. N Engl J Med 2020;382(13):1219–31.

[72] Viprakasit V, Tyan P, Rodmai S, Taher AT. Identification and key management of non-transfusion-dependent thalassaemia patients: not a rare but potentially under-recognised condition. Orphanet J Rare Dis 2014;9:131.

[73] Lawson SE, Roberts IA, Amrolia P, Dokal I, Szydlo R, Darbyshire PJ. Bone marrow transplantation for beta-thalassaemia major: the UK experience in two paediatric centres. Br J Haematol 2003;120(2):289–95.

[74] La Nasa G, Argiolu F, Giardini C, Pession A, Fagioli F, Caocci G, Vacca A, De Stefano P, Piras E, Ledda A, Piroddi A, Littera R, Nesci S, Locatelli F. Unrelated bone marrow transplantation for beta-thalassemia patients: the experience of the Italian Bone Marrow Transplant Group. Ann N Y Acad Sci 2005;1054:186–95.

[75] Pinto FO, Roberts I. Cord blood stem cell transplantation for haemoglobinopathies. Br J Haematol 2008;141(3):309–24.

7

Disorders of Hemostasis and Thrombosis

Angela C. Weyand[1], Jordan A. Shavit[2], David Ginsburg[3]

[1]Department of Pediatrics, Division of Hematology/Oncology, University of Michigan Medical School, Ann Arbor, Michigan, United States

[2]Departments of Pediatrics and Human Genetics, University of Michigan Medical School, Ann Arbor, Michigan, United States

[3]Howard Hughes Medical Institute and Departments of Internal Medicine, Pediatrics, and Human Genetics, University of Michigan Medical School, Ann Arbor, Michigan, United States

7.1 OVERVIEW OF HEMOSTASIS AND THROMBOSIS

Maintenance of the integrity of the vascular tree is critical to all higher organisms with closed circulatory systems. A complex, highly regulated system has evolved for this purpose, consisting of extensive interactions between the endothelial cell lining of the blood vessel, the blood platelet, and an intricate cascade of plasma proteins [1,2]. This complex system can be conceptually divided into three major limbs, as illustrated in Fig. 7.1. Although depicted as discrete compartments, multiple connections link these pathways to each other and to other homeostatic processes, including inflammation and tissue remodeling.

The first response to vessel injury often involves the platelet limb of hemostasis (Fig. 7.1). Exposure of specific ligands in the injured vessel wall leads to the adhesion of blood platelets at the site of injury, subsequent platelet activation recruiting additional platelets, and activation of the plasma coagulation cascade. The coagulation cascade can also be triggered directly by tissue injury, exposing tissue factor to the plasma milieu (Fig. 7.2).

These processes result in a temporary plug to the injured vessel composed of a mixture of fibrin meshwork and activated platelets. A complex response then ensues, resulting in breakdown and removal of the blood clot and repair of the injured tissue. The final regulatory cascade limb of hemostasis, referred to as the fibrinolytic system, serves to dissolve the blood clot, limiting the spread of the thrombotic process and speeding its resolution. The fibrinolytic and coagulation cascades are both characterized by sequential interactions between a series of plasma proteases and their specific cofactors and inhibitors. A number of regulatory feedback loops within the system serve to dampen the coagulation cascade, most notably the protein C anticoagulant pathway (Fig. 7.3).

A shift in the delicate balance between the complex interacting limbs of the clotting cascade can result in pathologic thrombosis or hemorrhage. Considerable insight into the functions of the various components of this system has come from characterization of the associated human genetic diseases. In addition, subtle variation at several of the genetic loci discussed in this chapter can also contribute to susceptibility to atherosclerosis,

Emery and Rimoin's Principles and Practice of Medical Genetics and Genomics. https://doi.org/10.1016/B978-0-12-812534-2.00003-5

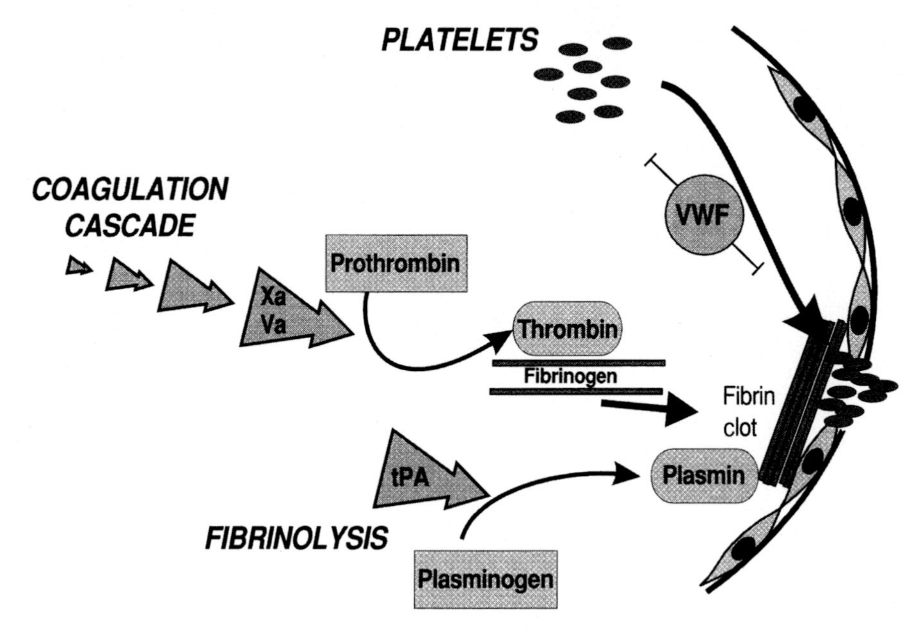

Figure 7.1 Overview of hemostasis. The platelet, coagulation cascade, and fibrinolysis limbs of hemostasis are illustrated schematically. *Xa*, factor Xa; *Va*, factor Va; *VWF*, von Willebrand factor; *tPA*, tissue plasminogen activator. See text for description.

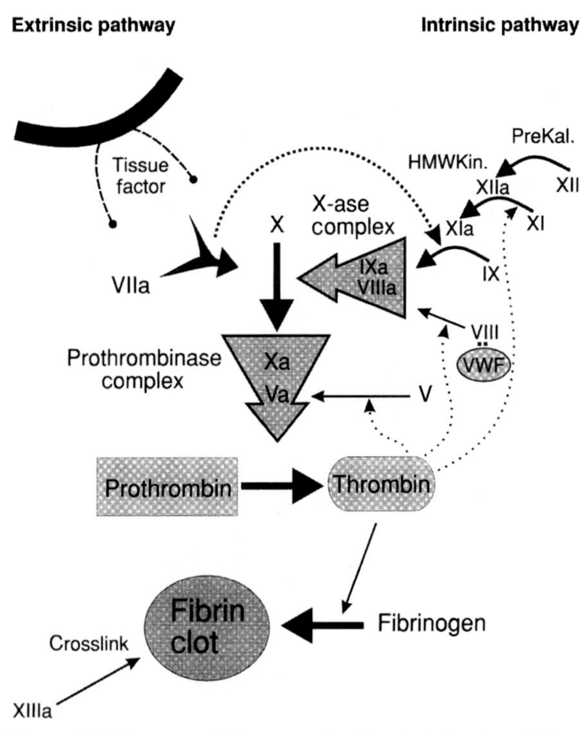

Figure 7.2 The coagulation cascade. See text for description. *HMWKin*, high-molecular-weight kininogen; *PreKal*, prekallikrein; *VWF*, von Willebrand factor. Other clotting factors are indicated by their Roman numeral designations.

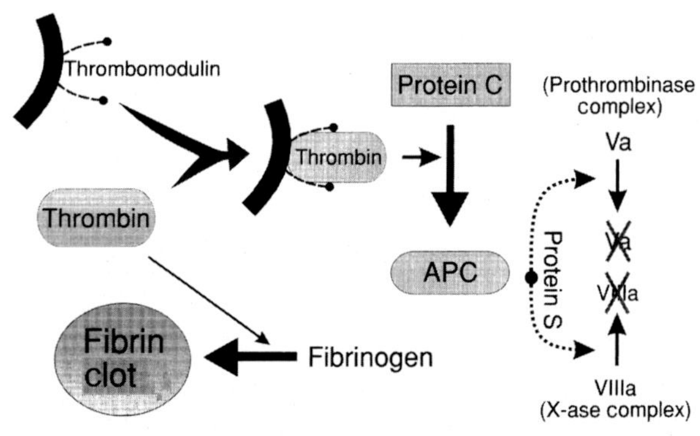

Figure 7.3 Protein C anticoagulant pathway. When thrombin is bound to thrombomodulin, its protease specificity is altered and no longer efficiently generates a fibrin clot from fibrinogen. Instead, thrombomodulin-bound thrombin now specifically cleaves protein C to generate activated protein C (APC). APC proteolytically inactivates factor Va and factor VIIIa, aided by its cofactor, protein S. *Va*, active form of factor V; *VIIIa*, active form of factor VIII. Factors Va and VIIIa are critical components of the prothrombinase and X-ase complexes, respectively.

arterial thrombosis, and venous thromboembolism, the leading causes of death and morbidity in developed countries. This chapter briefly reviews the biology of the plasma coagulation cascade, platelets, and the fibrinolytic system, as background for a discussion of the inherited bleeding and thrombotic disorders resulting from genetic defects in the components of these systems.

7.2 THE COAGULATION CASCADE

The coagulation cascade was historically one of the first biologic systems associated with human disease to be studied in detail at the biochemical level [3]. This was due in large part to the ready availability of the relevant proteins through the blood. Investigations over many years identified an ordered cascade consisting of a plasma protease activating an inactive zymogen target to an active protease form, which subsequently acts on the next step in the cascade (Fig. 7.2). The components of the coagulation cascade were historically assigned Roman numerals, generally in the reverse order of their activation in the cascade (Table 7.1). Factors V through XIII are commonly referred to by their Roman numeral, whereas factors I through IV have more common names (fibrinogen, prothrombin, tissue thromboplastin (a combination of tissue factor (TF) and phospholipids), and calcium, respectively). The term "factor VI" was abandoned when it was shown to simply represent the activated form of factor V (factor Va).

TABLE 7.1 Components of the Classic Coagulation Cascade

Factor Number	Synonym	Chromosomal Localization
I	Fibrinogen	4q28
II	Prothrombin	11p11-q12
III	Tissue thromboplastin (TF and phospholipids)	1p21.3 (TF)
IV	Calcium	-
V	Proaccelerin	1q23
VI	Activated form of factor V (FVa; FVI no longer used)	-
VII	Proconvertin	13q34
VIII	Antihemophilic factor	Xq28
IX	Christmas factor	Xq27.1-q27.2
X	Stuart–Prower factor	13q34
XI	Plasma thrombo-plastin antecedent	4q35
XII	Hageman factor	5q33-qter
XIII	Fibrin stabilizing factor,	6p25-p24 (A subunit)
	Plasma transglu-taminase	1q31-q32.1 (B subunit)

Although the clotting cascade was traditionally divided into the "intrinsic" and "extrinsic" pathways, the biologic significance of this distinction has been questioned. The extrinsic pathway is now viewed as the major mechanism of coagulation activation under most circumstances. The intrinsic pathway (Fig. 7.2) begins with the activation of factor XII (FXII) to FXIIa on surfaces such as glass, a property exploited in several laboratory tests. FXIIa, in the presence of high-molecular-weight kininogen, then activates FXI to FXIa. FXIa activates FIX, feeding into the common portion of the coagulation cascade. The absence of bleeding symptoms in patients with deficiencies of FXII or high-molecular-weight kininogen, despite marked prolongation of clotting times ex vivo, suggests that these factors are not important physiologic activators of the coagulation cascade. By contrast, deficiency of FXI, the next step in the intrinsic pathway, is clearly associated with significant bleeding. This paradox appears to be explained by the demonstration of FXI activation by thrombin and autoactivation by FXIa. These reactions are thought to form an amplification loop, leading to additional activation of FIX by FXIa [4,5].

The coagulation cascade is thought to be activated primarily through the so-called extrinsic pathway (often termed "initiation" [6]), binding of FVIIa to tissue factor on a cell membrane surface, with subsequent activation of FIX to IXa or of FX directly. FIXa and FXa are serine proteases, each of which forms a distinct membrane-bound complex with a specific cofactor, FVIIIa in the case of FIXa and FVa in the case of FXa. Factors V and VIII are highly homologous, nonenzymatic proteins that serve as essential cofactors for their cognate proteases. The FVIIIa/IXa complex, sometimes referred to as the X-ase (or tenase) complex, specifically cleaves FX to generate its active form, FXa. In turn, FXa interacts with FVa to form an enzyme complex referred to as "prothrombinase," which specifically activates prothrombin to thrombin. Initiation generates a small amount of activated thrombin that, through feedback, activates the intrinsic pathway (often termed "amplification" and "propagation" [6]) by activating factors XI, VIII, and V, leading to a "thrombin burst." This excess of thrombin, the final protease in the cascade, acts directly on fibrinogen to generate the fibrin polymer that forms the structural basis of the blood clot.

Tissue factor expression is induced on endothelial cells and monocytes by inflammatory mediators such as tumor necrosis factor and endotoxin. In addition, cells in the subendothelium and tissues surrounding blood vessels express large amounts of tissue factor. Thus, vascular injury initiates the cascade by exposing the coagulation proteins, including FVIIa, to tissue factor. Although a low level of FVII is constitutively present in the circulation, it is inactive until it encounters tissue factor. In the prothrombin time (PT) assay, coagulation is initiated by addition of a crude brain extract that contains large amounts of tissue factor on a lipid membrane surface. This assay is most sensitive to abnormalities in the vitamin-K-dependent enzymes including thrombin, FVII, FIX, and FX. These factors share a common posttranslational processing event that modifies selected glutamic acid residues to γ-carboxyglutamic acid. Vitamin K is an essential cofactor for this reaction, and thus vitamin K deficiency, or antagonism of vitamin K by warfarin, leads to bleeding as a result of the deficient function of these proteins.

7.2.1 Protein C Anticoagulant Pathway and Other Inhibitors of Coagulation

Fig. 7.3 illustrates a negative feedback regulatory loop in the coagulation cascade, based on the activation of the anticoagulant serine protease, protein C, by thrombin. Abnormalities in this pathway are associated with several common inherited thrombotic disorders (see below). Although thrombin is normally a procoagulant protein, when bound to the transmembrane protein thrombomodulin, thrombin is converted from a procoagulant to an anticoagulant by changing its protease target specificity. Thrombomodulin-bound thrombin cleaves protein C to form activated protein C (APC). Accelerated by its cofactor protein S, APC proteolytically inactivates FVa and FVIIIa, thereby damping down the clotting cascade. Thus, activated thrombin can feed back on itself to turn off the activation of additional thrombin. As one might predict, deficiencies of protein C, protein S, and thrombomodulin are all associated with inherited thrombotic disorders. A remarkably common variant in FV (factor V Leiden) renders FVa resistant to inactivation by APC, also resulting in a mild prothrombotic disorder.

Several important inhibitors also contribute to the regulation of the coagulation cascade, including tissue factor pathway inhibitor, protein C inhibitor, and two specific inhibitors of thrombin, antithrombin III and heparin cofactor II. Antithrombin III also inhibits FXa and several other clotting cascade proteases. The inhibitory activity of antithrombin III is dramatically stimulated by heparin, providing the primary mechanism for the latter's anticoagulant function. Inherited deficiency of antithrombin III is associated with thrombosis (see below).

7.2.2 Platelets

Defects affecting the platelet limb of hemostasis can also result in abnormal bleeding. Platelets provide the most immediate response to vascular injury by forming a cellular plug at the site. Deficiencies in platelet number (thrombocytopenia) or platelet function (qualitative platelet disorders) are associated with abnormal bleeding. The major ligand facilitating binding of the platelet to the vessel wall is the multimeric plasma protein, von Willebrand factor (VWF) (Fig. 7.1). Defects in VWF result in a platelet-like bleeding disorder, von Willebrand disease (VWD) (see below).

7.2.3 The Fibrinolytic System

The final limb of the hemostasis tree depicted in Fig. 7.1 is provided by the fibrinolytic system. This proteolytic cascade results in dissolution of the fibrin clot. The final enzyme in this pathway is plasminogen, a serine protease zymogen that in its active form (plasmin) specifically degrades fibrin and may also contribute to tissue remodeling through proteolysis of other matrix components. Plasminogen is activated by plasminogen activators (PA), which include urokinase-type and tissue-type PA (uPA and tPA, respectively). Plasmin, as well as both plasminogen activators, is inhibited by specific serine protease inhibitors, α_2-antiplasmin and plasminogen activator inhibitor-1, respectively. Deficiencies of these inhibitors are associated with abnormal bleeding, whereas deficiency of plasminogen has recently been uncovered as the explanation for a rare autosomal recessive disorder, ligneous conjunctivitis (see below).

7.2.4 Inherited Disorders of the Coagulation Cascade

Inherited bleeding disorders due to deficiencies of factors within the coagulation cascade (Table 7.1) generally result in similar phenotypes. Hemorrhage into deep tissues, particularly the joints (hemarthroses), is characteristic, as is increased bleeding following surgery or trauma. The pattern of bleeding can often be distinguished clinically from that associated with defective

TABLE 7.2 Clinical Findings in Inherited Bleeding Disorders

	Platelet-type Defects	Coagulation Cascade Defects
Timing of bleeding sites	Early after trauma or spontaneous	Delayed
	Skin and mucous membranes, petechiae, ecchymoses	Deep tissue hematomas, including joint, muscle and retroperitoneum
Inherited disorders	von Willebrand disease, Glanzmann thrombasthenia, Bernardsoulier syndrome, other inherited platelet defects	Hemophilia a and B, factor V deficiency, deficiency of factors XI, VII, II, or X; afibrinogenemia

platelet function, as outlined in Table 7.2. The bleeding associated with coagulation cascade disorders is generally delayed compared to that of platelet defects. The latter is more often from mucosal surfaces, particularly the nose, oral cavity, and gastrointestinal tract, in contrast to the deep tissue hemorrhage characteristic of abnormalities in the coagulation cascade.

7.2.5 Hemophilias

The term *hemophilia* is generally reserved for two specific inherited X-linked disorders with nearly indistinguishable phenotypes, FVIII deficiency (hemophilia A or classic hemophilia) and FIX deficiency (hemophilia B or Christmas disease), although FXI deficiency is sometimes referred to as hemophilia C. Hemophilia A is the most common inherited severe bleeding disorder with a frequency in most populations of approximately 1:5000 males [7]. Its X-linked pattern of inheritance was first recognized by Jewish scholars in the second century AD who exempted a male infant from circumcision if his mother and mother's sisters had other sons who had died of bleeding following their circumcisions [3,8]. The syndrome was rediscovered in the late 18th and early 19th centuries and is often referred to as "classical hemophilia" [8–10]. A number of cases of hemophilia occurred among several branches of the royal families of Europe, and for many years, it was unclear whether these represented hemophilia A or B [11]. However, advanced technologies including next-generation sequencing led to the identification of a causative pathogenic variant in the *FIX* locus [12].

FVIII is a nonenzymatic cofactor that is assembled on membrane surfaces together with FIXa to form the "X-ase" complex (Fig. 7.2) that subsequently activates the zymogen FX to the active protease FXa. The central nature of this step in the cascade and the critical role of FVIII were largely revealed by the severe clinical manifestations of FVIII deficiency. Hemophilia B is also an X-linked recessive disorder, but with the defect due to deficiency of coagulation factor IX (FIX). The FIX gene is also located on the long arm of the X chromosome, approximately 15 megabases (Mb) from the FVIII gene. Hemophilia B, sometimes referred to as Christmas disease, is approximately one-fifth to one-tenth as common as hemophilia A, with an estimated prevalence of approximately 1:30,000 males [7].

Clinical Features. The clinical manifestations of hemophilia vary considerably from a severe bleeding disorder presenting at birth to a very mild condition that may be asymptomatic or only diagnosed late in life [7,9,13]. The severity of the disease can be predicted from the level of residual FVIII or FIX activity (Table 7.3). Patients with very low or no factor activity (less than 1%) are affected with severe hemophilia, whereas those with factor levels of 1%–5% have moderate and those with 5%–25% mild disease. Factor levels above 25% are generally associated with a normal phenotype (Table 7.3). Hemophilia should be suspected in any male with a severe congenital bleeding disorder and also in older males with mild bleeding, and the diagnosis can usually be readily established by screening tests, followed by specific factor assays. Although carriers usually demonstrate reduced plasma levels of FVIII or IX (in the range of 50%), these values are unreliable for carrier detection given wide variation in normal factor levels. More accurate classification of carriers and prenatal diagnosis in an "at risk" pregnancy can usually be provided by linkage analysis or direct

TABLE 7.3 Clinical Classification of Hemophilia A and B

Classification	FVIII Or FIX Activity	Clinical Manifestations
Severe	<1%	Spontaneous hemorrhage beginning, in early infancy. Frequent hemarthroses; hemarthroses and other serious hemorrhages requiring factor replacement
Moderate	1%–5%	Hemorrhaging following trauma; occasional spontaneous hemorrhage
Mild	5%–25%	Bleeding generally only following significant trauma or surgery
	>25%	No significant bleeding

mutation detection. Where such testing is not possible, definitive prenatal diagnosis can be established by fetal blood sampling.

The typical clinical presentation in hemophilia patients includes joint and muscle hemorrhages primarily in the large joints, easy bruising, and excessive, sometimes fatal hemorrhage after trauma or surgery. Although joint hemorrhage is generally triggered by stress or trauma, the precipitating event may not be recognized. Hemorrhage into joints is unusual until children begin ambulation, and the disease may go undiagnosed until that time. However, many severe hemophiliacs are diagnosed around the time of birth due to umbilical hemorrhage or following circumcision.

Hemarthroses (joint hemorrhages) generally begin with pain and limitation of motion. After several hours, pain will increase and joint swelling and warmth will be noted. Damage to the joint results from a chronic inflammatory reaction, and the resulting synovitis renders the joint more susceptible to repeat hemorrhage and accumulated damage (Figs. 7.4 and 7.5). The radiographic findings associated with the hemophilic joint are characteristic, but not significantly different between hemophilia A and B. Hemophilic arthropathy often leads to permanent disability and may require joint replacement in selected cases.

Other forms of internal bleeding may be more immediately life-threatening. Intramuscular hematomas can result in compression and compartment syndromes. Bleeding into other sensitive structures can lead to airway obstruction or other complications. The most urgent type of bleeding is intracranial, accounting for about 25% of deaths among hemophiliacs prior to acquired immunodeficiency syndrome (AIDS).

Modern treatment with factor replacement has resulted in a dramatic improvement in life expectancy from 11.4 years during the early 1900s to between 60 and 70 years in 1980. The AIDS epidemic had a major impact on the hemophilia community and for a period during the 1980–1990s it was a central clinical manifestation of this disease. Most hemophiliacs receiving blood products prior to 1984 were exposed to the human immunodeficiency virus (HIV) and became infected, [7,9,10,13]. Many patients developed overt AIDS, approximately two-thirds of whom died. With improved screening and treatment of blood-derived factor, as well as recombinant factor products, this is no longer a major clinical issue.

Genetics. The FVIII gene is located near the tip of the long arm of the X chromosome (Xq28). It is a large gene spanning 186 kilobases (kb), constituting 0.1% of the X chromosome, and is dispersed across 26 exons corresponding to a 9-kb mRNA [14]. The FVIII protein sequence demonstrates a repeated structure, with three A domains, duplicated C domains, and a large central B domain, the latter encoded in a single exon [15–17]. The FVIII sequence and gene structure are highly homologous to FV [18]. The B domain in both proteins appears to be a dispensable connector, since recombinant FVIII or FV from which the B domain has been deleted still maintains relatively normal function [19–21]. The B domain is excised during FVIII activation resulting in a heavy chain, consisting of the A1 and A2 domains, and a light chain containing the A3, C1, and C2 domains, held together in a calcium-dependent fashion. Cleavage at arginine 1689 by thrombin liberates FVIII from VWF, allowing FVIII to bind to a phospholipid surface and interact with FIXa to form the X-ase complex [10].

A broad spectrum of genetic defects within the FVIII gene has been defined, resulting in a range of hemophilia phenotypes determined solely by the amount of residual FVIII activity [10,22]. As predicted by Haldane,

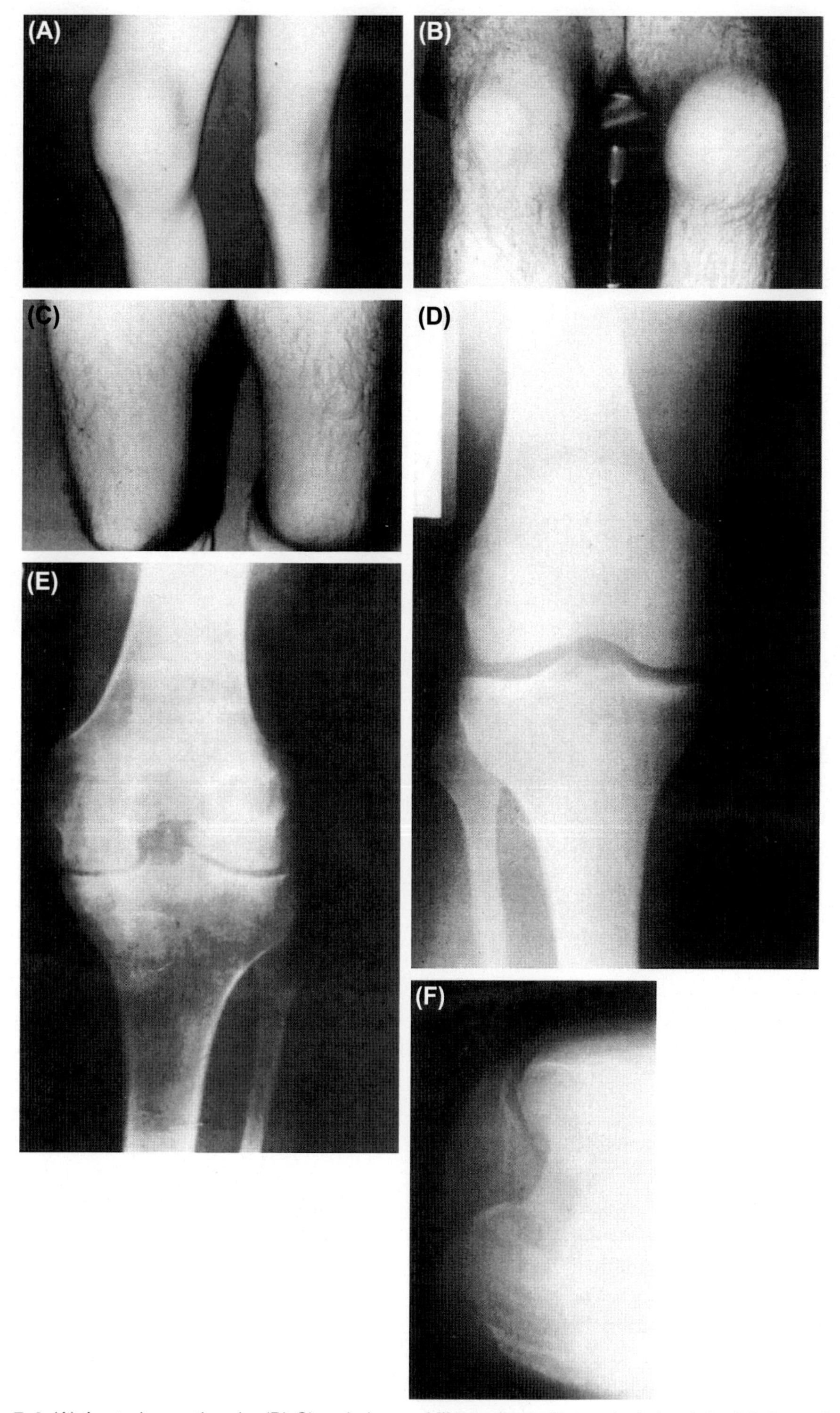

Figure 7.4 (A) Acute hemarthrosis. (B) Chronic hemophiliac arthropathy particularly of the left knee demonstrating enlargement of femoral condyles with associated quadriceps wasting (C). Radiographs of right (D) and left (E) knees of the same individual illustrating rarefaction, enlargement of condyles and intercondylar notch, and loss of cartilage along with degeneration of the patellofemoral surface (F).

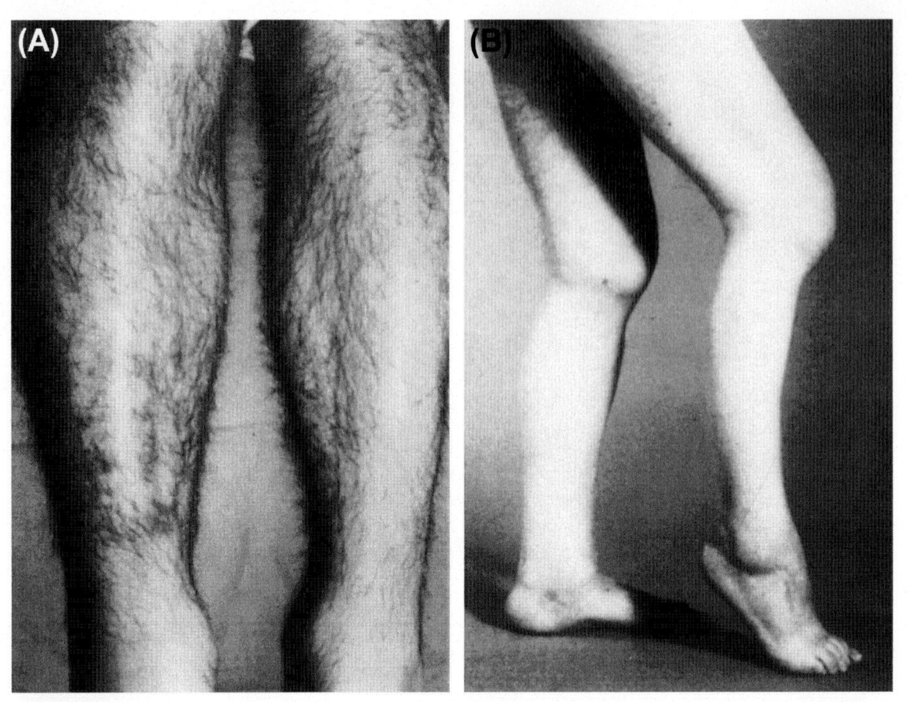

Figure 7.5 (A) Hematoma in calf. If inadequately treated, ischemic necrosis and subsequent fibrosis will ensue to give equinus deformity (B).

about one-third of cases appear to be new mutations, although this will likely decrease as availability of factor products leads to survival well beyond the reproductive years. Homozygous hemophilia A in females is predicted to occur at a frequency of less than 1:25,000,000. Female hemophilia A may also result from hemizygosity for an FVIII pathogenic variant as a result of extreme inequality of X-chromosome inactivation, an unbalanced translocation, or other X-chromosome defect interfering with X inactivation.

A large number of hemophiliacs have been studied at the DNA level [23], and databases of known pathogenic variants have been available over the years, maintained by individual investigators. With the advent of next-generation sequencing and accumulation of increasing quantities of variant data, these have largely accumulated in centralized databases such as ClinVar, which is maintained at the National Institutes of Health, https://www.ncbi.nlm.nih.gov/clinvar/ [24].

Approximately 5% of patients carry deletions within the FVIII gene, generally leading to complete loss of protein expression. Other defects, including frameshift, nonsense and missense variants, have also been described. A novel mutation mechanism accounts for a large proportion of severe hemophilia A patients. This inversion occurs via a recombination between a small gene called gene A, located within intron 22 and one of two additional copies of gene A located upstream of the gene, toward the tip of the X chromosome. The resulting inversion (Fig. 7.6) disrupts the FVIII gene and removes the C-terminus of the protein encoded by exons 23–26. Two mRNAs are transcribed from the inversion, one beginning in gene A and including exons 23–26, and the derivative of the original FVIII transcript, now containing only exons 1–22. Remarkably, this recurrent mutation is responsible for approximately 45% of severe hemophilia A [26]. A very similar recurrent mutation mechanism involves repeated sequence in intron 1 of the factor VIII gene [27], resulting in a large gene inversion and accounting for about 5% of severe cases.

Twenty to thirty percent of hemophilic patients develop clinically significant antibodies against FVIII (FVIII inhibitors), which severely complicate therapy. Inhibitor development is more frequent among patients with large gene deletions, or the factor VIII gene inversion, compared to other types of variants. Patients with missense variants that result in the synthesis of residual endogenous protein, even though nonfunctional, may develop immune tolerance, accounting for considerably lower inhibitor prevalence [28]. There appears to be a genetic predisposition to inhibitor development and studies have included examination of

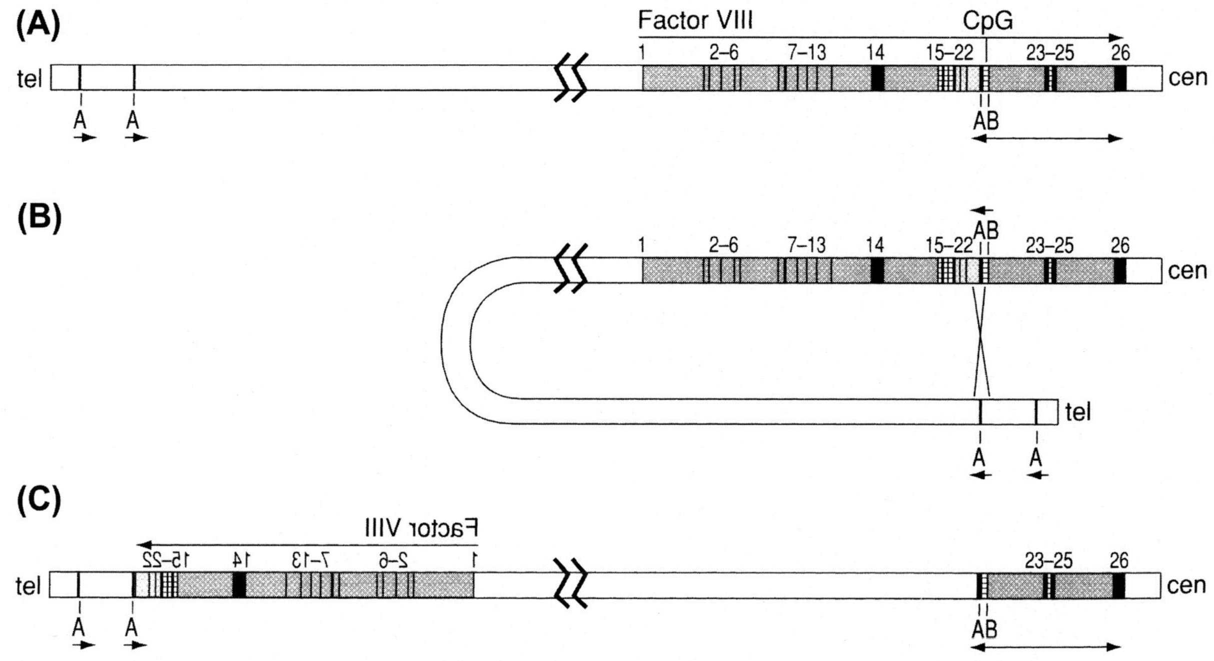

Figure 7.6 Model of factor VIII gene inversion. (A) Structure of the normal factor VIII gene. tel, telomere, cen, centromere. A, the two copies of the *A* gene upstream of the factor VIII gene and the single copy within intron 22. B, another small locus that shares the *A* gene promoter. The arrows indicate the directions of transcription. (B) Model for homologous recombination between the intron 22 copy of the *A* gene and one of the two upstream copies. The indicated crossover results in an inversion of the sequence between the two recombined *A* genes (C). (From Ref. Lakich D, Kazazian HH, Jr, Antonarakis SE, Gitschier J. Inversions disrupting the factor VIII gene are a common cause of severe haemophilia A. Nat Genet 1993;5:236–241, with permission.)

associations with various unlinked polymorphisms, HLA genotype, and *FVIII* variants [29,30].

Hemophilia B also demonstrates typical X-linked recessive inheritance. Rare cases of females affected with this disease generally demonstrate extreme X chromosome inactivation or an X chromosome abnormality. The FIX gene is 34 kb in length and contains eight exons. The gene is located approximately 15 Mb toward the centromere from the FVIII gene. The gene encodes a 461-amino acid precursor protein. FIX is a member of the serine proteinase family and shows significant homology to other members. A signal peptide is followed by a propeptide domain that is posttranslationally modified by the addition of γ-carboxyl groups to glutamic acid, a step dependent on vitamin K and required for function. FIX shares this feature with a group of vitamin-K-dependent coagulation enzymes, including factors II (prothrombin), VII, and X, as well as protein C, protein S, and protein Z [31]. The regulation of FIX gene transcription has been extensively studied. A unique variant of hemophilia B has provided important insights into the regulation of FIX expression. This variant termed FIX Leyden is associated with very low levels of FIX and severe hemophilia, but promptly improves at puberty with a rise of FIX levels to near the normal range [32]). A number of single-nucleotide pathogenic variants in the promoter region of the FIX gene, most within a 40-bp segment surrounding the major transcription start site, have been identified in these patients [32–35].

A large number of gross gene deletions, small insertions/deletions, and single-nucleotide pathogenic variants have been identified in thousands of hemophilia B patients. Complete databases of known mutations have been maintained by a consortium of investigators, but as for factor VIII, the most up to date and reliable data are likely to be found in larger repositories, such as ClinVar.

As for FVIII, inhibitor antibody formation is observed primarily in patients with large gene deletions, to a lesser extent with nonsense variants, and rarely with missense variants [28]. Approximately 1/3 of FIX gene mutations have been shown to occur at CpG dinucleotides [36] and approximately 1/3 of severe patients appear to represent new mutations, as also observed in hemophilia A and consistent with the original hypothesis of Haldane for X-linked lethal disorders.

Clinical Management. Treatment for hemophilia historically relied on intravenous replacement of the deficient factor activity, with highly purified factor concentrates becoming the mainstays of therapy over the past four decades. Unfortunately, the purification processes used until the mid-1980s failed to inactivate common viral contaminants, including hepatitis B and C, cytomegalovirus (CMV), and HIV. As a result, all these infections were nearly uniform among patients heavily treated during that period, but all currently available FVIII concentrates are thought to be free of most viral hazards, though there remains concern about slow viruses or other unknown infectious agents [7,9,10]. Recombinant FVIII production was a major goal driving the cloning of the *FVIII* gene in 1984 [16,37,38], finally reaching the clinic with FDA approval in 1994.

Prophylaxis is the standard approach for the treatment of newly diagnosed cases in most centers. Patients with mild or moderate hemophilia A can also be treated with desmopressin (DDAVP), which will result in a two- to fivefold increase in FVIII and VWF levels that can last a number of hours [96]. Oral ε-aminocaproic acid (EACA) and local methods are often adequate for superficial procedures, but major surgery or trauma typically requires factor concentrates.

Patients are generally managed with recombinant factor or another highly purified plasma-derived product, with most clinicians preferring the former. This viewpoint changed somewhat with the results of the SIPPET study, which demonstrated a lower risk of inhibitors with plasma derived products [39]. More recently, recombinant factor products with extended half-lives have become available. Half-life extension is achieved with protein conjugation, chemical modification through PEGylation, protein sequence modification, or expression in human cell lines, resulting in 2.4–4.8-fold extension over standard half-life and the ability to do weekly infusions for hemophilia B patients [40]. Despite these innovations, the half-life extension of FVIII has only reached approximately 1.5-fold longer than standard half-life products [41]. This is likely due to altered FVIII continuing to complex with VWF and being limited to the latter's half-life. A new product has demonstrated sustained elevation of FVIII levels in clinical trials, with a fourfold extension in half-life [42]. This was achieved through fusion of the VWF D'D3 domain to an extended FVIII product, which might enhance stability [43] and also interfere with binding to endogenous VWF [42].

A new category of nonfactor products has emerged, which serve as substitution or hemostasis rebalancing therapy, with several in various stages of development [44]. Emicizumab, a humanized bispecific antibody that bridges FX and FIXa, effectively substituting for FVIIIa, is approved for use in hemophilia A patients with and without inhibitors. Its subcutaneous mode of administration and weekly to monthly dosing provides patients with efficacious prophylaxis. Another treatment paradigm is modulation of natural anticoagulants to restore the hemostatic equilibrium in patients with bleeding disorders. Current therapies targeting the natural anticoagulants antithrombin, tissue factor pathway inhibitor (TFPI), protein C, and protein S are in various stages of development [40,44].

Gene therapy may offer an improved treatment for hemophilia in the future. Tolerance of wide levels of factor without major negative sequelae and the need for only small increases in activity to achieve major clinical improvement made hemophilia an excellent candidate for a gene therapy approach. Initial studies used various approaches, but more recently there has been a shift to nonintegrating recombinant adeno-associated viral (AAV) vectors. Due to the limited packaging capacity of AAV vectors, initial studies focused on hemophilia B. The first successful AAV-FIX gene therapy study demonstrated stable FIX expression out to 8 years and no late toxicity [45]. More recent trials have utilized the naturally occurring gain of function single-nucleotide variation (FIX-Padua) resulting in higher levels of FIX expression [46,47].

Gene therapy for hemophilia A has lagged behind B due to the size of the FVIII gene. Deletion of the nonfunctional B-domain has allowed incorporation into AAV vectors, and the first successful trial has been published recently [48]. Preexisting immunity to AAV subtypes was a contraindication to gene therapy in early trials, but this is changing. The primary toxicity observed has been in the liver, which is effectively treated with corticosteroids. However, since hepatocytes are the target for AAV, this is often accompanied by loss of transgene expression. Given existing immunity in many patients and potential loss of expression, other vectors are also being evaluated. Gene editing approaches using CRISPR/Cas9, zinc finger nucleases (ZFN), and TALENs are ongoing. Use of ZFN targeted insertion of the FIX gene is the most advanced, with a phase 1 study currently in recruitment [49].

The modern molecular genetics of hemophilia provide powerful tools for genetic counseling and prenatal diagnosis. Diagnosis of the common FVIII inversion is now routinely performed in a number of laboratories. Of particular importance for genetic counseling, it appears that the FVIII inversion generally only occurs during male meiosis, possibly due to inhibition of inversion by the pairing of Xq homologues, which occurs during a female meiosis [26,50]. Thus, if a new hemophiliac has the inversion, the mother is generally a carrier, with the event having occurred in the germline of the maternal grandfather. Somatic mosaicism is surprisingly frequent in hemophilia A, occurring in up to 13% of families, although more common in families with single-nucleotide pathogenic variants, particularly CpG transitions [51].

Efficient screening of all 26 FVIII exons and associated splice junctions can identify the molecular defect in the vast majority of hemophilia A cases not due to FVIII gene inversions. Such direct DNA sequence screening is available through several commercial laboratories. In rare patients for whom the precise variant cannot be identified by direct DNA analysis, accurate genetic diagnosis usually can still be achieved by linkage analysis. In questionable or uninformative cases, a fetal blood sample can be collected and plasma FVIII activity measured directly, providing definitive prenatal diagnosis.

Although no common recurrent mutation mechanism (as occurs in hemophilia A) has been identified in hemophilia B, DNA testing can be used for prenatal diagnosis or to definitively establish carrier status in at-risk females and is also available through clinical laboratories. In addition, DNA testing for a known familial variant or for linkage analysis can be used for prenatal diagnosis or to definitively establish carrier status in females.

7.2.5.1 von Willebrand Disease (VWD)

von Willebrand factor (VWF) is a multimeric plasma glycoprotein that forms an adhesive link between platelets and the blood vessel wall at sites of vascular injury (Fig. 7.1) and serves as the carrier for FVIII in plasma [52,53]. The major cause of bleeding in mild VWD is related to the deficiency in platelet adhesive function. This bleeding resembles that associated with thrombocytopenia or a mild platelet functional defect. In contrast to the deep tissue bleeding of hemophilia, VWD patients suffer prolonged cutaneous or mucosal bleeding, such as spontaneous nosebleeds or gastrointestinal bleeding. More serious bleeding is unusual but can complicate major surgery or trauma. Heavy menstrual bleeding is also a frequent complication in female patients. VWD is due to either a quantitative or a qualitative defect in VWF. In contrast to the hemophilias, where straightforward correlation between factor activity and clinical phenotype can be made (Table 7.3), the clinical manifestations of VWD are complex, with extensive phenotypic and genotypic heterogeneity. Although VWD is more frequent than hemophilia, the most common forms are mild and frequently undiagnosed.

Clinical Features. Multiple clinical subtypes of VWD have been described. Traditionally, the most common variant of VWD, type 1, is associated with simple quantitative deficiency of VWF, with levels in the range of 3%–50% of normal. Type 1 VWD appears to account for about 70% of clinically significant VWD [54], and inheritance is generally autosomal dominant with incomplete penetrance. A multidisciplinary guideline panel [55] recommended that a diagnosis of VWD generally be restricted to patients with VWF levels <30% of normal or <50% with bleeding symptoms.

The standard laboratory workup for VWD consists of a series of coagulation tests, often referred to as the "VWD workup" [56]. This includes FVIII activity, VWF antigen level (previously referred to as FVIII-related antigen or FVIIIR:Ag), and a ristocetin cofactor assay of VWF functional activity. All three of these tests are generally proportionately decreased in typical type 1 VWD. The ristocetin cofactor assay is the most widely used measure of VWF function but due to wide variability and poor sensitivity, multiple new assays have been developed. The GP1bM assay also evaluates platelet dependent VWF activity but measures VWF activity without the need for ristocetin, resulting in superior precision and sensitivity [57]. VWF:CB is a collagen-binding assay typically evaluated using ELISA methods [58]. VWF:FVIIIB is another ELISA assay, which evaluates the capacity of VWF to bind exogenous FVIII and is useful in differentiating between mild hemophilia A and type 2N VWD ([60], see below). To properly subclassify the type of VWD, an additional test must be performed to examine the multimeric structure of plasma VWF. VWF multimer analysis is generally performed by agarose gel electrophoresis (Fig. 7.7). In type 1 VWD, although VWF is quantitatively decreased, the multimer pattern is normal.

A number of less common variants of VWD have been distinguished, including types 1C, 2A, 2B, 2N, and 2M [61]. Their features are summarized in Table 7.4. Type 1C VWD is associated with accelerated VWF clearance, resulting in a markedly elevated VWF propeptide to VWF antigen ratio. Type 2A VWD is associated with selective loss of large and intermediate-size VWF multimers, the forms of VWF that generally exhibit the greatest adhesive function. Type 2B VWD is associated with spontaneous or increased binding of VWF to platelets, with subsequent clearance from circulation of the largest multimers, as well as platelets, resulting in thrombocytopenia and a multimer pattern similar to type 2A VWD. Type 2M VWD is often misdiagnosed as type 1 VWD, which has resulted in patient reclassification after analysis by *VWF* gene sequencing [62]. In contrast to type 1 VWD, which has a VWF activity to antigen ratio close to 1, type 2M is characterized by a ratio less than 0.5–0.7, with normal multimers [63].

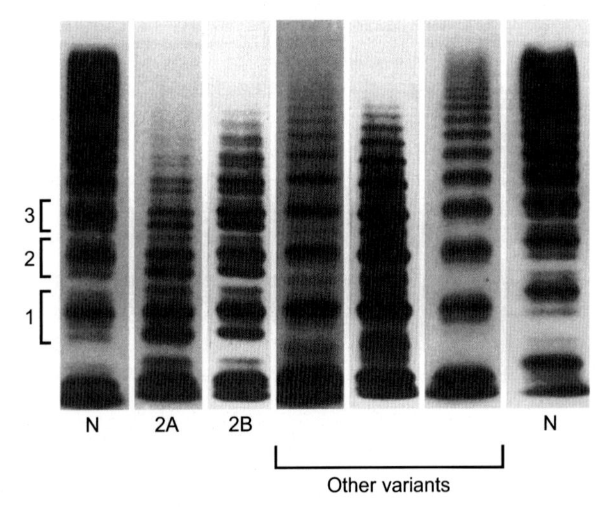

Figure 7.7 VWF multimer analysis. VWF multimers from plasma of patients with various qualitative variants of VWD are shown here, as analyzed by agarose gel electrophoresis. The brackets to the left encompass three individual multimer subunits including the main band and its associated satellite bands. The finer subclassification of VWD variants is often based on variations in these satellites. N, normal sample, 2A and 2B, type 2A and 2B VWD, respectively. (From Ref. Berkowitz SD, Ruggeri ZM, Zimmerman TS. von Willebrand disease. In: Zimmerman TS, Ruggeri ZM, editors. Coagulation and bleeding disorders. The role of factor VIII and von Willebrand factor. New York: Marcel Dekker, Inc.; 1989. p. 215–259, with permission.)

Type 2N VWD, or VWD Normandy, is a unique variant associated with decreased VWF binding to FVIII [56,67], but normal platelet binding function. Homozygotes or compound heterozygotes can have moderately decreased FVIII levels resembling hemophilia A, but are distinguished by autosomal recessive inheritance. Unlike those with hemophilia, type 2N VWD patients require VWF replacement and respond poorly to purified FVIII concentrates, thus differentiation is of particular clinical importance.

Type 3 VWD, the most severe form, is associated with extremely low or undetectable levels of VWF and low levels of FVIII [56]. This disorder is typically autosomal recessive, but in some families one or both parents may be affected with characteristic type 1 VWD. Type 3 VWD is associated with severe bleeding, both the mucosal type typical of VWD as well as deep tissue bleeding characteristic of FVIII deficiency. Thus, VWD testing needs to be included in the analysis of all patients with severe FVIII deficiency.

Genetics. The *VWF* gene is composed of 52 exons spanning approximately 180 kb on the short arm of chromosome 12 [68]. There is also a nonprocessed partial pseudogene, localized to chromosome 22, which duplicates approximately the middle third of the VWF gene (exons 23 to 34). *VWF* encodes a 2813-amino acid protein composed of a number of internally repeated, homologous domains (Fig. 7.8). As for the hemophilias, a large number of other variants have been reported. A database of known VWD pathogenic variants and polymorphisms has been maintained by various consortia over the years, but given the increasing identification of pathogenic variants by next-generation sequencing, ClinVar is likely to be the best source of information going forward.

TABLE 7.4 Summary of VWD Subtypes

Subtype	Frequency	Clinical Features	Diagnosis	Molecular Basis
Type 1	1–10:1000; most common VWD variant (>70% of VWD)	Mild to moderate bleeding; autosomal dominant; incomplete penetrance (approximately 60%)	VWF:Ag, VWF:RCo, and FVIII all proportionately decreased (20–50 U/dL); normal multimer distribution	Some cases are heterozygotes for type 3 VWF Causative *VWF* mutation identified in ~65% of patients with higher detection in patients with lower VWF levels [64] Unlinked loci, in addition to the *VWF* gene, likely to contribute to phenotype
Type 1C	~15% of type 1 [65]	Mild to moderate bleeding; autosomal dominant inheritance	VWF: Ag, VWF: RCo, and FVIII all proportionately decreased; normal multimer distribution, increased clearance on desmopressin trial [55]	VWF Vicenza, R1205H (prototypic clearance mutant), mutations in the D3 region of VWF
Type 3	1-5:10^6	Severe bleeding disorder; autosomal recessive inheritance	Markedly decreased or undetectable VWF:Ag, VWF:RCo, and FVIII	VWF gene deletions; nonsense mutations; frameshift mutations; other cis-defects in mRNA expression; some cases are homozygous for type 1 defect
Type 2A	Approximately 10%–15% of clinical VWD cases	Mild to moderate bleeding disorder; autosomal dominant, more complete penetrance than type 1; generally poor response to DDAVP	Variably decreased VWF:Ag, VWF:RCo, and FVIII; absent high and intermediate size VWF multimers with prominent satellite bands	Missense mutations clustered within VWF A2 domain; two subgroups: Group 1, defect in intracellular transport; group 2, proteolysis in plasma after secretion
Type 2B	Uncommon variant (<5% of clinical VWD)	Mild to moderate bleeding disorder; autosomal dominant, more complete penetrance than type 1; ? DDAVP contraindicated	Variably decreased VWF:Ag, VWF:RCo, and FVIII; loss of large multimers; enhanced RIPA; thrombo cytopenia	Missense mutations clustered in VWF A1 domain resulting in increased or spontaneous binding to platelet GPIb

Continued

TABLE 7.4 **Summary of VWD Subtypes—cont'd**

Subtype	Frequency	Clinical Features	Diagnosis	Molecular Basis
Type 2M	Uncommon variants but may be underdiagnosed due to difficulty in proper identification with commonly used assays [66]	Mild to moderate bleeding disorder; autosomal dominant; may respond to DDAVP	Variably decreased VWF:Ag, VWF:RCo, and FVIII; disproportionately low VWF:RCo relative to VWF:Ag; normal multimers	Missense mutations clustered in VWF A1 domain resulting in decreased platelet adhesion (most commonly) Missense mutations clustered in VWF A3 domain resulting in selective deficiency of VWF binding to collagen.
Type 2N (FVIII binding defects; VWD Normandy)	Allele frequency may be as high as 1% in some populations	Variable bleeding disorder; homozygotes (or compound heterozygotes) may present as autosomal hemophilia A	Variable VWF:Ag and VWF:RCo; disproportionately low FVIII; generally normal multimers; decreased or absent VWF binding to FVIII	Missense mutations within the N-terminus of mature VWF which interfere with FVIII binding
Platelet-type VWD ("Pseudo-VWD")	Rare (case reports)	Similar to type 2B VWD	Can be distinguished from type 2B by mixing studies with normal platelets and plasma	Missense mutation within GPIb α-chain resulting in increased or spontaneous binding to VWF

GP, glycoprotein; *RIPA*, ristocetin-induced platelet aggregation; *VWD*, von Willebrand disease; *VWF*, von Willebrand factor; *VWF:Ag*, VWF antigen; *VWF:RCo*, VWF ristocetin cofactor activity. Modified from Ref. Ginsburg D, Bowie EJW. Molecular genetics of von Willebrand disease. Blood 1992;79:2507–2519, with permission.

Deletion appears to be uncommon but can encompass the entire *VWF* locus. Gene deletions and nonsense and frameshift variants have been identified in a number of cases of type 3 VWD. A frameshift in exon 18 is common in Scandinavia and has been shown to be the defect in the original family described by von Willebrand [70,71]. Several large-scale, multicenter sequencing efforts have provided new insight into the molecular genetics of type 1 VWD [72–74]. *VWF* gene sequencing in approximately 330 type 1 families identified candidate variants in 67% of index cases. Seventy-five percent were missense variants scattered throughout the gene, and the rest were small insertions/deletions, splicing variants, nonsense variants, or promoter variants (summarized in Ref. [75]). Using more stringent criteria (VWF levels <30% on two occasions), a study of 556 subjects found genetic defects in 91% of Type 1 subjects, with a total of 155 novel variants, highlighting the high rate of variability in the VWF gene [76].

Pathogenic variants have also been identified accounting for the majority of type 2A VWD, restricted primarily to the A2 repeat within exon 28 (Fig. 7.7). These variants have been shown to produce type 2A VWD via two distinct mechanisms [77]. In group 1, a variant results in the misfolding of the protein and retention in the ER, with multimerization accounting for the dominant negative effect. In group 2, multimers containing abnormal subunits are secreted but show increased susceptibility to proteolysis in plasma by ADAMTS13 (see below), resulting in degradation of the largest multimers to smaller forms.

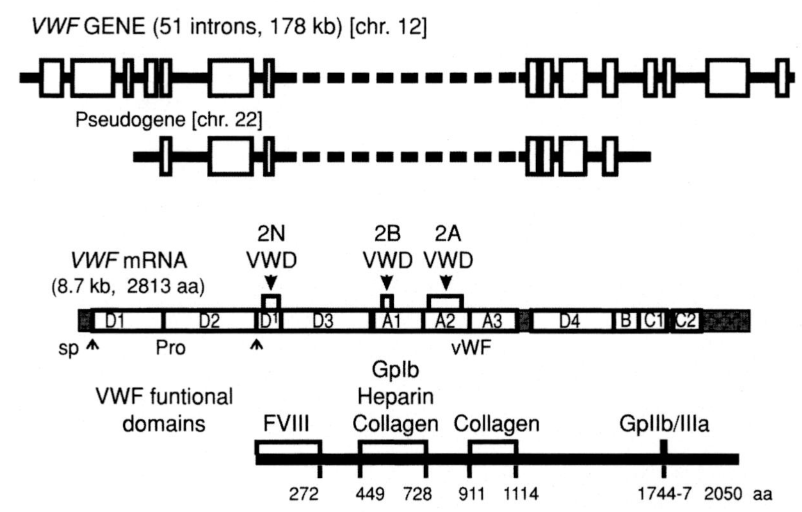

Figure 7.8 Structure of *VWF* gene, mRNA, and protein. The *VWF* gene and pseudogene are depicted at the top with boxes representing exons and the solid black line, introns. The *VWF* mRNA encoding the full prepro-VWF subunit is depicted in the middle as the stippled bar and lettered boxes. The locations of signal peptide (sp) and propeptide (pro-) cleavage sites are indicated by arrowheads below the bar, and the lettered boxes denote regions of internally repeated sequence. The clusters of variants responsible for type 2A VWD, type 2B VWD, and FVIII binding defects (type 2N) are indicated above the bar. The approximate localizations for known VWF functional domains within the mature VWF sequence are indicated at the bottom. Numbers underneath the domains refer to amino acid residues within the mature VWF subunit. *aa*, amino acids; *chr.*, chromosome; *kb*, kilobase pairs. (From Ginsburg D, Bowie EJW. Molecular genetics of von Willebrand disease. Blood 1992;79:2507–2519, with permission.)

Type 2B VWD has been shown to be due to a limited number of pathogenic variants confined to the A1 repeat of VWF, corresponding to the binding domain for its major platelet receptor, GPIb (Fig. 7.7). All these variants appear to result in "gain of function" producing spontaneous or increased binding of VWF to platelets. Finally, a group of variants has also been identified near the N-terminus of VWF, a region previously shown to contain the FVIII binding domain, which results in the type 2N variant of VWD, associated with decreased FVIII binding [56].

Type 2M is usually due to pathogenic variants in exon 28, which includes the A1 domain, the primarily functional domain of VWF necessary for the ristocetin cofactor and GP1bM assays. These are usually missense variants or in-frame deletions and are autosomal dominant with a high degree of penetrance [78].

Although several examples have been reported [79], the frequency of new mutation in VWD appears to be low, consistent with the presumably modest decrease in reproductive fitness. It has also been proposed that mild VWD might actually provide some protection against pathologic thrombosis, including myocardial infarction [80]. Consistent with this idea, high VWF levels are associated with venous and arterial thrombosis, at least in part through elevated FVIII levels [81–83]. Furthermore, large genome-wide association studies have strongly implicated ABO blood group as a risk locus [84–86], which is believed to be due to the ABO genotype dependent modification of VWF plasma levels (see below).

Penetrance is decreased in type 1 VWD (approximately 60%) [87]and appears to be more complete in type 2A and 2B VWD. A number of factors contribute to the variable penetrance and expressivity of VWD. ABO blood group has been shown to modify VWF levels significantly, accounting for ~30% of genetic variance in VWF antigen. VWF levels in type O individuals are on average 30%–40% lower than other blood types [88,89]. Families with higher heritability and lower VWF levels are more likely to have identifiable *VWF* gene pathogenic variants [90].

Environmental effects also modify VWF levels, including estrogens, stress, underlying vascular disease, and thyroid dysfunction.

The true prevalence of VWD is controversial. Large screening studies have suggested that mild VWD may affect about 1% of individuals in several populations [91,92]. However, there is clearly overlap in VWF levels between normal and VWD patients [93]. Prospective analyses of large primary care populations suggest that the prevalence of symptomatic disease is closer to 0.1% [94,95].

Management. The treatment of choice for mild type 1 VWD is DDAVP. Administration of DDAVP results in a two- to threefold increase in VWF levels, lasting 6–12 h [52,96]. Although tachyphylaxis can occur, administration every 12–24 h is frequently sufficient. DDAVP is generally ineffective in type 2A VWD, although some patients have been noted to respond, and is often considered contraindicated in type 2B VWD, since use may lead to worsening thrombocytopenia. Response is variable in patients with type 2M and 2N and generally ineffective in type 3. Patients with type 1C initially respond to DDAVP, but this response is short-lived due to rapid clearance. Minor procedures can often be managed with oral EACA.

In patients unresponsive to DDAVP or with contraindications, factor replacement is used. Several specific FVIII concentrate preparations containing large quantities of VWF of normal multimer structure have been shown to be therapeutically effective [52]. A recombinant VWF product is also now available [97].

Prenatal diagnosis is not generally performed in families with mild VWD, although sequencing is available.

7.2.6 Pseudo-von Willebrand Disease

Pseudo- or platelet-type VWD is a platelet disorder that closely mimics type 2B VWD [98]. It is characterized by spontaneous interaction between plasma VWF and platelets, clearance of the complexes from plasma, and results in thrombocytopenia and loss of large VWF multimers. However, mixing studies with normal platelets and plasma reveal that the defect in this disorder resides in the platelets rather than in VWF. The defect has been shown to be a gain of function variant in the platelet GPIb receptor for VWF [99,100]. Replacement of VWF is ineffective in this disorder and treatment with DDAVP would appear to be contraindicated, thus platelet transfusion is the treatment of choice for prophylaxis or significant bleeding.

7.2.7 Factor XI Deficiency

After the classic hemophilias, autosomal recessive factor XI deficiency is the most common inherited defect in the coagulation cascade [101]. It is particularly common among Ashkenazi Jews, accounting for half of all identified cases, and a heterozygote frequency of 9%. A number of missense variants in the FXI gene have been identified in affected patients, with two variants accounting for >90% of disease alleles among Ashkenazi Jews [102]. Detailed haplotype analysis demonstrated an ancient founder allele for the common Glu117Stop variant in this population (allele frequency 0.022), which apparently arose before the separation of Ashkenazi and non-Ashkenazi Jews. The Phe283Leu variant, also common in this population (allele frequency equals 0.025), appears to have arisen more recently [103]. A Cys38Arg founder variant has been identified in the French Basque population (allele frequency of 0.01) [104].

In contrast to classical hemophilia, the plasma level of FXI does not correlate well with bleeding severity, but there does appear to be genotype–phenotype correlation. Significant bleeding complications are most often associated with FXI levels <10%. Bleeding is most common after surgical procedures and hemarthroses are rare. Minor bleeding can generally be managed with antifibrinolytic agents, but when necessary, replacement is with fresh frozen plasma (FFP) or factor concentrate if available.

7.2.8 Deficiency in Contact Activation Factors

Inherited deficiency of FXII, prekallikrein, and high-molecular-weight kininogen has been reported. From a coagulation standpoint, these disorders are primarily a laboratory curiosity, as they are abnormal in clinical assays that are also used to screen for deficiencies of coagulation factors such as factors VIII and IX, but are generally not associated with clinically significant bleeding. Recent data suggest the involvement of this pathway in pathologic thrombosis after exposure of blood to artificial surfaces (e.g., medical devices), with the possibility of targeting this pathway as a novel therapeutic [105].

7.2.9 Disorders of Fibrinogen

The three fibrinogen genes, α, β, and γ, are located in a cluster on chromosome 4. Defects in any one of the three chains can result in markedly decreased fibrinogen (hypofibrinogenemia) or absent fibrinogen (afibrinogenemia) [106,107]. Inheritance of afibrinogenemia is usually autosomal recessive, with consanguinity noted in many affected families. Over 150 families with congenital afibrinogenemia have been reported, and a recent report suggests that >80% of the pathogenic variants lie in the fibrinogen α-gene. The spectrum of variants includes a large gene deletion eliminating most of the fibrinogen α-gene, as well as splice, frameshift, and nonsense variants [106]. Despite the central position of fibrinogen within the hemostatic system, symptoms can be surprisingly mild. Although afibrinogenemic blood is unclottable ex vivo, bleeding generally appears to be less severe than in hemophilia A, a puzzling observation, given its downstream location in the cascade. Hemarthroses, mucosal, and deep tissue bleeding occur, but intracranial hemorrhage is the leading cause of death. Pregnancies in afibrinogenemic females generally result in fetal loss. Reports of paradoxical thromboembolism have been noted in patients following fibrinogen replacement. Hypofibrinogenemic patients rarely have significant bleeding, unless the fibrinogen level is considerably <50 mg/dL. Therapy has generally been infusion of plasma, though a fibrinogen concentrate is also available [108].

Qualitative abnormalities in fibrinogen also occur and are termed "dysfibrinogenemia," with defects in all three chains reported. Inheritance is generally autosomal dominant, indicating a dominant negative effect of the mutant variant, since hypofibrinogenemic patients with 50% normal levels are generally asymptomatic. Clinical symptoms are varied, and ~40% of identified patients are asymptomatic, with 30% demonstrating hemorrhage and 10%–15% thrombotic complications. A few patients have exhibited both hemorrhage and thrombosis, usually mild, but fatal complications can occur [109,110]. Inherited abnormalities in fibrinogen must be distinguished from a wide variety of acquired disorders that alter fibrinogen levels.

7.2.10 Factor XIII Deficiency

Factor XIII (FXIII) is a plasma and platelet-associated protein that cross-links the fibrin clot, significantly enhancing its stability. Without cross-linking, clots are more easily lysed by the fibrinolytic system, resulting in bleeding. Homozygous FXIII deficiency is rare with frequency estimated at $<1:10^6$. Inheritance is autosomal recessive with pathogenic variants either in the A subunit on chromosome 6 or in the B subunit on chromosome 1, the vast majority in the former [107,111]. Plasma levels in patients are generally <1%, and the disorder frequently presents with umbilical cord bleeding after birth. Soft tissue hemorrhage and postsurgical traumatic bleeding are common and are classically described as being delayed by 1–2 days, although early bleeding can also be seen. Poor wound healing and abnormal scar formation are often noted. Intracranial hemorrhage is a particularly common complication and has been reported to occur in up to 25% of patients. All routine laboratory clotting tests are normal. The diagnosis is established by a urea solubility test as FXIII-deficient clots are readily lysed in 5M urea, whereas normal clots are resistant. Treatment was traditionally with FFP or cryoprecipitate, which can be given as infrequently as every 4–6 weeks, due to a long half-life (9–19 days). Since concentrates of FXIII and recombinant FXIII-A are now available, these are the preferred therapies.

7.2.11 Defects in Other Coagulation Cascade Proteins

Prothrombin (factor II, FII) deficiency is rare, with only ~40 cases reported [107,112,113]. Inheritance is generally autosomal recessive; defects in the prothrombin gene (chromosome 11), affecting most of the molecule, have been identified. Severity of bleeding is correlated with the level of prothrombin activity. Residual activity has been seen in all patients leading to speculation that complete deficiency is incompatible with life. Compared to the much milder presentation of afibrinogenemia, these observations are consistent with the critical role of thrombin in platelet activation as well as fibrinogen cleavage. Replacement is through infusion of prothrombin complex concentrates or FFP.

Factor V (FV) deficiency is a rare autosomal recessive disorder, with an estimated frequency of 1:1,000,000 [107,113–115]. The clinical presentation is very similar to hemophilia, and it was originally referred to as "parahemophilia." FV is found both in plasma and platelets and the relative contributions of these two pools to hemostasis are unclear. Some investigators have suggested a close correlation of bleeding symptoms with the platelet level of FV, and platelet transfusions have appeared to be effective in some patients. However, standard therapy is FFP, as FV-containing concentrates are not currently available. Clinical symptoms are generally associated with levels <1%–20%. Residual low levels of plasma FV activity can be detected in nearly all FV-deficient patients, although a few examples of apparently complete deficiency have been reported, including a patient who is apparently homozygous for a frameshift mutation [116]. An unusual autosomal dominant variant of FV deficiency leading to moderate bleeding, known as FV Quebec, is associated with decreased platelet FV and a normal plasma pool [117]. However, subsequent studies in these patients indicate that the reduced platelet FV is the result of a generalized accelerated proteolysis of α-granule contents that is not restricted to FV alone [118], due to increased levels of urokinase-type plasminogen activator (PLAU) [119]. These patients were subsequently shown to have a tandem duplication of the *PLAU* gene [120].

FVII deficiency has an incidence of 1:500,000 and has been associated with missense variants in the FVII gene (chromosome 13) [107,113,121]. Although hemorrhage is most likely with FVII of <10%, correlation with levels is variable. Levels of <1% are associated with a bleeding pattern similar to classic hemophilia. Plasma or prothrombin complex concentrates are effective, though the treatment of choice is recombinant FVIIa.

Factor X deficiency is a rare disorder with an incidence estimated at 1:1,000,000 [122]. Analyses of patient DNAs have identified specific missense variants and deletions within the FX gene, located on chromosome 13 adjacent to F7 [107,113,122]. Inheritance is autosomal recessive and the clinical presentation in severely deficient patients (activity less than 1%) is very similar to hemophilia. As in hemophilia, significant bleeding is rare with levels >10–15%. Treatment is with plasma-derived factor X concentrate, fresh frozen plasma, or prothrombin complex concentrates.

A number of rare cases of multiple clotting factor deficiencies have been reported. Inherited defects in the γ-carboxylase or vitamin K reductase pathways have been shown to account for combined defects in the synthesis of all the vitamin-K-dependent factors, including FVII, FIX, FX, and prothrombin, which are dependent on vitamin-K-mediated carboxylation for activity [123]. Similar patterns are seen with vitamin K deficiency or with warfarin anticoagulant treatment. Several missense variants in the γ-glutamylcarboxylase gene have now been identified in pedigrees with this disorder [123,124]. This latter disorder is termed vitamin K clotting factor deficiency (VKCFD) with pathogenic variants in the γ-carboxylase gene (GGCX) termed VKCFD1. A second subtype of this disorder, VKCFD2, has been shown to be due to variants in the vitamin K2,3-epoxide reductase (VKOR) enzyme, which was first identified through study of these patients [125]. The same gene (VKORC1) was also independently identified through a direct expression cloning approach [126] and was also discovered to be responsible for warfarin resistance in mice and rats. Polymorphic variations in VKORC1, as well as a cytochrome P450 gene, have also been identified as important determinants of warfarin sensitivity in human patients, accounting for 30%–40% of the variance in warfarin dosing [127,128].

An autosomal recessive disorder resulting in combined deficiency of FV and FVIII is particularly prevalent in non-Ashkenazi Jews [129] but has also been identified in other populations. The gene responsible for this disorder was identified by positional cloning [130] and shown to encode the ER/Golgi intermediate compartment protein LMAN1 (also known as ERGIC-53). Complete deficiency for LMAN1 results in a coordinate decrease in both FV and VIII via partial block to the export of these specific proteins from the ER. Over 100 cases have been identified with ~3/4 of patients found to be homozygous or compound heterozygous for null mutations in *LMAN1*. The remaining subset of patients have a variant in a second gene, *MCFD2* [131]. MCFD2 and LMAN1 form a complex, which appears to function as a specific cargo receptor for FV and FVIII.

7.2.12 Inherited Disorders of Platelet Function

A number of inherited disorders lead to congenital bleeding by interfering with the platelet limb of coagulation. All these disorders present similar clinical phenotypes, characterized by predominantly mucosal, cutaneous, and prolonged bleeding from minor injuries (Table 7.2). Petechiae and purpura are often seen. Deep tissue bleeding and hemarthroses are less common than in disorders of the coagulation cascade, although intracranial hemorrhage can occur, particularly subarachnoid bleeding. By far the most common genetic cause of platelet-type bleeding is VWD, with the remaining inherited platelet disorders uncommon.

7.2.13 Bernard–Soulier Syndrome

Bernard–Soulier syndrome is a rare disorder associated with platelet-type bleeding, unusually large platelets, and thrombocytopenia. The main functional defect is loss of VWF-dependent platelet adhesion. Platelets in these patients have been shown to be missing the GPIb/V/IX complex on the platelet surface, the major receptor for VWF. Inheritance is generally autosomal recessive and frequently associated with consanguinity. Most patients appear to have variants within the GPIbα gene, although variants in the GpIbβ and GpIX genes have also been identified [132,133].

7.2.14 Glanzmann Thrombasthenia

Glanzmann thrombasthenia is autosomal recessive and along with Bernard–Soulier syndrome, is the most clearly defined of the platelet functional defects at the molecular level [133,134]. Platelets show a profound and characteristic defect in platelet aggregation, although the clinical presentation is similar to other platelet functional defects. It is due to absence or dysfunction of the GPIIb-IIIa ($\alpha_{IIb}\beta_{III}$) integrin receptor on the platelet surface. This receptor plays a major role in platelet aggregation by binding to RGDS sites in fibrinogen and possibly other plasma proteins, including VWF. Inheritance is autosomal recessive; a variety of defects have been identified in both the *GPIIb* (*ITGA2B*) and *GPIIIa* (*ITGB3*) genes [133,134]. Glanzmann thrombasthenia is divided into type I, characterized by complete absence of surface GPIIb/IIIa; type II, with GPIIbIIIa antigen decreased to 10%–20% of normal; and type III, with normal GPIIb/IIIa antigen but abnormal function. Type III appears to be due primarily to missense mutations in GPIIIa, whereas a variety of deletions and point mutations (many resulting in frameshifts or aberrant splicing) have been identified in type I patients, involving either the *GPIIb* or *GPIIIa* genes [135]. Interestingly, variants in the *ITGA2B* gene have also been linked to autosomal dominant thrombocytopenia [136].

7.2.15 Other Platelet Disorders

A large number of congenital platelet disorders can be grouped together under the designation of storage pool deficiencies. A deficiency in any of several specific types of platelet storage granules can be seen. The gray platelet syndrome (GPS) results from the absence of normal α-granules in megakaryocytes and platelets, which can be visualized using light microscopy. The inheritance is autosomal recessive, with only a few cases reported. Pathogenic variants in *NBEAL2* were identified as the cause of GPS [137–139].

Deficiency of dense bodies is also seen in some hereditary platelet disorders, but these often require electron microscopy for diagnosis. Several of these conditions, such as Hermansky–Pudlak syndrome, are associated with other clinical findings, such as oculocutaneous albinism. This disorder is particularly frequent in a restricted region of Puerto Rico. Abnormalities of dense granules are also seen in other genetic disorders, including Chediak–Higashi syndrome, thrombocytopenia and absent radii (TAR) syndrome, and Wiskott–Aldrich syndrome. Rare patients have also been described with defects in platelet activation, affecting either the arachidonate or cyclooxygenase pathways. The precise molecular defects and the genetics of these latter disorders have not been clearly defined [133]. A number of other rare disorders exist, both with and without constitutional phenotypes [140].

Treatment of qualitative platelet disorders is primarily supportive, with platelet transfusions as necessary. Antiplatelet drugs such as aspirin and nonsteroidal anti-inflammatory drugs (NSAIDs) should be avoided. Bleeding symptoms associated with selective serotonin reuptake inhibitors may be secondary to depletion of this mediator from platelet granules, resulting in platelet dysfunction, and may act synergistically with NSAIDs [141]. Hormonal therapy is helpful for menstrual bleeding. Dental hygiene is important and antifibrinolytics are often useful for minor procedures. Iron deficiency anemia frequently develops from chronic low-grade mucosal blood loss. DDAVP may be of benefit in some patients, although the mechanism is not well understood. Given its relative lack of toxicity, a therapeutic trial is reasonable.

7.2.16 Defects of the Fibrinolytic System

Review of the complex coagulation regulatory system depicted in Fig. 7.1 would suggest that abnormal overactivity of the fibrinolytic system could also result in pathologic bleeding [142]. Although relatively rare, hereditary deficiencies of the fibrinolytic inhibitors α_2-antiplasmin [143,144] and plasminogen activator inhibitor-1 (PAI-1) [145,146] have been reported. These disorders result in overactivity of endogenous plasmin, with accelerated clot lysis. Inheritance is autosomal recessive and is associated with increased fibrinolytic activity and a mild to moderate bleeding defect, usually following minor trauma or surgery. Description of these defects provided the first strong indication that the major function of the plasminogen activation system is in the regulation of blood coagulation. Treatment is generally with antifibrinolytic agents that inhibit plasmin function, such as EACA or tranexamic acid. Rare patients with abnormally elevated levels of tissue plasminogen activator (tPA) associated with a bleeding tendency have been reported [147], although some of these cases may represent deficiency of PAI-1 or another primary vascular defect [142].

Mild bleeding is also found as a component of other genetic disorders. Abnormal bleeding into the skin is commonly observed in several types of Ehlers–Danlos syndrome. Hereditary hemorrhagic telangiectasia frequently is associated with epistaxis and gastrointestinal blood loss.

7.2.17 α_1-Antitrypsin Pittsburgh

α_1-Antitrypsin Pittsburgh is an unusual disease described in only two patients [148,149], but is a highly instructive example of a dominant "gain of function" as a mechanism for human disease. This disorder is due to a single missense variant within the α_1-antitrypsin gene resulting in substitution of an arginine for methionine at the P1 position (Met358). α_1-Antitrypsin is a highly abundant plasma serine protease inhibitor (serpin) that inhibits the serine protease elastase. Deficiency of α_1-antitrypsin results in a common genetic disease associated with emphysema and liver disease. The effect of the α_1-antitrypsin Pittsburgh variant is to convert α_1-antitrypsin from an inhibitor of elastase to an efficient inhibitor of thrombin, an activity not exhibited in native α_1-antitrypsin. The abnormal α_1-antitrypsin thus resembles antithrombin III (AT3) in its activity. AT3 is the primary physiologic regulator of thrombin activity and deficiency results in a hereditary prothrombotic condition (see below). Since α_1-antitrypsin is present in considerably higher plasma concentrations, the α_1-antitrypsin Pittsburgh variant results in very high levels of AT3activity. In addition, α_1-antitrypsin is an acute-phase reactant and is markedly elevated during inflammatory challenges. The clinical presentation is a severe bleeding disorder that occurs with infectious illness, due to the resulting induction of α_1-antitrypsin and associated pathologic AT3 activity. The original patient died as a result of such a fatal hemorrhagic episode [148]. The second patient enjoyed a considerably milder clinical course, possibly due to a counterbalancing inhibition of protein C [149]. Although this variant has not been transmitted, inheritance would be assumed to be a complete autosomal dominant, given the unique "gain-of-function" effect. Few human genetic diseases so clearly and elegantly demonstrate this important molecular mechanism.

7.3 INHERITED DISORDERS PREDISPOSING TO THROMBOSIS

Pathologic thrombosis is a major cause of morbidity and mortality in the developed world. In addition, inherited abnormalities in coagulation balance may contribute to the development of atherosclerosis, leading to stroke and myocardial infarction.

Venous thrombosis, most commonly in the lower extremities, affects approximately 1–2 in 1000 individuals and is responsible for up to 100,000 deaths annually (cdc.gov). The occurrence of venous thrombosis in patients under the age of 45, recurrent unexplained thromboses, and positive family history, are all suggestive of an inherited predisposition to thrombosis or "`thrombophilia." Indeed, a positive family history can be elicited in ~40% of young patients [150,151]. Five well-defined genetic risk factors attributable to coagulation protein loss or gain of function have been associated with inherited thrombophilia: AT3 deficiency, protein C deficiency, protein S deficiency, the prothrombin G20210A variant, and factor V Leiden [152]. The latter variant is particularly common and is found in 20%–50% of patients with spontaneous venous thrombosis. However, no genetic susceptibility can be identified in greater than 50% of patients, and it is likely that a number of additional genetic risk factors remain to be discovered.

7.3.1 Antithrombin III Deficiency

AT3 is a member of the serpin gene family and a potent inhibitor of several serine proteases in the coagulation cascade including thrombin, factors IX, X, XI, and XII [1,2]. The anticoagulant heparin acts through potentiation of antithrombin III activity against FIX, FX, and thrombin. Hemostatic balance is very sensitive to variations in AT3. Levels in the range of 50% of normal are associated with a significantly increased risk of thrombosis, explaining the autosomal dominant inheritance. Although AT3 deficiency is generally defined as a level of <50% of normal, significant overlap can be observed between genetically deficient patients and normal individuals. The prevalence of AT3 deficiency is estimated at approximately one in 500–5000 [150,153,154].

7.3.2 Clinical Features

AT3 deficiency is generally subdivided into two types. Type I, or classic, AT3 deficiency is due to an absolute quantitative deficiency of the protein with corresponding decreases in antigen levels as well as functional activity. Type II deficiency is associated with normal levels of antigen but decreased functional activity, generally due to missense variants within specific important functional domains such as the heparin binding domain. Penetrance of thrombosis is incomplete. Although thrombosis can be spontaneous, it is often associated with other known risk factors such as pregnancy, contraceptive use, surgery, or trauma. Recurrent thrombotic episodes are common, as is progression to pulmonary embolism. The typical age of onset of thrombosis is in the third to fourth decade of life. Although presentation in childhood and infancy has been reported, thrombosis before puberty is unusual.

A variety of acquired disorders can also result in decreased AT3 levels and should be distinguished from the inherited disorder. These include acute thromboses, chronic liver disease, nephrotic syndrome, and disseminated intravascular coagulation (DIC). Estrogens can also mildly reduce AT3 levels. Acute treatment with heparin also results in decreased AT3, presumably due to accelerated clearance. Thus, patients presenting with thrombosis should not undergo laboratory evaluation while on heparin. Because of these confounding factors, an abnormal AT3 level should be confirmed with at least one additional measurement before establishing the diagnosis. Evaluation of other family members can also be helpful in supporting the diagnosis. The degree of thrombosis is often milder in patients with type II defects, presumably due to residual activity. Rare individuals with homozygous type II AT3 deficiency have been reported [150,153], but homozygous type I is thought to be incompatible with life.

As noted above, the penetrance of thrombosis in heterozygous AT3 deficiency is incomplete. A review of published cases in the literature suggests penetrance of approximately 50%, although this figure may still represent an overestimate, due to greater ascertainment of more highly penetrant families. In contrast, careful analysis of a single large pedigree suggested that an unbiased penetrance figure is probably closer to 20% [155]. This issue is of major importance in deciding whether to treat asymptomatic patients. Although the risk of thrombosis in asymptomatic AT3 carriers is probably greater than that in protein C, protein S, and factor V Leiden (discussed below), the risk still is probably less than the major complication rate of long-term anticoagulation with warfarin (which may be as high as 1% per year). Most experts do not recommend routine long-term anticoagulation of asymptomatic carriers [150,153,156–158]. One study suggests a risk for thrombosis of approximately 1.5% per year in AT3 carriers [156], 10–20 fold higher than the corresponding risk in normal individuals [157]. However, overall, heterozygous AT3 deficiency is associated with a normal survival and a low risk of fatal thromboembolic events [158]. For this reason, anticoagulant prophylaxis is generally reserved only for individuals with additional significant risk factors or in particularly high-risk situations.

Although acute thromboses may be treated with thrombolytic therapy using guidelines similar to those for other cases of thrombosis, treatment is generally with standard or low-molecular-weight heparin [159]. Difficulty in achieving adequate heparinization, or heparin resistance, may occur as a result of decreased AT3 and may initially suggest the diagnosis.

Standard long-term management of symptomatic AT3 deficiency generally relies on anticoagulation with warfarin or a direct oral anticoagulant (DOAC), small molecules that inhibit either thrombin or FXa at fixed doses without a need for routine laboratory monitoring [160], and exhibit fewer drug interactions than warfarin. Asymptomatic patients are generally not treated with anticoagulants, given the significant long-term hemorrhagic risk that is greater than the penetrance of thrombosis, except at times of risk, e.g., surgery or immobilization.

7.3.3 Genetics

A single copy of the *AT3* gene is located on human chromosome 1. A large number of molecular defects scattered throughout the gene have been identified [24] including deletion, frameshift, and nonsense variants all resulting in type I deficiency, as well as missense variants leading to type II. Structure/function studies of type II variants have often demonstrated the molecular mechanism to be a specific defect in heparin binding or disruption of the active site, interfering with protease inhibition.

Although DNA diagnosis can be easily obtained through commercial clinical testing laboratories, this is not routinely performed, given the large number of different variants that have been identified and the lack of clinical utility given the overall low penetrance. Although prenatal diagnosis by linkage or direct variant detection is possible, it is generally not indicated, given the usually mild nature of this disorder.

7.3.4 Protein C Deficiency

Griffin and coworkers [161] first identified protein C deficiency in 1981 in a family with recurrent thrombosis. Affected family members exhibited protein C antigen levels of <50% of normal. A large number of protein C-deficient patients have since been reported and their defects biochemically characterized.

Protein C is a protease zymogen that is activated when cleaved by thrombin bound to thrombomodulin on the endothelial cell surface. Activated protein C (APC) proteolytically inactivates factors Va and VIIIa on the platelet surface, thus downregulating thrombin generation by turning off the prothrombinase and X-ase complexes. In this way, thrombin autoregulates its own activity by activating the protein C anticoagulant pathway (Fig. 7.3). The presence of thrombomodulin on the surface of vascular endothelial cells may serve to ensure vascular patency through this mechanism. Protein C deficiency removes this important balance, resulting in hypercoagulability.

7.3.5 Clinical Features

Protein C deficiency is an autosomal dominant disorder with prevalence estimated at ~1:500 [150,153,162,163]. Estimates of the penetrance of protein C deficiency vary from 20% to 75%. In addition, the same variant may vary widely in its expression in different families. Heterozygous deficiency results in protein C levels in the range of ~50% of control, although there is considerable overlap with the lower portion of the normal range [163]. Clinical presentation is similar to AT3 deficiency, with venous thrombosis usually presenting in the third or fourth decade, and rarely before puberty. A number of cases of homozygous protein C deficiency have been reported, associated with severe neonatal purpura fulminans, which is fatal if not treated promptly [150,153].

7.3.6 Genetics

A large number of pathogenic variants in the protein C gene on chromosome 2 have been reported [164]. Like AT3 deficiency, heterozygous protein C deficiency is divided into type I, or classic, associated with both low protein C antigen as well as functional activity, and type II, associated with normal antigen and decreased functional activity. The diagnosis of protein C deficiency is generally based on functional and antigen tests, rather than DNA diagnosis, given the large number of variants that have been identified and the lack of clinical utility due to incomplete penetrance (as described for AT3). Most type I patients have nonsense or frameshift variants or, in some cases, missense variants.

The wide range in the degree of symptoms, even among families segregating the same variant, may be partially explained by cosegregation of another prothrombotic mutation (see Factor V Leiden below). Protein C is a vitamin-K-dependent protease that requires posttranslational modification by γ-carboxylation at its N-terminus for full functional activity. Thus, warfarin treatment reduces functional activity, and to a lesser extent immunologic activity, making it difficult to diagnosis this disorder during warfarin therapy.

7.3.7 Management

Anticoagulant management of protein C deficiency is similar to that described above for AT3 deficiency [150,153]. However, it is important to pretreat for sufficient time with full heparinization before beginning warfarin anticoagulation and to avoid large loading doses of warfarin. Abrupt initiation of treatment with a vitamin K antagonist has been associated with a dramatic thrombotic complication in the skin, referred to as "warfarin-induced skin necrosis." This problem is thought to result from an imbalance between the anticoagulant effect of warfarin and the earlier, more rapid, inhibition of protein C. The latter effect results in an initial worsening of the relative protein C deficiency and hypercoagulability until sufficient inhibition of the coagulation factors II, VII, IX, and X has occurred, producing a net anticoagulant effect. If this complication should develop, treatment with FFP or protein C concentrate is indicated.

7.3.8 Protein S Deficiency

Protein S is a cofactor for the anticoagulant activity of APC (Fig. 7.3). Thrombophilia due to heterozygous deficiency appears to be similar in frequency to protein C deficiency, and severe thrombosis has been reported in a few patients with apparent homozygous deficiency [150]. Approximately 60% of protein S antigen in plasma is in a complex with complement C4B-binding protein, but only the free fraction is functionally active as a cofactor. This interaction complicates the interpretation of protein S levels, as most screening tests measure only total protein S antigen. Type I deficiency is associated with a decrease in both total and free protein S antigen. Type II deficiency is characterized by normal free protein S antigen, but low protein S activity (APC cofactor activity). Type III is associated with a selective decrease in free protein S. The protein S gene and a homologous pseudogene are both located on human chromosome 3. Although a number of specific variants leading to protein S deficiency have been identified, and DNA testing can be obtained commercially, diagnosis does not usually impact clinical management due to incomplete penetrance. A study of a large cohort of well-established protein S-deficient patients and first degree relatives demonstrated that free protein S level is the most reliable screening test for protein S deficiency [165].

Clinical features of heterozygous protein S deficiency are similar to those of AT3 and protein C deficiency, with onset typically in the third or fourth decade. Acquired abnormalities in protein S can be associated with oral contraceptives, pregnancy, DIC, and acute thrombosis. In addition, C4B-binding protein is an acute phase reactant and elevation in inflammatory states can be associated with a decline in free protein S activity. It is also important to note that protein S measurements are uninterpretable in patients on warfarin, since protein S also requires γ-carboxylation. Long-term treatment is similar to that described above for AT3 and protein C deficiency.

7.3.9 Factor V Leiden

Until the mid-1990s, deficiencies of AT3, protein C, and protein S were the only known thrombophilic risk factors. However, measurement of these factors generally detected an abnormality in only 5%–10% of patients admitted with unexplained thrombosis [166]. Dahlbäck and colleagues first reported a poor anticoagulant response to APC in three families with familial thrombophilia and developed an assay to detect "APC resistance" [167,168]. Affected individuals and their families showed resistance to this effect of APC. Inheritance of this APC resistance appeared to be autosomal dominant.

The responsible pathogenic variant was subsequently identified as a G- > A substitution at nucleotide position 1691 of FV [169], resulting in the substitution of glutamine for arginine 506 (R506Q), a predicted protein C cleavage site (Fig. 7.9). The R506Q variant, commonly denoted as factor V Leiden, is found in 85%–100% of individuals with APC resistance, depending on the specific assay used. APC resistance in the absence of factor V Leiden also appears to confer an increased risk for thrombosis, due in some cases to a lupus anticoagulant or other acquired disorder, though the cause often remains unidentified. Variants at a second APC cleavage site within Factor V (R306) have also been reported, including factor V Hong Kong (R306G) [170] and factor V Cambridge (R306T) [171]. The R306G variant appears to have a prevalence of ~4% in parts of Asia [170]. Analysis of recombinant factor VIII indicates that mutations at the homologous APC cleavage sites in factor VIII do not produce APC resistance in the standard

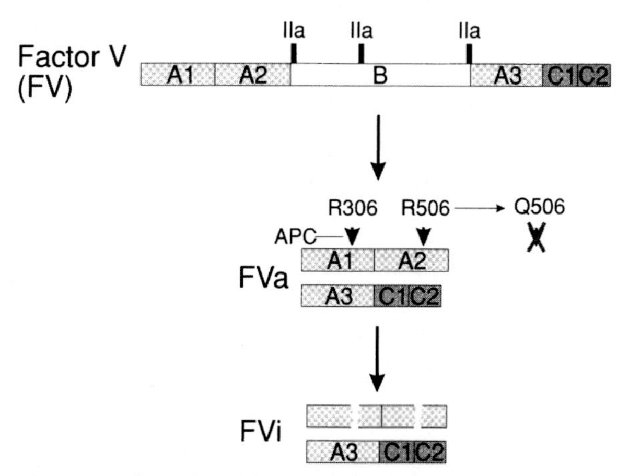

Figure 7.9 Domain structure and processing of factor V. Thrombin cleavage sites (IIa) are indicated. FVa, the active form of factor V; FVi, factor V which has been inactivated through cleavage by activated protein C (APC). The APC cleavage sites at Arg306 (R306) and Arg506 (R506) are indicated. The common factor V Leiden variant, substitution of glutamine for Arg506 (Q506), results in APC resistance.

clinical assay [172], and no such patients have been reported.

Factor V Leiden has been identified in diverse populations, although it is particularly prevalent in Europe. Allele frequency among Europeans ranges from 2% to 7%, with the highest frequency observed in Greece. Haplotype analysis suggests a single founder for this variant, who was estimated to have lived 21,000–34,000 years ago [173]. This variant appears to be quite uncommon in African and Asian populations [174].

Individuals homozygous for factor V Leiden have a high risk of thrombosis, with lifetime penetrance estimated at ~80%, compared to 10% in heterozygotes [150,153,175]. Oral contraceptives and hormone replacement therapy both confer an increased risk of thrombosis in factor V Leiden patients [150,176], though there are insufficient data to support routine genetic screening prior to initiating treatment [177]. Together with the prothrombin G2021A variant (see below), factor V Leiden is a major risk factor for thromboembolic complications during pregnancy and was observed among 44% of such patients in one study [178]. The presence of both variants resulted in a marked increase in the risk of thrombosis. Overall, the risk of thrombosis during pregnancy among factor V Leiden carriers was estimated at 0.2%, 0.5% for prothrombin G20210A, and 4.6% among patients with both variants. However, no clinical benefit has yet been demonstrated for routine variant screening [179].

Given the remarkably high prevalence of factor V Leiden, its relatively low penetrance for thrombosis, and lack of a significant effect on life expectancy [180] coupled with the significant morbidity of long-term anticoagulation with currently available drugs, no specific treatment can be recommended at this time for the asymptomatic patient, even those homozygous for the variant. Patients who develop symptomatic thrombosis are currently treated using the same guidelines applied to the general patient population [181–183]. Indeed, current evidence suggests that the risk of recurrent deep venous thrombosis is similar among carriers of factor V Leiden and patients without this variant, as well as in patients with the prothrombin G20210A, or even patients with both FV Leiden and prothrombin G20210A [184]. Some experts recommend more aggressive anticoagulation of patients carrying more than one prothrombotic mutation or in patients also exposed to another risk factor such as pregnancy [185,186]. However, this approach is controversial [181]. It is likely that the approach to treatment will be refined over the coming years with the identification of additional high-risk groups and further definition of interactions between genetic and environmental risk factors.

In addition to the APC resistance assay, direct DNA testing for the factor V Leiden variant is clinically available. Given the lack of any specific changes in therapy, routine screening is not currently recommended (2011). The remarkably high prevalence of factor V Leiden in European populations and its origin from a single founder variant argue strongly against a significantly adverse effect on life expectancy and also suggest the existence of a positive evolutionary selective advantage for factor V Leiden carriers. A significant survival advantage for factor V Leiden patients in the setting of severe sepsis has been reported [187], suggesting that an infectious pathogen may provide balancing selection accounting for the exceptionally high prevalence of this polymorphism in some human populations. A reduced risk of intrapartum bleeding complications in factor V Leiden patients might also confer a survival advantage [188].

7.3.10 The Prothrombin G20210A Variant

Analysis of the prothrombin gene as a candidate for venous thrombosis predisposition identified a common variant in the 3' untranslated region (G20210A) with an allele frequency of ~1% in European populations. This variant appears to be associated with elevated prothrombin levels, although elevated prothrombin level itself has not been found to be a risk factor for venous thrombosis [189]. The G20210A variant is at the 3′ terminus of the 3′ untranslated region and has been reported to increase the efficiency of prothrombin mRNA 3′ end processing [190], though subsequent studies suggest other mechanisms for mRNA accumulation, including linkage disequilibrium to another nearby SNP altering splicing efficiency [191]. The G20210A variant leads to an approximately 1.6-fold increased risk of thrombosis. As in factor V Leiden, the thrombotic risk seems to be primarily restricted to venous thrombosis.

7.3.11 Other Miscellaneous Prothrombotic Disorders

A number of other conditions have been associated with thrombosis but are not generally considered for routine evaluation. These include rare qualitative abnormalities in fibrin function referred to as dysfibrinogenemias (see above). A high level of factor XI has been reported as a risk factor for thrombosis, with doubling of the risk at levels present in the upper 10% of the population. Although the genetic basis for this variation in factor XI remains unknown, if confirmed, this additional factor could be an important genetic contribution to thrombophilia [192]. A similar association of elevated factor IX levels with thrombosis risk has also been reported [193], and a novel FIX variant(FIX Padua, R338L) was discovered in a patient with thrombosis who had an eightfold increase in FIX activity [194]. In several studies, plasma factor VIII level has been found to be an important determinant of thrombosis risk. In one analysis, elevated factor VIII was calculated to account for 16% of the population-wide attributable risk for thrombosis, compared to 25% for factor V Leiden, 4% for prothrombin 20,210A, 2% for protein C deficiency, and <1% for protein S and antithrombin III deficiency [195]. Family studies suggest that elevated factor VIII levels are at least in part genetically determined [196,197] and do not appear to be linked to the factor VIII locus itself.

Mild homocysteinemia may also be an independent risk factor for venous thrombosis [198], though this hypothesis is controversial since homocysteine lowering therapy has not been effective for prevention of recurrent thrombosis [199,200]. A very common polymorphism in the MTHFR gene (C677T), leading to a thermolabile enzyme and possible increased homocysteine levels, exhibits an allele frequency as high as 38% in some populations [201]. Several older single gene association studies suggested that this is variant is a risk factor for thrombosis. However, a large population-based case-control study of over 4000 subjects found no evidence of even a small risk [202], and multiple large GWAS did not detect a signal either, ruling out a significant contribution from this variant [84–86].

7.3.12 Plasminogen Deficiency

Inherited defects of fibrinolysis are a rare cause of familial thrombosis. Although a number of cases of partial plasminogen deficiency have been reported, the association with thrombosis is controversial [142]. Plasminogen deficiency has been classified as type I deficiency, with reduced antigen and functional activity and type II deficiency in which antigen is normal, but functional activity is proportionately decreased. Type II deficiency due to an Ala600Thr variant is very common in the Japanese population with an allele frequency of ~1%. This variant was initially identified in a patient with thrombosis [203], but other family members were unaffected. Although a number of other missense variants within the plasminogen gene have been identified, partial deficiency of plasminogen does not appear to be a significant risk factor for thrombosis, with similar prevalence observed among thrombosis patients and controls [204,205].

Homozygosity for null variants in the plasminogen gene has been shown to be the molecular basis for a rare ophthalmologic disorder, ligneous conjunctivitis [206] (reviewed in Ref. [142]. This disorder usually presents first in early infancy as a chronic, pseudomembranous conjunctivitis, although symptoms can be delayed into adulthood. The membranes that form over the eyes have a woody consistency and may follow minor trauma or infection (Fig. 7.10). These membranes can occur on other mucosal surfaces in the oropharynx, trachea, or female genital tract. Inheritance is autosomal recessive. Treatment of one of the original patients, who was homozygous for a nonsense variant (Glu460Ter), with a plasminogen concentrate prepared from human plasma produced a dramatic clinical response [207]. Remarkably, there have been no reports of thromboembolic complications in ligneous conjunctivitis patients or their families, suggesting that the primary role of plasminogen in humans is in the clearance of extravascular fibrin.

No genetic deficiencies of tPA or uPA have been reported in humans. Several families have been described with reduced fibrinolytic potential that could be due to either over production of PAI-1 or deficiency in tPA [150]. Although laboratory tests for plasminogen are available, routine screening in families with thrombosis is not generally indicated.

7.3.13 Thrombotic Thrombocytopenic Purpura (TTP)

TTP is a catastrophic, multisystem disorder characterized by the formation of platelet and VWF-rich

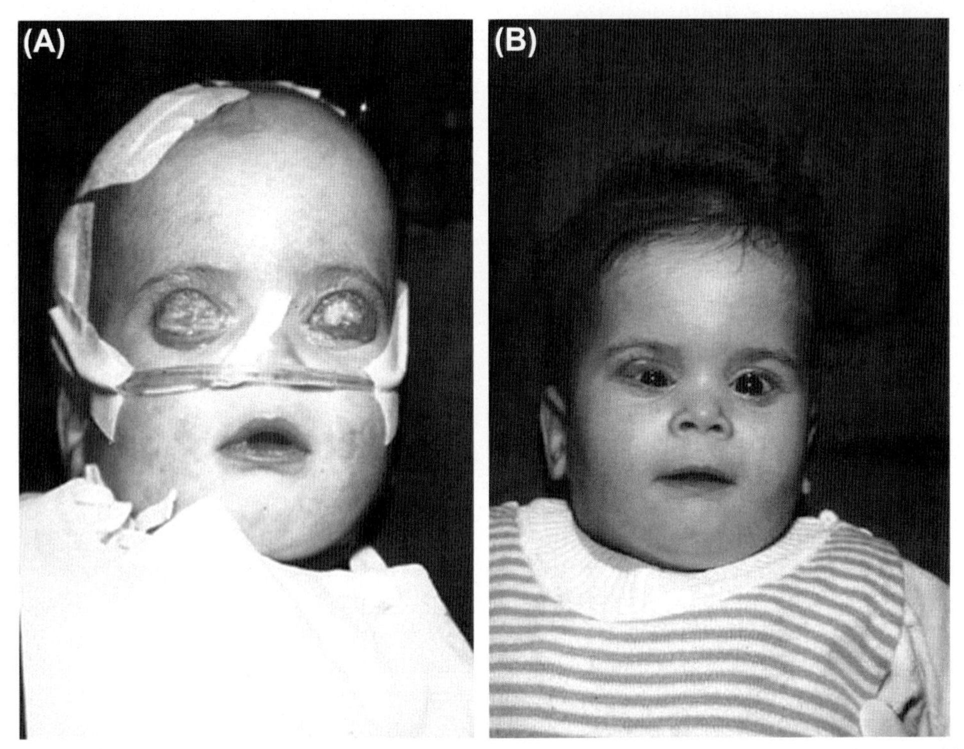

Figure 7.10 Ligneous conjunctivitis. (A) The initial appearance of a child with ligneous conjunctivitis resulting from homozygous plasminogen deficiency. (B) The same child after 7 months of replacement with Lys-plasminogen. (Adapted from Ref. Schott D, Dempfle CE, Beck P, Liermann A, Mohr-Pennert A, Goldner M, Mehlem P, Azuma H, Schuster V, Mingers AM, Schwarz HP, Kramer MD. Therapy with a purified plasminogen concentrate in an infant with ligneous conjunctivitis and homozygous plasminogen deficiency. N Engl J Med 1998;339:1679–1686 with permission.)

microthrombi in vessels of multiple organs, leading to the classic pentad of microangiopathic hemolytic anemia, thrombocytopenia, fever, and varying degrees of renal and neurological dysfunction [208,209]. Not all of these manifestations are present in all patients, although thrombocytopenia and hemolytic anemia are nearly universal. Typical age of onset is in the second or third decade, although this disorder can be seen at any age, and there is usually no clear precipitating event. Untreated, mortality is >90%, though with plasma exchange this figure falls to 10%–20% [209,210]. Most adult cases are sporadic and associated with the presence of acquired autoantibodies against a plasma metalloprotease that cleaves VWF [211,212]. Absence of this VWF-cleaving protease results in accumulation of unusually large multimeric forms of VWF, which are thought to trigger the pathologic platelet thrombi responsible for TTP [208,209]. A rare familial form of TTP, also referred to as Upshaw–Schulman syndrome, typically presents at birth or in early childhood and is usually highly responsive to the infusion of normal plasma [209]. The gene responsible for this rare autosomal recessive disorder was identified by positional cloning as the metalloprotease, ADAMTS13 [213]. It was independently identified at the same time by others as the VWF-cleaving protease previously associated with the sporadic form [214–216]. Pathogenic or likely pathogenic variants scattered throughout the ADAMTS13 gene have now been identified and appear to account for all cases of familial TTP, with no evidence to date for locus heterogeneity [208,213]. Patients with the familial form of TTP can generally be managed with simple plasma infusion, which presumably replaces the missing enzyme. Plasma exchange in adult patients with acquired disease may serve to partially remove autoantibodies, in addition to protein replacement [209,210]. Caplacizumab, an anti-von Willebrand factor humanized bivalent single domain nanobody, was recently approved for acquired TTP and has demonstrated efficacy in clinical trials [39].

7.4 INTERACTIONS AMONG MULTIPLE GENETIC DEFECTS

The discovery of the very common factor V Leiden mutation has also shed light on the penetrance of other less common familial hypercoagulable states. The risk of thrombosis varies considerably among protein C-deficient families. In some families, inheritance appears to be recessive and in others, dominant. In some cases, the same variant has been found in families of both types. Koeleman and coworkers [217] assessed the effect of factor V Leiden on the penetrance of protein C deficiency-related thrombosis. They observed a 19% prevalence of the factor V Leiden variant among symptomatic protein C-deficient patients, compared to a 2%–4% incidence in the control population. In six large pedigrees, 73% of patients inheriting both gene defects experienced thrombosis, compared to 31% with only protein C deficiency, and 13% with only factor V Leiden. Similar interaction of the factor V Leiden variant with protein S deficiency has also been noted [218], as has an increased risk of thrombosis associated with coinheritance of factor V Leiden and the prothrombin G20210 variant (see above).

Thus, venous thrombosis, like many other common, complex disorders, can be viewed as the product of interactions between a number of genetic and environmental factors. Factor V Leiden, the prothrombin 22,010 mutation, and likely a number of other common, yet to be identified genetic factors combine with environmental factors such as pregnancy, immobility, and underlying malignancy to determine the overall thrombosis risk for each individual. The strong foundation already in place for the basic biochemistry and physiology of blood coagulation should greatly facilitate deciphering the interactions between these multiple factors, eventually leading to accurate predictions of risk and individually tailored therapy. In this way, the treatment of thrombosis may serve as a valuable paradigm for other complex genetic disorders.

REFERENCES

[1] Colman RW, Clowes AW, George JN, Hirsh J, Marder VJ. Overview of hemostasis. In: Colman RW, Hirsh J, Marder VJ, Clowes AW, George JN, editors. Hemostasis and thrombosis basic principles and clinical practice, Vol. 4th. Philadelphia: Lippincott Williams & Wilkins; 2001. p. 3–20.

[2] Jenny NS, Mann KG. Coagulation cascade: an overview. In: Loscalzo J, Schafer AI, editors. Thrombosis and hemorrhage, vol. 3. Philadelphia: Lippincott Williams & Wilkins; 2003. p. 1–21.

[3] Ratnoff OD. Evolution of knowledge about hemostasis. In: Ratnoff OD, Forbes CD, editors. Disorders of hemostasis, Vol. Second. Philadelphia: W. B. Saunders Company; 1991. p. 1–17.

[4] Gailani D, Broze Jr GJ. Factor XI activation in a revised model of blood coagulation. Science 1991;253:909–12.

[5] Naito K, Fujikawa K. Activation of human blood coagulation factor XI independent of factor XII. Factor XI is activated by thrombin and factor XIa in the presence of negatively charged surfaces. J Biol Chem 1991;266:7353–8.

[6] Hoffman M, Monroe 3rd DM. A cell-based model of hemostasis. Thromb Haemostasis 2001;85:958–65.

[7] Mannucci PM, Tuddenham EGD. Medical progress - the hemophilias - from royal genes to gene therapy. N Engl J Med 2001;344:1773–9.

[8] Ingram GI. The history of haemophilia. J Clin Pathol 1976;29:469–79.

[9] Furie B, Limentani SA, Rosenfield CG. A practical guide to the evaluation and treatment of hemophilia. Blood 1994;84:3–9.

[10] Hoyer LW. Hemophilia A. N Engl J Med 1994;330:38–47.

[11] Stevens RF. The history of haemophilia in the royal families of Europe. Br J Haematol 1999;105:25–32.

[12] Rogaev EI, Grigorenko AP, Faskhutdinova G, Kittler EL, Moliaka YK. Genotype analysis identifies the cause of the "royal disease". Science 2009;326:817.

[13] Jones PK, Ratnoff OD. The changing prognosis of classic hemophilia (factor VIII "deficiency"). Am Intercult Mag 1991;114:641–8.

[14] Gitschier J, Wood WI, Goralka TM, Wion KL, Chen EY, Eaton DH, Vehar GA, Capon DJ, Lawn RM. Characterization of the human factor VIII gene. Nature 1984;312:326–30.

[15] Kaufman RJ. Biological regulation of factor VIII activity. ARM 1992;43:325–39.

[16] Toole JJ, Knopf JL, Wozney JM, Sultzman LA, Buecker JL, Pittman DD, Kaufman RJ, Brown E, Shoemaker C, Orr EC, Amphlett GW, Foster WB, Coe ML, Knutson GJ, Fass DN, Hewick RM. Molecular cloning of a cDNA encoding human antihaemophilic factor. Nature 1984;312:342–7.
[17] Vehar GA, Keyt B, Eaton D, Rodriguez H, O'Brien DP, Rotblat F, Oppermann H, Keck R, Wood WI, Harkins RN, Tuddenham EGD, Lawn RM, Capon DJ. Structure of human factor VIII. Nature 1984;312:337–42.
[18] Nicolaes GAF, DahlbÖck B. Factor V and thrombotic disease - description of a janus-faced protein. Arterioscler Thromb Vasc Biol 2002;22:530–8.
[19] Eaton DL, Wood WI, Eaton D, Hass PE, Hollingshead P, Wion K, Mather J, Lawn RM, Vehar GA, Gorman C. Construction and characterization of an active Factor VIII variant lacking the central one-third of the molecule. Biochemistry 1986;25:8343–7.
[20] Pittman DD, Marquette KA, Kaufman RJ. Role of the B-domain for factor VIII and factor V expression and function. Blood 1994;84:4214–25.
[21] Toole JJ, Pittman DD, Orr EC, Murtha P, Wasley LC, Kaufman RJ. A large region (~95 kDa) of human factor VIII is dispensable for in vitro procoagulant activity. Proc Natl Acad Sci U S A 1986;83:5939–42.
[22] Antonarakis SE, Kazazian Jr HH, Tuddenham EGD. Molecular etiology of factor VIII deficiency in Hemophilia A. Hum Mutat 1995;5:1–22.
[23] Kemball-Cook G, Tuddenham EGD. The factor VIII mutation database on the world wide web: the haemophilia A mutation, search, test and resource site. HAMSTeRS update (version 3.0). Nucleic Acids Res 1997;25:128–32.
[24] Landrum MJ, Lee JM, Benson M, Brown GR, Chao C, Chitipiralla S, Gu B, Hart J, Hoffman D, Jang W, Karapetyan K, Katz K, Liu C, Maddipatla Z, Malheiro A, McDaniel K, Ovetsky M, Riley G, Zhou G, Holmes JB, Kattman BL, Maglott DR. ClinVar: improving access to variant interpretations and supporting evidence. Nucleic Acids Res 2018;46. PubMed PMID: 29165669.
[25] Lakich D, Kazazian Jr HH, Antonarakis SE, Gitschier J. Inversions disrupting the factor VIII gene are a common cause of severe haemophilia A. Nat Genet 1993;5:236–41.
[26] Antonarakis SE, Rossiter JP, Young M, Horst J, de Moerloose P, Sommer SS, Ketterling RP, Kazazian Jr HH, N, grier C, Vinciguerra C, Gitschier J, Goossens M, Girodon E, Ghanem N, Plassa F, Lavergne JM, Vidaud M, Costa JM, Laurian Y, Lin SW, Lin SR, Shen MC, Lillicrap D, Taylor SAM. Factor VIII gene inversions in severe hemophilia A: results of an international consortium study. Blood 1995;86:2206–12.
[27] Bagnall RD, Waseem N, Green PM, Giannelli F. Recurrent inversion breaking intron 1 of the factor VIII gene is a frequent cause of severe hemophilia A. Blood 2002;99:168–74.
[28] Oldenburg J, Schroder J, Hermann BH, Muller-Reible C, Schwaab R, Tuddenham E. Environmental and genetic factors influencing inhibitor development. Semin Hematol 2004;41:82–8.
[29] Gouw SC, van den Berg HM. The multifactorial etiology of inhibitor development in hemophilia: genetics and environment. Semin Thromb Hemost 2009;35:723–34.
[30] Viel KR, Ameri A, Abshire TC, Iyer RV, Watts RG, Lutcher C, Channell C, Cole SA, Fernstrom KM, Nakaya S, Kasper CK, Thompson AR, Almasy L, Howard TE. Inhibitors of factor VIII in black patients with hemophilia. N Engl J Med 2009;360:1618–27.
[31] Broze Jr GJ. Protein Z-dependent regulation of coagulation. Thromb Haemostasis 2001;86:8–13.
[32] Briet E, Bertina RM, van Tilburg NH, Veltkamp JJ. Hemophilia B Leyden: a sex-linked hereditary disorder that improves after puberty. N Engl J Med 1982;306:788–90.
[33] Crossley M, Ludwig M, Stowell KM, De Vos P, Olek K, Brownlee GG. Recovery from hemophilia B Leyden: an androgen-responsive element in the factor IX promoter. Science 1992;257:377–9.
[34] Kurachi S, Deyashiki Y, Takeshita J, Kurachi K. Genetic mechanisms of age regulation of human blood coagulation factor IX. Science 1999;285:739–43.
[35] Kurachi S, Huo JS, Ameri A, Zhang K, Yoshizawa AC, Kurachi K. An age-related homeostasis mechanism is essential for spontaneous amelioration of hemophilia B Leyden. Proc Natl Acad Sci U S A 2009;106(19):7921–6. https://doi.org/10.1073/pnas.0902191106. Epub 2009 Apr 28. PMID: 19416882; PMCID: PMC2674395.
[36] Koeberl DD, Bottema CDK, Ketterling RP, Bridge PJ, Lillicrap DP, Sommer SS. Mutations causing hemophilia B: direct estimate of the underlying rates of spontaneous germ-line transitions, transversions, and deletions in a human gene. Am J Hum Gen 1990;47:202–17.
[37] Brownlee GG, Rizza C. Clotting factor VIII cloned. Nature 1984;312:307.
[38] Wood WI, Capon DJ, Simonsen DL, Eaton DL, Gitschier J, Keyt B, Seeburg PH, Smith DH, Hollingshead P, Wion KL, Delwart E, Tuddenham EGD, Vehar GA, Lawn RM. Expression of active human factor VIII from recombinant DNA clones. Nature 1984;312:330–7.

[39] Peyvandi F, Cannavò A, Garagiola I, Palla R, Mannucci PM, Rosendaal FR, sippet study group. Timing and severity of inhibitor development in recombinant versus plasma-derived factor VIII concentrates: a SIPPET analysis. J Thromb Haemost 2018;16(1):39–43. https://doi.org/10.1111/jth.13888. Epub 2017 Nov 16. PMID: 29080391.
[40] Weyand AC, Pipe SW. New therapies for hemophilia. Blood 2019;133(5):389–98. https://doi.org/10.1182/blood-2018-08-872291. Epub 2018 Dec 17. PMID: 30559264.
[41] Sankar AD, Weyand AC, Pipe SW. The evolution of recombinant factor replacement for hemophilia. Transfus Apher Sci 2019;58(5):596–600. https://doi.org/10.1016/j.transci.2019.08.010. Epub 2019 Aug 9. PMID: 31421983.
[42] Konkle BA, Shapiro AD, Quon DV, Staber JM, Kulkarni R, Ragni MV, Chhabra ES, Poloskey S, Rice K, Katragadda S, Fruebis J, Benson CC. BIVV001 fusion protein as factor VIII replacement therapy for hemophilia A. N Engl J Med 2020;383(11):1018–27. https://doi.org/10.1056/NEJMoa2002699. PMID: 32905674.
[43] Yee A, Gildersleeve RD, Gu S, Kretz CA, McGee BM, Carr KM, Pipe SW, Ginsburg D. A von Willebrand factor fragment containing the D'D3 domains is sufficient to stabilize coagulation factor VIII in mice. Blood 2014;124(3):445–52. https://doi.org/10.1182/blood-2013-11-540534. Epub 2014 May 21. PMID: 24850761; PMCID: PMC4102715.
[44] Zhao Y, Weyand AC, Shavit JA. Novel treatments for hemophilia through rebalancing of the coagulation cascade. Pediatr Blood Cancer 2021;68(5):e28934. https://doi.org/10.1002/pbc.28934. Epub 2021 Feb 12. PMID: 33577709.
[45] Nathwani AC, Reiss U, Tuddenham E, et al. Adeno-associated mediated gene transfer for hemophilia B: 8 year follow up and impact of removing "empty viral particles" on safety and efficacy of gene transfer. Blood 2018;132(Suppl. 1):491.
[46] George LA, Sullivan SK, Giermasz A, et al. Hemophilia B gene therapy with a high-specific-activity factor IX variant. N Engl J Med 2017;377(23):2215–27.
[47] Von Drygalski A, Giermasz A, Castaman G, et al. Phase 2b trial of AMT-061 (AAV5-Padua hFIX): translation into humans of an enhanced gene transfer vector for adults with severe or moderate-severe hemophilia B. Haemophilia 2019;25:30.
[48] Nathwani AC. Gene therapy for hemophilia. Hematol Am Soc Hematol Educ Prog 2019;2019(1):1–8. https://doi.org/10.1182/hematology.2019000007. PMID: 31808868; PMCID: PMC6913446.
[49] Batty P, Lillicrap D. Advances and challenges for hemophilia gene therapy. Hum Mol Genet 2019;28(R1):R95–101. https://doi.org/10.1093/hmg/ddz157. PMID: 31332444.
[50] Rossiter JP, Young M, Kimberland ML, Hutter P, Ketterling RP, Gitschier J, Horst J, Morris MA, Schaid DJ, de Moerloose P, Sommer SS, Kazazian Jr HH, Antonarakis SE. Factor VIII gene inversions causing severe hemophilia A originate almost exclusively in male germ cells. Hum Mol Genet 1994;3:1035–9.
[51] Leuer M, Oldenburg J, Lavergne JM, Ludwig M, Fregin A, Eigel A, Ljung R, Goodeve A, Peake IR, Olek K. Somatic mosaicism in hemophilia A: a fairly common event. Am J Hum Gen 2001;69:75–87.
[52] Mannucci PM. Treatment of von Willebrand's disease. N Engl J Med 2004;351:683–94.
[53] Ruggeri ZM, Ware J, Ginsburg D. von Willebrand factor. In: Loscalzo J, Schafer AI, editors. Thrombosis and hemorrhage, vol. 3. Philadelphia, PA: Lippincott Williams & Wilkins; 2003. p. 246–65.
[54] James P, Goodeve A. von Willebrand disease. Genet Med 2011;13:365–76. https://doi.org/10.1097/GIM.0b013e3182035931.
[55 James PD, Connell NT, Ameer B, Di Paola J, Eikenboom J, Giraud N, Haberichter S, Jacobs-Pratt V, Konkle B, McLintock C, McRae S, R Montgomery R, O'Donnell JS, Scappe N, Sidonio R, Flood VH, Husainat N, Kalot MA, Mustafa RA. ASH ISTH NHF WFH 2021 guidelines on the diagnosis of von Willebrand disease. Blood Adv 2021;5(1):280–300. https://doi.org/10.1182/bloodadvances.2020003265. PMID: 33570651; PMCID: PMC7805340.
[56] Nichols WC, Cooney KA, Ginsburg D, Ruggeri ZM. von Willebrand disease. In: Loscalzo J, Schafer AI, editors. Thrombosis and hemorrhage, vol. 3. Philadelphia: Lipincott Williams & Wilkins; 2003. p. 539–59.
[57] Flood VH, Gill JC, Morateck PA, Christopherson PA, Friedman KD, Haberichter SL, Hoffmann RG, Montgomery RR. Gain-of-function GPIb ELISA assay for VWF activity in the zimmerman program for the molecular and clinical biology of VWD. Blood 2011;117(6):e67–74. https://doi.org/10.1182/blood-2010-08-299016. Epub 2010 Dec 10. PMID: 21148813; PMCID: PMC3056647.
[58] Ng CJ, Di Paola J. von Willebrand disease: diagnostic strategies and treatment options. Pediatr Clin 2018;65(3):527–41. https://doi.org/10.1016/j.pcl.2018.02.004. PMID: 29803281.
[59] Berkowitz SD, Ruggeri ZM, Zimmerman TS. von Willebrand disease. In: Zimmerman TS, Ruggeri ZM,

editors. Coagulation and bleeding disorders. The role of factor VIII and von Willebrand factor. New York: Marcel Dekker, Inc; 1989. p. 215–59.

[60] Baronciani L, Peyvandi F. How we make an accurate diagnosis of von Willebrand disease. Thromb Res 2020 Dec;196:579–89. https://doi.org/10.1016/j.thromres.2019.07.010. Epub 2019 Jul 16. PMID: 31353031.

[61] Sadler JE, Budde U, Eikenboom JC, Favaloro EJ, Hill FG, Holmberg L, Ingerslev J, Lee CA, Lillicrap D, Mannucci PM, Mazurier C, Meyer D, Nichols WL, Nishino M, Peake IR, Rodeghiero F, Schneppenheim R, Ruggeri ZM, Srivastava A, Montgomery RR, Federici AB. Working Party on von Willebrand Disease Classification. Update on the pathophysiology and classification of von Willebrand disease: a report of the Subcommittee on von Willebrand Factor. J Thromb Haemost 2006;4(10):2103–14. https://doi.org/10.1111/j.1538-7836.2006.02146.x. Epub 2006 Aug 2. PMID: 16889557.

[62] James PD, Notley C, Hegadorn C, Poon MC, Walker I, Rapson D, Lillicrap D. Challenges in defining type 2M von Willebrand disease: results from a Canadian cohort study. J Thromb Haemostasis 2007;5:1914–22.

[63] Nichols WL, Hultin MB, James AH, Manco-Johnson MJ, Montgomery RR, Ortel TL, Rick ME, Sadler JE, Weinstein M, Yawn BP. von Willebrand disease (VWD): evidence-based diagnosis and management guidelines, the National Heart, Lung, and Blood Institute (NHLBI) Expert Panel report (USA). Haemophilia 2008;14:171–232.

[64] Rao ES, Ng CJ. Current approaches to diagnostic testing in von Willebrand Disease. Transfus Apher Sci 2018;57(4):463–5. https://doi.org/10.1016/j.transci.2018.07.005. Epub 2018 Jul 29. PMID: 30064913.

[65] James PD, Lillicrap D. The molecular characterization of von Willebrand disease: good in parts. Br J Haematol 2013;161(2):166–76. https://doi.org/10.1111/bjh.12249. Epub 2013 Feb 14. PMID: 23406206; PMCID: PMC3934371.

[66] Favaloro EJ, Pasalic L, Curnow J. Type 2M and Type 2A von Willebrand Disease: Similar but Different. Semin Thromb Hemost 2016;42(5):483–97. https://doi.org/10.1055/s-0036-1579641. Epub 2016 May 5. PMID: 27148841.

[67] Mazurier C, Meyer D. Factor VIII binding assay of von Willebrand factor and the diagnosis of type 2N von Willebrand disease - results of an international survey. On behalf of the Subcommittee on von Willebrand Factor of the Scientific and Standardization Committee of the ISTH. Thromb Haemostasis 1996;76:270–4.

[68] Sadler JE. Biochemistry and genetics of von Willebrand factor. Annu Rev Biochem 1998;67:395–424.

[69] Ginsburg D, Bowie EJW. Molecular genetics of von Willebrand disease. Blood 1992;79:2507–19.

[70] Mohlke KL, Nichols WC, Rehemtulla A, Kaufman RJ, FagerstrÖm HM, Ritvanen KLA, Kekom„ki R, Ginsburg D. A common frameshift mutation in von Willebrand factor does not alter mRNA stability but interferes with normal propeptide processing. Br J Haematol 1996;95:184–91.

[71] Zhang ZP, Blombäck M, Nyman D, Anvret M. Mutations of von Willebrand factor gene in families with von Willebrand disease in the Aland Islands. Proc Natl Acad Sci U S A 1993;90:7937–40.

[72] Cumming A, Grundy P, Keeney S, Lester W, Enayat S, Guilliatt A, Bowen D, Pasi J, Keeling D, Hill F, Bolton-Maggs PH, Hay C, Collins P. An investigation of the von Willebrand factor genotype in UK patients diagnosed to have type 1 von Willebrand disease. Thromb Haemostasis 2006;96:630–41.

[73] Goodeve A, Eikenboom J, Castaman G, Rodeghiero F, Federici AB, Batlle J, Meyer D, Mazurier C, Goudemand J, Schneppenheim R, Budde U, Ingerslev J, Habart D, Vorlova Z, Holmberg L, Lethagen S, Pasi J, Hill F, Hashemi Soteh M, Baronciani L, Hallden C, Guilliatt A, Lester W, Peake I. Phenotype and genotype of a cohort of families historically diagnosed with type 1 von Willebrand disease in the European study, Molecular and Clinical Markers for the Diagnosis and Management of Type 1 von Willebrand Disease (MCMDM-1VWD). Blood 2007;109:112–21.

[74] James PD, Notley C, Hegadorn C, Leggo J, Tuttle A, Tinlin S, Brown C, Andrews C, Labelle A, Chirinian Y, O'Brien L, Othman M, Rivard G, Rapson D, Hough C, Lillicrap D. The mutational spectrum of type 1 von Willebrand disease: results from a Canadian cohort study. Blood 2007;109:145–54.

[75] Goodeve A. Genetics of type 1 von Willebrand disease. Curr Opin Hematol 2007;14:444–9.

[76] Borras N, Batlle J, Perez-Rodriguez A, Lopez-Fernandez MF, Rodriguez-Trillo A, Loures E, et al. Molecular and clinical profile of von Willebrand disease in Spain (PCM-EVW-ES): comprehensive genetic analysis by next-generation sequencing of 480 patients. Haematologica 2017;102(12):2005–14.

[77] Lyons SE, Bruck ME, Bowie EJW, Ginsburg D. Impaired intracellular transport produced by a subset of type IIA von Willebrand disease mutations. J Biol Chem 1992;267:4424–30.

[78] Goodeve AC. The genetic basis of von Willebrand disease. Blood Rev 2010;24:123–34.

[79] Murray EW, Giles AR, Lillicrap D. Germ-line mosaicism for a valine-to-methionine substitution at residue 553 in the glycoprotein Ib-binding domain of von

Willebrand factor, causing Type IIB von Willebrand disease. Am J Hum Gen 1992;50:199–207.
[80] Badimon L, Badimon JJ, Chesebro JH, Fuster V. von Willebrand factor and cardiovascular disease. Thromb Haemostasis 1993;70:111–8.
[81] Guella I, Duga S, Ardissino D, Merlini PA, Peyvandi F, Mannucci PM, Asselta R. Common variants in the haemostatic gene pathway contribute to risk of early-onset myocardial infarction in the Italian population. Thromb Haemostasis 2011;106.
[82] Martinelli I. von Willebrand factor and factor VIII as risk factors for arterial and venous thrombosis. Semin Hematol 2005;42:49–55.
[83] Vischer UM. von Willebrand factor, endothelial dysfunction, and cardiovascular disease. J Thromb Haemostasis 2006;4:1186–93.
[84] Germain M, Chasman DI, de Haan H, Tang W, Lindström S, Weng LC, de Andrade M, de Visser MC, Wiggins KL, Suchon P, Saut N, Smadja DM, Le Gal G, van Hylckama Vlieg A, Di Narzo A, Hao K, Nelson CP, Rocanin-Arjo A, Folkersen L, Monajemi R, Rose LM, Brody JA, Slagboom E, Aïssi D, Gagnon F, Deleuze JF, Deloukas P, Tzourio C, Dartigues JF, Berr C, Taylor KD, Civelek M, Eriksson P, Cardiogenics Consortium, Psaty BM, Houwing-Duitermaat J, Goodall AH, Cambien F, Kraft P, Amouyel P, Samani NJ, Basu S, Ridker PM, Rosendaal FR, Kabrhel C, Folsom AR, Heit J, Reitsma PH, Trégouët DA, Smith NL, Morange PE. Meta-analysis of 65,734 individuals identifies TSPAN15 and SLC44A2 as two susceptibility loci for venous thromboembolism. Am J Hum Genet 2015;96(4):532–42. https://doi.org/10.1016/j.ajhg.2015.01.019. Epub 2015 Mar 12. PMID: 25772935; PMCID: PMC4385184.
[85] Herrera-Rivero M, Stoll M, Hegenbarth JC, Rühle F, Limperger V, Junker R, Franke A, Hoffmann P, Shneyder M, Stach M, Nowak-Göttl U. Single- and multimarker genome-wide scans evidence novel genetic risk modifiers for venous thromboembolism. Thromb Haemostasis 2021. https://doi.org/10.1055/s-0041-1723988. Epub ahead of print. PMID: 33592630.
[86] Klarin D, Emdin CA, Natarajan P, Conrad MF, INVENT Consortium, Kathiresan S. Genetic analysis of venous thromboembolism in UK biobank identifies the ZFPM2 locus and implicates obesity as a causal risk factor. Circ Cardiovasc Genet 2017;10(2):e001643. https://doi.org/10.1161/CIRCGENETICS.116.001643. PMID: 28373160; PMCID: PMC5395047.
[87] Miller CH, Graham JB, Goldin LR, Elston RC. Genetics of classic von Willebrand's disease. I. Phenotypic variation within families. Blood 1979;54:117–45.
[88] Gill JC, Endres-Brooks J, Bauer PJ, Marks WJ, Montgomery RR. The effect of ABO blood group on the diagnosis of von Willebrand Disease. Blood 1987;69:1691–5.
[89] Souto JC, Almasy L, Soria JM, Buil A, Stone W, Lathrop M, Blangero J, Fontcuberta J. Genome-wide linkage analysis of von Willebrand factor plasma levels: results from the GAIT project. Thromb Haemostasis 2003;89:468–74.
[90] Collins PW, Cumming AM, Goodeve AC, Lillicrap D. Type 1 von Willebrand disease: application of emerging data to clinical practice. Haemophilia 2008;14:685–96.
[91] Rodeghiero F, Castaman G, Dini E. Epidemiological investigation of the prevalence of von Willebrand's disease. Blood 1987;69:454–9.
[92] Werner EJ, Broxson EH, Tucker EL, Giroux DS, Shults J, Abshire TC. Prevalence of von Willebrand disease in children: a multiethnic study. J Pediatr 1993;123:893–8.
[93] Sadler JE. Von Willebrand disease type 1: a diagnosis in search of a disease. Blood 2003;101:2089–93.
[94] Bowman M, Hopman WM, Rapson D, Lillicrap D, James P. The prevalence of symptomatic von Willebrand disease in primary care practice. J Thromb Haemostasis 2010;8:213–6.
[95] Bowman M, Hopman WM, Rapson D, Lillicrap D, Silva M, James P. A prospective evaluation of the prevalence of symptomatic von Willebrand disease (VWD) in a pediatric primary care population. Pediatr Blood Cancer 2010;55:171–3.
[96] Mannucci PM. Desmopressin (DDAVP) in the treatment of bleeding disorders: the first 20 years. Blood 1997;90(7):2515–21. PMID: 9326215.
[97] Franchini M, Mannucci PM. Von Willebrand factor (Vonvendi®): the first recombinant product licensed for the treatment of von Willebrand disease. Expet Rev Hematol 2016;9(9):825–30. https://doi.org/10.1080/17474086.2016.1214070. Epub 2016 Jul 28. PMID: 27427955.
[98] Othman M. Platelet-type Von Willebrand disease: three decades in the life of a rare bleeding disorder. Blood Rev 2011;25:147–53.
[99] Miller JL. Platelet-type von Willebrand disease. Thromb Haemostasis 1996;75:865–9.
[100] Murata M, Russell SR, Ruggeri ZM, Ware J. Expression of the phenotypic abnormality of platelet-type von Willebrand disease in a recombinant glycoprotein Ibà fragment. J Clin Invest 1993;91:2133–7.
[101] Walsh PN. Factor XI: a renaissance. Semin Hematol 1992;29:189–201.
[102] Asakai R, Chung DW, Ratnoff OD, Davie EW. Factor XI (plasma thromboplastin antecedent) deficiency in

Ashkenazi Jews is a bleeding disorder that can result from three types of point mutations. Proc Natl Acad Sci U S A 1989;86:7667–71.

[103] Peretz H, Mulai A, Usher S, Zivelin A, Segal A, Weisman Z, Mittelman M, Lupo H, Lanir N, Brenner B, Shpilberg O, Seligsohn U. The two common mutations causing factor XI deficiency in Jews stem from distinct founders: one of ancient Middle Eastern origin and another of more recent European origin. Blood 1997;90:2654–9.

[104] Salomon O, Seligsohn U. New observations on factor XI deficiency. Haemophilia 2004;10(Suppl. 4):184–7.

[105] Tillman B, Gailani D. Inhibition of factors XI and XII for prevention of thrombosis induced by artificial surfaces. Semin Thromb Hemost 2018;44(1):60–9. https://doi.org/10.1055/s-0037-1603937. Epub 2017 Sep 12. PMID: 28898903; PMCID: PMC5794506.

[106] Neerman-Arbez M, de Moerloose P, Bridel C, Honsberger A, SchÖnbÖrner A, Rossier C, Peerlinck K, Claeyssens S, Di Michele D, D'Oiron R, Dreyfus M, Laubriat-Bianchin M, Dieval J, Antonarakis SE, Morris MA. Mutations in the fibrinogen Aà gene account for the majority of cases of congenital afibrinogenemia. Blood 2000;96:149–52.

[107] Roberts HR, Escobar MA. Other coagulation factor deficiencies. In: Loscalzo J, Schafer AI, editors. Thrombosis and hemorrhage, vol. 3. Baltimore: Lippincott Williams & Wilkins; 2003. p. 575–99.

[108] Manco-Johnson MJ, Dimichele D, Castaman G, Fremann S, Knaub S, Kalina U, Peyvandi F, Piseddu G, Mannucci P. Pharmacokinetics and safety of fibrinogen concentrate. J Thromb Haemostasis 2009;7:2064–9.

[109] Casini A, Blondon M, Lebreton A, Koegel J, Tintillier V, de Maistre E, Gautier P, Biron C, Neerman-Arbez M, de Moerloose P. Natural history of patients with congenital dysfibrinogenemia. Blood 2015;125(3):553–61. https://doi.org/10.1182/blood-2014-06-582866. Epub 2014 Oct 15. PMID: 25320241; PMCID: PMC4296015.

[110] Casini A, Neerman-Arbez M, Ariëns RA, de Moerloose P. Dysfibrinogenemia: from molecular anomalies to clinical manifestations and management. J Thromb Haemostasis 2015;13(6):909–19. https://doi.org/10.1111/jth.12916. Epub 2015 May 2. PMID: 25816717.

[111] Muszbek L, Bagoly Z, Cairo A, Peyvandi F. Novel aspects of factor XIII deficiency. Curr Opin Hematol 2011;18:366–72.

[112] Lancellotti S, De Cristofaro R. Congenital prothrombin deficiency. Semin Thromb Hemost 2009;35:367–81.

[113] Mannucci PM, Duga S, Peyvandi F. Recessively inherited coagulation disorders. Blood 2004;104:1243–52.

[114] Asselta R, Peyvandi F. Factor V deficiency. Semin Thromb Hemost 2009;35:382–9.

[115] Tracy PB, Mann KG. Abnormal formation of the prothrombinase complex: factor V deficiency and related disorders. Hum Pathol 1987;18:162–9.

[116] Guasch JF, Cannegieter S, Reitsma PH, Van't Veer-Korthof ET, Bertina RM. Severe coagulation factor V deficiency caused by a 4 bp deletion in the factor V gene. Br J Haematol 1998;101:32–9.

[117] Tracy PB, Giles AR, Mann KG, Eide LL, Hoogendoorn H, Rivard GE. Factor V (Quebec): a bleeding diathesis associated with a qualitative platelet Factor V deficiency. J Clin Invest 1984;74:1221–8.

[118] Janeway CM, Rivard GE, Tracy PB, Mann KG. Factor V Quebec revisited. Blood 1996;87:3571–8.

[119] Kahr WH, Zheng S, Sheth PM, Pai M, Cowie A, Bouchard M, Podor TJ, Rivard GE, Hayward CPM. Platelets from patients with the Quebec platelet disorder contain and secrete abnormal amounts of urokinase-type plasminogen activator. Blood 2001;98:257–65.

[120] Paterson AD, Rommens JM, Bharaj B, Blavignac J, Wong I, Diamandis M, Waye JS, Rivard GE, Hayward CP. Persons with Quebec platelet disorder have a tandem duplication of PLAU, the urokinase plasminogen activator gene. Blood 2010;115:1264–6.

[121] Mariani G, Bernardi F. Factor VII deficiency. Semin Thromb Hemost 2009;35:400–6.

[122] Menegatti M, Peyvandi F. Factor X deficiency. Semin Thromb Hemost 2009;35:407–15.

[123] Zhang B, Ginsburg D. Familial multiple coagulation factor deficiencies: new biologic insight from rare genetic bleeding disorders. J Thromb Haemost 2004;2:1564–72.

[124] Brenner B, S nchez-Vega B, Wu SM, Lanir N, Stafford DW, Solera J. A missense mutation in g -glutamyl carboxylase gene causes combined deficiency of all vitamin K-dependent blood coagulation factors. Blood 1998;92:4554–9.

[125] Rost S, Fregin A, Ivaskevicius V, Conzelmann E, Hortnagel K, Pelz HJ, Lappegard K, Seifried E, Scharrer I, Tuddenham EG, Muller CR, Strom TM, Oldenburg J. Mutations in VKORC1 cause warfarin resistance and multiple coagulation factor deficiency type 2. Nature 2004;427:537–41.

[126] Li T, Chang CY, Jin DY, Lin PJ, Khvorova A, Stafford DW. Identification of the gene for vitamin K epoxide reductase. Nature 2004;427:541–4.

[127] Rieder MJ, Reiner AP, Gage BF, Nickerson DA, Eby CS, McLeod HL, Blough DK, Thummel KE, Veenstra DL, Rettie AE. Effect of VKORC1 haplotypes on transcriptional regulation and warfarin dose. N Engl J Med 2005;352:2285–93.

[128] Wang L, McLeod HL, Weinshilboum RM. Genomics and drug response. N Engl J Med 2011;364:1144–53.
[129] Seligsohn U, Zivelin A, Zwang E. Combined factor V and factor VIII deficiency among non-Ashkenazi Jews. N Engl J Med 1982;307:1191–5.
[130] Nichols WC, Seligsohn U, Zivelin A, Terry VH, Hertel CE, Wheatley MA, Moussalli MJ, Hauri HP, Ciavarella N, Kaufman RJ, Ginsburg D. Mutations in the ER-Golgi intermediate compartment protein ERGIC-53 cause combined deficiency of coagulation factors V and VIII. Cell 1998;93:61–70.
[131] Zhang B, Cunningham MA, Nichols WC, Bernat JA, Seligsohn U, Pipe SW, McVey JH, Schulte-Overberg U, de Bosch N, Ruiz-Saez A, White GC, Tuddenham EGD, Kaufman RJ, Ginsburg D. Bleeding due to disruption of a cargo-specific ER to Golgi transport complex. Nat Genet 2003;34:220–5.
[132] Lopez JA, Andrews RK, Afshar-Kharghan V, Berndt MC. Bernard-Soulier syndrome. Blood 1998;91:4397–418.
[133] Nurden AT. Inherited abnormalities of platelets. Thromb Haemostasis 1999;82:468–80.
[134] Coller BS, Seligsohn U, Peretz H, Newman PJ. Glanzmann thrombasthenia: new insights from an historical perspective. Semin Hematol 1994;31:301–11.
[135] Bray PF. Inherited diseases of platelet glycoproteins: considerations for rapid molecular characterization. Thromb Haemostasis 1994;72:492–502.
[136] Khoriaty R, Ozel AB, Ramdas S, Ross C, Desch K, Shavit JA, Everett L, Siemieniak D, Li JZ, Ginsburg D. Genome-wide linkage analysis and whole-exome sequencing identifies an ITGA2B mutation in a family with thrombocytopenia. Br J Haematol 2019;186(4):574–9. https://doi.org/10.1111/bjh.15961. Epub 2019 May 23. PMID: 31119735; PMCID: PMC6679728.
[137] Albers CA, Cvejic A, Favier R, Bouwmans EE, Alessi MC, Bertone P, Jordan G, Kettleborough RN, Kiddle G, Kostadima M, Read RJ, Sipos B, Sivapalaratnam S, Smethurst PA, Stephens J, Voss K, Nurden A, Rendon A, Nurden P, Ouwehand WH. Exome sequencing identifies NBEAL2 as the causative gene for gray platelet syndrome. Nat Genet 2011;43:735–7.
[138] Gunay-Aygun M, Falik-Zaccai TC, Vilboux T, Zivony-Elboum Y, Gumruk F, Cetin M, Khayat M, Boerkoel CF, Kfir N, Huang Y, Maynard D, Dorward H, Berger K, Kleta R, Anikster Y, Arat M, Freiberg AS, Kehrel BE, Jurk K, Cruz P, Mullikin JC, White JG, Huizing M, Gahl WA. NBEAL2 is mutated in gray platelet syndrome and is required for biogenesis of platelet alpha-granules. Nat Genet 2011;43:732–4.
[139] Kahr WH, Hinckley J, Li L, Schwertz H, Christensen H, Rowley JW, Pluthero FG, Urban D, Fabbro S, Nixon B, Gadzinski R, Storck M, Wang K, Ryu GY, Jobe SM, Schutte BC, Moseley J, Loughran NB, Parkinson J, Weyrich AS, Di Paola J. Mutations in NBEAL2, encoding a BEACH protein, cause gray platelet syndrome. Nat Genet 2011;43:738–40.
[140] Lambert MP, Poncz M. Inherited platelet disorders. In: Orkin SH, Nathan DG, Ginsburg D, Look AT, Fisher DE, Lux SE, editors. Nathan and Oski's hematology of infancy and childhood. 7th ed. 2009. p. 1463–86.
[141] Turner MS, May DB, Arthur RR, Xiong GL. Clinical impact of selective serotonin reuptake inhibitors therapy with bleeding risks. J Intern Med 2007;261 :205–13.
[142] Ginsburg D. Disorders of the fibrinolytic system. In: Scriver CR, Beaudet AL, Sly WS, Valle D, Childs B, Vogelstein B, editors. The metabolic and molecular bases of inherited disease, vol. 8. New York: McGraw Hill; 2001. p. 4505–16.
[143] Aoki N, Saito H, Kamiya T, Koie K, Sakata Y, Kobakura M. Congenital deficiency of à 2 -plasmin inhibitor associated with severe hemorrhagic tendency. J Clin Invest 1979;63:877–84.
[144] Holmes WE, Lijnen HR, Nelles L, Kluft C, Nieuwenhuis HK, Rijken DC, Collen D. à 2 -Antiplasmin Enschede: alanine insertion and abolition of plasmin inhibitory activity. Science 1987;238:209–11.
[145] Fay WP, Parker AC, Condrey LR, Shapiro AD. Human plasminogen activator inhibitor-1 (PAI-1) deficiency: characterization of a large kindred with a null mutation in the PAI-1 gene. Blood 1997;90:204–8.
[146] Fay WP, Shapiro AD, Shih JL, Schleef RR, Ginsburg D. Complete deficiency of plasminogen-activator inhibitor type 1 due to a frame-shift mutation. N Engl J Med 1992;327:1729–33.
[147] Aoki N. Hemostasis associated with abnormalities of fibrinolysis. Blood Rev 1989;3:11–7.
[148] Owen MC, Brennan SO, Lewis JH, Carrell RW. Mutation of antitrypsin to antithrombin: alpha1-antitrypsin Pittsburgh (358 Met-Arg), a fatal bleeding disorder. N Engl J Med 1983;309:694–8.
[149] Vidaud D, Emmerich J, Alhenc-Gelas M, Yvart J, Fiessinger JN, Aiach M. Met 358 to Arg mutation of alpha 1 -antitrypsin associated with protein C deficiency in a patient with mild bleeding tendency. J Clin Invest 1992;89:1537–43.
[150] Bauer KA. Inherited and acquired hypercoagulable states. In: Loscalzo J, Schafer AI, editors. Thrombosis and hemorrhage, vol. 3. Baltimore: Lippincott Williams & Wilkins; 2003. p. 648–84.

[151] Malm J, Laurell M, Nilsson IM, Dahlb„ck B. Thromboembolic disease- critical evaluation of laboratory investigation. Thromb Haemostasis 1992;68:7–13.

[152] Dahlback B. Advances in understanding pathogenic mechanisms of thrombophilic disorders. Blood 2008;112:19–27.

[153] Crowther MA, Kelton JG. Congenital thrombophilic states associated with venous thrombosis: a qualitative overview and proposed classification system. Ann Intern Med 2003;138:128–34.

[154] Patnaik MM, Moll S. Inherited antithrombin deficiency: a review. Haemophilia 2008;14:1229–39.

[155] Demers C, Ginsberg JS, Hirsh J, Henderson P, Blajchman MA. Thrombosis in antithrombin-III-deficient persons. Am Intercult Mag 1992;116:754–61.

[156] Sanson BJ, Simioni P, Tormene D, Moia M, Friederich PW, Huisman MV, Prandoni P, Bura A, Rejto L, Wells P, Mannucci PM, Girolami A, Büller HR, Prins MH. The incidence of venous thromboembolism in asymptomatic carriers of a deficiency of antithrombin, protein C, or protein S: a prospective cohort study. Blood 1999;94:3702–6.

[157] Van Boven HH, Vandenbroucke JP, Briet E, Rosendaal FR. Gene-gene and gene-environment interactions determine risk of thrombosis in families with inherited antithrombin deficiency. Blood 1999;94:2590–4.

[158] Van Boven HH, Vandenbroucke JP, Westendorp RGJ, Rosendaal FR. Mortality and causes of death in inherited antithrombin deficiency. Thromb Haemostasis 1997;77:452–5.

[159] Kearon C, Crowther M, Hirsh J. Management of patients with hereditary hypercoagulable disorders. Annu Rev Med 2000;51:169–85.

[160] Eikelboom JW, Weitz JI. New anticoagulants. Circulation 2010;121:1523–32.

[161] Griffin JH, Evatt B, Zimmerman TS, Kleiss AJ. Deficiency of protein C in congenital thrombotic disease. J Clin Invest 1981;68:1370–3.

[162] Koster T, Rosendaal FR, Briet E, Van der Meer FJM, Colly LP, Trienekens PH, Poort SR, Reitsma PH, Vandenbroucke JP. Protein C deficiency in a controlled series of unselected outpatients: an infrequent but clear risk factor for venous thrombosis (Leiden thrombophilia study). Blood 1995;85:2756–61.

[163] Miletich J, Sherman L, Broze Jr GJ. Absence of thrombosis in subjects with heterozygous protein C deficiency. N Engl J Med 1987;317:991–6.

[164] Reitsma PH, Bernardi F, Doig RG, Gandrille S, Greengard JS, Ireland H, Krawczak M, Lind B, Long GL, Poort SR, Saito H, Sala N, Witt I, Cooper DN. Protein C deficiency: a database of mutations. 1995 update: on behalf of the subcommittee on plasma coagulation inhibitors of the Scientific and Standardization Committee of the ISTH. Thromb Haemostasis 1995;73:876–89.

[165] Makris M, Leach M, Beauchamp NJ, Daly ME, Cooper PC, Hampton KK, Bayliss P, Peake IR, Miller GJ, Preston FE. Genetic analysis, phenotypic diagnosis, and risk of venous thrombosis in families with inherited deficiencies of protein S. Blood 2000;95:1935–41.

[166] Nachman RL, Silverstein R. Hypercoagulable states. Am Intercult Mag 1993;119:819–27.

[167] Dahlback B. Inherited thrombophilia: resistance to activated protein C as a pathogenic factor of venous thromboembolism. Blood 1995;85:607–14.

[168] Dahlback B, Carlsson M, Svensson PJ. Familial thrombophilia due to a previously unrecognized mechanism characterized by poor anticoagulant response to activated protein C: prediction of a cofactor to activated protein C. Proc Natl Acad Sci U S A 1993;90:1004–8.

[169] Bertina RM, Koeleman BPC, Koster T, Rosendaal FR, Dirven RJ, de Ronde H, van der Velden PA, Reitsma PH. Mutation in blood coagulation factor V associated with resistance to activated protein C. Nature 1994;369:64–7.

[170] Chan WP, Lee CK, Kwong YL, Lam CK, Liang R. A novel mutation of Arg306 of factor V gene in Hong Kong Chinese. Blood 1998;91:1135–9.

[171] Williamson D, Brown K, Luddington R, Baglin C, Baglin T. Factor V Cambridge: a new mutation (Arg 306->Thr) associated with resistance to activated protein C. Blood 1998;91:1140–4.

[172] Amano K, Michnick DA, Moussalli M, Kaufman RJ. Mutation at either Arg336 or Arg562 in factor VIII is insufficient for complete resistance to activated protein C (APC)-mediated inactivation: implications for the APC resistance test. Thromb Haemostasis 1998;79:557–63.

[173] Zivelin A, Griffin JH, Xu X, Pabinger I, Samama M, Conard J, Brenner B, Eldor A, Seligsohn U. A single genetic origin for a common caucasian risk factor for venous thrombosis. Blood 1997;89:397–402.

[174] Rees DC, Cox M, Clegg JB. World distribution of factor V Leiden. Lancet 1995;346:1133–4.

[175] Rosendaal FR, Koster T, Vandenbroucke JP, Reitsma PH. High risk of thrombosis in patients homozygous for factor V Leiden (activated protein C resistance). Blood 1995;85:1504–8.

[176] Vandenbroucke JP, Koster T, Bri‰t E, Reitsma PH, Bertina RM, Rosendaal FR. Increased risk of venous thrombosis in oral-contraceptive users who are carriers of factor V Leiden mutation. Lancet 1994;344:1453–7.

[177] Kupferminc MJ, Eldor A, Steinman N, Many A, Bar-Am A, Jaffa A, Fait G, Lessing JB. Increased frequency of genetic thrombophilia in women with complications of pregnancy. N Engl J Med 1999;1:9–13.
[178] Gerhardt A, Scharf RE, Beckmann MW, Struve S, Bender HG, Pillny M, Sandmann W, Zotz RB. Prothrombin and factor V mutations in women with a history of thrombosis during pregnancy and the puerperium. N Engl J Med 2000;342:374–80.
[179] Dizon-Townson D, Miller C, Sibai B, Spong CY, Thom E, Wendel Jr G, Wenstrom K, Samuels P, Cotroneo MA, Moawad A, Sorokin Y, Meis P, Miodovnik M, O'Sullivan MJ, Conway D, Wapner RJ, Gabbe SG. The relationship of the factor V Leiden mutation and pregnancy outcomes for mother and fetus. Obstet Gynecol 2005;106:517–24.
[180] Hille ETM, Westendorp RGJ, Vandenbroucke JP, Rosendaal FR. Mortality and causes of death in families with the factor V leiden mutation (resistance to activated protein C). Blood 1997;89:1963–7.
[181] Bauer KA. Role of thrombophilia in deciding on the duration of anticoagulation. Semin Thromb Hemost 2004;30:633–7.
[182] Kearon C, Kahn SR, Agnelli G, Goldhaber S, Raskob GE, Comerota AJ. Antithrombotic therapy for venous thromboembolic disease: American College of Chest Physicians evidence-based clinical practice guidelines (8th edition). Chest 2008;133:454S–545S.
[183] Middeldorp S, van Hylckama Vlieg A. Does thrombophilia testing help in the clinical management of patients? Br J Haematol 2008;143:321–35.
[184] Lijfering WM, Middeldorp S, Veeger NJ, Hamulyak K, Prins MH, Buller HR, van der Meer J. Risk of recurrent venous thrombosis in homozygous carriers and double heterozygous carriers of factor V Leiden and prothrombin G20210A. Circulation 2010;121:1706–12.
[185] Brill-Edwards P, Ginsberg JS, Gent M, Hirsh J, Burrows R, Kearon C, Geerts W, Kovacs M, Weitz JI, Robinson S, Whittom R, Couture G, The Recurrence of Clot in This Pregnancy Study G. Safety of withholding heparin in pregnant women with a history of venous thromboembolism. N Engl J Med 2000;20:1439–44.
[186] Seligsohn U, Lubetsky A. Medical progress: genetic susceptibility to venous thrombosis. N Engl J Med 2001;344:1222–31.
[187] Kerlin BA, Yan SB, Isermann BH, Brandt JT, Sood R, Basson BR, Joyce DE, Weiler H, Dhainaut JF. Survival advantage associated with heterozygous factor V Leiden mutation in patients with severe sepsis and in mouse endotoxemia. Blood 2003;102:3085–92.
[188] Lindqvist PG, Svensson PJ, Dahlb„ck B, Mars l K. Factor V Q 506 mutation (activated protein C resistance) associated with reduced intrapartum blood loss - a possible evolutionary selection mechanism. Thromb Haemostasis 1998;79:69–73.
[189] Poort SR, Rosendaal FR, Reitsma PH, Bertina RM. A common genetic variation in the 3'-untranslated region of the prothrombin gene is associated with elevated plasma prothrombin levels and an increase in venous thrombosis. Blood 1996;88:3698–703.
[190] Gehring NH, Frede U, Neu-Yilik G, Hundsdoerfer P, Vetter B, Hentze MW, Kulozik AE. Increased efficiency of mRNA 3 ' end formation: a new genetic mechanism contributing to hereditary thrombophilia. Nat Genet 2001;28:389–92.
[191] von Ahsen N, Oellerich M. The intronic prothrombin 19911A>G polymorphism influences splicing efficiency and modulates effects of the 20210G>A polymorphism on mRNA amount and expression in a stable reporter gene assay system. Blood 2004;103:586–93.
[192] Meijers JCM, Tekelenburg WLH, Bouma BN, Bertina RM, Rosendaal FR. High levels of coagulation factor XI as a risk factor for venous thrombosis. N Engl J Med 2000;342:696–701.
[193] Van HV, van dLI, Bertina RM, Rosendaal FR. High levels of factor IX increase the risk of venous thrombosis. Blood 2000;95:3678–82.
[194] Simioni P, Tormene D, Tognin G, Gavasso S, Bulato C, Iacobelli NP, Finn JD, Spiezia L, Radu C, Arruda VR. X-linked thrombophilia with a mutant factor IX (factor IX Padua). N Engl J Med 2009;361:1671–5.
[195] Rosendaal FR. High levels of factor VIII and venous thrombosis. Thromb Haemostasis 2000;83:1–2.
[196] Kraaijenhagen RA, Anker PSI, Koopman MMW, Reitsma PH, Prins MH, van d, Ende A, Buller HR. High plasma concentration of factor VIII is a major risk factor for venous thromboembolism. Thromb Haemostasis 2000;83:5–9.
[197] O'Donnell J, Mumford AD, Manning RA, Laffan MA. Elevation of FVIII: C in venous thromboembolism is persistent and independent of the acute phase response. Thromb Haemostasis 2000;83:10–3.
[198] den Heijer M, Koster T, Blom HJ, Bos GMJ, Briet E, Reitsma PH, Vandenbroucke JP, Rosendaal FR. Hyperhomocysteinemia as a risk factor for deep-vein thrombosis. N Engl J Med 1996;334:759–62.
[199] den Heijer M, Willems HP, Blom HJ, Gerrits WB, Cattaneo M, Eichinger S, Rosendaal FR, Bos GM. Homocysteine lowering by B vitamins and the secondary prevention of deep vein thrombosis and pulmonary embolism: a randomized, placebo-controlled, double-blind trial. Blood 2007;109:139–44.
[200] Lonn E. Homocysteine in the prevention of ischemic heart disease, stroke and venous thromboembolism:

therapeutic target or just another distraction? Curr Opin Hematol 2007;14:481–7.
[201] D'Angelo A, Selhub J. Homocysteine and thrombotic disease. Blood 1997;90:1–11.
[202] Bezemer ID, Doggen CJ, Vos HL, Rosendaal FR. No association between the common MTHFR 677C->T polymorphism and venous thrombosis: results from the MEGA study. Arch Intern Med 2007;167:497–501.
[203] Aoki N, Moroi M, Sakata Y, Yoshida N, Matsuda M. Abnormal plasminogen. A hereditary molecular abnormality found in a patient with recurrent thrombosis. J Clin Invest 1978;61:1186–95.
[204] Demarmels Biasiutti F, Sulzer I, Stucki B, Wuillemin WA, Furlan M, Lammle B. Is plasminogen deficiency a thrombotic risk factor? A study on 23 thrombophilic patients and their family members. Thromb Haemostasis 1998;80:167–70.
[205] Tait RC, Walker ID, Conkie JA, Islam SIAM, McCall F. Isolated familial plasminogen deficiency may not be a risk factor for thrombosis. Thromb Haemostasis 1996;76:1004–8.
[206] Schuster V, Mingers AM, Seidenspinner S, Nüssgens Z, Pukrop T, Kreth HW. Homozygous mutations in the plasminogen gene of two unrelated girls with ligneous conjunctivitis. Blood 1997;90:958–66.
[207] Schott D, Dempfle CE, Beck P, Liermann A, Mohr-Pennert A, Goldner M, Mehlem P, Azuma H, Schuster V, Mingers AM, Schwarz HP, Kramer MD. Therapy with a purified plasminogen concentrate in an infant with ligneous conjunctivitis and homozygous plasminogen deficiency. N Engl J Med 1998;339:1679–86.
[208] Levy GG, Motto DG, Ginsburg D. ADAMTS13 turns 3. Blood 2005;106:11–7.
[209] Moake JL. Thrombotic microangiopathies. N Engl J Med 2002;347:589–600.
[210] George JN. How I treat patients with thrombotic thrombocytopenic purpura-hemolytic uremic syndrome. Blood 2000;96:1223–9.
[211] Furlan M, Robles R, Galbusera M, Remuzzi G, Kyrle PA, Brenner B, Krause M, Scharrer I, Aumann V, Mittler U, Solenthaler M, L„mmle B. von Willebrand factor-cleaving protease in thrombotic thrombocytopenic purpura and the hemolytic-uremic syndrome. N Engl J Med 1998;339:1578–84.
[212] Tsai HM, Lian ECY. Antibodies to von Willebrand factor-cleaving protease in acute thrombotic thrombocytopenic purpura. N Engl J Med 1998;339:1585–94.
[213] Levy GG, Nichols WC, Lian EC, Foroud T, McClintick JN, McGee BM, Yang AY, Siemieniak DR, Stark KR, Gruppo R, Sarode R, Shurin SB, Chandrasekaran V, Stabler SP, Sabio H, Bouhassira EE, Upshaw Jr JD, Ginsburg D, Tsai HM. Mutations in a member of the ADAMTS gene family cause thrombotic thrombocytopenic purpura. Nature 2001;413:488–94.
[214] Fujikawa K, Suzuki H, McMullen B, Chung D. Purification of human von Willebrand factor-cleaving protease and its identification as a new member of the metalloproteinase family. Blood 2001;98:1662–6.
[215] Gerritsen HE, Robles R, Lammle B, Furlan M. Partial amino acid sequence of purified von Willebrand factor-cleaving protease. Blood 2001;98:1654–61.
[216] Soejima K, Mimura N, Hirashima M, Maeda H, Hamamoto T, Nakagaki T, Nozaki C. A novel human metalloprotease synthesized in the liver and secreted into the blood: possibly, the von Willebrand factor-cleaving protease? J Biochem 2001;130:475–80.
[217] Koeleman BPC, Reitsma PH, Allaart CF, Bertina RM. Activated protein C resistance as an additional risk factor for thrombosis in protein C-deficient families. Blood 1994;84:1031–5.
[218] Koeleman BPC, Van Rumpt D, Hamuly k K, Reitsma PH, Bertina RM. Factor V Leiden: an additional risk factor for thrombosis in protein S deficient families? Thromb Haemostasis 1995;74:580–3.

FURTHER READING

Antoni G, Oudot-Mellakh T, Dimitromanolakis A, Germain M, Cohen W, Wells P, Lathrop M, Gagnon F, Morange PE, Tregouet DA. Combined analysis of three genome-wide association studies on vWF and FVIII plasma levels. BMC Med Genet 2011;12:102.

Azzi A, De Santis R, Morfini M, Zakrzewska K, Musso R, Santagostino E, Castaman G. TT virus contaminates first-generation recombinant factor VIII concentrates. Blood 2001;98:2571–3.

Bates SM, Greer IA, Middeldorp S, Veenstra DL, Prabulos AM, Vandvik PO. VTE, thrombophilia, antithrombotic therapy, and pregnancy: antithrombotic therapy and prevention of thrombosis, 9th ed: American College of Chest Physicians evidence-based clinical practice guidelines. Chest 2012;141:e691S–736S.

Battinelli EM, Bauer KA. Thrombophilias in pregnancy. Hematol Oncol Clin N Am 2011;25:323–33 [viii].

Berger M, Mattheisen M, Kulle B, Schmidt H, Oldenburg J, Bickeboller H, Walter U, Lindner TH, Strauch K, Schambeck CM. High factor VIII levels in venous thromboembolism show linkage to imprinted loci on chromosomes 5 and 11. Blood 2005;105:638–44.

Blau HM, Springer ML. Gene therapy - a novel form of drug delivery. N Engl J Med 1995;333:1204–7.

Bounameaux H. Factor V Leiden paradox: risk of deep-vein thrombosis but not of pulmonary embolism. Lancet 2000;356:182–3.

Bugge TH, Flick MJ, Daugherty CC, Degen JL. Plasminogen deficiency causes severe thrombosis but is compatible with development and reproduction. Gene Dev 1995;9:794–807.

Cui J, O'Shea KS, Purkayastha A, Saunders TL, Ginsburg D. Fatal haemorrhage and incomplete block to embryogenesis in mice lacking coagulation factor V. Nature 1996;384:66–8.

de Haan M, Kamp JJ, Briet E, Dubbeldam J. Noonan syndrome: partial factor XI deficiency. Am J Med Genet 1988;29:277–82.

Dombroski BA, Mathias SL, Nanthakumar E, Scott AF, Kazazian Jr HH. Isolation of an active human transposable element. Science 1991;254:1805–8.

Drew AF, Kaufman AH, Kombrinck KW, Danton MJS, Daugherty CC, Degen JL, Bugge TH. Ligneous conjunctivitis in plasminogen-deficient mice. Blood 1998;91:1616–24.

Evaluation of Genomic Applications in Practice and Prevention (EGAPP) Working Group. Recommendations from the EGAPP Working Group: routine testing for Factor V Leiden (R506Q) and prothrombin (20210G>A) mutations in adults with a history of idiopathic venous thromboembolism and their adult family members. Genet Med 2011;13:67–76.

Franchini M, Lippi G. Von Willebrand factor-containing factor VIII concentrates and inhibitors in haemophilia A. A critical literature review. Thromb Haemostasis 2010;104:931–40.

Frederiksen J, Juul K, Grande P, Jensen GB, Schroeder TV, Tybjaerg-Hansen A, Nordestgaard BG. Methylenetetrahydrofolate reductase polymorphism (C677T), hyperhomocysteinemia, and risk of ischemic cardiovascular disease and venous thromboembolism: prospective and case-control studies from the Copenhagen City Heart Study. Blood 2004;104:3046–51.

Giannelli F, Anagnostopoulos T, Green PM. Mutation rates in humans. II. Sporadic mutation-specific rates and rate of detrimental human mutations inferred from hemophilia B. Am J Hum Gen 1999;65:1580–7.

Ginsburg D. Identifying novel genetic determinants of hemostatic balance. J Thromb Haemost 2005;3:1561–8.

Gong IY, Tirona RG, Schwarz UI, Crown N, Dresser GK, Larue S, Langlois N, Lazo-Langner A, Zou G, Roden DM, Stein CM, Rodger M, Carrier M, Forgie M, Wells PS, Kim RB. Prospective evaluation of a pharmacogenetics-guided warfarin loading and maintenance dose regimen for initiation of therapy. Blood 2011;118:3163–71.

Heit JA, Sobell JL, Li H, Sommer SS. The incidence of venous thromboembolism among factor V Leiden carriers: a community-based cohort study. J Thromb Haemost 2005;3:305–11.

High KA. Gene therapy for haemophilia: a long and winding road. J Thromb Haemostasis 2011;9(Suppl. 1):2–11.

Iannuzzi MC, Hidaka N, Boehnke ML, Bruck ME, Hanna WT, Collins FS, Ginsburg D. Analysis of the relationship of von Willebrand disease (vWD) and hereditary hemorrhagic telangiectasia and identification of a potential type IIA vWD mutation (IIe865 to Thr). Am J Hum Gen 1991;48:757–63.

Juul K, Tybjaerg-Hansen A, Schnohr P, Nordestgaard BG. Factor V Leiden and the risk for venous thromboembolism in the adult Danish population. Ann Intern Med 2004;140:330–7.

Karger R, Donner-Banzhoff N, Muller HH, Kretschmer V, Hunink M. Diagnostic performance of the platelet function analyzer (PFA-100) for the detection of disorders of primary haemostasis in patients with a bleeding history-a systematic review and meta-analysis. Platelets 2007;18: 249–60.

Kazazian Jr HH, Wong C, Youssoufian H, Scott AF, Phillips DG, Antonarakis SE. Hemophilia A resulting from de novo insertion of L1 sequences represents a novel mechanism for mutation in man. Nature 1988;332:164–6.

Keijzer MB, Borm GF, Blom HJ, Bos GM, Rosendaal FR, den Heijer M. No interaction between factor V Leiden and hyperhomocysteinemia or MTHFR 677TT genotype in venous thrombosis. Results of a meta-analysis of published studies and a large case-only study. Thromb Haemostasis 2007;97:32–7.

Kluijtmans LAJ, den Heijer M, Reitsma PH, Heil SG, Blom HJ, Rosendaal FR. Thermolabile methylenetetrahydrofolate reductase and factor V Leiden in the risk of deep-vein thrombosis. Thromb Haemostasis 1998;79:254–8.

Koster T, Rosendaal FR, Briet E, Vandenbroucke JP. John Hageman's factor and deep-vein thrombosis: Leiden thrombophilia study. Br J Haematol 1994;87:422–4.

Kyrle PA, Minar E, Hirschl M, Bialonczyk C, Stain M, Schneider B, Weltermann A, Speiser W, Lechner K, Eichinger S. High plasma levels of factor VIII and the risk of recurrent venous thromboembolism. N Engl J Med 2000;343:457–62.

Lemmerhirt HL, Broman KW, Shavit JA, Ginsburg D. Genetic regulation of plasma von Willebrand factor levels: quantitative trait loci analysis in a mouse model. J Thromb Haemostasis 2007;5:329–35.

Lemmerhirt HL, Shavit JA, Levy GG, Cole SM, Long JC, Ginsburg D. Enhanced VWF biosynthesis and elevated plasma VWF due to a natural variant in the murine Vwf gene. Blood 2006;108:3061–7.

Lenzini P, Wadelius M, Kimmel S, Anderson JL, Jorgensen AL, Pirmohamed M, Caldwell MD, Limdi N,

Burmester JK, Dowd MB, Angchaisuksiri P, Bass AR, Chen J, Eriksson N, Rane A, Lindh JD, Carlquist JF, Horne BD, Grice G, Milligan PE, Eby C, Shin J, Kim H, Kurnik D, Stein CM, McMillin G, Pendleton RC, Berg RL, Deloukas P, Gage BF. Integration of genetic, clinical, and INR data to refine warfarin dosing. Clin Pharmacol Ther 2010;87:572–8.

Li H, Haurigot V, Doyon Y, Li T, Wong SY, Bhagwat AS, Malani N, Anguela XM, Sharma R, Ivanciu L, Murphy SL, Finn JD, Khazi FR, Zhou S, Paschon DE, Rebar EJ, Bushman FD, Gregory PD, Holmes MC, High KA. In vivo genome editing restores haemostasis in a mouse model of haemophilia. Nature 2011;475:217–21.

Li W, Rusiniak ME, Chintala S, Gautam R, Novak EK, Swank RT. Murine Hermansky-Pudlak syndrome genes: regulators of lysosome-related organelles. Bioessays 2004b;26:616–28.

Lind SE. The bleeding time does not predict surgical bleeding. Blood 1991;77:2547–52.

Manco-Johnson MJ, Abshire TC, Shapiro AD, Riske B, Hacker MR, Kilcoyne R, Ingram JD, Manco-Johnson ML, Funk S, Jacobson L, Valentino LA, Hoots WK, Buchanan GR, DiMichele D, Recht M, Brown D, Leissinger C, Bleak S, Cohen A, Mathew P, Matsunaga A, Medeiros D, Nugent D, Thomas GA, Thompson AA, McRedmond K, Soucie JM, Austin H, Evatt BL. Prophylaxis versus episodic treatment to prevent joint disease in boys with severe hemophilia. N Engl J Med 2007;357:535–44.

Mandel H, Brenner B, Berant M, Rosenberg N, Lanir N, Jakobs C, Fowler B, Seligsohn U. Coexistence of hereditary homocystinuria and factor V Leiden - effect on thrombosis. N Engl J Med 1996;334:763–8.

Martinelli I, Battaglioli T, Razzari C, Mannucci PM. Type and location of venous thromboembolism in patients with factor V Leiden or prothrombin G20210A and in those with no thrombophilia. J Thromb Haemostasis 2007;5:98–101.

Mei B, Pan C, Jiang H, Tjandra H, Strauss J, Chen Y, Liu T, Zhang X, Severs J, Newgren J, Chen J, Gu JM, Subramanyam B, Fournel MA, Pierce GF, Murphy JE. Rational design of a fully active, long-acting PEGylated factor VIII for hemophilia A treatment. Blood 2010;116:270–9.

Mohlke KL, Purkayastha AA, Westrick RJ, Smith PL, Petryniak B, Lowe JB, Ginsburg D. Mvwf, a dominant modifier of murine von Willebrand factor, results from altered lineage-specific expression of a glycosyltransferase. Cell 1999;96:111–20.

Nathwani AC, Davidoff AM, Linch DC. A review of gene therapy for haematological disorders. Br J Haematol 2005;128:3–17.

Nathwani AC, Tuddenham EG, Rangarajan S, Rosales C, McIntosh J, Linch DC, Chowdary P, Riddell A, Pie AJ, Harrington C, O'Beirne J, Smith K, Pasi J, Glader B, Rustagi P, Ng CY, Kay MA, Zhou J, Spence Y, Morton CL, Allay J, Coleman J, Sleep S, Cunningham JM, Srivastava D, Basner-Tschakarjan E, Mingozzi F, High KA, Gray JT, Reiss UM, Nienhuis AW, Davidoff AM. Adenovirus-associated virus vector-mediated gene transfer in hemophilia B. N Engl J Med 2011;365:2357–65.

Naylor JA, Green PM, Rizza CR, Giannelli F. Factor VIII gene explains all cases of haemophilia A. Lancet 1992;340:1066–7.

Nichols WC, Ginsburg D. von Willebrand disease. Medicine 1997;76:1–20.

Nilsson IM, Berntorp E, Lofqvist T, Pettersson H. Twenty-five years' experience of prophylactic treatment in severe haemophilia A and B. J Intern Med 1992;232:25–32.

Ostergaard H, Bjelke JR, Hansen L, Petersen LC, Pedersen AA, Elm T, Moller F, Hermit MB, Holm PK, Krogh TN, Petersen JM, Ezban M, Sorensen BB, Andersen MD, Agerso H, Ahmadian H, Balling KW, Christiansen ML, Knobe K, Nichols TC, Bjorn SE, Tranholm M. Prolonged half-life and preserved enzymatic properties of factor IX selectively PEGylated on native N-glycans in the activation peptide. Blood 2011;118:2333–41.

Peake IR, Bowen D, Bignell P, Liddell MB, Sadler JE, Standen G, Bloom AL. Family studies and prenatal diagnosis in severe von Willebrand Disease by polymerase chain reaction amplification of a variable number tandem repeat region of the von Willebrand factor gene. Blood 1990;76:555–61.

Ploplis VA, Carmeliet P, Vazirzadeh S, Van Vlaenderen I, Moons L, Plow EF, Collen D. Effects of disruption of the plasminogen gene on thrombosis, growth, and health in mice. Circulation 1995;92:2585–93.

Quiroga T, Goycoolea M, Munoz B, Morales M, Aranda E, Panes O, Pereira J, Mezzano D. Template bleeding time and PFA-100 have low sensitivity to screen patients with hereditary mucocutaneous hemorrhages: comparative study in 148 patients. J Thromb Haemost 2004;2:892–8.

Rallapalli PM, Kemball-Cook G, Tuddenham EG, Gomez K, Perkins SJ. An interactive mutation database for human coagulation factor IX provides novel insights into the phenotypes and genetics of hemophilia B. J Thromb Haemostasis 2013;11(7):1329–40. https://doi.org/10.1111/jth.12276. PMID: 23617593.

Ridker PM, Hennekens CH, Lindpaintner K, Stampfer MJ, Eisenberg PR, Miletich JP. Mutation in the gene coding for coagulation factor V and the risk of myocardial infarction, stroke, and venous thrombosis in apparently healthy men. N Engl J Med 1995;332:912–7.

Roberts HR, Monroe DM, White GC. The use of recombinant factor VIIa in the treatment of bleeding disorders. Blood 2004;104:3858–64.

Rose EH, Aledort LM. Nasal spray desmopressin (DDAVP) for mild hemophilia A and von Willebrand disease. Am Intercult Mag 1991;114:563–8.

Rosendaal FR, Siscovick DS, Schwartz SM, Psaty BM, Raghunathan TE, Vos HL. A common prothrombin variant (20210 G to A) increases the risk of myocardial infarction in young women. Blood 1997;90:1747–50.

Rosove MH, Grody WW. Should we be applying warfarin pharmacogenetics to clinical practice? No, not now. Am Intercult Mag 2009;151:270–3. W295.

Saunders RE, O'Connell NM, Lee CA, Perry DJ, Perkins SJ. The factor XI deficiency database: an interactive web database of mutations, phenotypes and structural analysis tools. Hum Mutat 2005;26:192–8. PM:16086308.

Schuster V, Seidenspinner S, Zeitler P, Escher C, Pleyer U, Bernauer W, Stiehm ER, Isenberg S, Seregard S, Olsson T, Mingers AM, Schambeck C, Kreth HW. Compound-heterozygous mutations in the plasminogen gene predispose to the development of ligneous conjunctivitis. Blood 1999;93:3457–66.

Shavit JA, Manichaikul A, Lemmerhirt HL, Broman KW, Ginsburg D. Modifiers of von Willebrand factor identified by natural variation in inbred strains of mice. Blood 2009;114:5368–74.

Smith NL, Chen MH, Dehghan A, Strachan DP, Basu S, Soranzo N, Hayward C, Rudan I, Sabater-Lleal M, Bis JC, de Maat MP, Rumley A, Kong X, Yang Q, Williams FM, Vitart V, Campbell H, Malarstig A, Wiggins KL, Van Duijn CM, McArdle WL, Pankow JS, Johnson AD, Silveira A, McKnight B, Uitterlinden AG, Aleksic N, Meigs JB, Peters A, Koenig W, Cushman M, Kathiresan S, Rotter JI, Bovill EG, Hofman A, Boerwinkle E, Tofler GH, Peden JF, Psaty BM, Leebeek F, Folsom AR, Larson MG, Spector TD, Wright AF, Wilson JF, Hamsten A, Lumley T, Witteman JC, Tang W, O'Donnell CJ. Novel associations of multiple genetic loci with plasma levels of factor VII, factor VIII, and von Willebrand factor: the CHARGE (Cohorts for Heart and Aging Research in Genome Epidemiology) Consortium. Circulation 2010;121:1382–92.

Sun H, Yang TL, Yang AY, Wang X, Ginsburg D. The murine platelet and plasma Factor V pools are biosynthetically distinct and sufficient for minimal hemostasis. Blood 2003;102:2856–61.

Sun WY, Witte DP, Degen JL, Colbert MC, Burkart MC, Holmb„ck K, Xiao Q, Bugge TH, Degen SJF. Prothrombin deficiency results in embryonic and neonatal lethality in mice. Proc Natl Acad Sci U S A 1998;95:7597–602.

Tsai HM, Sussman II, Ginsburg D, Lankhof H, Sixma JJ, Nagel RL. Proteolytic cleavage of recombinant type 2A von Willebrand factor mutants R834W and R834Q: inhibition by doxycycline and by monoclonal antibody VP-1. Blood 1997;89:1954–62.

Turecek PL, Schrenk G, Rottensteiner H, Varadi K, Bevers E, Lenting P, Ilk N, Sleytr UB, Ehrlich HJ, Schwarz HP. Structure and function of a recombinant von Willebrand factor drug candidate. Semin Thromb Hemost 2010;36:510–21.

Verma D, Moghimi B, LoDuca PA, Singh HD, Hoffman BE, Herzog RW, Daniell H. Oral delivery of bioencapsulated coagulation factor IX prevents inhibitor formation and fatal anaphylaxis in hemophilia B mice. Proc Natl Acad Sci U S A 2010;107:7101–6.

Waters EK, Genga RM, Schwartz MC, Nelson JA, Schaub RG, Olson KA, Kurz JC, McGinness KE. Aptamer ARC19499 mediates a procoagulant hemostatic effect by inhibiting tissue factor pathway inhibitor. Blood 2011;117:5514–22.

Welch GN, Loscalzo J. Homocysteine and antherothrombosis. N Engl J Med 1998;338:1042–50.

White RA, Peters LL, Adkison LR, Korsgren C, Cohen CM, Lux SE. The murine pallid mutation is a platelet storage pool disease associated with the protein 4.2 (pallidin) gene. Nat Genet 1992;2:80–3.

Xue J, Wu Q, Westfield LA, Tuley EA, Lu D, Zhang Q, Shim K, Zheng X, Sadler JE. Incomplete embryonic lethality and fatal neonatal hemorrhage caused by prothrombin deficiency in mice. Proc Natl Acad Sci U S A 1998;95:7603–7.

Yang TL, Cui J, Taylor JM, Yang A, Gruber SB, Ginsburg D. Rescue of fatal neonatal hemorrhage in factor V deficient mice by low level transgene expression. Thromb Haemostasis 2000;83:70–7.

8

Amyloidosis and Other Protein Deposition Diseases

Merrill D. Benson[†]

Professor of Pathology and Laboratory Medicine, Professor of Medical and Molecular Genetics, and Professor of Medicine, Indiana University School of Medicine, Indianapolis, IN, United States

8.1 INTRODUCTION

Amyloidosis (AL) is the classic example of protein deposition disease. However, there are many forms of AL, and by definition, these represent only those diseases associated with extracellular deposits of β-structured protein fibrils [1—3]. We now know that there are diseases characterized by protein deposits that may be extracellular or intracellular, and the deposits may contain either fibrillar or nonfibrillar protein aggregates.

Each type of protein deposition disease, whether fibrillar or nonfibrillar, is characterized by the specific protein that is the major constituent of the pathologic deposits. In the case of the amyloidoses, there are at least 35 proteins that can form extracellular β-structured fibril deposits, and each represents a separate disease entity [4]. Some of these diseases are systemic (involving several organ systems) and some are localized (limited to one organ). The systemic forms of AL may be sporadic or acquired (e.g., immunoglobulin AL); secondary to other conditions (e.g., reactive AA AL); or hereditary (e.g., familial amyloidotic polyneuropathy/transthyretin AL). The localized forms of AL may also be sporadic or acquired (e.g., AL amyloid in the upper respiratory tract, urinary tract, or plasmacytomas and, most important, Alzheimer disease); secondary to other conditions (e.g., infectious Creutzfeld—Jakob disease); or hereditary (e.g., many of the corneal dystrophies, familial Alzheimer disease, Gerstmann—Sträussler—Shceinker prionosis). Amyloid deposits composed of several specific proteins can produce systemic or localized, sporadic or secondary, and hereditary or nonhereditary pathologic conditions. Each of these diseases probably shares certain pathogenic mechanisms that lead to β-fibril formation, but there must also be specific factors that dictate where, when, and how amyloid deposits occur in each of these disease entities. A similar story can be constructed for the nonfibrillary protein deposition conditions, which have only recently come to attention. These include systemic immunoglobulin light-chain (LC) deposition disease and the localized, mainly intracellular, protein deposits such as β-synuclein (Lewy bodies), tau (Pick bodies), huntingtin, and neuroserpin. All of these protein deposits are associated with one or more disease entities. However, they do not form the β-structured fibrils that we call amyloid.

In this chapter, we are concerned with those forms of AL and other protein deposition diseases that give hereditary syndromes. First, however, we should consider the more common types of AL, because they

† Deceased.

Emery and Rimoin's Principles and Practice of Medical Genetics and Genomics. https://doi.org/10.1016/B978-0-12-812534-2.00010-2

can tell us much about the mechanisms of fibrillogenesis, and they often must be included in the differential diagnosis of hereditary AL.

Immunoglobulin light chain (LC) AL is the most common type of amyloidosis. It is a sporadic (acquired) disease that increases in incidence with age. While it has no definite hereditary features other than increased incidence in families with a history of multiple myeloma, one family with autosomal dominant systemic amyloidosis due to mutation in the constant region of the kappa immunoglobulin LC gene has been described [5]. Sporadic AL amyloidosis is a monoclonal plasma cell disease in which proteolyzed fragments of immunoglobulin LC protein (κ or λ) are deposited as β-structured fibrils [6]. These deposits are extracellular and occur in major organs such as the kidney, heart, liver, and spleen, and in the blood vessel walls of most organs. This is the most aggressive form of amyloidosis, with median 1-year survival after diagnosis of approximately 18 months for nontreated patients, 5-year survival of 20%, and 10-year survival of only 5%. It may cause carpal tunnel syndrome and sensorimotor polyneuropathy and, therefore, is often mistaken for transthyretin (TTR) amyloidosis (familial amyloidotic polyneuropathy). The cardiomyopathy of some forms of hereditary apolipoprotein A-I (apo A-I) AL may also be mistaken for AL. The prognosis is markedly different.

Much is known about the pathogenesis of AL, and this is significant for the consideration of amyloid fibrillogenesis in general. For example, X-ray diffraction analysis has shown the high degree of β-pleated sheet structure in the immunoglobulin LC domains [7]. This would be expected to be an important factor in fibrillogenesis. However, the greater incidence of λ LC amyloidosis than κ LC amyloidosis suggests that there are more cryptic structural factors that influence fibrillogenesis than just the overall β structure that is shared by both κ and λ LCs. The selective deposition as fibrils of one specific monoclonal LC in the presence of a plethora of polyclonal immunoglobulins also testifies to the importance of structure in fibrillogenesis. There is tremendous interest in the physical properties of peptides that contribute to β structure and fibril formation. However, it should be remembered that, while structure of a peptide may be the basis for amyloid fibril formation, there must be specific in vivo metabolic processes that lead to the end result.

Reactive (secondary) AA AL appears to be much less common than it used to be. This may be the result of improved treatment of the infections (tuberculosis) and inflammatory (rheumatoid arthritis) diseases that predispose to this form of AL. Research on AA AL has shown that increased synthesis of the fibril precursor protein, serum amyloid A (SAA), and certain tissue factors (amyloid-enhancing factor) are important for fibrillogenesis. The primary structure of the precursor protein is also important and, in the case of inbred mice, explains the species specificity of amyloid fibril formation [8]. Of particular interest to the geneticist is the AA AL associated with autosomal recessive febrile illnesses, such as familial Mediterranean fever (FMF) [9], and autosomal dominant febrile syndromes, such as TRAPS (TNF-α receptor—associated periodic syndromes) [10], Muckle—Wells syndrome [11], and familial cold urticaria [12]. The amyloidosis is not inherited: it is a reactive form of amyloidosis, a result of chronic overproduction of the amyloid precursor protein SAA. The fact that different mutations in the FMF gene (MeV) [13,14], in the tumor necrosis factor-α (TNF-α) receptor gene [15], and in the CIAS1 gene [16] result in varying degrees of amyloid induction points to other, unknown factors in fibrillogenesis.

8.2 HEREDITARY SYSTEMIC AMYLOIDOSIS

8.2.1 Transthyretin Amyloidosis

First and foremost, among the genetically determined forms of amyloidosis and other protein deposition diseases are the TTR amyloidoses [17] (Table 8.1). This condition was originally named familial amyloidotic polyneuropathy (FAP) by Andrade, who in 1952 first described this peculiar syndrome of sensorimotor neuropathy with varying degrees of organ involvement (heart, kidney, eye) [18]. This form of AL is now known to be the most widespread and prevalent of the hereditary ALs. It is transmitted as an autosomal dominant trait owing to mutations in the plasma protein TTR (formerly named prealbumin) [19]. There are more than 120 mutations in TTR known to cause the disease, and all are single amino acid substitutions

TABLE 8.1 Transthyretin Amyloidoses[a]

Mutation	Clinical Features[b]	Geographic Kindreds
Cys10Arg	Heart, eye, PN	United States (PA)
Leu12Pro	LM	United Kingdom
Met13Lys		France
Val14Leu	PN	Italy
Asp18Glu	PN	South America, United States
Asp18Gly	LM	Hungary
Asp18Asn	Heart	United States
Ala19Asp	Heart	Caucasian
Val20Ile	Heart, CTS	Germany, United States
Arg21Gln	Heart, N	France
Ser23Asn	Heart, PN, eye	United States
Pro24Ser	Heart, CTS, PN	United States
Ala25Ser	Heart, CTS, PN	United States
Ala25Thr	LM, PN	Japan
Val28Met	PN, AN	Portugal
Val30Met	PN, AN, eye, LM	Portugal, Japan, Sweden, United States (FAP I)
Val30Ala	Heart, AN	United States
Val30Leu	PN, heart	Japan
Val30Gly	LM, eye	United States
Val32Ala	AN, heart, PN	China
Val32Gly	AN, PN	France
Phe33Ile	PN, eye	Israel
Phe33Leu	PN, heart	United States
Phe33Val	PN	United Kingdom, Japan
Phe33Cys	CTS, eye, heart, kidney	United States
Arg34Ser	PN	United States
Arg34Thr	PN, heart	Italy
Arg34Gly	Eye	United Kingdom
Lys35Asn	PN, AN, heart	France
Lys35Thr	Eye	United States
Ala36Asp	Heart, PN	Japan
Ala36Pro	Eye, CTS	United States
Asp38Ala	PN, heart	Japan
Asp38Val	Heart	Germany
Asp39Val	Heart	Germany
Thr40Asn		Germany
Trp41Leu	Eye, PN	United States
Glu42Gly	PN, AN, heart	Japan, United States, Russia
Glu42Asp	Heart	France
Phe44Leu	Heart	Caucasian
Phe44Ser	PN, AN, heart	United States
Pne44Tyr	Heart	France
Ala45Thr	Heart	United States
Ala45Gly	Heart	Dutch
Ala45Asp	Heart, PN	United States

Continued

TABLE 8.1 **Transthyretin Amyloidoses[a]—cont'd**

Mutation	Clinical Features[b]	Geographic Kindreds
Ala45Ser	Heart	Sweden
Ala45Val		France
Gly47Arg	PN, AN	Japan
Gly47Ala	Heart, AN	Italy, Germany
Gly47Val	CTS, PN, AN, heart	Sri Lanka
Gly47Glu	Heart, PN, AN	Turkey, United States, Germany
Thr49Ala	Heart, CTS	France, Italy
Thr49Ile	PN, heart	Japan
Thr49Pro	Heart	United States
Thr49Ser	PN	India
Ser50Arg	An, PN	Japan, France/Italy
Ser50Ile	Heart, PN, AN	Japan
Glu51Gly	Heart	United States
Ser52Pro	PN, AN, heart, kidney	England
Gly53Glu	LM, heart	Basque
Gly53Ala	LM	United Kingdom
Gly53Arg	LM	United States
Glu54Asp	Not listed	Germany
Glu54Gln	Heart, PN	Romania
Glu54Gly	PN, AN, eye	England
Glu54Lys	PN, AN, heart, eye	Japan
Glu54Leu	Heart	Belgium
Leu55Pro	Heart, AN, eye	United States, Taiwan
Leu55Arg	LM	Germany
Leu55Gln	Eye, PN	United States
Leu55Glu	Heart, PN, AN	Sweden
His56Arg	Heart	United States
Gly57Arg	Heart	Sweden
Leu58His	CTS, heart	United States (MD) (FAP II)
Leu58Arg	CTS, AN, eye	Japan
Thr59Arg	Heart	Japan
Thr59Lys	Heart, PN, AN	Italy
Thr60Ala	Heart, CTS	United States (Appalachian)
Glu61Lys	PN	Japan
Glu61Gly	Heart, PN	United States
Glu62Lys	Heart	Caucasian
Phe64Ala		Germany
Phe64Ile	Heart, PN, AN	Caucasian
Phe64Leu	PN, CTS, heart	United States, Italy
Phe64Ser	LM, PN, eye	Canada, England
Gly67Arg	Eye, AN	Bangladesh
Gly67Glu	Heart, PN	China
Ile68Leu	Heart	Germany
Tyr69His	Eye	United States
Tyr69Ile	Heart, CTS, AN	Japan

Continued

TABLE 8.1 Transthyretin Amyloidoses[a]—cont'd

Mutation	Clinical Features[b]	Geographic Kindreds
Lys70Asn	Eye, CTS, PN	United States
Val71Ala	PN, eye, CTS	France, Spain
Glu72Gly	Heart	Caucasian
Ile73Val	PN, AN	Bangladesh
Tyr75Ile	Heart	France
Ser77Tyr	Kidney	United States (IL, TX), France
Ser77Phe	PN, AN, heart	France
Tyr78Phe	PN, CTS, skin	Italy, France
Ala81Thr	Heart	United States
Ala81Val	Heart	United Kingdom
Gly83Arg	Eye	China
Ile84Ser	Heart, CTS, eye, LM	United States (IN), Hungary (FAP II)
Ile84Asn	Heart, eye	United States
Ile84Thr	Heart, PN	Germany, United Kingdom
His88Arg	Heart	Sweden
Glu89Gln	PN, heart	Italy
Glu89Lys	PN, heart	United States
His90Asp	Heart	United Kingdom
Ala91Ser	PN, CTS, heart	France
Glu92Lys	Heart	Japan
Val93Met	PN	Mali
Val94Ala	Heart, PN, AN	Germany, United States
Ala97Gly	Heart, PN	Japan
Ala97Ser	PN, heart, AN	Taiwan, United States
Arg103Ser	Heart	United States
Ile107Val	Heart, CTS, PN	United States
Ile107Met	PN, heart	Germany
Ile107Phe	PN, AN	United Kingdom
Ala109Ser	PN, AN	Japan
Leu111Met	Heart	Denmark
Ser112Ile	PN, heart	Italy
Pro113Thr	Heart	France
Tyr114Cys	PN, AN, eye, LM	Japan
Tyr114His	CTS	Japan
Tyr116Ser	PN, CTS, AN	France
Ala120Ser	Heart	Afro-Caribbean
Ala120Thr	PN, CTS	Japan
Val122Ile	Heart	United States
ΔVal122	Heart, PN	United States (Ecuador)
Val122Ala	Heart, eye, PN	United States

[a]Tabulation of most amyloid-associated gene mutations may be reviewed on a website maintained by the National Amyloidosis Center, London, UK (amyloidosismutations.com).[b]AN autonomic neuropathy; CTS, carpal tunnel syndrome; Eye, vitreous deposits; LM, leptomeningeal, PN, peripheral neuropathy.

except one, which is loss of one amino acid residue (ΔVal122) [17]. For this discussion, the position of mutations within a peptide uses the numbering of the mature circulating blood protein (excluding the 20 residue signal peptide). Most recent reports will present both the mature protein number and the genetically accepted (+20) number.

Transthyretin is a plasma transport protein for thyroxine and retinol-binding protein/vitamin A [20]. It is synthesized in the liver as a single polypeptide chain of 127 amino acid residues [21,22]. The gene has been localized to 18q23 [23]. A murine gene knockout model shows that TTR is not essential for life, although expression in embryos and adults suggests that it is normally important in both fetal development and adult life [24].

Mutations in TTR presumably alter the molecular structure, and this leads to aberrant degradation and aggregation of the protein as amyloid fibrils (Fig. 8.1). Transthyretin has extensive β-pleated sheet structure, and this makes it a good candidate for fibril formation [25]. Indeed, TTR amyloid fibril deposits are commonly found in the hearts of aged humans who do not have any mutant form of the protein [26,27]. It seems logical, therefore, that mutations in TTR are not necessary for amyloidosis but, instead, move the age of onset to earlier in adult life. Like most autosomal dominant diseases, TTR amyloidosis is an adult-onset disease in which the genetic mutation has been passed on to the next generation before serious clinical disease presents itself.

Clinically, TTR amyloidosis usually starts as a small fiber peripheral neuropathy affecting the extremities and autonomic functions [18,28–30]. It often presents as parasthesias or dysesthesia in the lower limbs, but may start as carpal tunnel syndrome, alternating diarrhea and constipation, impotence in males, restrictive cardiomyopathy, or, in rare forms, central nervous system (CNS) syndromes due to leptomeningeal amyloid infiltration [31–33]. Vitreous opacities due to amyloid fibril deposition are seen in one-quarter to one-third of kindreds with mutant forms of TTR [17,34] (Fig. 8.2). In fact, the presence of vitreous amyloid in a patient essentially limits the diagnosis to some form of TTR amyloidosis. Progression of the neuropathy is cephalad and may lead to sensory neuropathy over much of the body, plus loss of motor function that precludes ambulation. Even paralysis of the recurrent laryngeal nerve has been seen with this type of

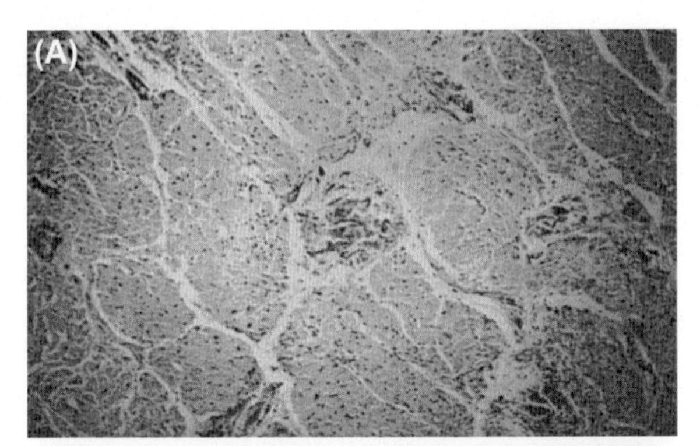

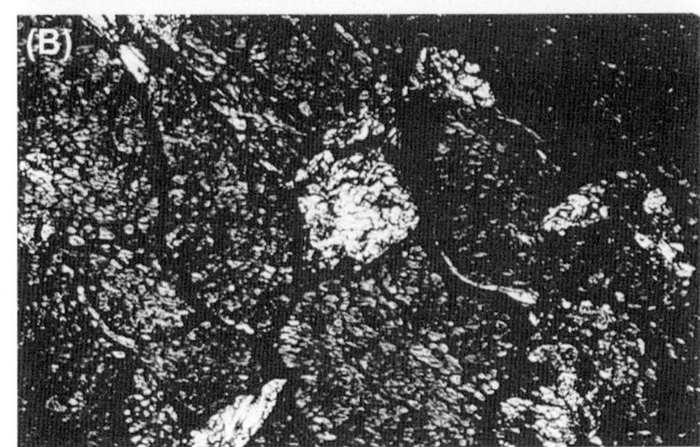

Figure 8.1 Amyloid deposits in the myocardium stain with Congo red (A) and give characteristic green birefringence when viewed by polarization microscopy (B). See Color Plate.

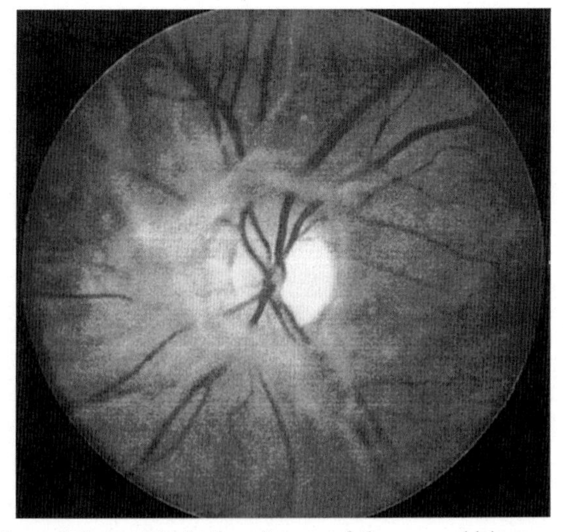

Figure 8.2 Amyloid in the vitreous of the eye, which causes progressive loss of vision, can be seen as fluffy deposits on funduscopic examination. See Color Plate.

amyloidosis. Death used to be the result of infection and inanition in many persons affected with FAP. Now, restrictive cardiomyopathy, and those syndromes associated with seriously compromised cardiac function (e.g., renal failure) are the most common cause of demise.

Disease onset varies from the third decade of life to advanced age, with examples of incomplete penetrance in a number of kindreds with different TTR mutations [35]. In general, it would seem that the disease is more rapidly progressive in individuals affected at an early age, with time from diagnosis to death being less than a decade. In older-onset syndromes (past age 50), it is not unusual to observe periods from diagnosis to death of 15–20 years. In these cases, the disease usually involves slowly progressive restrictive cardiomyopathy and lesser degrees of peripheral neuropathy.

The Val30Met form of TTR amyloidosis is the most common type of FAP. It was originally described in patients in northern Portugal, but has been found in kindreds in northern Sweden, Japan, Cypress, Turkey, and Greece, and, of course, in many American families of European descent [18,28,36,37]. Haplotype analysis has shown that the Portuguese, the Swedish, and many of the Japanese kindreds probably have the mutant gene that originated in northern Portugal [38]. Similar studies reveal that some English, Turkish, and Japanese kindreds probably have the Val30Met gene from other mutation events [39].

Other widespread forms of TTR amyloidosis include the Thr60Ala mutation, which was first discovered in a large kindred in the Appalachian region of the United States [40,41] but now has been shown to have originated in Ireland. There are kindreds of Irish ancestry with this mutation in Australia and New Zealand. The TTR Leu58His mutation was discovered in a number of kindreds in the eastern United States (Maryland and Pennsylvania) [29,42]. It has now been found in Germany in the birthplace of the ancestors of the American kindreds. Both the Thr60Ala and Leu58His mutations are associated with relatively late-onset disease with progressive cardiomyopathy.

The Ser77Tyr TTR AL has also been found in many countries, including the United States, Germany, and France [43,44]. Haplotype studies indicate that a French kindred probably had a separate mutation event [45]. Other TTR amyloidosis kindreds are relatively restricted geographically. The Indiana/Swiss Ile84Ser AL is present in mid-America, Switzerland, and one kindred in Hungary [30,34,46,47]. The Cys10Arg mutation has been found in one kindred in the United States [48]. The ΔVal122 TTR mutation, which was discovered in an Ecuadorian family, has now been found in Spain, perhaps indicating the country of origin [49].

Perhaps the greatest number of carriers of a TTR mutation that causes AL is represented by the Val122Ile mutation, which is present in 3%–4% of African Americans [50–52] Gly53Arg. This mutation, which has now been discovered in those parts of western Africa from which most American blacks originated, causes restrictive cardiomyopathy in later adult life, usually after age 60. There is often minor associated neuropathy, and the disease is frequently not recognized clinically. Instead, the patient is often thought to have heart failure due to previous hypertensive or coronary vascular disease.

8.3 CLINICAL VARIATIONS IN FAP

Although the TTR amyloidoses are all systemic diseases, the marked variation in organ system involvement leads to syndromes that obscure the underlying diagnosis (this is particularly true for certain mutations in TTR that are associated with relatively similar clinical presentations). The classic perception of FAP is of a syndrome of sensorimotor and autonomic neuropathy with malfunction of the gastrointestinal tract, orthostatic hypotension, varying degrees of restrictive cardiomyopathy, and nephropathy. This clinical presentation is typical of the Portuguese FAP, Val30Met [18], as well as Ser77Tyr [44] and a majority of the other mutations. In early studies, it was recognized that certain mutations resulted in disease with the initial presentation of carpal tunnel syndrome (FAP II), which is characteristic of the Leu58His mutation in the Maryland kindreds [29] and the Ile84Ser mutation in the Indiana kindred [30]. It has now been appreciated that peripheral neuropathy may be a minor aspect of the amyloidosis. This is particularly true of the

Thr60Ala [40], Ile68Leu, Ile84Ser [30], Ile84Asn, Ala97Gly, Ile107Val, and Val122Ile [50,52] mutations. The Leu111Met mutation originally described by Fredericksen and colleagues in Denmark is present in a kindred with restrictive cardiomyopathy and no signs of peripheral neuropathy [53]. Of particular interest are those TTR mutations that cause extensive leptomeningeal vascular amyloid deposition—Leu12Pro, Asp18Gly [54], Val30Gly [31,32], and Phe64Ser [33]—and are associated with intracranial hemorrhage, either subarachnoid or intracerebral, seizures, and hydrocephalus. In the case of two of these mutations (Asp18Gly and Val30Gly), neurodegeneration with dementia is a prominent feature of the syndrome [31,54]. Other clinical variations have been seen. In one kindred with Tyr69His, only vitreous amyloid was seen [55], whereas in another Tyr69His kindred, patients suffer with the consequences of leptomeningeal amyloidosis, and a Japanese kindred (Tyr114His) had only carpal tunnel syndrome as evidence of amyloidosis [56]. Thus, although multiorgan involvement with TTR amyloidosis may produce syndromes that are relatively easy to diagnose, the many variations in clinical presentation obviously make this group of diseases underappreciated.

8.4 GENETICS

The TTR ALs show an autosomal dominant inheritance. Most affected individuals are heterozygous for a point mutation, indicating that the disease is the result of "gain of function." This gain of function, however, would appear only to be the ability of the mutant TTR to participate in the synthesis of amyloid fibrils. A few TTR mutations have been found in the homozygous state. Homozygosity for TTR Val30Met is associated with clinical disease that is not distinguished from that seen in heterozygotes [35]. The same would appear to be true for the Val122Ile TTR mutation, which is typically expressed as restrictive cardiomyopathy in African Americans after age 60 [52,57]. Homozygosity for TTR Leu58His, however, has been reported to give earlier onset and more aggressive disease. Studies have suggested that inheritance of the mutant Val30Met gene from an affected mother results in earlier onset than inheritance from an affected father. Reduced penetrance of TTR AL is a confounding factor in diagnosis and has often resulted in a misdiagnosis of immunoglobulin amyloidosis, in which peripheral neuropathy and cardiomyopathy are often seen. Reduced penetrance is particularly high in the northern Sweden population with TTR Val30Met [35], but is also seen in the United States with the same mutation, with the Thr60Ala and Cys10Arg mutations, and with several other mutations [48]. On the other hand, extensive study of the Indiana/Swiss kindred with Ile84Ser has not identified an escapee over several generations.

Transthyretin mutations associated with amyloidosis have been found in most countries of the world (see Table 8.1). Although larger representations of mutations have been from a few countries, including Italy, Germany, Japan, France, and the United States, this is probably largely a reflection of scientific interest and medical sophistication in these countries. Since the disease has been studied at the molecular biology level, all of the mutations that have been discovered in the United States have been traced to immigrant populations. To date, no TTR mutation has been found in a Native American. No TTR mutation associated with amyloidosis has been found in Australian Aborigines or in Polynesians, and only the Val122Ile mutation has been described in individuals from Africa. This will undoubtedly change with the advancement and availability of DNA diagnostics. Several of the TTR mutations found in the United States have been traced to ancestors in Europe: the Leu58His TTR to Germany, the Val30Met to Portugal and Sweden [58], the Thr60Ala to Ireland, and the Ser77Tyr to Germany. The ΔVal122 TTR, originally discovered in a kindred in Ecuador, has now been reported from Spain [49].

Haplotype analysis based on polymorphisms in TTR has produced convincing evidence that the Val30Met mutation of the Portuguese was shared with the people of northern Sweden and with a number of kindreds in Japan; however, haplotype analysis indicates that at least two other mutation events occurred to produce the Val30Met mutation in English, Turkish, and Japanese families [38,39]. There are several CpG dinucleotide sequences in TTR, which may explain the increased incidence of some TTR mutations, including Val30Met and Val122Ile [38]. However, the large number of TTR mutations that have been discovered, and the fact that some mutations, such as Ser77Tyr, are seen on more than one haplotype and would not be predicted to have an increased mutation rate, indicate a high rate of retention of spontaneous mutations. Factors that might add to this observation include the following:

1. Transthyretin amyloidosis is an adult-onset disease and appears to have minimal effect on procreation.
2. Transthyretin, as shown by the murine knockout model, appears not to be necessary for survival and normal development [24].
3. Any loss of function by TTR mutation, such as reduced thyroxine binding or vitamin A transport, is compensated for by other biologic systems.
4. The identification of many TTR mutations may be enhanced by the fact that almost any alteration in structure of this heavily β-structured molecule will lead to abnormal metabolism and amyloid formation. There are only 12 reported mutations in TTR that have not been associated with the development of amyloidosis, whereas over 120 mutations can cause disease [17].

8.5 OTHER SYSTEMIC AMYLOIDOSES

8.5.1 Apolipoprotein A-I Amyloidosis

Apolipoprotein A-I AL was originally designated FAP III [59]. Although the first kindred with this type of amyloidosis had a clinical syndrome that included peripheral neuropathy and renal failure, the presence of hepatic amyloidosis and gastric ulcers in a large number of affected individuals indicated that it was different from FAP I and FAP II, which were subsequently shown to be related to mutations in TTR. This disparity was confirmed by the demonstration that the amyloid in FAP III was the result of a mutation in apo A-I in which an arginine was substituted for glycine at position 26 of the mature protein [60]. Subsequently, a number of mutations in apo A-I (at least 24) have been identified that are associated with systemic amyloidosis. Most are single nucleotide point mutations, but single nucleotide deletions with shift of reading frame and codon insertions have also been reported (Table 8.2). This is an autosomal dominant disease, with all affected individuals being heterozygous; however, the small numbers of kindreds with this disease have precluded an assessment of degree of penetrance. The gene for apo A-I is on chromosome 11 and codes for a mature protein of 243 amino acid residues that is the major protein constituent of plasma high-density lipoprotein. Amyloid deposits contain only the first 83–93 N-terminal residues of apo A-I and, as native apo A-I has been shown to have a predominantly α-helical structure, major rearrangement of the amino-terminal peptide must occur during catabolic processing so that β-pleated sheet amyloid fibrils are formed. One study has shown increased catabolism of mutant apo A-I Gly26Arg, and this may be a significant factor in the pathogenesis of the amyloidosis [61]. The first mutations in apo A-I associated with AL were found in the amino-terminal portion of the protein, the portion that forms amyloid fibrils. The clinical syndrome usually involves renal and hepatic amyloidosis. Only the Gly26Arg mutation is associated with neuropathy, and this has not been found in all kindreds with this mutation. Subsequently, mutations in the carboxyl-terminal portion of apo A-I were discovered and were shown to cause a different syndrome characterized more by cardiac, dermal, and laryngeal amyloid deposition (Fig. 8.3) [62–65]. These mutations indicate that incorporation into the amyloid fibril is not required to cause amyloid formation, and, indeed, alteration in metabolism is probably a major factor in pathogenesis.

8.5.2 Gelsolin-Related Amyloidosis

FAP IV was the designation given to the Finnish type of amyloidosis originally reported by Meretoja [66]. It is an adult-onset disease with lattice corneal dystrophy occurring as early as the second decade of life, followed by neuropathy of the cranial nerves, thickening and later laxity of facial skin, and a mild sensory neuropathy in the limbs. It is a systemic disease with deposits of amyloid also occurring in the heart and gastrointestinal tract. Analysis of the deposits has shown that the fibril protein is a 71–amino acid residue internal fragment of gelsolin. The disease in the original Finnish kindreds has been shown to be related to the substitution of asparagine for aspartic acid at position 187 of the mature protein [67,68]. The disease has now been reported in American families, in the Netherlands, and in Japan [69,70]. Most affected individuals are heterozygous for the Asp187Asn mutation, but a few homozygous individuals have been identified. A separate mutation at the same codon of gelsolin (Asp187Tyr) has been found in kindreds from Denmark and Czechoslovakia with similar syndromes, and more recently two mutations at other sites in gelsolin have been reported (see Table 8.2) [69].

TABLE 8.2 **Mutant Proteins Other Than Transthyretin Associated with Autosomal Dominant Systemic Amyloidosis**

Protein	Mutation	Clinical Features	Geographic Kindreds
Apolipoprotein A-I	Gly26Arg	PN[a], nephropathy	United States
	Glu34Lys	Nephropathy	Polish
	Trp50Arg	Nephropathy	United Kingdom
	Leu60Arg	Nephropathy	United Kingdom
	Leu64Pro	Nephropathy	United States, Italy
	Del 60–71 ins Val/Thr	Hepatic	Spain
	Del 70–72	Nephropathy	South Africa
	Asn74Lys(fs)[b]	Nephropathy	Germany
	Leu75Pro	Hepatic	Italy
	Leu90Pro	Cardiomyopathy, cutaneous, laryngeal	France
	Ala154(fs)[b]	Nephropathy	Germany
	His155Met(fs)[b]	Nephropathy, PN[a]	British
	Leu170Pro	Laryngeal	Germany
	Arg173Pro	Cardiomyopathy, cutaneous, laryngeal	United States
	Leu174Ser	Cardiomyopathy	Italy
	Ala175Pro	Laryngeal	United Kingdom
	Leu178His	Cardiomyopathy, laryngeal	France
Gelsolin	Gly167Arg	Nephropathy	United States
	Asn184Lys	Nephropathy	United States
	Asp187Asn	PN[a], lattice corneal dystrophy	Finland, United States, Japan
	Asp187Tyr	PN[a]	Denmark, Czech Republic
Cystatin C	Leu68Gln	Cerebral hemorrhage	Iceland
Fibrinogen Aα chain	Gly519Glu(fs)[b]	Nephropathy	France
	Phe521Leu(fs)[b]	Nephropathy	
	Val522Ala(fs)[b]	Nephropathy	United States
	Glu524Glu(fs)[b]	Nephropathy	United States
	Glu524Lys	Nephropathy	
	Thr525(fs)[b]	Nephropathy	China
	Glu526Val	Nephropathy	United States
	Glu526Lys	Nephropathy	Russia
	Thr538Lys	Nephropathy/Neuropathy	China
	Glu540Val	Nephropathy	Germany
	Pro552His	Nephropathy	Afro-Caribbean
	Arg554Leu	Nephropathy	Mexico
	Gly555Phe	Nephropathy	Norway
	Del 1636–50 ins CA 1649–50	Nephropathy	Korea

Continued

TABLE 8.2 Mutant Proteins Other Than Transthyretin Associated with Autosomal Dominant Systemic Amyloidosis—cont'd

Protein	Mutation	Clinical Features	Geographic Kindreds
Lysozyme	Tyr54Asn	Cardiomyopathy, gastrointestinal	United States
	Ile56Thr	Nephropathy, petechiae	United Kingdom
	Phe57Ile	Nephropathy	Canada
	Trp64Arg	Nephropathy	France
	Asp67Gly	Nephropathy	Rumania
	Asp67His	Nephropathy	United Kingdom
	Leu84Ser	Nephropathy	United States
	Trp112Arg	Nephropathy, gastrointestinal hemorrhage	Germany
Apolipoprotein A-II	Stop78Gly	Nephropathy	United States
	Stop78Ser	Nephropathy	United States
	Stop78Arg	Nephropathy	United States
	Stop78Leu	Nephropathy	Spain
Beta-2-microglobulin	Asp76Asn	Gastrointestinal	France
Apolipoprotein C-II	Glu69Val	Nephropathy	United States
Apolipoprotein C-III	Glu25Val	Nephropathy	France
Immunoglobulin K-LC	Cys33Ser	Nephropathy	United States

[a]PN, peripheral neuropathy.[b](fs), frame shift.

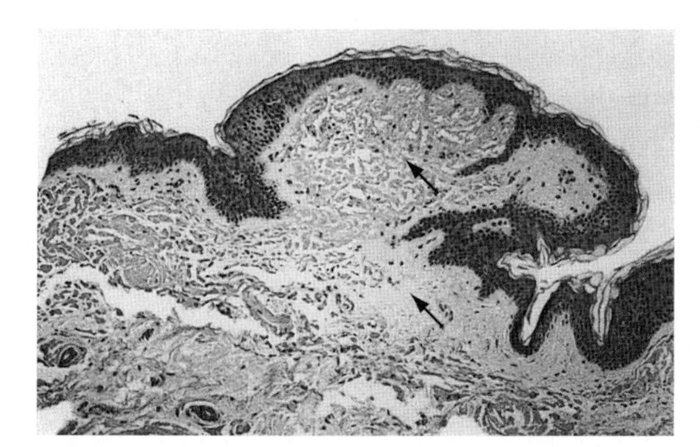

Figure 8.3 Apolipoprotein A-I amyloid deposits in the skin are characteristic of the Leu90Pro and Arg173Pro mutations. See Color Plate.

Gelsolin is a calcium-binding protein that is involved in the fragmentation of actin filaments [71,72]. It is coded by a single gene on chromosome 9 (9q32-q34), with two forms resulting from alternative splicing [71]. Cytoplasmic gelsolin has an important role in cytoskeletal reorganization, whereas plasma gelsolin binds actin and presumably functions to clear actin from the plasma. It is the plasma protein that is most likely the precursor of amyloid fibrils. In general, gelsolin amyloidosis is not associated with shortened longevity, although homozygosity for the Asp187Asn mutation is associated with severe renal disease [73]. The lattice corneal dystrophy is often not of great clinical significance and, if necessary, may be treated with corneal transplantation.

8.5.3 Lysozyme Amyloidosis

Ten mutations in lysozyme have been shown to cause systemic AL (see Table 8.2) [74–76]. In one family, a petechial skin rash was a presenting feature with subsequent renal failure (Ile56Thr), and in another family, petechial skin rash was associated with cardiomyopathy (Tyr54Asn). Other mutations—Asp67His, Trp64Arg, Phe57Ile, and Trp112Arg—are also associated with renal amyloidosis. Lysozyme is a bacteriolytic enzyme that is synthesized by polymorphonuclear leukocytes and macrophages. The gene on chromosome 12 codes for a 14,500-Da protein. The entire mature protein is incorporated into amyloid fibrils, but only the mutant form has been found in all tissues examined, which have been from heterozygous individuals [77].

8.5.4 Cystatin C Amyloidosis (Hereditary Cerebral Hemorrhage with Amyloidosis)

A mutant form of cystatin C, a serine protease inhibitor, has been shown to cause leptomeningeal vascular amyloidosis, which is associated with repeated intracranial hemorrhage [78]. The disease is systemic, with amyloid deposits in other organs such as the spleen. It is found principally in Iceland, is autosomal dominant, and usually occurs in the third or fourth decade of life. The amyloid deposits are composed of cystatin C lacking the first 10 amino acids of the mature protein [79]. A substitution of glutamine for leucine at position 58 of the amyloid subunit peptide determines the expression of the disease [80]. The gene for cystatin C is on chromosome 20.

8.5.5 Fibrinogen Aα Chain Amyloidosis

Eighteen mutations in the gene for fibrinogen Aαchain have been found to be associated with AL expressed as nephropathy (see Table 8.2) [81,82]. Five are point mutations that give single amino acid substitutions, and two are single nucleotide deletions that cause a shift in the reading frame of the messenger RNA and aberrant peptides with premature termination [83,84]. A deletion/insertion mutation associated with renal AL in a pediatric patient appears to have occurred de novo. All mutations are in the protease-sensitive carboxyl-terminal portion of the Aα chain and result in proteolytic peptides containing from 49 to 83 amino acid residues. The clinical syndrome is characteristic, with the development of hypertension followed by proteinuria and then progressive azotemia. Life can be prolonged for several years by dialysis, but hepatic and splenic involvement will occur. Renal transplantation has been shown to be followed by amyloid deposition in the transplanted organ within 1–10 years. Families in the United States, Canada, and Germany have been described with these mutations. The gene for fibrinogen Aα chain is on chromosome 4 and codes for a mature protein with a molecular weight of 66,000 Da [85]. Fibrinogen is synthesized exclusively by the liver and demonstrates a modest acute-phase response to injury. Since fibrinogen Aα protein coded by the normal allele in heterozygous individuals does not participate in amyloid fibril formation, liver transplantation prevents further production of amyloid deposition [86].

8.5.6 Apolipoprotein A-II Amyloidosis

In 1973, Weiss and Page [87] described the autopsy pathology of two sisters with renal amyloidosis. Subsequent study of affected offspring revealed that the amyloid was composed of apolipoprotein A-II that was 21 amino acid residues longer than the wild-type protein, the result of mutation in the A-II stop codon (Stop76Gly) [88]. The clinical syndrome is characterized by hypertension followed by slowly progressive renal failure. Other stop mutations (Stop76Ser, Stop76Arg, Stop76Leu) have now been described in families with similar clinical disease.

8.5.7 Apolipoprotein — CII Amyloidosis

Mutation in apolipoprotein-CII (Glu69Val) is associated with renal amyloidosis. Proteinuria and azotemia are features of this type of AL, which may occur in older individuals and be misdiagnosed as AL amyloidosis [89].

8.5.8 Apolipoprotein — CIII Amyloidosis

Mutation in apo-CIII (Glu25Val) is associated with renal amyloidosis [90]. Amyloid infiltration of salivary glands is a feature of this disease with sicca syndrome preceding the renal dysfunction, a feature that may distinguish this disease from the more common fibrinogen Aα-chain AL.

8.5.9 Immunoglobulin Kappa Light Chain Amyloidosis

One family with autosomal dominant inherited AL has been shown to have a mutation (Cys33Ser) in the constant region of their IG kappa LC gene, with renal involvement as the prominent feature of this disease [5]. Cardiac and GI amyloid deposition may occur.

8.5.10 Leukocyte Chemotactic Factor 2 Amyloidosis

Lect2 amyloidosis is a relatively frequent cause of renal amyloidosis in Mexicans although it has also been diagnosed in other ethnic groups [81]. No definite genetic pattern of inheritance has been discovered although homozygosity of the 40Val polymorphism of the Lect2 gene has been reported for most patients that have been studied. While renal involvement is the most common feature of ALect2, hepatic amyloid with a periportal globular pattern of deposition is characteristic of this disease.

8.6 DIAGNOSIS

The diagnosis of systemic hereditary AL is rarely made in a timely fashion. The systemic ALs can give a number of symptom complexes (congestive heart failure, renal failure, polyneuropathy, diarrhea) that reflect the clinical presentation of several more common and mundane diseases. Therefore, a heightened awareness of the various forms of hereditary AL is very important to early diagnosis and institution of appropriate therapies. There are a number of features of hereditary AL that should raise suspicion. A progressive small fiber sensorineuropathy with abnormalities of autonomic function is often seen with the hereditary TTR ALs. In such cases, a peripheral nerve biopsy should always be examined pathologically with the diagnosis of amyloidosis in mind, as amyloid deposits are often not appreciated on routine histologic stains (Fig. 8.4). The possible diagnosis of AL should be considered in any subject who presents with restrictive cardiomyopathy with or without orthostatic hypotension. Amyloid cardiomyopathy may give an anginal syndrome and often causes fatigue or dyspnea on exertion as signs of heart failure. Typical congestive features of heart failure, with peripheral edema, ascites, and pleural effusion, may not be seen until late in the course of the disease. The most

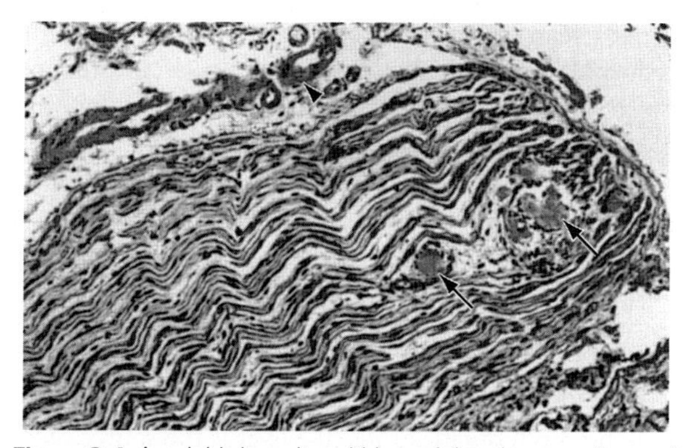

Figure 8.4 Amyloid deposits within peripheral nerves (*arrows*) are often not evenly distributed and may be missed on standard nerve biopsy. Amyloid in the walls of perineural blood vessels (*arrowhead*) may reveal the diagnosis. See Color Plate.

valuable tests for this condition are electrocardiography, which typically shows low voltage often with a pseudo-anteroseptal myocardial infarction pattern; and echocardiography or cardiac magnetic resonance imaging (MRI), which reveal a thickened interventricular septum and left ventricular posterior wall and enlarged left atrium, if the restrictive hemodynamics have been present for a long time. Chronic renal failure with or without heavy proteinuria should raise the suspicion of amyloidosis, as should unexplained impotence in males and chronic constipation alternating with diarrhea in either males or females. The presence of vitreous opacities affecting vision in any subject with heart failure or renal failure should raise the possibility of TTR AL [34]. As with most genetically determined diseases, family history can be very important in making the diagnosis. However, family history may be misleading, as the diagnosis of amyloidosis has often been confused with other diseases in previous generations, and many forms of the ALs, despite being autosomal dominant, fail to show complete penetrance. If a patient with AL has a family history of similar disease, or if there is a strong suspicion of hereditary AL, DNA analysis is usually the best way to confirm this suspicion [91]. The diagnosis of AL has first to be made by tissue biopsy of an affected organ, and although immunohistochemistry may be helpful in identifying the type of amyloid involved, the results of immunostaining are not always reliable. Most new patients who present with hereditary AL are not aware of the gene or specific mutation that is the cause of their disease, even if they have a positive family history. In this case, it is important for the diagnostician to narrow the field of differential diagnoses by analyzing the symptom complex, the ethnic origin of the patient or his or her ancestors, the age of onset of the disease, and the duration of disease before death. Table 8.1 lists the known disease-associated mutations in transthyretin and identifies where in the world significant kindreds have been identified. Unless a patient's syndrome matches that of a previously identified kindred with a specific mutation and geographic localization, it is probably not cost-effective to pursue direct DNA analysis for specific mutations. There are over 120 mutations in TTR that cause systemic amyloidosis, and although there is a polymerase chain reaction (PCR)-based test using specific restriction enzymes, allele-specific PCR, or induced mutation restriction analysis for each, it is not feasible to perform all of these tests to find the causative mutation. Instead, since automated

DNA sequencing of TTR exons is now commercially available, this form of DNA diagnosis has become the most efficient means to detect amyloid-associated mutations. Single-strand conformation polymorphism remains a useful screening procedure for new TTR mutations and for testing large numbers of subjects for a specific mutation. At the protein level, TTR variants can be detected by isoelectric focusing, but this technique is difficult and not generally available. The same is true for immunoprecipitation of TTR and the identification of mutant forms by mass spectrometry [92]. DNA testing for the other forms of hereditary AL, including apo A-I, lysozyme, fibrinogen Aα chain, and apo A-II, presents similar problems. None of these tests is readily available for the various mutations, and the clinician needs to consult one of the few laboratories dedicated to the study of amyloidosis. It is now relatively easy for the patient or physician to obtain the necessary information from the Internet (http://www.mayoclinic.com/health/amyloidosis/DS00431; http://www.iupui.edu/~amyloid; http://www.bu.edu/amyloid/; http://www.amyloidosissupport.com/index.html).

8.7 MANAGEMENT

There are three aspects to management of the patient with hereditary amyloidosis. First, the availability of DNA analysis allows the patient to have the benefit of a definitive diagnosis and avoid extensive, costly, and time-consuming medical evaluations searching for other diagnoses. Unfortunately, many patients with hereditary AL have already undergone cardiac catheterization, coronary angiography, multiple organ biopsy, and even psychiatric evaluation before the diagnosis of AL has been entertained. Once a mutated allele associated with a specific form of AL has been identified, DNA testing becomes available for at-risk individuals in the family, and these individuals may benefit from the results of this testing. Testing, of course, needs to be accompanied by proper counseling as to the genetics of the specific disease and the prognosis based on current knowledge. The identification of specific mutations associated with AL has made prenatal testing available; however, to date only prenatal testing for TTR mutations has been reported [93]. Either amniotic fluid or chorionic villus sampling is adequate for this first-trimester test, and results can be obtained within 48 h if the specific mutation in the parent is known.

A second aspect of the management of AL is the nonspecific therapies that might improve quality and length of life. For instance, a number of commonly used drugs aggravate the heart failure associated with restrictive cardiomyopathy. Treatment for chronic diarrhea, gastroparesis, urinary retention, and dysesthesias may help quality of life. Cardiac pacing and renal dialysis both may prolong life.

The third aspect of patient management centers on specific therapy or the lack thereof. In the last 20 years, approximately 2000 liver transplantations have been carried out as a specific treatment for TTR AL ([94], http://www.fapwtr.org/ram1.htm). Liver transplantation results in rapid loss of variant TTR from the plasma, as TTR is synthesized predominantly in the liver. The greatest success has been in patients with the Val30Met TTR mutation. The most immediate effect has been improvement of gastrointestinal function and possibly hypotension. There has been a report of improvement in nerve myelination in patients who have had liver transplantation early in the course of their disease. Unfortunately, there is now definite evidence that progression of systemic AL occurs in some patients who have had liver transplantation. This has been found to be true in patients with the Glu42Gly, Thr60Ala, Ile84Asn, and Cys10Arg mutations. Biochemical analysis of amyloid deposits in tissues of patients who died 3–5 years after liver transplantation have shown that normal TTR can continue to form amyloid after variant TTR has initiated the process but is no longer present [95,96]. Renal transplantation in patients with fibrinogen Aα chain AL usually results in amyloid deposition in the transplanted organ within 1–10 years. However, combined kidney and liver transplantation has shown good results, and it has been suggested that since normal fibrinogen is not found in the amyloid fibrils, liver transplantation may be a cure for the disease. Liver transplantation without renal transplant for one patient has shown no progression of disease and, actually, improvement in renal function [86]. A few liver transplantations have been done in patients with apo A-I AL and may delay progression of disease. Apolipoprotein A-I is synthesized in the liver and in the gastrointestinal tract, and it is not known whether both organs or just the liver is the source of the plasma precursor of the amyloid fibril deposits. A number of cardiac transplants have been done for patients with apo A-I cardiomyopathy and appear justified since this form of amyloidosis is very slowly progressive. Renal transplantations have also been done for patients with apo A-II amyloidosis since the renal disease is also slowly progressive [97].

To date there have been no specific therapies approved for the various forms of hereditary amyloidosis other than Tafamidis, which has been approved in the European Union for treatment of stage 1 TTR polyneuropathy [98]. This drug has now been the subject of a study to treat TTR cardiomyopathy and while preliminary reports are positive, it has not yet been approved by the FDA. Two drug trials testing the efficacy of inhibiting hepatic synthesis of TTR and thereby stopping progression of TTR amyloid polyneuropathy have been completed and report significant results [99,100].

Both an antisense oligonucleotide specific to TTR mRNA and an siRNA specific to TTR have shown very significant inhibition of progression of polyneuropathy in patients with hereditary TTR AL [101–103]. Both agents are presently under review by the FDA. These positive results suggest that TTR cardiomyopathy may also be inhibited but controlled studies have not been completed.

Other, rarer, forms of hereditary AL need similar attention by both basic science and clinical disciplines.

8.7.1 Hereditary Localized Amyloidosis

There are a number of localized forms of hereditary AL, each characterized by a specific protein that, because of a mutation in structure or a genetically determined alteration in metabolism, gains the features to participate in amyloid fibril formation. In most cases, the localized nature of the amyloid deposition is the result of synthesis of the fibril precursor protein that is restricted to the specific organ or tissue involved. There are, however, other factors necessary for the generation of amyloid deposits, and these must be present in proximity to the site of synthesis of the amyloid precursor protein. In most forms of localized amyloidosis, the fibril subunit protein is a proteolytic fragment of a larger precursor protein. Therefore, proteases that generate these fragments must be present and functional in the tissue where amyloid fibril deposition occurs. The same must be true for the various forms of proteoglycans, which are invariably part of amyloid deposits. As with the systemic forms of AL, once the amyloid fibrils are formed they must be resistant to removal by any localized tissue mechanisms.

8.8 ALZHEIMER DISEASE

By far the most common form of localized amyloidosis is seen in Alzheimer disease. Amyloid deposits, which are localized to cerebral cortical tissues and blood vessels, contain a 39– to 42–amino acid residue internal fragment from the C-terminal portion of a precursor protein expressed in the CNS [104]. This protein, β-amyloid precursor protein (β-APP), is coded by a single gene on chromosome 21 and has several alternatively spliced transcripts, the largest encoding a protein of 770 amino acid residues [105]. A few forms of familial Alzheimer disease are the result of point mutations in the β-APP gene that cause early-onset autosomal dominant dementia [106–109]. A larger number of familial forms of Alzheimer disease are the result of mutations in the presenilin genes (PS1 and PS2) localized to chromosomes 14 and 1 [110–113]. The majority of cases of Alzheimer disease, however, are age-related and do not show simple mendelian genetics. A current hypothesis for pathogenesis is that mutations associated with hereditary forms of Alzheimer disease accelerate the age of onset and progression of amyloid plaque formation, which is a normal aging phenomenon and may be modulated by such factors as apolipoprotein E.

8.9 GERSTMANN–STRÄUSSLER–SCHEINKER DISEASE

Gerstmann-Sträussler-Scheinker disease (GSS) is another hereditary form of amyloidosis localized to the CNS [114]. In this neurodegenerative disease, the amyloid deposits are derived from the prion protein, which is also associated with Creutzfeld–Jakob disease and other transmissible forms of spongiform encephalopathy [115,116]. Variant forms of GSS are associated with missense mutations or numbers of octapeptide repeats in the prion gene localized to chromosome 20 [117]. Pathogenesis is hypothesized to be related to transition in the tertiary structure of a protease-sensitive form of prion protein to a protease-resistant form that is capable of forming β-pleated sheet fibrils. Considerable variability in the clinical expression of the prion diseases is seen, and pathologic manifestations may include varying degrees of spongiform encephalopathy, cortical amyloid deposits, and amyloid angiopathy.

However, all are adult-onset diseases with autosomal dominant inheritance manifested as progressive neurodegeneration.

8.10 BRITISH DEMENTIA

Another form of localized amyloidosis limited to the CNS is the autosomal dominant familial British dementia, characterized by progressive dementia, spasticity, and cerebellar ataxia [118]. In this condition, cerebral amyloid angiopathy, nonneuritic and perivascular plaques, and neurofibrillary tangles represent the pathology. The disease is the result of a mutation in the stop codon for the BRI gene, which is localized to chromosome 13 [119]. Owing to a single base substitution in the stop codon, a larger 277-residue protein is produced. A carboxyl-terminal fragment containing 34 amino acid residues of the mutated protein is the ABRI amyloid subunit. A similar syndrome with dementia in a Danish kindred is associated with a 10-nucleotide duplication proximal to the stop codon of the BRI gene. This results in a new C-terminal peptide that is the precursor of amyloid fibrils [120].

8.11 CORNEAL DYSTROPHIES

A number of corneal dystrophies are examples of localized amyloidosis that are genetically determined [121] (Fig. 8.5). Because of the localized nature and small amount of amyloid in the corneal tissues, the chemical nature of the amyloid deposits remained unknown until recently, when genetic studies and molecular biology techniques identified causative genes and mutations. Lattice corneal dystrophy (type 2), which is related to mutations in the gelsolin gene, was the first corneal amyloid to be chemically characterized, but this is a systemic disease and the protein structure was determined from fibrils isolated from major organs [68]. A number of point mutations in the Big-h3 gene, localized to chromosome 5q31, have been found to cause varying phenotypes of corneal dystrophy [122,123]. Mutations in the M1S1 gene (chromosome 1p), which encodes a gastrointestinal tumor—associated antigen, have also been found to cause corneal amyloid deposition [124]. Although it is now possible to separate several of the various types of corneal dystrophy by DNA analysis, chemical characterization of the deposited amyloid fibril proteins has been delayed by the small amount and localized nature of the deposits. Given the unique environment of the cornea, characterization of the proteolytic processing and factors involved in deposition of amyloid fibrils in these diseases may be very revealing for amyloid fibrillogenesis in general.

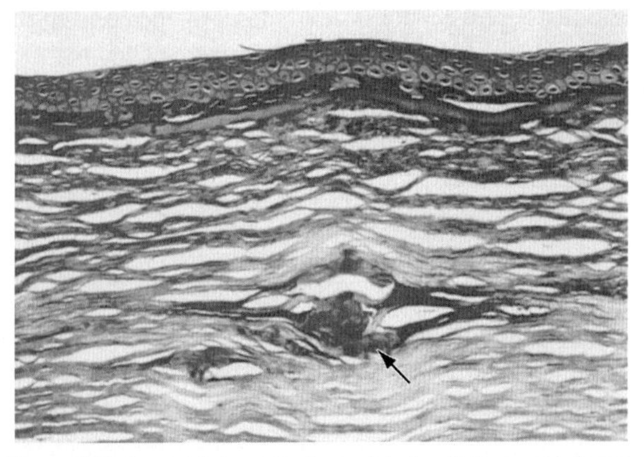

Figure 8.5 Amyloid deposits (*arrow*) in familial corneal dystrophy. See Color Plate.

8.12 OTHER LOCALIZED AMYLOIDOSES

There are numerous other forms of localized amyloidosis that are expressed in a familial manner. In two of these, the amyloid subunit protein has been identified. Isolated atrial amyloid contains a 28—amino acid residue carboxyl-terminal peptide of atrial natriuretic peptide [125,126]. In most cases, deposition of this form of amyloid in the cardiac atria is age-related and associated with chronic congestive heart failure. The familial atrial standstill syndrome is associated with this type of amyloid deposition and is inherited as an autosomal dominant condition [127]. Amyloid that is deposited in medullary carcinoma of the thyroid is derived from procalcitonin [128]. In the familial form of medullary carcinoma of the thyroid, an autosomal dominant condition that may be associated with other endocrinopathies, the amyloid deposition may be a result of overproduction of the amyloid precursor procalcitonin, as no structural defect or mutation in the amyloid precursor protein has been identified [129].

8.12.1 Hereditary Nonamyloid Protein Deposition Disease

While the number of amyloid proteins continues to grow, attention is now being directed toward diseases in which protein deposits occur intracellularly and the protein may or may not form fibrillar structures. Some of these diseases are hereditary and, as with most inherited forms of amyloidosis, are autosomal dominant. Included in this group are Huntington disease, frontotemporal dementia linked to chromosome 17, and familial Parkinson disease.

In Huntington disease, the disorder is associated with expanded CAG repeats in the Huntington gene (chromosome 4). Intranuclear deposits of the expressed protein with polyglutamine sequences are found in affected individuals, but the role of these protein deposits in the pathogenesis of the neurodegeneration is not known. Pick disease is characterized by presenile dementia with circumscribed cerebral atrophy. In this disease, protein deposits (Pick bodies) within the cytoplasm of cortical neurons contain abnormal forms of the microtubule-associated protein tau [130]. Familial forms of Pick disease are associated with mutations in the tau gene (chromosome 17) and appear to change the ratio of expressed isoforms of tau protein [112]. The role of protein deposition in the pathogenesis of the disease is still not known. Autosomal dominant Parkinson disease is associated with intracellular deposits that contain α-synuclein. In a few families with hereditary Parkinson disease, mutations in the α-synuclein gene (chromosome 4) have been found to be the cause of the disease [131].

Recently, families with progressive myoclonus epilepsy and dementia have been found to have intraneuronal cytoplasmic inclusions of neuroserpin, a serine protease inhibitor (Fig. 8.6). Two mutations in the neuroserpin gene (chromosome 3q26) have been found to cause these syndromes [132,133]. Mutation in the ferritin light polypeptide gene also causes intracellular protein deposits associated with dementia. Both of these syndromes are examples of pathology associated with protein deposition without actual amyloid β-fibril formation. Although the significance of such protein deposits in the pathogenesis of dementia is not known, identification of variant forms of intracellular proteins in these neurodegenerative diseases will certainly add to our knowledge of neuronal function.

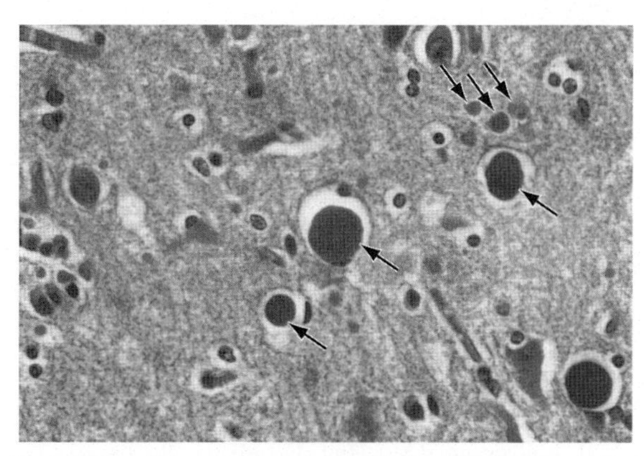

Figure 8.6 Intraneuronal nonamyloid deposits contain mutant neuroserpin in a patient with hereditary myoclonus epilepsy and dementia. See Color Plate.

8.13 CONCLUSION

The field of AL and other protein deposition diseases continues to evolve, not only in recognition of the genetic basis of previously enigmatic diseases but in understanding the processes by which normal biologic proteins transition from soluble to insoluble, and pathologic structure. This knowledge has, and certainly will continue, to provide the basis for new therapeutic interventions.

REFERENCES

[1] Bonar L, Cohen AS, Skinner MM. Characterization of the amyloid fibrils as a cross-B protein. Proc Soc Exp Biol Med 1969;131:1373–5.

[2] Cohen AS, Calkins E. Electron microscopic observation on a fibrous component in amyloid of diverse origins. Nature 1959;183:1202–3.

[3] Glenner GG. Amyloid deposits and amyloidosis: the β-fibrilloses. N Engl J Med 1980;302:1283–92. 1333–1343.

[4] Sipe JD, Benson MD, Buxbaum JN, Ikeda S, Merlini G, Saraiva MJ, Westermark P. Amyloid fibril proteins and amyloidosis: chemical identification and clinical classification International Society of Amyloidosis 2016 Nomenclature Guidelines. Amyloid 2016;23:209–13.

[5] Benson MD, Liepnieks JJ, Kluve-Beckerman B. Hereditary systemic immunoglobulin light-chain amyloidosis. Blood 2015;125:3281–6.

[6] Glenner GG, Terry W, Harada M, Isersky C, Page D. Amyloid fibril proteins: proof of homology with

immunoglobulin light chains by sequence analysis. Science 1971;172:1150–1.

[7] Poljak RJ, Anzel LM, Chen BL, Phizackerley RP, Saul F. The three-dimensional structure of the fab' fragment of a human myeloma immunoglobulin at 2.0 Å resolution. Proc Natl Acad Sci U S A 1974;71:3440–4.

[8] Hoffman JS, Ericsson LH, Eriksen N, Walsh KA, Benditt EP. Murine tissue amyloid protein AA NH2-terminal sequence identity with only one of two serum amyloid protein (ApoSAA) gene products. J Exp Med 1984;159:641–6.

[9] Heller H, Sohar E, Sherf L. Familial Mediterranean fever. Arch Intern Med 1958;102:50–71.

[10] Williamson LM, Hull D, Mehta R, Reeves WG, Robinson BH, Toghill PJ. Familial Hibernian fever. Q J Med 1982;51:469–80.

[11] Muckle TJ, Wells M. Urticaria, deafness and amyloidosis: a new heredo-familial syndrome. Q J Med 1962;31:235–48.

[12] Tindall JP, Beeker SK, Rosse WF. Familial cold urticaria: a generalized reaction involving leukocytosis. Arch Intern Med 1969;124:129–34.

[13] French FMF Consortium. A candidate gene for familial Mediterranean fever. Nat Genet 1997;17:25–31.

[14] International FMF Consortium. Ancient missense mutations in a new member of the RoRet gene family are likely to cause familial Mediterranean fever. Cell 1997;90:797–807.

[15] McDermott MF, Aksentijevich I, Galon J, McDermott EM, Ogunkolade BW, Centola M, Mansfield E, Gadina M, Karenko L, Pettersson T, McCarthy J, Frucht DM, Aringer M, Torosyan Y, Teppo AM, Wilson M, Karaarslan HM, Wan Y, Todd I, Wood G, Schlimgen R, Kumarajeewa TR, Cooper SM, Vella JP, Amos CI, Mulley J, Quane KA, Molloy MG, Ranki A, Powell RJ, Hitman GA, O'Shea JJ, Kastner DL. Germline mutations in the extracellular domains of the 55kDa TNF receptor, TNFR1, define a family of dominantly inherited autoinflammatory syndromes. Cell 1999;97:133–44.

[16] Hoffman HM, Muellor JL, Broide DH, Wanderer AA, Kolodner RD. Mutation of a new gene encoding a putative pyrin-like protein causes familial cold autoinflammatory syndrome and Muckle-Wells syndrome. Nat Genet 2001;29:301–5.

[17] Benson MD, Uemichi T. Review—transthyretin amyloidosis. Amyloid Int J Exp Clin Invest 1996;3:44–56.

[18] Andrade C. A peculiar form of peripheral neuropathy: familial atypical generalized amyloidosis with special involvement of the peripheral nerves. Brain 1952;75:408–28.

[19] Costa PP, Figuera AS, Bravo FR. Amyloid fibril protein related to prealbumin in familial amyloidotic polyneuropathy. Proc Natl Acad Sci USA 1978;75:4499–503.

[20] Robbins J. Thyroxine-binding proteins. Prog Clin Biol Res 1976;5:331–55.

[21] Costa RH, Lai E, Darnell JE. Transcriptional control of the mouse prealbumin (transthyretin) gene: both promotor sequences and a distinct enhancer are cell specific. Mol Cell Biol 1986;6:4697–708.

[22] Kanda Y, Goodman DS, Canfield RE, Morgan FJ. The amino acid sequence of human plasma prealbumin. J Biol Chem 1974;249:6796–805.

[23] Wallace MR, Naylor SL, Kluve-Beckerman B. Localization of the human prealbumin gene to chromosome 18. Biochem Biophys Res Commun 1985;129:753–8.

[24] Episkopou V, Maeda S, Nishiguchi S, Shimada K, Gaitanaris GA, Gottesman ME, Robertson EJ. Disruption of the transthyretin gene results in mice with depressed levels of plasma retinol and thyroid hormone. Proc Natl Acad Sci U S A 1993;90:2375–9.

[25] Blake CCF, Geisow MJ, Oatley SJ. Structure of prealbumin: secondary, tertiary and quaternary interactions determined by Fourier refinement at 1.8 Å. J Mol Biol 1978;121:339–56.

[26] Pitkanen P, Westermark P, Cornwell GG. Senile systemic amyloidosis. Am J Pathol 1984;117:391–9.

[27] Westermark P, Sletten K, Johansson B, Cornwell GG. Fibril in senile systemic amyloidosis is derived from normal transthyretin. Proc Natl Acad Sci U S A 1990;87:2843–5.

[28] Benson MD, Cohen AS. Generalized amyloid in a family of Swedish origin: a study of 426 family members in 7 generations of a new kinship with neuropathy, nephropathy and central nervous system involvement. Ann Intern Med 1977;86:419–24.

[29] Mahloudji M, Teasdall RD, Adamkiewicz JJ, Hartmann WH, Lambird PA, McKusick VA. The genetic amyloidoses: with particular reference to hereditary neuropathic amyloidosis, type II (Indiana or Rukavina type). Medicine 1969;48:1–37.

[30] Block WD, Carey JG, Curtis AC, Falls HF, Jackson CE, Rukavina JG. Primary systemic amyloidosis: a review and an experimental, genetic, and clinical study of 29 cases with particular emphasis on the familial form. Medicine 1956;35:239–334.

[31] Goren H, Steinberg MC, Farboody GH. Familial oculoleptomeningeal amyloidosis. Brain 1980;103:473–95.

[32] Petersen RB, Goren H, Cohen M, Richardson SL, Tresser N, Lynn A, Gali M, Estes M, Gambetti P. Transthyretin amyloidosis: a new mutation associated with dementia. Ann Neurol 1997;41:307–13.

[33] Uemichi T, Uitti RJ, Koeppen AH, Donat JR, Benson MD. Oculoleptomeningeal amyloidosis associated with a new transthyretin variant Ser64. Arch Neurol 1999;56:1152–5.
[34] Falls HF, Jackson J, Carey JH, Rukavina JG, Block WD. Ocular manifestations of hereditary primary systemic amyloidosis. Arch Ophthalmol 1955;54:660–4.
[35] Holmgren G, Bergstrom S, Drugge U, Lundgren E, Nording-Sikström C, Sandgren O, Steen L. Homozygosity for the transthyretin-Met30-gene in seven individuals with familial amyloidosis with polyneuropathy detected by restriction enzyme analysis of amplified genomic DNA sequences. Clin Genet 1992;41:39–41.
[36] Mita S, Maeda S, Shimada K, Araki S. Cloning and sequence analysis of cDNA for human prealbumin. Biochem Biophys Res Commun 1984;124:558–64.
[37] Sasaki H, Sakaki Y, Matsuo H, Goto I, Kuroiwa Y, Sahashi I, Takahashi A, Shinoda T, Isobe T, Takagi Y. Diagnosis of familial amyloidotic polyneuropathy by recombinant DNA techniques. Biochem Biophys Res Commun 1984;125:636–42.
[38] Yoshioka K, Furuya H, Sasaki H, Saraiva MJ, Costa PP, Sakaki Y. Haplotype analysis of familial amyloidotic polyneuropathy. Hum Genet 1989;82:9–13.
[39] Waits RP, Uemichi T, Benson MD. Haplotype analysis of the transthyretin gene: evidence for multiple recurrence of the Met30 mutation in the Caucasian population. Amyloid Int J Exp Clin Invest 1995;2:114–8.
[40] Benson MD, Wallace MR, Tejada E, Baumann H, Page B. Hereditary amyloidosis: description of a new American kindred with late onset cardiomyopathy. Arthritis Rheum 1987;30:195–200.
[41] Wallace MR, Dwulet FE, Conneally PM, Benson MD. Biochemical and molecular genetic characterization of a new variant prealbumin associated with hereditary amyloidosis. J Clin Invest 1986;78:6–12.
[42] Nichols WC, Liepnieks JJ, McKusick VA, Benson MD. Direct sequencing of the gene for Maryland/German familial amyloidotic polyneuropathy type II and genotyping by allele-specific enzymatic amplification. Genomics 1989;5:535–40.
[43] Satier F, Nichols WC, Benson MD. Diagnosis of familial amyloidotic polyneuropathy in France. Clin Genet 1990;38:469–73.
[44] Wallace MR, Dwulet FE, Williams EC, Conneally PM, Benson MD. Identification of a new hereditary amyloidosis prealbumin variant, Tyr-77, and detection of the gene by DNA analysis. J Clin Invest 1988;81:189–93.
[45] Zhao N, Aoyama N, Benson MD, Skinner M, Satier F, Sakaki Y. Haplotype analysis of His58, Ala60, and Tyr77 types of familial amyloidotic polyneuropathy. Amyloid Int J Exp Clin Invest 1994;1:75–9.
[46] Dwulet FE, Benson MD. Characterization of a transthyretin (prealbumin) variant associated with familial amyloidotic polyneuropathy type II (Indiana/Swiss). J Clin Invest 1986;78:880–6.
[47] Zólyomi Z, Benson MD, Halász K, Uemichi T, Fekete G. Transthyretin mutation (serine 84) associated with familial amyloid polyneuropathy in a Hungarian family. Amyloid Int J Exp Clin Invest 1998;5:30–4.
[48] Uemichi T, Murrell JR, Zeldenrust S, Benson MD. A new mutant transthyretin (Arg 10) associated with familial amyloid polyneuropathy. J Med Genet 1992;29:888–91.
[49] Uemichi T, Liepnieks JJ, Waits RP, Benson MD. A trinucleotide deletion in the transthyretin gene (ΔV122) in a kindred with familial amyloidotic polyneuropathy. Neurology 1997;48:1667–70.
[50] Gorevic PD, Prelli FC, Wright J, Pras M, Frangione B. Systemic senile amyloidosis. Identification of a new prealbumin (transthyretin) variant in cardiac tissue: immunologic and biochemical similarity to one form of familial amyloidotic polyneuropathy. J Clin Invest 1989;83:836–43.
[51] Jacobson DR, Pastore R, Pool S, Malendowicz S, Kane I, Shivji A, Embury SH, Ballas SK, Buxbaum JN. Revised transthyretin Ile122 allele frequency in African-Americans. Hum Genet 1996;98:236–8.
[52] Nichols WC, Liepnieks JJ, Snyder EL, Benson MD. Senile cardiac amyloidosis associated with homozygosity for a transthyretin variant (IIe-122). J Lab Clin Med 1991;117:175–80.
[53] Frederiksen T, Gotzsche H, Harboe N, Kiaer W, Mellemgaard K. Familial primary amyloidosis with severe amyloid heart disease. Am J Med 1962;33:328–48.
[54] Vidal R, Garzuly F, Budka H, Lalowski M, Linke RP, Brittig F, Frangione B, Wisniewski T. Meningocerebrovascular amyloidosis associated with a novel transthyretin mis-sence mutation at codon 18 (TTRD18G). Am J Pathol 1996;148:361–6.
[55] Zeldenrust SR, Skinner M, Harding J, Share J, Benson MD. A new transthyretin variant (His69) associated with vitreous amyloid in an FAP family. Amyloid 1994;1:17–22.
[56] Murakami T, Tachibana S, Endo Y, Kawai R, Hara M, Tanase S, Ando M. Familial carpal tunnel syndrome due to amyloidogenic transthyretin His114 variant. Neurology 1994;44:315–8.
[57] Jacobson DR, Gorevic PD, Buxbaum JN. A homozygous transthyretin variant associated with senile systemic amyloidosis: evidence for a late-onset

disease of genetic etiology. Am J Hum Genet 1990;47:127–36.
[58] Dwulet FE, Benson MD. Polymorphism of human plasma thyroxine binding prealbumin. Biochem Biophys Res Commun 1983;114:657–62.
[59] Van Allen MW, Frohlich JA, Davis JR. Inherited predisposition to generalized amyloidosis. Neurology 1969;19:10–25.
[60] Nichols WC, Gregg RE, Brewer Jr HB, Benson MD. A mutation in apolipoprotein A-I in the Iowa type of familial amyloidotic polyneuropathy. Genomics 1990;8:318–23.
[61] Rader DJ, Gregg RE, Meng MS, Schaefer JR, Zech LA, Benson MD, Brewer Jr HB, et al. In vivo metabolism of a mutant apolipoprotein, apoA-IIowa, associated with hypoalphalipoproteinemia and hereditary systemic amyloidosis. J Lipid Res 1992;33:755–63.
[62] Hamidi Asl K, Liepnieks JJ, Nakamura M, Parker F, Benson MD. A novel apolipoprotein A-1 variant, Arg 173Pro, associated with cardiac and cutaneous amyloidosis. Biochem Biophys Res Commun 1999;257:584–8.
[63] Hamidi Asl L, Liepnieks JJ, Hamidi Asl K, Uemichi T, Moulin G, Desjoyaux E, Loire R, Delpech M, Grateau G, Benson MD. Hereditary amyloid cardiomyopathy caused by a variant apolipoprotein A1. Am J Pathol 1999;154:221–7.
[64] Mendes de Sousa M, Vital C, Ostler D, Fernandes R, Pouget-Abadie J, Carles D, Saraiva MJ. Apolipoprotein AI and transthyretin as components of amyloid fibrils in a kindred with apoAI Leu178His amyloidosis. Am J Pathol 2000;156:1911–7.
[65] Obici L, Bellotti V, Mangione P, Stoppini M, Arbustini E, Verga L, Zorzoli I, Anesi E, Zanotti G, Campana C, Viganò M, Merlini G. The new apolipoprotein A-I variant Leu 174-Ser causes hereditary cardiac amyloidosis, and the amyloid fibrils are constituted by the 93-residue N-terminal polypeptide. Am J Pathol 1999;155:695–702.
[66] Meretoja J. Familial systemic paramyloidosis with lattice dystrophy of the cornea, progressive cranial neuropathy, skin changes and various internal symptoms. Ann Clin Res 1969;1:314–24.
[67] Levy E, Haltia M, Fernandez-Madrid I, Koivunen O, Ghiso J, Prelli F, Frangione B. Mutation in gelsolin gene in Finnish hereditary amyloidosis. J Exp Med 1990;172:1865–7.
[68] Maury CPJ. Isolation and characterization of cardiac amyloid in familial amyloid polyneuropathy type IV (Finnish): relation of the amyloid protein to variant gelsolin. Biochim Biophys Acta 1990;1096:84–6.
[69] De la Chapelle A, Tolvanen R, Boysen G, Santavy J, Bleeker-Wagemakers L, Maury CP, Kere J. Gelsolin-derived familial amyloidosis caused by asparagine or tyrosine substitution for aspartic acid at residue 187. Nat Genet 1992;2:157–60.
[70] Sunada Y, Shimizu T, Nakase H, Ohta S, Asaoka T, Amano S, Sawa M, Kagawa Y, Kanazawa I, Mannen T. Inherited amyloid polyneuropathy type IV (gelsolin variant) in a Japanese family. Ann Neurol 1993;33:57–62.
[71] Kwiatkowski DJ, Mehl R, Yin HL. Genomic organization and biosynthesis of secreted and cytoplasmic forms of gelsolin. J Cell Biol 1988;106:375–84.
[72] Yin HL, Kwiatkowski DJ, Mole JE, Cole FS. Structure and biosynthesis of cytoplasmic and secreted variants of gelsolin. J Biol Chem 1984;259:5271.
[73] Maury CPJ, Kere J, Tolvanen R, de la Chapelle A. Homozygosity for the Asn187 gelsolin mutation in Finnish-type familial amyloidosis is associated with severe renal disease. Genomics 1992;13:902–3.
[74] Pepys MB, Hawkins PN, Booth DR, Vigushin DM, Tennent GA, Soutar AK, Totty N, Nguyen O, Blake CCF, Terry CJ, Feest TG, Zalin AM, Hsuan JJ. Human lysozyme gene mutations cause hereditary systemic amyloidosis. Nature 1993;362:553–7.
[75] Valleix S, Drunat S, Philit J-B, Adoue D, Piette JC, Droz D, MacGregor B, Canet D, Delpech M, Grateau G. Hereditary renal amyloidosis caused by a new variant lysozyme W64R in a French family. Kidney Int 2002;61:907–12.
[76] Yazaki M, Farrell SA, Benson MD. A novel lysozyme mutation Phe57Ile associated with hereditary renal amyloidosis. Kidney Int 2003;63:1652–7.
[77] Booth DR, Sunde M, Bellotti V, Robinson CV, Hutchinson WL, Fraser PE, Hawkins PN, Dobson CM, Radford SE, Blake CC, Pepys MB. Instability, unfolding and aggregation of human lysozyme variants underlying amyloid fibrillogenesis. Nature 1997;385:787–93.
[78] Gudmundsson G, Hallgrimsson J, Jonasson TA, Bjarnason O. Hereditary cerebral hemorrhage with amyloidosis. Brain 1972;95:387–404.
[79] Cohen DH, Feiner H, Jensson O, Frangione B. Amyloid fibril in hereditary cerebral hemorrhage with amyloidosis (HCHWA) is related to gastroentero-pancreatic neuroendocrine protein, gamma trace. J Exp Med 1983;158:623–8.
[80] Abrahamson M, Jonsdottir S, Olafsson I, Jensson O, Grubb A. Hereditary cystatin C amyloid angiopathy: identification of the disease-causing mutation and specific diagnosis by polymerase chain reaction based analysis. Hum Genet 1992;89:377–80.

[81] Benson MD, Liepnieks J, Uemichi T, Wheeler G, Correa R. Hereditary renal amyloidosis associated with a mutant fibrinogen α-chain. Nat Genet 1993;3:252–5.
[82] Uemichi T, Liepnieks JJ, Benson MD. Hereditary renal amyloidosis with a novel variant fibrinogen. J Clin Invest 1994;93:731–6.
[83] Hamidi Asl L, Liepnieks JJ, Uemichi T, Rebibou JM, Justrabo E, Droz D, Mousson C, Chalopin JM, Benson MD, Delpech M, Grateau G. Renal amyloidosis with a frame shift mutation in fibrinogen Aα-chain gene producing a novel amyloid protein. Blood 1997;90:4799–805.
[84] Uemichi T, Liepnieks JJ, Yamada T, Gertz MA, Bang N, Benson MD. A frame shift mutation in the fibrinogen Aα-chain gene in a kindred with renal amyloidosis. Blood 1996;87:4197–203.
[85] Doolittle RF, Watt KWK, Cottrell BA, Strong DD, Riley M. The amino acid sequence of the α-chain of human fibrinogen. Nature 1979;280:464–8.
[86] Fix OK, Stock PG, Lee BK, Benson MD. Liver transplant alone without kidney transplant for fibrinogen Aα-chain (AFib) renal amyloidosis. Amyloid 2015;23:132–3.
[87] Weiss SW, Page DL. Amyloid nephropathy of Ostertag with special reference to renal glomerular giant cells. Am J Pathol 1973;72:447–60.
[88] Benson MD, Liepnieks JJ, Yazaki M, Yamashita T, Hamidi Asl K, Guenther B, Kluve-Beckerman B. A new human hereditary amyloidosis: the result of a stop-codon mutation in the apolipoprotein AII gene. Genomics 2001;72:272–7.
[89] Nasr SH, Dasari S, Hasadari L, Theis JD, Vrana JA, Gertz MA, Prasuna M, Zimmermann MT, Grogg KL, Dispenzieri A, Sethi S, Highsmith Jr WE, Merlini G, Leung N, Kurtin PJ. Novel type of renal amyloidosis derived from apolipoprotein-CII. J Am Soc Nephrol 2017;28:439–45.
[90] Valleix S, Verona G, Jourde-Chiche N, Nédelec B, Mangione PP, Bridoux F, Mange A, Dogan A, Goujon J-M, Lhomme M, Dauteuille C, Chabert M, Porcari R, Waudby CA, Relini A, Talmud PJ, Kovrov O, Olivecrona G, Stoppini M, Christodoulou J, Hawkins PN, Grateau G, Delpech M, Kontush A, Gillmore JD, Kalopissis AD, Bellotti V. D25V apolipoprotein C-III variant causes dominant hereditary systemic amyloidosis and confers cardiovascular protective lipoprotein profile. Nat Commun 2016;7:10353.
[91] Nichols WC, Benson MD. Hereditary amyloidosis: detection of variant prealbumin genes by restriction enzyme analysis of amplified genomic DNA sequences. Clin Genet 1990;37:44–53.
[92] Ranløv I, Ando Y, Ohlsson PI, Holmgren G, Ranløv PJ, Suhr OB. Rapid screening for amyloid-related variant forms of transthyretin is possible by electrospray ionization mass spectrometry. Eur J Clin Invest 1997;27:956–9.
[93] Morris M, Nichols WC, Benson MD. Prenatal diagnosis of hereditary amyloidosis in a Portuguese family. Am J Med Genet 1991;39:123–4.
[94] Holmgren G, Ericzon B-G, Growth C-G, Steen L, Suhr O, Andersen O, Wallin BG, Seymour A, Richardson S, Hawkins PN, Pepys MB. Clinical improvement and amyloid regression after liver transplantation in hereditary transthyretin amyloidosis. Lancet 1993;341:1113–6.
[95] Liepnieks JJ, Benson MD. Progression of cardiac amyloid deposition in hereditary transthyretin amyloidosis patients after liver transplantation. Amyloid 2007;14:277–82.
[96] Liepnieks JJ, Zhang LQ, Benson MD. Progression of transthyretin amyloid neuropathy after liver transplantation. Neurology 2010;75:324–7.
[97] Magy N, Liepnieks JJ, Yazaki M, Kluve-Beckerman B, Benson MD. Renal transplantation for apolipoprotein AII amyloidosis. Amyloid J Protein Fold Disord 2003;10:224–8.
[98] Cruz MW, Benson MD. A review of Tafamidis for the treatment of transthyretin-related amyloidosis. Neurol Ther 2015;4:61–79.
[99] Ackermann EJ, Guo S, Benson MD, Booten S, Freier S, Hughes SG, Kim TW, Jesse Kwo T, Matson J, Norris D, Yu R, Watt A, Monia BP. Suppressing transthyretin production in mice, monkeys and humans using 2nd-generation antisense oligonucleotides. Amyloid 2016;23:148–57.
[100] Benson MD, Kluve-Beckerman B, Zeldenrust SR. Targeted suppression of an amyloidogenic transthyretin with antisense oligonucleotides. Muscle Nerve 2006;33:609–18.
[101] Adams D, Gonzalez-Duarte A, O'Riordan W, et al. Patisiran, a RNAi Therapeutic, to improve outcomes in hereditary transthyretin-mediated (hATTR) amyloidosis. N Engl J Med 2018;379(1):11–21. PMID 29972753.
[102] Benson MD, Cruz WM, Berk JL, et al. Inotersen treatment for patiens with hereditary transthyretin amyloidosis. N Engl J Med 2018;379(1):22–31. PMID 2997257.
[103] Benson MD, Dasgupta NR, Rissing SM, Smith J, Feigenbaum H. Safety and efficacy of a TTR specific antisense oligonucleotide in patients with transthyretin amyloid cardiomyopathy. Amyloid 2017;24:219–25.

[104] Glenner GG, Wong CW. Alzheimer's disease: initial report of the purification and characterization of a novel cerebrovascular amyloid protein. Biochem Biophys Res Commun 1984;120:885–90.
[105] St George-Hyslop PH, Tanzi RE, Polinsky RJ, Haines JL, Nee L, Watkins PC, Myers RH, Feldman RG, Pollen D, Drachman D, Growdon J, Bruni A, Foncin J-F, Salmon D, Frommelt P, Amaducci L, Sorbi S, Piacentini S, Stewart GD, Hobbs WJ, Conneally PM, Gusella JF. The genetic defect causing familial Alzheimer's disease maps on chromosome 21. Science 1987;235:885–9.
[106] Chartier-Harlin M-C, Crawford F, Houlden H, Warren A, Hughes D, Fidani L, Goate A, Rossor M, Roques P, Hardy J, Mullan M. Early-onset Alzheimer's disease caused by mutations at codon 717 of the β-amyloid precursor protein gene. Nature 1991;353:844–6.
[107] Goate A, Chartier-Harlin M-C, Mullan M, Brown J, Crawford F, Fidani L, Giuffra L, Haynes A, Irving N, James L, Mant R, Newton P, Rooke K, Roques P, Talbot C, Pericak-Vance M, Roses A, Williamson R, Rossor M, Owen M, Hardy J. Segregation of a missense mutation in the amyloid precursor protein gene with familial Alzheimer's disease. Nature 1991;349:704–6.
[108] Haass C, Lemere CA, Capell A, Citron M, Seubert P, Schenk D, Lannfelt L, Selkoe DJ. The Swedish mutation causes early-onset Alzheimer's disease by β-secretase cleavage within the secretory pathway. Nat Med 1995;1:1291–6.
[109] Murrell J, Farlow M, Ghetti B, Benson MD. A mutation in the amyloid precursor protein associated with hereditary Alzheimer's disease. Science 1991;254:97–9.
[110] Rogaev EI, Sherrington R, Rogaeva EA, Levesque G, Ikeda M, Liang Y, Chi H, Lin C, Holman K, Tsuda T, Mar L, Sorbi S, Nacmias B, Piacentini S, Amaducci L, Chumakov I, Cohen D, Lannfelt L, Fraser PE, Rommens JM, St George-Hyslop PH. Familial Alzheimer's disease in kindreds with missense mutations in a novel gene on chromosome I related to the Alzheimer's disease type 3 gene. Nature 1995;376:775–8.
[111] Schellenberg GD. Genetic dissection of Alzheimer disease, a heterogeneous disorder. Proc Natl Acad Sci U S A 1995;92:8552–9.
[112] Spillantini MG, Bird T, Ghetti B. Frontotemporal dementia and parkinsonism linked to chromosome 17: a new group of tauopathies. Brain Pathol 1998;8:387–402.
[113] St George-Hyslop P, Haines J, Rogaev E, Mortilla M, Vaula G, Pericak-Vance M, Foncin JF, Montesi M, Bruni A, Sorbi S, Rainero I, Pinessi L, Pollen D, Polinsky R, Nee L, Kennedy J, Macciardi F, Rogaeva E, Liang Y, Alexandrova N, Lukiw W, Schlumpf K, Tanzi R, Tsuda T, Farrer L, Cantu JM, Duara R, Amaducci L, Bergamini L, Gusella J, Roses A, Crapper McLachlan D. Genetic evidence for a novel familial Alzheimer's disease locus on chromosome 14. Nat Genet 1992;2:330–4.
[114] Hsiao K, Dlouhy SR, Farlow MR, Cass C, Da Costa M, Conneally PM, Hodes ME, Ghetti B, Prusiner SB. Mutant prion proteins in Gerstmann-Sträussler-Scheinker disease with neurofibrillary tangles. Nat Genet 1992;1:68–71.
[115] Prusiner SB. Novel proteinaceous infectious particles cause scrapie. Science 1982;215:136–44.
[116] Prusiner SB, DeArmond SJ. Prion protein amyloid and neurodegeneration. Amyloid Int J Exp Clin Invest 1995;2:39–65.
[117] Ghetti B, Piccardo P. Amylose de la protéine prion. In: Grateau G, Benson MD, Delpech M, editors. Les amyloses. Paris: Médecine-Sciences/Flammarion; 2000. p. 523–45.
[118] Plant GT, Révész T, Barnard RO, Harding AE, Gautier-Smith PC. Familial cerebral amyloid angiopathy with nonneuritic plaque formation. Brain 1990;113:721–47.
[119] Vidal R, Frangione B, Rostagno A, Mead S, Révész T, Plant G, Ghiso J. A stop-codon mutation in the BRI gene associated with familial British dementia. Nature 1999;399:776–81.
[120] Vidal R, Révész T, Rostagno A, Kim E, Holton JL, Bek T, Bojsen-Møller M, Braendgaard H, Plant G, Ghiso J, Frangione B. A decamer duplication in the 3' region of the BRI gene originates an amyloid peptide that is associated with dementia in a Danish kindred. Proc Natl Acad Sci U S A 2000;97:4920–5.
[121] Klintworth GK. Advances in the molecular genetics of corneal dystrophies. Am J Ophthalmol 1999;128:747–54.
[122] Munier FL, Korvatska E, Djemaï A, Djemaï A, Le Paslier D, Zografos L, Pescia G, Schorderet DF. Kerato-epithelin mutations in four 5q31-linked corneal dystrophies. Nat Genet 1997;15:247–51.
[123] Stewart H, Black GCM, Donnai D, Bonshek RE, McCarthy J, Morgan S, Dixon MJ, Ridgway AA. A mutation within exon 14 of the TGFBI (BIGH3) gene on chromosome 5q31 causes an asymmetric, late-onset form of lattice corneal dystrophy. Ophthalmology 1999;106:964–70.
[124] Tsujikawa M, Kurahashi H, Tanaka T, Nishida K, Shimomura Y, Tano Y, Nakamura Y. Identification of

the gene responsible for gelatinous drop-like corneal dystrophy. Nat Genet 1999;21:420–3.
[125] Johansson B, Wernstedt C, Westermark P. Atrial natriuretic peptide deposited as atrial amyloid fibrils. Biochem Biophys Res Commun 1987;148:1087–92.
[126] Linke RP, Voigt C, Storkel FS, Eulitz M. N-terminal amino acid sequence analysis indicates that isolated atrial amyloid is derived from atrial natriuretic peptide. Virchows Arch B Cell Pathol 1988;55:125–7.
[127] Maeda S, Tanaka T, Hayashi T. Familial atrial standstill caused by amyloidosis. Br Heart J 1988;59:498–500.
[128] Sletten K, Westermark P, Natvig JB. Characterization of amyloid fibril proteins from medullary carcinoma of the thyroid. J Exp Med 1976;143:993–8.
[129] Keiser HR, Beaven MA, Doppman J, Wells Jr S, Buja LM. Sipple's syndrome: medullary thyroid carcinoma, pheochromocytoma, and parathyroid disease. Ann Intern Med 1973;78:561–79.
[130] Murrell JR, Spillantini MG, Zolo P, Guazzelli M, Smith MJ, Hasegawa M, Redi F, Crowther RA, Pietrini P, Ghetti B, Goedert M. Tau gene mutation G389R causes a tauopathy with abundant pick body-like inclusions and axonal deposits. J Neuropathol Exp Neurol 1999;58:1207–26.
[131] Polymeropoulos MH, Lavedan C, Leroy E, Ide SE, Dehejia A, Dutra A, Pike B, Root H, Rubenstein J, Boyer R, Stenroos ES, Chandrasekharappa S, Athanassiadou A, Papapetropoulos T, Johnson WG, Lazzarini AM, Duvoisin RC, Di Iorio G, Golbe LI, Nussbaum RL. Mutation in the α-synuclein gene identified in families with Parkinson's disease. Science 1997;276:2045–7.
[132] Davis RL, Shrimpton AE, Holohan PD, Bradshaw C, Feiglin D, Collins GH, Sonderegger P, Kinter J, Becker LM, Lacbawan F, Krasnewich D, Muenke M, Lawrence DA, Yerby MS, Shaw CM, Gooptu B, Elliott PR, Finch JT, Carrell RW, Lomas DA. Familial dementia caused by polymerization of mutant neuroserpin. Nature 1999;401:376–9.
[133] Yerby MS, Shaw C-M, Watson JMD. Progressive dementia and epilepsy in a young adult: unusual intraneuronal inclusions. Neurology 1986;36:68–71.

9

Leukemias, Lymphomas, and Plasma Cell Disorders

Jennifer J.D. Morrissette[1], Jacquelyn J. Roth[1], Selina M. Luger[2], Edward A. Stadtmauer[2]

[1]Division of Precision and Computational Diagnostics, Department of Pathology and Laboratory Medicine, Perelman School of Medicine, University of Pennsylvania, Philadelphia, PA, United States
[2]Division of Hematology-Oncology, Department of Medicine, Perelman School of Medicine, University of Pennsylvania, Philadelphia, PA, United States

9.1 INTRODUCTION

Hematological malignancies are cancers of the blood and lymphatic system and are increasingly characterized by acquired genetic and genomic abnormalities. Normal hematopoiesis is a highly regulated differentiation and maturation process initiating from hematopoietic progenitor stem cells evolving into highly specialized differentiated cells of myeloid or lymphoid lineage. Leukemia is characterized by a progressive expansion of mature (chronic leukemia) or immature (acute leukemia) myeloid or lymphoid cells in the bone marrow, whereas lymphoma is described by abnormal lymphoid proliferation, commonly in arising in the lymph node, but may also develop in the spleen, tonsils, or skin. In the earlier classifications, such as the French—American—British cooperative Group (FAB classification) in leukemia, and the European and American classification of lymphoid neoplasm (REAL) [4,5], hematopoietic malignancies were classified based on the morphologic features of the abnormal cell population in combination with the clinical presentation. With the expanding knowledge of oncogenesis, in particular the genetic mechanisms in leukemia and lymphoma, the World Health Organization (WHO) developed and revised a universally accepted classification system of myeloid and lymphoid neoplasms most recently in 2016 [6] according to the lineage of cell origin and the combination of morphologic, immunophenotypic, genetic, cytogenetic, and clinical criteria. Genetic abnormalities play an important role as diagnostic criteria for further subclassification of some neoplasms.

9.1.1 General Patterns of Chromosome Aberrations and Genomic Abnormalities in Hematological Malignancies

Detection of mutations and chromosome abnormalities is critical in the evaluation of most hematological malignancies. These abnormalities are present in the diagnostic specimen and can be used for treatment decisions and monitoring throughout the course of disease. For example, following treatment as the malignant clone responds to therapy the level of the acquired abnormalities contract, corresponding with the decreased burden of disease; at relapse the abnormalities detected in the initial study may reappear, or new

Emery and Rimoin's Principles and Practice of Medical Genetics and Genomics. https://doi.org/10.1016/B978-0-12-812534-2.00012-6

mutations may be observed due to clonal selection or evolution through therapy.

Cytogenetic evaluation of cells from patients with hematologic malignancies has resulted in major advances in our understanding of the specificity of some of the abnormalities observed in tumor cells [7]. Chromosomal abnormalities are nonrandom and represent a type of acquired clonal somatic mutation. Balanced or unbalanced chromosome translocations, inversion and insertion, loss and gain of chromosomes, and deletions and amplifications of parts of chromosomes are the common forms of chromosome abnormalities in cancer. The incidence of chromosome abnormalities varies based on disease status and types, from nearly 0% in Hodgkin lymphoma to 10%–25% in low-grade myelodysplastic syndrome (MDS) and myeloproliferative neoplasm (MPN) to 100% in chronic myeloid leukemia (CML) (by definition) [1,8–10]. Balanced chromosome translocations are common in de novo acute myeloid and lymphoid leukemia and in lymphoma, whereas losses or gains of chromosome materials are frequent in de novo MDS and in therapy-related MDS and AML. Balanced rearrangements can result in chimeric fusion genes, such as t(9;22) in CML resulting in the *BCR::ABL1* fusion or can lead to the juxtaposition of a strong enhancer/promoter to an oncogene or antiapoptosis gene, such as t(8;14) in Burkitt lymphoma resulting in the *IGH::MYC* rearrangement driving overexpression of intact *MYC* [7,11]. Chromosome translocations and/or inversions in myeloid and lymphoid leukemia and lymphoma frequently involve genes that are functionally involved in the transcriptional regulation, the cell cycle, apoptosis, and cytokine signaling pathways [12,13]. Loss or deletion of chromosome material may lead to the loss of relevant tumor suppressor genes, such as deletion of *TP53* through either monosomy 17 or deletion of the short arm of chromosome 17, whereas gain or duplication is associated with amplification of oncogenes, sometimes in the form of heterogeneous staining regions or double minutes (episomes), such as *MYC* amplification.

Chromosome studies miss many critical abnormalities, including submicroscopic imbalances and rearrangements and gene mutations [14,15]. Other methodologies, such as CMA, designed to detect genomic imbalances below the sensitivity of chromosome studies have identified novel genomic aberrations, such as copy-neutral loss of heterozygosity (CN-LOH) [16]. NGS can further delineate mutational profiles and, dependent on the test parameters, can combine copy number and mutation detection in a single assay. There are an increasing number of targetable mutations, such as FLT3 internal tandem duplications (ITDs) in AML and mutations that alter risk, such as *TP53* that require a more aggressive therapeutic approach. Additionally, the broad detection of mutations by NGS can infer clonal evolution and, in some cases, unrelated or branched clones such as multiple vertically complex (i.e., multiple variants occurring at the same genomic position) *NRAS* mutations in AML.

9.1.2 Significance of Detecting Acquired Chromosome and Gene Abnormalities in Hematological Malignancies

The close association of specific chromosome abnormalities with particular types of human leukemia and lymphoma has been well established in the past several decades and is very helpful in confirming the diagnosis [17]. For instance, the detection of t(9;22) by cytogenetic or fluorescence in situ hybridization (FISH) analysis will enable a precise diagnosis of CML and will rule out any other MPN, regardless of morphological features in bone marrow specimens [18]. In non-Hodgkin lymphomas (NHL), the detection of t(11;14) in lymph nodes, bone marrow, or peripheral blood samples strongly favors mantle cell lymphoma (MCL), rather than chronic lymphocytic leukemia (CLL). Because of its unique association with disease, detecting certain chromosome abnormalities greatly helps in selecting appropriate treatment, such as t(15;17) in acute promyelocytic leukemia (APL) and t(9;22) in CML. Once the t(15;17)/*PML::RARA* rearrangement is detected by karyotype, FISH, or PCR techniques, treatment using all-trans retinoic acid (ATRA) and arsenic trioxide will efficiently induce leukemia cells to differentiate and then apoptosis [19]. In CML with t(9;22), imatinib (also called Gleevec) treatment can induce most CML patients into clinical, hematological, cytogenetic, and molecular remission in a short period [20,21]. Moreover, chromosome abnormalities are the most significant prognostic factors for patients with acute leukemia and lymphoma. Patients with AML and t(8;21) and inv(16)/t(16;16), involving the alpha and beta subunits, respectively, of the core-binding factor

(CBF) have a very favorable prognosis, whereas patients with AML and a complex karyotype show a dismal clinical course [8,9,22].

Likewise, the detection of molecular abnormalities can alter the treatment regimens and prognosis for patients with hematological malignancies. The recent FDA approval of the FLT3 inhibitors, midostaurin and gilteritinib for FLT3 mutated AML in the untreated and refractory/relapsed settings, was based on studies demonstrating improved overall survival [23–25]. Additionally there are targeted therapies for *IHD1* and *IDH2* positive AMLs that improve overall survival [26,27]. For many malignancies, there is not a single abnormality and the detection of multiple abnormalities can be additive. For example, the presence of *TP53* mutation with a complex karyotype is more unfavorable than *TP53* mutation or complex karyotype alone [28]. Detection of residual genetic abnormalities prior to allogeneic bone marrow transplant portends an increased likelihood of relapse [29] and conversely detection of mutations 30 days postallogeneic transplant is associated with an increased likelihood for relapse [30]. Any abnormality noted at the time of diagnosis can be used as a biological marker to monitor the response to therapy or to detect residual disease in follow-up specimens.

It is worth noting that most hematologic malignancies are sporadic cancers. However, there is a growing literature involving cases associated with cancer predisposition genes, such as *DDX41* associated with familial MDS, and other rare genetic syndromes, such as the *FANC* genes associated with bone marrow failure and AML. Additionally, clonal hematopoiesis can predispose to myeloid malignancies and, surprisingly, cardiovascular events [31].

9.1.3 Application of Cytogenetic, FISH, Microarray, and Molecular Techniques in Diagnosis of Leukemia and Lymphoma

The principles and methods used for conventional cytogenetic analysis of malignant disease are very similar to those used for constitutional abnormalities, which have been outlined in detail in the Cancer cytogenetic analysis can be performed on almost any tissue with actively dividing (malignant) cells. For leukemia studies, the specimen is usually a bone marrow aspirate although a bone biopsy can also be processed successfully. Alternatively, in patients with circulating immature cells, a sample of peripheral blood can be analyzed. For lymphoma studies, the most appropriate tissue is a biopsy of the affected tissue such as lymph node, spleen, or tonsil. However, even when lymphoma cells are extensively present in the bone marrow, cytogenetic evaluation of a bone marrow aspirate rarely proves to be informative for the study of lymphoma, particularly in low-grade B-cell lymphoma, such as follicular lymphoma and CLL [32].

Cytogenetic analysis for detection of acquired abnormalities includes the complete analysis of at least 20 metaphase cells for a full analysis. However, this may not always be possible and depends on the cellularity, mitotic index, and quality of the specimen. An analysis of fewer than 20 cells is still informative when a clonal abnormality is detected, but may be suboptimal for detection of emerging clones. According to the International System of Human Chromosome Nomenclature [33], a chromosomal abnormality is considered to be clonal if a structural abnormality or gain of a chromosome is identified in two or more cells. Chromosome loss can occur as a technical artifact during metaphase cell preparation; thus, a loss of a chromosome is considered to be clonal when it occurs in three or more cells. Although chromosome analysis has the disadvantages of low resolution (about 10 MB), requirement for dividing cells, and is very labor-intensive, they are outweighed by the advantages, including detection of chromosome abnormalities in individual cells and novel disease-specific chromosome abnormalities. Additionally, the dependence of conventional cytogenetics studies on mitotic index rather than disease percentage allows for detection of low-level (emerging) clones.

The field of cytogenetic analysis was significantly advanced through the development of FISH in the 1980s [34,35]. FISH uses fluorescently labeled DNA probes that anneal to the complementary DNA sequences of specific regions of the genome. To perform FISH, tumor cells can be derived from fresh or fixed bone marrow or peripheral blood, touch preparations, or paraffin-embedded sectioned tissue. FISH can be used on metaphase or interphase cells, thus enhancing the detection of small nondividing tumor populations. The FISH probe can be gene or locus-specific, repetitive sequences specific for individual centromeres or telomeres of each chromosome. Because chromosome abnormalities are nonrandom in cancer, probes are designed to be able to identify specific rearrangements, generally designed as break-apart probes or dual-color/dual-fusion probes. Break-apart probes are engineered to show a disruption of a gene locus that rearranges with multiple partners, such as *IGH*, which can rearrange with *MYC, BCL2, CCND1*, and others. Break-apart probes do not identify the gene partner, so additional studies are necessary if that is clinically important. Dual-color dual fusion probes can detect specific gene rearrangements, such as *BCR::ABL1* or *IGH::MYC*. While these probes are specific, they will not detect specific breakpoints within a gene, which requires molecular methods. The high sensitivity and specificity of FISH and the speed with which the technique can be completed have made FISH a powerful tool with numerous applications, such as the rapid detection of numerical and structural chromosomal abnormalities in new acute leukemia cases [36], targetable rearrangements in ALL and CML, and in lymphoma specimens that are formalin-fixed and paraffin-embedded (FFPE). FISH can also be used to clarify ambiguous chromosomal or molecular abnormalities. One limitation of FISH is the specificity of the probes, such that only the region covered by the probe is analyzed, and other chromosome abnormalities may be missed.

Microarrays utilizing single-nucleotide polymorphisms (SNP arrays) are considered first tier testing for constitutional abnormalities, but are less common in the detection of acquired abnormalities in hematological malignancies. Microarray techniques provide genome-wide analysis of copy number and zygosity, allowing for detection of smaller regions that are gained and/or lost, as well as detection of copy number neutral loss of heterozygosity (cnnLOH) events that occur in hematological malignancies. The resolution of microarrays is much higher than chromosome analysis in detecting genomic and chromosomal abnormalities leading to copy number alterations in hematological neoplasms; however, one limitation of microarrays is that they will miss balanced rearrangements [15]. Various cryptic copy number imbalances and CN-LOH have been found in AML, MDS, MPN, ALL, and lymphoma [14,16,37].

Detection of molecular alterations in hematological malignancies is increasingly critical in the diagnosis and treatment of patients. The scope of the genomic testing has evolved from single-gene testing for specific abnormalities, such as JAK2 V617F in myeloproliferative neoplasms, FLT3-ITD in AML, and BRAF V600E in hairy cell leukemia (HCL) to broad molecular profiling at diagnosis, relapse, and remission. The determination regarding what gene(s) to sequence should be dependent on the clinical requirements. For example, targeted mutation testing for BRAF V600E would be appropriate for the diagnosis of HCL, since BRAF V600E is the causative mutation for HCL [38,39]. However, sequencing a single gene is insufficient for MDS and AML, as there are many mutations that are important in the diagnosis and treatment of these diseases [28,40]. Even after gene mutations have been detected in the diagnostic specimen, continued broad molecular profiling is important because therapy can result in clonal selection with different driver mutations. For example, patient with an FLT3-ITD at diagnosis and treated with an FLT3 inhibitor can relapse with an FLT3 negative clone [41], hence using an assay that will only identify FLT3 abnormalities would be insufficient to characterize the molecular profile of the new clone.

The detection of genome-wide abnormalities by chromosome analysis and/or CMA in addition to molecular testing by single gene or NGS leads to multiple tests ordered on patients with hematological malignancies, often present in the EMR as individual tests that, in some cases, appear contradictory. An advantage of whole-genome sequencing (WGS) is that it can theoretically simultaneously detect mutations across the genome as well genome-wide copy number changes and chromosome rearrangements. A recent paper using low-depth WGS sequencing was able to detect relevant mutations, chromosome rearrangements, and copy number changes in AML and MDS. The low-depth WGS identified nearly all of the abnormalities detected by chromosome and mutational analysis and even discovered other abnormalities missed by classic techniques [42]. It is reasonable to think that as improvements are made in the field of WGS, such as decreased cost, turnaround time, and improved bioinformatics pipelines, more major academic laboratories will move toward single comprehensive genomic assays.

9.2 MYELOPROLIFERATIVE NEOPLASMS

Myeloproliferative neoplasms (MPNs) are a group of rare hematological malignancies that include CML, chronic neutrophilic leukemia (CNL), primary myelofibrosis (PMF), polycythemia vera (PV), essential thrombocythemia (ET), and chronic eosinophilic leukemia. The MPNs are clonal disorders resulting from the proliferation of cells from one of the myeloid lineages (erythroid, granulocytic, and/or megakaryocytic) and are clinically characterized by abnormal blood counts, including high platelets, red blood cells and/or granulocytes, and resultant complications. Identification of pathognomonic mutations and fusion genes has altered the diagnosis and prognosis.

9.2.1 Chronic Myeloid Leukemia (CML)

CML is a clonal myeloproliferative disorder characterized by a recurrent gene fusion resulting in the juxtaposition of the *BCR* gene from chromosome 22 with the *ABL1* gene from chromosome 9 creating an oncogenic fusion gene classically residing on the derivative chromosome 22. The *BCR::ABL1* fusion typically results from a recurrent translocation, t(9;22) (q34;q11.2), which may include other translocations. CML has an incidence of 1–2/100,000 cases/year, with the incidence rising with age [43,44] and is responsible for 15%–20% of adult leukemias. CML is characterized by the proliferation of granulocytes and serves as a model for targeting genetic abnormalities in hematological malignancies [10,45–47].

9.2.1.1 Genetics and Mechanism of Disease

CML was the first cancer in which a unique chromosome abnormality, namely the Philadelphia (Ph) chromosome, was identified [1,48]. Improvements in chromosome analysis led to the discovery that the Philadelphia chromosome was part of a balanced translocation between chromosomes 9 and 22 [2]. The translocation was further characterized as a fusion gene, joining the 5′ end of *BCR* to the 3′ end of the *ABL1* gene, resulting in a chimeric protein with an activated ABL1 tyrosine kinase that was fundamental to the initiation of CML [49–52].

The t(9;22) is the hallmark of CML as recognized by the 2016 WHO classification of hematopoietic and lymphoid neoplasms and can be detected by conventional cytogenetic analysis in more than 90% of CML patients (Fig. 9.1A) [7,53]. It is present in all myeloid lineage cells, including granulocytes, monocytes, erythroid precursors,

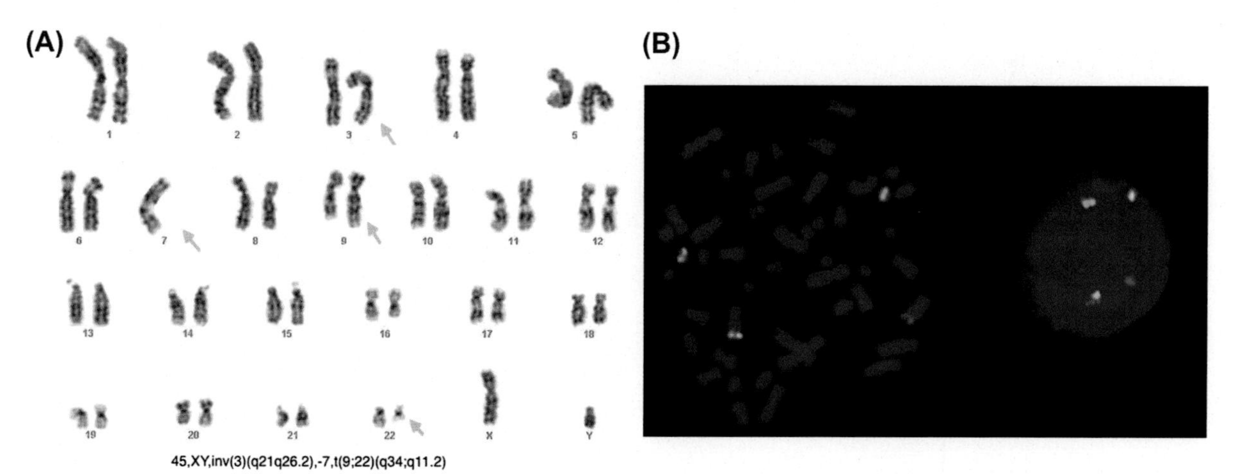

Figure 9.1 (A) Karyotype of a metaphase cell from a male patient with chronic myeloid leukemia (CML) in the blast phase. The karyotype is 45, XY, inv(3) (q21q26.2), −7, t(9;22)(q34;q11.2). Gain of additional chromosome aberrations, in addition to t(9;22), is frequently observed in CML during the accelerated and blast phases. (B) Fluorescence in situ hybridization (FISH) analysis of cells from the same patient reveals the BCR/ABL fusions in metaphase and interphase cells. The probes for ABL and BCR are labeled with SpectrumOrange and SpectrumGreen, respectively. Normal chromosomes 9 and 22 show a single *red* and *green* signal, whereas the derivative chromosomes 9 and 22 display a fusion (*yellow*) signal pattern containing the fused BCR and ABL probes. Abnormal interphase cells show a typical pattern with two fusions (*yellow*) and one *red* and one *green* signal.

megakaryocytes, and in all B and some T lymphocytes [54,55]. In the remaining 2%–10% of the patients with CML, a variant of the t(9;22) can be observed. These atypical cases can include complex chromosomal rearrangements, with the translocation involving additional chromosomes, or a cryptic rearrangement due to an insertion of the *BCR* gene into the *ABL1* gene or vice versa. The complementary use of chromosome analysis, FISH, and standard PCR studies can elucidate those rare complex rearrangements (Fig. 9.1B).

There are different breakpoints in *BCR* in patients with CML and other Philadelphia chromosome positive hematologic malignancies. The different *BCR* breakpoints result in distinct *BCR::ABL1* isoforms, with the most common breakpoint in CML joining *BCR* exon 13 or exon 14 (e13/e14) with *ABL1* exon 2 (a2), producing a fusion gene referred to as e13a2 (b2a2) or e14a2 (b3a2). This set of breakpoints result in a protein that has an apparent molecular weight of 210 kDa and is commonly referred to in the literature as p210. This is the fusion that has international standards (IS) for measurement of *BCR::ABL1* levels, which has established benchmarks for treatment response (Fig. 9.2) [3]. There are other, rarer breakpoints, including an isoform known as p190 results from a breakpoint joining exon 1 of *BCR* (e1) to exon 2 of *ABL1* (e1a2) [56]. The p190 isoform is more common in B cell ALL, but can be detected in CML. Another isoform exists joining *BCR* exon 19 (e19) with *ABL1* exon 2 (a2), producing a fusion gene referred to as e19a2 (b2a2) with a larger protein product of 230 kDa (p230). The p230 is typically associated with a more indolent disease course [57]. Both p210 and p190 isoforms are known to display dysregulated constitutive tyrosine kinase activity.

ABL1 is a member of the family of nonreceptor protein tyrosine kinases. The *BCR* gene is a member of the Rho family of small GTPases [58]. The fusion of the BCR coiled-coil motif with ABL1 results in constitutive activation of the ABL1 tyrosine kinase in the BCR::ABL1 fusion protein [59]. This allows for activation by homodimerization and tetramerization of BCR::ABL1 and results in cellular transformation. BCR::ABL (p210 and p190) is located exclusively in the cytoplasm and constitutively interferes with a variety of cytoplasmic and cytoskeletal signaling pathways [60]. The fusion protein BCR::ABL is also a powerful antiapoptotic molecule in mammalian cells and is directly

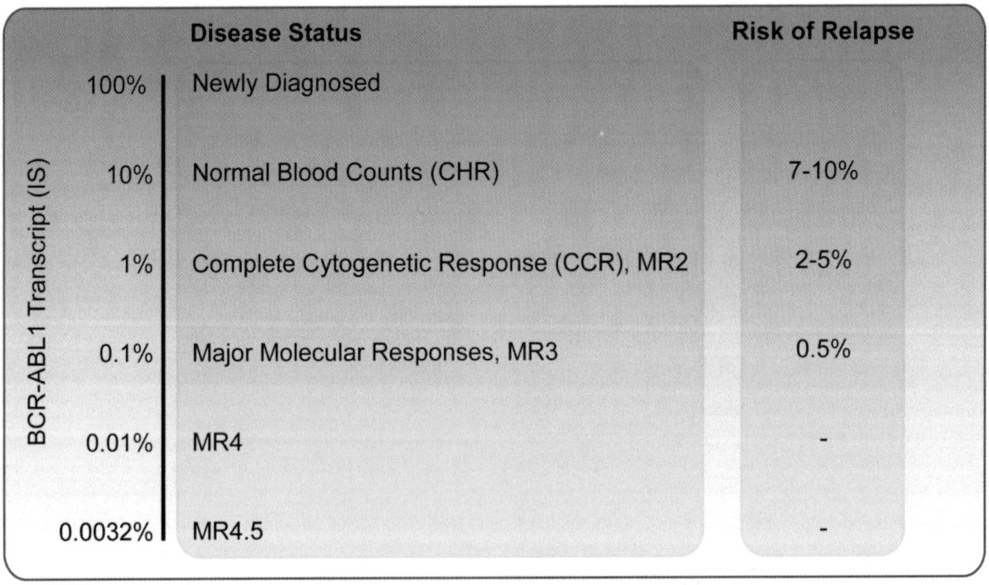

BCR-ABL1 Transcript (IS)	Disease Status	Risk of Relapse
100%	Newly Diagnosed	
10%	Normal Blood Counts (CHR)	7-10%
1%	Complete Cytogenetic Response (CCR), MR2	2-5%
0.1%	Major Molecular Responses, MR3	0.5%
0.01%	MR4	-
0.0032%	MR4.5	-

Figure 9.2 Monitoring of chronic-phase CML on BCR-ABL1 TKI therapy. *BCR-ABL1* transcript levels as measured by reverse-transcriptase PCR and their corresponding disease status. Risk of relapse per year numbers are derived from the original IRIS trial and subsequent European LeukemiaNet guidelines [386,387].

implicated in the defective responses of both immature and differentiated primary myeloid cells to growth factor deprivation [61,62].

Overall in CML, the appearance of additional abnormalities while on TKI therapy, such as +8 and i(17q), is concerning occult disease progression, which can present several months earlier than morphological evidence on bone marrow examination or clinical symptoms. The detection of additional chromosome abnormalities at diagnosis in Ph-positive CML, such as additional copies of the Philadelphia chromosome, monosomy 7, isochromosome 17q, and MECOM rearrangements in addition to the t(9;22), is associated with a poorer response to TKI therapy [63].

Risk factors for CML include genetic predisposition [64,65], older age, immunosuppression, and environmental exposures (radiation, smoking, organic solvents, herbicides, and pesticides for example) [66,67]. However, most cases of CML appear to be sporadic.

9.2.1.2 Therapeutic Implications

In its early chronic phase, CML typically presents with a proliferation of mature blood cells, an enlarged spleen, and/or lymphadenopathy. Without treatment CML will evolve, with a median time to progression of 3–4 years, into a life-threatening advanced hematological malignancy, which is characterized by increases in immature blood cells and associated cytopenias. Historical therapy involved control of blood counts with cytotoxic therapies or allogeneic bone marrow transplant during chronic phase, which prevented disease progression in the majority of patients, but was complicated by significant treatment-related mortality. After demonstration of the causative role of the *BCR::ABL1* rearrangement, imatinib was the first tyrosine kinase inhibitor designed to inhibit the ABL1 kinase and revolutionized the treatment of CML [3,21,68]. Treatment with imatinib has significantly improved the outcome of patients with CML, with 97% complete hematologic response and 87% complete cytogenetic remission in a 5-year follow-up study secondary to a decreased risk of disease progression [20].

In up to 15% of patients with CML on imatinib, additional acquired chromosomal abnormalities arise. In Ph-positive CML, the most common additional acquired chromosomal abnormalities are gain of chromosome 8 (33%), followed by an additional Ph chromosome (30%), i(17q) (20%), +19 (12%), loss of the Y chromosome (8% in XY individuals), trisomy 21 (7%), and loss of chromosome 7 (5%) [69,70]. These additional abnormalities may appear individually or in combination. Repeating conventional karyotype analysis at regular intervals is warranted to monitor CML patients for newly acquired changes [69,71]. The most commonly mutated gene in CML that is associated with progression is the *ABL1* gene itself. The acquisition of ABL1 kinase domain mutations can lead to failure of ABL1 kinase inhibition and aid in acquiring additional mutations. Most ABL1 resistance mutations occur in and around the imatinib binding site and have been shown to prevent or reduce imatinib binding while in an active conformation for the ABL1 kinase domain [3]. Second- and third-generation ABL1 TKIs have been developed, which overcome many of these mutations. One recurrent mutation that remains insensitive to multiple ABL1 inhibitors is the ABL1 T315I mutation. While a few patients have them at diagnosis, others develop resistance mutations during the course of treatment of their disease and are found to have a T315I or other mutations (Fig. 9.3) [3]. At the time of disease progression/loss of response to the current targeted therapy, or rising *BCR::ABL1* transcript, targeted ABL1 mutation testing should be performed to determine if there is a mutation that would direct therapy to a different TKI.

The *BCR::ABL1* fusion has been shown to be an early/initiating event and even in patients with long-term deep molecular response to targeted therapy, the *BCR::ABL1* fusion can often be detected at extremely low levels [72]. However, the STIM-Pilot/STIM1 trial demonstrated that some patients with undetectable *BCR::ABL1* transcripts could discontinue imatinib therapy with no disease relapse [73,74], with additional trials echoing these findings showing that approximately half of patients treated with imatinib and other ABL1 TKIs in a long-term MMR can discontinue therapy [3,75,76]. CML represents the model for detection of a targeted abnormality and subsequent treatment, with possible disease cure in many patients [21,77–79]. Furthermore, second-, third-, fourth-, and fifth-generation tyrosine kinase inhibitors have stronger ABL1 TK inhibition than imatinib and have showed deeper remission, which has allowed for attempts at treatment-free remission in significant numbers of patients [20,80].

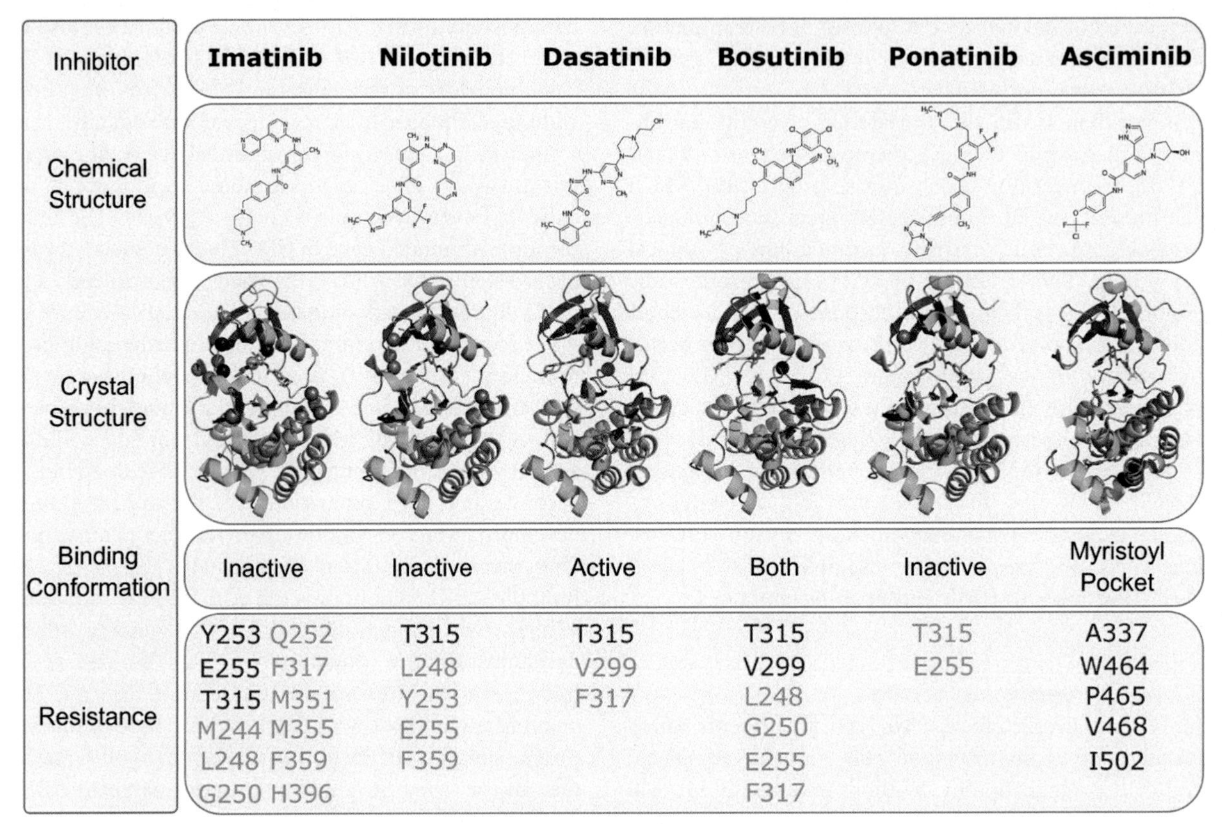

Figure 9.3 BCR-ABL1 tyrosine kinase inhibitors and resistance mechanisms. Chemical structures and published X-ray crystallographic structures of ABL1 complexed with kinase inhibitors are shown. Residues at which mutations are associated with strong resistance to a given TKI are indicated in red, while those associated with lesser degrees of resistance are listed in orange. Both T315 and E255 mutations do lead to an increase in the IC_{50} for ponatinib; however, they do not typically lead to clinical resistance in isolation, but do as a compound mutation. The structure of ABL1 complexed with asciminib shows nilotinib in the ATP-binding site for reference. T315I is indicated in purple for visual reference [388–393].

9.2.2 Other Myeloproliferative Neoplasms

In the 2016 WHO classification of leukemia and lymphoma, there are several additional MPN diseases, all with features of overproliferation of one or more myeloid lineages in the bone marrow and extramedullary areas [81]. CNL is a rare myeloproliferative disorder characterized by morphologic features. Three "classic" MPNs are PV, ET, and PMF. Although these hematological malignancies have some overlap, there are differences in overall survival (OS), with PMF having the shortest average survival of approximately 6 years OS, PV of approximately 14 years OS, and ET having a greater than 20-year OS.

9.2.2.1 Genetics and Mechanism of Disease

In CNL, mutations in colony-stimulating factor 3 receptor (*CSF3R*) are observed in approximately 80% of cases and have been included in the diagnostic criteria for this disorder. There are additional gene mutations detected in CNL, including *SETBP1* and *ASXL1* that are associated with a more adverse prognosis [82].

A recurrent mutation in *JAK2* that confers a gain of function was initially identified in PV, ET, and PMF [83–85]. JAK2 is a protein tyrosine kinase that is involved in signal transduction from cytokine receptors through the JAK-STAT signaling pathway [86]. The most common *JAK2* mutation is the dominant gain-of-function V617F mutation that disrupts the autoinhibitory JH2 kinase domain and leads to constitutive tyrosine phosphorylation activity and upregulation of STAT-mediated transcriptional activation [83]. The *JAK2* mutation is detected in almost 100% of patients with PV and in about 50% in patients with ET and PMF [18,84] The V617F mutation seems to occur exclusively in hematopoietic malignancies of the myeloid lineage. There is also an activating *JAK2* mutation associated with an exon 12 deletion that has less frequently been observed in PV.

Alternatively about 20%–25% of patients with PMF and ET have been found to have mutations in *CALR* in approximately 20%–25% of cases, with two recurrent hotspot mutations. The large 52 base pair deletion in the last exon of *CALR* (exon 9) is associated with a favorable outcome in PMF patients. In general, mutations in *CALR* are associated with less aggressive forms of NPM. Mutations in the *MPL* gene are observed in approximately 10% of PMF and 4% of ET, with the most common mutations in MPL being W515L/K [18]. Although these are the most common gene mutations, there are additional recurrent mutations described, most of which are commonly observed in myeloid malignancies (e.g., *TET2, ASXL1, DNMT3A*). Conventional cytogenetic studies continue to play a role in the MPNs. For example, additional copies of chromosome 9 are often associated with additional copies of mutated *JAK2*, and deletion 20q is associated with a favorable outcome.

9.2.2.2 Therapeutic Implications

The prognosis of these myeloproliferative disorders is related to increased risk of thrombotic events, a risk of progression to life-threatening myelofibrosis, and an aggressive myeloid malignancy. Small molecular kinase inhibitors target activating *JAK2* mutations and are associated with improvement in disease-related symptoms and manifestations but do not impact the risk of disease progression. Targeted treatments exist for some of the mutations identified in MPNs. The *CSF3R* mutations are targetable, with the common CSF3R T618I (exon 12) being sensitive to *JAK2* inhibitors and intracytoplasmic truncating mutations in exon 17 sensitive to dasatinib.

9.2.3 Myeloid/Lymphoid Neoplasms with Eosinophilia and Gene Rearrangement

The WHO 2016 [85] characterizes a subset of neoplasms with hypereosinophilia that classify disease subtypes based on recurrent gene rearrangements [86] rather than cell lineage. These abnormalities include rearrangements of the platelet-derived growth factor receptor alpha and beta (*PDGFRA* and *PDGFRB*) genes, fibroblast growth factor receptor 1 (*FGFR1*), and *PCM1::JAK2* rearrangement.

9.2.3.1 Genetics and Mechanisms of Disease

The most common rearrangement in this class involves *PDGFRA* rearrangement involving the locus at 4q12, with the rearrangement secondary to a cryptic interstitial deletion of the CHIC2 gene locus, resulting in the *FIP1L1::PDGFRA* fusion [87]. This rearrangement can be detected by FISH or through identification of the fusion transcript. In *PDGFRB* rearrangements, the most common rearrangement involving *ETV6* (12p13) and *PDGFRB* (5q33), but over 30 partners, has been characterized [88].

The *FGFR1* rearrangement involves the rearrangements of the short arm of chromosome 8p11 with a variety of partners. Of note, there have been reports of rearrangements involving *FGFR1* but with the band designation varying, likely due to poor banding, so FISH confirmation of rearrangements involving the short arm of chromosome 8 should be considered if clinically indicated [86]. Cases with *FGFR1* rearrangements tend to be aggressive and often also have mutated *RUNX1* [89]. Clinical presentation is highly variable ranging from mild cytopenias to acute leukemia. Presence of eosinophilia is similarly variable in those with FGFR1 rearrangements.

The *PCM1::JAK2* fusion, typically from a translocation between chromosomes 8p22 and 9p24, respectively, is a relatively new entity added to the WHO [85]. This translocation joins the coiled-coil domains of *PCM1*, which mediate oligomerization, with the tyrosine kinase domain of *JAK2*, resulting in a constitutively activated tyrosine kinase domain of *JAK2* [90] and is associated with a high risk of rapid disease progression.

9.2.3.2 Therapeutic Implications

Detection of *PDGFRA/B* rearrangements is critical because patients respond well to the treatment with imatinib and other tyrosine kinase inhibitors. Conversely, patients with *FGFR1* rearrangements have a high risk of rapid disease progression, and these patients are considered for early allogeneic stem cell transplant. The *PCM1::JAK2* rearrangement is rare, and while studies are ongoing with ruxolitinib, a JAK2 inhibitor, patients are typically offered allogeneic hematopoietic stem cell transplantation before there is evidence of disease progression [91].

9.3 MYELODYSPLASTIC SYNDROMES

The MDSs are a heterogeneous group of clonal hematological stem cell disorders demonstrating morphologic dysplasia and caused by an expansion of mutated hematopoietic stem cells. MDS is characterized by the inability of the bone marrow to produce mature peripheral blood cells leading to low blood counts, most commonly anemia, due to abnormalities in the affected hematopoietic cell line, and a variable risk of transformation to AML [6]. The number of blasts in the bone marrow and/or peripheral blood is by definition is below 20%. MDS may arise from precursor conditions such as clonal hematopoiesis of indeterminate potential (CHIP), secondary to exposures to cytotoxic agents, or germline mutations in cancer predisposition genes, or due to unknown etiologies. Diagnosis is made based on identification of dysplasia in affected cell lines. While MDS subtypes have traditionally been categorized by morphologic differences, there are now specific cytogenetic abnormalities and gene mutations that correlate with specific subtypes.

9.3.1 Genetics and Mechanism of Disease

The genetic abnormalities associated with MDS can largely be grouped into chromosome abnormalities and gene mutations. Clonal chromosome abnormalities in bone marrow samples can be detected in 40%–70% of patients with primary MDS at diagnosis and in up to 90% of therapy-related myeloid neoplasms. These cytogenetic changes are evaluated as a component of the International Prognostic Scoring System (IPSS-R) that when combined with other disease-related factors determines a risk score for survival and median time to 25% AML evolution [92]. Complex cytogenetics, defined as three or more chromosome abnormalities, are associated with a poor prognosis and an increased risk for transformation into acute myeloid leukemia (AML). While recurrent gene mutations in MDS have improved our understanding of MDS, new algorithms incorporating these changes into the prognostic calculation are now being developed [93].

Chromosomal changes in MDS nearly always display genomic imbalances, that is, gain or loss of whole chromosomes and/or chromosome regions with rare (less than 3%) of cases showing balanced rearrangements [93,94] The most common chromosomal abnormalities in MDS include del(5q)/-5, −7/del(7q), +8, and del(20q) (Table 9.1) [95] and are also

TABLE 9.1 Recurrent Chromosomal Abnormalities and their frequencies in MDS at Diagnosis

	FREQUENCY	
Chromosomal Abnormality	**MDS (%)**	**Therapy-Related MDS (%)**
Loss of chromosome 7 or del(7q)	10	50
del(5q)	10	40
Gain of chromosome 8	10	–
del(20q)	5–8	–
Loss of Y chromosome[a]	5	–
Isochromosome 17q or t(17p)	3–5	25–30
Loss of chromosome 13 or del(13q)	3	–
del(11q)	3	–
del(12p)or t(12p)	3	–

As a sole cytogenetic abnormality in the absence of morphological criteria, gain of chromosome 8, del(20q), and loss of Y chromosome are not considered definitive evidence of MDS.

[a]Sole loss of the Y chromosome is considered a malignancy-related abnormality when present in 75% or greater of the cells analyzed.

observed in therapy-related MDS. The detection of chromosome abnormalities is associated with overall survival and transformation of AML [96] (Fig. 9.4), [97–100]. These common chromosome abnormalities are not specific for MDS as they are also commonly detected in other myeloid disorders, that is, AML and MPN. However, the presence of some of these chromosome abnormalities, along with cytopenia, is considered MDS defining, even without morphological evidence of dysplasia in the bone marrow cells [101]. Recently, genomic microarray studies have discovered multiple lesions through the whole-genome analysis in MDS, including many small cryptic genomic imbalances, such as deletions and duplications, and CN-LOH in MDS patients with and without clonal cytogenetic abnormalities. Some of these novel genomic abnormalities detected by microarray in MDS have been implicated in disease progression and the pathobiology of MDS [102–105].

In addition to the recurrent chromosome abnormalities observed in the bone marrow of patients with MDS, there are recurrently mutated genes in MDS [40]. These genes fall into distinct pathways that include chromatin modification (e.g., *EZH2*, *ASXL1*), DNA methylation (e.g., *DNMT3A*, *TET2*, *IDH1/2*), signal transduction (e.g., *JAK2*, *KRAS*, *CBL*), RNA splicing (e.g., *SF3B1*, *U2AF1*, *SRSF2* and *ZRSR2*), transcriptional regulation (e.g., *EVI1*, *RUNX1*, *GATA2*), and tumor suppressors (e.g., TP53, CDKN2A) (Fig. 9.5) [106]. TP53 mutations are nearly always associated with a complex karyotype and a poor prognosis [107]. There are many subtypes of MDS and the frequency of specific gene mutations varies between these subtypes (Fig. 9.6), [40].

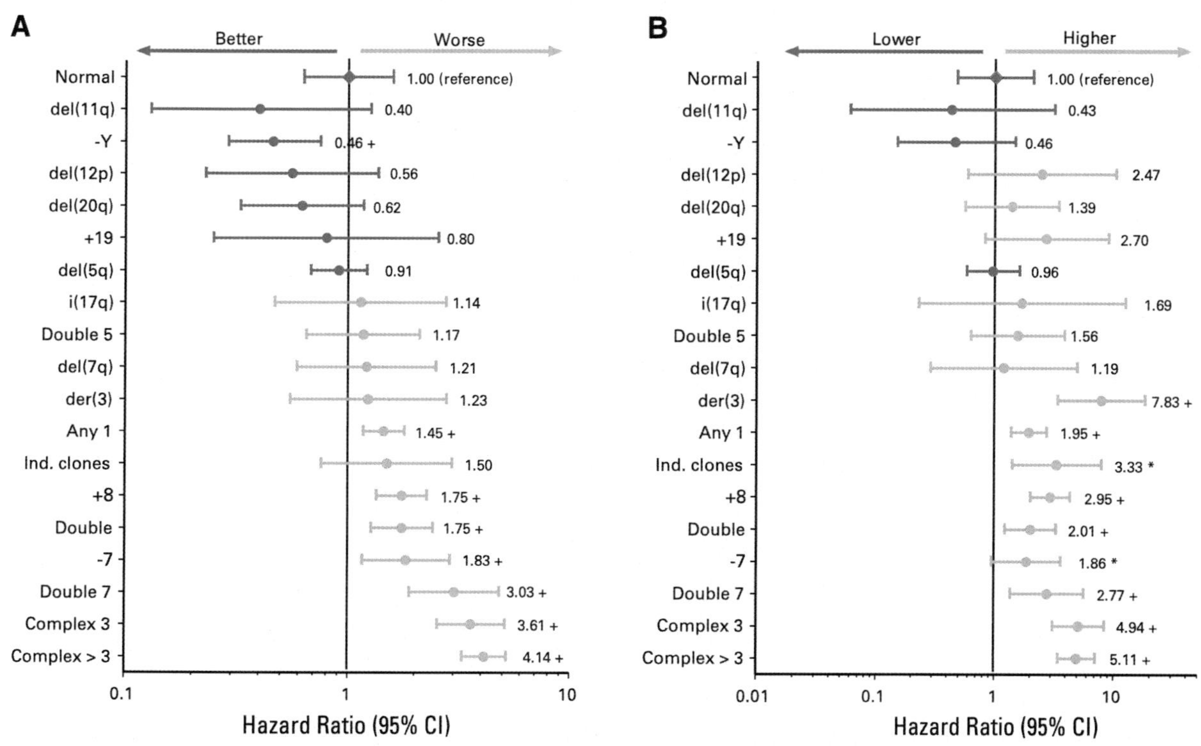

Figure 9.4 Overall survival and risk of acute myeloid leukemia transformation in distinct cytogenetic subgroups in MDS patients. (A) Overall survival ($n = 1893$) and (B) risk of acute myeloid leukemia transformation ($n = 1691$) in distinct cytogenetic subgroups (abnormalities with $n < 10$ combined as Any 1). Any 1, any other single abnormality; double 5, double abnormalities including del(5q); double, any other combination of two abnormalities; double 7, double abnormalities including −7/7q−; complex 3, three abnormalities; complex >3, four or more abnormalities; + indicates $P < .01$ (as compared with reference category); (*) indicates $P < .05$ (as compared with reference category). Ind., independent.

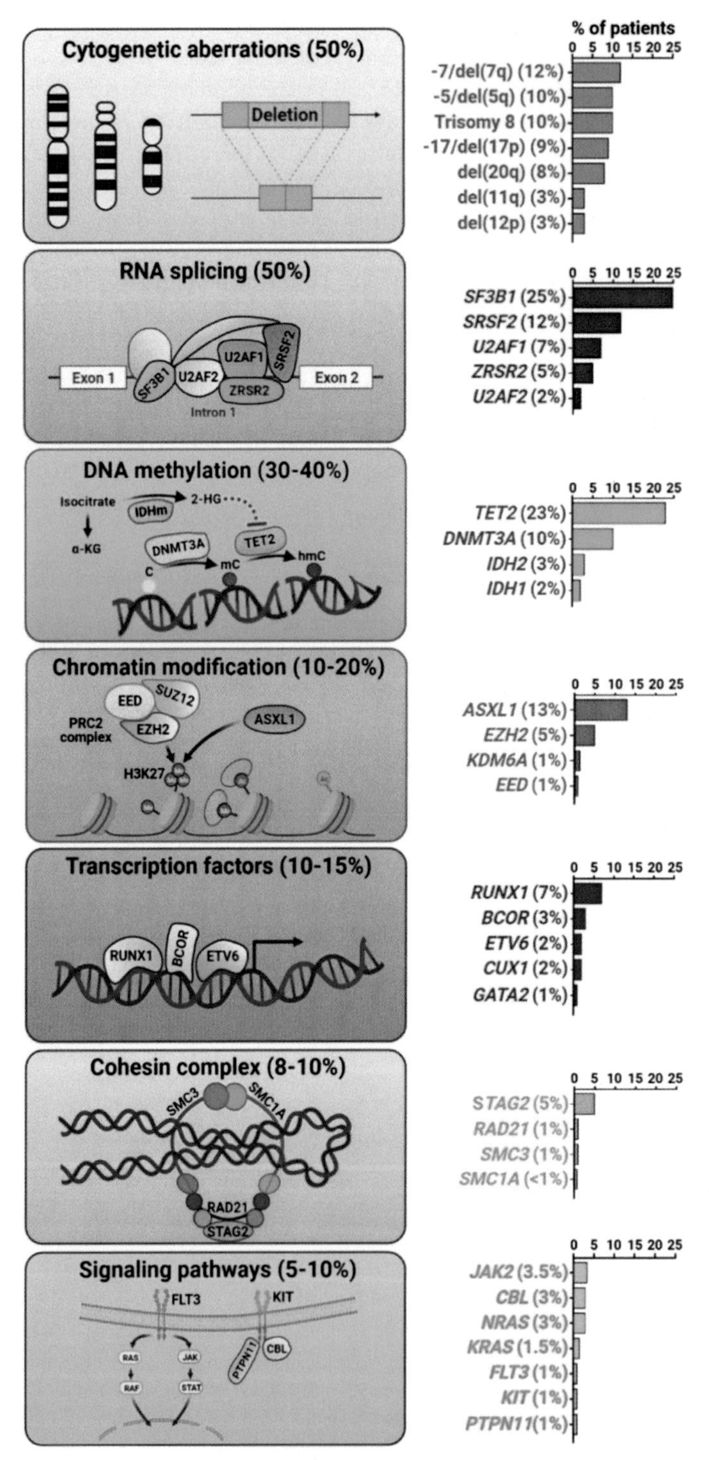

Figure 9.5 Somatic mutation landscape in myelodysplastic syndromes. Schemas highlight individual groups of mutations, and the associated bar graphs on the right provide the mutation percentages.

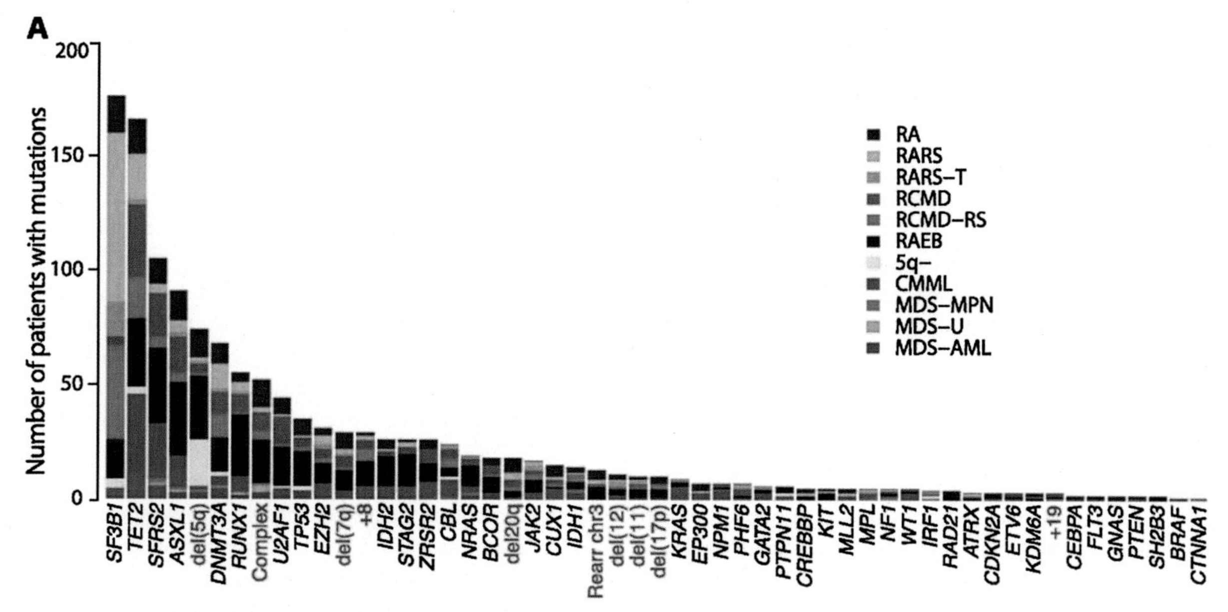

Figure 9.6 Commonly mutated genes and chromosome aberrations in MDS, with the proportions reflective of the MDS subtype. Genomic architecture of MDS. (A) Frequency of driver mutations identified in the sequencing screen or by cytogenetics in the cohort of 738 patients, broken down by MDS subtype.

The diagnosis of MDS is not always easily diagnosed on morphology alone. Cytogenetic analysis can lead to the confirmation of a diagnosis of MDS with detection of certain MDS-related chromosome abnormalities, when patients with refractory cytopenia lack definitive diagnostic morphologic features in bone marrow [101,108]. These unique chromosome aberrations include unbalanced abnormalities of loss or deletion of chromosome arms in 5q, 7q, 9q, 11q, 12p, 13q, 17p and idic(Xq), and balanced translocations and inversion, such as t(1;3), t(2;11), t(3;21), inv(3q), t(6;9), and t(11;16) (Table 9.1) [95,108]. Cytogenetic analysis while prognostic in all MDS patients can therefore also be diagnostic in some and is therefore considered an essential element of the evaluation of a patient with unexplained cytopenias.

Predisposition to myeloid malignancies due to underlying genetic abnormalities has been underappreciated, particularly in the adult population. For a comprehensive list of genes, please see Table 1 in Ref. [109]. There are several scenarios that may be suggestive that a patient should be further tested for a cancer predisposition. These include a cancer that arises at an earlier age than typical in the general population, a family history of malignances (including a history of hematological abnormalities, such as cytopenias), physical features associated with a germline predisposition syndrome, and the detection of certain gene mutations that are cancer predisposition genes. The detection of cancer predisposing germline variants is increasingly identified during routine genomic testing for hematological malignancies due to incorporation of massively parallel sequencing (MPS)/Next-Generation Sequencing (NGS) in routine clinical care.

Confirmation of germline cancer predisposition genetic variants can be difficult in hematological malignancies. Acquisition of normal tissue often requires an invasive procedure (e.g., skin biopsy that needs to be grown in culture). The age of presentation and whether the mutation is associated with a known syndrome may help ascertain whether a mutation detected in a hematological malignancy could represent a germline variant in a cancer predisposition gene. In the pediatric setting, many of these mutations are present in the setting of a bone marrow failure syndrome. The best recognized of these bone marrow failure syndromes is Fanconi anemia, with 80%—90% of cases caused by loss of function mutations in FANCA, FANCC, and

FANCG, with the other Fanconi anemia genes associated with the remaining cases. Germline gain-of-function mutations in the *SAMD9* and the *SAMD9L* genes are associated with MIRAGE syndrome, ataxia–pancytopenia syndrome, and myelodysplasia and leukemia syndrome with monosomy 7 syndrome. Inference of a germline variant may be elucidated if performing sequencing throughout the course of disease. This allows the comparison of variant allele fractions (VAF) over time and, reflected with disease state, can be used as a surrogate for germline testing (Fig. 9.7) [110].

9.3.2 Therapeutic Implications

As indicated above, chromosome abnormalities are incorporated into prognostic scoring systems such as the International Prognostic Scoring System (IPSS), the WHO Prognostic Scoring System, and the MD Anderson Cancer Center MDS model [111–115]. The Revised IPSS (IPSS-R) [92] stratifies MDS patients into five subgroups based on hematologic findings and cytogenetic abnormalities, with treatment based on separation into low risk (very low and low) and high risk (very high and high), with treatment of all patients including improvements of cytopenias. Treatment of patients with low-risk disease is additionally focused on supportive measures and quality of life, while in high-risk patients, the aim is to delay transformation to AML [116].

9.4 ACUTE MYELOID LEUKEMIA

AML is a clonal hematopoetic malignancy characterized by a block in the differentiation process that results in the accumulation of immature cells within one of the myeloid lineages. AML is a genetically heterogeneous disease, with hundreds of recurrent, structural cytogenetic, and molecular abnormalities reported in the literature. Morphological, immunophenotypic, cytogenetic, and molecular data are integrated into the final diagnosis, with cytogenetic and molecular abnormalities now essential in the appropriate classification of AML. Broad detection of cytogenetic and molecular abnormalities has allowed large multicenter studies to group patients into prognostic groups (Fig. 9.8) [28]. The development of targeted inhibitors for specific gene mutations has provided a leap forward in personalized diagnostics and personalized medicine in AML [117].

AML classification was traditionally based on the stage of maturation arrest using the FAB (French American British) system [4], but has been updated to include other features so that WHO 1016 includes categories such as: AML with recurrent genetic abnormalities, AML with myelodysplasia-related changes, therapy-related myeloid neoplasms and AML, not otherwise classified. In general, AML is diagnosed in patients having greater than 20% myeloid blasts in blood or marrow by morphologic assessment [6,101] with certain exceptions. These disease-defining exceptions include the presence of myeloid blasts with

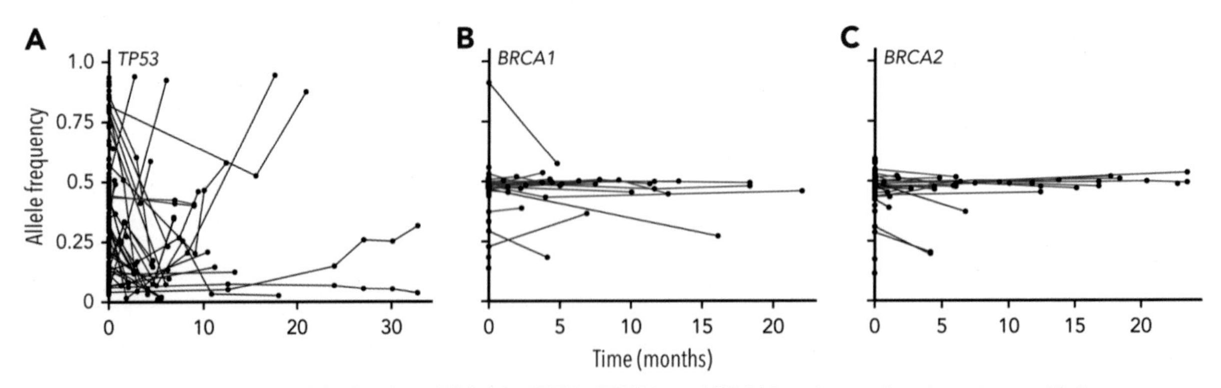

Figure 9.7 Variant allele fractions (VAFs) for TP53, BRCA1, and BRCA2 and over time in patients with hematopoietic malignancies. VAFs of DNA alterations in TP53 (A), BRCA1 (B), and BRCA2 (C) in individual patients at the University of Chicago are graphed over time. Each point indicates an individual variant identified in an in-house NGS assay, and *red* lines connect likely somatic variants; likely germline variants are shown in *blue*.

A

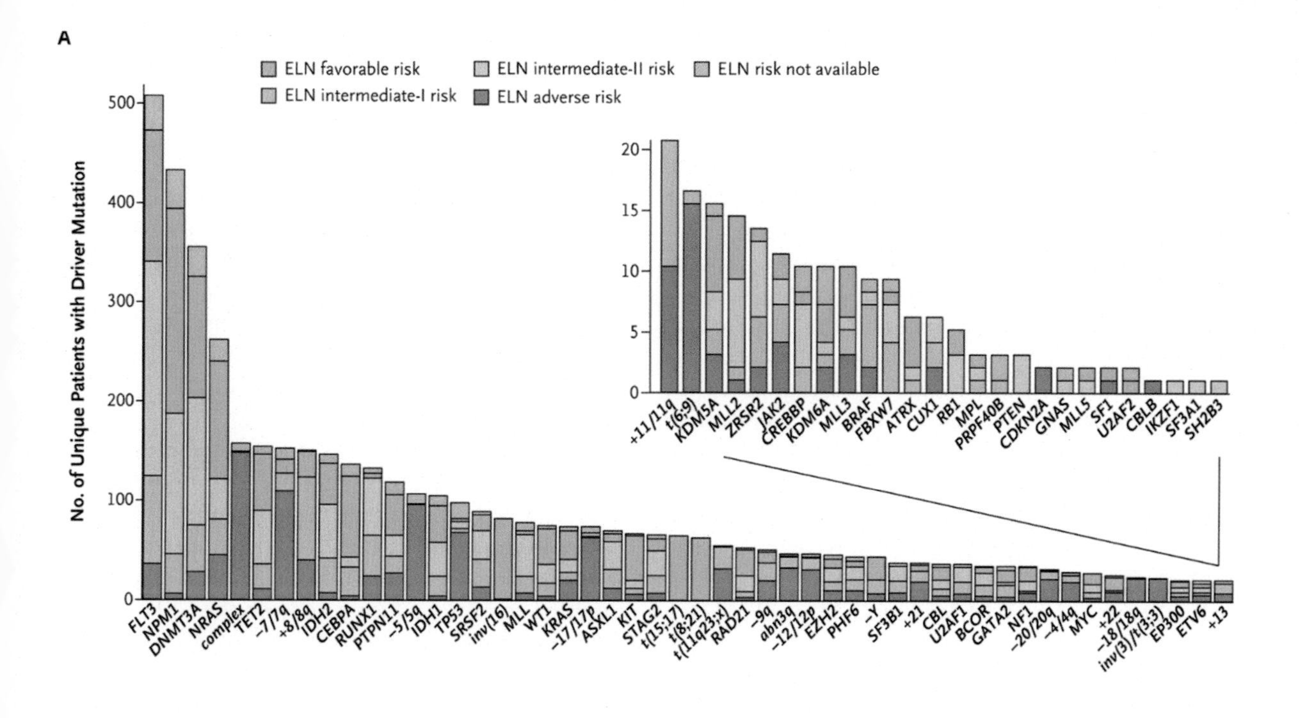

B

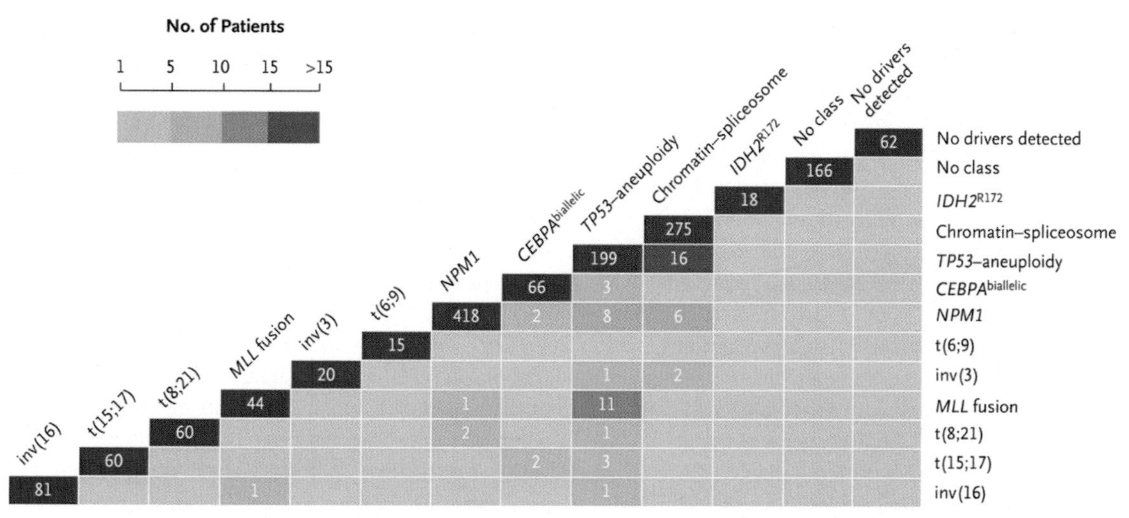

Figure 9.8 Landscape of driver mutations in acute myeloid leukemia (AML). Panel A shows driver events in 1540 patients with AML. Each bar represents a distinct driver lesion; the lesions include gene mutations, chromosomal aneuploidies, fusion genes, and complex karyotypes. The colors in each bar indicate the molecular risk according to the European LeukemiaNet (ELN) classification. Panel B shows the distribution of samples and overlap (*cross sections*) across molecular subgroups (*vertical bars*). Patients who had no driver mutations and those who had driver mutations but did not meet the criteria for any specific class are also included. The number at the top of each column is the number of patients assigned solely to the designated class; the numbers of patients meeting criteria for two or more classes are shown at the intersection of classes.

RUNX1T1::RUNX1 (t(8;21) (q21.3;q22), *CBFB::MYH11* (inv(16) (p13.1q22), t(16;16) (p13.1;q22), and *PML::RARA* (t(15;17) (q24;q22) rearrangements, or myeloid sarcoma (described below). This section will describe genetic abnormalities that are associated with AML and will highlight the clinical importance of the detection of these abnormalities.

Conventional cytogenetic studies remain an important part of the diagnostic work-up of AML patients, with clonal chromosome abnormalities detected in about 50%–80% of patients with AML [7,9] There are seven subtypes of AML collectively identified in the 2016 WHO classification system as AML with recurrent genetic abnormalities: *RUNX1T1::RUNX1*, *CBFB::MYH11*, and *PML::RARA*, as well as other rearrangements, *AF9::KMT2A* (previously known as *MLL*) (t(9;11) (p22;q23)), *DEK::NUP214* (t(6;9) (p23;q34)), *RPN1::MECOM* (inv(3) (q21q26.2)/t(3;3) (q21;q26.2)), and *RBM15::MKL1* (t(1;22) (p13;q13)).

9.4.1 Core Binding Factor AML

9.4.1.1 Genetics and Mechanism of Disease

Under normal circumstances, the RUNX1 and CBFB proteins heterodimerize to form the CBF (RUNX1/CBFB) transcription complex. The CBF (Core Binding Factor) transcription factor is critical to hematopoietic stem and progenitor cell growth, differentiation, and function. It binds directly to an enhancer core motif that is present in the transcriptional regulatory regions of a number of genes, including interleukin (*IL*)-*3*, *GM-CSF*, the *CSF1* receptor, myeloperoxidase, and neutrophil elastase [118,119]. CBF AML is characterized as containing either the *RUNX1T1::RUNX1* [t(8;21)] or *CBFB::MYH11* [inv(16)/t(16;16)] rearrangements, with the rearrangement of these genes leading to overexpression of these critical components of CBF.

The *RUNX1T1::RUNX1* rearrangement results in the juxtaposition of 5′ *RUNX1* with the 3′*RUNX1T1*, leading to a fusion gene and a chimeric protein [120,121], with the critical event containing the active fusion gene localized on the derivative chromosome 8. Development of AML associated with *RUNX1T1::RUNX1* results from both altered transcriptional regulation of normal *RUNX1* target genes and activation of new target genes that block programmed cell death and/or cellular differentiation pathways [122]. The breakpoints involved in this rearrangement cluster within a single intron of both genes, so that similar *RUNX1T1::RUNX1* chimeric transcripts are generated. These breakpoints colocalize with topoisomerase II cleavage sites and DNA hypersensitive sites [123] and may explain why this rearrangement can occur as a therapy-related malignancy in patients treated with topoisomerase II inhibitors.

The *RUNX1T1::RUNX1* rearranged disease identifies a morphologically and clinically distinct subset of AML [6]. The portion of *RUNX1* involved in the fusion contains the DNA-binding and heterodimerization motifs. Thus in *RUNX1T1::RUNX1* disease, the fusion protein contains not only these motifs but also most of the RUNX1T1 protein, and dimerization with CBFB is apparently unaffected.

The *CBFB* gene is located at 16q22 and is involved in the inv(16) (p13q22)/t(16;16) (p13;q22) resulting in the *CBFB::MYH11* fusion, and is the other CBF leukemia, which is observed in about 7% of newly diagnosed AML [124,125]. The pericentromeric inversion or translocation results in a fusion between the *CBFB* at 16q22 and the myosin heavy-chain gene (*MYH11*) at 16p13 [126–128]. The critical fusion is 5′ *CBFB*:3′ *MYH11*, present on the long arm of chromosome 16. As described earlier, CBFB forms a heterodimer with RUNX1, which directly binds to an enhancer core motif that is present in the transcriptional regulatory regions of a number of genes that are critical to myeloid cell growth, differentiation, and function. The CBFB protein increases the efficiency of heterodimerization and the affinity of RUNX1 binding to DNA. The CBFB:MYH11 fusion protein interferes and represses the function of the CBF transcription factor and, in cooperation with other genetic mutations, such as *KIT* and *RAS* mutations, leads to leukemia.

The detection of trisomy 22 in addition to *CBFB::MYH11* is associated with a more favorable outcome. As inv(16)/t(16;16) can be difficult to detect when chromosome banding is suboptimal, the observation of trisomy 22 should result in a review of the karyotype or FISH for *CBFB* rearrangement. A less favorable outcome is associated with trisomy 8, *FLT3* TKD, and *KIT* mutations. In contrast to *RUNX1T1::RUNX1*, the detection of *RAS* mutations does not appear to affect prognosis in the setting of *CBFB::MYH11*.

The detection of molecular abnormalities in addition to the *RUNX1T1::RUNX1* rearrangement can however modify the favorable outcome. Gain-of-function mutations of *KIT* have been detected in AML with t(8;21) and inv(16)/t(16;16), [129]. *KIT* is a member of the type III tyrosine kinase family and encodes a 145-kd transmembrane glycoprotein. Mutations of *KIT* have been shown to have adverse impact on the prognosis of CBF leukemia, with a significantly shorter event-free survival and overall survival rate in patients with *KIT* mutations compared with wild-type *KIT* patients [129]. *FLT3* tyrosine kinase domain (TKD) mutations are also recurrently observed and associated with a less favorable outcome. Detection of *NRAS* or *KRAS* mutations is associated with a favorable disease course.

9.4.1.2 Therapeutic Implications

Clinically, patients with *RUNX1T1::RUNX1* respond well to conventional treatment with high remission, low relapse, and high salvage remission rates so that this leukemia is characterized by a favorable prognosis in both adults and children with the overall 10-year survival of 61% and 80%, respectively [8,9,130]. Patients with *CBFB::MYH11* have a good response to standard chemotherapy with a rate of complete remission and 10-year overall survival of 92% and 55%, respectively [8]. As described above, gain-of-function mutations of *KIT* will negatively affect the prognosis of patients with inv(16)/t(16;16). As the choice of agents for standard chemotherapy differs for CBF leukemias than for other AMLs, it is important to have rapid FISH studies to identify these rearrangements prior to initiation of therapy.

9.4.2 Acute Promyelocytic Leukemia

9.4.2.1 Genetics and Mechanism of Disease

Acute promyelocytic leukemia (APL) is an aggressive subtype of AML, which is characterized by a structural rearrangement involving chromosomes 15 and 17 [2,131,132]. This rearrangement is unique to APL and is the basis for the diagnosis of APL in the WHO classification [6,101,108]. The translocation results in a fusion between the *PML* (promyelocytic leukemia) gene on chromosome 15 and the retinoic acid receptor alpha (*RARA*) gene on chromosome 17 [133]. The critical rearrangement is located on the abnormal chromosome 15, with the 5′ portion of *PML* fused to virtually the entire *RARA* gene, beginning with the *RARA* response element [134–137].

Under normal circumstances, RARA and retinoic receptor X form a heterodimer that binds to the retinoic acid response elements of target genes. In the absence of ligand, the target genes are in the inactive state and a complex of repressor proteins binds to the heterodimer preventing transcription. When ligand is present (e.g., retinoic acid), a conformational change in the heterodimer displaces the repressor complex, recruits transcriptional coactivators, resulting in expression of target genes and differentiation. The PML:RARA fusion protein is insensitive to the normal ligands leading to persistent transcriptional repression and preventing differentiation of promyelocytes. This maturation arrest can be overcome by treatment with all-trans retinoic acid (ATRA), which acts as a ligand for the fusion protein [133,138,139].

The *FLT3* internal tandem duplication (ITD) mutations are enriched in APL, observed in approximately 30%–40% of cases, and FLT3 TKD mutations are seen in approximately 10% of cases. The detection of FLT3-ITD does not alter the prognosis of patients with APL.

In addition to rearrangements with *PML*, *RARA* may be involved in other translocations forming fusions with genes including *PLZF* (11q23), *NPM1* (5q32), *NUMA1* (11q13), *FIP1L1* (4q12), and *STAT5B* (17q21) (Table 9.1) [95]. These variant translocations are associated with atypical APL. PLZF functions as a transcriptional repressor through its interaction with the same complex of repression proteins, and this interaction is not affected by ATRA [133,138].

9.4.2.2 Therapeutic Implications

With the introduction of ATRA for treatment, APL became the first malignancy in which genotype-specific therapy was used. This treatment leads to maturation of leukemic cells in virtually all *PML::RARA* patients and is combined with cytotoxic drugs to eradicate the leukemic clone. This leads to a remarkable improvement in the overall survival of APL patients. The addition of arsenic trioxide (ATO) to the ATRA and chemotherapy regimen achieves an overall survival rate of ~90%, in comparison to 60%–80% with ATRA and chemotherapy in a 5-year follow-up study [19]. Arsenic can directly bind to the PML protein in the PML:RARA fusion, resulting in homodimerization and multimerization facilitating subsequent degradation of the PML:RARA protein product via the proteasome [140]. In APL, the combination of ATO (targeting PML) and ATRA (targeting RARA) and chemotherapy has dramatically improved therapeutic efficacy and become a frontline treatment option in this otherwise very aggressive disease [141,142].

Mutations in *PML* and *RARA* have been associated with resistance to ATO and ATRA, respectively [143]. Additionally, some of the uncommon rearrangements of RARA with other partners do not respond to ATRA, and so the precise rearrangement at diagnosis is critical for appropriate therapy. Cells expressing PLZF::RARA fusion proteins are not sensitive to ATRA-induced differentiation, although they contain identical RARA domains to those of PML::RARA. Treatment with HDAC inhibitors, however, induces differentiation of *PLZF::RARA* cells in vitro.

Patient with APL have a high risk of a life-threatening complications at presentation due to a coagulopathy predisposing them to both life-threatening clots and hemorrhage. Cytotoxic chemotherapy can further increase the risk of coagulopathy if used. ATRA and ATO effectively decrease the coagulopathy but subject the patient to additional risk secondary to what is called APL differentiation syndrome. With patients who present with blasts, which morphologically appear concerning for APL, ATRA is often started emergently to decrease the risk of potential complications even before a diagnosis is confirmed. Depending on the clinical scenario, emergent testing for the presence of the 15; 17 translocation may be needed as the results may impact immediate clinical decisions.

9.4.3 KMT2A-Rearranged AML

9.4.3.1 Genetics and Mechanism of Disease

Rearrangements of the long arm of chromosome 11q23 are recurrent in AML and acute - leukemia (ALL), involving the KMT2A gene, previously called MLL for mixed-lineage leukemia, due to cases with phenotypic lineage switching (diagnosed with ALL and relapse with AML [144,145]. KMT2A is a promiscuous gene with over 85 distinct partner genes identified [146]. The most common translocations in AML are t(9;11), t(6;11), and t(11;19) (q23;p13.1); in ALL they are t(4;11) and t(11;19) (q23;p13.3) [17,200–204]). The t(9;11) is part of the diagnostic criteria for AML in the WHO 2016 under AML with recurring genetic abnormalities (Table 9.1) [6]. Patients with the t(9;11) (p22;q23) had rates of complete remission and 10-year survival of 84% and 39%, respectively, demonstrating this as an intermediate prognosis, with other rearrangements of KMT2A with poorer outcomes [8,22,147]. Translocations involving KMT2A account for about three-quarters of the chromosomal abnormalities observed in infant leukemia (AML or ALL) [148–150]. Analysis of genetic data in infant leukemias suggests that the leukemia is initiated in a hematopoietic progenitor cell (HPC) and that the KMT2A/MLL chimeric gene plays an early, dominant role in establishing leukemic cells [151]. This contrasts with pediatric and adult KMT2A/MLL leukemias that are thought to arise in more mature HPCs [8,22,152–155]. KMT2A translocations can be seen de novo or as therapy-related myeloid neoplasms in patients who previously received treatment with drugs targeting topoisomerase II [156–158].

KMT2A can also be mutated via a partial tandem duplication (PTD) within the *KMT2A* gene. This is cytogenetically invisible and cannot be detected by standard FISH studies, requiring specialized testing. KMT2A-PTD is detected in 90% of AML cases with trisomy 11 and in 11% of patients with AML and normal karyotype and is associated with a poor prognosis [159,160].

9.4.3.2 Therapeutic Implications

Regardless of myeloid or lymphoid lineage involvement, most leukemia patients with 11q23/MLL rearrangements have a very dismal prognosis. However, patients with t(9;11) (p22;q23) tend to have an intermediate response to standard therapy(Fig. 9.9), [8]. Patients with KMT2A rearrangements are often considered candidates for allogeneic bone marrow transplantation.

9.4.4 Additional Categories of AML

9.4.4.1 Genetics and Mechanism of Disease

The t(3;3) and inv(3) are abnormalities that are classified by 2016 WHO classification system as AML with recurrent genetic abnormalities and represent the rarely recurring chromosome abnormalities, occurring in only 1% of AML cases [8,161]. It is detected in de novo AML, CML in blast crisis, and in therapy-related MDS/AML (Table 9.1) [95]. Rearrangements involving 3q26.2 in AML are associated with thrombocytosis in the peripheral blood and increased atypical megakaryocytes in the bone marrow [161,162]. The inv(3)/t(3;3) rearrangements reposition the distal *GATA2* enhancer (3q21) to activate *MECOM* expression (3q26) [163,164]. MECOM encodes the transcriptional isoforms MDS1-EVI1 and EVI1, and the rearrangement results in the repositioned GATA2 enhancer to overexpress EVI1, but not MDS1-EVI1, and concurrently

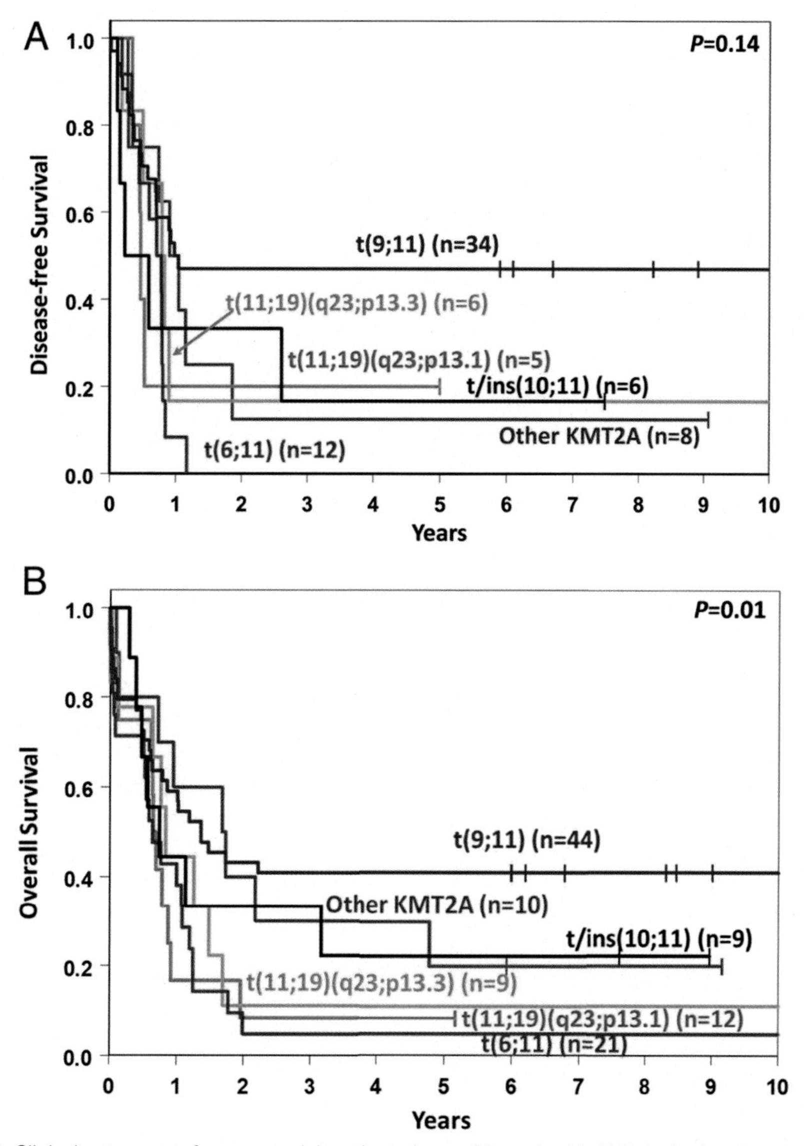

Figure 9.9 Clinical outcomes of younger adult patients (age <60 year) with AML harboring the most frequent recurring rearrangements involving 11q23/KMT2A. (A) DFS. (B) OS.

confers *GATA2* haploinsufficiency. As the most common additional aberration to inv(3q)/t(3;3), loss of chromosome 7 contributes to the adverse prognostic impact on relapse-free survival and overall survival [8]. There are rare, atypical rearrangements that involve the MECOM locus using enhancer hijacking from other genes that have the same downstream effects (EVI1 overexpression, absent *MDS1-EVI1* expression, and potential *GATA2* involvement) with the same poor prognosis [165], so that abnormalities involving 3q26 should be investigated for MECOM rearrangements.

The t(6;9) (p22;q34) is a distinct entity recognized by 2016 WHO classification system as "AML with recurrent genetic abnormalities" and results in the juxtaposition of *DEK* on chromosome 6 with *NUP214* (also known as *CAN*) on chromosome 9 and is observed in less than 1% of newly diagnosed AML. NUP214 is an essential component of the nuclear pore complex at the nuclear envelope and is required for proper nucleocytoplasmic transport. The DEK protein is a major component of the chromatin to modify the DNA structure by introducing supercoils. This results in the creation of a nucleoporin fusion protein that acts as a transcription factor and also alters nuclear transport. There is a high cooccurrence (70%) of FMS-like tyrosine kinase 3 (*FLT3*) gene internal tandem duplications (ITD) [6,8,22,166]. Patients with the t(6;9) (p23;q34) have a poor outcome with standard therapy. The poor prognosis is likely also due to the high incidence of *FLT3*-ITD mutations [166,167], but can be overcome by allogenic bone marrow transplantation [168].

The t(1;22) (p13;q13), involving the RNA-binding motif protein-15 (RBM15) at 1p13 and a DNA-binding motif protein known as megakaryocyte leukemia-1 (MKL1) at 22q13, is a rare entity (<0.5%) exclusively detected in newly diagnosed AML [8,169]. This is an entity that can be seen in neonatal leukemia (defined as observed in the first 28 days of life). It typically presents with megakaryoblastosis occurring in infants and is associated with acute megakaryoblastic leukemia in young children, most often without Down syndrome and prior MDS or transient myeloproliferation [170]. Overall, infant AML has a poor prognosis, so the specific prognostic significance of t(1;22) in AML treated with modern therapy is not clear [169].

In addition to the recurrent chromosomal rearrangements described above, other cytogenetic abnormalities are prognostic in AML, including the overall pattern of abnormalities [9], with the presence of complexity and/or loss of chromosomes 5, 7, and 17 associated with a very poor prognosis. The detection of a monosomal karyotype, defined as the presence of two or more autosomal monosomies or one monosomy plus at least one structural abnormality, was associated with a particularly poor prognosis, with the initial study showing a 4% 4-year overall survival (OS) [171]. Monosomal karyotype has been observed in 5%–10% of all AMLs and is enriched in patients over the age of 60 years (approximately 20%). A monosomal karyotype does not need to be complex. For example, monosomy 7 is seen in approximately 30% of cases with inv(3) rearrangements involving MECOM, meeting the criteria for a monosomal karyotype. More recent studies have shown that cases with monosomy 17, with loss of TP53, are an independent predictor for worse survival among monosomal karyotypes patients as well as those with highly complex karyotypes (i.e., those with greater than or equal to five abnormalities) [171]. One caveat for determining monosomal karyotype is that the presence of marker chromosomes and other material of unknown origin could result in the inappropriate calling of chromosome loss. This can be ascertained by FISH for centromeric material from chromosomes that are lost.

Although cytogenetic abnormalities were initially the defining entities for AML, the detection of gene mutations has filled in the gaps for those patients with normal karyotypes and has added additional information to karyotypically abnormal cases. The molecular landscape of mutations in AML is complex, with driver mutations described in over 75 genes, often with more than one driver gene detected [172]. The mutated genes can be subclassified based on function: tumor suppressor, transcription factor, spliceosome complex, signaling pathway, NPM1, DNA methylation, chromatin regulation, and cohesion complex, in addition to the transcription factor fusions, described above. The detection of gene mutations at diagnosis and throughout the course of disease has demonstrated the complexity of AML and how cooccurrence of mutations with or without cytogenetic abnormalities can influence outcome [28]. For example, the detection of TP53 mutation with a complex karyotype has an additive effect resulting in very unfavorable outcomes, worse than that observed with a single occurrence (e.g., TP53 mutation only, with a normal karyotype) [28]. The triple-mutation status of NPM1, DNMT3A, and FLT3-ITD in patients with karyotypically normal AML is associated with a poorer overall survival; however, newer chemotherapeutic regimens may change this metric [173].

FLT3 is one of the most commonly mutated genes in AML with two different types of mutations detected, internal tandem duplications (ITDs) and tyrosine kinase domain mutations. FLT3 ITDs are in frame mutations that typically occur in the juxtamembrane domain and result in the constitutive activation of the tyrosine kinase domains via autophosphorylation. FLT3 inhibitors have been developed that are used in addition to standard chemotherapy in the initial therapy of patients with FLT3 mutations. The identification of the targeted FLT3 therapy, gilteritinib, improved the survival of patients with relapsed or refractory FLT3-mutated AML compared with standard chemotherapy changing the risk profile for this common marker [174]. Resistance appears to occur either through the expansion of clones containing FLT3 mutations, D835 and F691, or through the expansion of clones with other drivers, such as NRAS [41].

The CCAAT/enhancer binding and protein alpha gene (*CEBPA*) encodes a key transcription factor that regulates myeloid cell differentiation and proliferation, with mutations detected in about 10% of normal karyotype AML (NK-AML), [175]. The detection of *CEBPA* biallelic mutations characterized by both frameshift mutations in the N-terminal and in-frame insertions/deletions (indels) in the C-terminal domain has defined a distinctive disease entity with a very favorable prognosis [176]. In 2016, the WHO designated CEBPA mutated AML as an entity only in the setting of biallelic mutations.

NPM1 encodes nucleophosmin, a nuclear-cytoplasmic shutting phosphoprotein with pleiotropic functions. *NPM1* mutations are detected in 25%–35% of AML, and in about 50% of NK-AML, and are associated with good prognosis, in particular in the absence of *FLT3*-ITD [177]. AMLs with *NPM1* are included as a disease entity in the 2016 WHO classification. The mutations in NPM1 occur in the terminal exon and are typically small insertions that alter protein localization, with mutation NPM1 protein sequestered in the cytoplasm. Although NPM1 is considered a favorable prognostic mutation, this can be mitigated by additional mutations, with about 40% of NPM1-mutated AMLs also harboring an FLT3-ITD. ELN has included the NPM1 and FLT3 status in its classification, with FLT3-ITD-negative/NPM1-positive considered favorable; FLT3-ITD-negative/NPM1-negative or FLT3-ITD-positive/NPM1-positive considered intermediate; and FLT3-ITD-positive/NPM1-negative associated with a poor prognosis. Moreover, studies have shown that molecular monitoring for absence of NPM1 mutations in NPM1 mutant AML after 2 cycles of therapy predicts for long-term disease-free survival.

9.4.4.2 Therapeutic Implications

Therapies directed toward the various subclasses of AML are based on the associated prognoses and risk factors described above. These therapies can range anywhere from systemic broad-range agents, such as chemotherapy, to targeted agents, such as those that target specific genetic alterations.

9.5 THERAPY-RELATED MYELOID NEOPLASMS

9.5.1 Genetics and Mechanism of Disease

Therapy-related myeloid neoplasms result from chemotherapy and/or radiation therapy exposure used in the treatment of both malignant and nonmalignant diseases [178,179]. Characteristic nonrandom chromosome abnormalities are commonly observed in the bone marrow cells of patients with t-MDS/t-AML/tMDS/MPN, with high-risk cytogenetic features common (with some exceptions noted below). Patients with tMN typically are associated either with previous treatment with alkylating agents and/or radiation or with topoisomerase II inhibitors. In patients with prior exposure to alkylating agents and/or radiation, the t-MN typically evolves 4–6 years after the treatment for primary disease and presents with an insidious MDS phase prior to t-AML. Abnormalities of chromosomes 5 and/or 7 are most common, occurring in almost 70% of t-MDS/t-AML patients [180,181]. The other subgroup of t-AML/MDS, cancer patients have been treated with drugs that inhibit the ligase function of topoisomerase II (epipodophyllotoxins, anthracyclines, etc.), resulting in DNA with double-strand breaks. These leukemias typically develop as acute disease with no previous MDS stage in these patients having a different karyotypic pattern, namely translocations involving the *KMT2A/MLL* gene at 11q23 and the *RUNX1* gene at 21q22 [97,178,182–186]. Virtually all patients with t(11;16)(q23;p13) that involves the *MLL* and *CREBBP* genes have treatment-related hematologic disorders induced by exposure to topoisomerase II inhibitors [157,187]. The genomic breakpoints in t-AML often coincide with topo II cleavage sites, DNase I-hypersensitive (HS) sites, and scaffold attachment regions, implying that these chromatin structural elements may influence the location of these translocation breakpoints [188–193]. Therapy-related APL with t(15;17) (q22;q21) is particularly associated with exposure to mitoxantrone and epirubicin; both preferentially induce topoisomerase II-mediated DNA damage in a "hotspot region" within the *PML* gene.

Although t-MN is associated with prior therapy, there is an increased incidence of more than one cancer prior to the development of the t-MN, with approximately 20% of patients with t-MN containing a mutation in a cancer germline predisposition gene [194].

There is evolving evidence that predisposition genes involved in the DNA damage response pathways may increase the susceptibility to cytotoxic chemotherapy and/or radiation, due to deficiencies in pathways associated with DNA repair, genomic instability, cell cycle arrest, and apoptosis (reviewed in Fig. 9.10) [194].

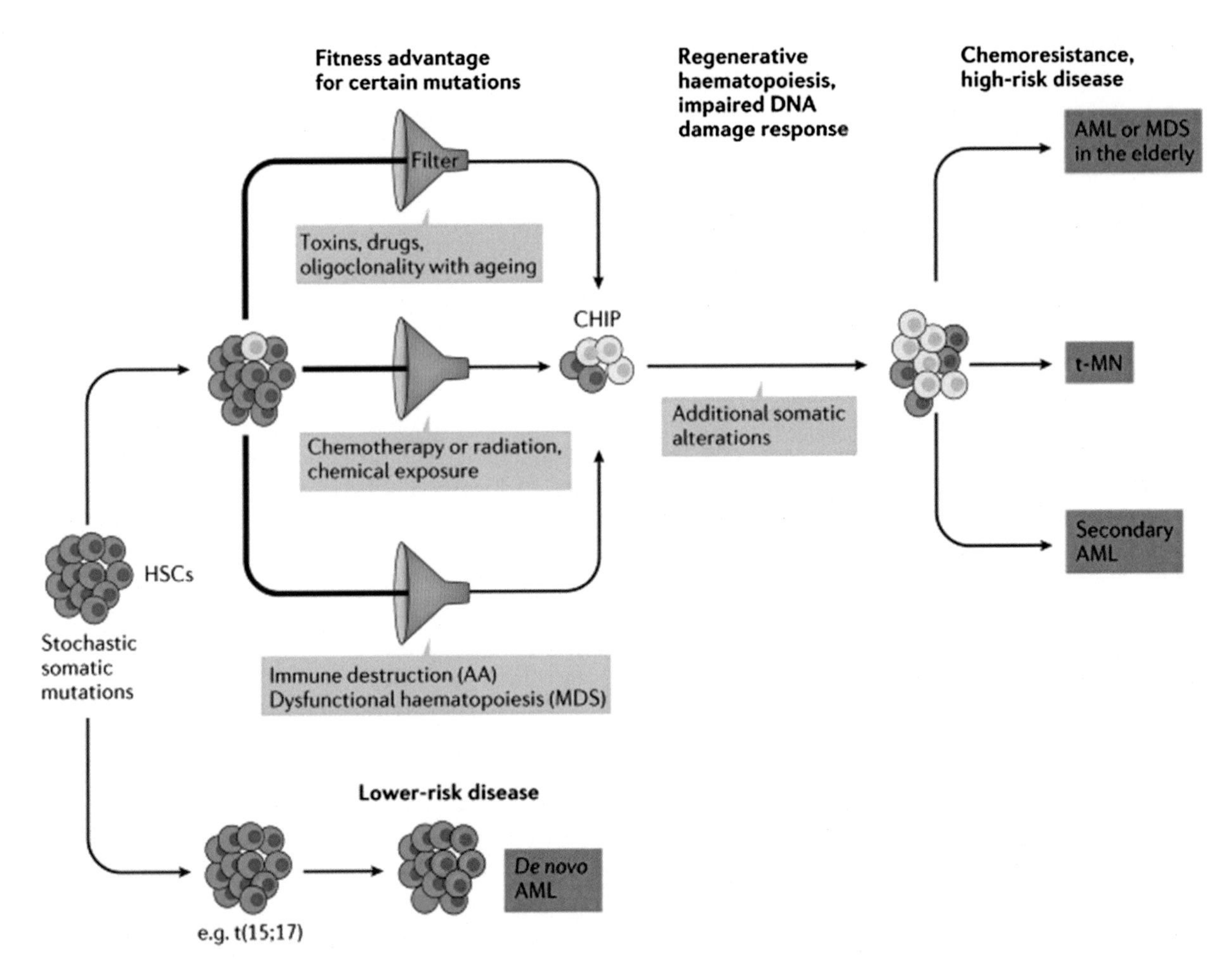

Figure 9.10 Model for the role of clonal selection in the etiology of high-risk myeloid neoplasms. Stochastic mutations occur in hematopoietic stem cells over time. Certain mutated genes provide a "fitness" advantage in the context of various competitive conditions. Competition filters for mutant HSCs at the expense of healthy HSCs, resulting in clonal hematopoiesis of indeterminate potential (CHIP). Over time, mutant clones can acquire additional mutations due to mutations in the DNA damage response genes and/or increased proliferation in the context of regenerative hematopoiesis. In hematopoiesis, competition arises in a number of different contexts. Loss of HSC diversity with age, and the cumulative lifetime exposure to toxins and drugs, among other factors, may select for mutant clones that ultimately give rise to myeloid neoplasms in the elderly. Chemotherapy and radiation exposure promote therapy-related myeloid neoplasms. Inflammation, immune destruction, and dysfunctional hematopoiesis (including niche-based effects) can give rise to secondary myeloid neoplasms arising from aplastic anemia or myelodysplastic syndrome. Regardless of the selective pressure, the competitive filter increases the likelihood that the malignant clones will have inherent therapy resistance. In contrast, a stochastic mutation that gives rise to AML without an antecedent clonal selection is more likely to be lower-risk disease.

9.5.2 Therapeutic Implications

Therapy-related myeloid neoplasms (tMN) are associated with an unfavorable prognosis, with a 5-year survival rate of ~10%, and treated based on clinical classification of the disease. The exceptions to this dismal prognosis are when the tMN contains the *PML::RARA* [t(15;17)] rearrangement, *RUNX1T1::RUNX1* [t(8;21)] rearrangement, or *CBFB* rearrangement. In cases with these rearrangements, the patient is treated using the standard chemotherapy for induction, assuming the that patient is an induction candidate.

9.6 CLONAL HEMATOPOIESIS

During the normal aging process, all tissues acquire somatic mutations, some of which can confer a growth advantage [195]. Clonal hematopoiesis (CH) is defined as the overrepresentation of cells derived from a hematopoietic progenitor containing one or more of these somatic genomic alterations.

9.6.1 Genetics and Mechanism of Disease

Myeloid malignancies are often associated with mutations in genes that involve DNA methylation, chromatin modification, RNA splicing, transcription regulation, cell growth signaling, and DNA damage response [28,40]. However, mutations are acquired across the genome as part of normal aging, and there is a distinct subset of genes that are associated with age-related clonal hematopoiesis (ARCH) most commonly *DNMT3A*, *TET2* (involved in DNA methylation), and *ASXL1* (involved in chromatin modification). Other leukemia driver genes are less frequently mutated in CH, but can occur in a hematopoietic stem cell in the absence of a diagnosis of a myeloid neoplasm. Clonal hematopoiesis is generally defined as having a mutational burden (variant allele fraction, VAF) of greater than 2%, but can be as high as 20%–30%. Sequential acquisition of somatic mutations occurs prior to the diagnosis of hematological malignancies, with mutations improving the fitness of those cells leading to clonal expansions and an increased likelihood of acquiring additional mutations.

Clonal hematopoiesis of indeterminate potential (CHIP) is defined by the presence of a somatic mutation (VAF $\geq$ 2%) in a leukemia-associated driver gene in the absence of persistent cytopenias and with no evidence of a myeloid marrow disorder morphologically. These patients have a slightly increased risk of progression to a myeloid disorder (0.5%–1%/year), but most importantly survival is decreased because of an increase in deaths from cardiac disease.

Clonal cytopenias of undetermined significance (CCUS) is defined by the presence of a somatic mutation in a leukemia-associated driver gene in the presence of a cytopenia that has been present for $\geq$4 months. These patients have a significant risk (50%–90%) of progression to MDS within 5 years dependent on the type, number, and VAF of mutations. Sequential biopsies are needed to identify progression of somatic mutations or progression to over myeloid malignancy.

The detection of clonal hematopoiesis in patients with other cancers is frequent and can be observed in pretreatment and posttreatment specimens [196]. In these settings, there are important implications regarding clone size and risk for treatment-related myeloid neoplasms. The specific mutations detected are dependent on exposures, with *ASXL1* mutations enriched in smokers and DNA response gene mutations (e.g., *TP53*) associated with topoisomerase II inhibitors. Importantly, the variant allele fractions (VAFs) of these mutated genes change during the course of therapy, most prominently observed with radiation therapy. Detection of *TP53* mutations in patients with CH at the time of solid tumor diagnosis was observed in patients who developed therapy-related myeloid neoplasms.

Surprisingly, the detection of clonal hematopoiesis has also been associated with an increased incidence of cardiovascular events [197]. CHIP carriers with mutations in *DNMT3A*, *TET2*, *ASXL1*, and *JAK2* were found to be associated with coronary heart disease and increased coronary artery calcification, with a 1.9-fold increased risk.

The incidence of CH increases with age and has associated risks of development of myeloid malignancies and cardiovascular events [197,198]. The risk for individuals carrying CH mutations to progress to AML or MDS is estimated to be 0.5%–1% per year and is enriched in individuals with a VAF of any single mutation being greater than 10%, mutations in *TP53* or spliceosome genes, and those with more than one mutation [198]. Although CH is associated with an increased risk of hematological neoplasms, most cases do not progress to overt disease and rather represent "benign" clones (Fig. 9.11) [199].

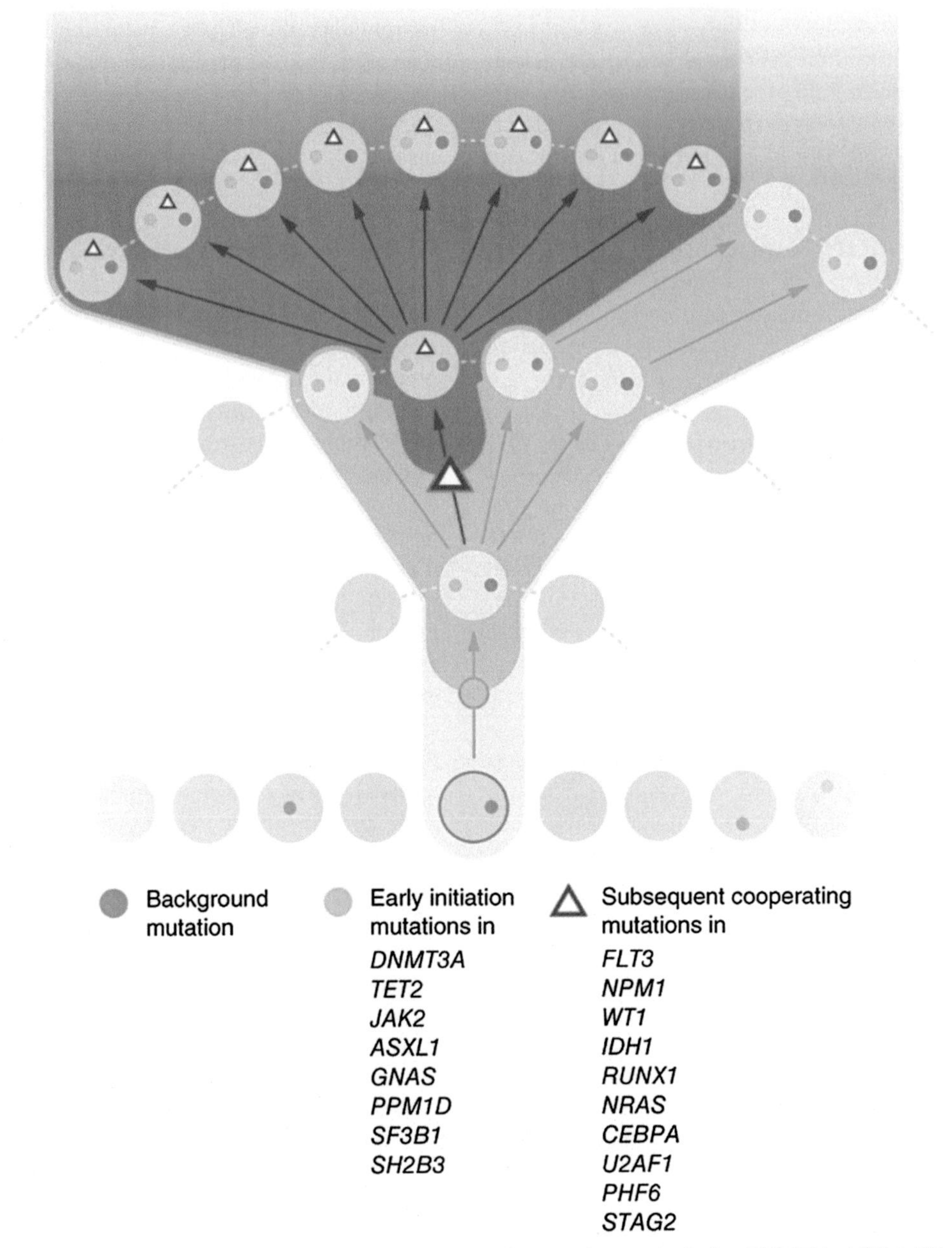

Figure 9.11 Clonal expansion model. The distinct roles of a set of genes including DNMT3A, ASXL1, TET2, GNAS, JAK2, PPM1D, IDH1, NRAS, NPM1, and FLT3 in the initiation of hematopoietic clonal expansion.

9.6.2 Therapeutic Implications

The fact that CH mutations can be found in the absence of other signs of hematologic disease suggests that the presence of mutations alone cannot be used to diagnose or treat a hematological malignancy, which requires additional morphologic findings for appropriate interpretation.

9.7 ACUTE LYMPHOBLASTIC LEUKEMIA

Acute Lymphoblastic Leukemia (ALL) is a cancer of the lymphoid lineage in which the differentiation process is blocked resulting in the accumulation of immature lymphoid cells (blasts) and can affect the B or T cell lineage. ALL is the most frequent leukemia in children (Fig. 9.12) [200], highest in the 1–4-year age range, and dropping through adulthood, with the majority (~85%) being of the B cell lineage.

Most cases of ALL occur spontaneously; however, mutations in nonsyndromic cancer predisposition genes can also predispose to the development of ALL, including mutations in *PAX5, IKZF1, ETV6, PTPN11,* CDKN2A, and CDKN2B [201]. Environmental exposures to pesticides, radiation, and infections have also been associated with risk for development of ALL. Cytogenetic analysis remains an important aspect in the diagnostic work-up of B cell ALL, with chromosome rearrangements and chromosomal alterations characteristic and providing prognostic information.

A better understanding of the complex interrelation of the disease and patient characteristics with therapeutic intervention changed the prognosis in children with ALL. Long-term event-free survival, which is virtually synonymous with cure, which before 1965 was only 5%, now exceeds 80% [22,202,203]. This is not the case in adult patients, where cure rates are only 20%–40%. The most useful prognostic indicators in ALL are age, white blood cell (WBC) count, immunophenotype, MRD detection, and karyotype [204–208]. The incidence of chromosome abnormalities is sharply different in childhood and adult ALL, with the chromosome abnormalities associated with prognosis. For example, t(4;11) is most common in infant (younger than 12 months) B ALL cases and is considered unfavorable; *ETV6::RUNX1*/t(12;21) and hyperdiploid karyotypes are most common in pediatric ALL and are considered favorable; and *BCR::ABL1*/t(9;22) is common in adult ALL [209–212].

In general, favorable factors in ALL include age (from 1 to less than 10 years of age), low blast count, B cell lineage, specific chromosome abnormalities (hyperdiploidy, *ETV6::RUNX1*/t(12;21) and *TCF3::PBX1*/t(1;19), *ERG* deletion), no CNS involvement, and low to no minimal residual disease following induction chemotherapy. Unfavorable factors include age (under 1 year or greater or equal to 10 years), high blast count, T cell lineage, specific chromosome abnormalities (hypodiploidy, *BCR::ABL1*, KMT2A/MLL rearrangements, TCF3:HLF rearrangement, complex karyotype, IKZF1 deletion, Ph-like rearrangements, MEF2D rearrangement), and persistence of disease of disease postinduction chemotherapy [213]. Some of the genetic factors are described in more detail below and summarized in Table 9.2 [213].

9.7.1 B-Cell Acute Lymphoblastic Leukemia

9.7.1.1 Genetics and Mechanism of Disease

High hyperdiploidy is a recurrent abnormality observed in patients with B ALL, characterized by chromosome numbers between modal chromosome numbers of 51 and 67. The chromosomal gains in B ALL associated high hyperdiploidy is nonrandom, with gains of chromosomes X, 4, 6, 10, 14, 17, 18, and 21, with gains of one or two copies of the chromosomes. Detection of the "triple trisomy," gains of chromosomes 4, 10, and 17 is associated with a very favorable prognosis [214,215] In general, patients with a hyperdiploid clone have a very favorable overall survival, which is often coincident with clinical features of a good prognosis, including age between 1 and 9 years, low WBC count, and favorable immunophenotypes (non-T or -B markers), that is early pre-B or pre-B. High hyperdiploidy is seen in 25%–35% of pediatric ALL cases and is much less frequent in the AYA and adult populations. Molecular profiling has identified mutations in receptor tyrosine kinases (e.g., FLT3), RAS pathway genes (e.g., NRAS, KRAS, PTPN11), and histone modifying genes (e.g., CREBBP, SETD2, EZH2) [214]. High hyperdiploidy needs to be carefully interpreted as a separate entity from duplication of near haploid or severe hypodiploid clones, which carry a very unfavorable prognosis. Endoreduplication of a near haploid karyotype can often be ascertained based on the number of each chromosome (2 or 4) and the chromosomes that have four chromosomes (usually chromosomes 14 and 21) [216].

The *ETV6::RUNX1* fusion gene is a recurrent abnormality in B cell ALL and is associated with a favorable prognosis [215]. This rearrangement is cytogenetically invisible, so needs to be detected by FISH or molecular techniques [217–225]. The t(12;21) is present in about 25% of childhood ALL [207]. The t(12;21) results in fusion of the 5′ helix–loop–helix (HLH) dimerization domain of ETV6 with the 3′ DNA-binding and transactivation domain of *RUNX1* [219]. The ETV6

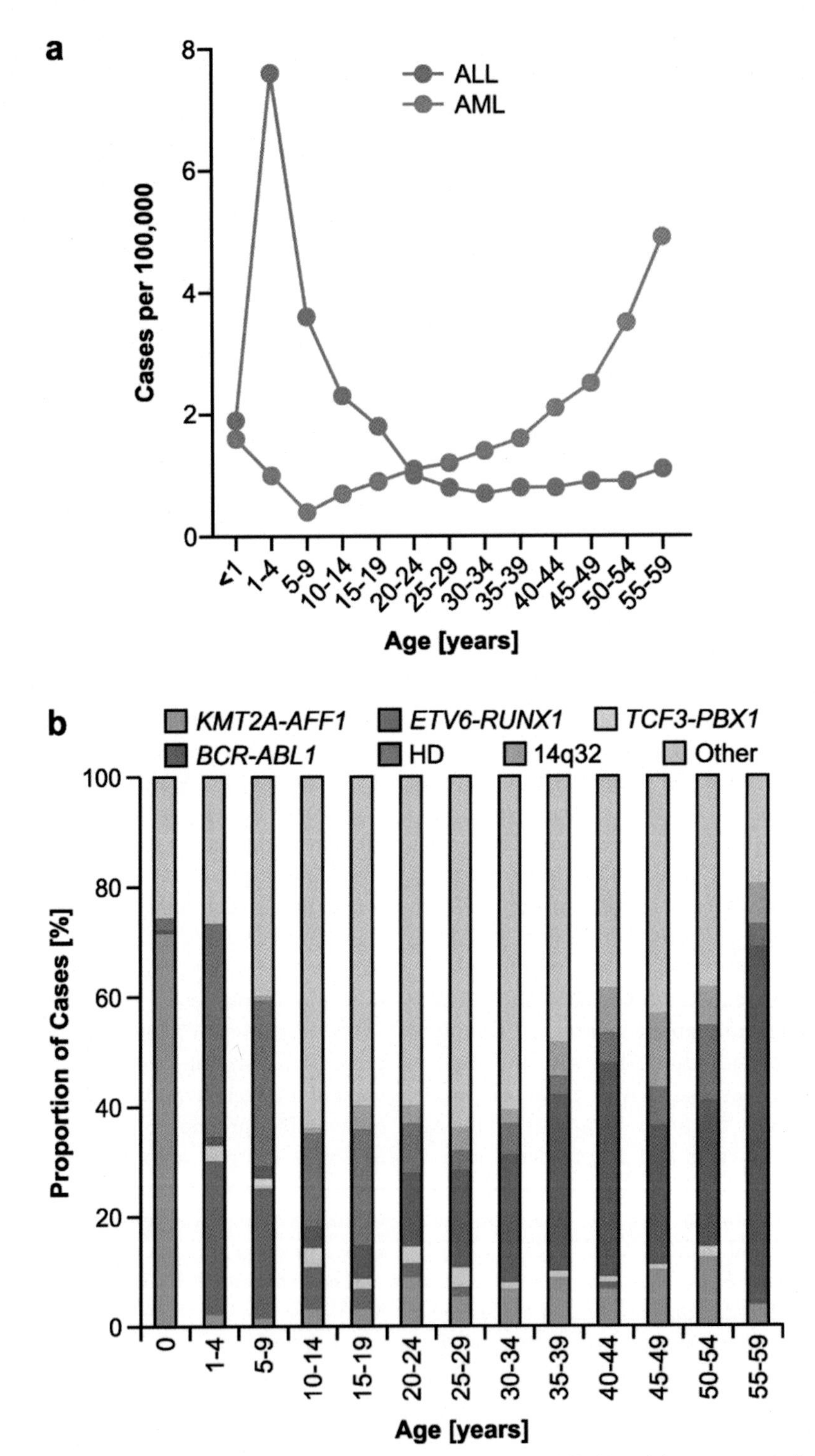

Figure 9.12 Differences in age at presentation between ALL and AML(A) and frequency of chromosome abnormalities in ALL based on age at presentation (B). Age distribution and major subtypes of ALL. (A) Age distribution of ALL and AML in the USA from 1975 to 2016. Cases per 100,000 are shown. ALL has a clear

protein functions as a transcriptional repressor [226,227]. RUNX1 heterodimerizes with CBFB to form a transcription factor, CBF [see AML with t(8;21) and inv(16)/t(16;16)]. The ETV6::RUNX1 fusion protein inhibits transactivation of gene expression by the normal RUNX1 protein, which requires the HLH domain of ETV6 [228,229]. In more than a half of ALL patients with the t(12;21), the ETV6 allele on the other chromosome 12 homolog is also deleted, typically observed by FISH, suggestive of a tumor suppressor effect by ETV6 [229].

The *ETV6::RUNX1* fusion was found at birth in six of nine patients with childhood B ALL diagnosed between 2 and 5 years of age [230]. These findings indicate that the translocations had occurred in utero, and the cells with the translocation had proliferated at birth to be detectable by RT-PCR. The observation that a significant temporal difference between detection of the *ETV6::RUNX1* rearrangement and development of disease suggests additional genetic abnormalities is for transformation to B ALL.

The t(1;19) (q23;p13.3) is another recurrent chromosome abnormality identified in about 5% of all childhood B ALL and in 30% of childhood ALL with pre-B cell phenotype [231,232]. The t(1;19) occurs either as a reciprocal translocation, t(1;19) (q23;p13.3) or, more often, as an unbalanced form containing only the derivative 19 from the rearrangement, der(19) t(1;19) (q23;p13.3) [233]. This rearrangement joins the *PBX* gene on chromosome 1 and the *TCF3* (*E2A*) gene on chromosome 19 at 19p13.3 [231,234,235]. The t(1;19) was initially associated with early treatment failure and poor prognosis, but can be overcome by more intensive chemotherapy, with, t(1;19), included in the favorable prognostic category with t(12;21) and hyperdiploidy [208].

The most common translocation involving the KMT2A/MLL gene in ALL is t(4;11) (q21;q23), observed in more than 60% of infants, 2% of children, and 3%–6% of adults with ALL and is associated with an unfavorable prognosis [236]. The t(4;11) results in a fusion between *KMT2A/MLL* on 11q23 with *AF4* at 4q21 [148,237] and can be detected by RT-PCR in virtually all t(4;11) patients [155]. The association of t(4;11) with neonatal or early-childhood ALL is particularly interesting in view of the low incidence of ALL in this age group [238,239]. Although the morphology often appears lymphoid, other features are more suggestive of a monocytic leukemia, suggesting that the t(4;11) in infant leukemia is derived from very early precursor cells that have dual-lineage capabilities. Studies using FISH probes for *KMT2A/MLL* to screen cells from infant leukemias found that 73%–80% have *KMT2A/MLL* rearrangements, including t(4;11), t(11;19), and t(9;11) [240]. The 5-year event-free survival for infants with these three *KMT2A/MLL* translocations was significantly lower than that of infants lacking *KMT2A/MLL* rearrangement or other *MLL* rearrangements: 22%–30% versus 53%–61% [152,155]. Children with a t(4;11) present with very high leukocyte counts, which is a poor prognostic factor [208,241].

The *BCR::ABL1* rearrangement, resulting in the Philadelphia chromosome (Ph, the derivative chromosome 22), is rare in children (2%–5%) and the most frequent rearrangement in adults (25%) [23,281]. Although this rearrangement has been considered a poor prognostic indicator, the addition of tyrosine kinase inhibitors that target the ABL1 kinase has improved the prognosis. The detection of mutation in IKZF1 in conjunction with *BCR::ABL1* is associated with a poor prognosis [242]. At the cytogenetic level, the breakpoints appear identical to those in CML (9q34 and 22q11.2). In approximately 30%–50% of adult patients, the molecular rearrangement is identical to that observed in CML, that is, the t(9;22) produces a chimeric gene that encodes for a 210-kd fusion protein (p210), and the Ph chromosome is present in both myeloid and lymphoid cells, suggesting that they actually are cases of CML in lymphoid blast crisis. In the remaining patients, the breakpoint in the BCR gene occurs in the m-bcr and results in smaller fusion proteins (190kd, p190) [243–245]. Both the p210 and p190

peak at ages 1–4 and 5–9, whereas AML rates rise with age. (B) Major subtypes of ALL divided by age groups. The KMT2A-AFF1 fusion is very prevalent in infants, ETV6-RUNX1 and high hyperdiploidy (HD) dominate childhood ALLs, and BCR-ABL1 is the most prevalent aberration in adults. Data for (A) taken from Nowell PC, Hungerford DA. Chromosome studies on normal and leukemic human leukocytes. J Natl Cancer Inst. July 1960;25:85–109. (B) Taken from Braun TP, Eide CA, Druker BJ. Response and resistance to BCR-ABL1-targeted therapies. Cancer Cell. April 13, 2020;37(4):530–42.

TABLE 9.2 Main genetic subtypes of B-cell acute lymphoblastic leukaemia

	Frequency	Mutations	Prognosis
High hyperdiploid (gain of ≥ 5 chromosomes)	25% children; 3% AYAs and adults	RTK-RAS signaling pathway, histone modifiers	Favourable
Near-haploid (24–31 chromosomes)	2% children; < 1% AYAs and adults	RAS-activating, *IKZF3*	Poor
Low-hypodiploid (32–39 chromosomes)	<1% children; 5% AYAs; >10% adults	*TP53, IKZF2, RB1*	Very poor
MLL (KMT2A) rearrangements	>80% infants; <1% children; 4% AYAs; 15% adults	*MLL (KMT2A)* rearrangement, few additional mutations (PI3K-RAS signaling pathway)	Very poor
ETV6-RUNX1 translocation, t(12;21) (q13;q22)	30% children; <5% AYAs and adults	*ETV6-RUNX1*	Favourable
TCF3-PBX1 translocation, t(1;19) (q23;p13)	5% children, AYAs and adults	*TCF3PBX1*	Favourable
TCF3-HLF variant of t(1;19) (q23;p13)	<1% acute lymphoblastic leukemia	*TCF3-HLF*	Poor
BCR—ABL1 philadelphia chromosomes, t(9;22) (q34:q11)	2%–5% children, 6% AYAs; >25% adults	*BCR—ABL1* fusion gene, common deletions of *IKZF1, CDKN2A CDKN2B,* and *PAX5*	Poor (improved with tyrosine kinase inhibitors)
Philadelphia chromosomes-like acute lymphoblastic leukemia	10% children; 25%–30% AYAs; 20% adults	Rearrangements of *CRLF2* (about 50%), *ABL*-class tyrosine kinase genes (12%) and *JAK2* (10%) mutations of *EPOR* (3%–10%); mutations activating JAK-STAT (10%) and RAS (2%–8%) signaling pathways	Poor
DUX4 and *ERG*-deregulated acute lymphoblastic leukemia	5–10% acute lymphoblastic leukemia	*DUX4* rearrangements and over expression. *ERG* deletions	Favourable, including if coexistence of *IKZF1* mutations (about 40% of patients)
MEF2D-rearranged acute lymphoblastic leukemia	4% children; 7% AYAs and adults	*MEF2D* is fused to *BCL9* (most frequent fusion event), *HNRNPUL1, SS18, FOXJ2, CSF1R,* or *DAZAP1*	Poor
ZNF384-rearranged acute lymphoblastic leukemia	5% children; 10% AYAs and adults	*ZNF384* rearranged with a transcriptional regulator or chromatin modifier (*EP300, CREBBP, TAF15, SYNRG, EWSR1,TCF3, ARID1B, BMP2K,* or *SMARCA2*)	Intermediate

AYAs, adolescents and young adults. Adapted from Malard F, Mohty M. Acute lymphoblastic leukaemia. Lancet. 2020 Apr 4;395(10230):1146–1162. https://doi.org/10.1016/S0140-6736(19)33018-1. PMID: 32247396.

fusion proteins participate in constitutive signaling via the RAS pathway of signal transduction.

A hypodiploid clone with fewer than 46 chromosomes can be detected in about 5%–6% of ALL patients. There are three distinct subgroups of hypodiploidy in ALL, that is, near haploidy (23–29 chromosomes), low hypodiploidy (33–39 chromosomes), and high hypodiploidy (42–45 chromosomes) [246]. The near-haploid clone is often accompanied by a doubling of the chromosome complement (endoreduplication) resulting in a hyperdiploid population with a gain of two copies of certain chromosomes, which is distinguishable from the more common hyperdiploid clone with a single gain of certain chromosomes. In some patients, cytogenetic analysis may detect the endoreduplicated clone only. ALL patients with hypodiploid clones generally have a poor prognosis, especially patients with near-haploid and low-hypodiploid clones [246]. In this regard, it is critical to avoid confusing the endoreduplicated (doubled) cases with a true hyperdiploid population, as the prognosis and treatment differ between these groups.

The Philadelphia chromosome–like acute lymphoblastic leukemia (Ph-like ALL) is a subtype of B ALL that displays a similar gene expression profile and signaling to Ph-positive B ALL without the presence of the 9;22 translocation [247] Ph-like B ALL is a heterogeneous disease characterized by a variety of genomic alterations. These rearrangements activate kinase and cytokine receptor signaling and can be grouped based on if the rearrangement results in a phenocopy of BCR:ABL1 rearrangement (rearrangements involving ABL1, ABL2, CSF1R, and PDGFRB) or if the rearrangement results in activation of the JAK/STAT pathway (rearrangements involving CRLF2, JAK2) and EPOR (typically rearrangements with IGH) [248]. The most common rearrangement involves the CRLF2 gene (cytokine receptor-like factor 2) localized on the short arm of the sex chromosomes in the pseudoautosomal region, Xp22.3 and Yp11.3 [248]. There are multiple rearrangements that result in the deregulation of the *CRLF2* gene in ALL, including overexpression due to a rearrangement with IGH on chromosome 14 [249], interstitial deletions involving the juxtaposition of CRLF2 to the promoter of the *P2RY8* gene (Xp22.3/Yp11.3) [250,251], and activating point mutations (F232C) in *CRLF2* [252]. Some of these mutations occur in pathways for which targeted agents are available so identification of Ph-like mutations can result in addition of targeted agents to standard therapy.

9.7.1.2 Therapeutic Implications

The use of tyrosine kinase (TK) inhibitors in Ph + ALL, including imatinib, dasatinib, and nilotinib, is part of the frontline treatment, in combination with chemotherapy, and has showed a significant improvement in complete remission (95%) and overall survival (over 50% with 3-year follow-up) [253]. Clinical trials are underway in children with Ph-like ALL to determine whether addition of TK therapy in combination with chemotherapy improves remission and overall survival in these patients.

9.7.2 T-Cell Acute Lymphoblastic Leukemia

9.7.2.1 Genetics and Mechanism of Disease

T-cell acute lymphoblastic leukemia (T-ALL) is an aggressive malignancy caused by the accumulation of genomic lesions that affect the development of T cells; T-ALL represents ~15% of pediatric and ~25% of adult of ALL cases. The different chromosome and genomic abnormalities observed in T-ALL likely reflect different pathways and their cooperation in leukemogenesis in T-ALL [254]. Compared with the high frequency (80%–90%) of chromosome abnormalities in B-cell ALL, an abnormal karyotype is present in about 50%–70% of patients with T-ALL [207,255]. Gene expression profiling and genomic SNP microarray studies have identified genomic alterations and defined several distinct genetic subgroups of T-ALL that correspond to T-cell development [256,257]. The most common recurrent cytogenetic abnormalities occur in various translocations with the alpha/delta TCR loci (*TRA/TRD*) at 14q11.2, the beta locus (*TRB*) at 7q34, and the gamma locus (*TRG*) at 7p14 (Table 9.3) [258]. The common partner genes are *TAL1/SCL* (1p32), *HOX11L2/TLX3* (5q35), *HOXA* (7p15), *MYC* (8q24.1), *TAL2* (9q31), *HOX11/TLX1*(10q24), *RBTN1/LMO1*(11q15), *RBTN2/LMO2* (11p13), *CCND2* (12p13), *TCL1* (14q32), and *LYL1* (19p13) (Table 9.2) [213,259–265]. With few exceptions, the involved gene on the partner chromosome encodes a cell cycle inhibitor or a transcription factor whose expression is deregulated or activated as a result of juxtaposition with the regulatory regions of one of these TCR loci. Genomic microdeletions involving the short arm of chromosome 9 at the *CDKN2A* tumor suppressor

TABLE 9.3 Genetic Events Underlying the Initiation and Progression of Myeloma to Plasma Cell Leukemia

Inherited variation

SNPs
- 2p: *DTNB* and *DNMT3A*
- 3p: *ULK4* and *TRAK1*
- 7p: *DNAH11* and *CDCA7L*

Primary genetic events (% of tumors)

IGH @ *translocations and genes affected*
- t(4;14): *FGFR3* and *MMSET* (11%)
- t(6;14):*CCND3* (< 1%)
- t(11;14):*CCND1* (14%)
- t(14;16):*MAF* (3%)
- t(14;20):*MAFB* (1.5%)

Hyperdiploidy (57%)
- Trisomies of chromosomes 3, 5, 79, 11, 15, 19 and 21

Secondary genetic events (% of tumors)

Gains
- 1q: *CKS1B* and *ANP32E* (40%)
- 12p: *LTBR*
- 17q: *NIK*

Secondary translocations
- t(8;14):*MYC*
- Other non—*IGH@* translocations

Epigenetic events
- Global hypomethylation (MGUS to myeloma)
- Gene-specific hypermethylation (myeloma to plasma cell leukemia)

Secondary genetic events (% of tumors) (continued)

Molecular hallmarks
- Immortalization
- G1/S abnormality (*CDKN2C*, *RB1* (3%), *CCND1* (3%) and *CDKN2A*)
- Proliferation (*NRAS* (21%), *KRAS* (28%), *BRAF* (5%) and *MYC* (1%))
- Resistance to apoptosis (*PI3K* and *AKT*)
- NF-kB pathway (*TRAF3* (3%),*CYLD* (3%) and I-kB)
- Abnormal localization and bone disease (*DKK1*, *FRZB* and *DNAH5* (8%))
- Abnormal plasma cell differentiation (*XBP1* (3%), *BLIMP1* (also known as *PRDM1*) (6%) and *IRF4* (5%))
- Abnormal DNA repair (*TP53* (6%), *MRE11A* (1%) and *PARP1*)
- RNA editing (*DIS3* (13%), *FAM46C* (10%) and *LRRK2* (5%))
- Epigenetic abnormalities (*KDM6A* (also known as *UTX*) (10%), *MLL* (1%) *MMSET* (8%), *HOXA9* and *KDM6B*)
- Abnormal immune surveillance
- Abnormal energy metabolism and ADME events

Deletions
- 1p: CDKN2C, FAF1 and FAM46C (30%)
- 6q (33%)
- 8p (25%)
- 11q: BIRC2 and BIRC3 (7%)
- 13: RB1 and DIS3 (45%)
- 14q: TRAF3 (38%)
- 16q: CYLD and WWOX (35%)
- 17p: TP53 (8%)

ADME, absorption, distribution, metabolism and excretion; *ANP32E*, acidic leucine-rich nuclear phosphoprotein 32 family, member E; *BIRC*, baculoviral IAP repeat-containing protein; *BLIMP1*, B lymphocyte-induced maturation protein 1; *CCND*, cyclin D; *CDCA7L*, cell division cycle-associated 7-like; *CDKN*, cyclin-dependent kinase inhibitor; *CKS1B*, CDC28 protein kinase 1B; *CYLD*, cylindromatosis; *DNAH*, dynein, axonemal, heavy chain; *DNMT3A*, DNA methyltranferase 3A; *DTNB*, dystrobrevin, beta; *FAF1*, FAS-associated factor 1; *FGFR3*, fibroblast growth factor receptor 3; *HOXA9*, homeobox A9; *IGH@*, immunoglobulin heavy chain locus; *IRF4*, Interferon regulatory factor 4; *I-kB*, inhibitor of nuclear factor-kB; *KDM*, lysine demethylase; *LRRK2*, leucine-rich repeat kinase 2; *LTBR*, lymphotoxin beta receptor; *MGUS*, monoclonal gammopathy of undermined significance; *MLL*, mixed-lineage leukemia; *MMSET*, multiple myeloma SET domain; *MRE11A*, meiotic recombination 11A; *NF-kB*, nuclear factor kB; *PARP1*, poly(ADP-ribose) polymerase 1; *TRAF3*, tumor necrosis factor receptor-associated factor 3; *TRAK1*, trafficking protein, kinesin-binding 1; *WWOX*, WW domain-containing oxidoreductase; *XBP1*, X box-binding protein 1.

locus containing the *p16/INK4a* and *p14/ARF* are recurrently observed in T ALL [266]. A recurrent cryptic rearrangement amplification event involving the *ABL1* gene fused with *NUP214* occurs in approximately 6% of cases [267]. The episomal amplification event is acentric, so the number of copies of this fusion varies from cell to cell. Another recurrent rearrangement involves the *TAL1* gene (1p32) and has been observed in ~25% of T-ALL cases. The most common rearrangement is a submicroscopic deletion, joining *TAL1* with *STIL* (also on chromosome 1), but a *TAL1::TRA/D* fusion from a t(1;14) (p32;q11) translocation has also been observed.

The prognostic relevance of *TAL1* rearrangement is unclear.

One of the most commonly mutated and deleted genes in T-ALL is *NOTCH1*, seen in ~50% of cases [268]. Although NOTCH1 mutations are seen in T-ALL across age groups, the prognostic implications appear to differ, with NOTCH1 mutations considered favorable in the pediatric population, but less so in adults [269]. Inactivation mutations in FBXW7, another gene in the NOTCH signaling pathway, are observed in ~14% of T-ALL cases [269]. Gene mutations that are associated with a more adverse clinical course include PTEN mutations and deletions, KRAS and NRAS mutations, IDH1/2 and DNMT3A (references).

9.7.2.2 Therapeutic Implications

The detection of cytogenetic and molecular abnormalities in T ALL is largely prognostic. The *ABL1::NUP214* amplification represents a target for ABL1 tyrosine kinase directed therapy [270], even though it is thought to be a weak oncogenic driver, requiring cooperation with other gene mutations.

9.7.3 Prenatal Origins of Childhood Leukemia

The detection of *KMT2A/MLL* rearrangements, *ETV6::RUNX1* fusion and hyperdiploid clones in ALL, and the *RUNX1T1::RUNX1* fusion in AML, in pediatric ALL, in blood spots on Guthrie cards from children who later developed leukemia arise demonstrate that the first hit for acute leukemias can occur prenatally [271,272]. The detection of *ETV6::RUNX1* and *RUNX1T1::RUNX1* fusion transcripts in cord blood samples suggests that the abnormalities arise during fetal hematopoiesis [273]. The frequency of detection of disease-associated transcripts was approximately 100-fold greater than the incidence for development of leukemia, suggesting that secondary, complementary mutations are required for transformation [273]. As mentioned above, similar findings were identified when *GATA1* was studied in fetal tissue [274]. Notably, preleukemia clones such as the *ETV6::RUNX1* fusion occur in utero at a high rate in excess of clinical leukemia, and the concordance of infant ALL in monozygotic twins is around 10%–20%, consistent with a multihit hypothesis [275,276]. Thus, all the data indicate that additional genetic mutations and a dysregulated immune system are required for the development of an overt leukemia (reviewed in Ref. [200]).

9.7.4 Down Syndrome and the Risk of Leukemia

Children with Down syndrome have a 20-fold increased risk of developing acute leukemia, 50% of which will be acute megakaryoblastic leukemia. About 10%–30% of infants with Down syndrome develop transient myeloproliferative disease (TMD) or transient abnormal myelopoiesis (TAM). TAM presents as a profound leukocytosis with blasts indistinguishable from leukemic megakaryoblasts, but usually disappears spontaneously during the first months of life. Patients who present with TAM have been shown to have an acquired truncating mutation in GATA1 in addition to their trisomy 21 [277]. Studies of fetal tissue demonstrated that *GATA1* mutations can occur prenatally and thought to be early events in leukemogenesis requiring additional genetic events and/or environmental exposures for the full development of leukemia [274].

However, ~20% of those who present with TMD will develop acute megakaryoblastic leukemia in the first 4 years of life [278]. Most Down syndrome children with AML have additional acquired chromosome abnormalities, most commonly trisomy 8 (25%) or, less commonly, trisomies 19 or 21%–13% and 17%, respectively [279]. The transition from TAM to a myeloid malignancy is associated with gain of mutations, typically in cohesion genes [280,281].

9.7.5 Lymphomas

Lymphomas represent multiple distinct entities of malignancies that involve the lymphatic system, which includes the lymph nodes, spleen, and thymus gland. Lymphomas can are generally grouped into three major categories: B-cell neoplasms, T-cell, and NK-cell neoplasms, which are generally grouped as non-Hodgkin lymphomas (NHLs) and Hodgkin lymphomas (HLs) [6].

9.7.6 Non-hodgkin Lymphomas

NHLs comprise approximately 90% of lymphomas and display a wide range of histologic morphologic features, which can make diagnosis difficult. Approximately 85%–90% of NHLs are B-cell in origin, with the remaining divided between T-cell and NK-cell origin. While most lymphomas can be diagnosed accurately based on morphologic and immunophenotypic findings, cytogenetic and molecular testing can assist in the classification and prognostic/risk stratification. NHLs can range

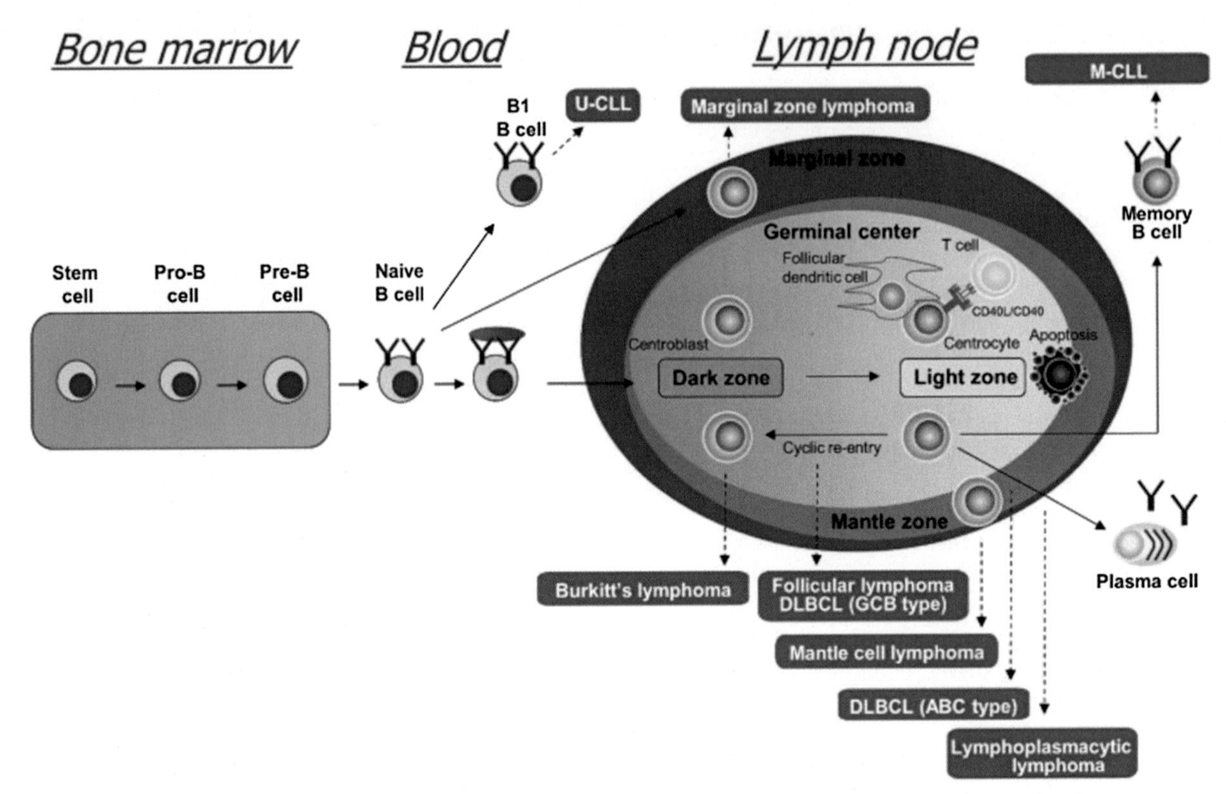

Figure 9.13 B cell development and cell of origin of B cell malignancies node. Lymphomas arise from different regions of the lymph.

from relatively indolent disease to aggressive lymphomas. The most common B-cell lineage NHLs include follicular lymphoma (FL), diffuse large B-cell lymphoma (DLBCL), and mantle cell lymphoma and often arise in the lymph node (Fig. 9.13) [282].

9.8 MATURE B CELL NEOPLASMS

9.8.1 Genetics and Mechanisms of Disease

FL is typically an indolent B-cell neoplasm and accounts for approximately 20%–25% of all NHLs, with an overall survival rate of greater than 10 years for ~80% of patients due to the advent of new therapies [6]. FL arises from precursor B-cells in the germinal center region of the lymph node and is graded from 1 to 3, with grade 1 being the most indolent disease and grade 3 having a higher risk of transformation into diffuse large B-cell lymphoma (DLBCL), a more aggressive disease [283].

The t(14;18) (q32;q21) translocation is recurrently observed in B-cell NHLs and is seen in greater than 85% of FL cases [284,285]. This rearrangement results in placing the *BCL2* gene under transcriptional control of the immunoglobulin heavy-chain (IGH) regulatory regions. The *IGH::BCL2* rearrangement leads to overexpression of *BCL2*, which acts to protect cells from apoptosis [286]. The t(14;18) can be detected by conventional cytogenetics or by FISH; as many lymphoma specimens are FFPE for histologic subtyping, effectively killing the cells, FISH is often the only technique able to detect this rearrangement and is superior to PCR methods due to the variability of the breakpoints in the rearranged genes. Although *BCL2* rearranges with the *IGH* locus in the majority of FL, there are cases where there is a rearrangement with other immunoglobulin genes, *IGK* (2p12) or *IGL* (22q11.2). The t(14;18) (q32;q21) is not disease defining for FL and can also be detected in approximately 20%–30% of cases of de novo DLBCLs.

The *EZH2* gene encodes an epigenetic regulator that is mutated in approximately 20% of patients with FL [287]. *EZH2*-targeted therapies are available with patients harboring *EZH2* mutations showing an improved response to those with the *EZH2* wild-type disease (69% response vs. 35% response, respectively. The response in the *EZH2* wild-type patients is thought to be due to other mutations in other genes in the same pathway (such as *CREBBP* and *KMT2D*) or other mechanisms [288]. These discoveries led to the approval by the US Food and Drug Administration (FDA) of tazemetostat for the treatment of relapsed/refractory FL [289]. As with most targeted therapies, acquired resistance mutations occur to circumvent *EZH2* activity and include the activation of prosurvival pathways (IGF-1R, PI3K, and MAPK) and the selection for resistance mutations in *EZH2* that prevent drug binding [290].

FL is a relatively indolent disease, but can transform into DLBCLs at a rate of ~2–3% per year [288]. Transformation is often associated with the acquisition of new abnormalities, including *MYC* rearrangement, *TP53* and *B2M* mutations, and deletion of *CDKN2A/CDKN2B* [291].

MYC rearrangements are recurrently observed in NHL, and the interpretation is dependent on the histologic findings and additional chromosomal abnormalities. The most common rearrangement involves a translocation involving *MYC* at 8q24 and *IGH* at 14q32 (t(8;14) (q24;q32)), although rearrangements with *IGK* or *IGL* can also be observed. All *IGH/K/L::MYC* rearrangements cause deregulation of the *MYC* gene. *MYC* translocations are observed in ~90% of Burkitt lymphoma, ~5–10% of DLBCL, and about 40% of the WHO category B-cell lymphoma, unclassifiable, with the prognosis dependent on the diagnosis. The presence of an *MYC* rearrangement in DLBCL predicts a poor response to first-line chemotherapy [292]. In contrast, *MYC* rearrangement in Burkitt lymphoma is highly curable in children and young adults, but less so in older adults due to the toxicity of the standard therapeutic options in this age group. The t(8;14) is found in 75%–85% of Burkitt lymphomas, and variant translocations, t(2;8) (p12;q24) and t(8;22) (q24;q11.2), are described in 15%–25% of cases (Table 9.2).

The breakpoints in 8q24 demonstrate considerable variability at the molecular level, extending from over 300 kb centromeric to at least the same distance telomeric to *MYC* [293]. The critical chromosome differs dependent on the *MYC* rearrangement, with a proximal *MYC* breakpoint leading to *MYC* translocating to chromosome 14 in the t(8;14) rearrangement, while for the t(2;8) and t(8;22), the breakpoint is distal to *MYC*, with the enhancers moving to chromosome 8 for those rearrangements. Interestingly, the location of breakpoints appears to differ in endemic, sporadic, and HIV-associated Burkitt lymphoma. Regardless of the molecular breakpoints, the consequence of these translocations is the deregulated expression of a full-length *MYC* protein. *MYC* acts as a transcription factor and plays a central role in a number of cellular processes, including proliferation and apoptosis [294]. *MYC* participates in regulating chromatin dynamics by directly binding to several proteins that control epigenetic modifications of both histone and the DNA methylation state and is also involved in the regulation of microRNA expression [295]. The t(8;14) and particularly other *MYC* abnormalities are also observed in other lymphoma, such as DLBCL with overlapping features of Burkitt lymphoma, and follicular lymphoma, MCL, and T-cell leukemia/lymphoma and CLL [296,297]. In all these diseases, *MYC* abnormalities are prognostic and associated with a more aggressive phenotype, resistance to standard chemotherapy, and a higher mortality [298]. In addition to cytogenetic abnormalities, there are many genes that are recurrently mutated in DLBCL (Fig. 9.14) [291]. These molecular subtypes are associated with distinct pathogenic mechanisms and outcomes, notable TP53 mutations identify a more aggressive disease [291].

The *BCL6* gene on chromosome 3q27 is one of the most common rearrangements in NHL, seen in approximately 10% of cases, and as high as 30% in DLBCL cases. Over 20 different partners have been identified in *BCL6* rearrangements, resulting in the dysregulation of *BCL6* by promoter substitution [291] leading to unregulated expression. Somatic hypermutation (SHM) is a separate mechanism associated with *BCL6* overexpression and has been shown to occur in about 15% of cases.

The detection of *MYC* rearrangements with additional rearrangements of *BCL2* and/or *BCL6* is considered "high grade B cell lymphomas" by the WHO, which are sometimes referred to as double-hit (or triple-hit) lymphomas (DHL), where an *MYC* rearrangement is seen with a second (or third) rearrangement of *BCL2* (85%), *BCL6* (5%), or all three

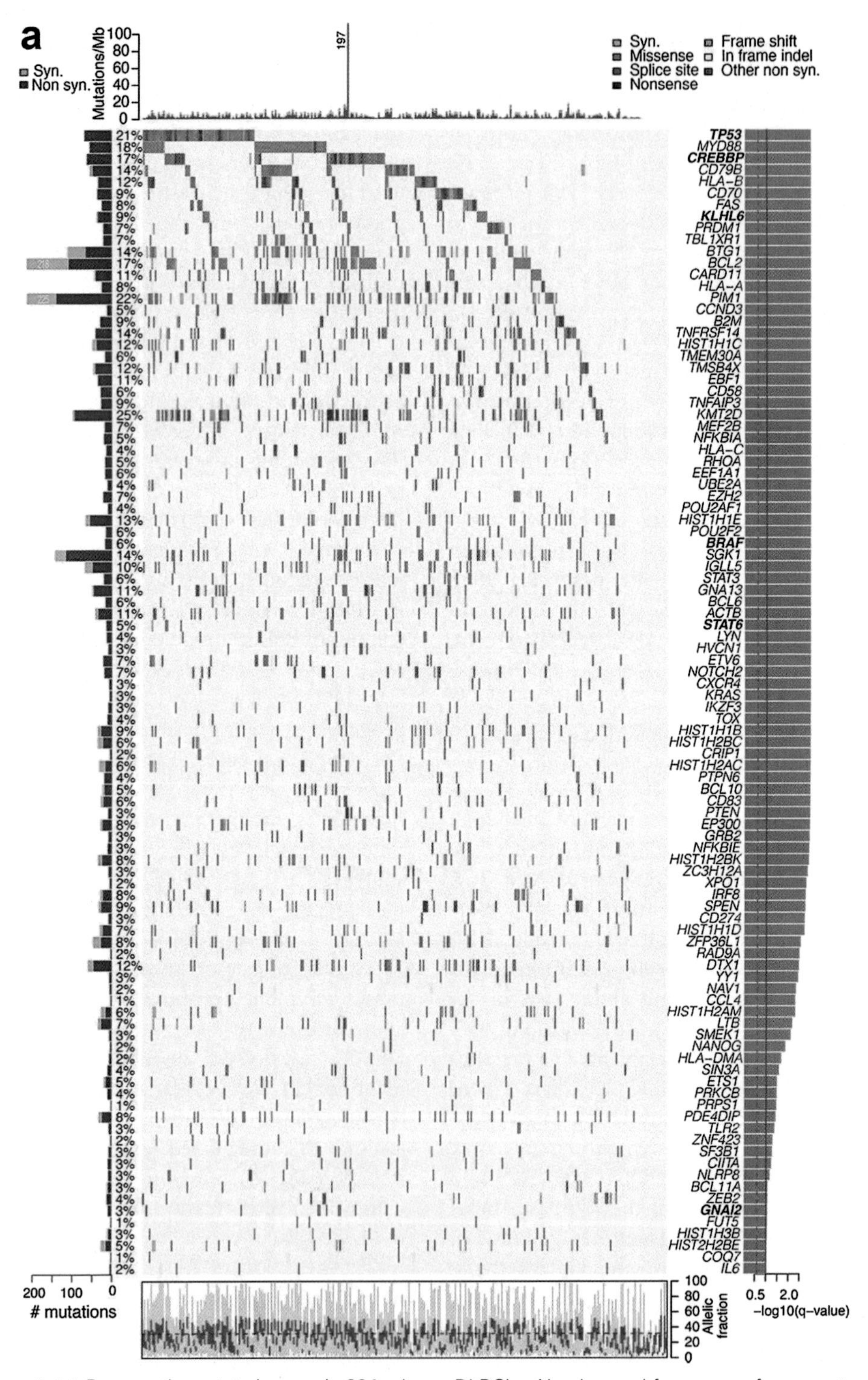

Figure 9.14 Recurrently mutated genes in 304 primary DLBCLs. Number and frequency of recurrent mutations (*left*), gene–sample matrix of recurrently mutated genes (color-coded by type, *center*), ranked by their significance (MutSig2CV q-value, *right*). Total mutation density across the cohort is shown at the *top*, allelic fraction of mutations at the *bottom*. *Asterisk* indicates hypermutator case.

(10%) [299]. These lymphomas often display a complex karyotype with multiple chromosomal abnormalities on cytogenetic analysis [300].

DHLs are observed in about 5%–10% of DLBCLs, but a higher percentage of these cases have overexpression of these genes without detection of the rearrangement. These rearrangements are most commonly detected by FISH or chromosome analysis, with FISH often representing the only available technology for detection, since lymph nodes are typically FFPE for morphologic assessment. The morphologic identification of aggressive appearing regions of the submitted sample can enrich detection of DHLs, so laboratory detection of abnormalities is directly related to pathologic assessment and communication of relevant areas of the specimen for study.

The dysregulation of *CCND1* by rearrangement with *IGH*, t(11;14) (q13;q32) or *CCND1::IGH*, which results in the overexpression of cyclin D1 [formerly referred to as *BCL1*], is detected in the majority of mantle cell lymphomas (MCL). Cyclin D1 activates cyclin-dependent kinases 4 and 6 (*CDK4* and *CDK6*), which inactivate *RB1*, allowing cells to progress in the cell cycle, leading to rapid cell proliferation [301]. MCL with *MYC* expression has been associated with intrinsic ibrutinib resistance [302]. There are two WHO recognized subtypes of MLC, conventional (cMCL) and nonnodal (nnMCL), which have the same t(11;14) rearrangement, but different recurrent chromosome and mutational profiles [303].

The t(11;14) rearrangement is not specific to MCL and is also common rearrangement in plasma cell neoplasms. FISH can detect t(11;14) in every patient with MCL, regardless of their morphological or clinical presentation, whereas RT-PCR and conventional cytogenetics are successful in 50%–60% and 70%–75%, respectively [304]. The detection of t(11;14) fusion is clinically significant because MCL has an aggressive clinical course, and the morphology and immunophenotypes of MCL cells are variable, making it difficult to separate from CLL and other low-grade B-cell lymphoma and leukemia.

Marginal zone lymphomas are considered an indolent NHL and are categorized as extranodal MZL of mucosa-associated lymphoid tissue (MALT lymphoma), splenic MZL, and nodal MZL. Subtyping of these entities can be enhanced by cytogenetic and molecular studies. Trisomy of chromosomes 3 and 18 and deletions at 6q23 are frequent events in all MZLs [305]. Extranodal marginal zone lymphomas (MALT lymphomas) are the only subtype characterized by translocations, including the t(11;18) (q21;q21) translocation that involves the *BIRC3* gene (an apoptosis inhibitor gene on chromosome 11) and the *MALT1* gene on chromosome 18, the t(14;18) (q32;q21) involving *IGH* and *MALT1*, which is seen in MALT lymphomas arising in the parotid, liver, and conjunctiva, and the t(3;14) (p14;q32) involving *FOXP1* and *IGH*, observed mainly in MALT lymphomas arising in the thyroid, ocular adnexa, and skin. Splenic marginal zone lymphomas are specifically characterized by deletions of chromosome 7q. Loss of *PTPRD* and a much higher prevalence of mutations affecting *KMT2D* (MLL2) are more characteristic of nodal marginal zone lymphomas.

The observation of familial risk for FL has shown a 2–4 fold increased risk for individuals with a family history of NHL or FL (respectively) [306,307], but these studies were not able to ascertain whether the risk was genetic or environmental.

9.8.1.1 Therapeutic Implications

There are many chemotherapies and immunotherapies used in the treatment of B cell malignancies, which are beyond the scope of this manuscript. However, there are several pan-B-cell malignancy treatments that are worth mentioning. Constitutive activation of B-cell receptor (BCR) signaling is a primary mechanism for disease progression in B cell lymphomas, driven through increased activation of Bruton tyrosine kinase (BTK). The BTK inhibitor ibrutinib is an FDA-approved agent for chronic lymphocytic leukemia (CLL), mantle cell lymphoma (MCL), and Waldenstrom's macroglobulinemia (WM) (including patients carrying del(17p) or *TP53* mutation), and clinical trials have shown response in DLBCL and FL [308,309]. The detection of acquired mutations in the *BTK* or *PLCG2* genes was found in 85% of patients who became resistant to ibrutinib, and these mutations will appear early, months before clinical relapse, and can potentially be used as an opportunity for early intervention when used as a biomarker for future relapse [310].

A cellular therapy targeting the CD19 protein, which is present on all B cells, represents a personalized

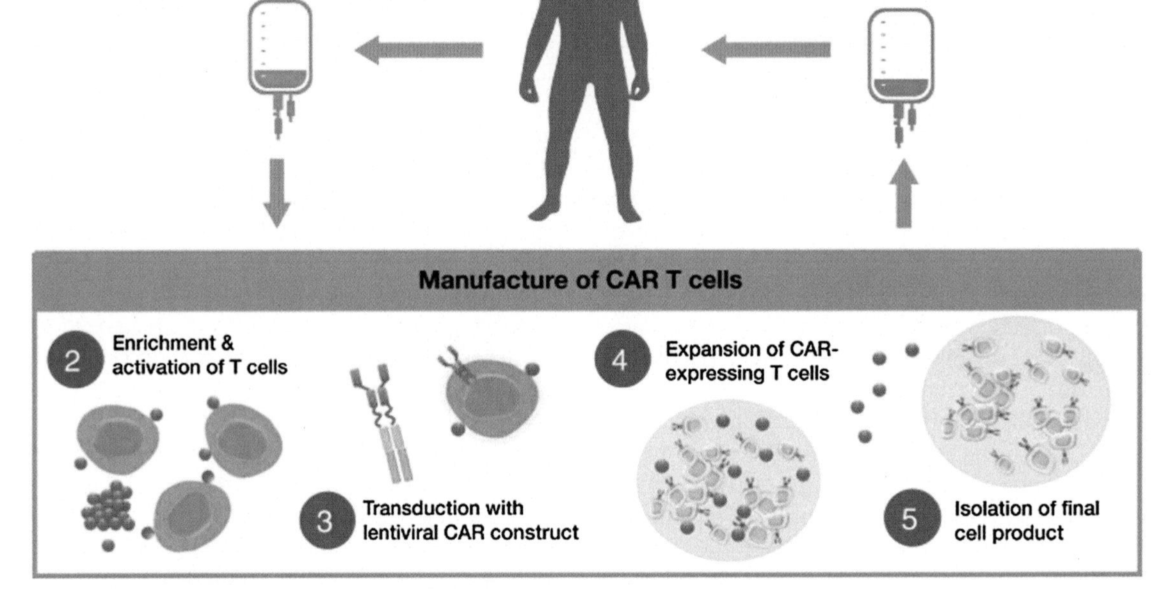

Figure 9.15 Diagram of CAR T cell treatment process. The treatment process for patients receiving CAR T cell therapy begins with leukapheresis of the patient's T cells. Once isolated, autologous T cells are sent for manufacturing to produce genetically modified CAR T cells, which are reprogrammed to facilitate targeted killing of CD19+ B cells. The treatment process is completed with intravenous infusion of CAR T cells back to the patient. CAR chimeric antigen receptor.

therapy for the treatment of B cell neoplasms, which include B cell ALL and B cell lymphomas. This therapy cultures and transduces the patient's T-cells ex vivo such that they express a synthetic chimeric antigen receptor (CAR) that targets CD19 expressing B cells. These cells are infused back into the patient and destroy the B cell neoplasm (Fig. 9.15) reviewed in Refs. [311,312]. Dramatic clinical responses and high rates of complete remission have been observed in the setting of CAR T-cell therapy of B-cell malignancies and have resulted in two recent FDA approvals for treatment of relapse/refractory B cell acute lymphoblastic leukemia and certain types of B-cell lymphomas.

9.8.2 Hodgkin Lymphomas

Hodgkin lymphomas (HLs) are most commonly a cancer of the lymph nodes and can be generally categorized as classic HL (cHL) or nodular lymphocyte predominant Hodgkin lymphoma (NLPHL) and comprise about 10% of all lymphomas. The neoplastic cells of interest are Reed Sternberg (RS) cells, which are rare and scattered within the background of more abundant mixed inflammatory nonneoplastic cells, making molecular characterization difficult. Due to this feature HL cytogenetic and molecular characterization is not usually part of the clinical work-up of this entity, although molecular and cytogenetic features have been described (reviewed in Ref. [313]). Of note, HLs that are refractory to treatment are more likely to contain *TP53* mutations [314].

9.8.3 Chronic Lymphocytic Leukemia

CLL is the most common chronic leukemia in the United States and Europe. It is virtually unknown in people less than 30 years old, but after the age of 70, its incidence is about 50 cases per 100,000. It is considered to be a monoclonal neoplastic proliferation of small lymphocytes that in 95% of cases are of B-cell origin. The disease has a prolonged natural history, sometimes measured in decades; cells presumably gradually acquire sequential genetic abnormalities, rendering them more malignant. The diagnosis of CLL does not depend on the detection of cytogenetic or molecular abnormalities; however, genetic studies are critical to determine risk assessment and may also guide therapeutic decision-making. Conventional cytogenetics, FISH, mutational analysis, and somatic hypermutation analysis of the variable region of the IGH gene (IGHV) are central components of the genetic assessment of CLL patients.

9.8.3.1 Genetics and Mechanism of Disease

CLL cells can be coaxed to grow in cell culture using a variety of mitogens that can provide abnormal karyotypic results in more than 80% of CLL bone marrow or peripheral blood samples [32,315,316]. There are recurrent abnormalities in CLL, which can be screened by FISH studies of interphase cells, with an abnormality rate of approximately 85% [317]. Current practice in the diagnosis of CLL now includes an interphase FISH analysis with a cocktail of probes for the following loci: *TP53* (17p13), *ATM* (11q22.3), 13q14.3, chromosome 12 centromere, and *CCND1::IGH* (summary in Fig. 9.16) [317,318].

Conventional cytogenetic and FISH studies have shown that the most common genetic abnormalities in CLL (up to 55% of all CLL cases) are deletions or translocations involving 13q14.3 [317]. The presence of del(13)(q14.3) as a sole abnormality is typically associated with a good prognosis (median survival 133 months) [317]. However, there are several other factors that may mitigate this favorable prognosis, including the size of the deletion, the percentage of positive cells, and heterozygous versus homozygous deletions. The size of 13q14 deletions has been grouped into those exclusive (type 1) versus inclusive (type 2) of *RB1*. The larger deletions inclusive of *RB1* are seen in ~20% of all CLLs and associated with a shorter survival [319]. High percentage of nuclei with a 13q deletion has been associated with a less favorable outcome, with the percentage varying in different studies [320,321]. Biallelic deletion of the 13q14 region can be seen, with the prognostic relevance of biallelic versus monoallelic 13q deletions suggestive of a less favorable outcome [322]. Trisomy 12, the most common abnormality detected by classic cytogenetic studies, is only present in 16% of cases when using FISH. The gain of chromosome 12 as a sole abnormality or in the presence of del 13q is associated with a good to moderate prognosis (median survival 114 mos) [323]. About 7% of CLL patients have deletions of the *TP53* gene at 17p13.3, and these have been associated with shortened time to treatment and poor survival (median survival 32 mos) and chemotherapy resistance [317,324]. Deletions of the *ATM* gene locus are detected in 18% of cases of CLL using FISH. Patients with *ATM* deletions often present with significant lymphadenopathy, and this deletion is associated with a moderate to poor outcome (median survival 79 mos) [317,325]. Notably, both deletions of *TP53* and *ATM* in CLL are often observed as secondary abnormalities during disease progression. In addition to deletions and trisomies, there are structural rearrangements that are seen in CLL, most involving the *IGH* promoter. FISH using an IGH break-apart probe can be useful for the detection of additional alterations. Of note, the *CCND1::IGH* rearrangement is often tested for new diagnosis in CLL patients to rule out mantle cell lymphoma (discussed below).

Combining conventional karyotype and FISH studies has shown that the combination of these methods can further prognosticate treatment naïve CLL patients (Fig. 9.17) [326]. The identification of complex karyotype (defined in CLL as five or more abnormalities) as an independent prognostic indicator, and associated with resistance to standard therapies, *BCR* inhibitors or venetoclax (*BCL2* selective inhibitor) has led to refine prognostic categories: 1. *TP53* loss and monosomy 15; 2. *ATM* loss and 14q32 (IGH) translocation; and 3. trisomy for chromosomes 12, 18, and 19 or t(14;18)(q32;q21), without deletion of *ATM* or *TP53*.

It is thought that the initial events in CLL are the chromosomal gains and losses detected by conventional cytogenetics and FISH studies, with additional molecular genetic alterations occurring as the disease develops. The most commonly mutated genes in CLL are *SF3B1*, *ATM*, *TP53*, *NOTCH1*, and *POT1* (Fig. 9.18) [327], and the intermediate and late driver mutations are acquired and often cooccur with the same early driver events. For example, patients with an initial deletion of 13q14 are likely to have *SF3B1* or *POT1* as the intermediate molecular events, with late drivers usually including *TP53* mutations and other chromosome abnormalities such as deletion of 6q21 [328].

CLL has one of the strongest inherited predispositions of all the hematological malignancies with an increased risk of 2.4–8.5-fold compared to the normal population [329] with approximately 10% of CLL patients having a family history of CLL.

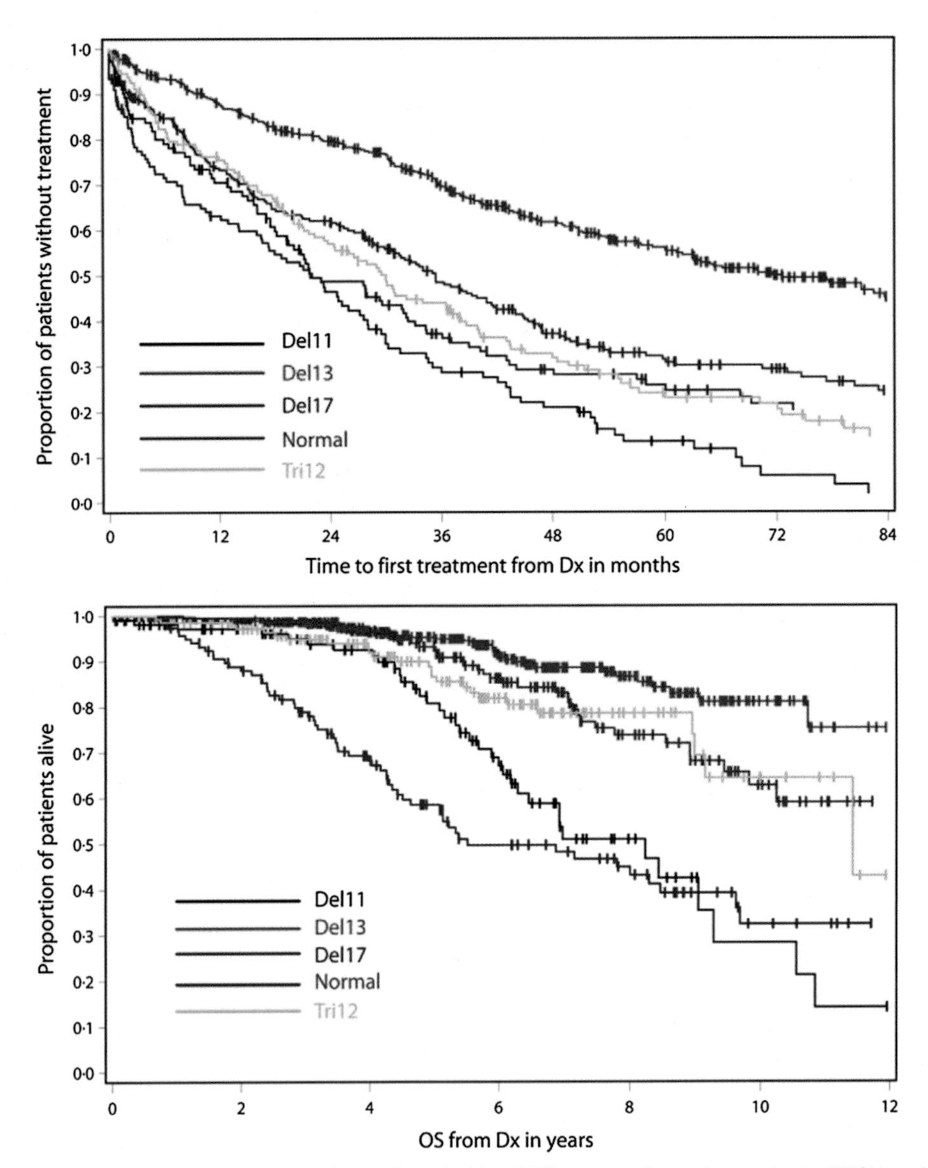

Figure 9.16 Time to first treatment and overall survival by FISH category for patients who had FISH analysis less than 4 years after diagnosis of chronic lymphocytic leukemia. *Dx*, diagnosis; *FISH*, fluorescence in situ hybridization.

9.8.3.2 Therapeutic Implications

Genetic abnormalities do not drive treatment in CLL, rather the time to first anti-CLL treatment is based on disease progression. *TP53* deletions and *TP53* mutations often are associated with short time to first treatment and generally poor overall survival. Treatment with venetoclax, a BCL2 inhibitor, and ibrutinib, a BTK inhibitor, has shown efficacy in treating *TP53* mutated CLL [330]. Chimeric antigen receptor T cells (CAR T cells) are a cellular therapy using the patient's T cells modified to recognize the cell surface marker CD19, present on B cells, and have been approved for chemotherapy-resistant CLL [331].

(A)

risk group according to CBA			Combined CBA and FISH risk	risk group according to FISH abnormalities: Idel13q14 (405 pts)	Trisomy12 (383 pts)	*ATM* loss (138 pts)	*TP53* loss (92 pts)
Normal CBA				48%	40%	2%	1%
abnormal CBA	non-complex CBA <3	single del13q		19%	48%	54%	30%
		Others*		26%			
	Complex CBA>=3			7%	12%	44%	69%

(B)

risk group according to CBA			Combined CBA and FISH risk	risk group according to FISH abnormalities: idel13q14	trisomy12	*ATM* loss	*TP53* loss
Normal CBA				low-risk	int-risk	high-risk	high-risk
abnormal CBA	non-complex CBA <3 *	single del13q		low-risk		high-risk	high-risk
		trisomy 12,18,19			low-risk		
		Others		int-risk	int-risk		
		t(8;14)(q24;q32) amp2p(NMYC)		high-risk	high-risk		
	complex CBA >=3	3<=CK<5		high-risk	high-risk	high-risk	very high-risk
		CK>=5		very high-risk	very high-risk	very high-risk	very high-risk

Figure 9.17 Risk categories for treatment naïve CLL patients combining chromosome analysis with FISH. (A) Combined CBA/FISH risk proposal according to our data. (B) Combined CBA/FISH risk proposal with additional cytogenetic data according to published studies. (*) some anomalies may refine prognostic, that is, combined trisomies 12, 18, and 19 as good risk factor, and t(8;14) (q24;q32) and amp2p(NMYC) as lower risk factor. Risk group color code: low-risk in *green*; int-risk in *blue*; high-risk in *orange*; and very high-risk in *red*.

9.8.4 Plasma Cell Neoplasms

Multiple myeloma (MM) is a malignant plasma cell disorder accounting for approximately 10% of all blood cancers. MM appears to evolve from an asymptomatic proliferation of clonal plasma cells designated "monoclonal gammopathy of undetermined significance" (MGUS). MGUS is present in up to 3% of the population over age 50 years. After a period of observation without progression, MGUS can still progress to a plasma cell malignancy at a rate of approximately 1% per year [332].

MM arises from the malignant transformation of postgerminal center plasma cells [333]. All steps of the pathogenesis are not yet known, but somatic alterations in the variable region of immunoglobulin genes resulting from an abnormal response to antigen stimulation appear most likely. The result is a malignant plasma cell clone producing monoclonal immunoglobulin or light chain in addition to other cytokines. Progression from MGUS to MM results from additional abnormalities in these genetically unstable cells interacting with the bone marrow microenvironment.

9.8.4.1 Genetics and Mechanisms of Disease

Genomic abnormalities can be detected in virtually all cases of MGUS and MM. Most MGUS is initiated by either chromosomal translocations involving the immunoglobulin heavy chain (IgH) (approximately

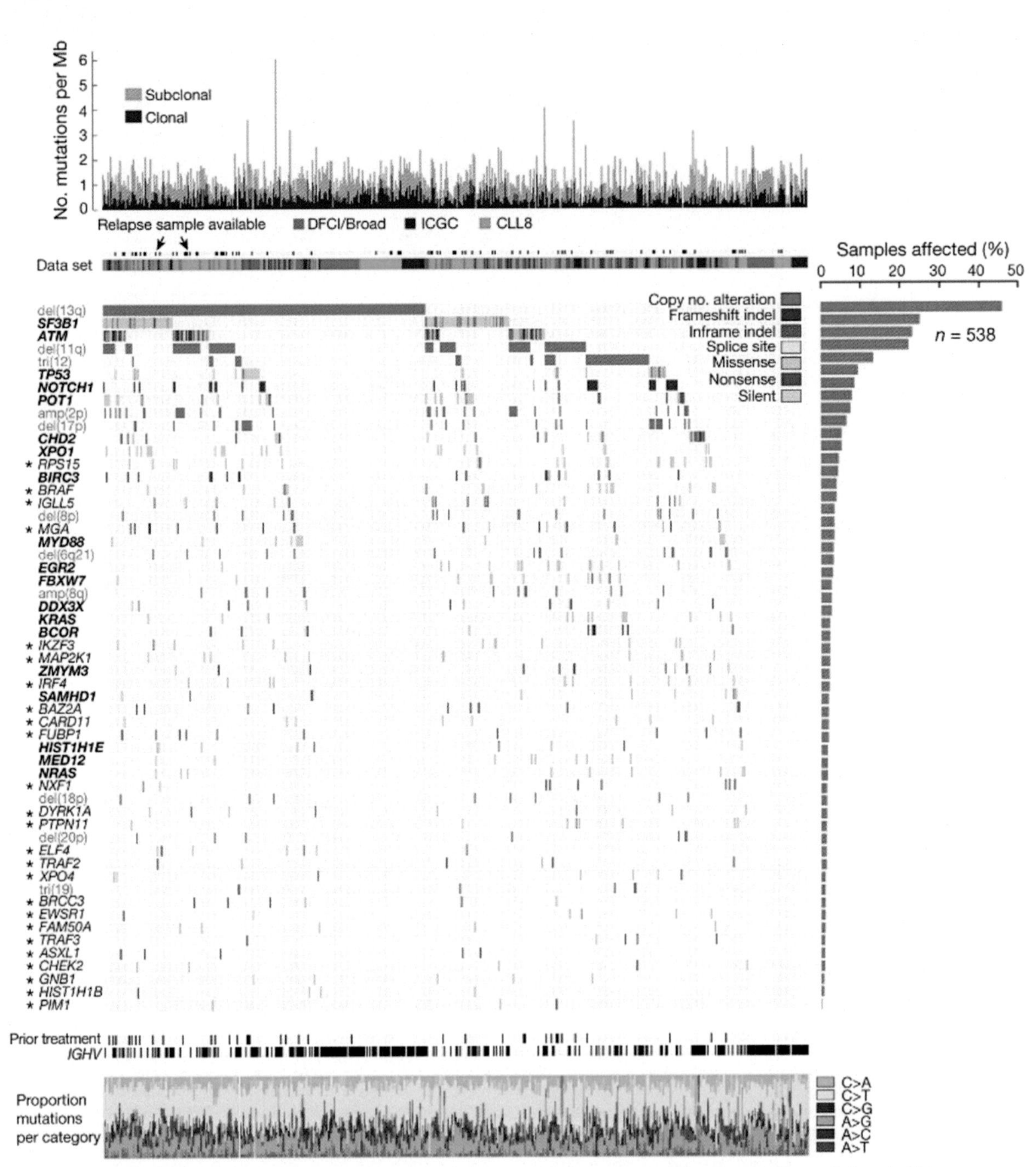

Figure 9.18 The landscape of putative driver mutations and recurrent somatic copy number variations in CLL. Somatic mutation information is shown across the 55 putative driver genes and recurrent sCNVs (*rows*) for 538 primary patient samples (from CLL8 [*green*], Spanish ICGC [*red*], DFCI/Broad [*blue*]) that underwent WES (*columns*). Blue labels-recurrent sCNVs; Bold labels-putative CLL cancer genes previously identified in Landau et al. [327]; *asterisked* labels—additional cancer genes identified in this study. Samples were annotated for IGHV status (*black*—mutated; *white*—unmutated; *red*—unknown), and for exposure to therapy prior to sampling (*black*—prior therapy; *white*—no prior therapy; *red*—unknown prior treatment status).

40%) or trisomies (also approximately 40%) or both (approximately 10%) [334,335]. Additionally, loss of chromosome 13 or a del(13q) is the most frequently observed chromosomal loss in MM. Deletions of 17p13, the TP53 locus, are found in 10% of MM and are associated with poor prognosis, including plasma cell leukemia, extramedullary disease [336].

The normal primary immune response to an antigen is the formation of IgM antibodies. When reexposed to the same antigen, there is frequently a class switch where the genes coding for the variable portion of IgH move from the IgM constant region to the IgG or IgA constant regions to create a secondary immune response. MGUS and MM transformative events seem to be frequently caused by erroneous translocations occurring during these switch recombinations. Recombination events involving the IgH locus, which is located on chromosome 14q32, result in overexpression of the translocated gene—a key step for the development of malignant transformation. IGH rearrangements are detectable by interphase FISH analysis in 50% of patients with MGUS, 60%–75% of patients with MM, and more than 80% of patients with plasma cell leukemia [337].

Five recurrent chromosomal translocations have been observed and are associated with various prognoses. The t(11;14) (q13;q32) is found in 16% of cases and results in the IGH/CCND1 (cyclin D1 gene) fusion. The t(4;14) (p16;q32) is noted in about 15% of patients and deregulates the expression of the fibroblast growth factor receptor 3 gene (FGFR3) and MMSET on the der(4) chromosome. The t(14;16) (q32;q23) is present in about 5% of cases and involves the juxtaposition of IGH and MAF (v-maf musculoaponeurotic fibrosarcoma oncogene homolog). The t(14;20) involves 20q11 and the musculoaponeurotic fibrosarcoma oncology family, protein B gene (MAFB) and is seen in approximately 2% of cases [258]. The t(6;14) (p21;q32) is detected in about 3% of MM patients and leads to CCND3 (cyclin D3) overexpression. The t(4;14) and t(14;16) are associated with a poor clinical outcome, whereas the t(11;14) and t(6;14) confer a more favorable prognosis [338,339].The non-IGH genes involved in these translocations are transcription factors, growth factor receptors, and cell cycle mediators that promote growth and replication. It should be noted that other chromosomal alterations, such as insertions, can produce the same gene rearrangements.

Hyperdiploidy, the gain of additional copies of chromosomes in the clonal cell populations, is common in malignant plasma cell disorders. Trisomies of one or more odd-numbered chromosomes are typically involved (often with the exception of chromosome 13). A hyperdiploid karyotype in MM is associated with longer survival as compared to translocations involving 14q32 [340]. Interestingly, samples with a hyperdiploid karyotype very rarely have rearrangements of IGH suggesting that there are two mutually exclusive pathways for the development of MM. Overexpression of genes on these chromosomes may promote growth and replication of the mutated clone [341].

RAS mutations in N-and K-RAS genes occur in up to 40% of MM and play a critical role, in cooperation with loss or deletion of chromosome 13, in disease progression, and are associated with worse survival [342,343]. Gene expression profiling analysis in MM showed dysregulated and/or increased expression of cyclin D1, D2, or D3 in all of the above genetic pathways virtually in all MM and MGUS patients [344]. Cyclin D1 is biallelically dysregulated in a majority of hyperdiploid myeloma patients. Increased expression of one of the cyclin D proteins facilitates activity of CDK4 or CDK6, which can facilitate G1 to S cell cycle progression.

Genetic abnormalities and bone marrow microenvironment-driven deregulations, such as increased levels of IL-6, IGF1, HGF, VEGF, CXCL12, TNFα, BAFF, APRIL, and CCL3 derived from stromal, bone, and immune cells result in the activation of multiple signaling pathways. They trigger tumor cell proliferation, survival, drug resistance, migration, secretion of humoral factors but also promote immunoparesis, bone marrow angiogenesis, and bone disease (Fig. 9.19) [345]. The pathways most prominently include the RAS/RAF/MEK/ERK-,PI3K/AKT-, NFκB-, and STAT-pathway but also the WNT-, Hedgehog-, and TNFα-pathway [335,346]. Of these pathways, only the RAS/RAF/MEK/ERK has been associated with frequent mutations in MM.

The RAS/RAF/MEK/ERK pathway is a pathway of intracellular kinases involved in proliferation, growth, adhesion, and apoptosis. Mutations of the MAPK pathway are among the most common mutations found in MM, with a prevalence of 43%–53% of patients [347]. The number of patients with mutations seems to be higher in relapsed disease, with up

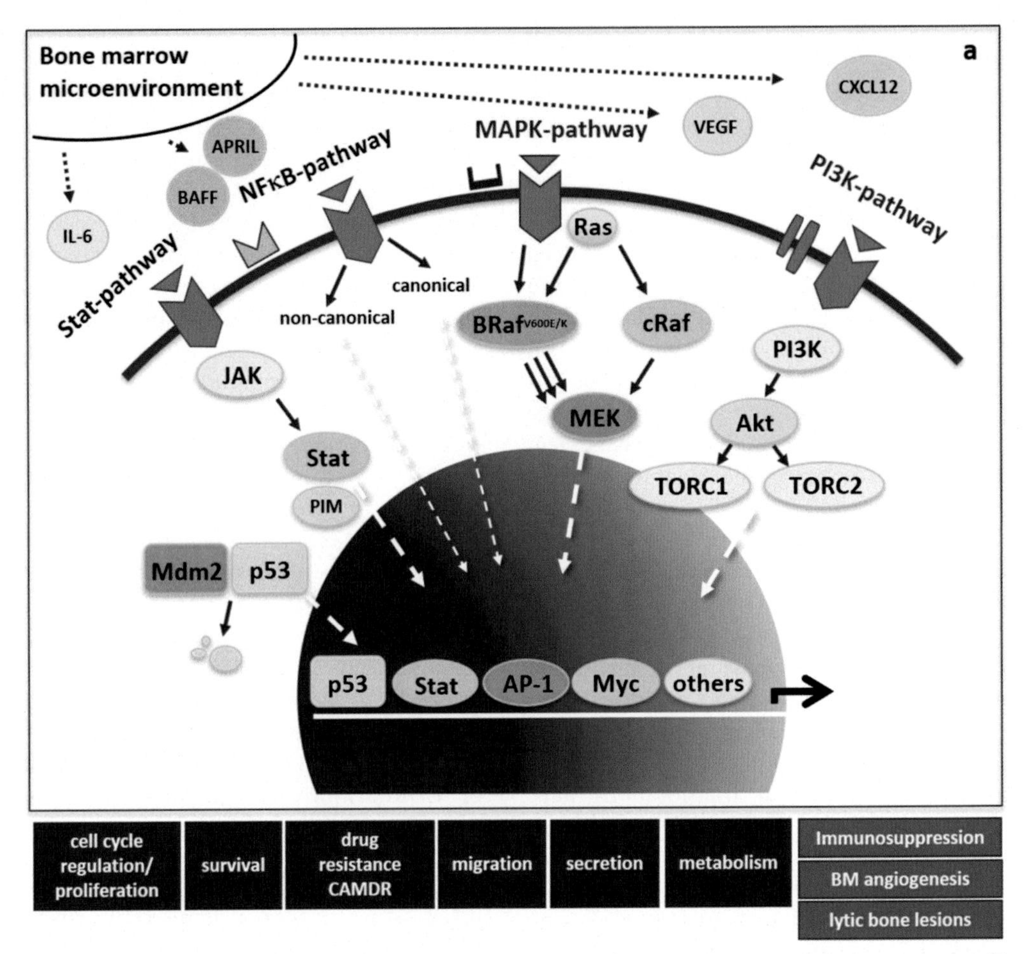

Figure 9.19 Deregulated signaling cascades and pathway-directed therapies in multiple myeloma. (A) Signaling pathways in multiple myeloma. In multiple myeloma, genetic abnormalities and bone marrow microenvironment-driven deregulations, such as increased levels of IL-6, IGF1, HGF, VEGF, CXCL12, TNFα, BAFF, APRIL, and CCL3 derived from stromal, bone, and immune cells result in the activation of multiple signaling pathways, most prominently including the RAS/RAF/MEK/ERK-, PI3K/AKT-, NFκB-, and STAT-pathway but also the WNT-, Hedgehog-, and TNFα-pathway. They trigger tumor cell proliferation, survival, drug resistance, migration, secretion of humoral factors but also promote immunoparesis, bone marrow angiogenesis, and bone disease.

to 72% of patients showing mutations in the r/r setting [348]. Only KRAS G12D and BRAFV600E consistently led to phosphorylation of downstream target ERK. Other mutations were associated with increased pERK-levels only in a small percentage of cases. This suggests that despite the high prevalence, many of the mutations in the RAS/RAF/MEK/ERK pathway may not lead to significant dysregulation of this pathway on their own [345]. The BRAF V600E mutation is also relatively common in MM patients, being present in 2%–4% of all newly diagnosed MM-patients with the prevalence of mutations increasing to about 8% in r/r patients and patients with extramedullary disease.

A significant proportion of patients with MGUS will not progress to MM, suggesting progression may require acquisition of additional genetic lesions. Secondary cytogenetic abnormities include further IgH translocations frequently involving the MYC oncogene (8q24), 6p25, 20q11, and 1q21 [349–351].

Risk factors for the development of MGUS include genetic predisposition, older age, immunosuppression, and environmental exposures (radiation, smoking, organic solvents, herbicides, and pesticides, for example) [352,353]. Most cases do not have a clear causal relationship, however.

9.8.4.2 Therapeutic Implications

The prognosis of patients with MM is dependent on four key factors: staging (disease characteristics), patient factors (host characteristics), disease biology (clinical characteristics, including plasmablastic morphology and the presence of extramedullary involvement), and response to therapy. The Revised International Staging System (R-ISS) has become the preferred staging system because of its simplicity and lack of subjectivity. The R-ISS uses serum B2M, serum albumin, serum lactate dehydrogenase (LDH), and bone marrow FISH results to stratify patients into three risk groups; Stage I—B2M < 3.5 mg/L, serum albumin ≥3.5 g/dL, normal LDH, and no TP53 deletion, t(4;14), or t(14;16) by FISH, Stage II—neither stage I nor stage III and Stage III—B2M ≥ 5.5 mg/L and elevated LDH and/or—TP53 deletion, t(4;14), or t(14;16) by FISH. The R-ISS was developed based on 3060 patients with newly diagnosed MM enrolled onto one of 11 international trials. R-ISS I ($n = 871$) showed estimated OS and PFS at 5 years were 82 and 55%, respectively; R-ISS II ($n = 1894$) showed estimated OS and PFS at 5 years were 62 and 36%, respectively, and R-ISS III ($n = 295$) showed estimated OS and PFS at 5 years were 40% and 24%, respectively [354] (Table 9.4).

TABLE 9.4 Standard risk factors for MM and the R-ISS

Prognostic Factor	Criteria
ISS stage	
I	Serum β_2-microglobulin <3.5 mg/L, serum albumin ≥3.5 g/dL
II	Not ISS stage I or II
III	Serum β_2-microglobulin ≥5.5 mg/L
CA by iFISH	
High risk	Presence of del(17p)and/or translocation t(4;14) and/or translocation t(14;16)
Standard risk	No high-risk CA
LDH	
Normal	Serum LDH < the upper limit of normal
High	Serum LDH < the upper limit of normal
A new model for risk stratification for MM	
R-ISS stage	
I	ISS stage I and standard-risk CA by iFISH and normal LDH
II	Not R-ISS stage I or III
III	ISS stage III and either high-risk CA by iFISH or high LDH

CA, chromosomal abnormalities; *iFISH*, interphase fluorescent in situ hybridization; *ISS*, international staging system; *LDH*, lactate dehydrogenase; *MM*, multiple myeloma; *R-ISS*, revised international staging system.

The sensitivity of metaphase cytogenetics is limited in comparison to FISH, and thus the approach has less overall utility. At present, FISH is used primarily for prognostic purposes. Patients with t(4;14), t(14;16), t(14;20), TP53 deletion, or gain 1q by FISH account for approximately 25% of MM and have a shortened median survival with standard therapy [355,356]. Due to their prognostic value and the widespread availability of probes for FISH testing, t(4;14), t(14;16), and TP53 deletion are considered high-risk cytogenetics in the R-ISS. While loss of chromosome 13 and hypodiploidy have been considered adverse prognostic factors when detected by conventional cytogenetics, these are not independent predictors of poor outcome when detected by FISH.

The t(4;14) was previously associated with a high rate of relapse, even in those undergoing high-dose chemotherapy followed by autologous HCT [357]. While t(4;14) is considered a high-risk genetic abnormality, there is evidence that bortezomib can overcome the adverse prognosis associated with this finding [358] TP53 deletion is highly associated with a low rate of complete response, rapid disease progression, plasma cell leukemia, and central nervous system involvement [359]. The presence of mutated TP53 on the other allele ("double hit") likely impacts the prognosis of patients with TP53 deletion [360].

Hyperdiploidy is a well-recognized predictor of favorable outcomes in MM. There are conflicting data on whether the presence of trisomies can ameliorate some of the adverse prognostic effects of high-risk cytogenetic abnormalities [361]. Risk assignment is not at this time downgraded for those with trisomies. Abnormalities in both the short and long arms of chromosome 1 have been associated with shorter survival [362,363]. In general, 1q gain (three copies), 1q amplification (four or more copies), and 1p deletion, and 1q21 aberrations are associated with both disease progression and poor prognosis [364,365]. Finally, translocations involving MYC were detected by FISH in 8% of newly diagnosed MM. MYC translocations were associated with a higher disease burden and shorter OS [365].

The t(11;14) translocation is so far the one potentially targetable translocation in MM. Dysregulation of the intrinsic apoptotic pathway contributes to hematologic malignancy pathogenesis. Members of the B-cell lymphoma 2 (BCL2) protein family function as key regulators of the intrinsic apoptosis pathway. Anti-apoptotic BCL2 family proteins, including BCL2, myeloid cell leukemia 1 (MCL1), and BCL-XL, contribute to development and progression of MM [366]. MM cells harboring t(11;14) express high levels of BCL2 relative to BCL-XL and MCL1. Venetoclax is a highly selective, potent, oral BCL2 inhibitor that restores apoptosis by direct, high-affinity binding to its prosurvival target, BCL2. Venetoclax, as a monotherapy, has demonstrated meaningful clinical activity in RRMM [367]. Therapeutic approaches that combine venetoclax with agents that increase BCL2 dependency or have complementary mechanisms of action such as the proteasome inhibitors bortezomib and carfilzomib further potentiate therapeutic efficacy in MM [368,369]. Pivotal studies are underway in t(11;14) translocated MM.

Aside from larger chromosomal rearrangements, targeting of smaller genomic variants in MM is now possible. There are several available and potent inhibitors specific for the BRAF V600 E/K mutation available, such as vemurafenib, encorafenib, and dabrafenib. All substances have an acceptable safety profile, with the most common side effects being blurred vision, macula edema, cramps, arthralgia, diarrhea, skin rash, decreased left ventricular function, anemia, and thrombocytopenia.

9.8.5 Waldenstrom Macroglobulinemia

WM is an uncommon B-cell malignancy characterized by lymphoplasmacytic lymphoma in the bone marrow and IgM monoclonal gammopathy in the blood or urine [370]. WM is frequently an indolent or low-grade lymphoma but considered incurable with current therapies. Wide clinical heterogeneity can lead to bulky lymphadenopathy, organomegaly, cytopenias, renal insufficiency, constitutional symptoms, neuropathy, and hyperviscosity from the excess IgM (macroglobulinemia). WM represents 1%–2% of hematologic cancers and is seen more frequently in males, with a median age at diagnosis of 63 years [371].

9.8.5.1 Genetics and Mechanisms of Disease

Myeloid differentiation primary response 88 (MYD88) and C-X-C chemokine receptor type 4 (CXCR4) somatic mutations are found in more than 90% and 30%–35% of WM patients, respectively, and have been shown to play a pivotal role in WM tumorigenesis [372]. Using NGS, recurring somatic mutations in MYD88, CXCR4, ARID1A, and CD79 are identified in WM lymphoplasmacytic lymphoma cells [373]. Copy number alterations, including those in chromosome 6q that affect regulatory genes for NFKB, Bruton tyrosine kinase (BTK), BCL2, and apoptotic signaling, are also identified. Most patients with WM (95%–97%) carry a point mutation in MYD88 that switches leucine to proline at amino acid position 265, those that are wild type for MYD88 show a more aggressive disease course [374].

Additionally, cytogenetic abnormalities identified in WM include deletion of the long arm of chromosome 6 (del6q) (20%–40%), del13q (10%–15%), trisomy 18 (18%) and deletion of the short arm of chromosome 17 (del17p) (8%) [375]. Limited information concerning clinical impact of these chromosomal abnormalities on clinical outcome in WM is available, though del17p appears to be associated with shorter progression-free survival [376].

MYD88 L265P is found to be expressed in 93%–97% of patients with WM and was identified in both sorted B cells and plasma cells that make up the malignant clone in WM [377]. Non-L265P MYD88 mutations have also been identified, although expression estimates for these variants are 1%–2% in WM [378]. Mutated MYD88 is also detectable in patients with IgM MGUS, suggesting an early oncogenic role for MYD88 in WM pathogenesis. Patients with IgM MGUS with detectable mutated MYD88 and patients with a higher mutated allele burden are at greater risk of progression to WM [379]. MYD88 is an adaptor protein that interacts with the Toll-like and interleukin (IL)-1 receptors and dimerizes upon receptor activation. Dimerization of MYD88 triggers downstream signaling, leading to nuclear factor-kB (NFKB) activation (Fig. 9.20) [377]. BTK triggers NFKB in mutated MYD88. Mutated MYD88 can also upregulate transcription of the SRC family member HCK that normally is downregulated in late stages of B-cell ontogeny and can transactivate HCK via IL-6. Activated HCK triggers prosurvival signaling of mutated MYD88 WM cells through BTK, PI3K/AKT, and MAPK/ERK1/2.

Somatic mutations involving the C-terminal domain of CXCR4 are unique to WM and present in up to 40% of patients with WM. Although CXCR4 variants almost always occur in those with MYD88 mutations, some

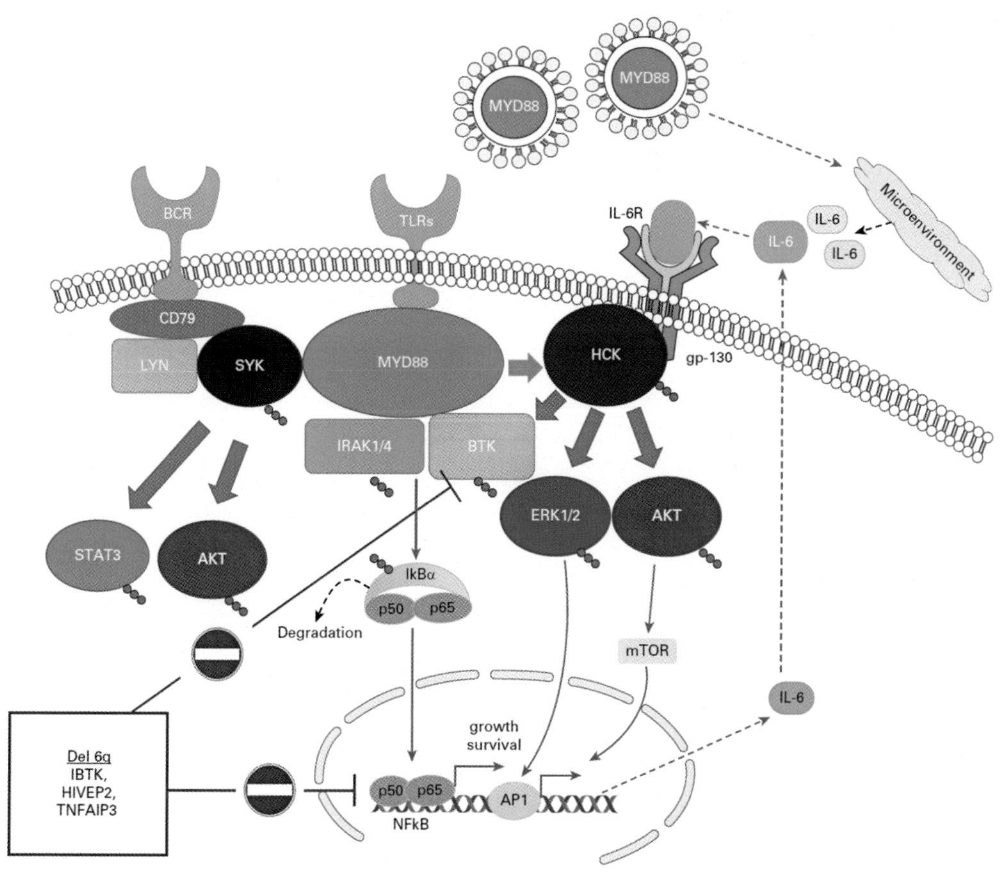

Figure 9.20 Prosurvival signaling mediated by mutated MYD88. Mutated MYD88 triggers assembly of the Myddosome , which includes activated BTK and IRAK4/IRAK1 that transactivate NFKB. Mutated MYD88 also transcriptionally upregulates and transactivates through interleukin (IL)-6 the SRC family member HCK, which triggers activation of BTK itself, as well as AKT and ERK. Mutated MYD88 also mediates cross talk through LYN-activated SYK, which activates STAT3 and AKT prosurvival signaling. Deletions in chromosome 6q result in the loss of important regulators of MYD88 signaling, including inhibitor of BTK (IBTK) and the NFKB regulators HIVEP2 and TNFAIP3. Waldenström macroglobulinemia cells also export mutated MYD88 via extracellular vesicles to induce signaling cascades in mast cells and macrophages that provide a supportive proinflammatory microenvironment. BCR, B-cell receptor; mTOR, mammalian target of rapamycin; TLRs, Toll-like receptors.

patients with wild-type MYD88 can also express CXCR4 mutations [380]. Mutations in the C-terminal domain of CXCR4 lead to loss of regulatory serines and promote sustained CXCL12-mediated activation of the AKT and ERK pathways and in vivo WM disease progression and dissemination in mice [381]. Despite the autonomous prosurvival signaling associated with CXCR4 mutations, inhibition of MYD88 triggers apoptosis in both wild-type and mutated CXCR4-expressing WM cells, consistent with a primary role for mutated MYD88 survival signaling in WM.

9.8.5.2 Therapeutic Implications

MYD88 and CXCR4 mutation status may be useful in treatment selection for symptomatic patients (Fig. 9.21) [377]. To minimize risks of short- and long-term treatment-related adverse effects associated with chemoimmunotherapy, including neuropathy, immunosuppression, and stem-cell damage, as well as secondary myelodysplasia and malignancies, the use of ibrutinib can be considered for treatment-naïve patients with WM who carry only the MYD88 mutation. Current nonrandomized comparisons of data from ibrutinib monotherapy and combined ibrutinib and rituximab studies show little difference in patients who only have MYD88 mutation; hence for whom no contraindication to ibrutinib exists [382]. Alternatives to ibrutinib that can be considered include bendamustine and rituximab (Benda-R) or PI-based therapy such as bortezomib, dexamethasone, and rituximab (BDR).

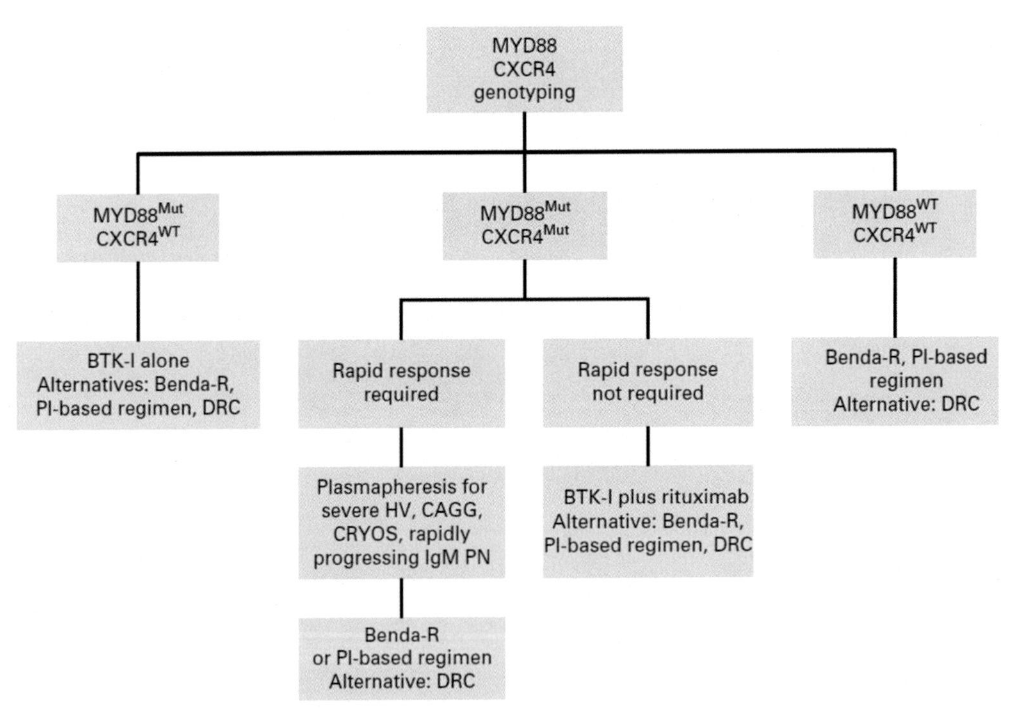

Figure 9.21 Genomic-based treatment algorithm for symptomatic, treatment-naïve patients with Waldenström macroglobulinemia. Rituximab should be held for serum immunoglobulin M (IgM) $\geq$ 4000 mg/dL to prevent symptomatic IgM flare. Bendamustine and rituximab (Benda-R) should be considered as primary choice in patients with bulky adenopathy or extramedullary disease. Proteasome inhibitor (PI)-based regimens should be considered in patients with symptomatic amyloidosis, with autologous stem-cell transplantation as possible consolidation. Rituximab alone or with ibrutinib if MYD88 is mutated or bendamustine may be considered in patients with IgM peripheral neuropathy (PN) depending on severity and pace of progression. Cyclophosphamide-based therapy such as dexamethasone, cyclophosphamide, and rituximab (DRC) represents an alternative to Benda-R but may be less effective. Maintenance rituximab may be considered in patients responding to rituximab-based regimens.*BTK-I*, bruton tyrosine kinase inhibitor; *CAGG*, cold agglutinemia; *CRYOS*, cryoglobulinemia; *HV*, hyperviscosity; *Mut*, mutated; *WT*, wild type (not mutated).

For patients with both MYD88 and CXCR4 mutations, the necessity to achieve a rapid response must be considered. Nonsense CXCR4-mutated patients in particular are at risk for presenting with symptomatic hyperviscosity requiring immediate disease control. Rituximab should be avoided in patients with WM with high IgM levels or symptomatic hyperviscosity as it can cause a flare of the disease [383]. For patients requiring more emergent responses, either bendamustine or a PI-based regimen may be considered. For patients without symptomatic hyperviscosity and emergent need for disease control, ibrutinib plus rituximab may be preferred versus ibrutinib alone because the median time to achieve a major response appears to be shorter with the combination [384]. For patients who have experienced multiple relapses and have been previously exposed to alkylators, PIs, and ibrutinib, the use of venetoclax may be considered. High overall response rates and durable responses are observed with venetoclax, including those who were CXCR4 mutated and previously exposed to a BTK inhibitor, although response depth was affected by these factors [385].

ACKNOWLEDGMENTS

The authors would like to thank Samantha DiPompeo for assistance with this chapter.

REFERENCES

[1] Nowell PC, Hungerford DA. Chromosome studies on normal and leukemic human leukocytes. J Natl Cancer Inst July 1960;25:85–109.

[2] Rowley JD. Letter: A new consistent chromosomal abnormality in chronic myelogenous leukaemia identified by quinacrine fluorescence and Giemsa staining. Nature June 1, 1973;243(5405):290–3.

[3] Braun TP, Eide CA, Druker BJ. Response and resistance to BCR-ABL1-targeted therapies. Cancer Cell April 13, 2020;37(4):530–42.

[4] Bennett JM, Catovsky D, Daniel MT, Flandrin G, Galton DA, Gralnick HR, et al. Proposals for the classification of the acute leukaemias. French-American-British (FAB) co-operative group. Br J Haematol August 1976;33(4):451–8.

[5] Harris NL, Jaffe ES, Stein H, Banks PM, Chan JK, Cleary ML, et al. A revised European–American classification of lymphoid neoplasms: a proposal from the international lymphoma study group. Blood September 1, 1994;84(5):1361–92.

[6] Campo E, Harris NL, Jaffe ES, Pileri SA, Stein H, Thiele J. WHO classification of tumours of haematopoietic and lymphoid tissues. In: World Health organization classification of tumours. Lyon, France: International Agency for Research on Cancer; 2017.

[7] Heim S, Mitelman F. Cancer cytogenetics. 3rd ed. New Jersey: John Wiley & Sons, Inc; 2009.

[8] Grimwade D, Hills RK, Moorman AV, Walker H, Chatters S, Goldstone AH, et al. Refinement of cytogenetic classification in acute myeloid leukemia: determination of prognostic significance of rare recurring chromosomal abnormalities among 5876 younger adult patients treated in the United Kingdom Medical Research Council trials. Blood 2010;116(3):354–65.

[9] Grimwade D, Walker H, Oliver F, Wheatley K, Harrison C, Harrison G, et al. The importance of diagnostic cytogenetics on outcome in AML: analysis of 1,612 patients entered into the MRC AML 10 trial. Blood 1998;92(7):2322–33.

[10] Rowley JD. The Philadelphia chromosome translocation a paradigm for understanding leukemia. Cancer 1990;65(10):2178–84.

[11] Rabbitts TH. Chromosomal translocations in human cancer. Nature 1994;372(6502):143–9.

[12] Rowley JD. The critical role OF chromosome translocations IN human leukemias. Annu Rev Genet 1998;32(1):495–519.

[13] Mullighan CG, Phillips LA, Su X, Ma J, Miller CB, Shurtleff SA, et al. Genomic analysis of the clonal origins of relapsed acute lymphoblastic leukemia. Science November 28, 2008;322(5906):1377–80.

[14] Gondek LP, Dunbar AJ, Szpurka H, McDevitt MA, Maciejewski JP. SNP array karyotyping allows for the detection of uniparental disomy and cryptic chromosomal abnormalities in MDS/MPD-U and MPD. PLoS One 2007;2(11):e1225.

[15] Maciejewski JP, Mufti GJ. Whole genome scanning as a cytogenetic tool in hematologic malignancies. Blood 2008;112(4):965–74.

[16] Gondek LP, Tiu R, O'Keefe CL, Sekeres MA, Theil KS, Maciejewski JP. Chromosomal lesions and uniparental disomy detected by SNP arrays in MDS, MDS/MPD, and MDS-derived AML. Blood February 1, 2008;111(3):1534–42.

[17] Bitter MA, Le Beau MM, Rowley JD, Larson RA, Golomb HM, Vardiman JW. Associations between morphology, karyotype, and clinical features in myeloid leukemias. Hum Pathol March 1987;18(3):211–25.

[18] Tefferi A, Skoda R, Vardiman JW. Myeloproliferative neoplasms: contemporary diagnosis using histology and genetics. Nat Rev Clin Oncol 2009;6(11):627–37.

[19] Estey E, Garcia-Manero G, Ferrajoli A, Faderl S, Verstovsek S, Jones D, et al. Use of all-trans retinoic acid plus arsenic trioxide as an alternative to chemotherapy in untreated acute promyelocytic leukemia. Blood 2006;107(9):3469–73.

[20] Druker BJ, Guilhot F, O'Brien SG, Gathmann I, Kantarjian H, Gattermann N, et al. Five-year follow-up of patients receiving imatinib for chronic myeloid leukemia. N Engl J Med 2006;355(23):2408–17.

[21] Druker BJ, Tamura S, Buchdunger E, Ohno S, Segal GM, Fanning S, et al. Effects of a selective inhibitor of the Abl tyrosine kinase on the growth of Bcr–Abl positive cells. Nat Med 1996;2(5):561–6.

[22] Harrison CJ, Hills RK, Moorman AV, Grimwade DJ, Hann I, Webb DK, et al. Cytogenetics of childhood acute myeloid leukemia: United Kingdom Medical Research Council Treatment trials AML 10 and 12. J Clin Oncol June 1, 2010;28(16):2674–81.

[23] Levis M, Perl AE. Gilteritinib: potent targeting of FLT3 mutations in AML. Blood Adv March 24, 2020;4(6):1178–91.

[24] Perl AE, Martinelli G, Cortes JE, Neubauer A, Berman E, Paolini S, et al. Gilteritinib or chemotherapy for relapsed or refractory FLT3-mutated AML. N Engl J Med October 31, 2019;381(18):1728–40.

[25] Pulte ED, Norsworthy KJ, Wang Y, Xu Q, Qosa H, Gudi R, et al. FDA approval summary: gilteritinib for relapsed or refractory acute myeloid leukemia with a *FLT3* mutation. Clin Cancer Res 2021;27(13):3515–21.

[26] DiNardo CD, Stein EM, de Botton S, Roboz GJ, Altman JK, Mims AS, et al. Durable remissions with ivosidenib in IDH1-mutated relapsed or refractory AML. N Engl J Med June 21, 2018;378(25):2386–98.

[27] Stein EM, DiNardo CD, Pollyea DA, Fathi AT, Roboz GJ, Altman JK, et al. Enasidenib in mutant IDH2 relapsed or refractory acute myeloid leukemia. Blood August 10, 2017;130(6):722–31.

[28] Papaemmanuil E, Gerstung M, Bullinger L, Gaidzik VI, Paschka P, Roberts ND, et al. Genomic classification and prognosis in acute myeloid leukemia. N Engl J Med 2016;374(23):2209–21.

[29] Gilleece MH, Labopin M, Yakoub-Agha I, Volin L, Socié G, Ljungman P, et al. Measurable residual disease, conditioning regimen intensity, and age predict outcome of allogeneic hematopoietic cell transplantation for acute myeloid leukemia in first remission: a registry analysis of 2292 patients by the acute Leukemia working party European society of blood and marrow transplantation. Am J Hematol 2018;93(9):1142–52.

[30] Zheng WS, Hu YL, Guan LX, Peng B, Wang SY. The effect of the detection of minimal residual disease for the prognosis and the choice of post-remission therapy of intermediate-risk acute myeloid leukemia without FLT3-ITD, NPM1 and biallelic CEBPA mutations. Hematology December 2021;26(1):179–85.

[31] Spitzer B, Levine RL. Toward more complete prognostication for patients with clonal hematopoiesis. Blood Cancer Discov 2021 -05;2(3):192–4.

[32] Larramendy ML, Siitonen SM, Zhu Y, Hurme M, Vilpo L, Vilpo JA, et al. Optimized mitogen stimulation induces proliferation of neoplastic B cells in chronic lymphocytic leukemia: significance for cytogenetic analysis.The Tampere Chronic Lympocytic Leukemia group. Cytogenet Cell Genet 1998;82(3–4):215–21.

[33] McGowan-Jordan J, Hastings RJ, Moore S. ISCN 2020 An international system for human cytogenomic nomenclature (2020). ISBN: 978-3-318-06706-4 e-ISBN: 978-3-318-06867-2 DOI: 10.1159/isbn.978-3-318-06867-2. Karger Publishers; 2020.

[34] Gozzetti A, Le Beau MM. Fluorescence in situ hybridization: uses and limitations. Semin Hematol October 2000;37(4):320–33.

[35] Pinkel D, Straume T, Gray JW. Cytogenetic analysis using quantitative, high-sensitivity, fluorescence hybridization. Proc Natl Acad Sci USA May 1986;83(9):2934–8.

[36] Nelson ND, McMahon CM, El-Sharkawy Navarro F, Freyer CW, Roth JJ, Luger SM, et al. Rapid fluorescence in situ hybridisation optimises induction therapy for acute myeloid leukaemia. Br J Haematol December 2020;191(5):935–8.

[37] Hosoya N, Sanada M, Nannya Y, Nakazaki K, Wang L, Hangaishi A, et al. Genomewide screening of DNA copy number changes in chronic myelogenous leukemia with the use of high-resolution array-based comparative genomic hybridization. Genes Chromosomes Cancer May 2006;45(5):482–94.

[38] Tiacci E, Trifonov V, Schiavoni G, Holmes A, Kern W, Martelli MP, et al. BRAF mutations in hairy-cell leukemia. N Engl J Med June 16, 2011;364(24):2305–15.

[39] Falini B, Martelli MP, Tiacci E. BRAF V600E mutation in hairy cell leukemia: from bench to bedside. Blood October 13, 2016;128(15):1918–27.

[40] Papaemmanuil E, Gerstung M, Malcovati L, Tauro S, Gundem G, Van Loo P, et al. Clinical and biological implications of driver mutations in myelodysplastic syndromes. Blood November 21, 2013;122(22). 3616,3627; quiz 3699.

[41] McMahon CM, Ferng T, Canaani J, Wang ES, Morrissette JJD, Eastburn DJ, et al. Clonal selection with RAS pathway activation mediates secondary

clinical resistance to selective FLT3 inhibition in acute myeloid leukemia. Cancer Discov August 2019;9(8):1050–63.
[42] Duncavage EJ, Schroeder MC, O'Laughlin M, Wilson R, MacMillan S, Bohannon A, et al. Genome sequencing as an alternative to cytogenetic analysis in myeloid cancers. N Engl J Med 2021;384(10):924–35.
[43] Hoffmann VS, Baccarani M, Hasford J, Lindoerfer D, Burgstaller S, Sertic D, et al. The EUTOS population-based registry: incidence and clinical characteristics of 2904 CML patients in 20 European Countries. Leukemia June 2015;29(6):1336–43.
[44] Höglund M, Sandin F, Simonsson B. Epidemiology of chronic myeloid leukaemia: an update. Ann Hematol April 2015;94(Suppl. 2):241.
[45] Daley GQ, Van Etten RA, Baltimore D. Induction of chronic myelogenous leukemia in mice by the P210bcr/abl gene of the Philadelphia chromosome. Science February 16, 1990;247(4944):824–30.
[46] Melo JV, Barnes DJ. Chronic myeloid leukaemia as a model of disease evolution in human cancer. Nat Rev Cancer June 2007;7(6):441–53.
[47] Sawyers CL. Chronic myeloid leukemia. N Engl J Med April 29, 1999;340(17):1330–40.
[48] Nowell PC, Hungerford DA. A minute chromosome in human chronicgranulocytic leukemia. Science 1960;142:1497.
[49] Klein A, Kessel AG, Grosveld G, Bartram CR, Hagemeijer A, Bootsma D, et al. A cellular oncogene is translocated to the Philadelphia chromosome in chronic myelocytic leukaemia. Nature 1982;300(5894):765–7.
[50] Groffen J, Stephenson JR, Heisterkamp N, de Klein A, Bartram CR, Grosveld G. Philadelphia chromosomal breakpoints are clustered within a limited region, bcr, on chromosome 22. Cell January 1984;36(1):93–9.
[51] Grosveld G, Verwoerd T, van Agthoven T, de Klein A, Ramachandran KL, Heisterkamp N, et al. The chronic myelocytic cell line K562 contains a breakpoint in bcr and produces a chimeric bcr/c-abl transcript. Mol Cell Biol 1986;6(2):607–16.
[52] Shtivelman E, Lifshitz B, Gale RP, Canaani E. Fused transcript of abl and bcr genes in chronic myelogenous leukaemia. Nature 1985;315(6020):550–4.
[53] Kamps MP. E2A-Pbx1 induces growth, blocks differentiation, and interacts with other homeodomain proteins regulating normal differentiation. Curr Top Microbiol Immunol 1997;220:25–43.
[54] Anastasi J, Feng J, Dickstein JI, Le Beau MM, Rubin CM, Larson RA, et al. Lineage involvement by BCR/ABL in Ph+ lymphoblastic leukemias: chronic myelogenous leukemia presenting in lymphoid blast vs Ph+ acute lymphoblastic leukemia. Leukemia May 1996;10(5):795–802.
[55] Golub TR, Barker GF, Lovett M, Gilliland DG. Fusion of PDGF receptor beta to a novel ets-like gene, tel, in chronic myelomonocytic leukemia with t(5;12) chromosomal translocation. Cell April 22, 1994;77(2):307–16.
[56] Mahon FX, Etienne G. Deep molecular response in chronic myeloid leukemia: the new goal of therapy? Clin Cancer Res January 15, 2014;20(2):310–22.
[57] Chissoe SL, Bodenteich A, Wang Y, Wang Y, Burian D, Clifton SW, et al. Sequence and analysis of the human ABL gene, the BCR gene, and regions involved in the Philadelphia chromosomal translocation. Genomics May 1, 1995;27(1):67–82.
[58] Pane F, Frigeri F, Sindona M, Luciano L, Ferrara F, Cimino R, et al. Neutrophilic-chronic myeloid leukemia: a distinct disease with a specific molecular marker (BCR/ABL with C3/A2 junction). Blood October 01, 1996;88(7):2410–4.
[59] Hotchin NA, Hall A. Regulation of the actin cytoskeleton, integrins and cell growth by the Rho family of SmallGTPases. Cancer Surv 1996;(27):311–22.
[60] McWhirter JR, Wang JY. Effect of Bcr sequences on the cellular function of the Bcr-Abl oncoprotein. Oncogene October 2, 1997;15(14):1625–34.
[61] Wetzler M, Talpaz M, Van Etten RA, Hirsh-Ginsberg C, Beran M, Kurzrock R. Subcellular localization of Bcr, Abl, and Bcr-Abl proteins in normal and leukemic cells and correlation of expression with myeloid differentiation. J Clin Invest 1993 -10;92(4):1925–39.
[62] Ohno T, Hada S, Sugiyama T, Mizumoto T, Furukawa H, Nagai K. Chronic myeloid leukemia with minor-bcr breakpoint developed hybrid type of blast crisis. Am J Hematol April 1998;57(4):320–5.
[63] Wark G, Heyworth CM, Spooncer E, Czaplewski L, Francis JM, Dexter TM, et al. Abl protein kinase abrogates the response of multipotent haemopoietic cells to the growth inhibitor macrophage inflammatory protein-1 alpha. Oncogene 1998;16(10):1319–24.
[64] Wang W, Cortes JE, Tang G, Khoury JD, Wang S, Bueso-Ramos CE, et al. Risk stratification of chromosomal abnormalities in chronic myelogenous leukemia in the era of tyrosine kinase inhibitor therapy. Blood June 02, 2016;127(22):2742–50.
[65] Ranjan A, Penninga E, Jelsig AM, Hasselbalch HC, Bjerrum OW. Inheritance of the chronic myeloproliferative neoplasms. A systematic review. Clin Genet February 2013;83(2):99–107.
[66] Landgren O, Goldin LR, Kristinsson SY, Helgadottir EA, Samuelsson J, Björkholm M. Increased risks of polycythemia vera, essential thrombocythemia,

and myelofibrosis among 24,577 first-degree relatives of 11,039 patients with myeloproliferative neoplasms in Sweden. Blood September 15, 2008;112(6):2199–204.
[67] Bizzozero OJ, Johnson KG, Ciocco A. Radiation-related leukemia in Hiroshima and Nagasaki, 1946–1964. I. Distribution, incidence and appearance time. N Engl J Med May 05, 1966;274(20):1095–101.
[68] Druker BJ, Sawyers CL, Kantarjian H, Resta DJ, Reese SF, Ford JM, et al. Activity of a specific inhibitor of the BCR-ABL tyrosine kinase in the blast crisis of chronic myeloid leukemia and acute lymphoblastic leukemia with the Philadelphia chromosome. N Engl J Med April 5, 2001;344(14):1038–42.
[69] Johansson B, Fioretos T, Mitelman F. Cytogenetic and molecular genetic evolution of chronic myeloid leukemia. Acta Haematol 2002;107(2):76–94.
[70] Sokal JE, Gomez GA, Baccarani M, Tura S, Clarkson BD, Cervantes F, et al. Prognostic significance of additional cytogenetic abnormalities at diagnosis of Philadelphia chromosome-positive chronic granulocytic leukemia. Blood July 1988;72(1):294–8.
[71] Baccarani M, Saglio G, Goldman J, Hochhaus A, Simonsson B, Appelbaum F, et al. Evolving concepts in the management of chronic myeloid leukemia: recommendations from an expert panel on behalf of the European LeukemiaNet. Blood September 15, 2006;108(6):1809–20.
[72] Eisterer W, Jiang X, Christ O, Glimm H, Lee KH, Pang E, et al. Different subsets of primary chronic myeloid leukemia stem cells engraft immunodeficient mice and produce a model of the human disease. Leukemia March 2005;19(3):435–41.
[73] Mahon F, Réa D, Guilhot J, Guilhot F, Huguet F, Nicolini F, et al. Discontinuation of imatinib in patients with chronic myeloid leukaemia who have maintained complete molecular remission for at least 2 years: the prospective, multicentre Stop Imatinib (STIM) trial. Lancet Oncol November 2010;11(11):1029–35.
[74] Rousselot P, Huguet F, Rea D, Legros L, Cayuela JM, Maarek O, et al. Imatinib mesylate discontinuation in patients with chronic myelogenous leukemia in complete molecular remission for more than 2 years. Blood January 01, 2007;109(1):58–60.
[75] Hochhaus A, Masszi T, Giles FJ, Radich JP, Ross DM, Gómez Casares MT, et al. Treatment-free remission following frontline nilotinib in patients with chronic myeloid leukemia in chronic phase: results from the ENESTfreedom study. Leukemia July 2017;31(7):1525–31.
[76] Mahon F, Boquimpani C, Kim D, Benyamini N, Clementino NCD, Shuvaev V, et al. Treatment-free remission after second-line nilotinib treatment in patients with chronic myeloid leukemia in chronic phase. Ann Intern Med 2018;168(7):461–70.
[77] Deininger MW, Goldman JM, Melo JV. The molecular biology of chronic myeloid leukemia. Blood November 15, 2000;96(10):3343–56.
[78] Mittelman F. The third international workshop on chromosomes in leukemia. Lund, Sweden, july 21-25, 1980. Introduction. Cancer Genet Cytogenet October 1981;4(2):96–8.
[79] Salesse S, Verfaillie CM. BCR/ABL: from molecular mechanisms of leukemia induction to treatment of chronic myelogenous leukemia. Oncogene December 9, 2002;21(56):8547–59.
[80] La Rosee P, Hochhaus A. Molecular pathogenesis of tyrosine kinase resistance in chronic myeloid leukemia. Curr Opin Hematol 2010;(17):91–6.
[81] Arber DA, Orazi A, Hasserjian R, Thiele J, Borowitz MJ, Le Beau MM, et al. The 2016 revision to the World Health Organization classification of myeloid neoplasms and acute leukemia. Blood May 19, 2016;127(20):2391–405.
[82] Maxson JE, Tyner JW. Genomics of chronic neutrophilic leukemia. Blood February 09, 2017;129(6):715–22.
[83] Schieber M, Crispino JD, Stein B. Myelofibrosis in 2019: moving beyond JAK2 inhibition. Blood Cancer J September 11, 2019;9(9):74.
[84] Vardiman JW, Brunning RD, Arber DA. Introduction and overview of the classification of the MyeloidNeoplasms. In: Swerdlow SH, Campo E, Harris NL, editors. WHO classification of tumors of hematopoietic and lymphoid tissues. WHO Press; 2008.
[85] Bain BJ, Horny HP, Arber DA. Myeloid/lymphoid neoplasms with eosinophilia and rearrangements of PDGFRA, PDGFRB or FGFR1, or with PCM1-JAK2. In: Swerdlow SH, Campo E, Harris NL, et al., editors. WHO classification of tumours of haematopoietic and lymphoid tissues; 2017.
[86] Pozdnyakova O, Orazi A, Kelemen K, King R, Reichard KK, Craig FE, et al. Myeloid/lymphoid neoplasms associated with eosinophilia and rearrangements of PDGFRA, PDGFRB, or FGFR1 or with PCM1-JAK2. Am J Clin Pathol 2021;155(2):160–78.
[87] Cools J, DeAngelo DJ, Gotlib J, Stover EH, Legare RD, Cortes J, et al. A tyrosine kinase created by fusion of the PDGFRA and FIP1L1 genes as a therapeutic target of imatinib in idiopathic hypereosinophilic syndrome. N Engl J Med 2003 -03-27;348(13):1201–14.
[88] Shomali W, Gotlib J. World Health Organization-defined eosinophilic disorders: 2019 update on diagnosis, risk stratification, and management. Am J Hematol 2019 -10;94(10):1149–67.

[89] Strati P, Tang G, Duose DY, Mallampati S, Luthra R, Patel KP, et al. Myeloid/lymphoid neoplasms with FGFR1 rearrangement. Leuk Lymphoma July 2018;59(7):1672–6.
[90] Lacronique V, Boureux A, Valle VD, Poirel H, Quang CT, Mauchauffé M, et al. A TEL-JAK2 fusion protein with constitutive kinase activity in human leukemia. Science November 14, 1997;278(5341):1309–12.
[91] Sun Y, Cai Y, Chen J, Cen J, Zhu M, Pan J, et al. Diagnosis and treatment of myeloproliferative neoplasms with PCM1-JAK2 rearrangement: case report and literature review. Front Oncol 2021;11:4139.
[92] Greenberg PL, Tuechler H, Schanz J, Sanz G, Garcia-Manero G, Solé F, et al. Revised international prognostic scoring system for myelodysplastic syndromes. Blood September 20, 2012;120(12):2454–65.
[93] Nazha A, Narkhede M, Radivoyevitch T, Seastone DJ, Patel BJ, Gerds AT, et al. Incorporation of molecular data into the revised international prognostic scoring system in treated patients with myelodysplastic syndromes. Leukemia November 2016;30(11):2214–20.
[94] Bejar R, Levine R, Ebert BL. Unraveling the molecular pathophysiology of myelodysplastic syndromes. J Clin Oncol February 10, 2011;29(5):504–15.
[95] Hosono N. Genetic abnormalities and pathophysiology of MDS. Int J Clin Oncol August 2019;24(8):885–92.
[96] Schanz J, Tüchler H, Solé F, Mallo M, Luño E, Cervera J, et al. New comprehensive cytogenetic scoring system for primary myelodysplastic syndromes (MDS) and oligoblastic acute myeloid leukemia after MDS derived from an international database merge. J Clin Oncol March 10, 2012;30(8):820–9.
[97] Pedersen-Bjergaard J, Andersen MK, Andersen MT, Christiansen DH. Genetics of therapy-related myelodysplasia and acute myeloid leukemia. Leukemia February 2008;22(2):240–8.
[98] Pozdnyakova O, Miron PM, Tang G, Walter O, Raza A, Woda B, et al. Cytogenetic abnormalities in a series of 1,029 patients with primary myelodysplastic syndromes: a report from the US with a focus on some undefined single chromosomal abnormalities. Cancer December 15, 2008;113(12):3331–40.
[99] Solé F, Espinet B, Sanz GF, Cervera J, Calasanz MJ, Luño E, et al. Incidence, characterization and prognostic significance of chromosomal abnormalities in 640 patients with primary myelodysplastic syndromes. Grupo Cooperativo Español de Citogenética Hematológica. Br J Haematol February 2000;108(2):346–56.
[100] Toyama K, Ohyashiki K, Yoshida Y, Abe T, Asano S, Hirai H, et al. Clinical implications of chromosomal abnormalities in 401 patients with myelodysplastic syndromes: a multicentric study in Japan. Leukemia April 1993;7(4):499–508.
[101] Vardiman JW, Thiele J, Arber DA, Brunning RD, Borowitz MJ, Porwit A, et al. The 2008 revision of the World Health Organization (WHO) classification of myeloid neoplasms and acute leukemia: rationale and important changes. Blood July 30, 2009;114(5):937–51.
[102] Boultwood J, Pellagatti A, Cattan H, Lawrie CH, Giagounidis A, Malcovati L, et al. Gene expression profiling of CD34+ cells in patients with the 5q-syndrome. Br J Haematol November 2007;139(4):578–89.
[103] Heinrichs S, Kulkarni RV, Bueso-Ramos C, Levine RL, Loh ML, Li C, et al. Accurate detection of uniparental disomy and microdeletions by SNP array analysis in myelodysplastic syndromes with normal cytogenetics. Leukemia 2009;23(9):1605–13.
[104] Moorman AV, Harrison CJ, Buck GA, Richards SM, Secker-Walker LM, Martineau M, et al. Karyotype is an independent prognostic factor in adult acute lymphoblastic leukemia (ALL): analysis of cytogenetic data from patients treated on the Medical Research Council (MRC) UKALLXII/Eastern Cooperative Oncology Group (ECOG) 2993 trial. Blood April 15, 2007;109(8):3189–97.
[105] Xu F, Li X, Wu L, Zhang Q, Yang R, Yang Y, et al. Overexpression of the EZH2, RING1 and BMI1 genes is common in myelodysplastic syndromes: relation to adverse epigenetic alteration and poor prognostic scoring. Ann Hematol June 2011;90(6):643–53.
[106] Saygin C, Godley LA. Genetics of myelodysplastic syndromes. Cancers July 06, 2021;13(14).
[107] Bernard E, Nannya Y, Hasserjian RP, Devlin SM, Tuechler H, Medina-Martinez J, et al. Implications of TP53 allelic state for genome stability, clinical presentation and outcomes in myelodysplastic syndromes. Nat Med 2020;26(10):1549–56.
[108] Jaffe ES, Harris NL, Stein H. Pathology and genetics of tumours of haematopoietic and lymphoid tissues. Lyon: IARC Press; 2001.
[109] Trottier AM, Godley LA. Inherited predisposition to haematopoietic malignancies: overcoming barriers and exploring opportunities. Br J Haematol August 2021;194(4):663–76.
[110] Kraft IL, Godley LA. Identifying potential germline variants from sequencing hematopoietic malignancies. Blood November 26, 2020;136(22):2498–506.

[111] Greenberg P, Cox C, LeBeau MM, Fenaux P, Morel P, Sanz G, et al. International scoring system for evaluating prognosis in myelodysplastic syndromes. Blood March 15, 1997;89(6):2079–88.

[112] Kantarjian H, O'Brien S, Ravandi F, Cortes J, Shan J, Bennett JM, et al. Proposal for a new risk model in myelodysplastic syndrome that accounts for events not considered in the original International Prognostic Scoring System. Cancer September 15, 2008;113(6):1351–61.

[113] Kita-Sasai Y, Horiike S, Misawa S, Kaneko H, Kobayashi M, Nakao M, et al. International prognostic scoring system and TP53 mutations are independent prognostic indicators for patients with myelodysplastic syndrome. Br J Haematol November 2001;115(2):309–12.

[114] Malcovati L, Germing U, Kuendgen A, Della Porta MG, Pascutto C, Invernizzi R, et al. Time-dependent prognostic scoring system for predicting survival and leukemic evolution in myelodysplastic syndromes. J Clin Oncol August 10, 2007;25(23):3503–10.

[115] Sanz GF, Sanz MA, Vallespí T, Cañizo MC, Torrabadella M, García S, et al. Two regression models and a scoring system for predicting survival and planning treatment in myelodysplastic syndromes: a multivariate analysis of prognostic factors in 370 patients. Blood July 1989;74(1):395–408.

[116] Scalzulli E, Pepe S, Colafigli G, Breccia M. Therapeutic strategies in low and high-risk MDS: what does the future have to offer? Blood Rev January 2021;45:100689.

[117] Döhner H, Wei AH, Löwenberg B. Towards precision medicine for AML. Nat Rev Clin Oncol 2021;18(9):577–90.

[118] Meyers S, Lenny N, Hiebert SW. The t(8;21) fusion protein interferes with AML-1B-dependent transcriptional activation. Mol Cell Biol April 1995;15(4):1974–82.

[119] Okuda T, Cai Z, Yang S, Lenny N, Lyu CJ, van Deursen JM, et al. Expression of a knocked-in AML1-ETO leukemia gene inhibits the establishment of normal definitive hematopoiesis and directly generates dysplastic hematopoietic progenitors. Blood May 1, 1998;91(9):3134–43.

[120] Nucifora G, Birn DJ, Erickson P, Gao J, LeBeau MM, Drabkin HA, et al. Detection of DNA rearrangements in the AML1 and ETO loci and of an AML1/ETO fusion mRNA in patients with t(8;21) acute myeloid leukemia. Blood February 15, 1993;81(4):883–8.

[121] Nucifora G, Larson RA, Rowley JD. Persistence of the 8;21 translocation in patients with acute myeloid leukemia type M2 in long-term remission. Blood 1993;82(3):712–5.

[122] Speck NA, Gilliland DG. Core-binding factors in haematopoiesis and leukaemia. Nat Rev Cancer July 2002;2(7):502–13.

[123] Zhang Y, Strissel P, Strick R, Chen J, Nucifora G, Le Beau MM, et al. Genomic DNA breakpoints in *AML1/RUNX1* and *ETO* cluster with topoisomerase II DNA cleavage and DNase I hypersensitive sites in t(8;21) leukemia. Proc Natl Acad Sci USA 2002;99(5):3070–5.

[124] Bitter MA, Le Beau MM, Larson RA. A morphologic and cytochemical study of acute myelomonocytic leukemia with abnormal marrow eosinophils associated within v(16)(p13q22). Am J Clin Pathol 1984;(81):733–41.

[125] Ferrando AA, Neuberg DS, Dodge RK, Paietta E, Larson RA, Wiernik PH, et al. Prognostic importance of TLX1 (HOX11) oncogene expression in adults with T-cell acute lymphoblastic leukaemia. Lancet February 14, 2004;363(9408):535–6.

[126] Liu P, Tarlé SA, Hajra A, Claxton DF, Marlton P, Freedman M, et al. Fusion between transcription factor CBF beta/PEBP2 beta and a myosin heavy chain in acute myeloid leukemia. Science August 20, 1993;261(5124):1041–4.

[127] Lutterbach B, Hou Y, Durst KL, Hiebert SW. The inv(16) encodes an acute myeloid leukemia 1 transcriptional corepressor. Proc Natl Acad Sci USA October 26, 1999;96(22):12822–7.

[128] Shurtleff SA, Meyers S, Hiebert SW, Raimondi SC, Head DR, Willman CL, et al. Heterogeneity in CBF beta/MYH11 fusion messages encoded by the inv(16)(p13q22) and the t(16;16)(p13;q22) in acute myelogenous leukemia. Blood June 15, 1995;85(12):3695–703.

[129] Pollard JA, Alonzo TA, Gerbing RB, Ho PA, Zeng R, Ravindranath Y, et al. Prevalence and prognostic significance of KIT mutations in pediatric patients with core binding factor AML enrolled on serial pediatric cooperative trials for de novo AML. Blood 2010;115(12):2372–9.

[130] Mrózek K, Bloomfield CD. Clinical significance of the most common chromosome translocations in adult acute myeloid leukemia. J Natl Cancer Inst Monogr 2008;(39):52–7.

[131] Larson RA, Kondo K, Vardiman JW, Butler AE, Golomb HM, Rowley JD. Evidence for a 15;17 translocation in every patient with acute promyelocytic leukemia. Am J Med May 1984;76(5):827–41.

[132] Rowley JD, Golomb HM, Vardiman J, Fukuhara S, Dougherty C, Potter D. Further evidence for a non-random chromosomal abnormality in acute promyelocytic leukemia. Int J Cancer December 15, 1977;20(6):869–72.

[133] Grignani F, De Matteis S, Nervi C, Tomassoni L, Gelmetti V, Cioce M, et al. Fusion proteins of the retinoic acid receptor-alpha recruit histone deacetylase in promyelocytic leukaemia. Nature February 19, 1998;391(6669):815–8.
[134] de Thé H, Chomienne C, Lanotte M, Degos L, Dejean A. The t(15;17) translocation of acute promyelocytic leukaemia fuses the retinoic acid receptor alpha gene to a novel transcribed locus. Nature October 11, 1990;347(6293):558–61.
[135] Greene ME, Mundschau G, Wechsler J, McDevitt M, Gamis A, Karp J, et al. Mutations in GATA1 in both transient myeloproliferative disorder and acute megakaryoblastic leukemia of Down syndrome. Blood Cells Mol Dis 2003;31(3):351–6.
[136] Melnick A, Licht JD. Deconstructing a disease: RARalpha, its fusion partners, and their roles in the pathogenesis of acute promyelocytic leukemia. Blood May 15, 1999;93(10):3167–215.
[137] Solomon E, Borrow J, Goddard AD. Chromosome aberrations and cancer. Science November 22, 1991;254(5035):1153–60.
[138] Lin RJ, Nagy L, Inoue S, Shao W, Miller Jr WH, Evans RM. Role of the histone deacetylase complex in acute promyelocytic leukaemia. Nature February 19, 1998;391(6669):811–4.
[139] Fazi F, Zardo G, Gelmetti V, Travaglini L, Ciolfi A, Di Croce L, et al. Heterochromatic gene repression of the retinoic acid pathway in acute myeloid leukemia. Blood 2007;109(10):4432–40.
[140] Zhang XW, Yan XJ, Zhou ZR, Yang FF, Wu ZY, Sun HB, et al. Arsenic trioxide controls the fate of the PML-RARalpha oncoprotein by directly binding PML. Science April 9, 2010;328(5975):240–3.
[141] Lo-Coco F, Avvisati G, Vignetti M, Thiede C, Orlando SM, Iacobelli S, et al. Retinoic acid and arsenic trioxide for acute promyelocytic leukemia. N Engl J Med July 11, 2013;369(2):111–21.
[142] Burnett AK, Russell NH, Hills RK, Bowen D, Kell J, Knapper S, et al. Arsenic trioxide and all-trans retinoic acid treatment for acute promyelocytic leukaemia in all risk groups (AML17): results of a randomised, controlled, phase 3 trial. Lancet Oncol 2015;16(13):1295–305.
[143] Madan V, Shyamsunder P, Han L, Mayakonda A, Nagata Y, Sundaresan J, et al. Comprehensive mutational analysis of primary and relapse acute promyelocytic leukemia. Leukemia 2016;30(8):1672–81.
[144] Stass S, Mirro J, Melvin S, Pui CH, Murphy SB, Williams D. Lineage switch in acute leukemia. Blood September 1984;64(3):701–6.
[145] Gagnon GA, Childs CC, LeMaistre A, Keating M, Cork A, Trujillo JM, et al. Molecular heterogeneity in acute leukemia lineage switch. Blood November 1, 1989;74(6):2088–95.
[146] Bill M, Mrózek K, Kohlschmidt J, Eisfeld A, Walker CJ, Nicolet D, et al. Mutational landscape and clinical outcome of patients with de novo acute myeloid leukemia and rearrangements involving 11q23/*KMT2A*. Proc Natl Acad Sci USA 2020;117(42):26340–6.
[147] Chen Y, Kantarjian H, Pierce S, Faderl S, O'Brien S, Qiao W, et al. Prognostic significance of 11q23 aberrations in adult acute myeloid leukemia and the role of allogeneic stem cell transplantation. Leukemia 2013;27(4):836–42.
[148] Cimino G, Lo Coco F, Biondi A, Elia L, Luciano A, Croce CM, et al. ALL-1 gene at chromosome 11q23 is consistently altered in acute leukemia of early infancy. Blood 1993;82(2):544–6.
[149] Downing JR, Head DR, Raimondi SC, Carroll AJ, Curcio-Brint AM, Motroni TA, et al. The der(11)-encoded MLL/AF-4 fusion transcript is consistently detected in t(4;11)(q21;q23)-containing acute lymphoblastic leukemia. Blood January 15, 1994;83(2):330–5.
[150] Sorensen PH, Chen CS, Smith FO, Arthur DC, Domer PH, Bernstein ID, et al. Molecular rearrangements of the MLL gene are present in most cases of infant acute myeloid leukemia and are strongly correlated with monocytic or myelomonocytic phenotypes. J Clin Invest 1994;93(1):429–37.
[151] Andersson AK, Ma J, Wang J, Chen X, Gedman AL, Dang J, et al. The landscape of somatic mutations in infant MLL-rearranged acute lymphoblastic leukemias. Nat Genet April 2015;47(4):330–7.
[152] Heerema NA, Sather HN, Ge J, Arthur DC, Hilden JM, Trigg ME, et al. Cytogenetic studies of infant acute lymphoblastic leukemia: poor prognosis of infants with t(4;11)—a report of the Children's Cancer Group. Leukemia May 1999;13(5):679–86.
[153] Balgobind BV, Raimondi SC, Harbott J, Zimmermann M, Alonzo TA, Auvrignon A, et al. Novel prognostic subgroups in childhood 11q23/MLL-rearranged acute myeloid leukemia: results of an international retrospective study. Blood September 17, 2009;114(12):2489–96.
[154] Hilden JM, Dinndorf PA, Meerbaum SO, Sather H, Villaluna D, Heerema NA, et al. Analysis of prognostic factors of acute lymphoblastic leukemia in infants: report on CCG 1953 from the Children's Oncology Group. Blood July 15, 2006;108(2):441–51.
[155] Hilden JM, Frestedt JL, Moore RO, Heerema NA, Arthur DC, Reaman GH, et al. Molecular analysis of

infant acute lymphoblastic leukemia: MLL gene rearrangement and reverse transcriptase-polymerase chain reaction for t(4;11)(q21;q23). Blood November 15, 1995;86(10):3876–82.
[156] Ziemin-van der Poel S, McCabe NR, Gill HJ, Espinosa 3 R, Patel Y, Harden A, et al. Identification of a gene, MLL, that spans the breakpoint in 11q23 translocations associated with human leukemias. Proc Natl Acad Sci USA 1991;88(23):10735–9.
[157] Bloomfield CD, Archer KJ, Mrózek K, Lillington DM, Kaneko Y, Head DR, et al. 11q23 balanced chromosome aberrations in treatment-related myelodysplastic syndromes and acute leukemia: report from an international workshop. Genes Chromosomes Cancer April 2002;33(4):362–78.
[158] Meyer C, Kowarz E, Hofmann J, Renneville A, Zuna J, Trka J, et al. New insights to the MLL recombinome of acute leukemias. Leukemia August 2009;23(8):1490–9.
[159] Caligiuri MA, Strout MP, Oberkircher AR, Yu F, de la Chapelle A, Bloomfield CD. The partial tandem duplication of ALL1 in acute myeloid leukemia with normal cytogenetics or trisomy 11 is restricted to one chromosome. Proc Natl Acad Sci U S A April 15, 1997;94(8):3899–902.
[160] Schichman SA, Caligiuri MA, Gu Y, Strout MP, Canaani E, Bloomfield CD, et al. ALL-1 partial duplication in acute leukemia. Proc Natl Acad Sci U S A 1994;91(13):6236–9.
[161] Bitter MA, Neilly ME, Le Beau MM, Pearson MG, Rowley JD. Rearrangements of chromosome 3 involving bands 3q21 and 3q26 are associated with normal or elevated platelet counts in acute nonlymphocytic leukemia. Blood December 1985;66(6):1362–70.
[162] Pintado T, Ferro MT, San Román C, Mayayo M, Laraña JG. Clinical correlations of the 3q21;q26 cytogenetic anomaly. A leukemic or myelodysplastic syndrome with preserved or increased platelet production and lack of response to cytotoxic drug therapy. Cancer February 1, 1985;55(3):535–41.
[163] Gröschel S, Sanders MA, Hoogenboezem R, de Wit E, Bouwman BAM, Erpelinck C, et al. A single oncogenic enhancer rearrangement causes concomitant EVI1 and GATA2 deregulation in leukemia. Cell 2014;157(2):369–81.
[164] Yamazaki H, Suzuki M, Otsuki A, Shimizu R, Bresnick EH, Engel JD, et al. A remote GATA2 hematopoietic enhancer drives leukemogenesis in inv(3)(q21;q26) by activating EVI1 expression. Cancer Cell 2014;25(4):415–27.
[165] Ottema S, Mulet-Lazaro R, Beverloo HB, Erpelinck C, van Herk S, van der Helm R, et al. Atypical 3q26/MECOM rearrangements genocopy inv(3)/t(3;3) in acute myeloid leukemia. Blood 2020;136(2):224–34.
[166] Chi Y, Lindgren V, Quigley S, Gaitonde S. Acute myelogenous leukemia with t(6;9)(p23;q34) and marrow basophilia: an overview. Arch Pathol Lab Med November 2008;132(11):1835–7.
[167] Visconte V, Shetty S, Przychodzen B, Hirsch C, Bodo J, Maciejewski JP, et al. Clinicopathologic and molecular characterization of myeloid neoplasms with isolated t(6;9)(p23;q34). Int Jnl Lab Hem 2017;39(4):409–17.
[168] Kayser S, Hills RK, Luskin MR, Brunner AM. Christine Terré, Jörg Westermann, et al. Allogeneic hematopoietic cell transplantation improves outcome of adults with t(6;9) acute myeloid leukemia: results from an international collaborative study. Haematol 2020;105(1):161–9.
[169] Duchayne E, Fenneteau O, Pages MP, Sainty D, Arnoulet C, Dastugue N, et al. Acute megakaryoblastic leukaemia: a national clinical and biological study of 53 adult and childhood cases by the Groupe Français d'Hématologie Cellulaire (GFHC). Leuk Lymphoma January 2003;44(1):49–58.
[170] Bernstein J, Dastugue N, Haas OA, Harbott J, Heere NA, Huret JL, et al. Nineteen cases of the t(1;22)(p13;q13) acute megakaryblastic leukaemia of infants/children and a review of 39 cases: report from a t(1;22) study group. Leukemia 2000;14(1):216–8.
[171] Strickland SA, Sun Z, Ketterling RP, Cherry AM, Cripe LD, Dewald G, et al. Independent prognostic significance of monosomy 17 and impact of karyotype complexity in monosomal karyotype/complex karyotype Acute myeloid leukemia: results from four ECOG-ACRIN prospective therapeutic trials. Leuk Res 2017;59:55–64.
[172] Cancer Genome Atlas Research Network, Ley TJ, Miller C, Ding L, Raphael BJ, Mungall AJ, Robertson A, Hoadley K, Triche Jr TJ, Laird PW, Baty JD, Fulton LL, Fulton R, Heath SE, Kalicki-Veizer J, Kandoth C, Klco JM, Koboldt DC, Kanchi KL, Kulkarni S, Lamprecht TL, Larson DE, Lin L, Lu C, McLellan MD, McMichael JF, Payton J, Schmidt H, Spencer DH, Tomasson MH, Wallis JW, Wartman LD, Watson MA, Welch J, Wendl MC, Ally A, Balasundaram M, Birol I, Butterfield Y, Chiu R, Chu A, Chuah E, Chun HJ, Corbett R, Dhalla N, Guin R, He A, Hirst C, Hirst M, Holt RA, Jones S, Karsan A, Lee D, Li HI, Marra MA, Mayo M, Moore RA, Mungall K, Parker J, Pleasance E, Plettner P, Schein J, Stoll D, Swanson L, Tam A, Thiessen N, Varhol R, Wye N, Zhao Y, Gabriel S, Getz G, Sougnez C, Zou L, Leiserson MD, Vandin F, Wu HT, Applebaum F, Baylin SB, Akbani R, Broom BM, Chen K, Motter TC, Nguyen K, Weinstein JN, Zhang N, Ferguson ML, Adams C,

Black A, Bowen J, Gastier-Foster J, Grossman T, Lichtenberg T, Wise L, Davidsen T, Demchok JA, Shaw KR, Sheth M, Sofia HJ, Yang L, Downing JR, Eley G. Genomic and epigenomic landscapes of adult de novo acute myeloid leukemia. N Engl J Med 2013;368(22):2059–74.
[173] Bezerra MF, Lima AS, Piqué-Borràs M, Silveira DR, Coelho-Silva JL, Pereira-Martins DA, et al. Co-occurrence of DNMT3A, NPM1, FLT3 mutations identifies a subset of acute myeloid leukemia with adverse prognosis. Blood 2020;135(11):870–5.
[174] Perl AE, Altman JK, Cortes J, Smith C, Litzow M, Baer MR, et al. Selective inhibition of FLT3 by gilteritinib in relapsed or refractory acute myeloid leukaemia: a multicentre, first-in-human, open-label, phase 1-2 study. Lancet Oncol August 2017;18(8):1061–75.
[175] Marcucci G, Haferlach T, Döhner H. Molecular genetics of adult acute myeloid leukemia: prognostic and therapeutic implications. J Clin Oncol February 10, 2011;29(5):475–86.
[176] Behdad A, Weigelin HC, Elenitoba-Johnson KS, Betz BL. A clinical grade sequencing-based assay for CEBPA mutation testing: report of a large series of myeloid neoplasms. J Mol Diagn January 2015;17(1):76–84.
[177] Cazzaniga G, Dell'Oro MG, Mecucci C, Giarin E, Masetti R, Rossi V, et al. Nucleophosmin mutations in childhood acute myelogenous leukemia with normal karyotype. Blood August 15, 2005;106(4):1419–22.
[178] Pedersen-Bjergaard J, Philip P. Balanced translocations involving chromosome bands 11q23 and 21q22 are highly characteristic of myelodysplasia and leukemia following therapy with cytostatic agents targeting at DNA-topoisomerase II. Blood August 15, 1991;78(4):1147–8.
[179] Pedersen-Bjergaard J, Rowley JD. The balanced and the unbalanced chromosome aberrations of acute myeloid leukemia may develop in different ways and may contribute differently to malignant transformation. Blood May 15, 1994;83(10):2780–6.
[180] Smith SM, Le Beau MM, Huo D, Karrison T, Sobecks RM, Anastasi J, et al. Clinical-cytogenetic associations in 306 patients with therapy-related myelodysplasia and myeloid leukemia: the University of Chicago series. Blood July 1, 2003;102(1):43–52.
[181] Zhao N, Stoffel A, Wang PW, Eisenbart JD, Espinosa 3rd R, Larson RA, et al. Molecular delineation of the smallest commonly deleted region of chromosome 5 in malignant myeloid diseases to 1-1.5 Mb and preparation of a PAC-based physical map. Proc Natl Acad Sci U S A June 24, 1997;94(13):6948–53.
[182] Ahuja HG, Felix CA, Aplan PD. The t(11;20)(p15;q11) chromosomal translocation associated with therapy-related myelodysplastic syndrome results in an NUP98-TOP1 fusion. Blood November 1, 1999;94(9):3258–61.
[183] Arai Y, Hosoda F, Kobayashi H, Arai K, Hayashi Y, Kamada N, et al. The inv(11)(p15q22) chromosome translocation of de novo and therapy-related myeloid malignancies results in fusion of the nucleoporin gene, NUP98, with the putative RNA helicase gene, DDX10. Blood June 1, 1997;89(11):3936–44.
[184] Libura J, Slater DJ, Felix CA, Richardson C. Therapy-related acute myeloid leukemia–like MLL rearrangements are induced by etoposide in primary human CD34+ cells and remain stable after clonal expansion. Blood 2005;105(5):2124–31.
[185] Pui CH, Relling MV. Topoisomerase II inhibitor-related acute myeloid leukaemia. Br J Haematol April 2000;109(1):13–23.
[186] Roulston D, Espinosa III R, Nucifora G, Larson RA, Le Beau MM, Rowley JD. CBFA2(AML1) translocations with novel partner chromosomes in myeloid leukemias: association with prior therapy. Blood 1998;92(8):2879–85.
[187] Rowley JD, Reshmi S, Sobulo O, Musvee T, Anastasi J, Raimondi S, et al. All patients with the T(11;16)(q23;p13.3) that involves MLL and CBP have treatment-related hematologic disorders. Blood 1997;90(2):535–41.
[188] Aplan PD, Chervinsky DS, Stanulla M, Burhans WC. Site-specific DNA cleavage within the MLL breakpoint cluster region induced by topoisomerase II inhibitors. Blood April 1, 1996;87(7):2649–58.
[189] Domer PH, Head DR, Renganathan N, Raimondi SC, Yang E, Atlas M. Molecular analysis of 13 cases of MLL/11q23 secondary acute leukemia and identification of topoisomerase II consensus-binding sequences near the chromosomal breakpoint of a secondary leukemia with the t(4;11). Leukemia August 1995;9(8):1305–12.
[190] Strissel PL, Strick R, Rowley JD, Zeleznik-Le NJ. An in vivo topoisomerase II cleavage site and a DNase I hypersensitive site colocalize near exon 9 in the MLL breakpoint cluster region. Blood November 15, 1998;92(10):3793–803.
[191] Strick R, Strissel PL, Borgers S, Smith SL, Rowley JD. Dietary bioflavonoids induce cleavage in the MLL gene and may contribute to infant leukemia. Proc Natl Acad Sci U S A April 25, 2000;97(9):4790–5.
[192] Strissel PL, Dann HA, Pomykala HM, Diaz MO, Rowley JD, Olopade OI. Scaffold-associated regions in the human type I interferon gene cluster on the short

arm of chromosome 9. Genomics January 15, 1998;47(2):217–29.

[193] Strissel PL, Strick R, Tomek RJ, Roe BA, Rowley JD, Zeleznik-Le NJ. DNA structural properties of AF9 are similar to MLL and could act as recombination hot spots resulting in MLL/AF9 translocations and leukemogenesis. Hum Mol Genet July 1, 2000;9(11):1671–9.

[194] McNerney ME, Godley LA, Le Beau MM. Therapy-related myeloid neoplasms: when genetics and environment collide. Nat Rev Cancer August 24, 2017;17(9):513–27.

[195] Alexandrov LB, Nik-Zainal S, Wedge DC, Aparicio SAJR, Behjati S, Biankin AV, et al. Signatures of mutational processes in human cancer. Nature August 22, 2013;500(7463):415–21.

[196] Bolton KL, Ptashkin RN, Gao T, Braunstein L, Devlin SM, Kelly D, et al. Cancer therapy shapes the fitness landscape of clonal hematopoiesis. Nat Genet November 2020;52(11):1219–26.

[197] Jaiswal S, Natarajan P, Silver AJ, Gibson CJ, Bick AG, Shvartz E, et al. Clonal hematopoiesis and risk of atherosclerotic cardiovascular disease. N Engl J Med July 13, 2017;377(2):111–21.

[198] Warren JT, Link DC. Clonal hematopoiesis and risk for hematologic malignancy. Blood October 01, 2020;136(14):1599–605.

[199] Xie M, Lu C, Wang J, McLellan MD, Johnson KJ, Wendl MC, et al. Age-related mutations associated with clonal hematopoietic expansion and malignancies. Nat Med December 2014;20(12):1472–8.

[200] Hein D, Borkhardt A, Fischer U. Insights into the prenatal origin of childhood acute lymphoblastic leukemia. Cancer Metastasis Rev 2020;39(1):161–71.

[201] Bloom M, Maciaszek JL, Clark ME, Pui CH, Nichols KE. Recent advances in genetic predisposition to pediatric acute lymphoblastic leukemia. Expet Rev Hematol January 2020;13(1):55–70.

[202] Larson RA, Dodge RK, Burns CP, Lee EJ, Stone RM, Schulman P, et al. A five-drug remission induction regimen with intensive consolidation for adults with acute lymphoblastic leukemia: cancer and leukemia group B study 8811. Blood 1995;85(8):2025–37.

[203] Williams DL, Look AT, Melvin SL, Roberson PK, Dahl G, Flake T, et al. New chromosomal translocations correlate with specific immunophenotypes of childhood acute lymphoblastic leukemia. Cell January 1984;36(1):101–9.

[204] Bloomfield CD, Goldman AI, Alimena G, Berger R, Borgström GH, Brandt L, et al. Chromosomal abnormalities identify high-risk and low-risk patients with acute lymphoblastic leukemia. Blood February 1986;67(2):415–20.

[205] Cavé H, van der Werff ten Bosch J, Suciu S, Guidal C, Waterkeyn C, Otten J, et al. Clinical significance of minimal residual disease in childhood acute lymphoblastic leukemia. European Organization for Research and Treatment of Cancer–Childhood Leukemia Cooperative Group. N Engl J Med August 27, 1998;339(9):591–8.

[206] Chessels JM, Swansbury GJ, Reeves B, Bailey CC, Richards SM. Cytogenetics and prognosis in childhood lymphoblastic leukaemia: results of MRC UKALL X. Medical research council working party in childhood leukaemia. Br J Haematol October 1997;99(1):93–100.

[207] Pui CH, Carroll WL, Meshinchi S, Arceci RJ. Biology, risk stratification, and therapy of pediatric acute leukemias: an update. J Clin Oncol February 10, 2011;29(5):551–65.

[208] Pui CH, Robison LL, Look AT. Acute lymphoblastic leukaemia. Lancet March 22, 2008;371(9617):1030–43.

[209] Gaynor J, Chapman D, Little C, McKenzie S, Miller W, Andreeff M, et al. A cause-specific hazard rate analysis of prognostic factors among 199 adults with acute lymphoblastic leukemia: the Memorial Hospital experience since 1969. J Clin Oncol June 1988;6(6):1014–30.

[210] Mancini M, Scappaticci D, Cimino G, Nanni M, Derme V, Elia L, et al. A comprehensive genetic classification of adult acute lymphoblastic leukemia (ALL): analysis of the GIMEMA 0496 protocol. Blood May 1, 2005;105(9):3434–41.

[211] Harvey RC, Mullighan CG, Chen IM, Wharton W, Mikhail FM, Carroll AJ, et al. Rearrangement of CRLF2 is associated with mutation of JAK kinases, alteration of IKZF1, Hispanic/Latino ethnicity, and a poor outcome in pediatric B-progenitor acute lymphoblastic leukemia. Blood July 1, 2010;115(26):5312–21.

[212] Moorman AV, Chilton L, Wilkinson J, Ensor HM, Bown N, Proctor SJ. A population-based cytogenetic study of adults with acute lymphoblastic leukemia. Blood January 14, 2010;115(2):206–14.

[213] Malard F, Mohty M. Acute lymphoblastic leukaemia. Lancet April 4, 2020;395(10230):1146–62.

[214] Paulsson K, Lilljebjörn H, Biloglav A, Olsson L, Rissler M, Castor A, et al. The genomic landscape of high hyperdiploid childhood acute lymphoblastic leukemia. Nat Genet June 2015;47(6):672–6.

[215] Mattano LA, Devidas M, Maloney KW, Wang C, Friedmann AM, Buckley P, et al. Favorable trisomies and ETV6-RUNX1 predict cure in low-risk B-cell acute lymphoblastic leukemia: results from children's oncology group trial AALL0331. J Clin Orthod 2021;39(14):1540–52.

[216] Enshaei A, Vora A, Harrison CJ, Moppett J, Moorman AV. Defining low-risk high hyperdiploidy in

patients with paediatric acute lymphoblastic leukaemia: a retrospective analysis of data from the UKALL97/99 and UKALL2003 clinical trials. Lancet Haematol November 2021;8(11):e828–39.
[217] Aguiar RC, Sohal J, van Rhee F, Carapeti M, Franklin IM, Goldstone AH, et al. TEL-AML1 fusion in acute lymphoblastic leukaemia of adults. M.R.C. Adult Leukaemia Working Party. Br J Haematol December 1996;95(4):673–7.
[218] Borkhardt A, Cazzaniga G, Viehmann S, Valsecchi MG, Ludwig WD, Burci L, et al. Incidence and clinical relevance of TEL/AML1 fusion genes in children with acute lymphoblastic leukemia enrolled in the German and Italian multicenter therapy trials. Associazione Italiana Ematologia Oncologia Pediatrica and the Berlin-Frankfurt-Münster Study Group. Blood July 15, 1997;90(2):571–7.
[219] Golub TR, Barker GF, Bohlander SK, Hiebert SW, Ward DC, Bray-Ward P, et al. Fusion of the TEL gene on 12p13 to the AML1 gene on 21q22 in acute lymphoblastic leukemia. Proc Natl Acad Sci U S A 1995;92(11):4917–21.
[220] Romana SP, Mauchauffé M, Le Coniat M, Chumakov I, Le Paslier D, Berger R, et al. The t(12;21) of acute lymphoblastic leukemia results in a tel-AML1 gene fusion. Blood June 15, 1995;85(12):3662–70.
[221] Rubnitz JE, Look AT. Molecular basis of leukemogenesis. Curr Opin Hematol July 1998;5(4):264–70.
[222] Shurtleff SA, Buijs A, Behm FG, Rubnitz JE, Raimondi SC, Hancock ML, et al. TEL/AML1 fusion resulting from a cryptic t(12;21) is the most common genetic lesion in pediatric ALL and defines a subgroup of patients with an excellent prognosis. Leukemia December 1995;9(12):1985–9.
[223] Lanza C, Volpe G, Basso G, Gottardi E, Perfetto F, Cilli V, et al. The common TEL/AML1 rearrangement does not represent a frequent event in acute lymphoblastic leukaemia occurring in children with Down syndrome. Leukemia 1997;11(6):820–1.
[224] Rubnitz JE, Downing JR, Pui CH, Shurtleff SA, Raimondi SC, Evans WE, et al. TEL gene rearrangement in acute lymphoblastic leukemia: a new genetic marker with prognostic significance. J Clin Oncol March 1997;15(3):1150–7.
[225] Rubnitz JE, Pui CH, Downing JR. The role of TEL fusion genes in pediatric leukemias. Leukemia January 1999;13(1):6–13.
[226] Lopez RG, Carron C, Oury C, Gardellin P, Bernard O, Ghysdael J. TEL is a sequence-specific transcriptional repressor. J Biol Chem 1999;274(42):30132–8.
[227] Roman-Gomez J, Jimenez-Velasco A, Agirre X, Castillejo JA, Navarro G, Calasanz MJ, et al. CpG island methylator phenotype redefines the prognostic effect of t(12;21) in childhood acute lymphoblastic leukemia. Clin Cancer Res 2006;12(16):4845–50.
[228] Ford AM, Palmi C, Bueno C, Hong D, Cardus P, Knight D, et al. The TEL-AML1 leukemia fusion gene dysregulates the TGF-beta pathway in early B lineage progenitor cells. J Clin Invest 2009;119(4):826–36.
[229] Stegmaier K, Pendse S, Barker GF, Bray-Ward P, Ward DC, Montgomery KT, et al. Frequent loss of heterozygosity at the TEL gene locus in acute lymphoblastic leukemia of childhood. Blood July 1, 1995;86(1):38–44.
[230] Wiemels JL, Greaves M. Structure and possible mechanisms of TEL-AML1 gene fusions in childhood acute lymphoblastic leukemia. Cancer Res August 15, 1999;59(16):4075–82.
[231] Devaraj PE, Foroni L, Janossy G, Hoffbrand AV, Secker-Walker LM. Expression of the E2A-PBX1 fusion transcripts in t(1;19)(q23;p13) and der(19) t(1;19) at diagnosis and in remission of acute lymphoblastic leukemia with different B lineage immunophenotypes. Leukemia May 1995;9(5):821–5.
[232] Hunger SP. Chromosomal translocations involving the E2A gene in acute lymphoblastic leukemia: clinical features and molecular pathogenesis. Blood February 15, 1996;87(4):1211–24.
[233] Pui CH, Raimondi SC, Hancock ML, Rivera GK, Ribeiro RC, Mahmoud HH, et al. Immunologic, cytogenetic, and clinical characterization of childhood acute lymphoblastic leukemia with the t(1;19) (q23;p13) or its derivative. J Clin Orthod 1994;12(12):2601–6.
[234] Foa R, Vitale A, Mancini M, Cuneo A, Mecucci C, Elia L, et al. E2A-PBX1 fusion in adult acute lymphoblastic leukaemia: biological and clinical features. Br J Haematol February 2003;120(3):484–7.
[235] Hunger SP, Galili N, Carroll AJ, Crist WM, Link MP, Cleary ML. The t(1;19)(q23;p13) results in consistent fusion of E2A and PBX1 coding sequences in acute lymphoblastic leukemias. Blood February 15, 1991;77(4):687–93.
[236] Taki T, Ida K, Bessho F, Hanada R, Kikuchi A, Yamamoto K, et al. Frequency and clinical significance of the MLL gene rearrangements in infant acute leukemia. Leukemia August 1996;10(8):1303–7.
[237] Chen CS, Sorensen PH, Domer PH, Reaman GH, Korsmeyer SJ, Heerema NA, et al. Molecular rearrangements on chromosome 11q23 predominate in infant acute lymphoblastic leukemia and are associated

with specific biologic variables and poor outcome. Blood May 1, 1993;81(9):2386–93.
[238] Rowley JD. Rearrangements involving chromosome band 11Q23 in acute leukaemia. Semin Cancer Biol December 1993;4(6):377–85.
[239] Gu Y, Nakamura T, Alder H, Prasad R, Canaani O, Cimino G, et al. The t(4;11) chromosome translocation of human acute leukemias fuses the ALL-1 gene, related to Drosophila trithorax, to the AF-4 gene. Cell November 13, 1992;71(4):701–8.
[240] Rubnitz JE, Link MP, Shuster JJ, Carroll AJ, Hakami N, Frankel LS, et al. Frequency and prognostic significance of HRX rearrangements in infant acute lymphoblastic leukemia: a Pediatric Oncology Group study. Blood July 15, 1994;84(2):570–3.
[241] Pui CH, Frankel LS, Carroll AJ, Raimondi SC, Shuster JJ, Head DR, et al. Clinical characteristics and treatment outcome of childhood acute lymphoblastic leukemia with the t(4;11)(q21;q23): a collaborative study of 40 cases. Blood February 1, 1991;77(3):440–7.
[242] van der Veer A, Zaliova M, Mottadelli F, De Lorenzo P, te Kronnie G, Harrison CJ, et al. IKZF1 status as a prognostic feature in BCR-ABL1–positive childhood ALL. Blood 2014;123(11):1691–8.
[243] Clark SS, McLaughlin J, Timmons M, Pendergast AM, Ben-Neriah Y, Dow LW, et al. Expression of a distinctive BCR-ABL oncogene in Ph1-positive acute lymphocytic leukemia (ALL). Science February 12, 1988;239(4841 Pt 1):775–7.
[244] Fainstein E, Marcelle C, Rosner A, Canaani E, Gale RP, Dreazen O, et al. A new fused transcript in Philadelphia chromosome positive acute lymphocytic leukaemia. Nature 1987;330(6146):386–8.
[245] Hermans A, Heisterkamp N, von Linden M, van Baal S, Meijer D, van der Plas D, et al. Unique fusion of bcr and c-abl genes in Philadelphia chromosome positive acute lymphoblastic leukemia. Cell October 9, 1987;51(1):33–40.
[246] Harrison CJ, Moorman AV, Broadfield ZJ, Cheung KL, Harris RL, Reza Jalali G, et al. Three distinct subgroups of hypodiploidy in acute lymphoblastic leukaemia. Br J Haematol June 2004;125(5):552–9.
[247] Roberts KG, Li Y, Payne-Turner D, Harvey RC, Yang Y, Pei D, et al. Targetable kinase-activating lesions in ph-like acute lymphoblastic leukemia. N Engl J Med 2014;371(11):1005–15.
[248] Tasian SK, Loh ML, Hunger SP. Philadelphia chromosome–like acute lymphoblastic leukemia. Blood 2017;130(19):2064–72.
[249] Russell LJ, Capasso M, Vater I, Akasaka T, Bernard OA, Calasanz MJ, et al. Deregulated expression of cytokine receptor gene, CRLF2, is involved in lymphoid transformation in B-cell precursor acute lymphoblastic leukemia. Blood September 24, 2009;114(13):2688–98.
[250] Cario G, Zimmermann M, Romey R, Gesk S, Vater I, Harbott J, et al. Presence of the P2RY8-CRLF2 rearrangement is associated with a poor prognosis in non-high-risk precursor B-cell acute lymphoblastic leukemia in children treated according to the ALL-BFM 2000 protocol. Blood July 1, 2010;115(26):5393–7.
[251] Mullighan CG, Collins-Underwood JR, Phillips LA, Loudin MG, Liu W, Zhang J, et al. Rearrangement of CRLF2 in B-progenitor-and Down syndrome-associated acute lymphoblastic leukemia. Nat Genet November 2009;41(11):1243–6.
[252] Yoda A, Yoda Y, Chiaretti S, Bar-Natan M, Mani K, Rodig SJ, et al. Functional screening identifies CRLF2 in precursor B-cell acute lymphoblastic leukemia. Proc Natl Acad Sci U S A 2010;107(1):252–7.
[253] Ravandi F, O'Brien S, Thomas D, Faderl S, Jones D, Garris R, et al. First report of phase 2 study of dasatinib with hyper-CVAD for the frontline treatment of patients with Philadelphia chromosome-positive (Ph+) acute lymphoblastic leukemia. Blood September 23, 2010;116(12):2070–7.
[254] Tosello V, Mansour MR, Barnes K, Paganin M, Sulis ML, Jenkinson S, et al. WT1 mutations in T-ALL. Blood July 30, 2009;114(5):1038–45.
[255] Heerema NA, Sather HN, Sensel MG, Kraft P, Nachman JB, Steinherz PG, et al. Frequency and clinical significance of cytogenetic abnormalities in pediatric T-lineage acute lymphoblastic leukemia: a report from the Children's Cancer Group. J Clin Oncol April 1998;16(4):1270–8.
[256] Garand R, Vannier JP, Béné MC, Faure G, Favre M, Bernard A. Comparison of outcome, clinical, laboratory, and immunological features in 164 children and adults with T-ALL. The Groupe d'Etude Immunologique des Leucémies. Leukemia 1990;4(11):739–44.
[257] McCaw BK, Hecht F, Harnden DG, Teplitz RL. Somatic rearrangement of chromosome 14 in human lymphocytes. Proc Natl Acad Sci U S A June 1975;72(6):2071–5.
[258] Morgan GJ, Walker BA, Davies FE. The genetic architecture of multiple myeloma. Nat Rev Cancer April 12, 2012;12(5):335–48.
[259] De Keersmaecker K, Graux C, Odero MD, Mentens N, Somers R, Maertens J, et al. Fusion of EML1 to ABL1 in T-cell acute lymphoblastic leukemia with cryptic t(9;14)(q34;q32). Blood June 15, 2005;105(12):4849–52.
[260] Erikson J, Williams DL, Finan J, Nowell PC, Croce CM. Locus of the alpha-chain of the T-cell receptor is split by chromosome translocation in T-cell leukemias. Science August 23, 1985;229(4715):784–6.

[261] McCormack MP, Rabbitts TH. Activation of the T-cell oncogene LMO2 after gene therapy for X-linked severe combined immunodeficiency. N Engl J Med February 26, 2004;350(9):913–22.

[262] Shima EA, Le Beau MM, McKeithan TW, Minowada J, Showe LC, Mak TW, et al. Gene encoding the alpha chain of the T-cell receptor is moved immediately downstream of c-myc in a chromosomal 8;14 translocation in a cell line from a human T-cell leukemia. Proc Natl Acad Sci U S A May 1986;83(10):3439–43.

[263] Strehl S, Nebral K, König M, Harbott J, Strobl H, Ratei R, et al. ETV6-NCOA2: a novel fusion gene in acute leukemia associated with coexpression of T-lymphoid and myeloid markers and frequent NOTCH1 mutations. Clin Cancer Res February 15, 2008;14(4):977–83.

[264] Uckun FM, Sensel MG, Sun L, Steinherz PG, Trigg ME, Heerema NA, et al. Biology and treatment of childhood T-lineage acute lymphoblastic leukemia. Blood February 1, 1998;91(3):735–46.

[265] Burnett RC, Thirman MJ, Rowley JD, Diaz MO. Molecular analysis of the T-cell acute lymphoblastic leukemia-associated t(1;7)(p34;q34) that fuses LCK and TCRB. Blood 1994;84(4):1232–6.

[266] Wang HP, Zhou YL, Huang X, Zhang Y, Qian JJ, Li JH, et al. CDKN2A deletions are associated with poor outcomes in 101 adults with T-cell acute lymphoblastic leukemia. Am J Hematol March 1, 2021;96(3):312–9.

[267] Graux C, Cools J, Melotte C, Quentmeier H, Ferrando A, Levine R, et al. Fusion of NUP214 to ABL1 on amplified episomes in T-cell acute lymphoblastic leukemia. Nat Genet October 2004;36(10):1084–9.

[268] Weng AP, Ferrando AA, Lee W, Morris 4th JP, Silverman LB, Sanchez-Irizarry C, et al. Activating mutations of NOTCH1 in human T cell acute lymphoblastic leukemia. Science October 8, 2004;306(5694):269–71.

[269] Girardi T, Vicente C, Cools J, De Keersmaecker K. The genetics and molecular biology of T-ALL. Blood 2017;129(9):1113–23.

[270] Stergianou K, Fox C, Russell NH. Fusion of NUP214 to ABL1 on amplified episomes in T-ALL—implications for treatment. Leukemia 2005;19(9):1680–1.

[271] Gale KB, Ford AM, Repp R, Borkhardt A, Keller C, Eden OB, et al. Backtracking leukemia to birth: identification of clonotypic gene fusion sequences in neonatal blood spots. Proc Natl Acad Sci U S A 1997;94(25):13950–4.

[272] Ford AM, Ridge SA, Cabrera ME, Mahmoud H, Steel CM, Chan LC, et al. In utero rearrangements in the trithorax-related oncogene in infant leukaemias. Nature May 27, 1993;363(6427):358–60.

[273] Mori H, Colman SM, Xiao Z, Ford AM, Healy LE, Donaldson C, et al. Chromosome translocations and covert leukemic clones are generated during normal fetal development. Proc Natl Acad Sci U S A June 11, 2002;99(12):8242–7.

[274] Taub JW, Mundschau G, Ge Y, Poulik JM, Qureshi F, Jensen T, et al. Prenatal origin of GATA1 mutations may be an initiating step in the development of megakaryocytic leukemia in Down syndrome. Blood 2004;104(5):1588–9.

[275] Greaves MF, Maia AT, Wiemels JL, Ford AM. Leukemia in twins: lessons in natural history. Blood October 1, 2003;102(7):2321–33.

[276] Greaves MF, Wiemels J. Origins of chromosome translocations in childhood leukaemia. Nat Rev Cancer September 2003;3(9):639–49.

[277] Wechsler J, Greene M, McDevitt MA, Anastasi J, Karp JE, Le Beau MM, et al. Acquired mutations in GATA1 in the megakaryoblastic leukemia of Down syndrome. Nat Genet September 2002;32(1):148–52.

[278] Zipursky A, Brown EJ, Christensen H, Doyle J. Transient myeloproliferative disorder (transient leukemia) and hematologic manifestations of Down syndrome. Clin Lab Med March 1999;19(1). 157,67, vii.

[279] Iselius L, Jacobs P, Morton N. Leukaemia and transient leukaemia in Down syndrome. Hum Genet October 1990;85(5):477–85.

[280] Nikolaev SI, Santoni F, Vannier A, Falconnet E, Giarin E, Basso G, et al. Exome sequencing identifies putative drivers of progression of transient myeloproliferative disorder to AMKL in infants with Down syndrome. Blood July 25, 2013;122(4):554–61.

[281] Labuhn M, Perkins K, Matzk S, Varghese L, Garnett C, Papaemmanuil E, et al. Mechanisms of progression of myeloid preleukemia to transformed myeloid leukemia in children with down syndrome. Cancer Cell August 12, 2019;36(2). 123,138.e10.

[282] Efremov DG, Turkalj S, Laurenti L. Mechanisms of B Cell receptor activation and responses to B cell receptor inhibitors in B cell malignancies. Cancers 2020;12(6).

[283] Carbone A, Roulland S, Gloghini A, Younes A, von Keudell G, López-Guillermo A, et al. Follicular lymphoma. Nat Rev Dis Prim 2019;5(1):83.

[284] Tsujimoto Y, Cossman J, Jaffe E, Croce CM. Involvement of the bcl-2 gene in human follicular lymphoma. Science June 21, 1985;228(4706):1440–3.

[285] Cleary ML, Sklar J. Nucleotide sequence of a t(14;18) chromosomal breakpoint in follicular lymphoma and

demonstration of a breakpoint-cluster region near a transcriptionally active locus on chromosome 18. Proc Natl Acad Sci U S A November 1985;82(21):7439–43.
[286] Vaux DL, Cory S, Adams JM. Bcl-2 gene promotes haemopoietic cell survival and cooperates with c-myc to immortalize pre-B cells. Nature September 29, 1988;335(6189):440–2.
[287] Morin RD, Johnson NA, Severson TM, Mungall AJ, An J, Goya R, et al. Somatic mutations altering EZH2 (Tyr641) in follicular and diffuse large B-cell lymphomas of germinal-center origin. Nat Genet 2010;42(2):181–5.
[288] Morschhauser F, Tilly H, Chaidos A, McKay P, Phillips T, Assouline S, et al. Tazemetostat for patients with relapsed or refractory follicular lymphoma: an open-label, single-arm, multicentre, phase 2 trial. Lancet Oncol 2020;21(11):1433–42.
[289] Morin RD, Arthur SE, Assouline S. Treating lymphoma is now a bit EZ-er. Blood Adv 2021;5(8):2256–63.
[290] Bisserier M, Wajapeyee N. Mechanisms of resistance to EZH2 inhibitors in diffuse large B-cell lymphomas. Blood 2018;131(19):2125–37.
[291] Chapuy B, Stewart C, Dunford AJ, Kim J, Kamburov A, Redd RA, et al. Molecular subtypes of diffuse large B cell lymphoma are associated with distinct pathogenic mechanisms and outcomes. Nat Med 2018;24(5):679–90.
[292] Liu Y, Barta SK. Diffuse large B-cell lymphoma: 2019 update on diagnosis, risk stratification, and treatment. Am J Hematol May 2019;94(5):604–16.
[293] Zeidler R, Joos S, Delecluse HJ, Klobeck G, Vuillaume M, Lenoir GM, et al. Breakpoints of Burkitt's lymphoma t(8;22) translocations map within a distance of 300 kb downstream of MYC. Genes Chromosomes Cancer April 1994;9(4):282–7.
[294] Pelengaris S, Khan M, Evan G. c-MYC: more than just a matter of life and death. Nat Rev Cancer October 2002;2(10):764–76.
[295] Janz S. Myc translocations in B cell and plasma cell neoplasms. DNA Repair September 8, 2006;5(9–10):1213–24.
[296] Lossos IS. Molecular pathogenesis of diffuse large B-cell lymphoma. J Clin Oncol September 10, 2005;23(26):6351–7.
[297] Sarris A, Ford R. Recent advances in the molecular pathogenesis of lymphomas. Curr Opin Oncol September 1999;11(5):351–63.
[298] Smith SM, Anastasi J, Cohen KS, Godley LA. The impact of MYC expression in lymphoma biology: beyond Burkitt lymphoma. Blood Cells Mol Dis December 15, 2010;45(4):317–23.
[299] Aukema SM, Siebert R, Schuuring E, van Imhoff GW, Kluin-Nelemans HC, Boerma EJ, et al. Double-hit B-cell lymphomas. Blood February 24, 2011;117(8):2319–31.
[300] Boerma EG, Siebert R, Kluin PM, Baudis M. Translocations involving 8q24 in Burkitt lymphoma and other malignant lymphomas: a historical review of cytogenetics in the light of todays knowledge. Leukemia February 2009;23(2):225–34.
[301] Dowdy SF, Hinds PW, Louie K, Reed SI, Arnold A, Weinberg RA. Physical interaction of the retinoblastoma protein with human D cyclins. Cell May 7, 1993;73(3):499–511.
[302] Lee J, Zhang LL, Wu W, Guo H, Li Y, Sukhanova M, et al. Activation of MYC, a bona fide client of HSP90, contributes to intrinsic ibrutinib resistance in mantle cell lymphoma. Blood Adv 2018;2(16):2039–51.
[303] Nadeu F, Martin-Garcia D, Clot G, Díaz-Navarro A, Duran-Ferrer M, Navarro A, et al. Genomic and epigenomic insights into the origin, pathogenesis, and clinical behavior of mantle cell lymphoma subtypes. Blood 2020;136(12):1419–32.
[304] Li JY, Gaillard F, Moreau A, Harousseau JL, Laboisse C, Milpied N, et al. Detection of translocation t(11;14)(q13;q32) in mantle cell lymphoma by fluorescence in situ hybridization. Am J Pathol May 1999;154(5):1449–52.
[305] Rinaldi A, Mian M, Chigrinova E, Arcaini L, Bhagat G, Novak U, et al. Genome-wide DNA profiling of marginal zone lymphomas identifies subtype-specific lesions with an impact on the clinical outcome. Blood 2011;117(5):1595–604.
[306] Linet MS, Vajdic CM, Morton LM, de Roos AJ, Skibola CF, Boffetta P, et al. Medical history, lifestyle, family history, and occupational risk factors for follicular lymphoma: the InterLymph Non-Hodgkin Lymphoma Subtypes Project. J Natl Cancer Inst Monogr August 2014;2014(48):26–40.
[307] Goldin LR, Björkholm M, Kristinsson SY, Turesson I, Landgren O. Highly increased familial risks for specific lymphoma subtypes. Br J Haematol 2009;146(1):91–4.
[308] Younes A, Sehn LH, Johnson P, Zinzani PL, Hong X, Zhu J, et al. Randomized phase III trial of ibrutinib and rituximab plus cyclophosphamide, doxorubicin, vincristine, and prednisone in non–germinal center B-cell diffuse large B-cell lymphoma. J Clin Orthod 2019;37(15):1285–95.
[309] Bartlett NL, Costello BA, LaPlant BR, Ansell SM, Kuruvilla JG, Reeder CB, et al. Single-agent ibrutinib in

relapsed or refractory follicular lymphoma: a phase 2 consortium trial. Blood 2018;131(2):182–90.
[310] Woyach JA, Ruppert AS, Guinn D, Lehman A, Blachly JS, Lozanski A, et al. BTKC481S-Mediated resistance to ibrutinib in chronic lymphocytic leukemia. J Clin Orthod 2017;35(13):1437–43.
[311] Hucks G, Rheingold SR. The journey to CAR T cell therapy: the pediatric and young adult experience with relapsed or refractory B-ALL. Blood Cancer J 2019;9(2):10.
[312] June CH, O'Connor RS, Kawalekar OU, Ghassemi S, Milone MC. CAR T cell immunotherapy for human cancer. Science March 23, 2018;359(6382):1361–5.
[313] Hurwitz SN, Bagg A. A 2020 vision into Hodgkin lymphoma biology. Adv Anat Pathol September 2020;27(5):269–77.
[314] Mata E, Fernández S, Astudillo A, Fernández R, García-Cosío M, Sánchez-Beato M, et al. Genomic analyses of microdissected Hodgkin and Reed-Sternberg cells: mutations in epigenetic regulators and p53 are frequent in refractory classic Hodgkin lymphoma. Blood Cancer J 2019;9(3):34.
[315] Decker T, Schneller F, Kronschnabl M, Dechow T, Lipford GB, Wagner H, et al. Immunostimulatory CpG-oligonucleotides induce functional high affinity IL-2 receptors on B-CLL cells: costimulation with IL-2 results in a highly immunogenic phenotype. Exp Hematol May 2000;28(5):558–68.
[316] Dicker F, Schnittger S, Haferlach T, Kern W, Schoch C. Immunostimulatory oligonucleotide-induced metaphase cytogenetics detect chromosomal aberrations in 80% of CLL patients: a study of 132 CLL cases with correlation to FISH, IgVH status, and CD38 expression. Blood November 1, 2006;108(9):3152–60.
[317] Döhner H, Stilgenbauer S, Benner A, Leupolt E, Kröber A, Bullinger L, et al. Genomic aberrations and survival in chronic lymphocytic leukemia. N Engl J Med 2000;343(26):1910–6.
[318] Van Dyke DL, Werner L, Rassenti LZ, Neuberg D, Ghia E, Heerema NA, et al. The Dohner fluorescence in situ hybridization prognostic classification of chronic lymphocytic leukaemia (CLL): the CLL Research Consortium experience. Br J Haematol 2016;173(1):105–13.
[319] Ouillette P, Collins R, Shakhan S, Li J, Li C, Shedden K, et al. The prognostic significance of various 13q14 deletions in chronic lymphocytic leukemia. Clin Cancer Res 2011;17(21):6778–90.
[320] Dal Bo M, Rossi FM, Rossi D, Deambrogi C, Bertoni F, Del Giudice I, et al. 13q14 Deletion size and number of deleted cells both influence prognosis in chronic lymphocytic leukemia. Gene Chromosome Cancer 2011;50(8):633–43.
[321] Hernández JÁ, Rodríguez AE, González M, Benito R, Fontanillo C, Sandoval V, et al. A high number of losses in 13q14 chromosome band is associated with a worse outcome and biological differences in patients with B-cell chronic lymphoid leukemia. Haematol 2009;94(3):364–71.
[322] Chena C, Avalos JS, Bezares RF, Arrossagaray G, Turdó K, Bistmans A, et al. Biallelic deletion 13q14.3 in patients with chronic lymphocytic leukemia: cytogenetic, FISH and clinical studies. Eur J Haematol August 2008;81(2):94–9.
[323] Döhner H, Stilgenbauer S, Döhner K, Bentz M, Lichter P. Chromosome aberrations in B-cell chronic lymphocytic leukemia: reassessment based on molecular cytogenetic analysis. J Mol Med (Berl) February 1999;77(2):266–81.
[324] Döhner H, Fischer K, Bentz M, Hansen K, Benner A, Cabot G, et al. p53 gene deletion predicts for poor survival and non-response to therapy with purine analogs in chronic B-cell leukemias. Blood March 15, 1995;85(6):1580–9.
[325] Stankovic T, Weber P, Stewart G, Bedenham T, Murray J, Byrd PJ, et al. Inactivation of ataxia telangiectasia mutated gene in B-cell chronic lymphocytic leukaemia. Lancet January 2, 1999;353(9146):26–9.
[326] Senouci A, Smol T, Tricot S, Bakala J, Moulessehoul S, Quilichini B, et al. Cytogenetic landscape in 1012 newly diagnosed chronic lymphocytic leukemia. Eur J Haematol December 2019;103(6):607–13.
[327] Landau DA, Tausch E, Taylor-Weiner A, Stewart C, Reiter JG, Bahlo J, et al. Mutations driving CLL and their evolution in progression and relapse. Nature 2015;526(7574):525–30.
[328] Hallek M, Shanafelt TD, Eichhorst B. Chronic lymphocytic leukaemia. Lancet April 14, 2018;391(10129):1524–37.
[329] Cerhan JR, Slager SL. Familial predisposition and genetic risk factors for lymphoma. Blood November 12, 2015;126(20):2265–73.
[330] Jain N, Keating M, Thompson P, Ferrajoli A, Burger J, Borthakur G, et al. Ibrutinib and venetoclax for first-line treatment of CLL. N Engl J Med 2019;380(22):2095–103.

[331] Porter DL, Levine BL, Kalos M, Bagg A, June CH. Chimeric antigen receptor-modified T cells in chronic lymphoid leukemia. N Engl J Med August 25, 2011;365(8):725–33.

[332] Landgren O, Kyle RA, Pfeiffer RM, Katzmann JA, Caporaso NE, Hayes RB, et al. Monoclonal gammopathy of undetermined significance (MGUS) consistently precedes multiple myeloma: a prospective study. Blood May 28, 2009;113(22):5412–7.

[333] Matsui W, Wang Q, Barber JP, Brennan S, Smith BD, Borrello I, et al. Clonogenic multiple myeloma progenitors, stem cell properties, and drug resistance. Cancer Res January 01, 2008;68(1):190–7.

[334] Fonseca R, Bergsagel PL, Drach J, Shaughnessy J, Gutierrez N, Stewart AK, et al. International Myeloma Working Group molecular classification of multiple myeloma: spotlight review. Leukemia December 2009;23(12):2210–21.

[335] Walker BA, Mavrommatis K, Wardell CP, Ashby TC, Bauer M, Davies FE, et al. Identification of novel mutational drivers reveals oncogene dependencies in multiple myeloma. Blood August 09, 2018;132(6):587–97.

[336] Reece D, Song KW, Fu T, Roland B, Chang H, Horsman DE, et al. Influence of cytogenetics in patients with relapsed or refractory multiple myeloma treated with lenalidomide plus dexamethasone: adverse effect of deletion 17p13. Blood July 16, 2009;114(3):522–5.

[337] Kuehl WM, Bergsagel PL. Multiple myeloma: evolving genetic events and host interactions. Nat Rev Cancer March 2002;2(3):175–87.

[338] Fonseca R, Bailey RJ, Ahmann GJ, Rajkumar SV, Hoyer JD, Lust JA, et al. Genomic abnormalities in monoclonal gammopathy of undetermined significance. Blood August 15, 2002;100(4):1417–24.

[339] Avet-Loiseau H, Facon T, Grosbois B, Magrangeas F, Rapp M, Harousseau J, et al. Oncogenesis of multiple myeloma: 14q32 and 13q chromosomal abnormalities are not randomly distributed, but correlate with natural history, immunological features, and clinical presentation. Blood March 15, 2002;99(6):2185–91.

[340] Smadja NV, Bastard C, Brigaudeau C, Leroux D, Fruchart C. Hypodiploidy is a major prognostic factor in multiple myeloma. Blood January 10, 2001;98(7):2229–38.

[341] Magrangeas F, Lodé L, Wuilleme S, Minvielle S, Avet-Loiseau H. Genetic heterogeneity in multiple myeloma. Leukemia February 2005;19(2):191–4.

[342] Kalakonda N, Rothwell DG, Scarffe JH, Norton JD. Detection of N-Ras codon 61 mutations in subpopulations of tumor cells in multiple myeloma at presentation. Blood January 09, 2001;98(5):1555–60.

[343] Mulligan G, Lichter DI, Di Bacco A, Blakemore SJ, Berger A, Koenig E, et al. Mutation of NRAS but not KRAS significantly reduces myeloma sensitivity to single-agent bortezomib therapy. Blood January 30, 2014;123(5):632–9.

[344] Bergsagel PL, Kuehl WM, Zhan F, Sawyer J, Barlogie B, Shaughnessy J. Cyclin D dysregulation: an early and unifying pathogenic event in multiple myeloma. Blood July 01, 2005;106(1):296–303.

[345] John L, Krauth MT, Podar K, Raab M. Pathway-directed therapy in multiple myeloma. Cancers 2021;13(7).

[346] Chapman MA, Lawrence MS, Keats JJ, Cibulskis K, Sougnez C, Schinzel AC, et al. Initial genome sequencing and analysis of multiple myeloma. Nature March 24, 2011;471(7339):467–72.

[347] Xu J, Pfarr N, Endris V, Mai EK, Md Hanafiah NH, Lehners N, et al. Molecular signaling in multiple myeloma: association of RAS/RAF mutations and MEK/ERK pathway activation. Oncogenesis May 15, 2017;6(5):e337.

[348] Kortüm KM, Mai EK, Hanafiah NH, Shi C, Zhu Y, Bruins L, et al. Targeted sequencing of refractory myeloma reveals a high incidence of mutations in CRBN and Ras pathway genes. Blood September 01, 2016;128(9):1226–33.

[349] Holien T, Våtsveen TK, Hella H, Waage A, Sundan A. Addiction to c-MYC in multiple myeloma. Blood September 20, 2012;120(12):2450–3.

[350] Hanamura I, Stewart JP, Huang Y, Zhan F, Santra M, Sawyer JR, et al. Frequent gain of chromosome band 1q21 in plasma-cell dyscrasias detected by fluorescence in situ hybridization: incidence increases from MGUS to relapsed myeloma and is related to prognosis and disease progression following tandem stem-cell transplantation. Blood September 01, 2006;108(5):1724–32.

[351] Misund K, Keane N, Stein CK, Asmann YW, Day G, Welsh S, et al. MYC dysregulation in the progression of multiple myeloma. Leukemia January 2020;34(1):322–6.

[352] Iwanaga M, Tagawa M, Tsukasaki K, Matsuo T, Yokota K, Miyazaki Y, et al. Relationship between monoclonal gammopathy of undetermined significance and radiation exposure in Nagasaki atomic bomb survivors. Blood February 19, 2009;113(8):1639–50.

[353] Pertesi M, Went M, Hansson M, Hemminki K, Houlston RS, Nilsson B. Genetic predisposition for multiple myeloma. Leukemia March 2020;34(3):697–708.

[354] Palumbo A, Avet-Loiseau H, Oliva S, Lokhorst HM, Goldschmidt H, Rosinol L, et al. Revised international staging system for multiple myeloma: a report from

international myeloma working group. J Clin Oncol September 10, 2015;33(26):2863–9.
[355] Avet-Loiseau H, Hulin C, Campion L, Rodon P, Marit G, Attal M, et al. Chromosomal abnormalities are major prognostic factors in elderly patients with multiple myeloma: the intergroupe francophone du myélome experience. J Clin Oncol August 01, 2013;31(22):2806–9.
[356] Sonneveld P, Avet-Loiseau H, Lonial S, Usmani S, Siegel D, Anderson KC, et al. Treatment of multiple myeloma with high-risk cytogenetics: a consensus of the International Myeloma Working Group. Blood June 16, 2016;127(24):2955–62.
[357] Chang H, Sloan S, Li D, Zhuang L, Yi Q, Chen CI, et al. The t(4;14) is associated with poor prognosis in myeloma patients undergoing autologous stem cell transplant. Br J Haematol April 2004;125(1):64–8.
[358] San Miguel JF, Schlag R, Khuageva NK, Dimopoulos MA, Shpilberg O, Kropff M, et al. Bortezomib plus melphalan and prednisone for initial treatment of multiple myeloma. N Engl J Med 2008;359(9):906–17.
[359] Drach J, Ackermann J, Fritz E, Krömer E, Schuster R, Gisslinger H, et al. Presence of a p53 gene deletion in patients with multiple myeloma predicts for short survival after conventional-dose chemotherapy. Blood August 01, 1998;92(3):802–9.
[360] Corre J, Perrot A, Caillot D, Belhadj K, Hulin C, Leleu X, et al. del(17p) without TP53 mutation confers a poor prognosis in intensively treated newly diagnosed patients with multiple myeloma. Blood March 04, 2021;137(9):1192–5.
[361] Pawlyn C, Melchor L, Murison A, Wardell CP, Brioli A, Boyle EM, et al. Coexistent hyperdiploidy does not abrogate poor prognosis in myeloma with adverse cytogenetics and may precede IGH translocations. Blood January 29, 2015;125(5):831–40.
[362] Hebraud B, Leleu X, Lauwers-Cances V, Roussel M, Caillot D, Marit G, et al. Deletion of the 1p32 region is a major independent prognostic factor in young patients with myeloma: the IFM experience on 1195 patients. Leukemia March 2014;28(3):675–9.
[363] Caltagirone S, Ruggeri M, Aschero S, Gilestro M, Oddolo D, Gay F, et al. Chromosome 1 abnormalities in elderly patients with newly diagnosed multiple myeloma treated with novel therapies. Haematologica October 2014;99(10):1611–7.
[364] Giri S, Huntington SF, Wang R, Zeidan AM, Podoltsev N, Gore SD, et al. Chromosome 1 abnormalities and survival of patients with multiple myeloma in the era of novel agents. Blood Adv May 22, 2020;4(10):2245–53.
[365] Abdallah N, Baughn LB, Rajkumar SV, Kapoor P, Gertz MA, Dispenzieri A, et al. Implications of MYC rearrangements in newly diagnosed multiple myeloma. Clin Cancer Res December 15, 2020;26(24):6581–8.
[366] Touzeau C, Maciag P, Amiot M, Moreau P. Targeting Bcl-2 for the treatment of multiple myeloma. Leukemia September 2018;32(9):1899–907.
[367] Kumar S, Kaufman JL, Gasparetto C, Mikhael J, Vij R, Pegourie B, et al. Efficacy of venetoclax as targeted therapy for relapsed/refractory t(11;14) multiple myeloma. Blood November 30, 2017;130(22):2401–9.
[368] Costa LJ, Davies FE, Monohan GP, Kovacsovics T, Burwick N, Jakubowiak A, et al. Phase 2 study of venetoclax plus carfilzomib and dexamethasone in patients with relapsed/refractory multiple myeloma. Blood Adv October 12, 2021;5(19):3748–59.
[369] Moreau P, Chanan-Khan A, Roberts AW, Agarwal AB, Facon T, Kumar S, et al. Promising efficacy and acceptable safety of venetoclax plus bortezomib and dexamethasone in relapsed/refractory MM. Blood November 30, 2017;130(22):2392–400.
[370] Swerdlow SH, Campo E, Pileri SA, Harris NL, Stein H, Siebert R, et al. The 2016 revision of the World Health Organization classification of lymphoid neoplasms. Blood May 19, 2016;127(20):2375–90.
[371] Braggio E, Philipsborn C, Novak A, Hodge L, Ansell S, Fonseca R. Molecular pathogenesis of Waldenström's macroglobulinemia. 1 September 01, 2012;97(9):1281–90.
[372] Yun S, Johnson AC, Okolo ON, Arnold SJ, McBride A, Zhang L, et al. Waldenström macroglobulinemia: review of pathogenesis and management. Clin Lymphoma, Myeloma Leukemia May 1, 2017;17(5):252–62.
[373] Hunter ZR, Yang G, Xu L, Liu X, Castillo JJ, Treon SP. Genomics, signaling, and treatment of Waldenström macroglobulinemia. J Clin Orthod March 20, 2017;35(9):994–1001.
[374] Treon SP, Gustine J, Xu L, Manning RJ, Tsakmaklis N, Demos M, et al. MYD88 wild-type Waldenstrom Macroglobulinaemia: differential diagnosis, risk of histological transformation, and overall survival. Br J Haematol February 2018;180(3):374–80.
[375] Krzisch D, Guedes N, Boccon-Gibod C, Baron M, Bravetti C, Davi F, et al. Cytogenetic and molecular abnormalities in Waldenström's macroglobulinemia patients: correlations and prognostic impact. Am J Hematol 2021;96(12):1569–79.

[376] Nguyen-Khac F, Lambert J, Chapiro E, Grelier A, Mould S, Barin C, et al. Chromosomal aberrations and their prognostic value in a series of 174 untreated patients with Waldenström's macroglobulinemia. 1 April 01, 2013;98(4):649–54.

[377] Treon SP, Xu L, Guerrera ML, Jimenez C, Hunter ZR, Liu X, et al. Genomic landscape of Waldenström macroglobulinemia and its impact on treatment strategies. J Clin Orthod 2020;38(11):1198–208.

[378] Treon SP, Xu L, Hunter Z. MYD88 mutations and response to ibrutinib in waldenström's macroglobulinemia. N Engl J Med August 06, 2015;373(6):584–6.

[379] Xu L, Hunter ZR, Yang G, Zhou Y, Cao Y, Liu X, et al. MYD88 L265P in Waldenström macroglobulinemia, immunoglobulin M monoclonal gammopathy, and other B-cell lymphoproliferative disorders using conventional and quantitative allele-specific polymerase chain reaction. Blood March 14, 2013;121(11):2051–8.

[380] Xu L, Hunter ZR, Tsakmaklis N, Cao Y, Yang G, Chen J, et al. Clonal architecture of CXCR4 WHIM-like mutations in Waldenström macroglobulinaemia. Br J Haematol 2016;172(5):735–44.

[381] Cao Y, Hunter ZR, Liu X, Xu L, Yang G, Chen J, et al. The WHIM-like CXCR4(S338X) somatic mutation activates AKT and ERK, and promotes resistance to ibrutinib and other agents used in the treatment of Waldenstrom's macroglobulinemia. Leukemia January 2015;29(1):169–76.

[382] Treon SP, Gustine J, Meid K, Yang G, Xu L, Liu X, et al. Ibrutinib monotherapy in symptomatic, treatment-naïve patients with Waldenström macroglobulinemia. J Clin Orthod September 20, 2018;36(27):2755–61.

[383] Treon SP, Cao Y, Xu L, Yang G, Liu X, Hunter ZR. Somatic mutations in MYD88 and CXCR4 are determinants of clinical presentation and overall survival in Waldenström macroglobulinemia. Blood May 1, 2014;123(18):2791–6.

[384] Dimopoulos MA, Trotman J, Tedeschi A, Matous JV, Macdonald D, Tam C, et al. Ibrutinib for patients with rituximab-refractory Waldenström's macroglobulinaemia (iNNOVATE): an open-label substudy of an international, multicentre, phase 3 trial. Lancet Oncol February 1, 2017;18(2):241–50.

[385] Davids MS, Roberts AW, Seymour JF, Pagel JM, Kahl BS, Wierda WG, et al. Phase I first-in-human study of venetoclax in patients with relapsed or refractory non-hodgkin lymphoma. J Clin Orthod 2017;35(8):826–33.

[386] Hughes TJ, Kaeda J, Branford S, Rudzki Z, Hochhaus A, Hensley ML, et al., International Randomised Study of Interferon versus STI571 (IRIS) Study Group. Frequency of major molecular responses to imatinib or interferon alfa plus cytarabine in newly diagnosed chronic myeloid leukemia. N Engl J Med 2003;349:1423–32.

[387] Marin D, Milojkovic D, Olavarria E, Khorashad JS, de Lavallade H, Reid AG, et al. European LeukemiaNet criteria for failure or suboptimal response reliably identify patients with CML in early chronic phase treated with imatinib whose eventual outcome is poor. Blood 2008;112:4437–44.

[388] Cowan-Jacob SW, Fendrich G, Floersheimer A, Furet P, Liebetanz J, Rummel G, et al. Structural biology contributions to the discovery of drugs to treat chronic myelogenous leukaemia. Acta Crystallogr D Biol Crystallogr 2007;63:80–93.

[389] Levinson NM, Boxer SG. Structural and spectroscopic analysis of the kinase inhibitor bosutinib and an isomer of bosutinib binding to the Abl tyrosine kinase domain. PLoS One 2012;7.

[390] O'Hare T, Shakespeare WC, Zhu X, Eide CA, Rivera VM, Wang F, et al. AP24534, a pan-BCR-ABL inhibitor for chronic myeloid leukemia, potently inhibits the T315I mutant and overcomes mutation-based resistance. Cancer Cell 2009;16:401–12.

[391] Tokarski JS, Newitt JA, Chang CY, Cheng JD, Wittekind M, Kiefer SE, et al. The structure of Dasatinib (BMS-354825) bound to activated ABL kinase domain elucidates its inhibitory activity against imatinib-resistant ABL mutants. Cancer Res 2006;66:5790–7.

[392] Weisberg E, Manley PW, Breitenstein W, Brüggen J, Cowan-Jacob SW, Ray A. Characterization of AMN107, a selective inhibitor of native and mutant Bcr-Abl. Cancer Cell. Cancer Cell 2005;7:129–41.

[393] Wylie AA, Schoepfer J, Jahnke W, Cowan-Jacob SW, Loo A, Furet P, et al. The allosteric inhibitor ABL001 enables dual targeting of BCR-ABL1. Nature 2017;543:733–7.

PART III

Immunologic Disorders

10

Inherited Complement Deficiencies

Kathleen E. Sullivan
Division of Allergy Immunology, The Children's Hospital of Philadelphia, University of Pennsylvania Perelman School of Medicine, Philadelphia, PA, United States

GLOSSARY

AH50 The assay is used to define the dilution of serum capable of lysing 50% of nonsensitized rabbit red cells. This assay measures the intactness of the alternative pathway through the terminal components.

Alternative Pathway Factor B, Factor D, properdin, and the terminal components.

Anaphylatoxins C3a, C4a, C5a. These are mediators of smooth muscle contraction, degranulation of mast cells, enhanced neutrophil aggregation, increased vascular permeability.

C3 tick-over This term is occasionally used to describe spontaneous C3 hydrolysis.

CH50 The assay is used to define the dilution of serum capable of lysing 50% of the sensitized sheep red cells. This assay measures the intactness of the classical pathway through the terminal components.

Classical Pathway C1, C4, C2, C3, and the terminal components.

Lectin activation pathway MBL, MASP1, MASP2, C3, and the terminal components.

Membrane Attack Complex (terminal components) C5, C6, C7, C8, C9.

Opsonization Renders a particle more easily phagocytosed.

10.1 INTRODUCTION

The complement system is a group of evolutionarily ancient proteins comprising the central cascade and multiple regulatory proteins. Seven receptors mediate many of the biological functions of the complement ligands. Most complement proteins are produced in the liver, although C1q, properdin, and C7 are produced predominantly by myeloid cells, and factor D is produced by adipocytes [1–5]. Many other cells produce small amounts of complement components after proinflammatory stimuli, and this is thought to magnify the local response to infection [6]. The major functions of the complement system are host defense, protection of endothelial surfaces, and waste clearance. Each of these functions depends on a slightly different subset of complement proteins. The roles of the specific proteins have been largely elucidated through the study of inborn errors affecting the individual components.

Complement nomenclature follows certain patterns with the classical pathway components generally indicated with an upper case C followed by a number that roughly correlates with the position in the cascade (C4 appears out of order). Alternative pathway members are generally referred to as a "factor" and are designated with a letter (factor B, factor D, factor H). As protein fragments are cleaved off, they are given lower case letter identifiers with the "a" most often designating the

Emery and Rimoin's Principles and Practice of Medical Genetics and Genomics. https://doi.org/10.1016/B978-0-12-812534-2.00001-1

smaller fragment (the exception is C2a which is larger than C2b). In some cases, the two fragments can be further cleaved, and those smaller fragments are named with additional lower case letters. When a cleavage product is inactive, it is proceeded by the letter "i". Protein complexes with enzymatic activity are termed convertases, i.e., C3 convertase and C5 convertase.

10.2 INTRODUCTION TO THE COMPLEMENT SYSTEM

The complement cascade originally evolved to opsonize pathogens and enhance their phagocytosis. The most evolutionarily ancient members of the complement cascade are factor B and C3, the minimal components needed to effect opsonization. As the adaptive immune system evolved, the complement system coevolved such that it now interfaces with nearly all aspects of host defense. A model for the organization of the complement cascade has three activation arms: the classical pathway, the lectin activation pathway, and the alternative pathway (Fig. 10.1). These three pathways converge at the central protein, C3, and allow it to bind to the nearest surface, usually a pathogen. The lectin activation pathway and the alternative pathway are truly part of the innate immune system, but the classical pathway is activated typically by antigen—antibody complexes and therefore is dependent on the adaptive immune system. C3 is appropriately central to any discussion of complement as it constitutes a powerful opsonin and markedly enhances phagocytosis of the pathogen. It also functions as an important costimulatory molecule for B cells. Cleavage of C3 leads to formation of a C5 convertase and activation of the terminal components, which catalyze the formation of a pore in suitable membranes.

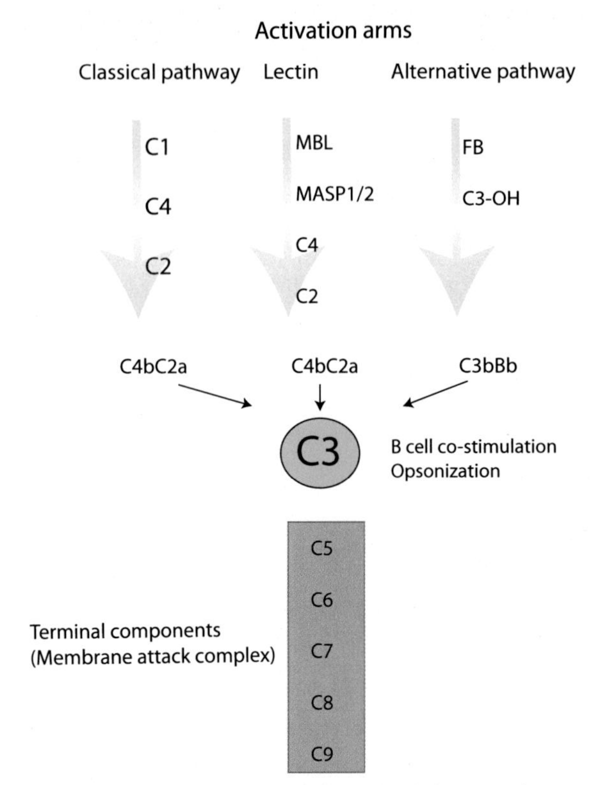

Figure 10.1 The three activation arms of the complement system. Activation of C3 is important for opsonization and B cell costimulation. These two major functions of host defense are dependent on C3 and its appropriate activation. The terminal components or membrane attack complex is required for lysis of bacteria.

10.3 THE CLASSICAL PATHWAY

The classical pathway is activated primarily by immune complexes. A conformational change occurs when antibody binds antigen, which renders the antibody molecule capable of interacting with C1 [7,8]. Only IgG and IgM activate complement and IgM is much more efficient than IgG. In addition, not all isotypes of IgG are equivalent. IgG3 is the most efficient followed by IgG1 and IgG2. IgG4 is not able to activate complement. A single molecule of IgM is sufficient to activate complement while many molecules of IgG bound to a particle are required to activate complement. In addition to immune complexes, the classical pathway is activated by apoptotic debris and a variety of other proteins and nucleic acids. Once activated, C1, C4, and C2 interact to produce a C3 convertase: C4aC2b. Cleavage products resulting from this activation pathway lead to inflammation with erythema (vasodilatation) and edema (vascular leak).

10.4 THE ALTERNATIVE PATHWAY

The alternative pathway is generally activated through the recognition of oligosaccharide and charge differences common to pathogens. The alternative pathway exploits the instability of the native C3 molecule and, on activator surfaces, nucleates a complex of C3(H_2O) Bb, which cleaves additional C3 to opsonize bacteria and initiate activation of the terminal components via the mature enzyme C3bBbC3b. Note that the classical pathway provides an important substrate for the alternative pathway, C3b.

The basis of this pathway relies on spontaneous hydrolysis of C3, which occurs in the serum at a rate of 0.2%–0.4% per hour [9]. The hydrolyzed C3 undergoes a conformational change that facilitates interaction with factor B. When factor B is bound to hydrolyzed C3, it is cleaved by factor D, which is in turn activated by MASP-3. C3bBb is the alternative pathway C3 converting enzyme, which is stabilized by properdin and cleaves additional C3 into C3b and C3a, leading to the C5 convertase C3bBbC3b [10]. The regulation of this pathway is largely through the effect of factor H on surfaces. A nonactivator surface (usually our own cells) binds factor H avidly due rich sialic acid residues [11–13]. In contrast, on activator surfaces (pathogens), factor H cannot displace factor B from C3b, and the alternative pathway activation is allowed to proceed. Activator surfaces are often coated with mannose or N-acetyl glucosamine.

10.5 THE LECTIN ACTIVATION PATHWAY

Components of the lectin pathway include mannose-binding lectin (MBL), H-ficolin (ficolin-3), L-ficolin (ficolin-2), M-ficolin (ficolin-1), and the serine proteases (MASP1, MASP2). Collectins can also interface with the lectin pathway. The pattern recognition components, MBL, ficolins, and collectins, recognize arrays of carbohydrates and acetyl groups on pathogens. MBL recognizes oligosaccharides specific to pathogens in a manner similar to that of factor H discussed above [14–16]. Mammalian glycoproteins are generally decorated with galactose and sialic acid, not recognized by MBL. In contrast, MBL avidly binds to oligosaccharides associated with bacteria, yeast, and parasites such as mannose, N-acetyl-glucosamine, fucose, and glucose [14]. MBL also binds to agalactosyl IgG with high affinity [17]. This unusual IgG is produced primarily at times of inflammation, and this antibody would therefore amplify the complement activation at sites of inflammation.

MBL undergoes a conformation change after engaging carbohydrate, leading to activation of MASP1 and MASP2. Both activated MASP1 and MASP2 can cleave C2, but C4 is cleaved only by MASP-2. The end result of the lectin activation pathway is the same C3 convertase as the classical pathway, namely C4bC2a.

10.6 THE MEMBRANE ATTACK COMPLEX

Once C3 is cleaved by any of the activation arms described above, it becomes a part of the next enzymatic complex and cleaves C5. The large cleavage fragment of C5 becomes attached to the surface of the pathogen. C5b binds to C6 and C7 and inserts into a lipid membrane [18,19]. C5b also binds directly to C8, which then becomes incorporated into the complex. Once this happens, the membrane integrity is compromised. The addition of C9 leads to the formation of true pore with enhanced stability [20].

10.7 REGULATION OF COMPLEMENT ACTIVATION

The regulators of complement are divided into fluid phase regulators and membrane-bound regulators (Table 10.1). C1 inhibitor is perhaps the most clinically important of the regulatory proteins, and it is a fluid phase regulator. C1 inhibitor is a serine protease, which inhibits the low-level autoactivation of C1 (Fig. 10.2) [21]. Immune complex activation of C1 is preserved. C1 inhibitor directly binds to C1s and C1r, leading to dissociation from C1q. C1 inhibitor has a similar function in the lectin activation pathway. Relevant for the clinical manifestations of C1 inhibitor deficiency are its roles in the coagulation pathway. C1 inhibitor has important roles inhibiting factor XII (Hageman factor) and prekallikrein [22,23].

The other important fluid phase regulators are C4-binding protein, which displaces C2a and inhibits cleavage of C3 [24,25]. Factor I and factor H regulate the alternative pathway (Fig. 10.3) [26–29]. Factor I inactivates C3b, and its activity is enhanced by factor H. Factor H identifies nonactivator surfaces though the recognition of mammalian oligosaccharides and displaces Bb from C3b on those surfaces. These two regulators are critical in the prevention of significant spontaneous activation of the alternative pathway. Factor I also acts to inhibit C4b from the classical pathway [29,30].

The membrane-bound regulators of complement consist of decay accelerating factor (DAF, CD55), membrane cofactor protein (MCP, CD46), and CD59 [29,31–36]. DAF dissociates the C3 convertase,

TABLE 10.1 Complement Regulatory Proteins

Protein	Gene	Localization	Function
C1 inhibitor	*SERPING1*	Serum	Binds to C1r and C1s and dissociates the C1 complex
C4 binding protein	*C4BPA*	Serum	Cofactor for factor I cleavage of C4b
Factor I	*CFI*	Serum	Cleaves C3b and C4b
Factor H	*CFH*	Serum	Defines activator surface
Decay accelerating factor (DAF)	*CD55*	Ubiquitous-cell membrane	Dissociates both C3 and C5 convertases
Membrane cofactor protein (MCP)	*CD46*	Hematopoietic cells except erythrocytes	Cofactor for C3b cleavage by factor I
CD59	*CD59*	Hematopoietic cells, endothelial cells, epithelial cells, glomerular cells	Inhibits the membrane attack complex

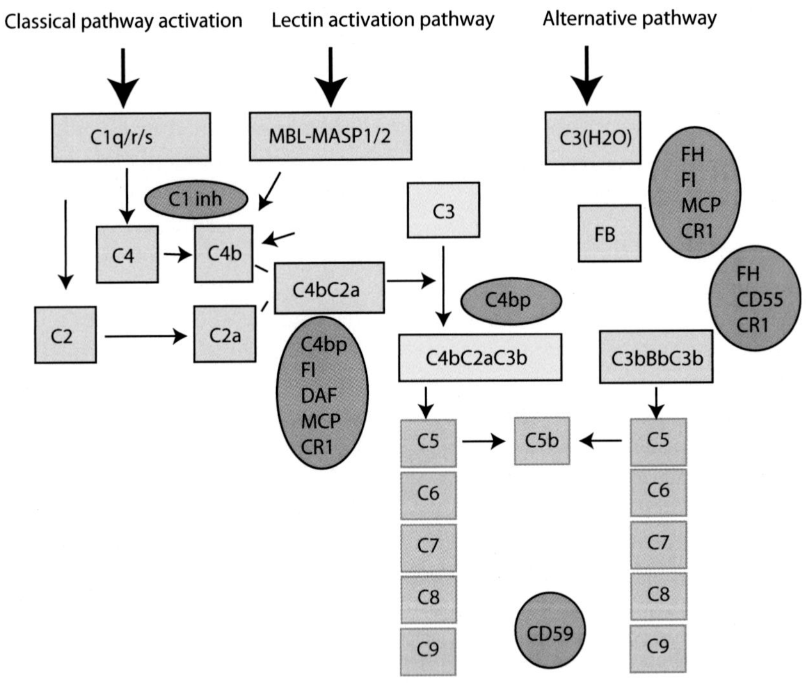

Figure 10.2 The regulation of the complement cascade. The activation pathways are schematized with the relevant regulatory proteins shown as red ovals.

and MCP serves as a cofactor for factor I cleavage of C3b and C4b. CD59 inhibits the membrane attack complex through inhibition of C9 binding. Several receptors terminate complement function and have a regulatory function such as CR1-binding C3b and C4b and serving as a cofactor for factor I mediated cleavage. CR2 has a similar role supporting cleavage of C3b.

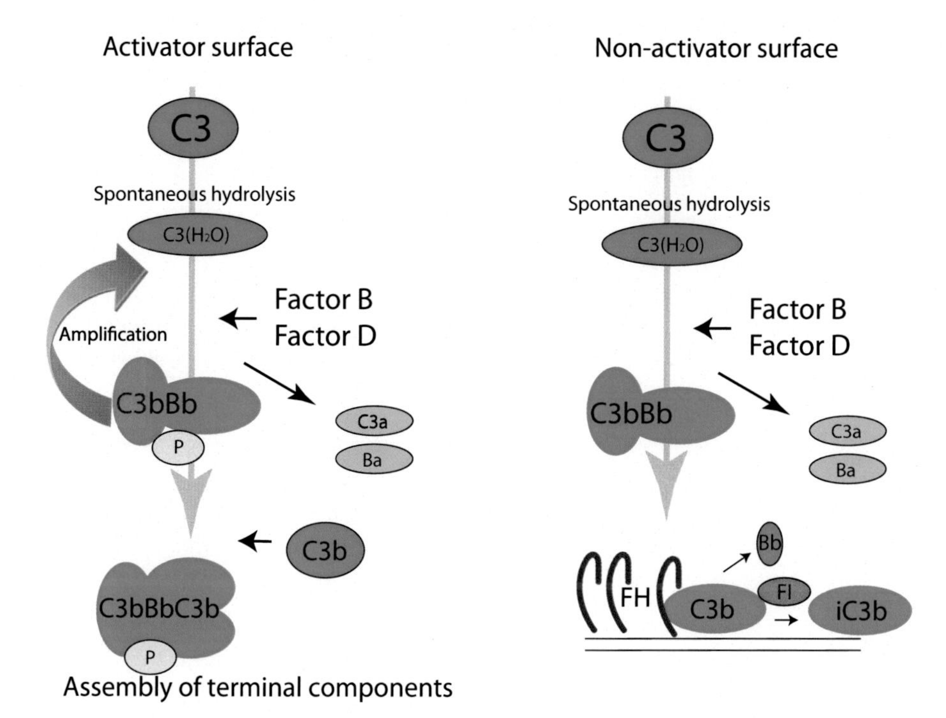

Figure 10.3 **The alternative pathway.** The alternative activation pathway is schematized showing the key difference between a nonactivator surface, i.e., human tissues, and an activator surface, i.e., pathogen surfaces. Factor H on nonactivator surfaces is central to prevention of C3 activation.

10.8 INHERITED COMPLEMENT DEFICIENCIES

10.8.1 Overview

Most inherited complement deficiencies are autosomal recessive and except where noted, the mutations are diverse (Table 10.2). Two key exceptions are properdin deficiency, which is X-linked and C1 inhibitor deficiency, which is autosomal dominant (hemizygous). The defects associated with susceptibility to atypical hemolytic uremic syndrome (aHUS) are particularly diverse, with activating and inactivating mutations as well as recessive and dominant patterns of inheritance. MBL mutations are also diverse and often additive. Inherited cascade component deficiencies are typically associated with a CH50 or AH50 of near zero, and identification of the specific deficiency can be defined with a serum mixing approach. Sequencing approaches are becoming more routine for cascade component deficiencies and have been used routinely for regulatory defects for the last decade. C1 inhibitor deficiency is an exception to this, as functional testing is widely available and is quite reliable.

10.8.1.1 Defects in Proteins Involved in the Activation of C3

10.8.1.1.1 C1 complex deficiencies. C1 is a multisubunit protein. C1q binds to antibody, and C1r and C1s are the enzymatic components, responsible for cleavage. C1q, C1r, and C1s deficiencies present nearly uniformly with early onset systemic lupus erythematosus (SLE) [37–39]. The SLE is typically severe and arises in childhood. The autoantibody profile is similar to that seen in other SLE patients although anti-dsDNA antibodies may be less common [40,41]. An additional phenotype is recurrent bacterial infections, most often systemic infections with encapsulated organisms, and infection is the most common cause of death. All three deficiencies are inherited in an autosomal recessive pattern. Bone marrow transplant has been used successfully as a treatment for C1q deficiency as this component is produced in significant amounts by myeloid cells [42].

TABLE 10.2 Inherited Complement Deficiencies[a]

Deficiency	Gene	Inheritance Pattern	Number of Cases Reported	Clinical Features, Diagnostic Strategy
C1q	*C1Q*	AR	10–100	SLE, infections, CH50 near zero
C1r	*C1R*	AR (AD GOF)	10–100	SLE, infections, CH50 near zero. GOF: periodonatal Ehlers Danlos
C1s	*C1S*	AR (AD GOF)	10–100	SLE, infections, CH50 near zero. GOF: periodontal Ehlers Danlos
C4	*C4A* + *C4B*	AR	10–100	SLE, infections, CH50 near zero
C2	*C2*	AR	Many	SLE, infections, some asymptomatic, CH50 near zero
C3	*C3*	AR (AD GOF)	10–100	Infections frequent and severe, glomerulonephritis, CH50 near zero. aHUS has low CH50
Factor D	*CFD*	AR	<10	Neisseria, AH50 near zero
Factor B	*CFB*	AR (AD GOF)	<10	Neisseria, AH50 near zero, C3 level. AHUS has low AH50/CH50
Properdin	*CFP*	XL	>100	Neisseria, AH50 diminished
C5	*C5*	AR	10–100	Neisseria, CH50 near zero
C6	*C6*	AR	>100	Neisseria, CH50 near zero
C7	*C7*	AR	>100	Neisseria, CH50 near zero
C8α–γ	*C8A*	AR	>100	Neisseria, CH50 near zero
C8β	*C8B*	AR		Neisseria, CH50 near zero
C9	*C9*	AR	Many	Neisseria, CH50 diminished
Factor I	*CFI*	AR and AD	10–100	Neisseria, low C3. aHUS, usually requires mutation analysis, some patients have inflammation
Factor H	*CFH*	AR and AD	10–100	Neisseria, low C3. aHUS, usually requires mutation analysis
MCP	*CD46*	AR	<10	aHUS, mutation analysis required
C1 inhibitor	*SERPING1*	AD	Many	Angioedema, C1 antigen and functional levels low
CR3/CR4	*ITGB2*	AR	>100	Leukocyte adhesion deficiency, flow cytometry
CD59	*CD59*	AR	<10	Hemolysis, flow cytometry

[a]MBL, MASP2, and ficolin are not reported in this table as their phenotype is not completely established.

Gain-of-function mutations in C1r and C1s have been associated with rapidly progressing periodontitis with detached gingiva and tooth loss by early adulthood, pretibial hyperpigmentation, vascular fragility, and leukodystrophy [43,44]. Typical for this Ehlers–Danlos-like phenotype, there is often joint hypermobility. The relationship of this phenotype to the traditionally recognized functions of complement is not understood.

10.8.1.1.2 C4 deficiency. There are two distinct genes for C4 termed *C4A* and *C4B*. They encode highly homologous proteins although C4A binds more avidly to protein while C4B binds more avidly to carbohydrate. Within each C4 locus, there can be deletions or duplications or simple inactivating mutations [45,46]. Partial C4 deficiencies are extremely common. 1%–2% of the general population and up to 15% of patients with SLE have complete C4A deficiency. 1%–2% of the population has complete C4B deficiency and up to 15% of patients with invasive bacterial disease are C4B-deficient [47]. Complete C4 deficiency due to four inactive alleles is quite rare, and the majority of patients with complete C4 deficiency have had SLE [48–50]. Some complete C4-deficient individuals can be asymptomatic but most have recurrent infections with encapsulated organisms. The SLE and infection phenotypes can be separate or occur together in an individual.

10.8.1.1.3 C2 deficiency. C2 deficiency is the most common of the inherited complement component deficiencies in Caucasians with a frequency of 1:10,000. The common Caucasian mutation is a 28bp deletion. Most C2-deficient individuals are asymptomatic although their susceptibility to infection is thought to be increased. In total, 20%–40% of C2-deficient individuals will develop SLE [51–53]. Anti-Ro antibodies are common in C2-deficient patients with SLE although anti-dsDNA antibodies are infrequent [54]. In spite of SLE being the most common phenotype, the most common cause of death among C2-deficient patients is sepsis [52]. Other systemic infections such as meningitis, pneumonia, epiglottitis, and peritonitis have been seen, and the most common organisms have been *S. pneumoniae* and *H. influenzae*.

10.8.1.1.4 C3 deficiency. C3 deficiency is the least common of the complement cascade protein deficiencies, and it has a severe phenotype [55–59]. All patients have a profound predisposition to infection, and the infections are characteristic of neutrophil dysfunction (abscesses), humoral deficiencies (sinopulmonary disease), and complement deficiencies (sepsis, meningitis). Membranoproliferative glomerulonephritis occurs in a third of the cases of C3 deficiency [51,53]. One other feature of C3 deficiency deserves mention. During infections, a vasculitic rash may appear and symptoms of serum sickness may occasionally be seen. These unusual findings are due to the lack of immune complex solubilization by C3. C3 deficiency is rare, with fewer than 30 cases reported in the literature. There is a founder effect among the Afrikaans-speaking population [55]. Hypomorphic C3 mutations have been identified in some patients with autoimmune disease, but the phenotype and prevalence are not clear [60,61].

Activating mutations of C3 can be associated with aHUS [62,63]. These patients have been infrequent but have all had mutations that lead to chronic activation and consumption of C3. These patients have had diminished, but not absent, serum C3.

10.8.1.1.5 MBL deficiency. MBL deficiency is common with 2%–7% of people affected [64] across all racial and ethnic backgrounds. There are common structural mutations, which destabilize the higher-order complexes [65] and several promoter mutations, which lead to impaired production [65]. Combinations of mild mutations can lead to complete loss of function. The consequences of MBL deficiency are controversial [65–68]. MBL deficiency may be a mild risk factor for infection but is unlikely to have a significant effect. Similarly, an effect on susceptibility to autoimmune disease has been proposed; however, the effect has not been consistently identified, and if there is an effect, it must be small.

10.8.1.1.6 MASP2 deficiency. MASP2 protein deficiency is seen in 5%–10% of the population [69]. The common polymorphism, D105G, leads to compromised interactions with MBL and ficolin. Like MBL deficiency, it may represent a risk for infection, but the effect is modest [67]. There is similarly no strong evidence for an association of MASP2 deficiency and autoimmune disease [70]. A separate set of variants, G634R and R203W, have been identified as being associated in one study with herpes simplex encephalitis [71]. The MASP2 gene is predicted to be tolerant of loss-of-function, and both variants have frequencies in the general population of $\sim 10^{-5}$.

10.8.1.1.7 MASP3 deficiency. MASP3, encoded by the *MASP1* gene as an isoform, is the activator of factor D. Deficiencies of MASP3 and the collectin, CK-L1, have been associated with the 3 MC syndrome [72,73]. This syndrome is associated with hypertelorism, blepharophimosis, arched eyebrows, cleft lip/palate, poor growth, developmental delay, and hearing loss. Surprisingly, CK-L1 was found to serve as a guide for neural crest cells in development. Infection has not been a prominent feature, and traditional studies of complement function in 3 MC syndrome have not been reported.

10.8.1.1.8 Factor B deficiency. A single case of mutation-validated complete factor B deficiency in a patient with recurrent infections with encapsulated organisms has been reported [74]. The disorder was suspected after studies demonstrated an absent AH50. An additional patient with vasculitis has been described, but there was no genetic analysis. Heterozygous gain-of-function mutations of factor B are found in approximately 2%–4% of patients with aHUS.

10.8.1.1.9 Factor D deficiency. Neisserial infections are the most common manifestation of factor D deficiency [51,53,75]. Factor D deficiency should be suspected when the AH50 activity is absent.

10.8.1.1.10 Properdin deficiency. Properdin deficiency is the only X-linked complement deficiency, and the identified mutations have been diverse, with some affecting protein function and some having a null phenotype. Approximately half of the properdin-deficient individuals present with meningococcal disease [51,53,76–79]. There is a high fatality rate in meningococcal disease in properdin-deficient patients. There may be a founder effect in Tunisian Jewish people; however, properdin deficiency is seen on all ethnic backgrounds.

10.8.1.2 Defects in Proteins in the Membrane Attack Complex

10.8.1.2.1 C5 deficiency. C5 deficiency is associated with susceptibility to meningococcal and gonococcal disease. C5 deficiency is found on a variety of ethnic and racial backgrounds, and the mutations are diverse.

10.8.1.2.2 C6 deficiency. C6 deficiency is one of the more common complement disorders and occurs more frequently in African Americans and in people of South Africa. C6 deficiency is associated with meningococcemia, meningococcal meningitis and disseminated gonococcal disease [51,53,80,81]. There are two notable mutations associated with variants of C6 deficiency. In one case, a splice defect leads to a smaller than usual protein, C6SD [82]. This protein functions less efficiently than wild-type C6; however, it is not clear whether bearing C6SD leads to compromised host defense. The other variation is combined C6 and C7 deficiency [83].

10.8.1.2.3 C7 deficiency. C7 deficiency is not particularly common, and the most common presentation has been neisserial disease [51,53,84]. There may be a founder affect among Moroccan Jewish kindreds [85], but generally the mutations are diverse [84].

10.8.1.2.4 C8 deficiency. C8 is composed of three chains: α, β, γ. The α and γ chains are covalently attached and bind to the β chain. C8β deficiency is more common in Caucasians while C8α-γ deficiency is more common among African Americans and Japanese [86–89]. The majority of the C8β mutations are due to a single base pair transition leading to a premature stop codon [86,90]. The majority of the C8α-γ mutations are due to a 10bp insertion leading to a stop codon [87]. All types of C8 deficiency are associated with susceptibility to neisserial disease [51,53]. Meningococcal meningitis, meningococcemia, and disseminated gonococcus have been seen.

10.8.1.2.5 C9 deficiency. C9 deficiency is seen with high frequency in Japan and Korea [91–93]. Approximately 0.05% of people in Japan are C9-deficient. It is more difficult to diagnose than most of the other complement cascade protein deficiencies because the CH50 is diminished but not absent [94]. As is true for the other terminal complement component deficiencies, C9 deficiency is associated with neisserial disease although the penetrance appears to be less than that seen in other terminal component deficiencies [95]. The common Japanese mutation is a nonsense mutation in exon 4 [96].

10.8.1.3 Defects in Regulatory Proteins

10.8.1.3.1 C1 inhibitor deficiency. C1 inhibitor deficiency is associated with recurrent episodes of submucosal or subcutaneous edema [97]. It is unusual in that it is most classically due to a heterozygous mutation and is inherited in an autosomal dominant fashion although homozygous deficient patients exist [22,98–101]. Serum levels are typically slightly less than 50% of normal due to accelerated consumption in affected individual and in those cases where serum protein levels are normal, the functional level is slightly less than 50% of normal. Approximately half of the patients present in childhood although rare asymptomatic adults have been identified. The frequency of angioedema episodes is highly variable as is the severity. The extremities, face, genitalia, and gastrointestinal tract are most often involved. Upper airway swelling can lead to respiratory arrest.

The angioedema typically progresses for 1–2 days and resolves in another 2–3 days, and common triggers are illness, hormonal fluctuations, trauma, and stress. The C1 inhibitor promoter is androgen-responsive, which is why men can be less severely affected than female patients [102–104]. It is also the mechanism underlying the oldest therapeutic modality, attenuated androgens.

10.8.1.3.2 C4 binding protein deficiency. A single kindred with C4-binding protein deficiency has been described [105] although no mutation was identified. Angioedema, vasculitis, and arthritis were the main features.

10.8.1.3.3 Factor I deficiency. Factor I deficiency has three phenotypes, and the genotype–phenotype relationship is not fully understood. When factor I is completely lacking, C3 cleavage occurs unchecked and a secondary deficit in C3 occurs. Both the CH50 and AH50 are depressed but not absent, and C3 antigen levels are low [106–109]. Neisserial disease and infections with *S. pneumoniae* and *H. influenzae* have been described [51,53]. The infection phenotype is most often associated with autosomal recessive mutations.

The second phenotype is renal disease, either aHUS or membranoproliferative glomerulonephritis II

[110–112]. These cases of factor I deficiency are difficult to identify because traditional complement studies are often normal. The factor I level is typically normal as the mutations inactivate critical binding sites. The fenestrated endothelium of the glomerulus is rich in polyanions, where complement can be activated in the absence of regulatory proteins. Mutations most often affect the catalytic domain [113].

Recently, an inflammatory phenotype has been described [114,115]. Patients with this phenotype have had partial loss-of-function, and the main inflammatory features were hemorrhagic leukoencephalopathy, cerebral edema, and meningoencephalitis. The mechanism of this phenotype is not understood.

10.8.1.3.4 Factor H deficiency. Infections, aHUS, and macular degeneration are the main phenotypes seen in factor H deficiency [116–119]. There is a strong genotype–phenotype correlation. The infections are due to consumption of C3 and a secondary partial C3 deficiency [120]. C3 levels are typically diminished and the CH50 and AH50 are low. Factor H levels may be diminished. The mutations associated with this phenotype are diverse. The other two phenotypes are less likely to be associated with low factor H levels and mutation testing it typically required. Kindreds with membranoproliferative glomerulonephritis have been identified with factor H deficiency, and an acquired phenocopy is due to auto-antibodies directed to factor H [121]. The most common phenotype is now known to be aHUS. Factor H defects were found to be the underlying basis for 15%–30% of patients with aHUS. Mutations typically occur at the C-terminus. Both autosomal recessive and heterozygous mutations have been seen. The disease is typically early onset and severe with recurrences in many cases [122].

A common (allele frequency 0.5–0.9 in different populations) polymorphism of factor H (the H402Y variant) was identified as a risk factor for macular degeneration with an Odds Ratio of 2–7 [117,123]. The retinal deposits called drusen contain factor H and terminal complement components. This polymorphism affects protection of the endothelium allowing smoldering complement activation and gradual damage.

10.8.1.3.5 Membrane cofactor protein (CD46) deficiency. Membrane cofactor protein (MCP) deficiency is associated with a later onset of aHUS compared to factor H and factor I deficiencies [122,124–126]. MCP mutations are thought to account for approximately 10% of all aHUS [127]. In contrast to factor H and factor I deficiencies, renal transplantation can be successful because MCP is expressed on the kidney; however, relapses have been described [128,129]. Traditional complement analyses are normal.

10.8.1.3.6 CD59 deficiency and paroxysmal nocturnal hemoglobinuria. CD59 deficiency is associated with chronic hemolytic anemia and recurrent stroke [130,131]. Early childhood strokes should prompt consideration of this condition. CD59 confers protection from intravascular complement-mediated lysis and is expressed on hematopoietic cells and endothelial cells.

More common than congenital absence of CD59 is paroxysmal nocturnal hemoglobinuria, which is due to somatic mutations of PIG-A [132]. This condition is associated with sudden, recurring episodes of dark urine triggered by stresses, such as infections or physical exertion. Pancytopenia and thrombosis are two of the major complications in this condition. PIG-A is required for GPI anchored proteins such as C8-binding protein, DAF, and CD59 [133]. The diagnosis of PNH is made by flow cytometry for CD59 or CD55 (DAF).

10.8.1.3.7 DAF deficiency (CD55). DAF deficiency is also termed the Inab blood group phenotype [134–136]. DAF deficiency is associated with a type of protein losing enteropathy due to unregulated complement deposition in the gastrointestinal tract [137,138].

10.8.1.3.8 CR3/CR4 deficiency. This disorder is a defect in the three β2 integrin adhesion molecules. This deficiency is more often called leukocyte adhesion deficiency type I (LAD type I). Mutations in the common β chain (CD18) lead to failure to express adequate α chains [139,140]. This disorder is discussed in more detail in the chapter on immunodeficiencies.

10.9 MANAGEMENT OF COMPLEMENT DEFICIENCIES

The management of complement deficiencies is dependent on the type of defect. With few exceptions, there are no trials supporting the management strategies offered here. The management approaches offered in this chapter represent possible interventions based on current literature. As this is a rapidly moving field, it is wise to seek out expert advice when confronted with a complement-deficient patient.

10.9.1 Early Classical Component Deficiencies

C1, C4, and C2 deficiencies are all associated with bacterial infections and SLE. The treatment for SLE in these patients does not differ from standard approaches, although the more severely affected patients could in theory be treated with a liver transplant. Uniquely for C1q deficiency, bone marrow transplant has been used successfully in several cases. The overall poor outcome for patients with these deficiencies supports novel strategies to improve quality of life and outcomes.

To mitigate the susceptibility to infection, patients are often vaccinated to raise the titers of antibodies to encapsulated organisms. For early complement component deficiencies, the major risks are *S. pneumoniae* and *H. influenzae* [51–53]. Vaccines to these entities exist, and there is reason to believe that having high titers of antibody may offer protection. Alternatively, prophylactic antibiotics may offer protection from serious infection. In one study, half of the C2-deficient patients had serious infections such as sepsis and meningitis. Infection was the leading cause of death, accounting for over 10% of the deaths in this cohort [52]. Patients on immunosuppressive medication for rheumatologic disorders may require yet more vigilance and protection from infection.

10.9.2 C3 Deficiency

The infections seen in C3-deficient patients are very severe and management must address loss of opsonization, loss of B cell costimulation, and loss of immune complex solubilization [120]. Immune globulin replacement is sensible to compensate for the compromised B cell function, and prophylactic antibiotics could ameliorate the risk of infection. The membranoproliferative glomerulonephritis seen in C3-deficient patients has no specific intervention although renal transplantation has been attempted. The recurrence risk has not been characterized; however, one would anticipate significant recurrence risk. Those patients with aHUS due to defects in regulatory proteins are often treated with eculizumab or ravulizumab, C5 inhibitors. This approach has not yet been investigated in C3 deficiency specifically but would likely offer benefit. Liver transplantation could be considered given the severity of the disease [141].

10.9.3 Factor D and Properdin Deficiency

Patients with factor D and properdin deficiency are susceptible to neisserial disease as well as *S. pneumoniae* and *H. influenzae*. Vaccination to achieve high titers of antibody to those entities could theoretically provide benefit. Traditionally, prophylactic antibiotics have been used for some patients in an effort to prevent infections [142–144].

10.9.4 Defects Associated With a Risk of Neisserial Infections

Terminal component deficiencies are all associated with an increased risk of neisserial disease. Meningococcal disease is by far the most common, but disseminated gonococcal infections have been described. Vaccination every 3 years with the meningococcal vaccine decreases the frequency of meningococcal episodes to 20% of what nonvaccinated individuals experience [145–148]. No study has examined prophylactic antibiotics, and it may be that careful monitoring and hypervaccination may be sufficient. Factor H and factor I deficiencies can also be associated with susceptibility to neisserial infections. There are no data to support a specific treatment strategy, but the interventions described for patients with terminal components deficiencies could be considered.

10.9.5 C1 Inhibitor Deficiency

Current management options include attenuated androgens to raise expression of the intact gene, purified C1 inhibitor, ecallantide (a kallikrein inhibitor), lanadelumab (a kallikrein inhibitor), and icatibant (a bradykinin B2 receptor antagonist). Each has demonstrated efficacy although some are approved for prophylaxis and some are approved for acute treatment. Treatment requires finding the optimal approach for the patient in terms of medical needs, convenience, cost, and availability of urgent care should airway involvement occur.

10.9.6 Defects Associated with aHUS

This phenotype is referred to as atypical HUS because it is not triggered by a bacterial toxin. As is done for typical HUS, some patients receive pheresis and fresh frozen plasma replacement [124,149,150]. For soluble protein defects, C5 inhibitors are now the mainstay of

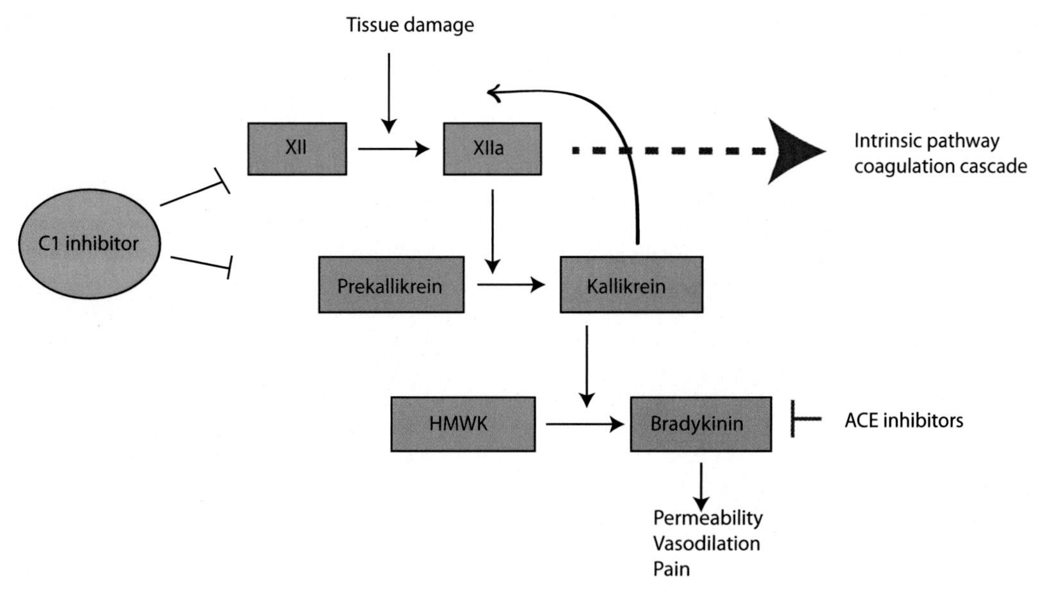

Figure 10.4 The role of C1 inhibitor. C1 inhibitor also functions in the complement cascade, but the main pathologic features in C1 inhibitor deficiency are due to compromised inhibition of this kallikrein—bradykinin pathway.

treatment with liver transplant performed in some cases. Fresh frozen plasma has been used in a prophylactic approach when C5 inhibitors are not available. For patients with end-stage renal disease, the recurrence of disease in patients with soluble factor deficiencies is unacceptably high, and renal transplantation is not recommended [128,129]. In contrast, renal disease in MCP typically does not recur in the transplanted kidney. Combined liver and kidney transplantation has been performed successfully [141] (Fig. 10.4).

REFERENCES

[1] Anthony R, el-Omar E, Lappin DF, MacSween RN, Whaley K. Regulation of hepatic synthesis of C3 and C4 during the acute-phase response in the rat. Eur J Immunol 1989;19:1405—12.

[2] Schwaeble W, Huemer HP, Most J, Dierich MP, Strobel M, Claus C, Reid KB, Ziegler-Heitbrock HW. Expression of properdin in human monocytes. Eur J Biochem 1994;219:759—64.

[3] Scoazec JY, Delautier D, Moreau A, et al. Expression of complement-regulatory proteins in normal and UW-preserved human liver. Gastroenterology 1994;107:505—16.

[4] Wilkison WO, Min HY, Claffey KP, Satterberg BL, Spiegelman BM. Control of the adipsin gene in adipocyte differentiation. Identification of distinct nuclear factors binding to single- and double-stranded DNA. J Biol Chem 1990;265:477—82.

[5] Ziccardi RJ. The first component of human complement (C1): activation and control. Springer Semin Immunopathol 1983;6:213—30.

[6] Passwell J, Schreiner GF, Nonaka M, Beuscher HU, Colten HR. Local extrahepatic expression of complement genes C3, factor B, C2, and C4 is increased in murine lupus nephritis. J Clin Invest 1988;82: 1676—84.

[7] Gal P, Cseh S, Schumaker VN, Zavodszky P. The structure and function of the first component of complement: genetic engineering approach (a review). Acta Microbiol Immunol Hung 1994;41:361—80.

[8] Sim RB, Reid KBM. C1: molecular interactions with activating systems. Immunol Today 1991;12:307—11.

[9] Pangburn MK, Muller-Eberhard HJ. Initiation of the alternative complement pathway due to spontaneous hydrolysis of the thioester of C3. Ann N Y Acad Sci 1983;421:291—8.

[10] Nolan KF, Schwaeble W, Kaluz S, Dierich MP, Reid KBM. Molecular cloning of the cDNA coding for properdin, a positive regulator of the alternative pathway of human complement. Eur J Immunol 1991;21:771—6.

[11] Rodriguez de Cordoba S, Esparza-Gordillo J, Goicoechea de Jorge E, Lopez-Trascasa M, Sanchez-Corral P. The human complement factor H: functional roles, genetic variations and disease associations. Mol Immunol 2004;41:355–67.
[12] Zipfel PF. Complement factor H: physiology and pathophysiology. Semin Thromb Hemost 2001;27:191–9.
[13] Zipfel PF, Skerka C, Hellwage J, et al. Factor H family proteins: on complement, microbes and human diseases. Biochem Soc Trans 2002;30:971–8.
[14] Childs RA, Drickamer K, Kawasaki T, Thiel S, Mizuochi T, Feizi T. Neoglycolipids as probes of oligosaccharide recognition by recombinant and natural mannose-binding proteins of rat and man. Biochem J 1989;262:1018–22.
[15] Matsushita M, Fujita T. Cleavage of the third component of complement (C3) by mannose-bindnign protein-associated serine protease (MASP) with subsequent complement activation. Immunobiology 1995;194:443–8.
[16] Sastry K, Herman GA, Day L, Deignan E, Bruns G, Morton CC, Ezekowitz RAB. The human mannose binding protein gene. J Exp Med 1989;170:1175–89.
[17] Malhotra R, Wormald MR, Rudd PM, Fischer PB, Dwek RA, Sim RB. Glycosylation changes of IgG associated with rheumatoid arthritis can activate complement via the mannose-binding protein. Nat Med 1995;1:237–43.
[18] Halperin JA, Taratuska A, Rynkiewicz M, Nicholson-Weller A. Transient changes in erythrocyte membrane permeability are induced by sublytic amounts of the complement membrane attack complex (C5b-9). Blood 1993;81:200–5.
[19] Salama A, Bhakdi S, Mueller-Eckhardt C, Kayser W. Deposition of the terminal C5b-9 complement complex on erythrocytes by human red cell autoantibodies. Br J Haematol 1983;55:161–9.
[20] Bhakdi S, Tranum-Jensen J. C5b-9 assembly: average binding of one C9 molecule to C5b-8 without poly-C9 formation generates a stable transmembrane pore. J Immunol 1986;136:2999–3005.
[21] Zahedi K, Prada AE, Davis 3rd AE. Structure and regulation of the C1 inhibitor gene. Behring Inst Mitt 1993:115–9.
[22] Cicardi M, Zingale L, Zanichelli A, Pappalardo E, Cicardi B. C1 inhibitor: molecular and clinical aspects. Springer Semin Immunopathol 2005;27:286–98.
[23] Davis 3rd AE. Biological effects of C1 inhibitor. Drug News Perspect 2004;17:439–46.
[24] Gronski P, Bodenbender L, Kanzy EJ, Seiler FR. C4-binding protein prevents spontaneous cleavage of C3 in sera of patients with hereditary angioedema. Complement 1988;5:1–12.
[25] Hessing M, van TVC, Bouma BN. The binding site of human C4b-binding protein on complement C4 is localized in the alpha'-chain. J Immunol 1990;144:2632–7.
[26] Discipio RG, Hugli TE. Circular dichroism studies of human factor H. A regulatory component of the complement system. Biochim Biophys Acta 1982;709:58–64.
[27] Isenman DE. Conformational changes accompanying proteolytic cleavage of human complement protein C3b by the regulatory enzyme factor I and its cofactor H. Spectroscopic and enzymological studies. J Biol Chem 1983;258:4238–44.
[28] Kinoshita T, Nussenzweig V. Regulatory proteins for the activated third and fourth components of complement (C3b and C4b) in mice. I. Isolation and characterization of factor H: the serum cofactor for the C3b/C4b inactivator (factor I). J Immunol Methods 1984;71:247–57.
[29] Masaki T, Matsumoto M, Nakanishi I, Yasuda R, Seya T. Factor I-dependent inactivation of human complement C4b of the classical pathway by C3b/C4b receptor (CR1, CD35) and membrane cofactor protein (MCP, CD46). J Biochem 1992;111:573–8.
[30] Hardig Y, Hillarp A, Dahlback B. The amino terminal module of the C4b-binding protein alph-chain is crucial for C4b binding and factor I-cofactor function. Biochem J 1997;323:469–75.
[31] Hansch GM, Weller PF, Nicholson-Weller A. Release of C8 binding protein (C8bp) from the cell membrane by phosphatidylinositol-specific phospholipase C. Blood 1988;72:1089–92.
[32] Lublin DM, Krsek-Staples J, Pangburn MK, Atkinson JP. Biosynthesis and glycosylation of the human complement regulatory protein decay-accelerating factor. J Immunol 1986;137:1629–35.
[33] Lublin DM, Lemons RS, Le Beau MM, et al. The gene encoding decay-accelerating factor (DAF) is located in the complement-regulatory locus on the long arm of chromosome 1. J Exp Med 1987;165:1731–6.
[34] Lublin DM, Liszewski MK, Post TW, et al. Molecular cloning and chromosomal localization of human membrane cofactor protein (MCP). Evidence for inclusion in the multigene family of complement-regulatory proteins. J Exp Med 1988;168:181–94.
[35] Rollins SA, Zhao J, Ninomiya H, Sims PJ. Inhibition of homologous complement by CD59 is mediated by a species-selective recognition conferred through binding to C8 within C5b-8 or C9 within C5b-9. J Immunol 1991;146:2345–51.

[36] Taguchi R, Funahashi Y, Ikezawa H, Nakashima I. Analysis of PI (phosphatidylinositol)-anchoring antigens in a patient of paroxysmal nocturnal hemoglobinuria (PNH) reveals deficiency of 1F5 antigen (CD59), a new complement-regulatory factor. FEBS (Fed Eur Biochem Soc) Lett 1990;261:142–6.
[37] Botto M, Walport MJ. C1q, autoimmunity and apoptosis. Immunobiology 2002;205:395–406.
[38] Hannema AJ, Kluin-Nelemans JC, Hack CE, Eerenberg-Belmer JM, Mallee C, vanHelden HPT. SLE syndrome and functional deficiency of C1q in members of a large family. Clin Exp Immunol 1984;55:106–14.
[39] Slingsby JH, Norsworthy P, Pearce G, et al. Homozygous hereditary C1q deficiency and systemic lupus erythematosus. Arthritis Rheum 1996;39:663–70.
[40] Bowness P, Davies KA, Norsworthy PJ, Athanassiou P, Taylor-Wiedeman J, Borysiewicz LK, Meyer PAR, Walport MJ. Hereditary C1q deficiency and systemic lupus erythematosus. Q J Med 1994;87:455–64.
[41] Walport MJ, Davies KA, Botto M. C1q and systemic lupus erythematosus. Immunobiology 1998;199:265–85.
[42] Arkwright PD, Riley P, Hughes SM, Alachkar H, Wynn RF. Successful cure of C1q deficiency in human subjects treated with hematopoietic stem cell transplantation. J Allergy Clin Immunol 2014;133:265–7.
[43] Kapferer-Seebacher I, Pepin M, Werner R, , et alMolecular Basis of Periodontal, E.D.S.C., Byers PH, Zschocke J. Periodontal Ehlers-Danlos syndrome is caused by mutations in C1R and C1S, which encode subcomponents C1r and C1s of complement. Am J Hum Genet 2016;99:1005–14.
[44] Kapferer-Seebacher I, Waisfisz Q, Boesch S, et al. Periodontal Ehlers-Danlos syndrome is associated with leukoencephalopathy. Neurogenetics 2019;20:1–8.
[45] Ballow M, McLean R, Einarson M. Hereditary C4 deficiency- genetic studies and linkage to the HLA. Transplant Proc 1979;11:1710–2.
[46] Fredrikson GN, Truedsson L, Kjellman M. DNA analysis in a MHC heterozygous patients with complete C4 deficiency-homozygosity for C4 gene deletion and C4 pseudogene. Exp Clin Immunogenet 1991;8:29–37.
[47] Bishof NA, Welch TR, Beischel LS. C4B deficiency: a risk factor for bacteremia with encapsulated organisms. JID (J Infect Dis) 1990;162:248–50.
[48] Colten HR. Navigating the maze of complement genetics: a guide for clinicians. Curr Allergy Asthma Rep 2002;2:379–84.
[49] Colten HR, Rosen FS. Complement deficiencies. Annu Rev Immunol 1992;10:809–34.
[50] Mascart-Lemone F, Hauptmann G, Goetz J, Duchateau J, Delespesse G, Vray B, Dab I. Genetic deficiency of C4 presenting with recurrent infections and a SLE-like disease. Genetic and immunologic studies. Am J Med 1983;75:295–304.
[51] Figueroa JE, Densen P. Infectious diseases associated with complement deficiencies. Clin Microbiol Rev 1991;4:359–95.
[52] Jonsson G, Truedsson L, Sturfelt G, Oxelius VA, Braconier JH, Sjoholm AG. Hereditary C2 deficiency in Sweden: frequent occurrence of invasive infection, atherosclerosis, and rheumatic disease. Medicine 2005;84:23–34.
[53] Ross SC, Densen P. Complement deficiency states and infection: epidemiology, pathogeneisis and consequences of neisserial and other infections in an immune deficiency. Medicine 1984;63:243–73.
[54] Vandersteen PR, Provost TT, Jordan RE, McDuffie FC. C2 deficient systemic lupus erythematosus. Arch Dermatol 1982;118:584–7.
[55] Botto M, Fong KY, So AK, Barlow R, Routier R, Morley BJ, Walport MJ. Homozygous hereditary C3 deficiency due to a partial gene deletion. Proc Natl Acad Sci 1992;89:4957–61.
[56] Botto M, Fong KY, So AK, Rudge A, Walport MJ. Molecular basis of hereditary C3 deficiency. J Clin Invest 1990;86:1158–63.
[57] Grumach AS, Vilela MM, Gonzalez CH, et al. Inherited C3 deficiency of the complement system. Braz J Med Biol Res 1988;21:247–57.
[58] Peleg D, Harit-Bustan H, Katz Y, Peller S, Schlesinger M, Schonfeld S. Inherited C3 deficiency and meningococcal disease in a teenager. Pediatr Infect Dis J 1992;11:401–4.
[59] Singer L, Colten HR, Wetsel RA. Complement C3 deficiency: human, animal, and experimental models. Pathobiology 1994;62:14–28.
[60] McLean RH, Bryan RK, Winkelstein J. Hypomorphic variant of the slow allele of C3 associated with hypocomplementemia and hematuria. Am J Med 1985;78:865–8.
[61] McLean RH, Weinstein A, Damajanov I, Rothfield N. Hypomorphic variant of C3, arthritis, and chronic glomerulonephritis. J Pediatr 1978;93:937–43.
[62] Fremeaux-Bacchi V, Miller EC, Liszewski MK, et al. Mutations in complement C3 predispose to development of atypical hemolytic uremic syndrome. Blood 2008;112:4948–52.
[63] Matsumoto T, Toyoda H, Amano K, et al. Clinical manifestation of patients with atypical hemolytic uremic syndrome with the C3 p.I1157T variation in the kinki region of Japan. Clin Appl Thromb Hemost 2018;24:1301–7.

[64] Thiel S, Frederiksen PD, Jensenius JC. Clinical manifestations of mannan-binding lectin deficiency. Mol Immunol 2006;43:86–96.
[65] Garred P, Larsen F, Seyfarth J, Fujita R, Madsen HO. Mannose-binding lectin and its genetic variants. Gene Immun 2006;7:85–94.
[66] Casanova JL, Abel L. Human mannose-binding lectin in immunity: friend, foe, or both? J Exp Med 2004;199:1295–9.
[67] Garcia-Laorden MI, Sole-Violan J, Rodriguez de Castro F, et al. Mannose-binding lectin and mannose-binding lectin-associated serine protease 2 in susceptibility, severity, and outcome of pneumonia in adults. J Allergy Clin Immunol 2008;122. 368–374, 374 e361–362.
[68] Kakkanaiah VN, Shen GQ, Ojo-Amaize EA, Peter JB. Association of low concentrations of serum mannose-binding protein with recurrent infections in adults. Clin Diagn Lab Immunol 1998;5:319–21.
[69] Lozano F, Suarez B, Munoz A, Jensenius JC, Mensa J, Vives J, Horcajada JP. Novel MASP2 variants detected among North African and Sub-Saharan individuals. Tissue Antigens 2005;66:131–5.
[70] Garcia-Laorden MI, Hernandez-Brito E, Munoz-Almagro C, et al. Should MASP-2 deficiency Be considered a primary immunodeficiency? Relevance of the lectin pathway. J Clin Immunol 2020;40:203–10.
[71] Bibert S, Piret J, Quinodoz M, et al. Herpes simplex encephalitis in adult patients with MASP-2 deficiency. PLoS Pathog 2019;15:e1008168.
[72] Rooryck C, Diaz-Font A, Osborn DP, et al. Mutations in lectin complement pathway genes COLEC11 and MASP1 cause 3MC syndrome. Nat Genet 2011;43:197–203.
[73] Urquhart J, Roberts R, de Silva D, et al. Exploring the genetic basis of 3MC syndrome: findings in 12 further families. Am J Med Genet 2016;170A:1216–24.
[74] Slade C, Bosco J, Unglik G, Bleasel K, Nagel M, Winship I. Deficiency in complement factor B. N Engl J Med 2013;369:1667–9.
[75] Kluin-Nelemans H, van Velzen-Blad H, van Helden HPT, Daha MR. Functional deficiency of complement factor D in a monozygous twin. Clin Exp Immunol 1984;58:724–30.
[76] Densen P, Weiler JM, Griffss JM, Hoffmann LG. Familial properdin deficiency and fatal meningococcemia: correction of the bactericidal defect by vaccination. N Engl J Med 1987;316:922–7.
[77] Fredrikson GN, Westburg J, Kuijper EJ, et al. Molecular genetic characterization of properdin deficiency type III: dysfunction due to a single point mutation in exon 9 of the structural gene causing a tyrosine to aspartic acid interchange. Mol Immunol 1996;33(Suppl. 1):1.
[78] Nielson HE, Kock C. Congenital properdin deficiency and meningococcal infection. Clin Immunol Immunopathol 1987;44:134–8.
[79] Sjoholm AG, Braconier JH, Soderstrom C. Properdin deficiency in a family with fulminant meningococcal infections. Clin Exp Immunol 1982;50:291–7.
[80] Nishizaka H, Horiuchi T, Zhu Z-B, et al. Molecular bases for inherited human complement component C6 deficiency in two unrelated individuals. J Immunol 1996;156:2309–15.
[81] Zhu ZB, Totemchokchyakarn K, Atkinson TP, Whiteley RJ, Volanakis JE. Molecular defects leading to human complement component C6 deficiency. FASEB J 1996;10:A1446.
[82] Wurzner R, Hobart MJ, Fernie BA, et al. Molecular basis of subtotal complement C6 deficiency. A carboxy-terminally truncated but functionally active C6. J Clin Invest 1995;95:1877–83.
[83] Fernie BA, Wurzner R, Morgan BP, Lachmann PJ, Hobart MJ. Molecular basis of combined C6 and C7 deficiency. Mol Immunol 1996;33(Suppl. 1):59.
[84] Nishizaka H, Horiuchi T, Zhu Z-B, Fukumori Y, Volanakis JE. Genetic bases of human complement C7 deficiency. J Immunol 1996;157:4239–43.
[85] Halle D, Elstein D, Geudalia D, Sasson A, Shinar E, Schlesinger M, Zimran A. High prevalence of complement C7 deficiency among healthy blood donors of Moroccan Jewish ancestry. Am J Med Genet 2001;99:325–7.
[86] Kaufmann T, Hansch G, Rittner C, Spath P, Tedesco F, Schneider PM. Genetic basis of human complement C8 beta deficiency. J Immunol 1993;150:4943–7.
[87] Kojima T, Horiuchi T, Nishizaka H, et al. Genetic basis of human complement C8 alpha-gamma deficiency. J Immunol 1998;161:3762–6.
[88] Komatsu M, Yamamoto K, Mikami H, Sodetz JM. Genetic deficiency of complement component C8 in the rabbit: evidence of a translational defect in expression of the alpha-gamma subunit. Biochem Genet 1991;29:271–4.
[89] Kotnik V, Luznik-Bufon T, Schneider PM, Kirschfink M. Molecular, genetic, and functional analysis of homozygous C8 b-chain deficiency in two siblings. Immunopharmacology 1997;38:215–21.
[90] Saucedo L, Ackerman L, Platonov AE, Gewurz A, Rakita RM, Densen P. Delineation of additional genetic bases for C8b deficiency. J Immunol 1995;155:5022–8.
[91] Hayama K, Sugai N, Tanaka S, et al. High-incidence of C9 deficiency throughout Japan: there are no

significant differences in incidence among eight areas of Japan. Int Arch Allergy Appl Immunol 1989;90:400–4.
[92] Kang HJ, Kim HS, Lee YK, Cho HC. High incidence of complement C9 deficiency in Koreans. Ann Clin Lab Sci 2005;35:144–8.
[93] Kira R, Ihara K, Watanabe K, et al. Molecular epidemiology of C9 deficiency heterozygotes with an Arg95Stop mutation of the C9 gene in Japan. J Hum Genet 1999;44:109–11.
[94] Hobart MJ, Fernie BA, Wurzner R, Oldroyd RG, Harrison RA, Joysey V, Lachmann PJ. Difficulties in the ascertainment of C9 deficiency: lessons to be drawn from a compound heterozygote C9-deficient subject. Clin Exp Immunol 1997;108:500–6.
[95] Fukumori Y, Yoshimura K, Ohnoki S, Yamaguchi H, Akagaki Y, Inai S. A high incidence of C9 deficiency among healthy blood donors in Osaka, Japan. Int Immunol 1989;1:85–9.
[96] Kira R, Ihara K, Takada H, Gondo K, Hara T. Nonsense mutation in exon 4 of human complement C9 gene is the major cause of Japanese complement C9 deficiency. Hum Genet 1998;102:605–10.
[97] Agostoni A, Aygoren-Pursun E, Binkley KE, et al. Hereditary and acquired angioedema: problems and progress: proceedings of the third C1 esterase inhibitor deficiency workshop and beyond. J Allergy Clin Immunol 2004;114:S51–131.
[98] Bissler JJ, Donaldson VH, Davis 3rd AE. Contiguous deletion and duplication mutations resulting in type 1 hereditary angioneurotic edema. Hum Genet 1994;93:265–9.
[99] Bissler JJ, Meng QS, Emery T. C1 inhibitor gene sequence facilitates frameshift mutations. Mol Med 1998;4:795–806.
[100] Blanch A, Roche O, Urrutia I, Gamboa P, Fontan G, Lopez-Trascasa M. First case of homozygous C1 inhibitor deficiency. J Allergy Clin Immunol 2006;118:1330–5.
[101] Lopez-Lera A, Favier B, de la Cruz RM, Garrido S, Drouet C, Lopez-Trascasa M. A new case of homozygous C1-inhibitor deficiency suggests a role for Arg378 in the control of kinin pathway activation. J Allergy Clin Immunol 2010;126. 1307–1310 e1303.
[102] Falus A, Feher K, Walcz E, Brozic M, Fust G, Hidvegi T, Feher T, Merety K. Hormonal regulation of complement biosynthesis in human cell lines I. Androgens and gamma interferon stimulate the biosynthesis and gene expression of C1 inhibitor in human cell lines U937 and HepG2. Mol Immunol 1990;27:191–5.
[103] Lener M, Vinci G, Duponchel C, Meo T, Tosi M. Molecular cloning, gene structure and expression profile of mouse C1 inhibitor. Eur J Biochem 1998;254:117–22.
[104] Prada AE, Zahedi K, Davis 3rd AE. Regulation of C1 inhibitor synthesis. Immunobiology 1998;199:377–88.
[105] Trapp RG, Fletcher M, Forristal J, West CD. C4 binding protein deficiency in a patient with atypical Behcet's disease. J Rheumatol 1987;14:135–8.
[106] Bonnin AJ, Zeitz HJ, Gewurz A. Complement factor I deficiency with recurrent aseptic meningitis. Arch Intern Med 1993;153:1380–3.
[107] Leitao MF, Vilela MM, Rutz R, Grumach AS, Condino-Neto A, Kirschfink M. Complement factor I deficiency in a family with recurrent infections. Immunopharmacology 1997;38:207–13.
[108] Morley BJ, Vyse TJ, Bartok I, et al. Molecular basis of hereditary factor I deficiency. Mol Immunol 1996;33(Suppl. 1):71.
[109] Vyse TJ, Spath PJ, Davies KA, et al. Hereditary complement factor I deficiency. Q J Med 1994;87:385–401.
[110] Fremeaux-Bacchi V, Dragon-Durey MA, Blouin J, Vigneau C, Kuypers D, Boudailliez B, Loirat C, Rondeau E, Fridman WH. Complement factor I: a susceptibility gene for atypical haemolytic uraemic syndrome. J Med Genet 2004;41:e84.
[111] Genel F, Sjoholm AG, Skattum L, Truedsson L. Complement factor I deficiency associated with recurrent infections, vasculitis and immune complex glomerulonephritis. Scand J Infect Dis 2005;37:615–8.
[112] Kavanagh D, Kemp EJ, Mayland E, et al. Mutations in complement factor I predispose to development of atypical hemolytic uremic syndrome. J Am Soc Nephrol 2005;16:2150–5.
[113] Saunders RE, Abarrategui-Garrido C, et al. The interactive Factor H-atypical hemolytic uremic syndrome mutation database and website: update and integration of membrane cofactor protein and Factor I mutations with structural models. Hum Mutat 2007;28:222–34.
[114] Broderick L, Gandhi C, Mueller JL, et al. Mutations of complement factor I and potential mechanisms of neuroinflammation in acute hemorrhagic leukoencephalitis. J Clin Immunol 2013;33:162–71.
[115] Haerynck F, Stordeur P, Vandewalle J, et al. Complete factor I deficiency due to dysfunctional factor I with recurrent aseptic meningo-encephalitis. J Clin Immunol 2013;33:1293–301.
[116] Dragon-Durey MA, Fremeaux-Bacchi V, et al. Heterozygous and homozygous factor h deficiencies associated with hemolytic uremic syndrome or membranoproliferative glomerulonephritis: report and

genetic analysis of 16 cases. J Am Soc Nephrol 2004;15:787–95.
[117] Klein RJ, Zeiss C, Chew EY, et al. Complement factor H polymorphism in age-related macular degeneration. Science 2005;308:385–9.
[118] Nielsen HE, Christensen KC, Koch C, Thomsen BS, Heegaard NH, Tranum-Jensen J. Hereditary, complete deficiency of complement factor H associated with recurrent meningococcal disease. Scand J Immunol 1989;30:711–8.
[119] Sanchez-Corral P, Perez-Caballero D, Huarte O, et al. Structural and functional characterization of factor H mutations associated with atypical hemolytic uremic syndrome. Am J Hum Genet 2002;71:1285–95.
[120] Reis ES, Falcao DA, Isaac L. Clinical aspects and molecular basis of primary deficiencies of complement component C3 and its regulatory proteins factor I and factor H. Scand J Immunol 2006;63:155–68.
[121] Dragon-Durey MA, Blanc C, Garnier A, Hofer J, Sethi SK, Zimmerhackl LB. Anti-factor H autoantibody-associated hemolytic uremic syndrome: review of literature of the autoimmune form of HUS. Semin Thromb Hemost 2010;36:633–40.
[122] Caprioli J, Peng L, Remuzzi G. The hemolytic uremic syndromes. Curr Opin Crit Care 2005;11:487–92.
[123] Haines JL, Hauser MA, Schmidt S, et al. Complement factor H variant increases the risk of age-related macular degeneration. Science 2005;308:419–21.
[124] Caprioli J, Noris M, Brioschi S, et al. Genetics of HUS: the impact of MCP, CFH, and IF mutations on clinical presentation, response to treatment, and outcome. Blood 2006;108:1267–79.
[125] Richards A, Kemp EJ, Liszewski MK, et al. Mutations in human complement regulator, membrane cofactor protein (CD46), predispose to development of familial hemolytic uremic syndrome. Proc Natl Acad Sci USA 2003;100:12966–71.
[126] Zimmerhackl LB, Besbas N, Jungraithmayr T, et al. Epidemiology, clinical presentation, and pathophysiology of atypical and recurrent hemolytic uremic syndrome. Semin Thromb Hemost 2006;32:113–20.
[127] Maga TK, Nishimura CJ, Weaver AE, Frees KL, Smith RJ. Mutations in alternative pathway complement proteins in American patients with atypical hemolytic uremic syndrome. Hum Mutat 2010;31:E1445–60.
[128] Loirat C, Fremeaux-Bacchi V. Hemolytic uremic syndrome recurrence after renal transplantation. Pediatr Transplant 2008;12:619–29.
[129] Loirat C, Niaudet P. The risk of recurrence of hemolytic uremic syndrome after renal transplantation in children. Pediatr Nephrol 2003;18:1095–101.
[130] Ben-Zeev B, Tabib A, Nissenkorn A, et al. Devastating recurrent brain ischemic infarctions and retinal disease in pediatric patients with CD59 deficiency. Eur J Paediatr Neurol 2015;19:688–93.
[131] Yamashina M, Ueda E, Kinoshita T, Tet al. Inherited complete deficiency of 20-kilodalton homologous restriction factor (CD59) as a cause of paroxysmal nocturnal hemoglobinuria. N Engl J Med 1990;323:1184–9.
[132] Shichishima T, Noji H. Heterogeneity in the molecular pathogenesis of paroxysmal nocturnal hemoglobinuria (PNH) syndromes and expansion mechanism of a PNH clone. Int J Hematol 2006;84:97–103.
[133] Shichishima T. Glycosylphosphatidylinositol (GPI)-anchored membrane proteins in clinical pathophysiology of paroxysmal nocturnal hemoglobinuria (PNH). Fukushima J Med Sci 1995;41:1–13.
[134] Hue-Roye K, Powell VI, Patel G, Lane D, Maguire M, Chung A, Reid ME. Novel molecular basis of an Inab phenotype. Immunohematol 2005;21:53–5.
[135] Reid ME. Cromer-related blood group antigens and the glycosyl phosphatidylinositol-linked protein, decay-accelerating factor DAF (CD55). Immunohematol 1990;6:27–9.
[136] Telen MJ, Green AM. The Inab phenotype: characterization of the membrane protein and complement regulatory defect. Blood 1989;74:437–41.
[137] Ozen A. CHAPLE syndrome uncovers the primary role of complement in a familial form of Waldmann's disease. Immunol Rev 2019;287:20–32.
[138] Ozen A, Comrie WA, Ardy RC, et al. CD55 deficiency, early-onset protein-losing enteropathy, and thrombosis. N Engl J Med 2017;377:52–61.
[139] Arnaout MA, Dana N, Gupta SK, Tenen TG, Fathallah DM. Point mutations impairing cell surface expression of the common b subunit (CD18) in a patient with leukocyte adhesion deficiency. J Clin Invest 1990;85:977–81.
[140] Kishimoto TK, Hollander N, Roberts TM, Anderson DC, Springer TA. Heterogeneous mutations in the beta subunit common to the LFA-1, Mac-1, and p150,95 glycoproteins cause leukocyte adhesion deficiency. Cell 1987;50:193–202.
[141] Saland JM, Ruggenenti P, Remuzzi G, Consensus Study G. Liver-kidney transplantation to cure atypical hemolytic uremic syndrome. J Am Soc Nephrol 2009;20:940–9.
[142] Biesma DH, Hannema AJ, van Velzen-Blad H, Mulder L, van Zwieten R, Kluijt I, Roos D. A family with complement factor D deficiency. J Clin Invest 2001;108:233–40.

[143] Hiemstra PS, Langeler E, Compier B, et al. Complete and partial deficiencies of complement factor D in a Dutch family. J Clin Invest 1989;84:1957–61.
[144] Sjoholm AG, Kuijper EJ, Tijssen CC, Jansz A, Bol P, Spanjaard L, Zanen HC. Dysfunctional properdin in a Dutch family with meningococcal disease. N Engl J Med 1988;319:33–40.
[145] Drogari-Apiranthitou M, Fijen CA, Van De Beek D, Hensen EF, Dankert J, Kuijper EJ. Development of antibodies against tetravalent meningococcal polysaccharides in revaccinated complement-deficient patients. Clin Exp Immunol 2000;119:311–6.
[146] Fijen CA, Kuijper EJ, Drogari-Apiranthitou M, Van Leeuwen Y, Daha MR, Dankert J. Protection against meningococcal serogroup ACYW disease in complement-deficient individuals vaccinated with the tetravalent meningococcal capsular polysaccharide vaccine. Clin Exp Immunol 1998;114:362–9.
[147] Platonov AE, Beloborodov VB, Pavlova LI, Vershinina IV, Kayhty H. Vaccination of patients deficient in a late complement component with tetravalent meningococcal capsular polysaccharide vaccine. Clin Exp Immunol 1995;100:32–9.
[148] Schlesinger M, Kayhty H, Levy R, Bibi C, Meydan N, Levy J. Phagocytic killing and antibody response during the first year after tetravalent meningococcal vaccine in complement-deficient and in normal individuals. J Clin Immunol 2000;20:46–53.
[149] Ariceta G, Besbas N, Johnson S, , et alEuropean Paediatric Study Group for, H.U.S. Guideline for the investigation and initial therapy of diarrhea-negative hemolytic uremic syndrome. Pediatr Nephrol 2009;24:687–96.
[150] Goodship TH. Factor H genotype-phenotype correlations: lessons from aHUS, MPGN II, and AMD. Kidney Int 2006;70:12–3.

11

Heritable and Polygenic Inflammatory Disorders*

Reed E. Pyeritz

Perelman School of Medicine at the University of Pennsylvania, Philadelphia, PA, United States

11.1 INTRODUCTION

Autoimmunity is a breakdown of the normal mechanisms that maintain immunologic homeostasis in the immune system response to specific antigens. The pathogenesis of autoimmune disease is multifactorial, including both genetic and environmental factors affecting the onset, maintenance, and progression of diseases. Autoimmune disorders result from the recognition of self-antigen(s) and the subsequent attack on self-tissues, usually by the adaptive immune response. Autoimmune pathology can be caused by T lymphocytes, natural killer (NK) cell, and/or antibodies and can be modulated by hormonal effects. One of the unifying themes of many autoimmune disorders is that they are associated with specific alleles and genotypes of the highly polymorphic major histocompatibility complex (MHC) where, in humans, the human leukocyte antigen (HLA) loci reside. In addition, a host of other genes can contribute to the onset, course, management, and response to therapy of specific diseases. This chapter reviews the components of the human immune system and how they interact. It deals with several prevalent autoimmune disorders, rheumatoid arthritis (RA) and ankylosing spondylitis, a variety of oligoarticular arthritides, and psoriatic arthritis. There are a host of other disorders that have a component of autoimmunity in their pathogenesis that are covered elsewhere in the volumes of this book, such as systemic lupus erythematosus (SLE), Beçhet syndrome, celiac disease, type 1 diabetes mellitus, myasthenia gravis, Sjogren syndrome, and narcolepsy.

11.2 AUTOIMMUNITY

Autoimmunity is a breakdown of the normal mechanisms that maintain immunologic homeostasis in the immune response to specific antigens. The pathology of autoimmune disease is multifactorial, including both genetic and environmental factors affecting the onset, maintenance, and progression of the disease [1]. Autoimmune diseases result from the recognition of self-antigen(s) and the subsequent attack of self-tissues, usually by the adaptive immune response. Autoimmune pathology can be caused by T lymphocytes, natural killer (NK) cells, antibodies, or both and can be modulated by hormonal effects. One of the unifying themes of many autoimmune disorders is that they are associated with specific alleles and genotypes of the highly polymorphic MHC where, in humans, the HLA loci reside. Since the completion of the human genome project (http://www.genome.gov/10001772), a plethora of new information about the genetics of autoimmune

* This chapter is a revision of the previous edition's chapters by Sarah Keidel, Catherine Swales, and Paul Wordsworth and by Nancy L Reinsmoen, Kai Cao and Chih-hung Lai © 2013, Elsevier Ltd.

Emery and Rimoin's Principles and Practice of Medical Genetics and Genomics. https://doi.org/10.1016/B978-0-12-812534-2.00008-4

disease has allowed the identification of a number of new genes involved in autoimmunity. The aim of this chapter is to present an update of recent genetic associations of autoimmunity with common genetic variants obtained through the genome-wide association studies (GWAS) and how these data have changed our knowledge of selected diseases, which historically showed an association with HLA [2].

11.3 THE IMMUNE RESPONSE

11.3.1 The Adaptive Immune System

In general, the mechanisms that produce tissue damage and clinical disease in autoimmune disorders are the same as those that mediate protection from pathogens. The immune system consists of a system of interacting molecules and cells that function to: 1. recognize intracellular and extracellular threats (primarily pathogens); 2. alert and activate lymphocytes, monocytes, and NK cells for specific response to these threats; 3. commit the responding cells to an appropriately regulated response to the threat; 4. amplify the response, if necessary, by recruitment of additional cells, including polymorphonuclear (PMN) leukocytes and macrophages, as well as additional lymphocytes and monocytes; and 5. ultimately, terminate the specific immune response after the threat has resolved.

The major components of the human adaptive immune response consist of antigen-presenting cells (APCs), T lymphocytes, B lymphocytes, HLA molecules, peptides, and costimulatory molecules. An immune response occurs when a peptide bound to an HLA molecule is recognized by a T-cell receptor (TCR), a phenomenon known as antigen presentation. The specific peptide, HLA molecule, and T cell together determine the nature of the response. There are two general categories of HLA molecules that perform this antigen presentation function: class I molecules (e.g., A, B, and C) and class II molecules (e.g., DR, DQ, and DP), each class having a different tissue distribution.

X-ray crystallographic structural studies have revealed that the outer domains of both class I alpha-chain molecules and the class II alpha—beta heterodimers form a peptide-binding groove, consisting of a beta-pleated sheet floor and two alpha-helical walls. Typically, the peptides bound in class I clefts are octamers or nanomers with the ends of the peptide buried within pockets of the cleft. CL-II peptides bound in the class II groove are somewhat longer, usually 12—14-mers, with the ends protruding from the cleft. The extensive polymorphism in HLA molecules lies primarily in the amino acids forming the floor and sides of the groove. The topography of the groove is unique for each allelic group of HLA molecules. Because only certain amino acid residues can "fit" into certain pockets based on size and charge, the peptides that can be bound by any given HLA molecule are limited. Some polymorphic residues within the HLA-binding groove bind to the peptide, whereas others bind to the T-cell receptor [3]. The structure and the capacity of HLA molecules to bind different peptides differ widely among different individuals owing to the extensive genetic polymorphism of the HLA genes, accounting for the association between genetic polymorphisms and variable immune response. Thus, it is the specific combination of a peptide—HLA complex that provides the genetically controlled determinant of antigen-specific recognition.

It is the complex of specific peptide bound within a given HLA antigen that is recognized by the TCR. The TCR of CD4 T cells generally recognizes peptide in the context of class II molecules to provide T-cell help (Th). These T helper cells can be of at least two different types. In general, Th1 cells provide help to other T cells (CD8) to affect cell killing. Th2 cells provide help to B cells in the production of antibody. By virtue of the cytokines they release, Th1 cells are predominantly inflammatory, whereas Th2 cells can often inhibit the inflammatory response. However, Th2 cells can contribute to an overproduction of antibody, which is a pathologic feature of some autoimmune diseases. In contrast, the TCR of CD8 T cells generally recognizes peptide in the context of class I molecules. These cells perform target cell killing and are known as cytotoxic T lymphocytes (CTLs). In addition, a subpopulation of both CD4 and CD8 T cells can carry out a regulatory (suppressive) role, in which case they are referred to as T-regulatory cells (TR or Tregs). These latter cells are CD4+CD25+FOXP3+ or CD8+CD25+/−FOXP3+. The Tregs (2%—10% of the circulating T cells) are essential in the maintenance of normal tolerance to self-antigens. Dysregulation of the Tregs can lead to profound autoimmunity [4—6].

In the adaptive immune system, a specific immune response initiates when a peptide bound to an HLA class II molecule on the surface of an APC, such as a macrophage or dendritic cell, is recognized by the T-cell receptor on a CD4 T lymphocyte, which is specific for the peptide—HLA molecule complex. T-cell activation results from this antigen-specific recognition event, provided a costimulatory interaction between B7 molecules (also known as CD80 and CD86) on the APC and CD28 on the T cell has also occurred.

11.3.2 The Innate Immune System

NK cells are one of the first lines of defense of the innate immune system, functioning before initiation of the adaptive immune response, as well as when the latter system is subverted, as happens in some viral infections and cancer and likely in autoimmunity. Regulation of NK cytolytic activity is a function of engagement of one or more NK surface receptors that may be activating or inhibitory. Two of the primary sets of receptors present on human NK cells are called the killer immunoglobulin-like receptor (KIR) and the CD94/NKG2 lectin-like receptors. They each comprised gene families. The KIR genes (15 genes, two pseudogenes) located on chromosome 19q13.4 have enormous genetic diversity generated by a combination of variable gene content on any given haplotype, alternative splicing, and allelic polymorphism, and they recognize a limited set of HLA Class I molecules or epitopes. The more ancestral CD94/NKG2 genes (four genes plus NKG2D) located on chromosome 12p12.3—p13.2 have less diversity but an unconventional array of HLA Class I-like ligands. The ligands for the KIR are the HLA classical Class I molecules A, B, C and the nonclassical Class I, HLA-G. HLA Class II does not appear to be involved. The CD94/NKG2 and NKG2D ligands are generally HLA Class 1b molecules such as HLA-E or stress-inducible nonclassical HLA Class I-like molecules such as MICA and MICB [MHC Class I-related chain] proteins. In the innate immune system, de novo recognition of self-antigens can occur when cryptic epitopes are exposed due to cell death/apoptosis [7] and can lead to autoimmune disease.

11.3.3 Autoimmune Pathology

Autoimmunity is thought to involve the activation of a self-peptide-specific T cell or NK cell, leading to the development of autoreactive CTLs, the inappropriate activation of macrophages, and the release of cytokines as well as inappropriate T-cell help, resulting in the formation of antibodies to self-antigens. Autoimmune tissue damage is mediated by autoantibodies, autoreactive T cells, or NK cells. Autoantibody-mediated disorders are distinguished by the different types of antigens that are recognized and the different classes of immunoglobulin that result in disease. Allergic (atopic) reactions are mediated by the production of IL-4 by CD4 Th2 cells, leading to the production of immunoglobulin E by B cells, in response to extrinsic (not self) antigens. The IgE can bind to and activate mast cells via the Fcε R1, releasing histamines, and other mediators and causing clinical symptoms, such as asthma or allergic rhinitis (hay fever). Some autoimmune diseases result from the production of IgG antibodies in response to self-antigens, such as cell-surface receptors. The autoantibody can disrupt the normal function of the receptor by causing uncontrolled activation (agonist) or by blocking receptor function (antagonist) and signaling. In Graves' disease, autoantibodies to the thyroid-hormone-stimulating receptor cause activation and hyperthyroidism, whereas in myasthenia gravis, autoantibodies to the acetylcholine receptor disrupt function and cause progressive muscle weakness. Some antibody-mediated autoimmune diseases, such as pemphigus vulgaris (PV), involve the production of IgG in response to cell- or matrix-associated antigens (the skin matrix protein, desmoglein 3 for PV), leading to tissue damage (blistering of skin and mucosal membranes for PV). Another form of antibody-mediated disease, such as SLE, reflects the production of antibodies to soluble antigens subsequent to exposure to cryptic autoantigens. In these diseases, the tissue damage is caused by responses triggered by immune complexes. A selected set of autoimmune diseases and their associated HLA alleles are listed in Table 11.1, and the putative autoantigens and pathogenic immunological mechanisms for some of these diseases are listed in Table 11.2.

In most autoimmune diseases, the environmental trigger or the eliciting antigen is not known. In some autoimmune diseases, infection with specific pathogens may represent an environmental trigger, eliciting an autoimmune response in genetically predisposed individuals. There are several hypothetical mechanisms by which infectious agents could break tolerance and induce autoreactive responses, such as molecular mimicry, release of cryptic self-antigen, and activation of non-tolerized cells, among others. Although epidemiologic studies may ultimately reveal

TABLE 11.1 **Selected Autoimmune Diseases and Associated HLA Alleles**

Disease	Associated HLA Alleles
Organ-Specific (Endocrine)	
Type 1 diabetes	DRB1*03:01–DQB1 * 02:01; DRB1*04-DQB1*03:02*[a]
	DRB1*03:01–DQB1*02:01; DRB1*04-DQB1*03:02*[a]
Graves' disease	DRB1*03:01–DQB1*02:01
Hashimoto's thyroiditis	DRB1*11–DQB1*03:01
Idiopathic Addison's	DRB1*03:01–DQB1*02:01 disease
Organ-Specific (Other)	
Inflammatory bowel	DRB1*01:03–DQB1*05:01; disease DRB1*01:03–DQB1*03:01
Crohn's disease	DRB1*11:01
Ulcerative colitis	DRB1*11:01
Myasthenia gravis	DRB1*03:01–DQB1*02:01
Multiple sclerosis	DRB1*15:01–DQB1*06:02
Psoriasis vulgaris	C*06
Pemphigus vulgaris	DRB1*04:02–DQB1*03:02; DRB1*14:01–DQB1*05:03
Narcolepsy	DQB1*06:02
Celiac disease	DQA1*05:01–DQB1*02:01 (in cis or in trans; see text)
Dermatitis herpetiformis	DQA1*05:01–DQB1*02:01
Rheumatologic Diseases	
Rheumatoid arthritis (RA)	DRB1*04–DQB1*03:02[b]; DRB1*01–
	DQB1*05:01; DRB1*10–
	DQB1*05:01
Pauciarticular juvenile RA	DRB1*08–DQB1*04:02; DRB1*11–
	DQB1*03:01, DPB1*02:01
Ankylosing spondylitis	B*27
Reactive arthropathy	B*27 including Reiter's disease
Systemic lupus erythematosus (C4A null allele)	DRB1*03:01-DQB1*02:01[c]
	DRB1*15:01-DQB1*06:02

[a]Not all DRB1*04 alleles confer equal risk for type 1 diabetes; DRB1*04:03 and *04:06 are protective, whereas the highest risk DRB1*04 alleles are 04:01,04:02, and 04:05. The highest risk genotype is DRB1*03:01– DQB1*02:01/DRB1*04–DQB1*03:02 (see text).[b]Not all DRB1*04 alleles confer equal risk for RA; DRB1*04:01 and 04:04 and 04:05 appear to confer the greatest risk. These alleles also appear to be associated with the severity of disease, unlike the disease-associated DRB*01 and DRB1*10 (see text).[c]The DRB1*03:01 association with SLE is likely due to linkage disequilibrium with the C4A null allele on B*08-DRB1*03:01 haplotypes.

the role of an infectious environmental trigger, direct evidence for the initiation of autoimmunity by an infectious agent is still lacking for most human autoimmune diseases.

In clinical practice, many patients do not satisfy the established criteria for a specific autoimmune disorder, such a PV, SLE, RA, and so forth. These patients are classified as having "undifferentiated connective tissue disease" [8]. Being a diagnosis of exclusion, undifferentiated connective tissue disease can be frustrating for the patient and clinician alike. As with other autoimmune conditions, coordination among the health professional team and ongoing follow-up are essential.

11.3.4 Autoreactive T Cells and the Failure of Tolerance

The maturation of T cells in the thymus involves positive selection as well as negative selection (deletion of self-reactive T-cell clones). The presence of autoreactive T cells in autoimmune disorders indicates that this latter process is not complete. Current consensus is that

TABLE 11.2 Selected Autoimmune Diseases and Autoantigens

Disease	Immunologic Mechanism	Putative Autoantigen	Clinical Consequences
Graves' disease	Antibody to cell surface receptor (agonist)	Thyroid stimulatory hormone receptor (THS)	Hyperthyroidism
Myasthenia gravis	Antibody to receptor (antagonist)	Acetylcholine receptor	Progressive muscle weakness
Pemphigus vulgaris	Antibody to skin protein	Epidermal cadherin (desmoglein 3)	Blistering of skin and mucosal membranes
Type 1 diabetes	T-cell mediated	Insulin, GAD (glutamic acid decarboxylase), insulin antibodies (IA-2), Islet cell antibodies (ICA)	β-cell destruction, insulin insufficiency
Rheumatoid arthritis	T-cell mediated	Unknown synovial antigen	Inflammation and destruction of joints
SLE (systemic lupus erythematosus)	Deposition of immune	DNA, histones, small nuclear ribonucleoproteins (SnRNP), small cytoplasmic ribonucleoproteins (ScRNP), ribosomes	Arthritis, glomerulonephritis, vasculitis
Narcolepsy	Autoantibodies	Hypocretin, Triblles Homolog 2 (TRIB2)	Excessive sleepiness

only the highest-affinity self-reactive T cells are deleted in the thymus. Thus, self-reactive T cells with lower affinities may escape negative selection, as well as T cells specific for self-peptides that are not expressed in the thymus. These autoreactive T cells may under normal conditions be prevented from mediating autoimmune disease by never encountering the relevant antigen or by being in a state of Treg-induced quiescence or by some other unknown regulatory mechanisms. But environmental triggers, such as infection, may disrupt these mechanisms in genetically predisposed individuals.

11.3.5 Thymic Education: Positive and Negative Selection

The diversity of TCR specificities creates a broad functional potential for antigenic recognition, an essential component of the adaptive immune response. By chance, TCR specificities, which are potentially reactive with self-antigens, are also created during random TCR gene rearrangement, creating the potential for autoreactive T cells. In the normal course of T-cell maturation, however, several TCR selection steps are designed to protect the individual from expression of such autoreactive T cells. This pathway of T-cell development is called thymic education because it occurs predominantly in the thymus, the site of T-cell maturation [9].

After expression of rearranged TCR genes, immature T lymphocytes in the thymus are capable of receiving antigen-specific signals from HLA-peptide complexes expressed on APCs. Early in the maturation process, such T cells encounter HLA–peptide complexes presented by thymic epithelial cells arrayed in the thymic cortex. In a process known as positive selection, T cells at this immature stage receive growth and differentiation signals as a result of TCR ligation. T cells with rearranged TCR capable of reacting with self-HLA and self-peptide complexes continue to develop and mature [10]. T cells with TCR that do not recognize self-HLA–peptide complexes die within the thymus. This positive selection of T cells based on HLA–peptide recognition is responsible for the phenomenon known as MHC restriction, wherein T cells specific for a given peptide recognize that peptide only in the context of a particular HLA molecule. Thus, following TCR alpha and beta gene rearrangement, CD4, CD8 double-positive, immature T cells are positively selected to become single positive and restricted to either specific class I molecules (CD8 T cells) or specific class II molecules (CD4 T cells).

After positive selection, the T cells traffic to the thymic medullary areas, where they encounter HLA–peptide complexes expressed on a different type of APC, namely, hematopoietically derived monocytes and dendritic cells similar to the types of mature APCs used by the peripheral immune system to respond to antigenic challenge. In the thymic medulla, these immature T cells can proceed through one of the four developmental routes. One, when they encounter very strong antigenic signals from the HLA–peptide complex, an apoptosis (programmed death) response is triggered that results in the deletion or negative selection of these strongly autoreactive cells. Two, they can undergo receptor editing to display a receptor that is not, or is less strongly, self-reactive. Three, they can become tuned or anergic, so they are less responsive to self-antigens. Four, T cells specific for self-peptides that are not expressed in the thymus can exit the thymus as immunologically ignorant but potentially autoreactive and approximately 20%–50% of the resulting TCRs are dangerously self-reactive, requiring peripheral control [10]. Experimental animal models of immune maturation have demonstrated that failure of this negative selection maturation step results in several types of organ-specific and systemic autoimmunity. A complex set of poorly characterized developmental regulatory genes controls this T-cell maturation pathway, some of which may contribute to human autoimmunity and aberrant immune response.

11.3.6 Breakdown of Immunologic Homeostasis or Equilibrium

The clonal deletion of autoreactive T cells in the thymus is only one of the mechanisms for preventing autoimmunity. In general, the adaptive immune response to a specific antigen is a dynamic and regulated process, reflecting the ability of the vertebrate immune system to maintain a "precarious equilibrium between the extremes of reactivity and quiescence." The immune system has several fail-safe mechanisms to prevent aberrant or unregulated TCR activation, as well as several feedback mechanisms designed to guide pathways of T-cell commitment to specific effector functions. As a counterbalance to immune activation, the immune system has several mechanisms for downregulating lymphocytes and terminating an ongoing immune response. The disruption of these terminator pathways can result in uncontrolled immune amplification and autoimmune disease. Some mechanisms for lymphocyte regulation are mediated through cell surface intermolecular interactions, analogous to pathways used for activation. One of the most important of these pathways is the interaction between CTLA-4 molecules expressed on T lymphocytes and the CD80/CD86 molecules (B7) expressed on antigen-presenting cells. The CTLA-4-B7 recognition system acts as a counterbalance to the CD28-B7 recognition system involved in T-cell activation. When the CTLA-4 protein is expressed on T lymphocytes and is activated by contact with the B7 molecule on APCs, it recruits phosphatases to the site of TCR activation complexes, which intersect and interrupt the kinase-mediated activation pathways, resulting in a negative, rather than a positive, signaling pathway. The lymphoid protein phosphatase (LYP), encoded by the gene PTPN22, is known to associate with the negative regulatory kinase Csk and to downregulate T-cell activation. The R620W polymorphic variant at this locus is associated with a number of autoimmune diseases.

Another form of immune downregulation is mediated by the TNF cytokine family and by the Fas-Fas ligand set of cell surface ligands, leading to death of the target cells. These terminator effector pathways induce apoptosis, by triggering signaling cascades that activate death-mediating molecules such as caspases and that inhibit protective molecules such as BCL-2. Yet another set of signal transducing cell surface molecules that can maintain homeostasis are the inhibitory and activating receptor systems on NK cells. In general, receptor systems on lymphoid and myeloid cells can result either in activation or inhibition depending on the structure of the cytoplasmic domain and the signals transduced when the receptor is engaged by the ligand. An appropriate balance between activation and inhibition signals seems to be required for normal immune modulation, by regulating initiation, amplification, and termination of specific responses. The absence of inhibitor signaling, as demonstrated in knockout mice with targeted disruption of inhibitory receptors, can result in autoimmune diseases and unregulated inflammatory responses.

11.4 GENETICS OF AUTOIMMUNE DISEASES

Familial clustering and comparisons of disease concordance among monozygotic and dizygotic twins clearly indicate that a variety of autoimmune diseases have a strong genetic component. Unlike the case of monogenic diseases caused by mutations, the alleles and genotypes that predispose to most autoimmune diseases are present in normal healthy individuals and generally only a minority of individuals with the high-risk genotypes actually gets the disease. Most autoimmune diseases are polygenic with a significant environmental component. In principle, polymorphism in any of the genes that encode elements in the T-cell activation pathways described earlier could play a role in genetic predisposition to specific autoimmune diseases. Linkage analysis studies have been used until recently to study possible transmission of two or more linked genes on a chromosome. Studies of autoimmune disease in families show some familial aggregation, but the linkage analysis is generally of a low power. A linkage analysis study is feasible only if the genetic component of the disease is strong and samples are available from relevant family members.

Several historic studies indicate many autoimmune disorders that show linkage to and/or association with genes in the MHC and specific alleles of the HLA loci. The HLA region on the short arm of chromosome 6 (6p21.3) spans 3.6 Mb and contains approximately 200 genes, many of which are involved in immune function. Both the HLA class I and class II genes encode highly polymorphic cell surface molecules that bind and present processed antigens in the form of peptides to T lymphocytes. Recognition by the T cell of the HLA-peptide complex, along with a costimulatory signal, results in T-cell activation. The class I molecules HLA-A, -B, and -C are found on the surface of most nucleated cells, presenting peptides primarily derived from endogenously synthesized proteins (e.g., self, viral, and tumor peptides) to CD8 T cells. These heterodimers consist of an HLA-encoded alpha chain associated with a monomorphic polypeptide, beta-2 microglobulin, encoded by *B2MR* on chromosome 15.

The HLA class II molecules consist of HLA-encoded alpha and beta chains associated as heterodimers on the cell surface of antigen-presenting cells such as B cells, macrophages, and dendritic cells. Class II molecules HLA-DR, DQ, and DP serve as receptors for processed peptides, derived predominantly from membrane and extracellular proteins (e.g., self and bacterial peptides) and are presented to CD4 T cells. Both the HLA-DQ and DP regions contain one functional gene for each of their alpha (HLA-DRA) and beta (*HLA-DPB1*) chains, as well as the pseudogenes *DQA2*, *DQB2*, *DPA2*, and *DPB2*. The HLA-DR region, however, contains one functional gene for the alpha chain (DRA), but either one or two functional genes for the beta chain, depending on the haplotype. All individuals express a DRB1 encoded polymorphic polypeptide that is found on the cell surface in association with the monomorphic alpha chain. The other functional class II DRB genes DRB3, DRB4, and DRB5 encode a beta chain that forms a second cell surface heterodimer with the DRA-encoded alpha chain. In general, the DRB3 locus is found on haplotypes where DRB1 is *03 (comprising the serologic subtypes DR17 and DR18), *11, *12, *13, or *14; the DRB4 locus is found on haplotypes where DRB1 is *04, *07, or *09, and the DRB5 locus is found on haplotypes where DRB1 is *15 or *16 (corresponding to the serological subtypes DR15 and DR16 of serologic type DR2). DRB1*01, *08, and *10 haplotypes typically have only the DRB1 locus.

Many other genes with important immune functions are found within the MHC. In the class III region, between the HLA-DR region (class II) and the HLA-B locus (class I), are found the complement genes encoding C2 and C4, as well as the TNF-alpha and -beta loci. The MICA and MICB loci are located just centromeric of HLA-B. The *TAP1* and *TAP2* loci are located between the *DQ* and *DM* loci. The TAP loci encode the peptide transporter molecules involved in loading the newly synthesized class I molecules in the ER with peptides derived from proteolysis of proteins in the cytoplasm by the proteasome. The *LMP2* and *LMP7* genes, located near the TAP loci, encode the γ-interferon-inducible subunits of the proteasome. HLA-E and HLA-G are class I loci, whose polymorphism and tissue distribution are both much more limited than those of the classical HLA-A, B, and C genes. The HLA-G genes appear to be particularly important in pregnancy, where they are expressed at the maternal—fetal interface of the cytotrophoblast. There are many other genes within the MHC, whose function has not yet been elucidated. In general, these loci are all much less polymorphic than the HLA class I and class II loci.

11.5 HLA ALLELIC DIVERSITY AND POPULATION GENETICS

The HLA class I and class II genes are the most polymorphic coding sequences in the human genome.

Virtually all this extensive sequence diversity is localized, for the class I loci, in the second and third exons, and for the class II loci, in the second exon. For the class I molecule, the peptide-binding groove is formed by a single chain with the beta-pleated sheet floor and two alpha-helical walls being encoded by the second and third exons. The class II polymorphic second exons encode the outer domains of the alpha and beta chains, which, together form the characteristic peptide-binding groove.

The patterns of allelic sequence diversity for both the class I and class II loci are highly unusual; some alleles differ in the polymorphic exons by as much as 15%, and the sequence variation is distributed as a patchwork of localized polymorphic sequence motifs. The allelic diversity at these loci is thought to have been generated by recombinational mechanisms such as gene-conversion-like events or, to a lesser extent, by reciprocal recombination. Thus, new alleles appear to have been created by shuffling these discrete polymorphic sequence motifs. In addition, point mutation has contributed to sequence diversity at the HLA loci.

Although a very large number of alleles (e.g., >400 for HLA-DRB1) can be found in the global human population, a much smaller number (e.g., 30–50 for HLA-DRB1) is present in most individual populations. Many populations who have gone through bottlenecks or founding events (e.g., Indigenous Americans) show more limited allelic diversity. In general, different populations tend to have different distributions of alleles as well as exhibit different patterns of linkage disequilibrium. Strong linkage disequilibrium is a striking feature of the genetics of the HLA region and can create difficulties in moving from an observed disease association with an allele at one locus, to identifying a causal locus. In all populations, particular haplotypes consisting of specific alleles at the linked HLA loci are found much more frequently than would be expected at random. The linkage disequilibrium for this haplotype extends to the DPB locus, about 3 Mb centromeric of HLA-A. Although a variety of evolutionary forces can create linkage disequilibrium, selection for particular combinations of HLA alleles has been suggested as the primary cause for these extended haplotypes. This strong linkage disequilibrium makes it difficult to assess which allele or which combination of alleles on a disease-associated HLA haplotype is responsible for observed correlations with disease. But, the alleles found in linkage disequilibrium differ among various populations, disease association studies in many different ethnic groups can prove valuable in identifying the contributions of individual alleles.

11.6 GENETIC SUSCEPTIBILITY TO AUTOIMMUNE DISEASE

Linkage of genes within the HLA region to several different diseases has been demonstrated by cosegregation studies in families or with nonparametric approaches such as haplotype sharing among affected sib pairs. The genetic region identified by linkage studies can be large, on the order of 5 cM (5% recombination). Higher-resolution mapping of disease genes provided by disease association studies because these depend on strong linkage disequilibrium between the genetic marker and the disease allele. In general, the association of a marker with a given disease implies that the genetic marker is significantly <1 cM away from the disease gene. In case-control association studies, the frequency of a given genetic marker among unrelated patients is compared with the frequency in matched controls. Although many parameters influence linkage disequilibrium, such as population history, population admixture, the age of the marker and of the disease alleles, and so on, the strength of linkage disequilibrium is generally inversely related to physical distance. Thus, a strong association with a given disease, particularly if it is found in different ethnic groups, suggests either that the associated marker locus is very close to the disease locus or that it may itself confer susceptibility to the disease.

HLA alleles that are positively associated with disease are referred to as susceptible, whereas negatively associated alleles are termed protective. In heterozygote conditions in which one haplotype is susceptible and one is protective, protection is generally dominant. Specific HLA alleles may enhance the autoimmune response by influencing immunogenicity, thereby predisposing to autoimmunity by influencing the expressed repertoire of T cells.

These disease associations with specific HLA alleles have been demonstrated in numerous case-control studies, as well as family studies in a large number of different populations. Most of these diseases are associated with alleles in the HLA class II region, but several diseases are associated with specific class I alleles. Some of the more common diseases associated with specific

HLA alleles are listed below along with newer data of other genes that may also play a role in the disease process.

11.7 HLA AND OTHER GENOTYPES

Inflammatory arthropathies are multifactorial polygenic disorders. GWAS led to identifying many of the genes involved. Associations with HLAs were first described in 1973 although the precise mechanisms by which these cause disease are still unclear. The association with protein tyrosine phosphatase PTPN22 is now well established as the second strongest with RA and is particularly a good example of an association that is common to several autoimmune conditions. There is now convincing evidence for the involvement of more than 30 genes in RA and at least 14 in ankylosing spondylitis (AS). Many of these are key factors in the regulation of inflammatory and immunological responses, as expected, but others have less-obvious biological expectations. The association of RA with HLA-DRB1 alleles is robust, and its interaction with smoking as an environmental factor is strongly suggested. The association of AS with several genes in the IL-17 producing (Th17) lymphocyte subset marked this as a potential therapeutic target. The highly significant association of AS with ERAP1 provided an important example of gene–gene interaction in susceptibility to a complex disease since the association is entirely restricted to those individuals with AS who also carry HLA-B*27 [1]. This association has also been evaluated at the level of protein structure and function. ERAP1 variants associated with protection against AS have reduced ability to trim peptide antigens to optimal length for binding to HLA class I molecules [11]. This raised the possibility that inhibitors of ERAP1 could be protective against AS.

As another example, SLE is associated with both major histocompatibility loci and non-MHC loci [12]. Studying the genetic basis of any autoimmune disorder requires updated and precise classification, including subtyping of individual disorders. This is exemplified by SLE [13].

11.8 RHEUMATOID ARTHRITIS

RA is the most common systemic inflammatory rheumatic disease and an archetypical multifactorial one. Genetic factors clearly are important, but Mendelian inheritance is rare. Rather multiple host factors—genetic, epigenetic, endocrine—interact with environmental factors—smoking, diet, microbiota, infectious agents, socioeconomics—to lead to RA [14]. Two classes occur based on the presence or absence of RA-associated antibodies, particularly rheumatoid factor (RF) and anticitrullinated protein antibodies. Distinguishing between patients is important to predict course and management. Antibody-negative RA tends to be diagnosed later, while antibody-positive RA has a more aggressive course if untreated [15].

Clinical descriptions of RA appeared fairly late in the medical literature, and it was only distinguished from gout and osteoarthritis by Garrod over 100 years ago. Classical RA is characterized by a destructive inflammatory arthritis affecting the synovial joints, but a wide spectrum of severity of the arthropathy exists. Additionally, a number of extraarticular manifestations are common. The substantial variation in clinical features and severity of the disease underlines the likelihood that its etiology is also heterogeneous, including the genetic factors that are involved. Familial clustering of cases is well recognized, with a generally accepted fivefold excess sibling recurrence risk over the general population risk ($\lambda s \sim 5$) although this varies with the severity of the disease in the proband.

One of the prominent features of RA is the presence of RF, which is an autoantibody to the Fc portion of IgG and present in most RA patients. Recently, antibodies to citrullinated proteins have been detected in about 80% of RA patients with 98% specificity. Citrullination is mediated by peptidylarginine deiminases, of which, *PADI4* has been identified as an RA-susceptibility gene [16]. Citrullinated peptides bind with high affinity to DRB1*04:01, perhaps explaining some of the HLA association seen with this disease.

When patients with citrullinated protein antigen positive forms of RA are studied, the top signal maps to *DQA1*03:01* and the top SMP is located ~2.5 kb upstream from HLA-DRA. Other independent SNP signals within the class II region are located between *HLA-DQA1* and *DQA2*; within the *HLA-DOB* locus, and in the intron of the *HLA-DRA* locus. Seropositive homozygous patients have an OR equal to 4.57 compared with heterozygotes. No association or interaction with *HLA-DRB1* has been seen, suggesting that these are independent susceptibility loci. MICA and

MICB have also been implicated in RA. Because they are present on the synovial cells, they can activate autologous T cells through their NKG2D ligand.

Specific HLA class II alleles are associated with RA, with risk in white populations increased 5–10-fold for individuals with specific *HLA-DRB1*04* genes, such as *DRB1*04:01* and *DRB1*04:04* and **04:05*. In addition, *DRB1*01* and **10* also show a weak association in some studies. However, the same alleles are prevalent in the normal population, so that screening for disease susceptibility based on the presence of these alleles is not highly predictive and is not clinically indicated. The absolute risk to an individual positive for one of these specific DR4 susceptibility alleles is, on average, only 8% (range 5%–12%), compared with a population prevalence of approximately 1% [17]. In addition, there is marked clinical heterogeneity in the spectrum of RA associated with these HLA genes. Although 80%–90% of patients with long-standing erosive forms of RA are positive for the *DR*4* susceptibility alleles, only 55%–65% of newly diagnosed RA patients carry the same alleles. This observation has been the subject of considerable clinical interest because it appears to indicate that patients diagnosed with RA represent a spectrum of clinical heterogeneity that is caused, in part, by an underlying genetic heterogeneity. Indeed, further studies have shown that the *DR4* susceptibility genes are most prevalent in patients with RF and severe erosive RA, and that other class II susceptibility genes, such as *DRB1*01:01*, are markers for a broader clinical spectrum that includes seronegative and nonerosive forms of RA.

This genetic distinction between severe erosive RA and other forms of polyarthritis has potential clinical utility for prognosis. When individuals newly diagnosed with RA are analyzed, the presence of *DRB1*04* RA susceptibility alleles (e.g., *DRB1*04:01* or **04:04*) predicts progression to severe erosive disease. In studies of patients who met clinical criteria for RA but did not yet have erosive disease (X-ray evidence of joint erosions), the presence of *DRB1*04* susceptibility alleles correlates with the onset of erosive disease within 2 years. Patients who lacked these DR4 susceptibility genes had a high frequency of nonerosive polyarthritis and a low frequency of erosive and progressive RA. In other studies, using unselected RA patients, there was little additional prognostic value shown for use of these genetic tests in patients who already had erosive disease, indicating that the positive predictive value of *DRB1*04* analysis is probably limited to use in early patients, who do not yet have severe disease [18].

Therefore, although the association between HLA-DR genes and RA suggests a role for HLA molecules in disease susceptibility, the current practical clinical utility lies in the association between specific RA susceptibility alleles and prognosis for progressive erosive forms of disease. HLA-DRB1**04* genetic analysis is potentially useful in this regard as a means to select patients for alternative forms of therapy. In a randomized clinical trial comparing an aggressive form of multidrug therapy for RA with a single-drug regimen, there was a marked distinction in clinical outcome that corresponded to the *HLA-DRB1*04* genetic analysis [19]. The DRB1*04-negative patients, as expected, had overall milder disease course over the 2-year clinical trial, and they responded well to both arms of the clinical trial. Patients treated with the single-drug regimen responded to therapy with an excellent clinical outcome (83% response). In contrast, the DRB1*04-positive patients who were randomized to the single-drug regimen did very poorly, with a marginal (32%) clinical response. The DRB1*04-positive patients who received the multidrug treatment, however, had a 94% response rate, comparable to the DR4-negative groups (88%). Thus, the use of HLA-DR4 genetic typing to influence the selection of "aggressive" versus standard therapies in RA for newly diagnosed Caucasoid patients who do not yet have joint erosions may be a valuable adjunct to medical management in this disease.

Essential to any drug therapy for immune-mediated inflammatory diseases is attention to monitoring patient responses [20,21].

Typically, RA presents as a distal small-joint arthropathy of the hands (Fig. 11.1) and feet that may initially involve relatively few joints. Subsequently, the more proximal load-bearing joints become involved, potentially leading to severe functional disability. Any synovial joint, including the temporomandibular and cricoarytenoid joints, may be affected. Overall, the disease is approximately three times more common in women than men, but this varies with the age of onset. Thus, at the age of 30 years, women are nearly 10 times more frequently affected than men, while there is no gender difference in incidence by the age of 65 years. RA may present in many different ways. In about two-thirds of cases, there is pauciarticular, insidious onset

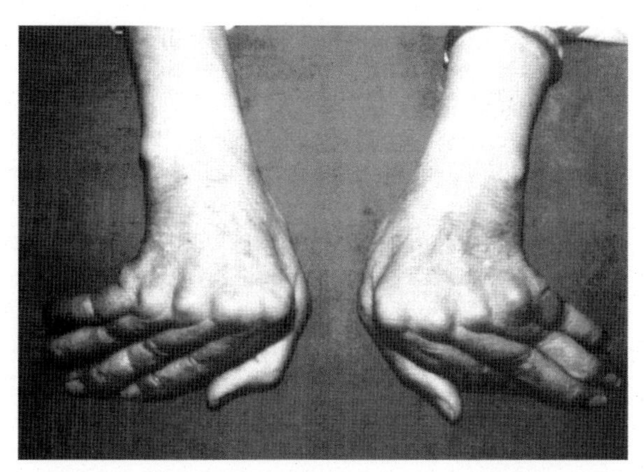

Figure 11.1 Classical rheumatoid deformity of the hands with ulnar deviation and subluxation at the metacarpophalangeal joints.

(typically in the hands and feet) with cumulative joint involvement over the course of months or years. This may be difficult to distinguish from other self-limiting causes of synovitis in the early stages of the disease. In contrast, the onset is explosive in a small minority of cases who develop widespread symmetrical polyarthritis over a few days or even overnight. Despite this dramatic onset, a proportion of such cases eventually shows complete resolution of synovitis and little joint destruction a year or two later. Systemic features, including fever, weight loss, and malaise, occasionally dominate the onset of RA, particularly in middle-aged men, prompting extensive investigation to exclude alternative causes, such as deep-seated infection or malignancy. Limb girdle symptoms similar to polymyalgia rheumatica may be prominent, particularly in those with later-onset disease. Occasionally, patients present with "palindromic rheumatism"; this consists of short-lived attacks (~48h) of mono-, oligo-, or polyarticular synovitis that initially resolve completely but recur at intervals and may eventually persist. Positive tests for RF and anticitrullinated protein antibodies (ACPAs) are predictive of progression to persistent rheumatoid disease.

Extraarticular features are common in RA. These include skin nodules, cachexia, and normochromic, normocytic anemia. Mild peripheral neuropathy with glove-and-stocking sensory loss is more common than mononeuritis multiplex caused by rheumatoid vasculitis of the vasa nervorum. Major rheumatoid vasculitis affects men more commonly than women and can cause life-threatening cutaneous ulceration, myocardial infarction, or bowel ischemia. Serositis is relatively common in RA; pericardial effusions are often asymptomatic and have been reported in up to 30% of those with RA on echocardiography. Rarely, constrictive pericarditis may result, and the heart valves are occasionally affected, particularly in those with nodular disease. Corticosteroids may be required to treat symptomatic pleurisy or pericarditis. About 20% of patients develop secondary Sjögren syndrome with keratoconjunctivitis sicca, xerostomia, or both. The eyes are commonly affected by episcleritis, which usually requires no treatment, but, less commonly, there may be scleritis, which may require systemic corticosteroids or immunosuppression. Lymphadenopathy is common, but <1% of patients develop Felty syndrome (RA, lymphadenopathy, neutropenia, and splenomegaly). This disorder rarely develops <10 years from the onset of RA. Patients with Felty syndrome probably suffer joint disease of similar severity to non-Felty patients but have more extraarticular manifestations. Familial recurrence of RA is more common where the proband has Felty syndrome.

Historically, despite active treatment with standard disease-modifying drugs, <50% of those with RA have been able to work full-time after 10 years of disease. Mortality is also increased by about 50%, reducing life expectancy by about 11 years (mainly due to increased cardiovascular disease and infections). New approaches to treatment with antitumor necrosis factor biologics have had a dramatic effect on joint disease in RA that may eventually be reflected in better mortality statistics.

11.8.1 Diagnostic Criteria

RA is regarded by many as a heterogeneous group of conditions with overlapping phenotypes. Over the years, many attempts have been made to define the disease more accurately using classification criteria, including those developed and routinely revised by the American College of Rheumatology. Comparisons between population studies are therefore somewhat complicated by differences in the diagnostic criteria that have been used. The 2010 American College of Rheumatology/European League Against Rheumatism classification criteria (Table 11.3) have a high degree of sensitivity and specificity. They have been modified particularly to detect early disease more effectively than

TABLE 11.3 2010 American College of Rheumatology/European League Against Rheumatism Classification Criteria for Rheumatoid Arthritis (From Reference xxx Aletaha)

Score ≥6/10 required for classification as definite RA: Add Scores A–D
A. Joint Involvement
1 large joint **0 points**; 2–10 large joints **1 point**; 1–3 small joints **2 points**; 4–10 small joints **3 points**; >10 joints including at least 1 small joint **5 points**.
B. Serology
Negative RF and negative ACPA **0 points**; low positive RF or low positive ACPA **2 points**; high positive (>3 times upper limit of normal) RF or high positive ACPA **3 points**.
C. Acute-Phase Reactants
Normal CRP and ESR **0 points**; abnormal ESR or CRP **1 point**.
D. Duration of Symptoms
<6 weeks **0 points**; ≥6 **1 point**.

the previous criteria because it is now recognized that early treatment of RA can prevent irreversible joint damage [22,23]. The diagnosis is not difficult in typical cases once the disease has been present for several months, but in early disease, a confident diagnosis may be difficult, particularly where there is palindromic onset, limited synovitis, or normal inflammatory markers. As many as one-fifth of patients presenting at an early synovitis clinic may turn out not to have RA after 1 year or so. The recognition that ACPA occurs with high frequency in RA has led to their development as diagnostic aids in early inflammatory arthritis. ACPAs have a similar sensitivity to RF for RA [~80%] but are more specific [~98%]. Both RF and ACPA may precede the development of clinical disease by months or years [24].

11.8.2 Differential Diagnosis

The diagnosis of RA is relatively straightforward except in the early phase, when distinction from self-limiting arthropathies (e.g., reactive arthritis, viral arthropathies, and crystal arthropathies) may be more difficult. Distinction from the seronegative arthropathies, notably psoriatic arthritis and reactive arthritis, may sometimes be problematic. Generalized nodal osteoarthritis is occasionally similar and may coexist with RA; typically, it affects the proximal and distal interphalangeal joints but usually spares the metacarpophalangeal joints and wrists. Osteoarthritis affecting the knees and hips is very common and is partly genetic in origin; large-joint osteoarthritis requiring joint replacement has an excess sibling recurrence risk of 2.3 [25].

Crystal arthritis is common and may be difficult to distinguish from RA, particularly in the elderly. For example, gout is polyarticular in onset in 10% of cases and quite commonly is not associated with elevated uric acid levels in the acute phase. On the other hand, its typical presentation with monoarticular lower limb arthropathy is quite distinct (50% of initial attacks involve the great toe, "podagra"). The diagnosis is established correctly by demonstrating intracellular needle-shaped crystals of monosodium urate in aspirates from affected tissues that are strongly negatively birefringent under polarized light. It is strongly familial. Pyrophosphate arthropathy may cause a "pseudo-rheumatoid" pattern of chronic arthropathy, as well as the more widely recognized acute episodes of "pseudogout," which can also punctuate the more chronic forms of pyrophosphate arthropathy. The diagnosis of crystal arthritis is best established by aspiration of the affected joint and demonstration of the relevant intracellular crystals under polarized light. Familial forms of chondrocalcinosis, characterized by early-onset pyrophosphate arthropathy, often have a dominant inheritance pattern. Gain-of-function mutations in *ANKH*, encoding a transmembrane transporter of inorganic pyrophosphate, have been described. Polymorphic variants of *ANKH* are also associated with sporadic forms of pyrophosphate arthritis [26]. Pyrophosphate arthropathy is also a classic component of hemochromatosis, in which abnormal iron handling causes its deposition in numerous tissues, including the liver, myocardium, and synovium. The defective *HFE* gene is a member of the immunoglobulin superfamily and has a very high pathogenic variant allele frequency (~0.1) in northern Europe. Premature degenerative arthritis, particularly in atypical sites, such as the metacarpophalangeal joints, wrists, and ankles, should raise suspicions, especially if articular calcification is present. The diagnosis is best established by demonstrating saturation of transferrin and elevated ferritin levels. Elevated ferritin levels alone may be spurious, since ferritin, an acute-phase reactant, is commonly elevated in inflammatory states including RA.

11.8.3 Pathology

Despite intensive research, the pathophysiology of RA remains incompletely understood. Animal models of the disease have provided some useful insights, including the central role of tumor necrosis factor α (TNFα) in rheumatoid inflammation. Such models are imperfect, however, and much useful information has been gleaned from careful studies of the disease in humans and from dissecting the genetic basis of the disease in GWAS.

There is convincing experimental and clinical evidence of involvement of many cell types, including macrophages, B lymphocytes, T lymphocytes, and synovial fibroblasts, all of which are overrepresented in RA synovium. In RA, the normal synovial lining expands from 1–3 to 10–15 cell layers, which is predominantly composed of macrophages and synovial fibroblasts; the additional influx and proliferation of immune and resident synovial cells also contribute to synovial hyperplasia. The most abundant stromal cells are the fibroblast-like synoviocytes (FLSs) that appear abnormally resistant to apoptosis and accumulate in the RA synovium. Activated FLSs are detectable early on in rheumatoid disease, and although multiple factors from both the innate (primarily TLR2) and adaptive (e.g., TNFα, IL-1, IL-6 and IL-17) immune systems support FLS activation, at least part of FLS activation appears to be independent of the surrounding inflammatory tissue. FLS activation leads to the upregulation of numerous chemokines (MIP-1α), cytokines, matrix metalloproteinases, and adhesion molecules (integrins, VCAM-1, and cadherins) required for the recruitment of inflammatory cells and their destructive effects in the joint (muller 6). In particular, the production of interleukin (IL)-15 by FLS induces the production of other cytokines, such as TNFα and IL-17, by T cells through direct cell–cell contact. This further stimulates the expression of IL-15 and IL-6 by FLS, creating a feedback loop that favors persistent inflammation. Microarray analysis has suggested two subgroups of FLS distinguished by their gene expression signatures. FLSs from highly inflamed areas have a TGFβ-activin A-inducible signature characteristic of myofibroblasts (which have a particular propensity for chemokine and cytokine production), while FLSs from less-inflamed synovial tissue have a predominance of insulin-like growth factor-regulated genes.

Classification of RA synovial tissue histomorphology has been attempted according to differential synovial infiltration by leukocyte populations. Most patients with RA have diffuse sublining infiltration consisting of scattered CD4 T lymphocytes and monocytes, but a significant proportion (up to 25% in some series [27]) develops more discrete synovial aggregates with T- and B-cell compartmentalization. Intriguingly, these aggregates can progress into ectopic lymphoid structures resembling germinal centers with characteristic follicular dendritic cell networks and proximity to high endothelial venules.

The hypertrophic synovium develops into invasive pannus, eroding the articular cartilage and bone, particularly at the points of synovial attachment, causing loss of joint space, instability, and deformity. Accumulation of osteoclasts at sites of bone erosion is characteristic. These multinucleated cells express receptor activator of nuclear factor-κB (RANK) and are derived from CD14-positive cells of monocyte/macrophage lineage under the influence of macrophage/monocyte colony-stimulating factor, RANK ligand (RANKL), and inflammatory cytokines. Increased numbers and activity of osteoclasts are both hallmarks of inflammatory bone loss. The inflammatory milieu within the RA joint also serves to augment not only osteoclast precursor recruitment from bone marrow but also their subsequent differentiation into mature osteoclasts. Mature osteoclasts secrete hydrochloric acid to dissolve inorganic bone matrix, while the bone matrix proteins are degraded by proteolytic enzymes (e.g., matrix metalloproteinases and cathepsin K). The main sources of RANKL are osteoblasts, FLSs, and activated T cells, and its expression in these cells is upregulated by proinflammatory cytokines in synovial tissue, including IL-1, IL-6, IL-17, and TNFα. Such cartilage and bone loss are manifested radiologically by juxtaarticular osteoporosis, narrowing of the joint space and the development of erosions and joint deformity (Fig. 11.2); however, the relationship between synovial inflammation and articular erosion is complex and variable. Occasionally, patients with prolonged synovitis do not erode, while erosions may be apparent in others at the time of presentation. Erosive disease is more common in those who are HLA-DRB1*04 positive.

Angiogenesis is a key event in the maintenance of synovial inflammation, delivering nutrients and immune cells to the site of inflammation. Despite new vessel formation, inflamed synovial tissue is invariably

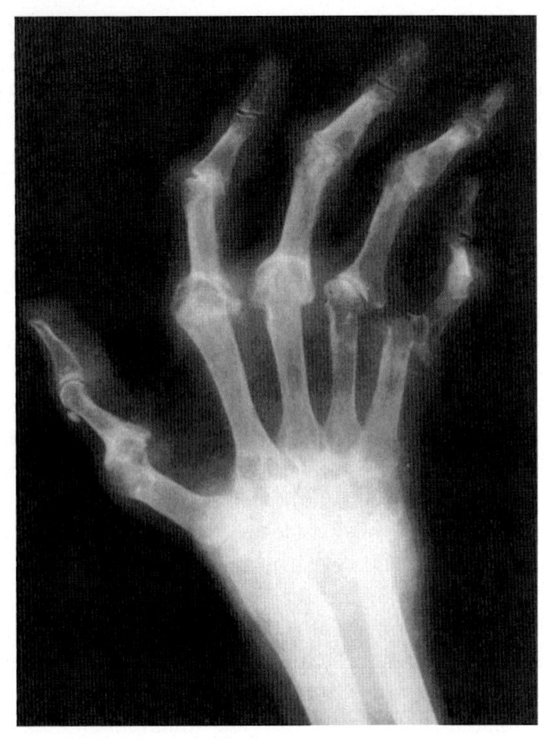

Figure 11.2 Radiographic appearances of advanced rheumatoid arthritis.

hypoxic: synovial proliferation leads not only to increased metabolic demand but also regional hypoperfusion as tissue hyperplasia increases the distance between proliferating cells and infiltrating vessels. The resultant low tissue oxygen tension drives the transcription of hypoxia-inducible factor (HIF) regulated genes, not least vascular endothelial growth factor (VEGF), the most potent proangiogenic molecule.

The complex pathophysiology of RA has been amply demonstrated by the efficacy of immunomodulatory drugs that target discrete elements of the immune system. These include biologic agents targeting cytokines, such as TNFα (therapeutic monoclonal antibodies and a recombinant TNF receptor/Ig fusion protein) and the IL6 receptor; T-cell activation (recombinant CTLA4/Ig fusion protein); and B cells (anti-CD20 agents).

11.8.4 Epidemiology

Depictions of RA in art and literature seem to be lacking until about 200 years ago in Europe. As a consequence, there has been speculation that the disease appeared relatively late in Europe and may even have been imported from the New World. RA has rarely been identified convincingly in the archeological record, but it may be mistaken for other diseases. Nonetheless, there is some evidence that the disease existed in Egypt in the third millennium B.C. and also in Roman Britain.

RA shows no preference by gender, ethnicity, or age, although the disease has a peak incidence in the age of 40–50 years. There is an excess of incidence in the puerperium, and while most women experience remission in pregnancy, recrudescence in the puerperium is almost invariable and may be severe.

The prevalence of RA in many populations worldwide is remarkably similar (~1%), but with some important exceptions that may give clues to the etiology of the disease. Environmental and/or genetic factors could account for the variations in the prevalence observed in certain ethnic groups [28]. For example, the disease is rare in most of sub-Saharan Africa (prevalence = 0.3%) but common in certain Native American groups (5%–7%). South African Blacks in a rural environment exhibit the same low prevalence of the disease as in most of sub-Saharan Africa. This contrasts starkly with the high frequency of the disease in their urban counterparts, similar to that in South Africans of European descent, which strongly suggests an environmental influence. By contrast, the high frequency of RA in the Chippewa and Yakima Native Americans of North America may be due to the high frequency of certain HLA class II alleles associated with RA: HLA-DRB1*04 and the rare HLA-DRB1*1402 allele, respectively. In most populations that have been studied, there are strong associations with various *HLA-DRB1*04* alleles (relative risk ~5) [21,24,25]. The *HLA-DRB1*04* series, originally defined serologically as the HLA-DR4 transplantation antigen, can be subdivided into numerous HLA-DRB1 alleles, not all of which are associated with RA.

11.8.5 Genetics

Concordance rates in monozygotic twins vary between studies from 12% to 30%. In the United Kingdom in the 1960s, concordance in identical twins with seropositive erosive disease was estimated as 30% [29]. This contrasts with a lower rate of 12% obtained in a nationwide survey of twins in Finland that included individuals with less-severe disease. Another UK study found 15% concordance in identical twins and demonstrated that this was highest in those twin pairs that were HLA-DRB1*04 positive [30]. In all of these studies, the rates for monozygotic twin concordance were four to five times higher than for dizygotic twins, highlighting the likely importance of genetic factors. There seems little doubt that nongenetic factors are important, not only in the development of RA but also in its progression. On the basis of these relatively small studies, broad sense heritability has been estimated to be about 55% [31,32].

Sib recurrence risk varies according to the severity of the disease in the proband. In one study in the 1960s [29], the excess risk to the sibs of individuals with mild nonerosive seronegative disease was barely greater than the general population frequency (λs 1.1). This is increased sixfold in the sibs of individuals with seropositive, erosive disease, where as many as 15% of individuals will have an affected first-degree relative [33]. Family recurrence risks of RA are also around seven times greater when the proband has Felty syndrome and are probably also increased when the index case has early-onset disease. Sibs of those with RA who share both HLA haplotypes identical by descent with the proband and are at particular risk [34] is especially interesting as this group may be more likely than others to be HLA-DRB1*04 homozygous [35]. The excess sibling recurrence risk over the general population risk (λs) is critically dependent on the criteria used in diagnosis. Thus mild, nonerosive disease is relatively common in the community at large, and sib recurrence risks are relatively low. In contrast, seropositive, erosive rheumatoid disease (that is typically followed up in hospital clinics) is much less common in the community and has higher sib recurrence risks. For this type of chronic, erosive disease, λs is probably nearer 12–14, and the genetic contribution is correspondingly higher.

RA is clearly linked to the MHC, as demonstrated in many populations (Table 11.4). HLA class I associations with RA are invariably weaker than those in the HLA class II region, particularly with *HLA-DRB1* alleles [36]. There are important associations not only with HLA-DRB1 but also with the HLA-B and HLA-DP loci [37].

11.8.6 Associations With the Classic Serologic HLA-DR (HLA-DRB1) Specificities

Although these associations were originally defined by serology, they are amply replicated at the DNA level. Associations with HLA-DRB1*04 are well established in most ethnic groups globally, typically with a relative risk of approximately 5. Some important exceptions to this general rule have been observed, particularly among Jews, some Indian populations, Chileans, and the Yakima Native Americans, in whom alternative associations have been found with *DRB1*01*, *DRB1*09*, and DR6 (*DRB1*1402*). The association with HLA-DRB1*09 originally described in Chileans is a minor association in the United Kingdom. Although the association with *DRB1*04* predominates in most populations, other weaker associations are not uncommon. Thus, associations with HLA-*DRB1**10 have been observed in Spaniards, Indians, and Jews, and with HLA-*DRB1*01* (DR1) in Indians, Jews, and southern European populations, in particular. While the strongest associations in the United Kingdom are with HLA-DRB1*04, weaker associations can also be detected with both *DRB1*01* and *DRB1*10* [36].

Not all DR4 haplotypes are equally associated with RA, despite the fact that SNP analysis indicates a high degree of conservation in the DNA flanking the HLA-DRB1 locus that encodes the polymorphic DRβ1 chain. Thus, HLA-DRB1*0401, *0404, *0408, and *0405 are positively associated with RA, but *0402, *0403, and *0407 are negatively associated. In the United Kingdom, there is a hierarchy of HLA-DRB1 susceptibility alleles (Table 11.3). The most widely held explanation for this observation (shared epitope hypothesis) is that there is a conserved functional epitope in the HLA-DR molecules positively associated with RA (encoded by *DRB1*0401, *0404, *0405, *0408, *0101, *0102, *10, and *1402*). In particular, a highly conserved sequence between amino acids 67 and 74 along the α-helix derived from the DRβ chain, which forms one side of the antigen-binding site of the DR molecule, is incriminated in susceptibility as shown in Fig. 11.3. All the rheumatoid-associated DR molecules share an identical or similar sequence, ^{67}LLEQRRAA^{74}; with the exception of *DRB1*0401*, where the substitution of a basic lysine for arginine at position 71 is a relatively conservative change. It has therefore been speculated that the capacity of the HLA-DR molecules to bind a potentially arthritogenic peptide may trigger an autoimmune response within the synovial joint, leading to the chronic process of inflammation and destruction that ensues. This shared epitope is associated predominantly with RF positivity. In addition, there is a

TABLE 11.4 HLA Sharing in Rheumatoid Arthritis—Affected Sib-Pairs

		HAPLOTYPE SHARING		
Ethnic Group	**Criteria**	**0**	**1**	**2**
French	1958 ARA C/D	0	1	0
Mixed	1958 ARA C/D	3	11	7
American	1958 ARA C/D	0	4	1
North Sweden	1958 ARA C	2	8	4
American	Seropositive erosive	2	0	1
Caucasoid	1958 ARA C/D	2	7	4
Italian	1958 ARA n/s	1	0	1
Canadians	1958 ARA C/D	1	0	2
Australasian	1958 ARA C/D	4	9	4
German	1958 ARA C/D	0	0	1
Norwegian	1958 ARA C/D	1	4	2
American	1958 ARA C/D	3	6	3
Dutch	1958 ARA C/D	1	8	0
North Sweden	1958 ARA C/D	1	1	0
UK white	1958 ARA n/s	5	20	17
Southern Irish	1958 ARA C/D/Pr/Po	6	12	16
UK white	1958 ARA D/Pr	6	24	23
Mixed	1958 ARA C/D	1	7	7
Dutch	1958 ARA D	0	0	2
American	1958 ARA C/D	4	8	4
North Indian, Hindu	n/s	0	3	6
English	1958 ARA C/D	3	6	6
American	1958 ARA C/D	4	15	18
Egyptian	1958 ARA n/s	0	7	10
German	1958 ARA C/D	7	4	1
Total		**57**	**165**	**140**
Percentage		**16**	**45**	**39**
Expected percentage		**25**	**50**	**25**

ARA, American Rheumatism Association; *C*, classical; *D*, definite; *ns*, not specified; *Po*, possible; *Pr*, probable. Original studies in this table are referenced elsewhere.

pronounced and dose-dependent risk of smoking in individuals carrying the shared epitope in developing RF-positive, but not RF-negative, RA [38]. This gene–environment interaction in RA may be explained by smoke-induced citrullination of pulmonary peptides resulting in increased binding affinity to shared epitope MHC class II receptors and thereby T-cell activation [38].

Associations with other MHC genes have been suggested by some studies. For example, TNF polymorphisms might influence responses to anti-TNF therapies [39]. It is extremely difficult to separate possible effects from the TNF locus and those arising from HLA-DRB1 haplotypes because of the linkage disequilibrium between the two loci [40]. To achieve this, association studies must be both adequately powered and also very carefully matched for controls (e.g., for HLA-DRB1 status). An extended TNF haplotype marking a 126-kb region centromeric to TNF was particularly interesting. This region contains *AIF1*, encoding allograft inflammatory factor that is implicated in inflammatory states [41]; however, such suggestions require very careful validation in carefully controlled, high-density genetic mapping. The longstanding suggestion that numerous HLA alleles, particularly in the MHC class II region, are involved in

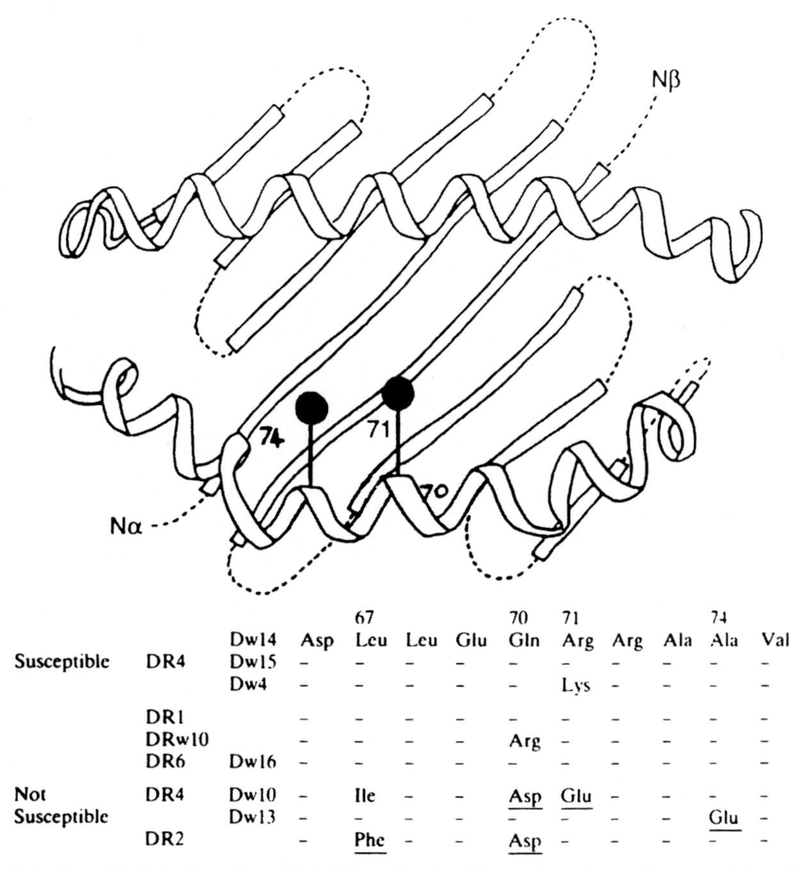

				67			70	71			74	
		Dw14	Asp	Leu	Leu	Glu	Gln	Arg	Arg	Ala	Ala	Val
Susceptible	DR4	Dw15	-	-	-	-	-	-	-	-	-	-
		Dw4	-	-	-	-	-	Lys	-	-	-	-
	DR1		-	-	-	-	-	-	-	-	-	-
	DRw10		-	-	-	-	Arg	-	-	-	-	-
	DR6	Dw16	-	-	-	-	-	-	-	-	-	-
Not Susceptible	DR4	Dw10	-	Ile	-	-	Asp	Glu	-	-	-	-
		Dw13	-	-	-	-	-	-	-	-	Glu	-
	DR2		-	Phe	-	-	Asp	-	-	-	-	-

Figure 11.3 Antigen-binding site of an HLA-DR molecule. Charged amino acids at position 71 in Dw10 (DRB1*0402) and 74 in Dw13 (DRB1*0403 and *0407) abolish susceptibility to rheumatoid arthritis; non-charged substitutions at position 71 in Dw4 (DRB1*0401) do not.

RA has been confirmed by recent high-density mapping studies of the MHC. It is now clear that contributions arise not only from HLA-DRB1 but also from HLA-B and HLA-DP [37]. It has also been possible to refine the precise epitopes within each of these HLA molecules that are associated with susceptibility. This is analogous to the complex HLA associations that have also been described with type 1 diabetes mellitus [42].

Despite the relatively strong association between RA and HLA-DRB1 alleles, at least 15% of patients do not carry the conserved DRB1 epitope. In these circumstances, some form of immunogenetic modulation may occur as a result of exposure of the host immune system to the noninherited HLA-DRB1*04 from the mother in utero [43]. HLA-DRB1*04 appears to be over-represented among noninherited maternal HLA antigens in patients with RA [44]. Noninherited maternal HLA-DR antigens can exert a protective effect. It could reflect an impact on the shaping of the T-cell receptor repertoire or an influence on tolerance mediated through exposure of the developing immune system to maternal antigens resulting from long-term microchimerism [45].

11.8.7 Pathogenesis

The shared epitope hypothesis put forward to explain the disparate HLA-DR associations with RA [46] implies the presentation of a single or limited range of potentially arthritogenic peptides by antigen-presenting cells within the joint to a subset of T lymphocytes capable of initiating a specific inflammatory response. This would most obviously produce a dominant model of susceptibility, but several studies have cast doubt on this, suggesting that the involvement of HLA genes fits best with a recessive model [47,48]. The hypothesis of HLA-DQ-mediated susceptibility would fit with this model as would the role of the HLA genes in shaping the T-cell receptor repertoire by positive and negative thymic selection. Another possibility invokes molecular mimicry between the shared epitope on rheumatoid-associated DR molecules with potential triggering pathogens although firm evidence for this has not been forthcoming, and such cross-reactive immune responses have not been observed routinely.

Early reports using PCR to detect expanded T-cell populations, particularly in the rheumatoid synovial compartment, were somewhat suggestive of an antigen- or superantigen-driven response. Although numerous subsequent studies have sought such oligoclonal expansions in patients with RA, there seems little consistency between individual studies to support such an idea [49].

The association with HLA-DRB1 variants and RA suggests that engagement of specific immune responses is important at some stage in the development of the disease. Additional signaling through costimulatory molecules in this "immunologic synapse" is critical to this process. Interactions between the CD80/86 molecules on antigen-presenting cells and their potential ligands on T cells (CD28 or cytoxic T lymphocyte associated antigen-4 (CTLA-4)) play an important role in determining activation. Variation in CTLA-4 is involved in susceptibility to autoimmune diabetes mellitus and thyroiditis. A large collaborative study from Sweden and North America suggests that there is also a weak association (OR 1.2) with RA [50]. It is certainly of interest in this context that the fusion protein CTLA4Ig has impressive disease-modifying effects in RA [51]. Other components of the immunologic synapse are potential candidates in the etiology of the disease. For example, CD28 is also associated with RA, and there is weak evidence that the inhibitory product of the programmed cell death 1 gene (PDCD1) may also be involved [52].

Major challenges remain in the identification of all the major genetic effects in RA, particularly those attributable to rare alleles. Even weak genetic influences could herald the identification of crucial pathogenic pathways and the means to modulate them.

11.8.8 Management

No single or combination of treatments cures RA. The earlier treatments begin, the better the chances are for a reduction in pain, joint deterioration, and evolution of systemic complications. Therefore, improved diagnostic criteria, better recognition of signs and symptoms by a variety of healthcare providers, intensive nonpharmacologic treatments such as physiotherapy, and judicious application of traditional and emerging targeted drugs have enhanced long-term outcomes and better quality of life [53]. About half of patients presenting with early undifferentiated polyarthritis have self-limiting disease, but only 15% of patients fulfilling criteria for RA go into remission [54]. In those with persistent arthritis, agents that modify the course of the inflammation are required, which are classified as disease-modifying antirheumatic drugs (DMARDs). These include sulfasalazine, methotrexate, antimalarials, gold salts, D-penicillamine, azathioprine, and leflunomide (often used in combination) as well as biologic DMARDs. Corticosteroids (either in low doses orally or as pulses) and nonsteroidal antiinflammatory drugs (NSAIDs) may be used at any time to suppress inflammation and erosive damage as second-line drugs are commenced. From the early 1990s, the treatment of the more severe forms of RA changed dramatically with the introduction of anti-TNF biologic agents [55,56]. In clinical trials recruiting patients with very active disease, these drugs routinely induced remission in a quarter of patients within 3 months. Further, they completely suppress the progression of erosions. Quite suddenly, it became unacceptable for rheumatologists to aim for anything less than complete suppression of the disease. This stimulated the use of combinations of the older disease modifying antiinflammatory drugs, such as methotrexate, sulfasalazine, and leflunomide, and if these failed to suppress the disease adequately, escalation to biologic therapies would follow. This has had a profound knock-on effect in the clinical practice of rheumatology with the development of sophisticated clinical and imaging tools to assess the degree of residual disease activity and the presence of potentially destructive synovitis. Patients must be treated aggressively early in the disease to prevent irreversible structural joint damage. Good responses to anti-TNF therapy can be anticipated in the majority of patients with ACR 20%, 50%, and 70% responses being achieved within 3 months by 60%, 40%, and 20% of patients with substantial disease activity before treatment (DAS28 $\geq$ 5.1). In those failing anti-TNF therapy, other options with broadly comparable efficacy include rituximab, an anti-CD20 monoclonal antibody B-cell depleting agent [57], tocilizumab (anti-IL6 receptor monoclonal antibody), and abatacept (recombinant CTLA4/immunoglobulin fusion protein).

Effective management of RA requires a coordinated approach by a multidisciplinary team because of the chronic nature of the condition. The primary objectives are the relief of pain and preservation of function. A combination of drugs, physical therapy, orthotics, and appliances to prevent or accommodate increasing long-term disability will be required. Surgical intervention is frequently needed at some stage in the course of the disease, but requires careful planning, particularly in patients with more severe forms of the disease, who may require many procedures.

The development of reliable methods of total joint replacement, particularly in the large weight-bearing joints of the hip and knee, has otherwise been the most important single development in the treatment of patients with RA. Good results from both hip and knee replacement can be anticipated in more than 90% of patients, and typically, the prosthesis will last for 15 years or more. Subluxation of the cervical spine is relatively common, particularly in patients with severe erosive disease, and may require surgical stabilization. The median standard mortality ratio for RA is approximately 1.5 compared with the general population [58], mainly from infection and cardiovascular disease. Only half of this can be attributed to classic risk factors such as smoking, hyperlipidemia, and hypertension, all of which should be tightly controlled.

11.9 SERONEGATIVE SPONDYLOARTHROPATHIES

The term seronegative spondyloarthropathy (SpA) refers to a group of inflammatory conditions characterized by inflammation of the entheses (sites of mechanical stress where ligaments or fibrocartilage interface with bone, e.g., sacroiliac joints) (Table 11.5). In contrast to RA, RF and ACPA are absent. These two characteristics differentiate these conditions from the more common RA, which is characterized pathologically by synovitis rather than enthesitis and in which RF is present in 85%.

Other features of SpA include an association with the MHC class I gene HLA-B*27; characteristic distribution of joint involvement with prominent axial, sacroiliac, and asymmetric lower limb peripheral large-

TABLE 11.5 Clinical Features of the Spondyloarthropathies

	Ankylosing Spondylitis	Reactive Arthritis	Psoriatic Spondyloarthritis	Enteropathic Spondyloarthritis
Sex	M > F	M > F	F > M	M = F
Age of onset (yr)	15–35	Any age	Any age	Any age
Uveitis	++	++	+	+
Conjunctivitis	—	++	—	—
Urethritis	—	++	—	—
Skin involvement	—	++	++	—
Mouth ulcers	—	++	—	+
Sacroiliitis	+++	++	++	++
Peripheral arthritis	Lower > upper	Lower > upper	Upper > lower	Lower > upper
Spinal symmetry	+++	+	+	++
Enthesopathy	++	++	++	++
Aortitis	+	+	?+	+
HLA-B*27 (%)	95	80	50	50
Risk for B*27-positive	2–8	10–20	Unknown	Unknown individual (%)
Self-limiting	—	++	—	—

—, rarely; ++, occasional; ++, frequent; ++++++, always.

TABLE 11.6 Revised New York Criteria for Ankylosing Spondylitis

Low back pain ≥3 months' duration (improved by exercise and not relieved by rest)
Limited back movement (sagittal and coronal)
Reduced chest expansion (compared to age- and sex-matched values)
Bilateral sacroiliitis (grade ≥2)
Unilateral sacroiliitis (grade ≥3)

Ankylosing spondylitis is diagnosed if significant radiographic evidence of sacroiliitis is present along with one or more clinical criteria.

joint arthritis; characteristic extraarticular features, particularly uveitis; and the formation of new bone at the site of inflammation, eventually leading to ankylosis. By contrast, seropositive RA is associated with the class II HLA-DRB1 genes, typically causes a small-joint symmetrical polyarthritis, has a different range of extraarticular features, and causes erosion of cartilage and bone rather than ankylosis.

Some patients with SpA fail to meet the criteria for any of these individual diseases and are said to have "undifferentiated spondyloarthritis" (Table 11.5).

11.9.1 Diagnostic Criteria

A variety of diagnostic criteria have been proposed for AS, reactive arthritis, and psoriatic arthritis. Criteria for AS have historically relied heavily on the presence of radiographic sacroiliitis for diagnostic specificity. As there is a mean delay of 9 years between the onset of symptoms and the development of radiographic changes, the sensitivity of these criteria in early disease is poor [59]. These criteria have been criticized for being too restrictive as they exclude a significant group of patients with clear features of SpA, but who do not clearly fit into any of the currently defined disease groups. In response to these shortcomings, the Assessment of Spondyloarthritis International Society (ASAS) has developed criteria based on clinical features and magnetic resonance imaging (MRI), which are much more sensitive at picking up early disease [60]. A combination of simple screening questions designed to identify inflammatory back pain combined with HLA-B*27 testing and the use of MRI to detect sacroiliitis can have a considerable impact on the diagnosis of early disease. Early diagnosis is of increasing importance since the advent of anti-TNF treatment, which can have life-changing efficacy in those with AS.

11.10 SPONDYLOARTHRITIDES

In this category are disorders with overlapping clinical features but distinct diagnostic criteria. Included are AS, psoriatic arthritis, inflammatory bowel disease–associated arthritis, and reactive arthritis [61]. They share a predisposition to having HLA-B27. Diagnosis often is delayed because of the diverse signs and symptoms some of which develop with age [62]. MRI is useful when routine radiography fails to detect areas of inflammation [14].

11.11 ANKYLOSING SPONDYLITIS

11.11.1 Epidemiology

The most widely used diagnostic criteria for AS are the modified New York criteria (Table 11.6). Those are heavily reliant on the presence of radiographic evidence of sacroiliitis, which may take several years to develop. MRI allows much earlier detection of sacroiliac joint abnormalities, particularly active inflammation, and is the imaging technique of choice for early diagnosis [63]. AS typically develops in early adulthood, with more than 90% of cases diagnosed before the age of 40 years. There is typically considerable delay in diagnosis, particularly in females among whom atypical patterns of joint involvement appear to be more common. Overall, men are also more commonly affected than women (ratio ~2.8:1). Estimates of the prevalence of AS in western Europe and North America vary between 0.05% and 1.4%, depending on study methodologies.

What is not in doubt is that the prevalence of AS roughly parallels the prevalence of the main susceptibility factor, HLA-B*27, in different populations. Thus, a high prevalence of AS is found in some populations, including Native Americans (B27 prevalence, 18%–50%), Norwegian Lapps (25%–30%), and Alaskan Eskimos (25%–40%). Populations with a low prevalence of HLA-B*27, including most sub-Saharan African ethnic groups, Australian Aborigines, and Indigenous South American, have correspondingly low prevalences of AS. However, the relationship is not simple since other environmental and genetic factors play a significant role in the etiology of the disease.

HLA-B*27 itself is polymorphic, and there is considerable interest in the possibility that allelic differences may be responsible for different degrees of disease susceptibility associated with the B27 subtypes [64].

11.11.2 Pathology

As in all spondyloarthropathies, the basic pathologic lesion of AS is enthesitis. Entheses represent specialized areas of bone adapted to cope with stress loading at interfaces with ligaments or fibrocartilage. Entheses have large vascular beds and are the site of relatively high connective tissue metabolic activity and are thus vulnerable to the effects of inflammation. Plasmacytic and lymphocytic infiltrates are seen. Localized osteitis and osteoporosis occur initially. Later granulation tissue forms, fibrosis occurs, and reactive new bone formation begins. This process may continue until ankylosis occurs across the involved joint. Analogous changes occur at the attachment of the joint capsule to periarticular bone.

Although entheses and fibrocartilagenous joints are primarily affected, inflammation sometimes also involves synovial joints. In some patients, AS may even present as peripheral arthritis involving synovial joints. Eventually approximately 20% of cases develop significant hip arthritis, and involvement of the zygapophyseal joints is universal. Synovial tissue from these joints shows changes similar to RA, although typically less severe. There is villous hypertrophy, synovial cell hyperplasia, and lymphocytic, plasmacytic, and histiocytic infiltration. The cellular infiltrate is diffuse, but also shows some perivascular aggregation. Although the disease is characterized systemically by raised immunoglobulin (Ig)A levels, plasma cells in the synovium secrete principally IgG. Attention has recently been drawn to the high proportion of CD4 T cells expressing the killer immunoglobulin-like receptor KIR3DL2 and their production of IL-17 [65]. The synovial fluid contains fewer polymorphs and more lymphocytes than rheumatoid synovial fluid.

The earliest changes detectable by imaging are bone marrow edema of the sacroiliac joints (Fig. 11.4) and osteitis at the vertebral corners of the spine (Fig. 11.5) on fat-suppressed MRI, which reflect underlying inflammation. On plain radiographs, the earliest changes are sclerosis and erosions in the juxtaarticular bone, but these may take years to develop. In the sacroiliac joints, erosions appear as irregular variations in the width of the sacroiliac joint space and loss of clear definition of the joint line most obvious in the inferior iliac aspect of the synovial component of the joint. Late disease is characterized by new bone formation underneath inflamed periosteum; this is reflected on plain radiographs by "squaring" of the vertebral bodies (loss of the normal concave anterior surface) and the presence of syndesmophytes. In the sacroiliac joints, radiographs demonstrate periarticular

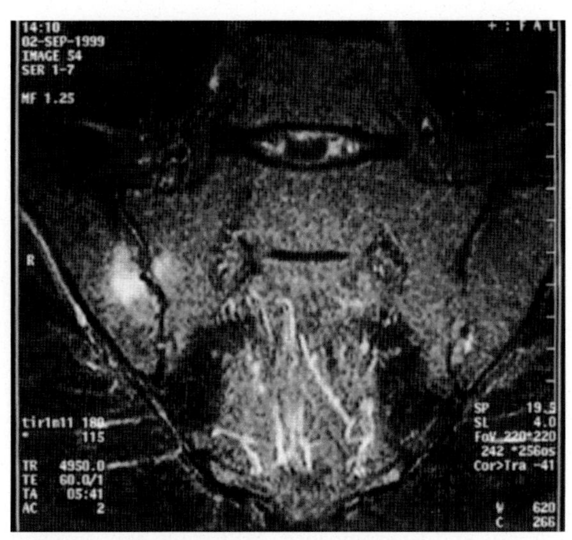

Figure 11.4 Tilted coronal STIR MRI of sacroiliac joints showing bone marrow edema (high signal) before (A) and after (B) anti-TNF therapy.

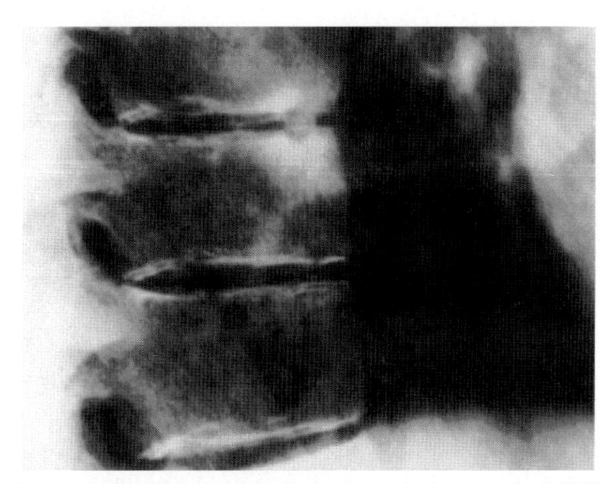

Figure 11.5 "Shiny corners" demonstrated on sagittal STIR MRI of the thoracolumbar spine due to osteitis in ankylosing spondylitis.

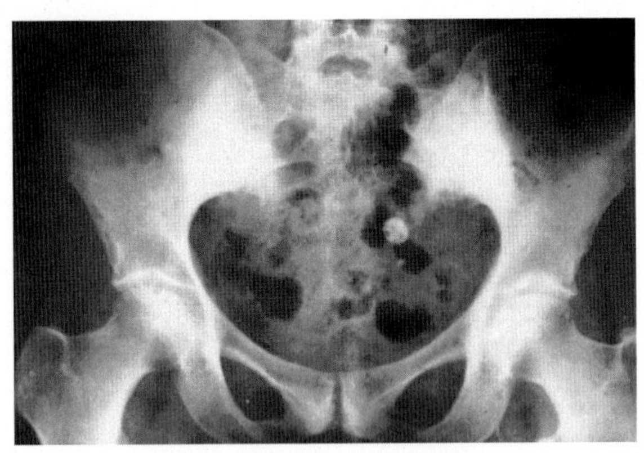

Figure 11.6 Florid syndesmophytes contributing to the appearance of a "bamboo spine" in long-standing AS.

sclerosis and later ankylosis. MRI scans may also show fibrosis and fatty replacement of the periarticular bone marrow but are inferior to plain radiographs in demonstrating the extent of new bone growth (syndesmophytes) (Fig. 11.6).

11.11.3 Other Clinical Features

Osteoporosis is common in AS, often from an early stage. Spinal fractures are common in advanced disease, reflecting both increased bone fragility and reduced flexibility. The eye is the most common extraskeletal site of involvement in AS. Anterior uveitis (involving the structures anterior to the lens—the iris and ciliary body) occurs in around 40% of cases of AS and 5%–10% of cases of reactive arthritis. It runs a relapsing/remitting course, is usually easily treated with cycloplegics and topical steroids, and is rarely sight-threatening unless neglected. Idiopathic anterior uveitis is also strongly associated with HLA-B*27 (95%) independent of AS.

Other uncommon overt extraskeletal manifestations of AS include cardiac and proximal aortic involvement by fibrosis and endarteritis [65]. This rarely results in clinically significant reduction in left ventricular function. Heart block and aortic regurgitation due to dilation of the aortic valve ring may be dramatic in onset. Standardized mortality rates are increased by about 50% in AS, much of which is due to the increased risk of cardiovascular disease. Management involves reducing standard risk factors such as smoking. Pulmonary involvement is predominantly mechanical because of the fusion of the costovertebral joints and thoracic kyphosis, reducing vital capacity. AS is occasionally complicated by upper lobe pulmonary cavitation or fibrosis (~1% of cases), which is usually of little clinical significance but may need to be distinguished from pulmonary tuberculosis or aspergillosis.

Juvenile SA begins by age 16 and typically persists into adult life [66].

11.11.4 Genetic Studies

A major role for genetic factors in the etiology of AS is emphasized by the high recurrence risks for AS among close relatives of patients [67]. Several relatively small twin studies have suggested a significant genetic component to AS. The largest study, undertaken in the United Kingdom, estimated monozygotic twin concordance at 75% compared to only 12.5% of dizygotic twins. Even in B27-concordant dizygotic twins, concordance for AS was only 27%. Variance modeling using these data suggests that broad sense heritability for AS is in excess of 92%, and that HLA-B*27 accounts for less than half of the genetic contribution [68]. Environmental factors are probably important but are very common or ubiquitous, therefore playing little part in determining population variance. It has become increasingly apparent that the genetic contribution to AS is polygenic. At least two HLA genes have already been clearly demonstrated to increase susceptibility to AS. *HLA-B*27* was the first in 1973 [69]. Subsequently, *HLA-B*60* has also been shown to be associated with AS in both B27-positive and -negative individuals [70], although this has not been a universal finding. HLA-B*60 is associated with a two- to threefold excess risk.

In the United Kingdom, more than 90% of patients with AS carry HLA-B*27 (relative risk ~160), but only 2%–8% of B27-positive individuals develop AS. The incidence of reactive arthritis following bacterial enteritis may be as high as 20% in B27-positive individuals, who are also at increased risk of subsequent AS. The fact that relatively few B27-positive individuals develop B27-related diseases probably reflects the role of other genes, rather than the environmental exposure.

There are at least 70 different HLA-B*27 alleles, differing from one another by between one and eight amino acids, which have evolved from the ancestral HLA-B*2705 subtype [70]. Different ethnic groups have distinct subtype distributions. In most white populations, more than 90% of B27 alleles are *HLA-B*2705* and the remainder almost entirely *HLA-B*2702*.

Association between AS and most HLA-B*27 subtypes has been described, at least sporadically (particularly HLA-B*2702, *2704, and *2705), although systematic studies have not always been possible because of the rarity of some subtypes, such as HLA-B*2701, *2707, and *2708. There has been particular interest in two subtypes (B*2703 and *2709) that appeared to be less obviously associated with AS. Early suggestions that the HLA-B*2703 allele might not be associated with AS were based on the relatively high frequency of this allele in Gambia where, nonetheless, AS is extremely rare, although sporadic cases of the disease in B*2703-positive individuals have been reported in neighboring Senegal. Approximately 50% of the B27-positive individuals in Gambia carry the most common white subtype, HLA-B*2705, which is positively associated with disease in all other populations studied so far. Despite this, AS is vanishingly rare in Gambia, perhaps suggesting the influence of other protective genetic or environmental effects in this population. In Europe there is good evidence from Sardinia that HLA-B*2709 is protective against AS.

As a diagnostic test, HLA-B*27 has a low positive predictive value, unless the patient already has a moderately high prior probability of disease. When the pretest probability of spondylitis is 0.5, the presence of B27 increases the likelihood of the disease to 0.92, whereas a negative result reduces it to 0.08. When the prior probability is low (e.g., in population screening programs), the main use of the test is its negative predictive value. In the clinical situation of early SpA with a suspicious history and examination but normal radiographs, a positive B27 test increases the probability from as low as 0.12 to only slightly greater than chance (0.62), adding very little to the clinical decision-making process [71]. Rudwaleit et al. have developed a very useful algorithm for the diagnosis of early AS, based on simple clinical questions to identify inflammatory back pain, HLA-B*27 typing, and MRI of the spine (Table 11.7) [60]. The other setting in which B27 tests may be useful is in determining the likelihood of the offspring of patients developing AS themselves. The pretest probability here is only 0.1, and positive testing for B27 increases this to only 0.2. A negative result makes it highly unlikely that the individual will develop disease and may in some circumstances be reassuring and justified.

Many theories have been proposed to explain the association with HLA-B*27, but none is universally accepted [72]. Doubts about whether HLA-B*27 itself was directly involved or whether it was a marker for a nearby linked gene have been resolved.

TABLE 11.7 **Assessment of Spondylo Arthritis International Society (ASAS) Classification Criteria for Axial Spondyloarthropathy 2009 (Rudwaleit 152).**

In patients with age at onset of symptoms <45 years and back pain ≥3 months
Either
1. Sacroiliitis (active inflammation on MRI/X-ray changes per modified New York criteria)
2. **PLUS** at least one SpA feature from:
a. Inflammatory back pain
b. Arthritis
c. Heel enthesitis
d. Uveitis
e. Dactylitis
f. Psoriasis
g. Inflammatory bowel disease
h. Good response to NSAIDs
i. Family history of SpA
j. HLA-B*27
k. Elevated CRP
Or
1. 1.Positive for *HLA-B2*7* **PLUS** at least 2 SpA features

The possibility that other genes within the MHC, in addition to HLA-B*27 and HLA-B*60, play a role remains somewhat controversial. Weak associations with HLA-DRB1*01 (OR 1.5; 95% confidence interval, 1.1–2.0) and the possibility that there is an extended B*27 haplotype containing another relevant locus have been suggested [73]; however, subsequent data from large-scale GWAS do not lend much support.

11.11.5 Other Genes Involved in AS

Since genetic predisposition to AS is polygenic and B27 accounts for less than half of heritability, other loci contribute to susceptibility [74]. In 1998, an initial genome screen on 120 affected sib-pairs in 105 families confirmed linkage to HLA but six other regions outside the MHC showed some evidence of linkage, with LOD scores of >1.00 [76]. The peak non-MHC linkage was on chromosome 16q (LOD = 2.6). Nominal linkage to the CYP2D6 microsatellite on chromosome 22 was found, and subsequent linkage and association studies with intragenic CYP2D6 markers have also proved positive [75]. This gene (otherwise known as debrisoquine hydroxylase) is one of the cytochrome P-450 enzyme complex genes. The poor metabolizer phenotype is found in 6% of the UK population, but how this might contribute to the etiology of AS is unclear.

Initial attempts to identify non-MHC genes using linkage analysis of affected sib-pair pedigrees met with only limited success. Subsequent meta-analysis of all published studies of linkage analysis was undertaken, which confirmed highly significant linkage to the MHC ($p < 3 \times 10^{-7}$) but also suggested linkage ($p < 7.4 \times 10^{-4}$) to regions on chromosome 6q, 10q, and 16q. Several other regions including the IL-1 gene cluster on chromosome 2 showed moderate evidence of linkage. The λ (LOCUS) value for the MHC was 4.5, suggesting that it accounted for around 34% of the heritability of AS.

First GWAS and then whole exome sequencing rapidly supplanted linkage analysis in identifying genetic effects in AS, building on the work of the WTCCC in other polygenic diseases including RA. In AS, two major international consortia have been established for these studies with sufficient power to reliably identify genes increasing the risk of AS by around 5%–10%. First, the Triple A (Australo-Anglo-American) Spondyloarthritis Consortium (TASC) and second, the International Genetics of AS (IGAS) consortium have generated samples of up to 13,000 cases for analysis. This has resulted in major genetic discoveries.

In 2007, TASC and WTCCC reported the results of a limited gene-targeted GWAS of 15,000 nonsynonymous single-nucleotide polymorphisms (ns SNPs) in 1000 cases [76]. The association with the MHC was confirmed ($OR \sim 120$, $p < 10^{-120}$) as expected. Outside the MHC, the most striking association was with ERAP1, encoding the endoplasmic reticulum aminopeptidase 1 ($p \sim 10^{-6}$), which was subsequently replicated in an independent dataset. Numerous SNPs in ERAP1 were associated with AS; rs30187 encoding K528R was associated with an OR of 1.3 ($p \sim 10^{-9}$). This association has been robustly replicated in several further studies [64,77,78], including a relatively large GWAS in Han Chinese ($n = 3937$) in which the strongest association was also with rs30187 ($p < 7 \times 10^{-4}$) [79]. In this study, the other interesting observation was of weak association ($p < 1.7 \times 10^{-3}$) with IL23R, encoding the IL23 receptor expressed by a subset of CD4 lymphocytes producing the cytokine IL17, known as T helper 17 (Th17) cells. IL23R is of particular interest because it is also associated with Crohn's disease and psoriasis, two conditions that are known to be overrepresented in patients with AS. This association was replicated in an independent dataset and subsequently in an independent UK study and meta-analysis of previously published studies [64,79]. The coding SNP rs11209026 is most strongly associated with AS ($OR = 0.61$, $p < 10^{-10}$) as in Crohn disease.

A subsequent GWAS reported by TASC in 2010 confirmed the association with the MHC, ERAP1, and IL23R but, in addition, identified several new associations [64]. This study was powered more heavily than previous studies, comparing 2053 unrelated subjects with AS with 5140 ethnically matched controls. Cases were genotyped for 370,000 SNPs on the Illumina HumHap 370 genotyping platform, and independent replication of positive results was undertaken in a further 898 cases and 1518 controls. Strong association was observed with two "gene deserts" on chromosomes 2p15 ($p < 10^{-19}$) and 21q22 ($p < 10^{-9}$) and also the genes *ILIR2* (the decoy receptor for IL1) and *ANTXR2* (encoding a receptor for anthrax toxin, but also functioning as a capillary growth factor). The region of chromosome 2p15 was again replicated in Han Chinese [79] with the marker rs10865331 most strongly associated ($p = 2 \times 10^{-8}$).

Several major breakthroughs in our understanding of the genetic basis of AS have recently been made. These relate in particular to the likely role of ERAP1, the key role of the Th17 lymphocyte development pathway, and the overlap between AS and other inflammatory disorders, including psoriasis and inflammatory bowel disease. In 2011, the TASC and WTCCC published the results of another large GWAS, in which a discovery set of 3023 cases of European ancestry was genotyped on the Illumina 660W-Quad microarray and compared with 8779 controls. A limited replication study of the top 50 "hits" was then undertaken in an independent set of 2111 cases and 4483 controls [77]. Of particular interest was the observation that the association with ERAP1 was limited to the B27-positive subset of patients, contrasting with the B27-negative minority (~9%) in whom no such association was apparent. Other associations, such as IL23R, were not influenced by HLA-B*27 status of the patients. It is therefore clear that there is synergy between HLA-B*27 and ERAP1 in susceptibility to AS. This most likely reflects a role for ERAP1 in trimming peptide antigens in the endoplasmic reticulum to optimal length (eight or nine amino acids) for binding to HLA class I molecules, including HLA-B27. The solution of the crystal structure of ERAP1 at around the same time presented a structural basis for the understanding of this function of ERAP1. Furthermore, we were able to show that particular variants associated with protection against AS, such as K528R (rs30187), showed significantly reduced peptide processing characteristics [11]. At least in theory, this would suggest that small-molecule inhibitors of ERAP1 could be of value in the treatment or prevention of AS.

The role of Th17 lymphocytes in AS is highlighted by the associations of a number of genes involved in this pathway [80]. A strong association exists between polymorphisms in *IL1* [81]. These include not only IL23R (expressed on Th17 lymphocytes) but also IL12B (encoding the common p40 chain of IL23 and IL12) and STAT3 (a key transcription factor for Th17 lymphocytes). Similar associations are seen in both psoriasis and inflammatory bowel disease, highlighting the shared genetic predisposition of these conditions. Other associations include the transcription factor RUNX3, which is involved in lymphocyte differentiation, which is driven by IL-7. In this respect, it is interesting that we have also observed a moderate association with the IL-7 receptor (8×10^{-5}). The latest TASC/WTCCC study also shows a strong association with PTGER4 (prostaglandin E receptor 4), which is also associated with Crohn disease. Prostaglandin E2 acts through this receptor to induce the production of IL23 and IL17, both of which are elevated in AS. Synovial levels of PTGER4 are increased in spondyloarthropathies. Of interest, β-glucan, a component of certain fungal and bacterial cell walls, induces PGE2 through interactions with components of the innate immune system, including CARD9 that is also associated with AS [82]. This would be consistent with the hypothesis that AS might be induced by ubiquitous organisms carrying β-glucan by stimulation of the Th17 pathway. PTGER4 is also part of the anabolic bone response to stress. Its expression is increased in the synovium of those with spondyloarthropathies, and this could potentially explain the new bone formation at sites of mechanical stress (entheses) in AS. These associations and doubtless many more will be further refined by the current IGAS consortium "ImmunoChip" experiment nearing completion in 2012. This will study known and candidate genes of inflammatory/immunological importance in 13,000 individuals globally and is likely to increase the number of target genes and pathways for study substantially.

11.11.6 Management

Pain relief and minimization of disability are the major aims of treatment. Physical exercises to maintain flexibility of the axial skeleton and good posture must be undertaken regularly, if necessary, under the supervision of a physiotherapist, since the benefit of intensive exercise regimens in hospital has been well recorded. Because of the likelihood of pronounced costovertebral joint involvement, lung function is often markedly reduced, and smoking should be strongly discouraged. NSAIDs are usually highly effective although there is only very weak evidence to suggest that they have any effect on the underlying rate of paraspinal ossification. Continuous treatment may be more effective at preventing new bone formation than on-demand therapy [83]. Sulfasalazine and methotrexate are valuable for patients with active peripheral joint involvement but have no discernible effect on the spinal disease. Likewise, corticosteroids administered systemically are usually ineffective for spinal disease. In contrast, local injection of corticosteroids may be very valuable for the management of peripheral joint synovitis and occasionally for costovertebral joint disease and sacroiliitis.

A plethora of agents targeted at R- and B-cells, proinflammatory mediators, and signaling molecules are new targets for therapy [84]. Management of active axial inflammation refractory to NSAIDs was revolutionized by the demonstration that anti-TNF agents are highly effective in this condition [85]. Whether these expensive drugs also retard the progression of joint ankylosis is unknown. Promising results were demonstrated in early trials of anti-IL17 therapy in active AS, but larger studies are awaited. Uveitis is common (up to 40%) and is usually responsive to topical corticosteroids and mydriatics. Osteoporosis is a relatively early complication of AS and may contribute to spinal deformity if wedge fractures occur. Bone densitometry measurements in the lumbar spine may be misleading because of the presence of syndesmophytes, which may cause spuriously high values. Joint surgery, particularly hip and knee arthroplasty, may be required in about 10% of patients but is particularly likely in those with juvenile-onset disease. Finally, a range of aids for daily living may be required, ranging from prismatic spectacles to allow severely kyphotic individuals to see where they are going, through to adaptations to cars (particularly additional mirrors for all-around vision) and wheelchairs for those with extreme disability and handicap.

11.12 REACTIVE ARTHRITIS (PREVIOUSLY REITER SYNDROME)

Classic reactive arthritis (previously Reiter syndrome) represents the triad of arthritis, conjunctivitis, and urethritis, developing after a triggering bacterial infection; however, incomplete forms of this triad are more common.

11.12.1 Clinical Features

Reactive arthritis usually develops between 1 and 3 weeks after a precipitating infection. The arthritis typically affects a few large lower limb joints asymmetrically. Diffuse swelling of fingers and toes, referred to as "sausage digits," is typical of reactive or psoriatic arthritis. Axial skeletal involvement (particularly sacroiliitis) affects up to 50% of those affected, particularly those who are HLA-B*27 positive. Plantar fasciitis and Achilles tendonitis are common. Mild conjunctivitis commonly precedes the arthritis. Mucocutaneous involvement may include mouth ulcers, sterile urethritis, circinate balanitis, and keratoderma blennorrhagica (a rash resembling pustular psoriasis). The illness can be very severe and associated with marked systemic features of fever, malaise, and weight loss. It runs a variable course depending on the nature of the triggering organism and host genetic factors. Many cases are extremely mild and last only a few days. For those with more sustained synovitis, the average duration is < 5 months but about 15% proceed to chronic disease and about one quarter may develop recurrent episodes. Chronic arthritis is much more likely in patients who are HLA-B*27 positive, 20% of whom may ultimately develop AS. Patients with concurrent human immunodeficiency virus (HIV) infection tend to develop more severe arthritis and are probably at greater risk of disease because of increased exposure to the relevant genital tract pathogens.

11.12.2 Epidemiology and Etiology

While environmental triggers and genetic susceptibility contribute to reactive arthritis, it is not known how these interact to cause arthritis. Between 68% and 85% of cases of reactive arthritis occur in B27-positive individuals, and, in epidemics of arthritogenic bacterial infections, approximately 20% of B27-positive individuals will develop reactive arthritis.

Although the triggering bacteria cannot be cultured from the affected joints, there is reasonable evidence for the presence of bacterial proteins and nucleic acid in the joints [86]. Cellular immunity to triggering bacteria has been demonstrated among both CD4-and CD8-positive B27-restricted T cells. These T cells are of the helper T cell type 1 subset of lymphocytes, which are thought to protect against intracellular infection. This seems appropriate as most arthritogenic pathogens are intracellular organisms, and it is consistent with theories suggesting a primary role for persistence of bacterial fragments within joints as the cause of reactive arthritis. The occurrence of reactive arthritis in face of CD4 lymphocyte depletion in HIV infection suggests that these cells are not critical in the development of the illness, in contrast with the situation in RA in which the disease often improves as the CD4 lymphocyte count falls. Other theories implicate B27 itself as the autoimmune target, due to either molecular mimicry or loss of tolerance resulting from chemical alteration of the molecule. Evidence that B27 may be associated with lesser protection against invasion by, and less-efficient clearance of, enteric bacteria provides a further possible explanation for B27-related arthritis.

11.12.3 Management

Pain relief from the musculoskeletal manifestations of reactive arthritis is the primary objective in nonpersistent disease. NSAIDs are commonly used but have only a moderate beneficial effect. Local steroid injections are appropriate for peripheral arthritis and enthesopathy and may need frequent repetition. Sulfasalazine, methotrexate, and azathioprine have demonstrated beneficial effects and are appropriate to use in cases in which a protracted or recurrent course seems likely. Anti-TNF therapy may be effective in refractory cases. In cases of presumed urogenitally acquired reactive arthritis, appropriate cultures should be taken for *Chlamydia* and *Neisseria gonorrhoeae*, as well as syphilis and HIV serology. Short courses of antibiotics are appropriate to clear any triggering infection, but longer courses of antibiotics may improve the late outcome of the arthritis [87].

11.13 ENTEROPATHIC ARTHRITIS

Arthritis may complicate up to 10% of cases of inflammatory bowel disease. It may affect the axial skeleton and/or peripheral joints and is approximately twice as common in Crohn disease as in ulcerative colitis. Axial arthritis occurs in 10%–15% of individuals with inflammatory bowel disease, but the association of sacroiliitis with HLA-B*27 may be weaker (50%–60%) in this group of patients than in those suffering from AS without inflammatory bowel disease or psoriasis (~95%). These figures are heavily dependent on the means of ascertainment. For example, using a UK database of patients with known AS, the prevalence of HLA-B*27 is 83% in those with spondylitis and bowel disease. Limited sacroiliitis (frequently asymptomatic) may be relatively common in inflammatory bowel disease, but HLA-B*27 increases the likelihood of developing more extensive axial disease.

Peripheral arthritis exists in two distinct clinical forms that are also immunologically distinct. Both forms occur about twice as commonly with Crohn disease as ulcerative colitis. The type 1 arthropathy (self-limiting, pauciarticular, and associated with flares of the bowel disease) is positively associated with HLA-B*27, HLA-B*35, and HLA-DRB1*0103 but the type 2 arthropathy (polyarticular, symmetrical, persistent, and not related to activity of the bowel disease) is associated with HLA-B*44 [88]. Since AS has a high frequency of subclinical ileitis, peripheral arthropathy may predate the onset of clinically apparent inflammatory bowel disease by months or occasionally years. In type 1 arthropathy, the knees and ankles are most frequently affected, whereas in type 2 arthropathy, the metacarpophalangeal joints, wrists, and knees are most commonly affected.

11.14 PSORIASIS AND PSORIATIC ARTHRITIS

Arthritis associated with psoriasis exhibits a considerable variety of clinical patterns and can be distinguished as a separate entity [89,90]. The arthropathy ranges through forms resembling RA to classical SpA. The extent to which these different expressions of psoriatic arthritis are under genetic controls is debatable since they do not appear to breed true in families with psoriatic arthritis [91].

11.14.1 Clinical Manifestations

Classically, psoriatic arthritis has been divided into five subsets depending on clinical and radiographic criteria: (1) "classical" psoriatic arthritis, confined to the distal interphalangeal joints of the hands and feet (5%); 2) arthritis mutilans (5%); (3) symmetrical polyarthritis, resembling RA (15%); 4) asymmetrical oligoarthritis (70%); and 5) spondyloarthritis (5%). The pattern of onset of peripheral arthritis does not predict outcome in most cases, supporting the development of classification criteria, in which all patients with peripheral arthritis are pooled together [92].

Dactylitis (sausage digits) is characteristic of reactive or psoriatic arthritis, occurring in more than one-third of patients. Achilles tendonitis and plantar fasciitis (both forms of enthesitis) are also common. Dystrophic nail changes, including pitting, ridging, discoloration, and onycholysis, are associated particularly with distal interphalangeal joint disease. Overall, approximately 80% of patients with psoriatic arthritis have nail changes, compared with only 20% of psoriatics without arthritis.

Psoriatic arthritis is often a benign illness, but severe joint damage is not uncommon. Patients may present with severe acute monoarthritis and marked constitutional symptoms, mimicking sepsis or reactive arthritis. Typically, an increasing number of joints become involved with time although there is little relationship between this and the severity of the skin disease. Arthritis mutilans, although rare, causes severe joint destruction and disability. Axial disease is common but highly variable and tends to cause less functional impairment than in idiopathic AS. Compared to the "pure" form of AS, psoriatic spondyloarthritis also tends to be less symmetrical; radiographs may show nonmarginal origin of syndesmophytes (origin not from the anterolateral border of the vertebral end plate); radiographic changes may skip vertebral segments; atlantoaxial involvement is more common; and unilateral sacroiliitis is not unusual. In reality these divisions into subgroups are somewhat arbitrary; of those patients with peripheral joint disease, probably one-third have evidence of axial disease on MRI, although this is commonly asymptomatic [93].

11.14.2 Epidemiology

Psoriasis itself occurs in 1%–2% of whites with an equal gender frequency. Common forms of arthritis may thus occur fortuitously with psoriasis, without there being any causal relationship. Psoriatic arthritis typically occurs in 7% of patients with psoriasis although frequencies of up to 42% have been reported in some series [94]. Arthritis is particularly common where there is nail involvement, but it may predate the onset of skin lesions. Although it usually starts in the third or fourth decade, pediatric onset is also quite common.

11.14.3 Pathogenesis

Inappropriate activation of the immune system appears to underlie both the skin and joint disease of psoriasis. In the psoriatic epidermis, keratinocytes proliferate and mature rapidly, leading to deficient adhesion of corneocytes that causes the characteristic scales of the psoriatic skin plaques. Significant infiltration of both the skin and joint synovium by T cells suggests a role for specific antigen presentation. Exacerbation of the skin lesions by CD4 depletion and HIV and/or acquired immunodeficiency syndrome has been observed, while specific T-cell-targeted therapy (e.g., cyclosporin A) may be effective. Furthermore, the efficacy of T-cell-targeted biologic therapies such as alefacept directed against CD2 in around 50% of patients supports this concept. An important role for TNFα in skin and joint lesions is emphasized by the efficacy of anti-TNF biologics although this is a complex relationship as anti-TNF agents have also been reported to precipitate flares of skin disease.

In the synovial membrane, there is evidence of new blood vessel growth and substantial increase in the amounts of transforming growth factor-β, platelet-derived growth factor, vascular endothelial growth factor, and angiopoietins.

11.14.4 Genetics

The heritability of psoriatic arthritis is substantial with a recurrence risk among first-degree relatives ($\lambda 1 = 30-55$) and siblings ($\lambda s = 30$) estimated to be at least threefold higher than in psoriasis [95,96]. Evidence from linkage and association studies led to the designation of the major psoriasis and psoriatic arthritis susceptibility locus within the MHC as PSORS1 (MIM*177,900) and is primarily associated with type 1 psoriasis (onset <40 years of age), as no association was found in patients with psoriatic arthritis and late-onset psoriasis [97]. The precise location of the PSORS1 risk allele is difficult to establish because of the extensive linkage disequilibrium within the MHC region; possibilities include *HLA-Cw*0602* and the SNP *rs10484554*T*.

Several other HLA associations have been reported in the literature; however, dissecting the contribution to psoriatic arthritis risk from the contribution to psoriasis risk is complex. Alleles specifically associated with arthritis independently of cutaneous disease included HLA-B*27, B*08, and B*38. Subgroup analysis demonstrated association of HLA-C*06 and HLA-B*27 with peripheral arthritis. HLA-B*27 is strongly associated with the development of sacroiliitis, and B*08, B*38, and B*39 are also associated with spondyloarthritis. Sacroiliitis is a common finding in patients with psoriatic arthritis, but these are frequently asymptomatic [93]. In psoriatic spondyloarthritis, the association with HLA-B*27 is less strong (50%–60%) than in the idiopathic forms of AS (95%) and may correlate with more severe forms of spinal involvement.

Non-HLA genes within the MHC region, which may contribute to psoriatic arthritis risk include MICA*002, and the TNF promoter polymorphisms TNF-238A and TNF-857T, although the evidence is contradictory [95,96].

Non-MHC gene loci including *IL23R*, *IL12B*, *TNIP1*, and *TRAF3IP2* [95,97] are associated. Significant associations have also been linked with regions on chromosome 4q27 containing the IL2 and IL21 genes and to chromosome 15q21 [98]. These associations hint at pathogenic mechanisms underlying PsA as they include genes encoding Th2 (IL13) and Th17 (IL12B, IL23R) cytokines and signaling pathways including multiple associations with NF-κB signaling (TNIP1, TRAF3IP2, NFKBIA, TNFAIP3, NOS2, FBXL19).

Children of affected fathers are more likely to develop psoriasis or psoriatic arthritis (16.2%) than children of affected mothers (8.3%) [99]. Evidence for imprinting is supported by the association of psoriatic arthritis with a region on chromosome 16q, which only confers risk when conditioned on paternal inheritance. Prognostically, males with axial psoriatic arthritis are more severely affected than females, in the absence of any differences in HLA distribution. A higher risk of progressive or erosive arthritis is found in patients carrying HLA-B*39, HLA-DRB1*04, HLA-DQ3 without HLA-DR7, HLA-B*27 plus HLA-DR7, or the AA variant of the IL4 SNP *rs1805010*, while patients with the combination HLA-*Cw*6* plus HLA-*DRB1*07* have a better prognosis [100].

Approximately 75% of the genetic risk of psoriatic arthritis remains unidentified, and further studies are required to identify risk alleles with smaller effect sizes, or indeed rare genetic variants conferring high risk [101].

11.14.5 Management

The primary target of therapy is remission or suppression of activity. The approach should be guided by the 2018 ACR/National Psoriasis Foundation Guideline for the Treatment of Psoriatic Arthritis [102]. NSAIDs similar to those used in RA are helpful for the peripheral joint arthritis. Intraarticular corticosteroids can be useful in peripheral arthritis and enthesitis, but oral corticosteroids have been associated with severe flares of cutaneous disease on withdrawal and are therefore not widely used. Anti-TNF biologic agents are highly effective in both the peripheral and axial joint manifestations of the disease although etanercept is considerably less effective at treating the skin manifestations [103,104].

Slow-acting disease-modifying antirheumatic drugs, such as sulfasalazine, methotrexate, and leflunomide, may be effective in some cases, but cyclosporin A, despite its proven efficacy in skin disease, is relatively ineffective on the joints. In contrast to their helpful effects in peripheral arthritis, these agents are generally ineffective for psoriatic spondylitis.

11.15 JUVENILE IDIOPATHIC ARTHRITIS

Historically, this has been a somewhat ill-defined entity, a situation exacerbated by the differences in nomenclature in Europe and North America. The term juvenile chronic arthritis was used in Europe, whereas in North America, these disorders were often called juvenile chronic RA (in contrast to the use of this term in Europe specifically for seropositive juvenile-onset RA). Further confusion may arise from the use of the eponymous term Still disease for juvenile idiopathic arthritis (JIA). In 1897, Sir George Frederick Still distinguished juvenile forms of arthritis from adult RA, commenting on the fever found in the systemic form. Subsequently, in 1959, Ansell and Bywaters distinguished these childhood forms on the basis of their mode of onset. Thereafter, in general, "Still disease" was reserved for the systemic onset form of the disease although it has also been sometimes loosely applied to juvenile arthritis overall. These problems of classification have made comparisons between studies difficult. It has also complicated analysis of the genetic component of these disorders since many studies have not adequately distinguished between the various forms.

Childhood arthropathy has many potential causes, and it is important to exclude sepsis and viral infection, in particular, before concluding that the child has JIA. A World Health Organization/International League Against Rheumatism report proposed a widely accepted classification based on clinical patterns of disease [105] that defines seven subtypes of JIA:

- Systemic onset (11%)
- Oligoarthritis and extended oligoarthritis (50%)
- RF-positive polyarthritis (3%)
- RF-negative polyarthritis (17%)
- Enthesitis-related arthritis (10%)
- Psoriatic arthritis (7%).

Specific diseases associated with joint inflammation, such as systemic lupus erythematosis, rheumatic fever, septic arthritis, and neoplasia, are excluded from this classification.

11.16 SYSTEMIC-ONSET JIA (STILL DISEASE)

Still disease is the form of childhood arthropathy that carries the most adverse prognosis, with a significant mortality of 10%. It is characterized by systemic features including quotidian fevers, evanescent rash, lymphadenopathy, hepatosplenomegaly, arthropathy, and polyserositis. These features, coupled with a pronounced neutrophilia ($>13 \times 10^9$/L in 75% or more of patients), may suggest infection, particularly when systemic features predate the arthropathy, which may happen by weeks or months. Other causes of fever and serositis should be considered, such as familial Mediterranean fever, particularly in patients of the appropriate ethnic background. Other important differential diagnoses include infection and malignancy. The disease has been reported in most populations worldwide and affects the sexes equally. The age of onset is variable but is usually between 4 and 6 years. It can also occur in adults, in whom it frequently causes diagnostic problems, particularly if systemic features predominate.

While some patients exhibit complete remission within 2 years of onset and others have repeated cycles of activity, the majority follow a chronic course [106]. Some of the systemic features, including the fever and malaise, may respond well to short-term NSAIDs. For patients with more severe disease or poor prognostic signs such as active fever and high physician's global score, the systemic features may require moderate- to high-dose corticosteroids (including intravenous methylprednisolone). Anti-IL1 biologic therapy (e.g., Anakinra) may also be indicated in these patients and is frequently dramatically effective [107].

Multiple intraarticular injections of corticosteroids can be very effective in managing mild joint disease, but in patients with persistent or severe polyarticular involvement, disease-modifying antirheumatic drugs may be necessary. Weekly methotrexate (either orally or subcutaneously) is particularly beneficial compared to other second-line antirheumatic agents (e.g., gold salts, D-penicillamine, sulfasalazine), which have, in general, been disappointing. For resistant articular disease, biologic therapies, including anti-TNF, but particularly anti-IL1 and anti-IL6 agents, may be effective [107].

11.17 OLIGOARTICULAR JIA

Oligoarticular JIA affects young girls at least six times more frequently than boys, with a peak incidence at under 3 years of age, although the disease may also present for the first time in adolescence. The prevalence is between 20 and 30 per 100,000 and most ethnic groups are affected. By definition children with this form of JIA have four or fewer joints affected within the first 6 months of disease, although in as many as one-third, the disease may subsequently extend to polyarticular involvement.

11.17.1 Clinical Features

Constitutional symptoms are not prominent in contrast to Still disease. Although joint pain is usually an obvious feature, the presentation is sometimes less obvious. Thus, a parent may notice a swollen joint in the absence of symptoms or observe nonspecific features, such as poor behavior or cessation from walking. Almost two-thirds of patients have only one joint involved, and more than 90% have no more than two joints involved, in the first 6 months. Those children who remain oligoarticular for 5 years are unlikely subsequently to develop more extensive disease.

Levels of acute-phase reactants, such as C-reactive protein, may be normal or only slightly risen. If there is marked elevation, a search for alternative explanations is merited. This group of patients is classically associated with the presence of antinuclear antibodies in the serum (40%–75%), usually in low titer (<1:640). Positive antinuclear antibodies indicate a higher risk of developing chronic anterior uveitis, which is the most serious potential complication of early-onset oligoarticular JIA, although all children with JIA are at risk. It causes blindness in up to 10% of patients, but this is avoidable if the appropriate screening and treatment steps are exercised. Ocular disease affects up to 57% of patients with oligoarticular JIA, usually with anterior uveitis although isolated posterior disease rarely may be present. Only a few of those with iritis (5%) have polyarticular disease at onset. JIA-related uveitis is usually asymptomatic and the onset of ocular features may be delayed. Therefore, slit-lamp examination is mandatory in all patients at diagnosis and regularly for many years thereafter. The uveitis is chronic, lasting between 2 and 15 years, and most commonly affects both eyes.

The joint disease associated with oligoarticular JIA typically resolves within 5 years, but chronic arthritis may sometimes recur, even many years later. Those children whose arthropathy extends to become polyarticular (about 20% overall) account for the majority exhibiting significant functional handicap at 15 years.

11.17.2 Management

It is important to keep the joints as normal as possible while the arthropathy is in its active phase. This can usually be accomplished by the use of NSIDs and intraarticular corticosteroids (best administered under general anesthetic in young patients). In general, there is a good response, and second-line drugs, such as methotrexate, are reserved for patients with persistent or severe disease. Physical methods of treatment to prevent contracture and preserve muscle strength are beneficial. Surgery is rarely needed except in those cases in which severe contractures develop or abnormalities of bone growth occur secondary to epiphyseal involvement. Anti-TNF biologics are potentially very effective in resistant cases. Topical corticosteroids and mydriatics are effective in 40% of patients with uveitis, but intraocular steroids or systemic corticosteroids are frequently required to prevent the formation of posterior synechiae between the lens and the iris. In some cases, immunosuppressive drugs such as azathioprine, chlorambucil, and cyclosporin A may be necessary to control the uveitis. The ocular disease is potentially sight-threatening and demands regular ophthalmologic review.

11.18 POLYARTICULAR JIA

Polyarticular JIA is much less clearly defined than the oligoarticular and systemic-onset forms. It is apparent that some patients whose disease begins with limited joint involvement subsequently progress to a more widespread form of arthropathy with a correspondingly poorer outcome (these patients are included under the extended oligoarticular disease subset). If this occurs, it is invariably within the first 5 years of disease. Other patients have polyarticular symptoms from the outset, but these appear to represent a relatively heterogeneous group with the exception of the subgroup with juvenile-onset seropositive RA (3% of all JIA). The prognosis of this specific subset is similar to the adult form of RA, with similar immunogenetic associations. Indeed, the HLA associations are even more striking in the juvenile form of the disease, and the association with the HLA-DRB1*0401/*0404 genotype is particularly striking.

11.18.1 Genetics of JIA

It has been estimated that JIA has a prevalence of between 20 and 120 per 100,000 and an annual incidence of 10–20 per 100,000. The prevalence appears to be similar in Europe and North America in general although a threefold excess risk to Indigenous American children compared with their white counterparts in British Columbia has been reported. No differences have been observed in the prevalence in North American Black and white children, but, by contrast, the condition appears to be rare in China. The available family studies provide support for a genetic component to JIA, but many of these studies predate the more accurate classification of JIA into various subgroups. Moroldo et al. [108] studied 71 sib-pairs affected by juvenile chronic arthritis and found substantial concordance (70%–80%) for the type of disease between the sibs. There is a slight excess recurrence in sibs for systemic-onset JIA, and a genetic component to the disease arising from HLA-linked genes has been suggested although these results are inconsistent.

HLA-DR alleles have been estimated to confer almost 20% of the total sibling recurrence risk in JIA although the risks associated with HLA alleles differ between the JIA subtypes. HLA-DRB1*08 (a rare allele in most white populations) is found consistently in between 25% and 50% of patients with oligoarticular JIA (relative risk ~12). HLA-DRB1*11 (relative risk ~5) is also increased, and some studies have suggested a weak association with HLA-DRB1*13. HLA-DRB1*13 is most closely associated with the presence of uveitis (OR 3.4) in oligoarticular disease. HLA-DRB1*04, which is associated with RA in adults is associated with juvenile seropositive polyarthritis (OR 3.2) but is protective against many other JIA subtypes (220). DRB1*07 is associated with decreased risk of JIA [110]. There is considerable evidence for a marked compound heterozygote effect with predisposing DRB1 alleles [110]. Gender has a significant influence on the age that the HLA risk is conferred. HLA analysis has been particularly useful in distinguishing a subset of children with pauciarticular disease, typically boys older than 9 years, in whom limited joint involvement with an associated enthesopathy is the first sign of spondyloarthritis. This variant is strongly associated with HLA-B*27 and is known to be associated with an excess of family members with sacroiliitis. Many of these children go on to develop AS although others appear to continue with a more peripheral form of enthesopathy without AS.

Other HLA alleles also contribute to risk. HLA-*DQA1*0103* is associated with oligoarticular disease and HLA-*DQA1**05 with oligoarticular and systemic-onset arthritis. HLA-DQA1*02 is protective against oligoarticular and seropositive polyarthritis [109]. DPB1*0201 confers risk of early-onset arthritis [110]. Previously reported HLA–DRB1/DQB1 risk associations were found to be solely due to the effect of the DRB1 locus [110].

Particular combinations of *HLA-A*2*, *DPB1*0201*, and other *DRB1* susceptibility alleles (*08,*11,*13) are associated with onset of oligoarticular disease under 3 years of age. The *DRB1*1501–DQA1*0102–DQB1*0602* haplotype protects against oligo- and polyarticular JIA.

Conducting studies of sufficient power to examine the risk contribution of non-HLA alleles is difficult because of the rarity of JIA and the phenotypic heterogeneity of the disease. While studies encompassing JIA as an umbrella disease may mask genetic associations with individual JIA subsets, subset subanalysis may fail to detect genetic risk alleles with small effects. Therefore, approaches to identifying candidate genes have included investigating loci associated with other autoimmune diseases [111].

More than 100 non-MHC risk alleles have been examined, and although many were found to be positively associated, only a small number of these have been replicated. The *PTPN22* and *IL2RA* gene associations have reached genome-wide significance. PTPN22 is involved with T-cell signaling and confers risk of oligo- and polyarticular arthritis, while IL2RA, which codes for the high-affinity IL2 receptor alpha chain, is particularly associated with ANA positivity, female gender, and oligoarthritis. Other significant associations include genes implicated in T-cell signaling and activation (*STAT4*, *VTCN1*), innate immunity (*TNFA*, *TNFAIP3*, *TRAF1/C5*, *MIF*, *SCL11A1*), and cartilage homeostasis (*WISP3*). Many of these genes are associated with risk of other autoimmune conditions such as RA [112]. For example, the *ERAP1* and *IL23R* gene variants, which are associated with AS in adults, increase risk of enthesitis-related arthritis [113] and juvenile psoriatic arthritis, respectively. These findings await replication.

The strong association of oligo- and polyarticular JIA with HLA class II alleles suggests pathogenic involvement of CD4 T cells. No etiologic agents have yet been identified in this disorder although autoantigens present in the eyes and the joints are strongly suggested as being responsible for the chronic inflammatory process in oligoarticular disease. The non-HLA associations implicate Tregs, which may influence the balance between pro- and antiinflammatory processes, as well as innate immune cells, which may contribute to cytokine production and CD4 T-cell activation. The distinct HLA and non-HLA genetic association between the JIA subsets implies differences in pathogenic mechanisms.

REFERENCES

[1] Invernizi P, Gershwin ME. The genetics of human autoimmune disease. J Autoimmun 2009;33:290–9.

[2] Stahl EA, et al. Genome wide association study meta-analysis identifies seven new rheumatoid arthritis risk loci. Nat Genet 2010;42(6):508–14.

[3] Garboczi DN, et al. Structure of the complex between human T-Cell receptor, viral peptide and HLA-A2. J Immunol 2010;85(11):6394–401.

[4] Kronenberg M, Rudensky A. Regulation of immunity by self-reactive T cells. Nature 2005;435:598–604.

[5] Rioux JD, Abbas AK. Paths to understanding the genetic basis of autoimmune disease. Nature 2005;435:584–9.

[6] Sakaguchi S. Naturally arising CD4? regulatory T cells for immunologic self-tolerance and negative control of immune responses. Annu Rev Immunol 2004;22:531–62.

[7] Rifkin IR, et al. Toll-like receptors, endogenous ligands, and systemic autoimmune disease. Immunol Rev 2005;20:427–42.

[8] Antunes M, et al. Undifferentiated connective tissue disease: state of the art on clinical practice guidelines. RMD Open 2019;26(4 Suppl. 1):e000786. https://doi.org/10.1136/rmdopen-2018-000786.

[9] Parham P. MHC class 1 molecules and KIRs in human history, health and survival. Nat Rev Immunol 2005;5:201–14.

[10] Goodnow CC, et al. Cellular and genetic mechanisms of self tolerance and autoimmunity. Nature 2005;2435:590–7.

[11] Kochan G, et al. Crystal structures of the endoplasmic reticulum aminopeptidase-1 (ERAP1) reveal the molecular basis for N-terminal peptide trimming. Proc Natl Acad Sci USA 2011;108(19):7745–50.

[12] Ha E, et al. Recent advances in understanding the genetic basis of systemic lupus erythematosus. Semin Immunopathol 2021. https://doi.org/10.1007/s00281-021-00900-w.

[13] Lythgoe H, et al. Classification of systemic lupus erythematosus in children and adults. Clin Immunol 2021;234:108898.

[14] Romao VC, Foseca JE. Etiology and risk factors for rheumatoid arthritis: a state-of-the-art review. Front Med 2021;i:6899698.

[15] DeStefano L, et al. The genetic, environmental, and immunopathologic complexity of autoantibody-negative rheumatoid arthritis. Int J Mol Sci 2021;22:12386.

[16] Suzuki A, et al. Functional haplotypes of PADI4, encoding citrullinating enzyme peptidylarginine deiminase 4, are associated with rheumatoid arthritis. Nat Genet 2003;34(4):395–402.

[17] Nepom GT, et al. Prognostic implications of HLA genotyping in the early assessment of patients with rheumatoid arthritis. J Rheumatol Suppl 1996:445–9.

[18] Seidl C, et al. Association of (Q)R/KRAA positive HLA-DRB1 alleles with disease progression in early active and severe rheumatoid arthritis. J Rheumatol 1999;26(4):773–6.
[19] O'Dell JR, et al. HLA-DRB1 typing in rheumatoid arthritis: predicting response to specific treatments. Ann Rheum Dis 1998;57(4):209–13.
[20] Syversen SW, et al. Effect of therapeutic drug monitoring vs standard therapy during maintenance infliximab therapy on disease control in patients with immune-mediated inflammatory diseases: a randomized clinical trial. JAMA 2021;326. https://doi.org/10.1001/jama.2021.21316.
[21] Wallace ZS, Sparks JA. Therapeutic drug monitoring for immune-mediated inflammatory diseases. JAMA 2021;326:2370–2.
[22] Vonkeman HE, van de Laar MAF. The new European League against Rheumatism/American College of Rheumatology diagnostic criteria for rheumatoid arthritis: how are they performing? Curr Opin Rheumatol 2013;24:354–9.
[23] Aletah D, et al. 2010 rheumatoid arthritis classification criteria: an American College of Rheumatology/European League against rheumatism collaborative initiative. Ann Rheum Dis 2010;69(9):1580–8.
[24] Kroot EJ, et al. The prognostic value of anti-cyclic citrullinated peptide antibody in patients with recent-onset rheumatoid arthritis. Arthritis Rheum 2000;4(8):1831–5.
[25] Chitnavis J, et al. Genetic influences in end-stage osteoarthritis. sibling risks of hip and knee replacement for idiopathic osteoarthritis. J Bone Jt Surg Br 1997;79(4):660–4.
[26] Zhang Y, et al. Association of sporadic chondrocalcinosis with a -4-basepair G-to-A transition in the 5'-untranslated region of ANKH that promotes enhanced expression of ANKH protein and excess generation of extracellular inorganic pyrophosphate. Arthritis Rheum 2005;52(4):1110–7.
[27] Klimiuk PA, et al. Tissue cytokine patterns distinguish variants of rheumatoid synovitis. Am J Pathol 1997;151(5):1311–9.
[28] Wordsworth BP, Bell JI. The immunogenetics of rheumatoid arthritis. Springer Semin Immunopathol 1992;14:59–78.
[29] Lawrence JS. Heberden Oration, 1969. Rheumatoid arthritis—nature or nurture? Ann Rheum Dis 1970;29(4):357–79.
[30] Brown MA, et al. Susceptibility to ankylosing spondylitis in twins: the role of genes, HLA, and the environment. Arthritis Rheum 1997;40(10):1823–8.
[31] MacGregor A, et al. HLA-DRB1*0401/0404 genotype and rheumatoid arthritis: increased association in men, young age at onset, and disease severity. J Rhematol 1995;22(6):1032–6.
[32] MacGregor AJ, et al. Characterizing the quantitative genetic contribution to rheumatoid arthritis using data from twins. Arthritis Rheum 2000;43:30–7.
[33] Wolfe F, Kleinheksel SM, Khan MA. Prevalence of familial occurrence in patients with rheumatoid arthritis. Br J Rheumatol 1988;27(Suppl. 2):150–2.
[34] Deighton CM, et al. Both inherited HLA-haplotypes are important in the predisposition to rheumatoid arthritis. Br J Rheumatol 1993;32(10):893–8.
[35] Wordsworth P, et al. HLA heterozygosity contributes to susceptibility to rheumatoid arthritis. Am J Hum Genet 1992;51(3):58–591.
[36] Hall FC, et al. Influence of the HLA-DRB1l locus on susceptibility and severity in rheumatoid arthritis. QJM 1996;89(11):821–9.
[37] Raychaudhuri S, et al. Five amino acids in three HLA proteins explain most of the association between MHC and seropositive rheumatoid arthritis. Nat Genet 2012;44(3):291–6.
[38] Klareskog A, et al. A new model for an etiology of rheumatoid arthritis: smoking may trigger HLA-DR (shared epitope)-restricted immune reactions to autoantigens modified by citrullination. Arthritis Rheum 2006;54(1):38–46.
[39] Kang CP, et al. The influence of a polymorphism at position-857 of the tumour necrosis factor alpha gene on clinical response to etanercept therapy in rheumatoid arthritis. Rheumatology 2005;44(4):547–52.
[40] Mattey DL, et al. Interaction between tumor necrosis factor microsatellite polymorphisms and the HLA-DRB1 shared epitope in rheumatoid arthritis - influence on disease outcome. Arthritis Rheum 1999;42(12):2698–704.
[41] Harney SMJ, et al. Fine mapping of the MHC class III region demonstrates association of AIF1 and rheumatoid arthritis. Rheumatology 2008;47(12):1761–7.
[42] Nejentsev S, et al. Localization of type 1 diabetes susceptibility to the MHC class I genes HLA-B and HLA-A. Nature 2007;450(7171):887–92.
[43] van Rood JJ, Claas F. Both self and non-inherited maternal HLA antigens influence the immune response. Immunol Today 2000;21:269–73.
[44] Harney S, et al. Non-inherited maternal HLA alleles are associated with rheumatoid arthritis. Rheumatology 2003;42(1):171–4.
[45] Feitsma AL, et al. Protective effect of noninherited maternal HLA-DR antigens on rheumatoid arthritis

development. Proc Natl Acad Sci U S A 2007;104(50):19966–70.

[46] Gregersen PK, Silver J, Winchester RJ. The shared epitope hypothesis. an approach to understanding the molecular genetics of susceptibility to rheumatoid arthritis. Arthritis Rheum 1987;30(11):1205–13.

[47] Rigby S, et al. Investigating the HLA component in rheumatoid arthritis: an additive (dominant) mode of inheritance is rejected, a recessive mode is preferred. Genet Epidemiol 1991;8(3):153–75.

[48] Dizier MH, et al. Investigation of the HLA component involved in rheumatoid arthritis (RA) by using the marker association-segregation chi-square (MASC) method: rejection of the unifying-shared-epitope hypothesis. Am J Hum Genet 1993;53(3):715–21.

[49] Wordsworth P. T cell genetics and rheumatoid arthritis (RA). Clin Exp Immunol 1998;111(3):469–71.

[50] Plenge RM, et al. Replication of putative candidate-gene associations with rheumatoid arthritis in >4,000 samples from North America and Sweden: association of susceptibility with PTPN22, CTLA4, and PADI4. Am J Hum Genet 2005;77(6):1044–60.

[51] Kremer JM, et al. Treatment of rheumatoid arthritis by selective inhibition of T-cell activation with fusion protein CTLA4Ig. N Engl J Med 2003;349(20):1907–15.

[52] James ES, et al. PDCD1: a tissue-specific susceptibility locus for inherited inflammatory disorders. Gene Immun 2005;6(5):430–7.

[53] Radu A-F, Bungau G. Management of rheumatoid arthritis: an overview. Cells 2021;10:2857.

[54] Harrison B, Symmons D. Early inflammatory polyarthritis: results from the Norfolk Arthritis Register with a review of the literature. II. outcome at three years. Rheumatology 2000;39(9):939–49.

[55] Maini R, et al. Infliximab (chimeric anti-tumour necrosis factor alpha monoclonal antibody) versus placebo in rheumatoid arthritis patients receiving concomitant methotrexate: a randomised phase III trial. ATTRACT Study Group. Lancet 1999;354(9194):1932–9.

[56] Moreland LW, et al. Etanercept therapy in rheumatoid arthritis. a randomized, controlled trial. Ann Intern Med 1999;130(6):478–86.

[57] Edwards JC, et al. Efficacy of B-cell-targeted therapy with rituximab in patients with rheumatoid arthritis. N Engl J Med 2004;350(25):2572–81.

[58] Sokka T, Abelson B, Pincus T. Mortality in rheumatoid arthritis: 2008 update. Clin Exp Rheumatol 2008;26(5 Suppl. 51):S35–61.

[59] Gran JT, Husby G. The epidemiology of ankylosing spondylitis. Semin Arthritis Rheum 1993;22(5):319–34.

[60] Rudwaleit M, et al. The development of assessment of Spondyloarthritis International Society Classification Criteria for Axial Spondyloarthritis (part I): classification of paper patients by expert opinion including uncertainty appraisal. Ann Rheum Dis 2009;68(6):770–6.

[61] Guilino GR, et al. Cellular and molecular diversity in spondyloarthritis. Sem Immnol 2021:101521. https://doi.org/10.1016/j.smim.2021.101521.

[62] Taitt HA, Blakrishan R. Spondyloarthridites. Emerg Med Clin N Am 2022;40:159–78.

[63] Braun J, Bollow M, Sieper J. Radiologic diagnosis and pathology of the spondyloarthropathies. Rheum Dis Clin N Am 1998;24(4):697–735.

[64] Reveille J, et al. Genome-Wide Association Study of Ankylosing Spondylitis identifies non-MHC susceptibility loci. Nat Genet 2010;42(2):123–7.

[65] Toussirot E. The risk of cardiovascular diseases in axial sponyloarthritis. Current insights. Front Med 2021;8. https://doi.org/10.3389/fmed.2021.782150.

[66] Srinivasalu H, et al. Recent updates in juvenile spondyloarthritis. Rheum Dis Clin N Am 2021;47:565–83.

[67] Wordsworth P. Genes in the spondyloarthropathies. Rheum Dis Clin N Am 1998;24:845–63.

[68] Schlosstein L, et al. High association of an HL-A antigen, W27, with ankylosing spondylitis. N Engl J Med 1973;288(14):704–6.

[69] Wei J, et al. HLA-B60 and B61 are strongly associated with ankylosing spondylitis in HLA-B27-negative Taiwan Chinese patients. Rheumatology 2004;43(7):839–42.

[70] Khan MA. Update: the twenty subtypes of HLA-B27. Curr Opin Rheumatol 2000;12:235–8.

[71] Baron M, Zendel I. HLA-B27 testing in ankylosing spondylitis: an analysis of the pretesting assumptions. J Rheumatol 1989;16(5):631–4. discussion 634-6.

[72] Alvarez I, Lopez de Castro JA. HLA-B27 and immunogenetics of spondyloarthropathies. Curr Opin Rheumatol 2000;12(4):248–53.

[73] Sims AM, et al. Non-B27 MHC associations of ankylosing spondylitis. Gene Immun 2007;8(2):115–23.

[74] Mousavi MJ, et al. Association of the genetic polymorphisms in inhibiting and activating molecules of immune system with rheumatoid arthritis: a systematic review and meta-analysis. J Res Med Sci 2021;26:22.

[75] Brown MA, et al. Polymorphisms of the CYP2D6 gene increase susceptibility to ankylosing spondylitis. Hum Mol Genet 2000;9(11):1563–6.

[76] Burton MA, et al. Ankylitis in West Africans—evidence for a non-HLA-B27 protective effect. Ann Rheum Dis 1997;56:68–70.

[77] Evans M, et al. Interaction between ERAP1 and HLA-B27 in ankylosing spondylitis implicates peptide handling in the mechanism for HLA-B27 in disease susceptibility. Nat Genet 2011;4(8):761–7.

[78] Harvey D, et al. Investigating the genetic association between ERAP1 and ankylosing spondylitis. Hum Mol Genet 2009;18(21):4204–12.
[79] Lin Z, et al. A genome-wide association study in Han Chinese identifies new susceptibility loci for ankylosing spondylitis. Nat Genet 2012;44(1):73–7.
[80] Danoy P, et al. Association of variants at 1q32 and STAT3 with ankylosing spondylitis suggests genetic overlap with Crohn's disease. PLoS Genet 2010;6(12):e1001195.
[81] Gao M, et al. Relationship between IL1 gene polymorphism and susceptibility to ankylosing spondylitis: and update and supplemented meta-analysis. Biochem Genet 2021. https://doi.org/10.1007/s10528-021.
[82] Pointon J, et al. Elucidating the chromosome 9 association with AS; CARD9 is a candidate gene. Gene Immun 2010;11(6):490–6.
[83] Wanders A, et al. Nonsteroidal antiinflammatory drugs reduce radiographic progression in patients with ankylosing spondylitis: a randomized clinical trial. Arthritis Rheum 2005;52(6):1756–65.
[84] Payandeh Z, et al. The role of cell organelles in rheumatoid arthritis with focus on exosomes. Biol Proced Online 2021;23:20.
[85] Baraliakos X, et al. Clinical response to discontinuation of anti-TNF therapy in patients with ankylosing spondylitis after 3 years of continuous treatment with infliximab. Arthritis Res Ther 2005;7(3):R439–44.
[86] Inman RD, et al. Chlamydia and associated arthritis. Curr Opin Rheumatol 2000;12(4):254–62.
[87] Yli-Kerttula T, et al. Effect of a three month course of ciprofloxacin on the late prognosis of reactive arthritis. Ann Rheum Dis 2003;62(9):880–4.
[88] Orchard TR, et al. Clinical phenotype is related to HLA genotype in the peripheral arthropathies of inflammatory bowel disease. Gastroenterology 2000;118(2):274–8.
[89] McGonagle D, Conaghan PG, Emery P. Psoriatic arthritis: a unified concept twenty years on. Arthritis Rheum 1999;42(6):1080–6.
[90] FitzGerald O, et al. Psoriatic arthritis. Nat Rev Dis Prim 2021;7(1):59.
[91] Myers A, et al. Recurrence risk for psoriasis and psoriatic arthritis within sibships. Rheumatology 2005;44(6):773–6.
[92] Jones SM, et al. Psoriatic arthritis: outcome of disease subsets and relationship of joint disease to nail and skin disease. Br J Rheumatol 1994;33(9):834–9.
[93] Williamson L, et al. Clinical assessment of sacroiliitis and HLA-B27 are poor predictors of sacroiliitis diagnosed by magnetic resonance imaging in psoriatic arthritis. Rheumatology 2004;43(1):85–8.
[94] Stern RS. The epidemiology of joint complaints in patients with psoriasis. J Rheumatol 1985;12(2):315–20.
[95] Rahmati S, et al. Complexities in genetics of psoriatic arthritis. Curr Rheumatol Rep 2020;22(4):10. https://doi.org/10.1007/s11926-020-0886-x.
[96] Chandran V. The genetics of psoriasis and psoriatic arthritis. Clin Rev Allergy Immunol 2013;44:149–56.
[97] Ho PY, et al. Investigating the role of the HLA-Cw*06 and HLA-DRB1 genes in susceptibility to psoriatic arthritis: comparison with psoriasis and undifferentiated inflammatory arthritis. Ann Rheum Dis 2008;67(5):677–82.
[98] Liu Y, et al. A genome-wide association study of psoriasis and psoriatic arthritis identifies new disease loci. PLoS Genet 2008;4:e1000041.
[99] Rahman P, et al. Excessive paternal transmission in psoriatic arthritis. Arthritis Rheum 1999;42(6):1228–31.
[100] Rahman P, Elder JT. Genetics of psoriasis and psoriatic arthritis: a report from the GRAPPA 2010 annual meeting. J Rheumatol 2012;39(2):431–3.
[101] O'Rielly DD, Rahman P. Genetics of susceptibility and treatment response in psoriatic arthritis. Nat Rev Rheumatol 2011;7(12):718–32.
[102] Singh JA, et al. Special article: 2018 American College of Rheumatology/National Psoriasis Foundation guideline for the treatment of psoriatic arthritis. Arthritis Rheumatol 2019;71:5–32.
[103] Gottlieb AB, Antoni CE. Treating psoriatic arthritis: how effective are TNF antagonists? Arthritis Res Ther 2004;6(Suppl. 2):S31–5.
[104] Mease PJ, et al. Etanercept in the treatment of psoriatic arthritis and psoriasis: a randomised trial. Lancet 2000;356(9227):38–390.
[105] Fink CW, Fernandez-Vina M, Stastny P. Clinical and genetic evidence that juvenile arthritis is not a single disease. Pediatr Clin N Am 1995;42(5):1155–69.
[106] Dewitt EM, et al. Consensus treatment plans for new-onset systemic juvenile idiopathic arthritis. Arthritis Care Res 2012;64(7):1001–10.
[107] Beukelman T, et al. 2011 American College of Rheumatology recommendations for the treatment of juvenile idiopathic arthritis: initiation and safety monitoring of therapeutic agents for the treatment of arthritis and systemic features. Arthritis Care Res 2011;63(4):465–82.
[108] Moroldo MB, et al. Juvenile rheumatoid arthritis in affected sibpairs. Arthritis Rheum 1997;40(11):1962–6.
[109] Thomson W, et al. Juvenile idiopathic arthritis classified by the ILAR criteria: HLA associations in UK patients. Rheumatology 2002;41(10):1183–9.

[110] Hollenbach JA, et al. Juvenile idiopathic arthritis and HLA class I and class II interactions and age-at-onset effects. Arthritis Rheum 2010;62(6):1781–91.
[111] Hinks A, et al. Investigation of rheumatoid arthritis susceptibility loci in juvenile idiopathic arthritis confirms high degree of overlap. Ann Rheum Dis 2012;71(7):1117–21.
[112] Hinks A, et al. Overlap of disease susceptibility loci for rheumatoid arthritis and juvenile idiopathic arthritis. Ann Rheum Dis 2010;69(6):1049–53.
[113] Hinks A, et al. Subtype specific genetic associations for juvenile idiopathic arthritis: ERAP1 with the enthesitis related arthritis subtype and IL23R with juvenile psoriatic arthritis. Arthritis Res Ther 2011;13(1):R12.

INDEX

Note: Page numbers followed by "f" indicate figures and "t" indicate tables.

G

H

N

O

P

Printed in the United States
by Baker & Taylor Publisher Services